ARTIFICIAL LIFE IV

Complex Adaptive Systems

John H. Holland, Christopher Langton, and Stewart W. Wilson, advisors

Adaptation in Natural and Artificial Systems: An Introductory Analysis with Applications to Biology, Control, and Artificial Intelligence
John H. Holland

Toward a Practice of Autonomous Systems: Proceedings of the First European Conference on Artificial Life
edited by Francisco J. Varela and Paul Bourgine

Genetic Programming: On the Programming of Computers by Means of Natural Selection
John R. Koza

From Animals to Animats 2: Proceedings of the Second International Conference on Simulation of Adaptive Behavior
edited by Jean-Arcady Meyer, Herbert L. Roitblat, and Stewart W. Wilson

Intelligent Behavior in Animals and Robots
David McFarland and Thomas Bösser

Advances in Genetic Programming
edited by Kenneth E. Kinnear, Jr.

Genetic Programming II: Automatic Discovery of Reusable Programs
John R. Koza

Turtles, Termites, and Traffic Jams: Explorations in Massively Parallel Microworlds
Mitchel Resnick

From Animals to Animats 3: Proceedings of the Third International Conference on Simulation of Adaptive Behavior
edited by Dave Cliff, Philip Husbands, Jean-Arcady Meyer, and Stewart W. Wilson

Artificial Life IV: Proceedings of the Fourth International Workshop on the Synthesis and Simulation of Living Systems
edited by Rodney A. Brooks and Pattie Maes

ARTIFICIAL LIFE IV

Proceedings of the Fourth International Workshop on the Synthesis and Simulation of Living Systems

edited by Rodney A. Brooks and Pattie Maes

A Bradford Book

The MIT Press
Cambridge, Massachusetts
London, England

Third printing, 1996

©1994 Massachusetts Institute of Technology

This book was printed and bound in the United States of America.

Library of Congress Cataloging-in-Publication Data

International Workshop on the Synthesis and Simulation of Living Systems (4th : 1994 : Massachusetts Institute of Technology)
  Artificial life IV : proceedings of the Fourth International Workshop on the Synthesis and Simulation of Living Systems / edited by Rodney A. Brooks and Pattie Maes.
      p.   cm. — (Complex adaptive systems)
  "A Bradford book."
  Includes bibliographical references and index.
  ISBN 0-262-52190-3
  1. Biological systems—Computer simulation—Congresses. 2. Biological systems—Simulation methods—Congresses. I. Brooks, Rodney Allen. II. Maes, Pattie, 1961–   . III. Title. IV. Title: Artificial life four. V. Title: Artificial life 4. VI. Series.
QH324.2.A75 1994
574'.01'13—dc20   94-26555
                                                                                    CIP

CONTENTS

Preface   xi

PLENARY TALKS

Evolutionary Systems for Brain Communications—Towards an Artificial Brain   3
    Katsunori Shimohara

Emergent Functionality in Robotic Agents through On-Line Evolution   8
    Luc Steels

LONG PAPERS

Artificial Fishes with Autonomous Locomotion, Perception, Behavior, and
Learning in a Simulated Physical World   17
    Demetri Terzopoulos, Xiaoyuan Tu, and Radek Grzeszczuk

Evolving 3D Morphology and Behavior by Competition   28
    Karl Sims

Altruism in the Evolution of Communication   40
    David H. Ackley and Michael L. Littman

Evolution of Metabolism for Morphogenesis   49
    Hiroaki Kitano

Competition, Coevolution and the Game of Tag   59
    Craig W. Reynolds

In Praise of Interactive Emergence, or Why Explanations Don't Have to
Wait for Implementations   70
    Horst Hendriks-Jansen

Spencer and Dewey on Life and Mind   80
    Peter Godfrey-Smith

Crossovers Generate Non-Random Recombinants under Darwinian Selection   90
    Gene Levinson

Steps Towards Co-Evolutionary Classification Neural Networks   102
    Jan Paredis

Self-Organisation in a System of Binary Strings   109
Wolfgang Banzhaf

Effects of Tree Size on Travelband Formation in Orang-Utans: Data
Analysis Suggested by a Model Study   119
Irenaeus J. A. te Boekhorst and Pauline Hogeweg

A Biologically Inspired Immune System for Computers   130
Jeffrey O. Kephart

Egrets of a Feather Flock Together   140
Yukihiko Toquenaga, Isamu Kajitani, and Tsutomu Hoshino

A Model of the Effects of Dispersal Distance on the Evolution of
Virulence in Parasites   152
C. C. Maley

Innate Biases and Critical Periods: Combining Evolution and Learning
in the Acquisition of Syntax   160
John Batali

Dynamics of Self-Assembling Systems—Analogy with Chemical Kinetics   172
Kazuo Hosokawa, Isao Shimoyama, and Hirofumi Miura

From Local Actions to Global Tasks: Stigmergy and Collective Robotics   181
R. Beckers, O. E. Holland, and J. L. Deneubourg

How to Evolve Autonomous Robots: Different Approaches in Evolutionary
Robotics   190
Stefano Nolfi, Dario Floreano, Orazio Miglino, and Francesco Mondada

Evolving Visual Routines   198
Michael Patrick Johnson, Pattie Maes, and Trevor Darrell

Evolving Sensors in Environments of Controlled Complexity   210
Filippo Menczer and Richard K. Belew

Traffic at the Edge of Chaos   222
Kai Nagel and Steen Rasmussen

A Phase Transition in Random Boolean Networks   236
James F. Lynch

Toward an Evolvable Model of Development for Autonomous Agent
Synthesis   246
Frank Dellaert and Randall D. Beer

Bifurcation Structure in Diversity Dynamics   258
Mark A. Bedau and Alan Bahm

On Modelling Life   269
  Chris Adami

SHORT PAPERS

Genes, Phenes and the Baldwin Effect: Learning and Evolution in a
Simulated Population   277
  Robert M. French and Adam Messinger

Evolving Multi-Cellular Artificial Life   283
  Kurt Thearling and Thomas S. Ray

Meshing of Engineering Domains by Meitotic Cell Division   289
  Kazuhiro Saitou and Mark J. Jakiela

Simulating Natural Spacing Patterns of Insect Bristles Using a Network
of Interacting Celloids   295
  Hiroaki Inayoshi

Character Recognition Agents   301
  Lijia Zhou and Stan Franklin

The Building Behavior of Lattice Swarms   307
  Eric Bonabeau, Guy Theraulaz, Eric Arpin, and Emmanuel Sardet

Modeling Adaptive Self-Organization   313
  Jari Vaario

Robot Herds: Group Behaviors for Systems with Significant Dynamics 319
  Jessica K. Hodgins and David C. Brogan

A Futures Market Simulation with Non-Rational Participants   325
  Michael de la Maza and Deniz Yuret

Evolutionary Differentiation of Learning Abilities—A Case Study on
Optimizing Parameter Values in Q-Learning by a Genetic Algorithm   331
  Tatsuo Unemi, Masahiro Nagayoshi, Nobumasa Hirayama,
  Toshiaki Nade, Kiyoshi Yano, and Yasuhiro Masujima

Exploring the Foundations of Artificial Societies: Experiments in
Evolving Solutions to Iterated N-Player Prisoner's Dilemma   337
  Steve Bankes

Evolutionary Dynamics of Altruistic Behavior in Optional and
Compulsory Versions of the Iterated Prisoner's Dilemma   343
  John Batali and Philip Kitcher

Evolving Cooperation in the Non-Iterated Prisoner's Dilemma: The
Importance of Spatial Organization 349
  Michael Oliphant

An Alternate Interpretation of the Iterated Prisoner's Dilemma and the
Evolution of Non-Mutual Cooperation 353
  Peter J. Angeline

Asymmetric Mutations Due to Semiconservative DNA Replication:
Double-Stranded DNA Type Genetic Algorithms 359
  Hirofumi Doi, Ken-nosuke Wada, and Mitsuru Furusawa

Embryological Development on Silicon 365
  P. Marchal, C. Piguet, D. Mange, A. Stauffer, and S. Durand

Development and Evolution of Hardware Behaviors 371
  Hitoshi Hemmi, Jun'ichi Mizoguchi, and Katsunori Shimohara

Evolutionary Learning in the 2D Artificial Life System "Avida" 377
  Chris Adami and C. Titus Brown

Asynchrony Induces Stability in Cellular Automata Based Models 382
  Hugues Bersini and Vincent Detours

Evolutionary Automata 388
  Murray Shanahan

Non-Uniform Cellular Automata: Evolution in Rule Space and Formation
of Complex Structures 394
  Moshe Sipper

Evolutionary Robots: Our Hands in Their Brains? 400
  James V. Stone

Universality Without Matter? 406
  Alvaro Moreno, Arantza Etxeberria, and Jon Umerez

Emergent Phenomena and Complexity 411
  Vince Darley

Autonomy vs. Environmental Dependency in Neural Knowledge
Representation 417
  Markus F. Peschl

Adiversity: Stepping Up Trophic Levels 424
  Takuya Saruwatari, Yukihiko Toquenaga, and Tsutomu Hoshino

Artificial Culture 430
  Nicholas Gessler

Explorations in the Emergence of Morphology and Locomotion Behavior in
Animated Characters   436
  Jeffrey Ventrella

An Instance of a Parasitic Replicator   442
  Alun Rhys Jones and Adrian J. West

Author Index   443

PREFACE

The term Artificial Life, or ALIFE, was coined by Chris Langton to cover a range of computational ideas concerned with attempts to synthesize phenomena normally associated with natural living systems. The media for these synthesis experiments include computers, robots, and (bio)chemical soups. The idea of the ALIFE workshops is to bring together researchers from diverse backgrounds who share a common intellectual starting point about the intrinsic interestingness of such endeavors.

Chris organized the first three groundbreaking ALIFE workshops in Santa Fe, New Mexico. But now the field which was his invention, has grown and matured to the point where it no longer belongs just to Santa Fe, but instead must go on the road around the world. The fourth ALIFE workshop was held at MIT, Cambridge, Massachussetts from July 6th to July 8th, 1994. This volume is the proceedings of that workshop. The fifth ALIFE workshop will be held in Japan in 1996.

The committee which selected the papers in this volume included the two editors along with:

David Ackley	Bellcore
Richard Belew	University of California at San Diego
David Jefferson	University of California at Los Angeles
Gerald Joyce	Research Institute of Scripps Clinic
Christopher Langton	Santa Fe Institute
Michael Littman	Brown University
Melanie Mitchell	Santa Fe Institute
Steen Rasmussen	Los Alamos National Laboratory and the Santa Fe Institute
Karl Sims	Thinking Machines Corporation
Charles Taylor	University of California at Los Angeles
Peter Todd	Rowland Institute

The papers are organized in this volume to reflect their presentation at the conference. There were four invited plenary speakers: Chris Langton, Katsunori Shimohara, Jack Szostak, and Luc Steels. Two of those speakers chose to provide written records for the proceedings. There were five sessions of long papers; a total of twenty five papers in all, recorded here in the order in which they were presented. The program committee felt that these papers represented mature and completed work of a level suitable for journal publication. There were four sessions of short papers; a total of twenty eight papers in all, again recorded here in the order in which they were presented. The program committee felt that these papers represented solid pieces of work that should be immediately available to other researchers in the field. Somewhere along the way a parasite seems to have infested the proceedings also.

Besides the papers in these proceedings, there were a large number of additional presentations at ALIFE IV. There was a large poster session on the afternoon of Thursday July 7th, where approximately fifty researchers presented their work in progress and held informal feedback sessions with browsing conference attendees. There were a number of demonstrations of systems, robots, and artistic endeavors, and there were special talks on the anthropology of the field, and the need for digital game reserves. In a special evening invited session, there were talks on the major efforts in existence to search for extra-terrestial intelligence (SETI). The conference thus spanned all of life as we know it, all that we could think of to create, and all that might exist out there elsewhere in the universe.

The material covered in these proceedings is diverse. As should be expected, the evolution of content matter visible in the proceedings of the earlier workshops continues with this volume. In particular, we see two shifts in emphasis. In the first shift, the sophistication of artificial worlds where evolving populations are studied, has increased dramatically, and now includes models of physical dynamics. In the second shift, there is more connection to real systems, exemplified by the following four trends:

- There is a new emphasis on agents and what it means to be an artificial lifeform beyond being able to merely reproduce; the questions here are how to organize the many concurrent phenomena that make up a life-like embodient in the world.

- There are new thrusts in the modelling of real biological systems.

- There is significant progress in using artificial life techniques in understanding non-biological phenomena.

- There are the beginnings of using evolutionary techniques for building programs that might potentially control physical agents in a way that is better than the best that can be done 'by hand' programming.

We hope that these proceedings, besides being an archival reference, will be a place to fossick around for old and new ideas, and that the work reported here will inspire much further work in the area of ALIFE.

The existence of these proceedings is due in large part to the enormous amount of work done on all aspects of them by Annika Pfluger of the MIT AI Lab, when she could have been having much more fun playing the cello. We thank her sincerely for her efforts. Harry Stanton and Teri Mendelsohn at the MIT Press have invisibly handled all the 'stuff' that happens in order to turn a camera ready pile of pages into a real physical publication that is distributed throughout the world.

Rodney Brooks, MIT Artificial Intelligence Lab
Pattie Maes, MIT Media Lab

Cambridge, MA, May, 1994.

There were a number of generous sponsors for the meeting at MIT. They included:

- The MIT Artificial Intelligence Laboratory

- The Santa Fe Institute

- The MIT Press

- Mitsubishi Corporation

- The Nippon Signal Co., Ltd.

- Uchidate Co., Ltd.

- Applied AI Systems, Inc.

PLENARY TALKS

Evolutionary Systems for Brain Communications
— Towards an Artificial Brain —

Katsunori Shimohara
Evolutionary Systems Department,
ATR Human Information Processing Research Laboratories,
2-2 Hikaridai, Seika-cho, Soraku-gun, Kyoto, 619-02, JAPAN
e-mail: katsu@hip.atr.co.jp

Abstract

The Evolutionary Systems Department at ATR aims to create new information processing systems, which are rich in autonomy and creativity. For this purpose, we introduce the concepts of "Evolution and Emergence" from Artificial Life into the modeling of the functions of the brain (the central organ for communications), such as information understanding and generation. This paper describes the research concepts we use toward our goal of creating an artificial brain as an evolutionary system, that will be able not only to develop new functionality spontaneously but also to grow and evolve its own structure autonomously.

1 Introduction

Computers that Can Generate Information

Computers that can spontaneously generate information may no longer be a dream, if one uses the ideas and methodologies of Artificial Life. Nowadays computers are indispensable tools in our productive and intellectual lives, as well as many other applications. However, in the sense that they only perform their programmed instructions, they are, in effect, passive slaves. If a computer could be given greater autonomy, it could become our partner, which we human beings can interact with for our thinking, e.g., it could adapt to us, give us ideas, gather information related to the items we wanted, and so forth. Will it be possible to create such a computer that can make its own judgments and generate information? Interactions between us and such a computer will make our ideas and imagination much richer, and thus amplify our creativity and productivity. In the new age of the "information super highway", where computers communicate with each other autonomously, such a creative world generated by cooperation between us and computers becomes increasingly possible.

At the Evolutionary Systems Department, ATR Human Information Processing Research Laboratories, we are researching into the possibilities of building such "computers that can generate information", that is, information processing systems, rich in autonomy and creativity, similar to the human brain.

Toward an Artificial Brain as an Evolutionary System

Let us call such information processing systems with autonomy and creativity "Artificial Brains". We do not intend to merely mimic artificially an biological brain in its function and structure, but wish to create an artificial brain that will be superior to the biological brain in certain respects.

In the mammalian brain, an enormous quantity of neurons are produced both before and after birth, many of which die, leaving many billions of neurons. In the infant brain, remaining neurons grow by spreading and lengthening their dendrites and axons, gradually forming networks of neurons by combining axons and dendrites through synapses. From then on, neurons continue to die until the death of the individual. Human infants are very flexible and creative in their thinking, because correspondingly the structure of their brains is growing and flexible. The plasticity of synapses allows them to be influenced by experience and to learn by modifying themselves little by little. Finally the structure of the networks is fixed. Ultimately the brain dies when the body dies. People can only leave their genes to their offspring no matter how hard they study or how much they experience. The evolution of the biological brain has followed this repetitive pattern of production, growth, and modification for countless generations.

However, if an artificial brain can be built, it should be possible to make part or all of it return to an infant brain state whenever flexibility is needed. That means it could restructure its neural networks and increase the number of its neurons, whenever necessary. While the biological brain is limited to a given size so that a body can support it, an artificial brain need not be limited in size. Also an artificial brain would not have to die! Therefore, the artificial brain could evolve a new part leaving the results of learning and experiences intact, and could add the new part to itself. It might be possible to build artificial brains which self-replicate and which can evolve separately, thus forming a "society of artificial brains."

Artificial brains have the potential to transcend the limits imposed on biological brains by their biological nature.

Our goal is to create such an artificial brain, that is,

an information processing system, rich in autonomy and creativity, based on an evolutionary system that will enable the spontaneous development of new functionality. This artificial brain will evolve its own hardware structure. For this purpose, we introduce the ideas and methodologies of Artificial Life as a new paradigm into the modeling of information processing in the brain.

2 Paradigm Shift to Life-Like and Society-Like Information Processing

Artificial Life is a challenging new research paradigm which investigates the concept of "life-as-it-could-be" as well as understanding "life-as-we-know-it", by synthesizing life-like phenomena in artificial media, using informational and computational techniques. From an engineering viewpoint, ALife can be regarded as a possible framework to achieve artificial systems which possess the advantages of living systems, such as autonomy, adaptation, evolution, self-replication, and self-reparation. A typical example is the epoch-making work of Tom Ray who showed that computer programs can evolve in the virtual world of the computer(Ray 1991). This encourages us to explore the possibility that a computer itself might evolve.

The ALife methodology can be explained as an information mechanism in the following way: Firstly, prepare a mass of elements and some framework to make them interact with one another. Some of these elements are activated by a stimulus or information from the environment, which triggers their interaction. Through such interactions, a kind of "whole", whether in terms of organization, structure, order, network, or global state, etc., spontaneously emerges. This "whole" can change due to its activation of other elements. Also, it is important to provide some mechanism whereby elements themselves can make changes.

At ATR, we intend to use the above-mentioned information processing mechanisms. Based on the above, we have adopted two approaches: a) life-like modeling and b) social modeling, in addition to the conventional learning model used in nervous systems, e.g. neural networks. In life-like modeling, the system should have a self assembling (embryological) capability similar to living systems in nature, in order to make its structure and components change, and to make it complex. In social modeling, the system should be regarded as a dynamical process where the global or macro-scopic order/state emerges through the behavior and the interaction of the micro-scopic components. In turn, the behavior of the micro-scopic components is influenced by the macro-scopic state. Thus, two directions of interaction—bottom-up, from micro to macro, and top-down, from macro to micro—make it possible that systems change interdependently.

In summary, we propose here life-like and society-like information processing schemes to build an information processing system rich in autonomy and creativity. It can be described by shifts in design principles, as follows;

- From centralized to DECENTRALIZED:
  The behavior of the system will not be centrally con-

trolled but will emerge due to the interaction of local components.

- From optimized to
  COLLECTIVE and REDUNDANT:
  Functions should be generated and realized as situated and self-organized collective behaviors of the system.

- From fixed to FLUID:
  Processing elements should emerge/disappear, replicate/decrease, and combine/separate, and then autonomously change through such mechanisms as metabolism, and/or natural selection.

3 Evolution and Emergent Mechanisms for Brain Communication

In order to give autonomy and creativity to an information processing system, it is vital that the system itself should have some mechanism to spontaneously generate change in its function and structure. We use two such mechanisms, "Evolution and Emergence", which we define as follows: "Evolution" is defined as a mechanism for generating change, and "Emergence" is defined as a mechanism for adjusting and integrating the changes into a system and for self-organizing the system. In addition, another key concept, called "Micro-Macro Dynamics" is proposed to work as a kind of a framework within which evolution and emergent mechanisms can be embedded.

The first step toward our goal of building an artificial brain is, therefore, to create evolutionary and emergent mechanisms, based on the the structure of micro-macro informational loops, in the brain, e.g., information processing and generation. One idea is to think of communication in the brain as a dynamical process in which information "macro" structures spontaneously and interdependently emerge from the interaction between "micro" informational elements, and from changes based on "micro-macro interactions" in an information system. Figure 1 illustrates such dynamics.

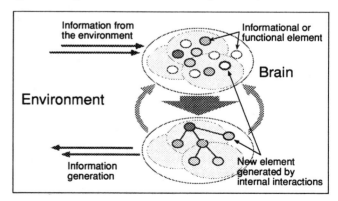

Figure 1: Information Processing Model within the Brain in Communication

Let us take an example concerning our understanding of what happens in a brain communication: First assume that we have a lot of knowledge and informational elements in the brain. At the beginning of a communication, as shown in figure 1, interactions between informational elements activated by a piece of information from the environment occur, which trigger an emerging process from a micro to a macro level in which elements self-organize into a kind of network. As the process unfolds from a macro to the micro level, the network itself influences the emerging process from the micro to the macro level by activating other elements, generating new ones or eliminating old ones, combining or separating some of them, and so on. Then, eventually, a form of informational structure of networking elements emerges.

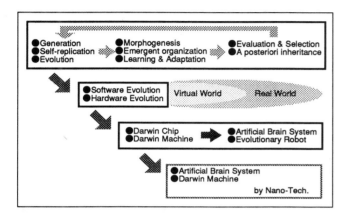

Figure 2: Research Topics and Sub-Goals

Figure 2 shows the research topics and sub-goals of our project toward building artificial brains as evolutionary systems. The box at the top of figure 2 represents the functions we should achieve. From a computer engineering viewpoint, research will be needed to develop new software and hardware technologies that will embody evolutionary and emergent computation methods. We will then implement these technologies as a "Darwin Chip" and "Darwin Machine", and hopefully achieve artificial brain systems and autonomous evolutionary robots with an artificial brain in the real world.

Longer term, we will be extending our research to the implementation of Darwin Machines and artificial brain systems using nano-technologies, as well as state-of-the-art nano- electronics and mechanics, because nano-technology will dominate 21st century science and technology. We believe that real evolvable hardware will only achieve its full potential once nano-technology is a practical reality.

Figure 3 shows the approaches we are taking to our research into evolutionary systems.

4 Current Research Projects at ATR

The research projects currently conducted in the Evolutionary Systems Department, ATR, are as follows:

- Evolutionary model of functionality—Software evolution

- Evolutionary model of structure—Hardware evolution

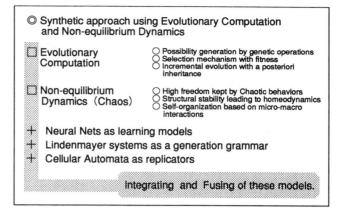

Figure 3: Approaches

- Evolutionary model of adaptive behaviors

- Emergent intelligence model

- Computational evolutionary biology and immune system modeling

- Dynamic information processing model

In the following, some of these research projects are briefly explained. Since we are now seeking and clarifying various possibilities for artificial brains as evolutionary systems, we don't yet have any specific application goals for the near future. However, some possible later applications are shown in figure 4.

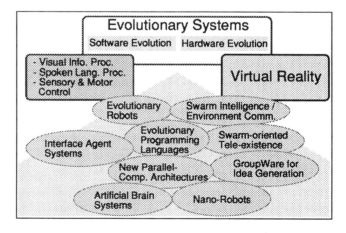

Figure 4: Possible Applications

4.1 Software Evolution

This project attempts to use evolution by natural selection in the medium of the digital computer to generate complex and intelligent software. The fundamental elements of evolution are self-replication with heritable variation. This can be implemented in a computer by writing self-replicating machine language programs and running them on a computer that makes mistakes.

These experiments result in rapid diversification of digital organisms. From a rudimentary ancestral self-replicating creature entire ecological communities emerge spontaneously. Natural evolution in this artificial system illustrates well established principles of evolutionary and ecological theory, and allows an experimental approach to the study of evolution, as well as observations of macro-evolutionary processes as they occur.

It is also a possible means of harnessing the evolutionary process for the production of complex computer software. The evolutionary process will be introduced into the context of the massively parallel computer to evolve complex MIMD parallel processes(Thearling and Ray 1994).

4.2 Hardware Evolution

Hardware evolution or evolvable hardware is one of the most challenging and significant areas of research in the use of evolutionary computation and methods. In order to create adaptive complex systems that are like living systems, we believe that not only software evolution but also hardware evolution is indispensable. New hardware architectures and devices, which will allow hardware to be evolved are needed.

It is also essential to create an evolutionary framework and/or computational mechanisms to guide hardware evolution and to clarify applicability and feasibility.

4.2.1 CAM-BRAIN: Cellular Automata Machine-Brain

In order to implement evolutionary computation or methods in hardware, a special framework is needed where structures built on reconfigurable hardware can be closely related to software functions realized on the structure.

Using special hardware called Cellular Automata Machines (CAMs), in this project, artificial neural networks as models of nervous systems will be grown inside the CAMs.

The aim of the CAM-Brain Project is to build (i.e. grow/evolve) an artificial brain by the year 2001. This artificial brain should initially contain thousands of interconnected artificial neural network modules, and be capable of controlling approximately 1000 "behaviors" in a "robot kitten". Using a family of CAMs, each with its own processor to measure the performance quality or fitness of the evolved neural circuits, will allow the neural modules and their interconnections to be grown and evolved at electronic speeds(de Garis 1993) (de Garis 1994).

4.2.2 Development and Evolution of Hardware Behaviors

Production Genetic Algorithms (PGAs) enable grammatical structures as well as HDL (Hardware Description Language) programs to evolve, toward an automated hardware design system using an evolutionary process.

In this PGA system, hardware specifications, which produce circuit behaviors, are automatically generated as HDL programs according to the grammar defined as

a rewriting system, and then evolve through PGAs. The PGAs introduce new chromosome representations and genetic operators to create self-generating mechanisms, like living systems in hardware design.

Experimental results show that using an evolutionary process, based on PGAs, a hardware specification program expands its circuit scale and as a result increases its functionality(Mizoguchi et al. 1994) (Hemmi et al. 1994).

4.3 Self-Organization and Evolution of Adaptive Behavior

The aim of this research project is to model biological adaptation that consists of self-organization and evolution processes. For evolution, the creation of new forms is essential, and for creation of new forms a self-organization process is necessary.

Thus we view evolutionary computation as "computational evolution" where the emphasis is on the construction of new forms that lead to the creation of new behaviors. To elaborate this fact, simple artifacts are modeled, starting from a single cell, that results, through multiple cell divisions and differentiation, in a complete organism capable of behaving in its environment. The point is to study the effect of variation in the environmental factors and genetic information, in order to model biological adaptation.

Furthermore, we will apply the same principles to engineering applications. As an approach to study the above processes, we have implemented a modeling language that is capable of modeling multilevel interactions within the environment, and within the objects themselves. These interactions are considered as the basic elements for self-organization and evolution, and they describe the physical interactions found in biological systems(Vaario 1994).

4.4 Genetic Mechanisms and Evolutionary System Theory

This project introduces an "Evolutionary System Theory" as a unified approach applicable to real world problems such as real-time adaptive control and mass genetic information analysis.

It also proposes a "Disparity Hypothesis" on the gene replication mechanism of an organism with double stranded DNA that postulates the disparity of fidelity between replication error on the leading strand and the lagging strand. The assertion is that by taking advantage of the disparity, an organism can evolve rapidly and acquire robustness even under fluctuating environmental conditions.

By applying the disparity model to a two dimensional maximum search problem with multi-peaks and diastrophism, it is possible to show that conservative and radical offspring can be bred together, and that the population can dynamically adapt(Wada, Wada, Doi, Tanaka, and Furusawa 1994).

5 Conclusion

We have shown the concepts of our research which aims to create an information processing system, rich in autonomy and creativity, for brain communication. Also, based on ALife ideas, a scheme of life-like and society-like information processing was proposed as a new engineering principle. "Evolution and emergence" and "micro-macro dynamics" are key concepts for evolutionary systems that will be able not only to develop new functionality as software but also to grow and evolve autonomously its own structure as hardware.

The goals are to create "evolutionary" mechanisms to generate change, and to create "emergent" mechanisms to self-organize a system, based on the structure of micro-macro informational loops, as well as to seek and clarify various possibilities for artificial nervous systems with these mechanisms as evolutionary systems. Artificial brains which will implement such evolutionary technologies as evolutionary systems should transcend the limitations of biological brains.

The ideas and approaches inspired by ALife increase in importance and are expected to have a major impact on the areas of communication, robotics, and computer engineering. One of the most stimulating results in this respect has shown us the possibility of creating autonomy and creativity in computers. This fact should remind us of the importance to control such autonomy and creativity of artificial creatures.

References

de Garis, H. (1993). Evolvable hardware: Genetic programming of Darwin machines. In *International Conference on Neural Networks and Genetic Algorithms*, Lecture Notes in Computer Science. Springer-Verlag.

de Garis, H. (1994). Cam-brain issues: Implementation and performance- scaling issues concerning the genetic programming of a cellular automata based artificial brain. In *IEEE Conference on Evolutionary Computation*.

Hemmi, H., J. Mizoguchi, and K. Shimohara (1994). Development and evolution of hardware behaviors. In R. Brooks and P. Maes (Eds.), *Artificial Life IV*. MIT Press.

Mizoguchi, J., H. Hemmi, and K. Shimohara (1994). Production genetic algorithms for automated hardware design through an evolutionary process. In *IEEE Conference on Evolutionary Computation*.

Ray, T. S. (1991). An approach to the synthesis of life. In C. G. Langton, C. Taylor, J. D. Farmer, and S. Rasmussen (Eds.), *Artificial Life II*, Volume X of *Santa Fe Institute Studies in the Sciences of Complexity*, Redwood City, CA, pp. 371–408. Addison-Wesley.

Thearling, K. and T. S. Ray (1994). Evolving a digital cambrian explosion: Some experiments with multi-cellular artificial organisms. In R. Brooks and P. Maes (Eds.), *Artificial Life IV*. MIT Press.

Vaario, J. (1994). Modeling adaptive self-organization. In R. Brooks and P. Maes (Eds.), *Artificial Life IV*. MIT Press.

Wada, K., Y. Wada, H. Doi, S. Tanaka, and M. Furusawa (1994). Evolutionary system: Structure and functions – disparity hypothesis and gene duplication. In *IEEE Conference on Evolutionary Computation*.

Emergent functionality in robotic agents through on-line evolution.

Luc Steels

Artificial Intelligence Laboratory

Vrije Universiteit Brussel

Pleinlaan 2, B-1050 Brussels, Belgium

E-mail: steels@arti.vub.ac.be

Abstract

The paper proposes an architecture for the on-line evolution of new behavioral competences on a robotic agent. Some experimental results for evolving a set of primitive behaviors are presented.

Introduction

A central question in ALife research is how new complexity and new functionality may emerge [Steels1994]. Selectionism and self-organisation have so far been put forward as the key explanatory principles [Langton1989]. These principles have been been applied at many level of biological systems, from the chemical reactions that explain the origin of life [Kaufmann1993] to the interaction between individuals in societies [Deneubourg1993]. This paper explores in how far selectionism and self-organisation may lead to the build up of behavioral complexity in animals. In the tradition of Alife research, this exploration takes place by building artificial systems, i.c. robotic agents.

There has already been a large amount of work attempting to use selectionist techniques for evolving behavioral competences. Holland for example has developed classifier systems and used genetic algorithms to evolve them [Holland1975]. Koza has shown how the reactive finite state machines proposed by Brooks [Brooks1991] can be evolved using genetic programming techniques [Koza1991]. The Sussex group [Cliffs,et.al.1993] has proposed an experimental environment for using genetic techniques based on real sensory data. Selectionist mechanisms have also been proposed by neurobiologists, notably Edelman, as an alternative explanation to the inductive or associative mechanisms dominating the literature on neural networks [Edelman1987].

All this work is extremely valuable and has provided inspiration and techniques for the work reported here. Our own approach differs however in the following respects:

- *Subsymbolic vs symbolic.* The classifier systems of Holland are in the tradition of symbolic AI. They assume that the world needs to be categorised in terms of predicates like 'object in left center field of vision' and that behavior must be decomposed into actions like 'cause eyes to look left'. We want to stay at the sub- or presymbolic level in which the dynamics of the world is directly coupled to an internal dynamics without prior segmentation or categorisation. This implies that we must evolve dynamical systems instead of symbolic computation rules.

- *Cooperation vs Subsumption.* Many architectures for robotic agents, such as the subsumption architecture [Brooks1991], allow different behavioral modules to inhibit one another. We want instead to create a 'level playing field' in which one behavioral module cannot inhibit another one. Different modules must cooperate or compete with each other in order to achieve a coherent behavior. Genuine conflicts are handled by motivational systems, as studied extensively in ethology [McFarland1992]. This 'level playing field' appears crucial to apply selectionism properly because otherwise some behavior systems get an unfair advantage.

- *On-line vs off-line.* Most of the genetic experiments so far have been performed in a simulation setting. Instead, we want to work on real robots in view of the large discrepancies between simulations and artificial systems and the complexity of building realistic simulators [Brooks1991]. Moreover we want the agent to remain viable as a task-achieving agent instead of running the genetic algorithm off-line and then transplanting a solution for testing on the robot, as in [Cliffs,et.al.1993]. Running genetic mechanisms on-line puts heavy constraints on the selectionist mechanisms that can be used but it brings the experimental conditions closer to real autonomous robotic agents.

- *Open vs. closed functionality* Most genetic experiments so far assume a fixed desired functionality so that the fitness function can be provided as input by the designer. We want instead an evolving open functionality with no *a priori* fitness function. To achieve this, we have created an experimental setup in which there are a large number of real world constraints acting on the agents (Steels1994b). The constraints range from internal constraints (limited decision time, limited memory) to external constraints (pressure to obtain enough energy in time, pressure to make other robots survive, pressure to avoid bodily damage). This way we obtain a situation in which there is natural selection as opposed to artificial selection, much in the same way the computational constraints in the Tierra system [Ray1992] constitute a natural selection environment for evolving copying programs.

This paper reports on progress towards the ambitious objectives stated above. Our explorations are far from finished but we have achieved already the following results: (1) a selectionist architecture has been designed, (2) it has been shown experimentally that a primary repertoire of behavioral competences can be evolved, (3) an implementation has been constructed on real robots and experimental results with this implementation have been obtained. Much further work needs to be done, particularly towards the evolution of secondary repertoires and towards the mathematical investigation of the genetic mechanisms that have been proven to be successful so far in experiments. Moreover many variants of the architecture have not been explored yet.

The rest of the paper is in four parts. First we discuss the target of evolution, which are dynamical systems running on a cooperative dynamics architecture. Then the proposed selectionist mechanism, known as the *selectron* is described. Third some experimental results are reported. The paper concludes with some indications how we are tackling the evolution of a secondary repertoire.

The PDL Robot Architecture

Before we concentrate our efforts on evolving behavioral competences we must decide in which form this competence will be implemented on the robot. At the same time, the choice of robot architecture is influenced by whether it supports selectionist mechanisms. We have adopted a cooperative dynamics architecture which is implemented with an associated language called PDL [Steels1993]. This architecture assumes that the sensors deliver a continuous (discretised) stream of data and that there is a continous (discretised) stream of parameters flowing to the effectors (e.g. the speed or acceleration of the motors). The time-varying data are made available as the values of *quantities* (sensory quantities and action parameters) which are stored in an array. There is also a set of internal quantities which are used to keep track of motivational states and world states, and to support internal bookkeeping. The architecture also features *processes*. A process establishes a dynamical relation between a set of quantities but is not in full control of the exact value. For example, there is no assignment. If a process wants a particular quantity to have a certain value, then it must drive the dynamics such that this quantity progressively reaches that value. A process is mathematically defined as a differential equation. It is discretised to a difference equation and then implemented by simulated parallellism similar to the way cellular automata approximate continuous dynamical systems.

The third basic unit in the architecture is called a *behavior system*. A behavior is a regularity in the interaction dynamics between an agent and the environment

(for example a certain distance is maintained from the wall) [Smithers1992]. A behavior system is the set of internal processes that are active when the regularity is observed. Each behavior system can be said to establish a particular *condition*. Examples of conditions are: approaching zero translation speed (halting behavior), approaching maximum left photo and right photo sensing (orientating towards a light source), approaching minimum infrared reflection (turning away from an obstacle), etc. Often the robot can monitor the satisfaction of a condition if it can sense the behavioral regularity.

The PDL architecture is guided by two principles which are both adapted to create a 'level playing field':

- All behavior systems (or more precisely all the processes of all the behavior systems) are active at the same time, i.e. there are no subsumption relations between behavior systems. This means concretely that, for example, the forward movement behavior system and the backward movement behavior system are both operational. The ultimate direction taken depends on which behavior system influences the overall behavior in the strongest way.

- The influences of the different behavior systems are summed. Although a process can (and usually does) internally perform a non-linear mapping, the additive combination guarantees that no process is viewed as more important than another one.

Given a set of behavior systems, each consisting of a set of processes, then the overall execution algorithm is defined by the following procedure:

1. All quantities are frozen

2. All processes are executed and their influences combined.

3. All quantities are changed based on the influences.

4. The action parameter quantities are sent to the effectors.

5. The latest sensory quantities are read in.

Then the procedure starts again from 1. This cycle takes place on our current robots with PC-level processors at a speed of at least 40 cycles per second, ensuring a very reactive behavior (Figure 1.).

To regulate the interaction between different behavior systems motivations and behavioral tendencies have been introduced. Both terms are used in their ethological sense [McFarland1992], although we introduce them to constrain the internal architecture of the robot not to explain empirically observed behavior as is usually done in ethology. A motivation is a quantity which reflects the cost of reaching a particular state. A typical example is a quantity *EnergyNeed* which is inversely proportional to the level of the battery (Figure 2.): the less energy there is, the higher the quantity EnergyNeed. Certain behavior systems are influenced by motivational quantities. For example, because running out of energy is fatal to the robot, the behavior of moving towards the charging station should be stronger as energy becomes

Figure 1: Typical example of robotic agents we use in our experiments. The body has been built with LegoTechnics^{TM}. The main processor is a pocket PC computer inserted in the robot body. We use a custom-made sensory-motor board to buffer and preprocess the sensors and regulate the flow of action parameters to the actuators.

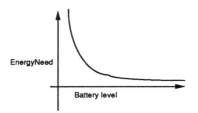

Figure 2: The graph plots a typical dependency of a motivational quantity on an external quantity, in this case EnergyNeed (y-axis) versus battery level (x-axis). As a lethal region is approached, EnergyNeed goes up exponentially.

low. The behavioral tendency is a function (usually multiplicative) of the motivational and cue strength. The same strength in behavioral tendency will be observed with low motivation but high cue strength as with high motivation but low cue strength. This is often shown by drawing the isoclines of the behavioral tendencies. (Figure 3.)

Motivations and behavioral tendencies are well established notions in ethology [McFarland1992] and have been suggested as applicable to robotics [McFarlandBoesser1994]. We incorporate them here as explicit internal quantities. There are processes which determine the level of motivations as well as the levels of behavioral tendencies. The influence imposed by a behavior system is always a (multiplicative) function of the tendency associated with the behavior.

The behavior of the robot is constrained by a set of built-in motivations and processes computed based on internal and external sensing. These built-in motivations set the physical boundaries of the robot (e.g. avoid

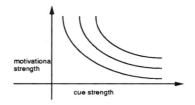

Figure 3: This graph plots the isoclines of the behavioral tendency as a function of the motivational strength (y-axis) and the cue strength (x-axis).

overcharging the batteries, avoid running out of energy) and are also a way to (indirectly) specify a mission to the robot. For example, if the robot must perform measurements, then its motivational strength to collect more measurements depends on the amount of measurements it has been able to gather.

The selectionist mechanism

The task of the selectionist mechanism that we seek can now be defined more clearly: While remaining viable in the environment, the robot must evolve an appropriate set of behavior systems which will ultimately guarantee its survival. This means in the first place maintaining adequate energy levels (assuming recharging opportunities). We make a distinction between the acquisition of a primary repertoire of basic functionalities such as forward movement, turning away from obstacles, orientating towards a light source, etc. and a secondary repertoire which combines primary (or secondary) behaviors by manipulating their behavioral tendencies. So far we have only reached solid results for deriving the primary repertoire, which is however often viewed as the most difficult part.

The basic idea of our approach centers around conditions. As stated earlier, conditions are dynamical properties of the environment or the internal state of the agent that a particular behavior system attempts to satisfy. Conditions are related to the notion of goals in classical control theory, but not to the notion of goal in symbolic AI, where goals are predicates on symbolically described world states.

A typical example of a condition is: Wheelcounter should approach zero. This is satisfied in the case of a halting behavior. To develop the primary repertoire, the robot generates conditions by combining randomly operators, constants, and quantities from a limited repertoire. For example, the condition "WheelCounter should approach the maximum value (which is 255)." could be generated. For each condition, a quantity corresponding to a behavioral-tendency is introduced. The higher this quantity, the faster we expect the condition to see satisfied. At the moment, the robot then concentrates on evolving a behavior system for the first condition generated, before moving on to the next condition. Evolving a behavior system means to derive a set of processes that are capable together to cause the condition to be satisfied.

There are many variants of selectionist mechanisms, but they typically involve the following steps [Koza1991]:

1. An initial population is generated.

2. The elements of the population are assigned a fitness value.

3. The constellation of the population is changed using copy and mutate operations whose probability of occurrence is a function of the fitness of the elements engaged in the operations.

We explore a particular variant of this general class of mechanisms which is adapted to the task at hand. We call this variant the *selectron*.

The initial population

In the present case, the elements of a population are processes, such as 'increase the translation', or 'decrease the rotation'. The impact of these processes is always a (multiplicative) function of the behavioral tendency associated with the behavior system. The initial population may contain multiple copies of a process (in which case the effect is enforced) or completely opposing processes (for example some pushing the speed of the translation motors up and some others pushing the speed down).

Assigning a fitness value

To determine the fitness, all processes in the population are allowed to run for a particular length of time which is called the *selection window* W. Typical lengths in our experiments are at least 10 cycles, i.e. 0.5 to 1 seconds. This is based on the experience that the quantities fluctuate too much for a step by step evaluation due to the tight interaction with the environment. After the test period the fitness is determined by testing how far the conditions are satisfied in relation to the level of their corresponding behavioral tendency. The test yields a vector indicating in how far the actual value deviates from the desired value (positive or negative). The test itself also takes place in a PDL process and is recorded in a quantity called *satisfaction* s which gets higher when the current state is further away from being satisfied. The average satisfaction during the time window W is equal to $S(T + W) = \sum_{j=1,W} s(t + j)/W$.

As processes are run, the system keeps track of the role of each process for the overall behavior. A process p_i has an impact v_i on a quantity q at time t. For all processes, $\Delta_q^+ = \sum v_i$ iff $v_i > 0$, and $\Delta_q^- = \sum v_i$ iff $v_i < 0$. The fraction d_i of a process p_i at time t is equal to $d_i = v_i/\Delta_q^+$ if $v_i > 0$ or $d_i = v_i/\Delta_q^-$ if $v_i < 0$. The majority direction at time t is $M = +1$ if $\Delta_q^+ > \Delta_q^-$, $M = -1$ otherwise. The role of a process p_i at time t is then defined as $r_i = Md_i$. The average role of a process p_i in the time period $[t, t + W]$ is equal to $a_i = \sum r_i/W$.

Changing the population

There are two ways in which the population changes: by copying which ensures that a process stays in the population or by mutation/creation which ensures that new processes enter. The latter is particularly needed if a needed candidate died off prematurely.

A process has a certain probability of being copied and/or surviving into the next generation (at time $t + W$). This probability is proportionate to the change in average satisfaction between t and $t + W$:

$$e(t + W) = \begin{cases} +1 & \text{if } S(t) < S(t + W) \\ -1 & \text{if } S(t) > S(t + W) \\ rand1, -1 & \text{if } S(t) = S(t + W) \end{cases} \quad (1)$$

The random choice between -1 and 1 helps to bring the system out of local minima. The probability of copying or deleting is such that the following constraints are satisfied:

$$\sharp P_m(t + W) = \gamma \, a_i(t + W) \, e(t + W) + \sharp P_m(t) \quad (2)$$

$$\sharp P_m(t + W) >= 0 \quad (3)$$

where $p_i \in P_m$. γ is a constant which is in our current experiments set equal to 10.

The mutation/creation rate is influenced by the amount of processes in the population. This amount is kept constant. In our current experiments mutation is not yet related to fitness although we plan to do so in the future.

All the necessary data to execute the selectionist procedures are collected at every time step and in a parallel fashion (e.g. every process maintains itself what its role was). Moreover the overhead in temporal and spatial complexity for performing selection is small enough to incorporate selectionism as part of the robot's normal operation in the environment. We currently let the behavior systems in the primary repertoire evolve one by one, although the selectron algorithm does not prescribe that only one condition is worked on at the time. Experiments with multiple behavioral tendencies remain future work.

Experiments

The following examples illustrate the behavior of the selectron and provide at the same time experimental evidence for the viability of the approach. We concentrate first on evolving behavior systems for moving forward and backward and for halting using only one sensory quantity (WheelCounter) and one action parameter (Translation speed of the motors). We first look at forward movement with tendency Tendency1. The quantities involved are WheelCounter, translation speed of the left and right motors, Tendency1, and satisfaction-1. The latter quantities correspond respectively to the behavioral tendency for engaging in forward behavior and the satisfaction with which this behavior is actually observed.

The initial population is as follows. The number of copies of a process in the population is indicated in parentheses behind the name.

```
process-0 (16):
  Translation <= Tendency1 *
  approach(-255,WheelCounter)/1000
process-1 (14):
  Translation <= Tendency1 *
  approach(0,translation)/1000
```

```
process-2 (15):
 Translation <= Tendency1 *
 approach(255,WheelCounter)/1000
process-3 (18):
 Translation <= Tendency1 *
 approach(-255,translation)/1000
process-5 (18):
 Translation <= Tendency1 *
 approach(255,WheelCounter)/1000
process-7 (15):
 Translation <= Tendency1 *
 approach(0,WheelCounter))/1000
```

Here are some steps in the execution of the algorithm. For each step the average satisfaction at time t and $t+W$ are displayed as well as the direction e. For each process, the sum of the influences r_i during the time period $[t, t+W]$ and the number of processes in the population after copying and elimination are indicated.

```
Step 1. Satisf: 25.800=>26.000;e=-1
process-0   1.94: 14
process-1  -0.03: 14
process-2  -1.79: 15
process-3   2.06: 14
process-5  -2.16: 19
process-7  -0.02: 15
Step 2. Satisf: 26.000=>23.800;e=1
process-0  -0.06: 13
process-1   0.07: 14
process-2  -0.03: 14
process-3  -0.10: 13
process-5   0.01: 19
process-7   0.12: 15
Step 3. Satisf: 23.800=>23.600;e=1
process-0  -0.01: 12
process-1   0.01: 14
process-2  -0.00: 13
process-3  -0.02: 12
process-5   0.00: 19
process-7   0.02: 15
Step 4. Satisf: 23.600=>23.000;e=1
process-0   1.79: 13
process-1   0.20: 14
process-2  -1.63: 11
process-3   1.78: 13
process-5  -2.37: 16
process-7   0.22: 15
```

We observe during this short sequence a gradual improvement in the fitness of the population (because satisfaction evolves to zero). Usually there is a period of heavy competition until the winning processes finally dominate. This happens quite rapidly once these processes gain a slight upperhand, similar to phase transitions. The same data as above is displayed graphically but for a complete evolutionary sequence in Figure 4. We see that the satisfaction (directly related to fitness) evolves towards zero and that two processes dominate. These processes cause the wheel counter to approach the maximum value (255). The final population has the following constellation:

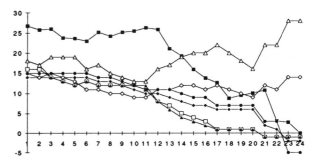

Figure 4: Evolution of the population and of the fitness function. After a competition, the 'winning' processes (process 2 and 5) take over quickly as in a phase transition.

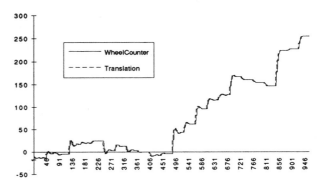

Figure 5: Evolution of the WheelCounter and translation quantity during the evolution of a forward movement behavior system.

```
process-0:  0
process-1:  0
process-2: 14
process-3:  0
process-5: 28
process-7:  0
```

The evolution of the WheelCounter (which is a sensory quantity) and the translational quantity (which is an action parameter) during the same time period are given in Figure 5. They indeed both reach their maximum values at the end of the evolutionary sequence.

Another typical run, now for acquiring the halting behavior, is shown in Figure 6. The evolution of WheelCounter and Translation speed as given in Figure 7. reflects the oscillation before the robot comes to a stand still. The processes at the end are:

```
process-0 (0):
 translation <= tendency-2 *
 approach (255, WheelCounter)/1000
process-1 (42):
 translation <= tendency-2 *
 approach (0, WheelCounter)/1000
process-4 (0):
 translation <= tendency-2 *
 approach (255, translation)/1000
```

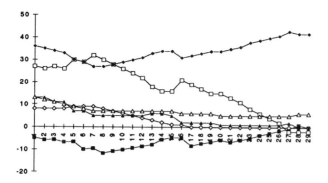

Figure 6: Competition between different processes attempting to establish a halting behavior.

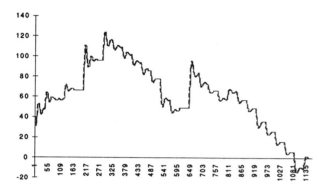

Figure 7: Evolution of the WheelCounter and translation quantities as the robot is evolving a halting behavior.

```
process-5 (0):
 translation <= tendency-2 *
 approach (-255, translation)/1000
process-6 (6):
 translation <= tendency-2 *
 approach (0, translation)/1000
```

Process-1, which moves Translation towards a state where the WheelCounter is zero dominates, and process-6, which indirectly contributes, both contain a non-empty population. Often we note a quasi-periodic oscillation before the dynamics settles on the winning processes. This is illustrates in the results of another experiment displayed in Figure 8.

In yet another evolutionary run the following solution was obtained for the halting problem:

```
process-1 (11):
 translation <= tendency-1 *
 approach(-255,translation)/1000
process-2 (11):
 translation <= tendency-1 *
 approach(255,WheelCounter)/1000
process-0 (10):
 translation <= tendency-1 *
 approach(0,translation)/1000
process-12 (7):
 translation <= tendency-1 *
```

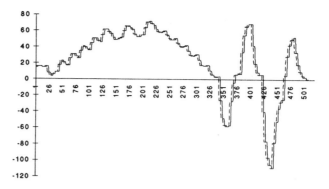

Figure 8: Evolution of the WheelCounter and translation quantities as the robot is evolving a halting behavior. We observe typically oscillations before the system settles in a solution.

```
 approach(0,WheelCounter)/1000
process-5 (6):
 translation <= tendency-1 *
 approach(-255,WheelCounter)/1000
process-6 (6):
 translation <= tendency-1 *
 approach(255,translation)/1000
```

Here we see a dynamic equilibrium between processes pushing towards the minimum (process-1 and process-5) and those pushing towards the maximum (process-2 and process-6) so that the net effect is to cause halting behavior.

We have consistently observed rapid success in generating such primitive behavioral competences where only few quantities are involved. In about 20 percent of the cases, 'good' processes are eliminated before a solution is found and then the system must rely on mutations to regenerate lost processes. But also in those cases adequate solutions are found.

Conclusion

The paper has discussed a selectionist mechanism, called the selectron, which has been designed to evolve new behavioral competences for a robotic agents. The mechanism satisfies the following constraints: (1) it can be used on-line, (2) it can be used on a real robot as it is operating in a dynamically changing real world environment, (3) there is no fixed fitness function imposed from the outside by the designer.

The selectron is based on a cooperative dynamics architecture which assumes that behavioral competence is decomposed into behavior systems. Each behavior system consists of a set of processes that establish a dynamical relationship between a set of sensory and action parameter quantities and a set of internal quantities. A behavior system, as it operates, satisfies a particular dynamic condition, for example approaching a zero wheel counter (halting behavior). To evolve a new behavior system, an initial population of processes is generated at random. These processes operate in parallel and determine the behavior of the robot in interaction with the environment. At regular time intervals, the evolution of the satisfaction of the condition is calculated and this is used to vary the population using fitness proportionate copy and elimination operations on processes. Initial experimental results demonstrate the viability of the approach.

There are many variations of the selectron algorithm which we have not investigated yet and experiments are in progress for deriving a secondary repertoire. Also a mathematical investigation remains future work. Much further work needs to be done to understand how the power of evolution can be unleashed on the development of behavior. But at the same time first results obtained so far are highly encouraging.

Acknowledgements

A lot of the work discussed in this paper is based on other work of members of the autonomous agents group. In particular, contributions by Filip Vertommen on the PDL architecture implementation on different robot platforms, Danny Vereertbrugghen for construction of the hardware, and Peter Stuer for implementation of the overall self-sufficiency experiments has proven highly valuable. Christophe Wauters has performed the experimentation on physical robots. Many discussions with David McFarland have helped to shape my ideas on useful mechanisms in ethology. Also thanks to Walter Van de Velde for discussions and comments, and for his help in preparing the final copy of this paper. The writing of this paper was partly sponsored by the IUAP Action of the Belgian Science Ministry. I am indebted to Rodney Brooks for inviting me to the conference and thus stimulating the writing of this paper.

References

[Brooks1991] Brooks, R. (1991b) Challenges for Complete Creature Architectures. In: Meyer, J-A., and S.W. Wilson (1991) From Animals to Animats. Proceedings of the First International Conference on Simulation of Adaptive Behavior. MIT Press/Bradford Books. Cambridge Ma. p. 434-443.

[Cliffs,et.al.1993] Cliff, D., P. Husbands, and I. Harvey (1993) Evolving Visually Guided Robots. In: Meyer, J-A., H.L. Roitblatt, and S.W. Wilson (1993) From Animals to Animats2. Proceedings of the Second International Conference on Simulation of Adaptive Behavior. MIT Press/Bradford Books, Cambridge Ma. p. 374-383.

[Deneubourg1993] Deneubourg, J-L, et.al. (1993) Self-organisation and life: from simple rules to global complexity. Proceedings of the Second European Conference on Artificial Life. ULB, Brussels.

[Edelman1987] Edelman, G. (1987) Neural Darwinism: The Theory of Neuronal Group Selection. Basic Books, New York.

[Holland1975] Holland, J.H. (1975) Adaptation in Natural and Artificial Systems. The University of Michigan Press, Ann Arbor, Michigan.

[Kaufmann1993] Kauffman, S.A. (1993) The origins of order: self organization and selection in evolution. Oxford University Press, Oxford.

[Koza1991] Koza, J. (1991) Evolving Emergent Wall Following Robotic Behavior sing the Genetic Programming Paradigm. In: Varela, F.J. and P. Bourgine (eds.) (1992) Toward a Practice of Autonomous Systems. Proceedings of the First European Conference on Artificial Life. MIT Press/Bradford Books, Cambridge Ma. p. 110-119.

[Langton1989] Langton, C.G. (1989) Artificial Life. Santa Fe Institute Studies in the Sciences of Complexity. Proc. Vol VI. Addison-Wesley, Reading Ma.

[McFarland1992] McFarland, D. (1992) Animals as cost-based Robots. International Studies in the Philosophy of Science, Vol 6, 2. p. 133-153.

[McFarlandBoesser1994] McFarland, D. and T. Boesser (1994) Intelligent Behavior in Animals and Robots. MIT Press/Bradford Books, Cambridge Ma.

[Ray1992] Ray, T. (1992) An Approach to the Synthesis of Life. In: Langton, C.G., C. Taylor, J.D. Farmer, and S. Rasmussen (1992) Artificial Life II. Proceedings of the Workshop on Artificial Life Held February, 1990 in Santa Fe, New Mexico. p. 325-371.

[Smithers1992] Smithers, T. (1992) Taking Eliminative Materialism Seriously: A Methodology for Autonomous Systems Research. In Varela, F.J. and P. Bourgine (eds.) (1992) Toward a Practice of Autonomous Systems. Proceedings of the First European Conference on Artificial Life. MIT Press/Bradford Books, Cambridge Ma. p. 31-40.

[Steels1993] Steels, L. (1993) Building Agents with Autonomous Behavior Systems. In: Steels, L. and R. Brooks (eds.) (1993) The 'artificial life' route to 'artificial intelligence'. Building situated embodied agents. Lawrence Erlbaum Associates, New Haven.

[Steels1994] Steels, L. (1994) The Artificial Life roots of Artificial Intelligence. Journal of Artificial Life. MIT Press. 1,1/2. p. 89-125.

[Steels1994b] Steels, L. (1994b) A case study in the behavior-oriented design of autonomous agents. Proceedings of the Simulation of Adaptive Behavior Conference. Brighton. Cambridge: MIT Press.

LONG PAPERS

Artificial Fishes with Autonomous Locomotion, Perception, Behavior, and Learning in a Simulated Physical World

Demetri Terzopoulos, Xiaoyuan Tu, and Radek Grzeszczuk

Department of Computer Science, University of Toronto
10 King's College Road, Toronto, Ontario, M5S 1A4, Canada
e-mail: {dt|tu|radek}@cs.toronto.edu

Abstract

We have developed artificial life patterned after natural organisms as evolved as those within the superclass Pisces. Our algorithms aspire to emulate not only the appearance, locomotion, and behavior of individual animals, but also the complex group behaviors evident in certain aquatic ecosystems. We model each animal holistically as an autonomous agent situated in its simulated physical domain. We develop a virtual marine world inhabited by realistic artificial fishes. Artificial fishes are able to learn how to control internal muscles in order to locomote hydrodynamically. They exhibit a repertoire of realistic behaviors that rely on their perception of their dynamic habitat.

1 Introduction

Imagine a virtual marine world inhabited by a variety of realistic fishes.[1] In the presence of underwater currents, the fishes employ their muscles and fins to gracefully swim around immobile obstacles and among moving aquatic plants and other fishes. They autonomously explore their dynamic world in search of food. Large, hungry predator fishes hunt for smaller prey fishes. Prey fishes swim around contentedly until they see a predator, at which point they take evasive action. When a predator appears in the distance, similar species of prey form schools to improve their chances of escape. When a predator approaches a school, the fishes scatter in terror. A chase ensues in which the predator selects victims and consumes them until satiated. Some species of fishes seem untroubled by predators. They find comfortable niches and forage on floating plankton when they are hungry. When compelled by their libidos, they engage in elaborate courtship rituals to secure mates.

The above scenario represents a formidable challenge in the quest for artificial life. We propose in this paper an approach which is capable of achieving the level of complexity described in the scenario, and hopefully beyond. Our approach is to create fully functional artificial animals—in this instance, artificial fishes. Artificial fishes are autonomous agents whose appearance, motivations, and complicated group interactions aspire to be as faithful as possible to nature's own. To this end, we pursue a bottom-up, compositional approach in which we

model not just form and superficial appearance, but also the basic physics of the animal within its environment, its means of locomotion, its perception of its world, its behavior, and its ability to learn. The holistic nature of our approach to synthesizing artificial fishes is crucial to achieving realism.

The long-term goal of our research is a computational theory that can potentially account for the interplay of physics, locomotion, perception, behavior, and learning in higher animals. A good touchstone of such a theory is its ability to produce visually convincing results in the form of realistic computer animation. Indeed, our original motivation was the development of algorithms that can produce realistic animation of animals with minimal intervention from an animator. For example, our animation "Go Fish!" [Tu *et al.*, 1993] shows a colorful variety of artificial fishes feeding in translucent water. A sharp hook on a line descends towards the hungry fishes and attracts them. A hapless fish, the first to bite the bait, is caught and dragged to the surface. The color plates are stills from our recent animation "The Undersea World of Jack Cousto." Plate 1a shows a variety of animated artificial fishes. The redish fish are engaged in a mating ritual, the greenish fish is a predator hunting for small prey, the remaining fishes are foraging on plankton (white dots). Note the dynamic seaweeds growing from the ocean bed. In Plate 1b, the large male in the foreground is courtship dancing with the female (top). Note the school of prey fish in the background. Schooling behavior is a common subterfuge for avoiding predators. Plate 1c shows a shark stalking the school. The detailed motions of the artificial fishes in the animations emulate the complexity and unpredictability of movement of their natural counterparts, and this enhances the beauty of the animation.

1.1 Background

Our holistic approach to developing artificial fishes is compatible with the "animat" approach proposed by Wilson [1991]. Artificial fishes are animats of unprecedented sophistication. They are virtual robots in a continuous 3D virtual world. Their physics-based modeling, motor control, perception, and behavioral simulation present challenges paralleling those encountered in building physical autonomous agents that couple perception to action through behavior (see, e.g., the compilation [Maes, 1991]). This sentiment is also expressed in the work of Beer [1990] (also Beer *et al.* in [Maes, 1991]) who have synthesized a virtual insect, a cockroach, with several behaviors in its 2D world. The paper by Brooks in [Maes,

---

[1] "Fish" is both singular and plural; when plural, it refers to more than one fish *within* the same species. The plural "fishes" is used when two or more species are involved [Wilson and Wilson, 1985].

1991] describes a physical insect robot "Genghis" that can locomote over irregular terrain.

Our work aims at animals more highly evolved and complex than insects. We attempt to emulate convincingly the external appearance of the animal as well so that the computational model will be visually convincing. The behavioral repertoire of the artificial fish is much more extensive than the schooling behaviors that were simulated in the behavioral animation work of Reynolds [1987] (recently Matarić [1994] has demonstrated flocking behaviors with physical robots). To deal with the broader behavioral repertoire, we exploit ideas from classical ethology [Tinbergen, 1950, Lorenz, 1973, Mcfarland, 1985, Adler, 1975]. Tinbergen's landmark studies of the three-spined stikleback highlight the great diversity of piscatorial behavior, even within a single species. We achieve the nontrivial patterns of behavior outlined in the introductory paragraph of this paper in stages. First, we implement primitive reflexive behaviors, such as obstacle avoidance, that directly couple perception to action [Braitenberg, 1984]. Then we combine the primitive behaviors into motivational behaviors whose activation depends also on the artificial fish's mental state, including hunger, libido, and fear.

Useful behavior is supported by perception of the environment as much as it is by action. Reynold's "boids" maintained flocking formations through perception of other nearby "boids" [Reynolds, 1987]. Our artificial fishes are currently able to sense their world through simulated visual perception within a deliberately limited field of view. Subject to the natural limitations of occlusion, they can sense lighting patterns, determine distances to objects, and identify objects. Furthermore, they are equipped with secondary nonvisual modalities, such as the ability to sense the local water temperature.

At its lowest level, our work relies upon computational physics. We model the biomechanics of a broad class of fishes and their muscle-based locomotion abilities that exploit the physics of their liquid medium [Alexander, 1992]. The mechanical model that we develop is inspired by the simple but surprisingly effective computer graphics model of snake and worm dynamics proposed by Miller [1988].

We have provided artificial fish with algorithms that enable them to learn automatically from first principles how to achieve hydrodynamic locomotion by controlling their internal muscle actuators. The locomotion learning algorithm that we describe is more continuous and closer connected to actuation than most of the animat "behavior learning" algorithms surveyed in [Meyer and Guillot, 1991]. Our multilevel reinforcement learning procedure first performs a global search for actuator activation functions that produce efficient locomotion. The process then abstracts these activation functions into a highly compact representation. The representation emphasizes the natural periodicities of the derived muscle actions and makes explicit the coordination among multiple muscles that leads to effective locomotion. Finally, the artificial fish can put into practice the compact, efficient controllers that it has learned. The learning technique enables it to accomplish higher level tasks guided by sensory perception—for example, it can maneuver to reach a visible target.

1.2 Functional Overview of the Artificial Fish

Fig. 1 shows an overview of an artificial fish situated in its world, illustrating the motor, perception, and behavior subsystems.

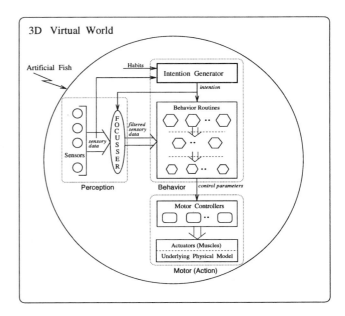

Figure 1: Control and information flow in artificial fish.

The motor system, comprising the actuators and a set of motor controllers (MCs), drives the dynamic model of the fish. We have crafted a mechanical model that represents a good compromise between anatomical consistency, hence realism, and computational efficiency. Our model is rich enough so that we can build MCs by gleaning information from the animal biomechanics literature [Alexander, 1992]. The MCs are parameterized procedures. Each is dedicated to carrying out a specific motor function, such as "swim forward" or "turn left." They translate natural control parameters such as the forward speed or angle of the turn into detailed muscle actions.

The perception system relies on a set of on-board virtual sensors to provide sensory information about the dynamic environment. The system includes a perceptual attention mechanism which allows the artificial fish to train its sensors at the world in a task-specific way, hence filtering out sensory information superfluous to its current behavioral needs. For example, the artificial fish attends to sensory information about nearby food sources when foraging.

The behavior system of the artificial fish mediates between its perception system and its motor system. An intention generator, the fish's "cognitive" center, harnesses the dynamics of the perception-action cycle. The innate character of the fish is established by a set of habit parameters that determine whether or not it is male/female, likes darkness, etc. The intention generator combines the habits with the incoming stream of sensory information to generate dynamic goals for the fish, such as to hunt and feed on prey. It ensures that goals have some persistence by exploiting a single-item memory. The intention generator also controls the perceptual attention mechanism. At every simulation time step, the intention generator activates behavior routines that attend to sensory information and compute the appropriate motor control parameters to carry the fish one step closer to fulfilling its current intention. Primitive behavior routines, such as obstacle avoidance, and more sophisticated motivational behavior routines, such as mating, implement the artificial fish's repertoire of behaviors.

2 Realistic Modeling of Form and Appearance

To achieve realism, our artificial fish model must first represent the form and appearance of real fishes with reasonable fidelity. To this end, we have perused photographs of real fishes, such as those shown in Fig. 2(a), and have built 3D geometric models of several different species using NURBS surfaces (Fig. 2(b)).

The next step is to map realistic textures onto the geometric fish model (Fig. 2(e)). We extract natural textures from digital images of the fish photos, employing a "snake-grid" tool to determine appropriate texture map coordinates in the image. Snakes [Kass et al., 1987] are interactive deformable contours that are subject to a force field derived from an image. The force field attracts them towards interesting image features such as intensity edges. A collection of mutually constrained snakes forms a deformable grid (Fig. 2(c–d)). The snake-grid floats freely over an image and it can be pulled into position using the mouse. When its border approaches the intensity edges that demarcate the fish from its background in the image, the border snakes lock on and adhere to these edges. The remaining snakes in the grid relax elastically to cover the imaged fish body with a favorable, nonuniform texture map coordinate system (Fig. 2(d)). The snake crossing points serve as texture map coordinates for the NURBS surfaces.

3 Physics-Based Fish Model and Locomotion

Studies into the dynamics of fish locomotion show that most fishes use their caudal fin as the primary motivator [Webb, 1989]. Caudal swimming normally uses posterior muscles on either side of the body, while turning normally uses anterior muscles. To synthesize realistic fish locomotion we have designed a dynamic fish model consisting of 23 nodal point masses and 91 springs. The spring arrangement maintains the structural stability of the body while allowing it to flex. Twelve of the springs running the length of the body also serve as simple muscles (Fig. 3).

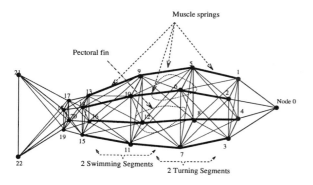

Figure 3: Dynamic fish model. Nodes are mass points. Lines are springs (at their rest lengths). Bold lines are muscle springs.

3.1 Mechanics

The mechanics of the spring-mass model are specified as follows: Let node i have mass m_i, position $\mathbf{x}_i(t) =$

$[x_i(t), y_i(t), z_i(t)]$, velocity $\mathbf{v}_i(t) = d\mathbf{x}_i/dt$, and acceleration $\mathbf{a}_i(t) = d^2\mathbf{x}_i/dt^2$. Let spring S_{ij} connect node i to node j. Denote its spring constant as c_{ij} and natural, rest length as l_{ij}. Its deformation is $e_{ij}(t) = \|\mathbf{r}_{ij}\| - l_{ij}$, where $\mathbf{r}_{ij} = \mathbf{x}_j(t) - \mathbf{x}_i(t)$. The force S_{ij} exerts on node i is $\mathbf{f}_{ij}^s = c_{ij}e_{ij}(t)\mathbf{r}_{ij}/\|\mathbf{r}_{ij}\|$ (and it exerts the force $-\mathbf{f}_{ij}^s$ on node j). The Lagrange equations of motion of the dynamic fish are:

$$m_i \frac{d^2\mathbf{x}_i}{dt^2} + \rho_i \frac{d\mathbf{x}_i}{dt} - \mathbf{w}_i = \mathbf{f}_i^w; \qquad i = 0, ..., 22, \quad (1)$$

where ρ_i is the damping factor, $\mathbf{w}_i(t) = \sum_{j \in N_i} \mathbf{f}_{ij}^s(t)$ is the net internal force on node i due to springs connecting it to nodes $j \in N_i$, where N_i is the index set of neighboring nodes. Finally, $\mathbf{f}_i^w$ is the external (hydrodynamic) force on node i.

To integrate the differential equations of motion, we employ a numerically stable, implicit Euler method [Press et al., 1986]. The method assembles the sparse stiffness matrix for the spring-mass system in "skyline" storage format. The matrix is factorized once at the start of the simulation and then resolved at each time step.[2]

We couple the control points of the aforementioned texture mapped NURBS body model to the time-varying positions of the mass points (the nodes in Fig. 3(a)), such that the fish body deforms in accordance with the simulated dynamics of the actuated spring-mass system.

3.2 Swimming Using Muscles and Hydrodynamics

The artificial fish moves as a real fish does, by contracting its muscles. If S_{ij} is a muscle spring, it is contracted by decreasing the rest length l_{ij}. For convenience, we assign a minimum contraction length l_{ij}^{min} to the muscle spring and express the contraction factor as a number in the range $[0, 1]$. The characteristic swinging of the fish's tail can be achieved by periodically contracting the swimming segment springs on one side of the body while relaxing their counterparts on the other side.

When the fish's tail swings, it sets in motion a volume of water. The inertia of the displaced water produces a reaction force normal to the fish's body proportional to the volume of water displaced per unit time, which propels the fish forward (Fig. 4(a)). Under certain assumptions, the instantaneous force on the surface S of a body due to a viscous fluid is approximately proportional to $-\int_S (\mathbf{n} \cdot \mathbf{v})\mathbf{n}\, dS$, where $\mathbf{n}$ is the unit outward normal function over the surface and $\mathbf{v}$ is the relative velocity function between the surface and the fluid. For efficiency, we triangulate the surface of the dynamic fish model between the nodes and approximate the force on each planar triangle as $\mathbf{f} = \min[0, -A(\mathbf{n} \cdot \mathbf{v})\mathbf{n}]$, where A is the area of the triangle and $\mathbf{v}$ is its velocity relative to the water. The $\mathbf{f}_i^w$ variables at each of the three nodes defining the triangle are incremented by $\mathbf{f}/3$.

---

[2]In our simulation: $m_i = 1.1$ for $i = 0$ and $13 \le i \le 19$; $m_i = 6.6$ for $1 \le i \le 4$ and $9 \le i \le 12$; $m_i = 11.0$ for $5 \le i \le 8$, and $m_i = 0.165$ for $i = 21, 22$. The cross springs (e.g., c_{27}) which resist shearing have spring constants $c_{ij} = 38.0$. The muscle springs (e.g., c_{26}) have spring constants $c_{ij} = 28.0$, and $c_{ij} = 30$ for the remaining springs. The damping factor $\rho_i = 0.05$ in (1) and the time step used in the Euler time-integration procedure is 0.055.

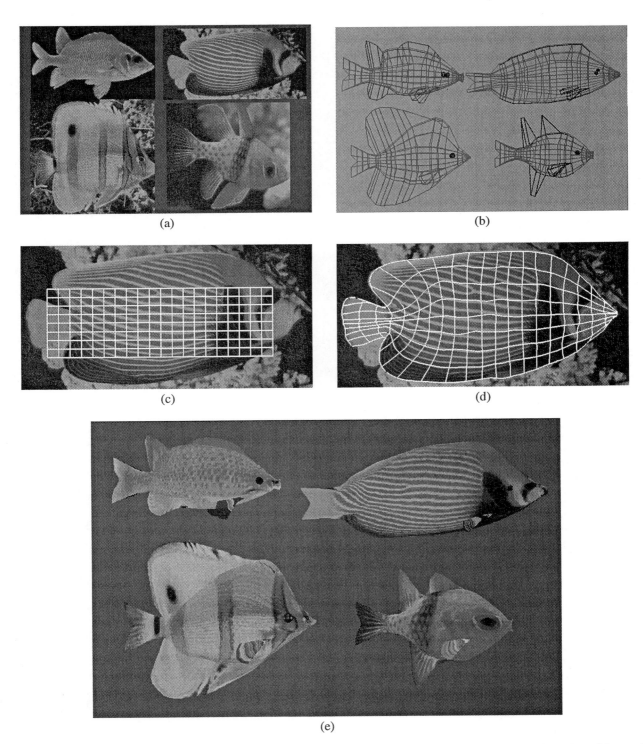

Figure 2: (a) Digitized images of fish photos. (b) 3D NURBS surface fish bodies. Initial (c) and final (d) snake-grid covering an imaged fish body. (e) Texture mapped 3D fish models.

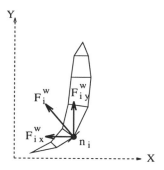

Figure 4: Hydrodynamic locomotion. (a) With tail swinging towards positive X axis, reaction force $\mathbf{F}_i^w$ at point n_i acts along the inward normal. Component $\mathbf{F}_{ix}^w$ resists the lateral movement, while $\mathbf{F}_{iy}^w$ is forward thrust. Aggregate thrust propels fish towards positive Y axis.

3.3 Motor Controllers

Currently the artificial fish has three MCs. The swim-MC produces straight swimming, while the left-turn-MC and right-turn-MC execute turns. The MCs prescribe muscle contractions to the mechanical model. The swimming MC controls the swimming segment muscles (see Fig. 3), while the turning MCs control the turning segment muscles.

According to [Webb, 1989], the swimming speed of most fishes is roughly proportional to the amplitude and frequency of the periodic lateral oscillation of the tail, below certain threshold values. Our experiments with the mechanical model agree well with these observations. Both the swimming speed and the turn angle of the fish model are approximately proportional to the contraction amplitudes and frequencies/rates of the muscle springs.

The swim-MC (swim-MC(speed) $\mapsto \{r_1, s_1, r_2, s_2\}$) converts a swim speed parameter into contraction amplitude and frequency control parameters for the anterior (r_1, s_1) and posterior (r_2, s_2) swim segments. One pair of parameters suffices to control each of the two swim segments because of symmetry—the four muscle springs have identical rest lengths and minimum contraction lengths, identical spring constants, and the contractions of the muscle spring pairs on opposite sides are exactly out of phase. Moreover, the swim-MC produces periodic muscle contractions in the posterior swim segment which lag 180 degrees behind those of the anterior swim segment; hence the mechanical model displays a sinusoidal body shape as the fish swims (see [Webb, 1989]).

By experimenting, we have found a set of four maximal parameters, $\hat{r}_1$, $\hat{s}_1$, $\hat{r}_2$ and $\hat{s}_2$, which produce the fastest swimming speed. The swim-MC generates slower swim speeds by specifying parameters that have values between 0 and the maximal parameters. For example, $\{0.8\hat{r}_1, \hat{s}_1, 0.7\hat{r}_2, \hat{s}_2\}$ results in a slower-swimming fish.

As mentioned earlier, most fishes use their anterior muscles for turning, and the turn angle is approximately proportional to the degree and speed of the anterior bend, up to the limit of the fish's physical strength [Webb, 1989]. The artificial fish turns by contracting and expanding the springs of the turning segments (Fig. 3) in similar fashion. For example, a left turn is achieved by quickly contracting the left side springs of the segments and relaxing those on the right side. This effectively deflects the fish's momentum and brings it into the

desired orientation. Then the contracted springs are restored to their rest lengths at a slower rate, so that the fish regains its original shape with minimal further change in orientation.

Similarly, the left and right turn MCs (turn-MC(angle) $\mapsto \{r_0, s_0, r_1, s_1\}$) convert a turn angle to control parameters for the anterior and posterior turning segments to execute the turn (note that the posterior turning segment also serves as the anterior swim segment). Through experimentation, we established 4 sets of parameter values $P_i = \{r_0^i, s_0^i, r_1^i, s_1^i\}$ which enable the fish to execute natural looking turns of approximately 30, 45, 60, and 90 degrees. By interpolating the key parameters, we define a steering map that allows the fish to generate turns of approximately any angle up to 90 degrees. Turns greater than 90 degrees are composed as sequential turns of lesser angles.

3.4 Pectoral Fins

On most fishes, the pectoral fins control pitching (the up-and-down motion of the body) and yawing (the side-to-side motion). The pectorals can be held close to the body to increase speed by reducing drag or they can be extended to serve as a brake by increasing drag [Wilson and Wilson, 1985]. Many reef fishes use a pectoral swimming style to achieve very fine motion control when foraging, including backwards motions, by keeping their bodies still and using their pectorals like oars.

The artificial fish has a pair of pectoral fins which enable it to navigate freely in its 3D world. The pectoral fins function in a similar, albeit simplified, manner to those on real fishes. Instead of creating a detailed physics-based model of the pectoral fins, we are content to simulate only their dynamic effect on the locomotion of the fish. This is because for our purposes the detailed movement of the pectoral fins is of lesser interest than the movement of the fish body. Furthermore, we wish to simplify the fish model and its numerical solution.

The pectoral fins (Fig. 5) work by applying reaction forces to nodes in the midsection, i.e. nodes $1 \leq i \leq 12$ (see Fig. 3).

Figure 5: The pectoral fins

The pectoral fins are analogous to the airfoils of an airplane. Pitch, yaw, and roll control stems from changing their orientations relative to the body; i.e., the angle $\pi/4 \leq \gamma \leq \pi$. Assuming that a fin has an area A, surface normal $\mathbf{n}$ and the fish has a velocity $\mathbf{v}$ relative to the water (Fig. 5), the fin force is $F_f = -A(\mathbf{n} \cdot \mathbf{v})\mathbf{n} = -A(\|\mathbf{v}\| \cos\gamma)\mathbf{n}$ which is distributed equally to the 12 midsection nodes. When the leading edge of a fin is elevated, a lift force is imparted on the body and the fish ascends, and when it is depressed a downward force is exerted and the fish descends. When the fin angles differ the fish yaws and rolls. The artificial fish can produce a braking effect by angling its fins to decrease its forward speed (i.e. $\gamma = \pi$). This motion control is useful, for instance, in maintaining schooling patterns.

4 Learning Muscle-Based Locomotion

We have discussed above how locomotion controllers may be carefully hand crafted using knowledge gleaned from the piscatorial biomechanics literature and long hours of experimentation. In this section, we consider the following general question: Given a physics-based model of an animal with internal muscle actuators capable of producing locomotion, such as the fish model of Fig. 3, is it possible for the model to learn from first principles how to control its actuators in order to locomote in a natural fashion? Furthermore, can it put into practice the controllers it has learned so as to accomplish higher level tasks guided by sensory perception? We demonstrate affirmative answers to both questions by applying a form of reinforcement learning to our model.

4.1 Learning Strategy

We formulate a bottom-up strategy for learning muscle controllers. At the initial stage of the learning process, the artificial fish has a fully functional body, but no "brain" — i.e., it does not know how best to contract its muscles in order to locomote. Through repeated forward simulations with perturbed muscle activation functions, the artificial fish learns how to locomote with increasing efficiency, by preserving those changes that improve its locomotion. Repeated improvements eventually produce locomotion that is very efficient.

When an adequate degree of optimization has been achieved in the low-level learning phase, the learning strategy then abstracts the activation functions into a highly compact representation. The representation drastically reduces the dimensionality of the learning problem by using basis functions that make explicit the natural periodicities of the derived muscle actions and the coordination among multiple muscles that yields effective locomotion. The artificial fish thus abstracts the low-level muscle activation functions that it has learned and associates them with specific higher-level tasks that it needs to perform. Finally, it can put into practice the compact, efficient controllers that it has learned, to accomplish higher level tasks guided by sensory perception—for example, it can locomote and execute turning maneuvers to reach a visible target. The learned controllers can subsequently be used by higher-level behaviors such as hunting.

4.2 Low-Level Learning

At the foundation of our approach lies the notion that natural motion patterns are energetically efficient. This allows us to reduce the problem of learning realistic locomotion into a problem of optimizing an objective function, for which various solution techniques are available.

The objective function takes the form

$$E(\mathbf{u}(t)) = \int_{t_0}^{t_1} \left(\mu_1 E_u(\mathbf{u}(t)) + \mu_2 E_v(\mathbf{v}(t)) \right) \, dt, \quad (2)$$

a weighted sum, with weighting variables μ_1, μ_2, of a term E_u that evaluates the vector of muscle actuator control functions $\mathbf{u}(t)$ and a term E_v that evaluates the resulting trajectory $\mathbf{v}(t)$ of the artificial fish. The actuation functions $\mathbf{u}(t)$ dictate the lengths l_{ij} of muscle springs in the dynamic model (see Sec. 3.2). Note that $\mathbf{v}(t)$ (and hence E) is computed through forward simulation of the dynamic model over a time interval $t_0 \le t \le t_1$ with the actuation function inputs $\mathbf{u}(t)$.

We may wish to guide the optimization by encouraging the smoothness of $\mathbf{u}$ through the term E_u. The rationale is that energy efficient actuations are usually not chaotic. We use two functions to encourage smoothness:

$$E_{u1} = -\frac{1}{2} \left| \frac{d\mathbf{u}}{dt} \right|^2 \quad \text{or} \quad E_{u2} = -\frac{1}{2} \left| \frac{d^2\mathbf{u}}{dt^2} \right|^2. \quad (3)$$

These terms are potential energy densities of linear and cubic splines in time, respectively. The former penalizes muscle effort much more than the latter.

The criterion E_v for a good trajectory that we used most often in our learning experiments was the final distance to a target location. Depending on the task, other possible criteria are: the closeness of match to a given speed, the distance it moves away from a target, etc.

Learning low level control involves the application of simulated annealing to optimize (2) [Press *et al.*, 1986]. Simulated annealing is applied after discretizing the actuator control functions $\mathbf{u}(t)$ in the time interval under consideration to obtain the set of discrete control points $\mathbf{u}_i = \mathbf{u}(t_i)$ for $1 \le i \le N$ (typically, N is set to 15 time samples, and the continuous $\mathbf{u}(t)$ is recovered through linear or cubic spline interpolation of the $\mathbf{u}_i$). The annealing algorithm repeatedly perturbs the $\mathbf{u}_i$ to modify the actuator activation functions that control the muscles in the fish. It retains those perturbations that produce increasingly better locomotion as measured by the objective function E, and sometimes accepts those that don't to escape local minima. Note that after performing a forward simulation using $\mathbf{u}(t)$, the artificial fish can evaluate E using its on-board sensors, so learning proceeds autonomously.

Fig. 6 shows a race between six identical artificial leopard sharks. The upper shark has completed only 90 annealing steps, which results in muscle control functions that are essentially random and achieve negligible locomotion. Sharks below it have learned for progressively longer periods of time (1350 more annealing steps). After learning for 6840 annealing steps, the bottommost shark locomotes the best and wins the race.

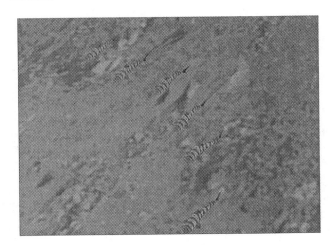

Figure 6: Race between sharks that have learned locomotion.

4.3 Abstraction of High-Level Controllers

The abstraction of higher level muscle controllers is a dimensionality reducing change of representation. More specifically, it seeks to compress the information content of the many control points to a compact form in terms of a few global basis functions; i.e., it tries to represent the control functions as accurately as possible in the form $\mathbf{u}(t) \approx \sum_{i=0}^{M} \alpha_i \mathbf{B}_i(t)$, where $\mathbf{B}_i$ are basis functions, α_i are scalar quantities, and M is a small number.

Since natural locomotion patterns are generally periodic [Pearson, 1991], the Fourier basis is a reasonable choice. We employ the FFT to perform the change in basis. If the Fourier space is a suitable representation that captures the temporal structure of the control functions, the dimensionality reduction can be achieved trivially by eliminating all basis functions whose coefficients α_i in the above approximation formula are negligible. This will result in a small set of M significant basis functions, usually 1 or 2, with associated coefficients that constitute the abstracted controller.

The artificial fish can now be trained to perform several locomotion tasks, such as swimming forward at different speeds and executing turns of different radii. After it has abstracted controllers for these tasks, it can construct a speed and steering map by interpolating across amplitudes, frequencies, and phases of the basis functions in the set abstracted controllers. Finally, it can put these learned abstractions into practice to accomplish higher level tasks, such as target tracking. Fig. 7 shows a shark model that has been trained to swim from target to target (balls); it has just swum from the left to the far target and has now turned to proceed to the near target.

Figure 7: Trained shark swimming between targets.

5 Sensory Perception

The perception system of the artificial fish, illustrated in Fig. 1, comprises a set of virtual on-board sensors and a perceptual focusser. Currently the artificial fish is equipped with two sensors that provide information about the dynamic environment—a temperature sensor that measures the ambient (virtual) water temperature at the center of the artificial fish's body and a cyclopean vision sensor.

5.1 Vision Sensor

We have not attempted to emulate the highly evolved vision system of real fishes. Instead, we have incorporated a simple cyclopean vision sensor into the fish model. The cyclopean vision sensor has a 300 degree spherical angle field of view extending to an effective radius V_r appropriate to the visibility of the translucent water (Fig.8). An object is "seen" if any part of it enters this view volume and it is not fully occluded by another object.

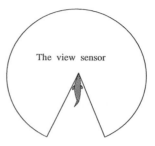

Figure 8: Vision sensor. Visual perception is limited to 300 degree solid angle.

The artificial fish's vision sensor has access to the geometry, material property, and illumination information that is available to the graphics pipeline for rendering purposes. In addition, the vision sensor can interrogate the object database to identify nearby objects and interrogate the physical simulation to obtain information such as the instantaneous velocities of objects of interest. In this way, the vision sensor extracts from the 3D virtual world only some of the most useful information that piscatorial visual processes can provide real fishes about their world, such as overall brightness, and the colors, sizes, distances, and identities of some objects.

A more realistic emulation of piscatorial visual processes would involve the application of various computer vision algorithms [Horn, 1986] to extract information from "retinal" images (and associated z-buffers) of the 3D world rendered from the vantage point of the artificial fish's cyclopean vision sensor. Currently, the fish determines the overall brightness of its environment by computing the mean intensity of the retinal image. Fig. 9 shows examples of retinal images acquired by a fish "witnessing" another fish being baited by a fishing line.

6 Behavioral Modeling

The artificial fish's behavior system runs continuously within the simulation loop. At each time step the intention generator issues an intention based on the fish's habits, mental state, and incoming sensory information. It then chooses and executes a behavior routine which in turn runs the appropriate motor controllers. It is important to note that the behavior routines are incremental by design. Their job is to get the artificial fish one step closer to fulfilling the intention during the current time step. The intention generator employs a memory mechanism to avoid dithering.

6.1 Habits and Mental State

The innate character of the fish is determined by a set of habit parameters that determine whether or not it likes brightness, darkness, cold, warmth, schooling, or is a male/female, etc.

The artificial fish has three mental state variables, hunger H, libido L, and fear F. The range of each variable is $[0, 1]$, with higher values indicating a stronger urge to eat, mate and

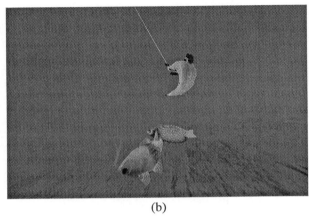

Figure 9: Fisheye view of the world showing fishing line (a) and hooked fish (b).

avoid danger, respectively. The variables are calculated as follows:

$$H(t) = \min[1 - n^e(t)R(\Delta t^H)/\alpha, 1],$$
$$L(t) = \min[s(\Delta t^L)(1 - H(t)), 1],$$
$$F(t) = \min\left[\sum_i F^i, 1\right], \text{ where } F^i = \min[D_0/d^i(t), 1];$$

where t is time, $n^e(t)$ is the amount of food consumed as measured by the number of food particles or prey fishes eaten, $R(x) = 1 - p_0 x$ with constant p_0 is the digestion rate, Δt^H is the time since the last meal, α is a constant that dictates the appetite of the fish (bigger fishes have a larger α), $s(x) = p_1 x$ with constant p_1 is the libido function, Δt^L is the time since the last mating, $D_0 = 100$ is a constant, and F^i and d^i are, respectively, the fear of and distance to sighted predator i. Nominal constants are $p_0 = 0.00067$ and $p_1 = 0.0025$. Certain choices can result in ravenous fishes (e.g, $p_0 = 0.005$) or sexual mania (e.g., $p_1 = 0.01$).

6.2 Intention Generator

Fig. 10 illustrates the generic intention generator which is responsible for the goal-directed behavior of the artificial fish in its dynamic world.

The intention generator first checks the sensory information stream to see if there is any immediate danger of collision.

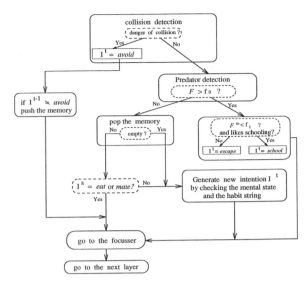

Figure 10: Generic intention generator (simplified). Set of intentions: { *avoid*, *escape*, *school*, *eat*, *mate*, *leave*, *wander* }. f_0 and f_1 are thresholds with $f_0 < f_1$.

If any object penetrates the fish's collision sensitivity region (a bounding box) then the intention I generated is to *avoid* collision. A large sensitivity region results in a 'timid' fish that takes evasive action to avoid a potential collision well in advance, while a tight sensitivity region yields a 'courageous' fish that takes evasive action only at the last second.

If there is no immediate danger of collision, the neighborhood is searched for predators, the fear state variable F and the most dangerous predator m for which $F^m \geq F^i$ are calculated. If the total fear $F > f_0$ (where $0.1 \leq f_0 \leq 0.5$ is a threshold value) evasive action is to be taken. If the most dangerous predator is not too threatening (i.e. $F^m < f_1$ where $f_1 > f_0$) and the fish has a schooling habit, then the *school* intention is generated, otherwise the *escape* intention is generated.

If fear is below threshold, the hunger and libido mental state variables H and L are calculated. If the greater of the two exceeds a threshold $0 < r < 0.5$, the intention generated will be to *eat* or *mate* accordingly.

If the above test fails, the intention generator accesses the ambient light and temperature information from the perception system. If the fish's habits dictate contentment with the ambient conditions, the intention generated will be to *wander* about, otherwise it will be to *leave* the vicinity.

Note that after the intention generator chooses an intention, it invokes the perceptual focus mechanism. For example, when the *avoid* intention is generated, the perception focusser is activated to locate the positions of the obstacles, paying special attention to the most dangerous one, generally the closest. Then the focusser computes qualitative constraints, such as *obstacle to the left* $\Rightarrow$ *no left turn*. The focusser passes only the position of the most dangerous obstacle along with these constraints to the behavior routines. When the intention of a male fish is to *mate*, the focusser targets the most desirable female fish; when the intention is to *escape* from predators, only the information about the most threatening predator is passed to the next layer; etc.

In a complex dynamic world, the artificial fish should have some persistence in its intentions, otherwise it will tend to dither, perpetually switching goals. If the current behavior is interrupted by a high priority event, the intention generator is able to store, in a single-item short term memory, the current intention and some associated information that may be used to resume the interrupted behavior. Persistence is particularly important in making long duration behaviors such as feeding, schooling, and mating more robust. Suppose, for example, that the current behavior is mating and an imminent collision is detected with another fish. This causes an *avoid* intention and the storage of the *mate* intention (we refer to the stored intention as I^s) along with the identity of the mating partner. After the obstacle has been cleared, the intention generator commands the focusser to generate up-to-date heading and range information about the mating partner, assuming it is still in viewing range.

Our design of the intention generator and focusser simplifies the modification of existing personalities and behaviors and the addition of new ones. For example, we can create artificial fishes with different persistences by augmenting the focusser with a new positive threshold. Suppose the current intention of a predator fish is to *eat* and let the distance to some currently targeted prey be l_c and the distance to some other prey be l_n. If $l_c - l_n$ is greater than the threshold, the fish will target the new prey. Varying the threshold will vary the fish's level of persistence. The same heuristic can be applied to mates when the fish is trying to *mate*. One can make the fish 'fickle' by setting the value of the threshold close to zero or make it 'devoted' by setting a large value.

6.3 Behavior Routines

Once the intention generator selects an intention it attempts to satisfy the intention by passing control to a behavior routine along with the data from the perception focusser. The artificial fish currently includes eight behavior routines: *avoiding-static-obstacle*, *avoiding-fish*, *eating-food*, *mating*, *leaving*, *wandering*, *escaping*, and *schooling* which serve the obvious purposes. The behavior routine uses the focused perceptual data to select an MC and provide it with the proper motor control parameters. We now briefly describe the function of the routines.

The *avoiding-static-obstacle* and *avoiding-fish* routines operate in similar fashion. Given the relative position of the obstacle, an appropriate MC (e.g. *left-turn-MC*) is chosen and the proper control parameters are calculated subject to the constraints imposed by other surrounding obstacles. For efficiency the *avoid-fish* routine treats the dynamic obstacle as a rectangular bounding box moving in a certain direction. Although collisions between fishes cannot always be avoided, bounding boxes can be easily adjusted such that they almost always are, and the method is very efficient. An enhancement would be to add collision resolution.

The *eating-food* routine tests the distance d from the fish's mouth to the food (see Fig. 5). If d is greater than some threshold value, the subroutine *chasing-target* is invoked.[3] When d is less than the threshold value the subroutine *suck-in* is activated where a "vacuum" force (to be explained in

---

[3]The *chasing-target* subroutine guides a fish as it swims towards a goal. It plays a crucial role in several behavior routines, but we cannot give the details for lack of space.

Sec. 6.1) is calculated and then exerted on the food.

The *mating* routine invokes four subroutines: *looping*, *circling*, *ascending* and *nuzzling* (see Sec. 7.3 for details). The *wandering-about* routine sets the fish swimming at a certain speed by invoking the swim-MC, while sending random turn angles to the turn-MCs. The *leaving* routine is similar to the *wandering-about* routine. The *escaping* routine chooses a suitable MC according to the relative position, orientation of the predator to the fish. The *schooling* routine will be discussed in Sec. 7.2.

7 Artificial Fish Types

The introductory paragraph of the paper described the behavior of three types of artificial fishes—predators, prey, and pacifists. This section presents their implementation details.

7.1 Predators

Fig. 11 is a schematic of the intention generator of a predator, which is a specialized version of Fig. 10. To simplify matters, predators currently are not preyed upon by other predators, so they perform no predator detection, and *escape*, *school*, and *mate* intentions are disabled ($F = 0$, $L = 0$). Since predators cruise perpetually, the *leave* intention is also disabled.

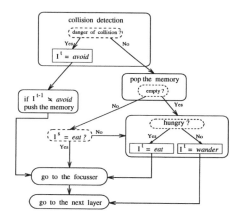

Figure 11: The intention generator of a Predator

Generally prey is in less danger of being hunted when it is far away from the predator, or is in a school, or is behind the predator. A predator chases prey k if the cost $C_k = d_k(1 + \beta_1 S_k + \beta_2 E_k/\pi)$ of reaching it is minimal. Here, d_k is the distance between the mouth of the predator and the center of prey k's body, $S_k = 1$ if prey k is in a school of fishes, otherwise $S_k = 0$, and the angle $E_k \in [0, \pi]$ (Fig. 5) measures the turning cost. β_1 and β_2 are parameters that tune the contributions of S_k and E_k. We use $\beta_1 = 0.5$ and $\beta_2 = 0.2$ in our implementation of the focusser. Plate 1c shows a shark predator stalking a school of prey fish.

Most teleost fishes do not bite on their victims like sharks do. When a fish is about to eat it swims close to the victim and extends its protrusile jaw, thus creating a hollow space within the mouth. The pressure difference between the inside and the outside of the mouth produces a vacuum force that sucks into the mouth the victim and anything else in the nearby water. The predator closes its mouth, expels the water through the gills, and grinds the food with pharyngeal jaws [Wilson and Wilson, 1985]. We simulate this process by enabling the artificial fish to open and close its mouth kinematically. To suck in prey, it opens its mouth and, while the mouth is open, exerts vacuum forces on fishes (the forces are added to external nodal forces $\mathbf{f}_i$ in equation (1) and other dynamic particles in the vicinity of the open mouth, drawing them in (Fig. 12).

Figure 12: Predator ingesting prey.

7.2 Prey

The intention generator of a prey fish is given by specializing the generic intention generator of Fig. 10.

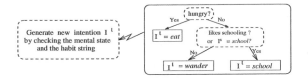

Figure 13: Modified portion of intention generator for prey.

Schooling and evading predators are the two distinct behaviors of prey. We briefly describe the implementation of the *schooling* behavior. Schooling is a complex behavior where all the fishes swim in generally the same direction. Each fish constantly adjusts its speed and direction to match those of other members of the school. They establish a certain distance from one another, roughly one body length from neighbors, on average [Wilson and Wilson, 1985]. Each member of a school of artificial fish acts autonomously, and the schooling behavior is achieved through sensory perception and locomotion. An inceptive school is formed when a few fish swim towards a lead fish. Once a fish is in some proximity to some other schooling fish, the *schooling* behavior routine outlined in Fig. 14 is invoked.

The intention generator prevents schooling fish from getting too close together, because the *avoid* collision intention has highest precedence. To create more compact schools, the collision sensitivity region of a schooling fish is decreased, once it gets into formation. When a large school encounters an obstacle, the autonomous behavior of individual fishes trying to avoid the obstacle may cause the school to split into

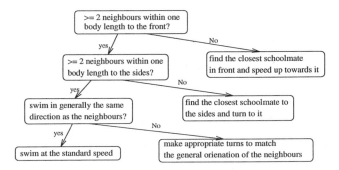

Figure 14: Schooling behavior routine.

two groups and rejoin once the obstacle is cleared and the *schooling* behavior routine regains control (Fig. 15).

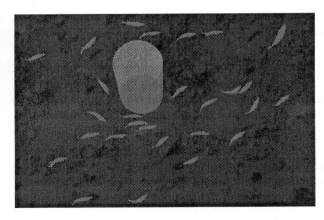

Figure 15: School of fish swimming past cylindrical obstacle (from lower left to upper right). Schooling behavior is interrupted by collision avoidance behavior and then resumed.

7.3 Pacifists

The intention generator of a pacifist differs from that of prey in that intention *mate* is activated and *escape* and *school* are deactivated.

Piscatorial mating behaviors show great interspecies and intraspecies diversity [Thresher, 1984]. However, two behaviors are prevalent: (i) nuzzling, where typically the male approaches the female from underneath and nudges her abdomen repeatedly until she is ready to spawn, and (ii) spawning ascent, where in its simplest form, the female rapidly swims towards the surface pursued by the male and releases gametes at the peak of her ascent. Moreover, courtship dancing is common in many species, albeit with substantial variation. Two frequently observed patterns are looping, in which the male swims vigorously up and down in a loop slightly above and in front of the female, and circling, in which the male and female circle, seemingly chasing each other's tails.

We have implemented a reasonably elaborate *courtship* behavior routine which simulates courtship dancing, circling, spawning ascent, and nuzzling behavior patterns in sequence (Plate 1b). A male fish selects a mating partner based on the following criteria: a female of the same species is more attractive than one of different species, and closer females are more attractive than ones further away. A female selects a partner similarly, but shows preference over the size of the male fish (stronger, more protective) rather than its distance.

Once fish i has selected a potential partner j based on the above criteria, it sends a signal to fish j, and there are three possibilities: *Case 1*: If fish j's intention is not to *mate*, fish i approaches j and follows it around using *chasing-target* with the center of j's body as target. *Case 2*: If fish j's intention is to *mate* but its intended partner is not fish i. In this case, if i is male it will perform a *looping* behavior in front of j for a certain amount of time. If j is impressed and selects i during this time limit, then the courtship sequence continues, otherwise i will discontinue *looping* and leave j to find a new potential partner. Otherwise, if i is female it will choose another potential male. *Case 3*: If fish j's intention is to *mate* and its intended partner is fish i, the *courtship* behavior starts with the male looping in front of the female while she hovers and bobs her head. Looping is simulated by invoking *chasing-target* at a point in front of the female's head which moves up and down at a certain frequency. The female's hovering and head bobbing is accomplished through motor control of her pectoral fins (i.e., parameter γ in Fig. 5).

The male counts the number of times his mouth reaches the vicinity of the moving point, and when the count exceeds a set threshold (currently 6) he makes a transition from *looping* to *circling* behavior. Although the threshold count is fixed, the actual motions and duration of looping is highly unpredictable for any number of reasons, including the fact that looping may be temporarily interrupted to handle high priority events such as potential collisions between the pair or with other fishes that may pass by.

Before the transition to *circling*, the female fish may reject her initial partner and turn to a new larger male fish if the latter joins in the *looping* display. At this point the initially engaged male turns away as in case 2 described above. *Circling* is achieved when the fishes use *chasing-target* to chase each other's tail.

The *circling* routine ends and the spawning *ascending* routine begins after the female has made a fixed number of turns during circling. The female fish ascends quickly through fast swimming followed by hovering. The male fish uses *chasing-target* to follow the abdomen of the female. The *nuzzling* routine requires the male to approach the her abdomen from below. Once his mouth touches her abdomen, the male backs off for a number of time steps. This procedure repeats, until the male successfully touches the female 3 times. To permit the mating pair to come close together, the regions of sensitivity are set very tightly to their bodies. It is intriguing to watch some of the male artificial fish's attempts fail because of an inappropriate approach angle which triggers the *avoiding-fish* response. The male turns away to avoid the collision and tries again.

8 Conclusion

We have demonstrated realistic looking artificial fishes that are capable of some astonishingly lifelike behaviors. To give an example of running times, our implementation can simulate 10 fishes, 15 food particles, and 5 static obstacles at about 4 frames/sec (including wireframe rendering time) on a Silicon Graphics R4400 Indigo2 workstation. The easy extensibility of our approach is made most evident by the complex patterns of mating behavior that we have been able to implement to date.

Our work opens up many avenues of research. For example, we may be within reach of computational models that can imitate the spawning behaviors of the female (release of gametes) and the male (fertilization), hence the evolution of new varieties of artificial fishes through simulated sexual reproduction. Interestingly, Pokhilko, Pajitnov, *et al.*, have already demonstrated the simulated breeding of fish models much simpler than ours using genetic algorithms, and this idea has resulted in the game "El-Fish" [Corcoran, 1992].

References

[Adler, 1975] H. E. Adler. *Fish Behavior: Why Fishes do What They Do*. T.F.H Publications, Neptune City, NJ, 1975.

[Alexander, 1992] R.M. Alexander. *Exploring Biomechanics*. Scientific American Library, New York, 1992.

[Beer, 1990] R. Beer. *Intelligence as Adaptive Behavior*. Academic press, NY, 1990.

[Braitenberg, 1984] V. Braitenberg. *Vehicles, Experiments in Synthetic Psychology*. MIT Press, Cambridge, MA, 1984.

[Corcoran, 1992] E. Corcoran. One fish, two fish: How to raise a school of tempting software toys. *Scientific American*, July 1992.

[Horn, 1986] B. K. P. Horn. *Robot Vision*. MIT Press, Cambridge, MA, 1986.

[Kass *et al.*, 1987] M. Kass, A. Witkin, and D. Terzopoulos. Snakes: Active contour models. *Int. J. of Computer Vision*, 1(4):321–331, 1987.

[Lorenz, 1973] K. Lorenz. *Foundations of Ethology*. Springer-Verlag, New York, 1973.

[Maes, 1991] P. Maes, editor. *Designing Autonomous Agents*. MIT Press, Cambridge, MA, 1991.

[Matarić, 1994] M. J. Matarić. *Interaction and Intelligent Behavior*. PhD thesis, Dept. of EECS, MIT, Cambridge, MA, May 1994.

[Mcfarland, 1985] D. Mcfarland. *Animal Behaviour*. Pitman, 1985.

[Meyer and Guillot, 1991] J.-A. Meyer and A. Guillot. Simulation of adaptive behavior in animats: Review and prospect. In J.-A. Meyer and S. Wilson, editors, *From Animals to Animats*, pages 2–14. MIT Press, Cambridge, MA, 1991.

[Miller, 1988] G. S. P. Miller. The motion dynamics of snakes and worms. *Computer Graphics*, 22(4):169–177, 1988.

[Pearson, 1991] K. G. Pearson. Sensory elements in pattern-generating networks. In *Making Them Move*, pages 111–127. Morgan Kaufmann, San Mateo, California, 1991.

[Press *et al.*, 1986] W. Press, B. Flannery, S. Teukolsky, and W. Vetterling. *Numerical Recipes: The Art of Scientific Computing*. Cambridge University Press, Cambridge, England, 1986.

[Reynolds, 1987] C. W. Reynolds. Flocks, herds, and schools: A distributed behavioral model. *Computer Graphics*, 21(4):25–34, 1987.

[Thresher, 1984] R. E. Thresher. *Reproduction in Reef Fishes*. T.F.H. Publications, Neptune City, NJ, 1984.

[Tinbergen, 1950] N. Tinbergen. *The Study of Instinct*. Clarendon Press, Oxford, England, 1950.

[Tu *et al.*, 1993] X. Tu, D. Terzopoulos, and E. Fiume. Go Fish! ACM SIGGRAPH Video Review Issue 91: SIGGRAPH'93 Electronic Theater, 1993.

[Webb, 1989] P. W. Webb. Form and function in fish swimming. *Scientific American*, 251(1), 1989.

[Wilson and Wilson, 1985] R. Wilson and J. Q. Wilson. *Watching Fishes*. Harper and Row, New York, 1985.

[Wilson, 1991] S. W. Wilson. The animat path to AI. In J.-A. Meyer and S. Wilson, editors, *From Animals to Animats*, pages 15–21. MIT Press, Cambridge, MA, 1991.

Evolving 3D Morphology and Behavior by Competition

Karl Sims

Thinking Machines Corporation
245 First Street, Cambridge, MA 02142

Abstract

This paper describes a system for the evolution and co-evolution of virtual creatures that compete in physically simulated three-dimensional worlds. Pairs of individuals enter one-on-one contests in which they contend to gain control of a common resource. The winners receive higher relative fitness scores allowing them to survive and reproduce. Realistic dynamics simulation including gravity, collisions, and friction, restricts the actions to physically plausible behaviors.

The morphology of these creatures and the neural systems for controlling their muscle forces are both genetically determined, and the morphology and behavior can adapt to each other as they evolve simultaneously. The genotypes are structured as directed graphs of nodes and connections, and they can efficiently but flexibly describe instructions for the development of creatures' bodies and control systems with repeating or recursive components. When simulated evolutions are performed with populations of competing creatures, interesting and diverse strategies and counter-strategies emerge.

1 Introduction

Interactions between evolving organisms are generally believed to have a strong influence on their resulting complexity and diversity. In natural evolutionary systems the measure of fitness is not constant: the reproducibility of an organism depends on many environmental factors including other evolving organisms, and is continuously in flux. Competition between organisms is thought to play a significant role in preventing static fitness landscapes and sustaining evolutionary change.

These effects are a distinguishing difference between natural evolution and optimization. Evolution proceeds with no explicit goal, but optimization, including the genetic algorithm, usually aims to search for individuals with the highest possible fitness values where the fitness measure has been predefined, remains constant, and depends only on the individual being tested.

The work presented here takes the former approach. The fitness of an individual is highly dependent on the specific behaviors of other individuals currently in the population. The hope is that virtual creatures with higher complexity and more interesting behavior will evolve than when applying the selection pressures of optimization alone.

Many simulations of co-evolving populations have been performed which involve competing individuals [1,2]. As examples, Lindgren has studied the evolutionary dynamics of competing game strategy rules [14], Hillis has demonstrated that co-evolving parasites can enhance evolutionary optimization [9], and Reynolds evolves vehicles for competition in the game of tag [19]. The work presented here involves similar evolutionary dynamics to help achieve interesting results when phenotypes have three-dimensional bodies and compete in physically simulated worlds.

In several cases, optimization has been used to automatically generate dynamic control systems for given two-dimensional articulated structures: de Garis has evolved weight values for neural networks [6], Ngo and Marks have applied genetic algorithms to generate stimulus-response pairs [16], and van de Panne and Fiume have optimized sensor-actuator networks [17]. Each of these methods has resulted in successful locomotion of two-dimensional stick figures.

The work presented here is related to these projects, but differs in several respects. Previously, control systems were generated for fixed structures that were user-designed, but here entire creatures are evolved: the evolution determines the creature morphologies as well as their control systems. The physical structure of a creature can adapt to its control system, and vice versa, as they evolve together. Also, here the creatures' bodies are three-dimensional and fully physically based. In addition, a developmental process is used to generate the creatures and their control systems, and allows similar components including their local neural circuitry to be defined once and then replicated, instead of requiring each to be separately specified. This approach is related to L-systems, graftal grammars, and object instancing techniques [8,11,13,15,23]. Finally, the previous work on articulated structures relies only on optimization, and competitions between individuals were not considered.

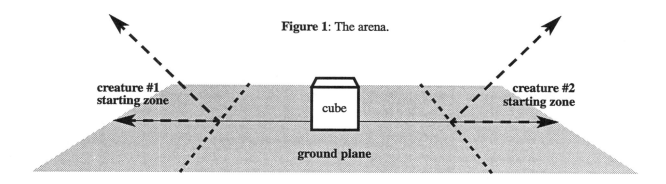

Figure 1: The arena.

A different version of the system described here has also been used to generate virtual creatures by optimizing for specific defined behaviors such as swimming, walking, and following [22].

Genotypes used in simulated evolutions and genetic algorithms have traditionally consisted of strings of binary digits [7,10]. Variable length genotypes such as hierarchical Lisp expressions or other computer programs can be useful in expanding the set of possible results beyond a predefined genetic space of fixed dimensions. Genetic languages such as these allow new parameters and new dimensions to be added to the genetic space as an evolution proceeds, and therefore define rather a *hyperspace* of possible results. This approach has been used to genetically program solutions to a variety of problems [3,12], as well as to explore procedurally generated images and dynamical systems [20,21].

In the spirit of unbounded genetic languages, *directed graphs* are presented here as an appropriate basis for a grammar that can be used to describe both the morphology and neural systems of virtual creatures. The level of complexity is variable for both genotype and phenotype. New features and functions can be added to creatures or existing ones removed, as they evolve.

The next section of this paper describes the environment of the simulated contest and how the competitors are scored. Section 3 discusses different simplified competition patterns for approximating competitive environments. Sections 4 and 5 present the genetic language that is used to represent creatures with arbitrary structure and behavior, and section 6 summarizes the physical simulation techniques used. Section 7 discusses the evolutionary simulations including the methods used for mutating and mating directed graph genotypes, and finally sections 8 and 9 provide results, discussion, and suggestions for future work.

2 The Contest

Figure 1 shows the arena in which two virtual creatures will compete to gain control of a single cube. The cube is placed in the center of the world, and the creatures start on opposite sides of the cube. The second contestant is initially turned by 180 degrees so the relative position of the cube to the crea-

ture is consistent from contest to contest no matter which starting side it is assigned. Each creature starts on the ground and behind a diagonal plane slanting up and away from the cube. Creatures are wedged into these "starting zones" until they contact both the ground plane and the diagonal plane, so taller creatures must start further back. This helps prevent the inelegant strategy of simply falling over onto the cube. Strategies like this that utilize only potential energy are further discouraged by relaxing a creature's body before it is placed in the starting zone. The effect of gravity is simulated until the creature reaches a stable minimum state.

At the start of the contest the creatures' nervous systems are activated, and a physical simulation of the creatures' bodies, the cube, and the ground plane begins. The winner is the creature that has the most control over the cube after a certain duration of simulated time (8 seconds were given). Instead of just defining a winner and loser, the margin of victory is determined in the form of a relative fitness value, so there is selection pressure not just to win, but to win by the largest possible margin.

The creatures' final distances to the cube are used to calculate their fitness scores. The shortest distance from any point on the surface of a creatures's parts to the center of the cube is used as its distance value. A creature gets a higher score by being closer to the cube, but also gets a higher score when its opponent is further away. This encourages creatures to reach the cube, but also gives points for keeping the opponent away from it. If d_1 and d_2 are the final shortest distances of each creature to the cube, then the fitnesses for each creature, f_1 and f_2, are given by:

$$f_1 = 1.0 + \frac{d_2 - d_1}{d_1 + d_2}$$

$$f_2 = 1.0 + \frac{d_1 - d_2}{d_1 + d_2}$$

This formulation puts all fitness values in the limited range of 0.0 to 2.0. If the two distances are equal the contestants receive tie scores of 1.0 each, and in all cases the scores will average 1.0.

Credit is also given for having "control" over the cube, beyond just as measured by the minimum distance to it. If both creatures end up contacting the cube, the winner is the one that surrounds it the most. This is approximated by further decreasing the distance value, as used above, when a creature is touching the cube on the side that opposes its center of mass. Since the initial distances are measured from the center of the cube they can be adjusted in this way and still remain positive.

During the simulated contest, if neither creature shows any movement for a full second, the simulation is stopped and the scores are evaluated early to save unnecessary computation.

3 Approximating Competitive Environments

There are many trade-offs to consider when simulating an evolution in which fitness is determined by discrete competitions between individuals. In this work, pairs of individuals compete one-on-one. At every generation of a simulated evolution the individuals in the population are paired up by some pattern and a number of competitions are performed to eventually determine a fitness value for every individual. The simulations of the competitions are by far the dominant computational requirement of the process, so the total number of competitions performed for each generation and the effectiveness of the pattern of competitions are important considerations.

In one extreme, each individual competes with all the others in the population and the average score determines the fitness (figure 2a). However, this requires $(N^2 - N)/2$ total competitions for a single-species population of N individuals. For large populations this is often unacceptable, especially if the competition time is significant, as it is in this work.

In the other extreme, each individual competes with just a single opponent (figure 2b). This requires only $N/2$ total competitions, but can cause inconsistency in the fitness values since each fitness is often highly dependent on the specific individual that happens to be assigned as the opponent. If the pairing is done at random, and especially if the mutation rate is high, fitness can be more dependent on the luck of receiving a poor opponent than on an individual's actual ability.

One compromise between these extremes is for each individual to compete against several opponents chosen at random for each generation. This can somewhat dilute the fitness inconsistency problem, but at the expense of more competition simulations.

A second compromise is a tournament pattern (figure 2c) which can efficiently determine a single overall winner with $N - 1$ competitions. But this also does not necessarily give all individuals fair scores because of the random initial opponent assignments. Also, this pattern does not easily apply to multi-species evolutions where competitions are not performed between individuals within the same species.

A third compromise is for each individual to compete once per generation, but all against the same opponent. The individual with the highest fitness from the previous generation is chosen as this one-to-beat (figure 2d). This also requires $N - 1$ competitions per generation, but effectively gives fair relative fitness values since all are playing against the same opponent which has proven to be competent. Various interesting instabilities can still occur over generations

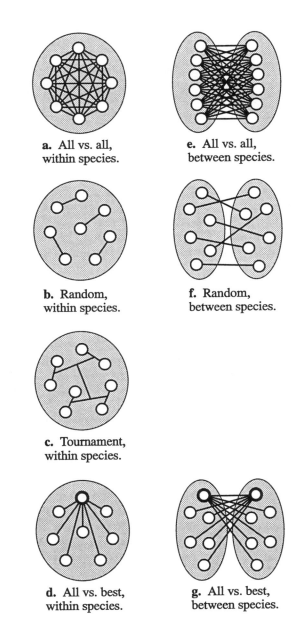

a. All vs. all, within species.

e. All vs. all, between species.

b. Random, within species.

f. Random, between species.

c. Tournament, within species.

d. All vs. best, within species.

g. All vs. best, between species.

Figure 2: Different pair-wise competition patterns for one and two species. The gray areas represent species of interbreeding individuals, and lines indicate competitions performed between individuals.

however, since the strategy of the "best" individual can change suddenly between generations.

The number of species in the population is another element to consider when simulating evolutions involving competition. A species may be described as an interbreeding subset of individuals in the population. In single-species evolutions individuals will compete against their relatives, but in multi-species evolutions individuals can optionally compete only against individuals from other species. Figure 2 shows graphical representations of some of the different competition patterns described above for both one and two species.

The resulting effects of using these different competition patterns is unfortunately difficult to quantify in this work, since by its nature a simple overall measure of success is absent. Evolutions were performed using several of the methods described above with both one and two species, and the results were subjectively judged. The most "interesting" results occurred when the all vs. best competition pattern was used. Both one and two species evolutions produced some intriguing strategies, but the multi-species simulations tended to produce more interesting interactions between the evolving creatures.

Genotype: directed graph.   **Phenotype**: hierarchy of 3D parts.

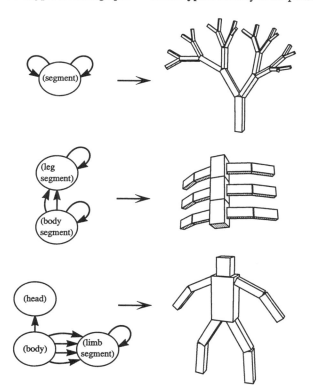

Figure 3: Designed examples of genotype graphs and corresponding creature morphologies.

4 Creature Morphology

In this work, the phenotype embodiment of a virtual creature is a hierarchy of articulated three-dimensional rigid parts. The genetic representation of this morphology is a directed graph of nodes and connections. Each graph contains the developmental instructions for growing a creature, and provides a way of reusing instructions to make similar or recursive components within the creature. A phenotype hierarchy of parts is made from a graph by starting at a defined *root-node* and synthesizing parts from the node information while tracing through the connections of the graph. The graph can be recurrent. Nodes can connect to themselves or in cycles to form recursive or fractal like structures. They can also connect to the same child multiple times to make duplicate instances of the same appendage.

Each node in the graph contains information describing a rigid part. The *dimensions* determine the physical shape of the part. A *joint-type* determines the constraints on the relative motion between this part and its parent by defining the number of degrees of freedom of the joint and the movement allowed for each degree of freedom. The different joint-types allowed are: *rigid, revolute, twist, universal, bend-twist, twist-bend,* or *spherical*. *Joint-limits* determine the point beyond which restoring spring forces will be exerted for each degree of freedom. A *recursive-limit* parameter determines how many times this node should generate a phenotype part when in a recursive cycle. A set of local *neurons* is also included in each node, and will be explained further in the next section. Finally, a node contains a set of *connections* to other nodes.

Each connection also contains information. The placement of a child part relative to its parent is decomposed into *position, orientation, scale,* and *reflection*, so each can be mutated independently. The position of attachment is constrained to be on the surface of the parent part. Reflections cause negative scaling, and allow similar but symmetrical sub-trees to be described. A *terminal-only* flag can cause a connection to be applied only when the recursive limit is reached, and permits tail or hand-like components to occur at the end of chains or repeating units.

Figure 3 shows some simple hand-designed graph topologies and resulting phenotype morphologies. Note that the parameters in the nodes and connections such as *recursive-limit* are not shown for the genotype even though they affect the morphology of the phenotype. The nodes are anthropomorphically labeled as "body," "leg segment," etc. but the genetic descriptions actually have no concept of specific categories of functional components.

5 Creature Behavior

A virtual "brain" determines the behavior of a creature. The brain is a dynamical system that accepts input sensor values and provides output effector values. The output values are applied as forces or torques at the degrees of freedom of the body's joints. This cycle of effects is shown in Figure 4.

Sensor, effector, and internal neuron signals are represented here by continuously variable scalars that may be positive or negative. Allowing negative values permits the implementation of single effectors that can both push and pull. Although this may not be biologically realistic, it simplifies the more natural development of muscle pairs.

5.1 Sensors

Each sensor is contained within a specific part of the body, and measures either aspects of that part or aspects of the world relative to that part. Three different types of sensors were used for these experiments:

1. *Joint angle sensors* give the current value for each degree of freedom of each joint.

2. *Contact sensors* activate (1.0) if a contact is made, and negatively activate (-1.0) if not. Each contact sensor has a sensitive region within a part's shape and activates when any contacts occur in that area. In this work, contact sensors are made available for each face of each part. No distinction is made between self-contact and environmental contact.

3. *Photosensors* react to a global light source position. Three photosensor signals provide the coordinates of the normalized light source direction relative to the orientation of the part. Shadows are not simulated, so photosensors continue to sense a light source even if it is blocked. Photosensors for two independent colors are made available. The source of one color is located in the desirable cube, and the other is located at the center of mass of the opponent. This effectively allows evolving nervous systems to incorporate specific "cube sensors" and "opponent sensors."

Other types of sensors, such as accelerometers, additional proprioceptors, or even sound or smell detectors could also be implemented, but these basic three are enough to allow some interesting and adaptive behaviors to occur.

5.2 Neurons

Internal neural nodes are used to give virtual creatures the possibility of arbitrary behavior. They allow a creature to have an internal state beyond its sensor values, and be affected by its history.

In this work, different neural nodes can perform diverse functions on their inputs to generate their output signals. Because of this, a creature's brain might resemble a dataflow computer program more than a typical artificial neural network. This approach is probably less biologically realistic than just using sum and threshold functions, but it is hoped that it makes the evolution of interesting behaviors more likely. The set of functions that neural nodes can have is: *sum, product, divide, sum-threshold, greater-than, sign-of, min, max, abs, if, interpolate, sin, cos, atan, log, expt, sigmoid, integrate, differentiate, smooth, memory, oscillate-wave,* and *oscillate-saw.*

Some functions compute an output directly from their

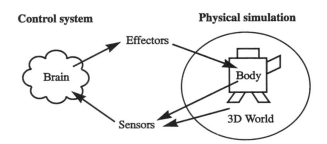

Figure 4: Cycle of effects between brain, body and world.

inputs, while others such as the oscillators retain some state and can give time varying outputs even when their inputs are constant. The number of inputs to a neuron depends on its function, and here is at most three. Each input contains a connection to another neuron or a sensor from which to receive a value. Alternatively, an input can simply receive a constant value. The input values are first scaled by weights before being operated on. The genetic parameters for each neural node include these weights as well as the function type and the connection information.

For each simulated time interval, every neuron computes its output value from its inputs. In this work, two brain time steps are performed for each dynamic simulation time step so signals can propagate through multiple neurons with less delay.

5.3 Effectors

Each effector simply contains a connection from a neuron or a sensor from which to receive a value. This input value is scaled by a constant weight, and then exerted as a joint force which affects the dynamic simulation and the resulting behavior of the creature. Different types of effectors, such as sound or scent emitters, might also be interesting, but only effectors that exert simulated muscle forces are used here.

Each effector controls a degree of freedom of a joint. The effectors for a given joint connecting two parts, are contained in the part further out in the hierarchy, so that each non-root part operates only a single joint connecting it to its parent. The angle sensors for that joint are also contained in this part.

Each effector is given a *maximum-strength* proportional to the maximum cross sectional area of the two parts it joins. Effector forces are scaled by these strengths and not permitted to exceed them. This is similar to the strength limits of natural muscles. As in nature, mass scales with volume but strength scales with area, so behavior does not always scale uniformly.

5.4 Combining Morphology and Control

The genotype descriptions of virtual brains and the actual phenotype brains are both directed graphs of nodes and connections. The nodes contain the sensors, neurons, and effectors, and the connections define the flow of signals between these nodes. These graphs can also be recurrent, and as a result the final control system can have feedback loops and cycles.

However, most of these neural elements exist within a specific part of the creature. Thus the genotype for the nervous system is a nested graph: the morphological nodes each contain graphs of the neural nodes and connections. Figure 5 shows an example of an evolved nested graph which describes a simple three-part creature as shown in figure 6.

When a creature is synthesized from its genetic description, the neural components described within each part are generated along with the morphological structure. This causes blocks of neural control circuitry to be replicated along with each instanced part, so each duplicated segment or appendage of a creature can have a similar but independent local control system.

These local control systems can be connected to enable the possibility of coordinated control. Connections are allowed between adjacent parts in the hierarchy. The neurons and effectors within a part can receive signals from sensors or neurons in their parent part or in their child parts.

Creatures are also allowed a set of neurons that are not associated with a specific part, and are copied only once into the phenotype. This gives the opportunity for the development of global synchronization or centralized control. These neurons can receive signals from each other or from sensors or neurons in specific instances of any of the creature's parts, and the neurons and effectors within the parts can optionally receive signals from these unassociated-neuron outputs.

In this way the genetic language for morphology and control is merged. A local control system is described for each type of part, and these are copied and connected into the hierarchy of the creature's body to make a complete distributed nervous system. Figure 6a shows the creature morphology resulting from the genotype in figure 5. Again, parameters describing shapes and weight values are not shown for the genotype even though they affect the pheno-

Figure 6a: The phenotype morphology generated from the evolved genotype shown in figure 5.

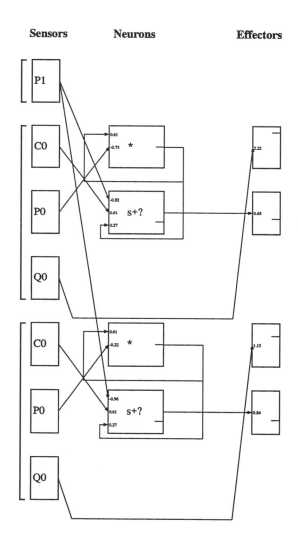

Sensors   Neurons   Effectors

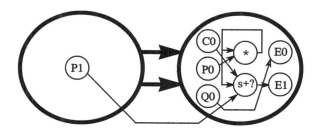

Figure 5: Example evolved nested graph genotype. The outer graph in bold describes a creature's morphology. The inner graph describes its neural circuitry. C0, P0, P1, and Q0 are contact and photosensors, E0 and E1 are effector outputs, and those labeled "*" and "s+?" are neural nodes that perform *product* and *sum-threshold* functions.

Figure 6b: The phenotype "brain" generated from the evolved genotype shown in figure 5. The effector outputs of this control system cause the morphology above to roll forward in tumbling motions.

type. Figure 6b shows the corresponding brain of this creature. The brackets on the left side of figure 6b group the neural components of each part. Two groups have similar neural systems because they were synthesized from the same genetic description. This creature can roll over the ground by making cyclic tumbling motions with its two arm-like appendages. Note that it can be difficult to analyze exactly how a control system such as this works, and some components may not actually be used at all. Fortunately, a primary benefit of using artificial evolution is that understanding these representations is not necessary.

6 Physical Simulation

Dynamics simulation is used to calculate the movement of creatures resulting from their interaction with a virtual three-dimensional world. There are several components of the physical simulation used in this work: articulated body dynamics, numerical integration, collision detection, and collision response with friction. These are only briefly summarized here, since physical simulation is not the emphasis of this paper.

Featherstone's recursive O(N) articulated body method is used to calculate the accelerations from the velocities and external forces of each hierarchy of connected rigid parts [5]. Integration determines the resulting motions from these accelerations and is performed by a Runge-Kutta-Fehlberg method which is a fourth order Runge-Kutta with an additional evaluation to estimate the error and adapt the step size. Typically between 1 and 5 integration time steps are performed for each frame of 1/30 second.

The shapes of parts are represented here by simple rectangular solids. Bounding box hierarchies are used to reduce the number of collision tests between parts from $O(N^2)$. Pairs whose world-space bounding boxes intersect are tested for penetrations, and collisions with a ground plane are also tested. If necessary, the previous time-step is reduced to keep any new penetration depths below a certain tolerance. Connected parts are permitted to interpenetrate but not rotate completely through each other. This is achieved by using adjusted shapes when testing for collisions between connected parts. The shape of the smaller part is clipped halfway back from its point of attachment so it can swing freely until its remote end makes contact.

Collision response is accomplished by a hybrid model using both impulses and penalty spring forces. At high velocities, instantaneous impulse forces are used, and at low velocities springs are used, to simulate collisions and contacts with arbitrary elasticity and friction parameters.

It is important that the physical simulation be reasonably accurate when optimizing for creatures that can move within it. Any bugs that allow energy leaks from non-conservation, or even round-off errors, will inevitably be discovered and exploited by the evolving creatures. Although this can be a lazy and often amusing approach for debugging a

physical modeling system, it is not necessarily the most practical.

7 Creature Evolution

An evolution of virtual creatures is begun by first creating an initial population of genotypes. Seed genotypes are synthesized "from scratch" by random generation of sets of nodes and connections. Alternatively, an existing genotype from a previous evolution can be used to seed an initial population.

Before creatures are paired off for competitions and fitness evaluation, some simple viability checks are performed, and inappropriate creatures are removed from the population by giving them zero fitness values. Those that have more than a specified number of parts are removed. A subset of genotypes will generate creatures whose parts initially interpenetrate. A short simulation with collision detection and response attempts to repel any intersecting parts, but those creatures with persistent interpenetrations are also discarded.

A *survival-ratio* determines the percentage of the population that will survive each generation. In this work, population sizes were typically 300, and the survival-ratio was 1/5. If the initially generated population has fewer individuals with positive fitness than the number that should survive, another round of seed genotypes is generated to replace those with zero fitness.

For each generation, creatures are grown from their genotypes, and their fitness values are measured by simulating one or more competitions with other individuals as described. The individuals whose fitnesses fall within the survival percentile are then reproduced, and their offspring fill the slots of those individuals that did not survive. The number of offspring that each surviving individual generates is proportional to its fitness. The survivors are kept in the population for the next generation, and the total size of the population is maintained. In multi-species evolutions, each sub-population is independently treated in this way so the number of individuals in each species remains constant and species do not die out.

Offspring are generated from the surviving creatures by copying and combining their directed graph genotypes. When these graphs are reproduced they are subjected to probabilistic variation or mutation, so the corresponding phenotypes are similar to their parents but have been altered or adjusted in random ways.

7.1 Mutating Directed Graphs

A directed graph is mutated by the following sequence of steps:

1. The internal parameters of each node are subjected to possible alterations. A mutation frequency for each parameter type determines the probability that a mutation will be applied to it at all. Boolean values are mutated by simply flipping their state. Scalar values are mutated by adding several random numbers to them for a Gaussian-like distribution

so small adjustments are more likely than drastic ones. The scale of an adjustment is relative to the original value, so large quantities can be varied more easily and small ones can be carefully tuned. A scalar can also be negated. After a mutation occurs, values are clamped to their legal bounds. Some parameters that only have a limited number of legal values are mutated by simply picking a new value at random from the set of possibilities.

2. A new random node is added to the graph. A new node normally has no effect on the phenotype unless a connection also mutates a pointer to it. Therefore a new node is always initially added, but then garbage collected later (in step 5) if it does not become connected. This type of mutation allows the complexity of the graph to grow as an evolution proceeds.

3. The parameters of each connection are subjected to possible mutations in the same way the node parameters were in step 1. With some frequency the connection pointer is moved to point to a different node which is chosen at random.

4. New random connections may be added and existing ones may be removed. In the case of the neural graphs these operations are not performed because the number of inputs for each element is fixed, but the morphological graphs can have a variable number of connections per node. Each existing node is subject to having a new random connection added to it, and each existing connection is subject to possible removal.

5. Unconnected elements are garbage collected. Connectedness is propagated outwards through the connections of the graph, starting from the root node of the morphology, and from the effector nodes of the neural graphs. Although leaving the disconnected nodes for possible reconnection might be advantageous, and is probably biologically analogous, at least the unconnected newly added ones are removed to prevent unnecessary growth in graph size.

Since mutations are performed on a per element basis, genotypes with only a few elements might not receive any mutations, where genotypes with many elements would receive enough mutations that they would rarely resemble their parents. This is compensated for by scaling the mutation frequencies by an amount inversely proportional to the size of the current graph being mutated, such that on the average at least one mutation occurs in the entire graph.

Mutation of nested directed graphs, as are used here to represent creatures, is performed by first mutating the outer graph and then mutating the inner layer of graphs. The inner graphs are mutated last because legal values for some of their parameters (inter-node neural input sources) can depend on the topology of the outer graph.

7.2 Mating Directed Graphs

Sexual reproduction allows components from more than one parent to be combined into new offspring. This permits features to evolve independently and later be merged into a sin-

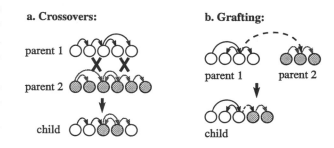

Figure 7: Two methods for mating directed graphs.

gle individual. Two different methods for mating directed graphs are used in this work.

The first is a *crossover* operation (figure 7a). The nodes of two parents are each aligned in a row as they are stored, and the nodes of the first parent are copied to make the child, but one or more crossover points determine when the copying source should switch to the other parent. The connections of a node are copied with it and simply point to the same relative node locations as before. If the copied connections now point out of bounds because of varying node numbers they are randomly reassigned.

A second mating method *grafts* two genotypes together by connecting a node of one parent to a node of another (figure 7b). The first parent is copied, and one of its connections is chosen at random and adjusted to point to a random node in the second parent. Newly unconnected nodes of the first parent are removed and the newly connected node of the second parent and any of its descendants are appended to the new graph.

A new directed graph can be produced by either of these two mating methods, or asexually by using only mutations. Offspring from matings are sometimes subjected to mutations afterwards, but with reduced mutation frequencies. In this work a reproduction method is chosen at random for each child to be produced by the surviving individuals using the ratios: 40% asexual, 30% crossovers, and 30% grafting. A second parent is chosen from the survivors if necessary, and a new genotype is produced from the parent or parents.

After a new generation of genotypes is created, a phenotype creature is generated from each, and again their fitness values are evaluated. As this cycle of variation and selection continues, the population is directed towards creatures with higher fitness.

7.3 Parallel Implementation

This process has been implemented to run in parallel on a Connection Machine® CM-5 in a master/slave message passing model. A single processing node contains the population and performs all the selection and reproduction operations. It farms out pairs of genotypes to the other nodes to be fitness tested, and gathers back the fitness values after they have been determined. The fitness tests each include a dynamics simulation for the competition and although many can execute in nearly real-time, they are still the dominant computational requirement of the system. Performing a fitness test per processor is a simple but effective way to parallelize this process, and the overall performance scales quite linearly with the number of processors, as long as the population size is somewhat larger than the number of processors.

Each fitness test takes a different amount of time to compute depending on the complexity of the creatures and how they attempt to move. To prevent idle processors from just waiting for others to finish, the slowest few simulations at the end of a generation are suspended and those individuals are removed from the population by giving them zero fitness. With this approach, an evolution with population size 300, run for 100 generations, might take about four hours to complete on a 32 processor CM-5.

8 Results and Discussion

Many independent evolutions were performed using the "all vs. best" competition pattern as described in section 3. Some single-species evolutions were performed in which all individuals both compete and breed with each other, but most included two species where individuals only compete with members of the opponent species.

Some examples of resulting two-species evolutionary dynamics are shown in Figure 8. The relative fitness of the best individuals of each species are plotted over 100 generations. The rate of evolutionary progress varied widely in different runs. Some species took many generations before they could even reach the cube at all, while others discovered a fairly successful strategy in the first 10 or 20 generations. Figure 8c shows an example where one species was successful fairly quickly and the other species never evolved an effective strategy to challenge it. The other three graphs in figure 8 show evolutions where more interactions occurred between the evolving species.

A variety of methods for reaching the cube were discovered. Some extended arms out onto the cube, and some reached out while falling forward to land on top of it. Others could crawl inch-worm style or roll towards the cube, and a few even developed leg-like appendages that they used to walk towards it.

The most interesting results often occurred when both species discovered methods for reaching the cube and then further evolved strategies to counter the opponent's behavior. Some creatures pushed their opponent away from the cube, some moved the cube away from its initial location and then followed it, and others simply covered up the cube to block the opponent's access. Some counter-strategies took advantage of a specific weakness in the original strategy and could be easily foiled in a few generations by a minor adaptation to the original strategy. Others permanently defeated the original strategy and required the first species to evolve another level of counter-counter-strategy to regain the lead.

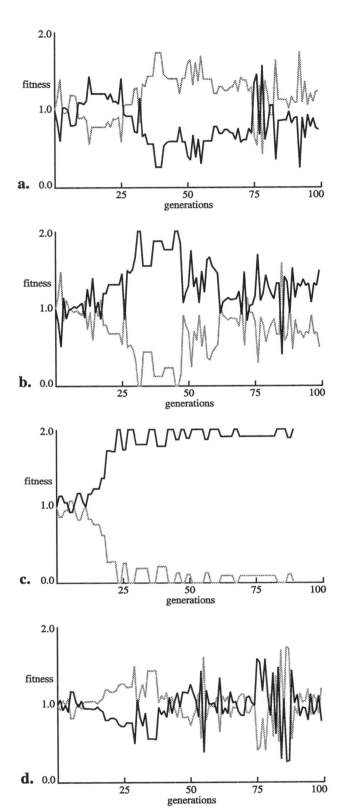

Figure 8: Relative fitness between two co-evolving and competing species, from four independent simulations.

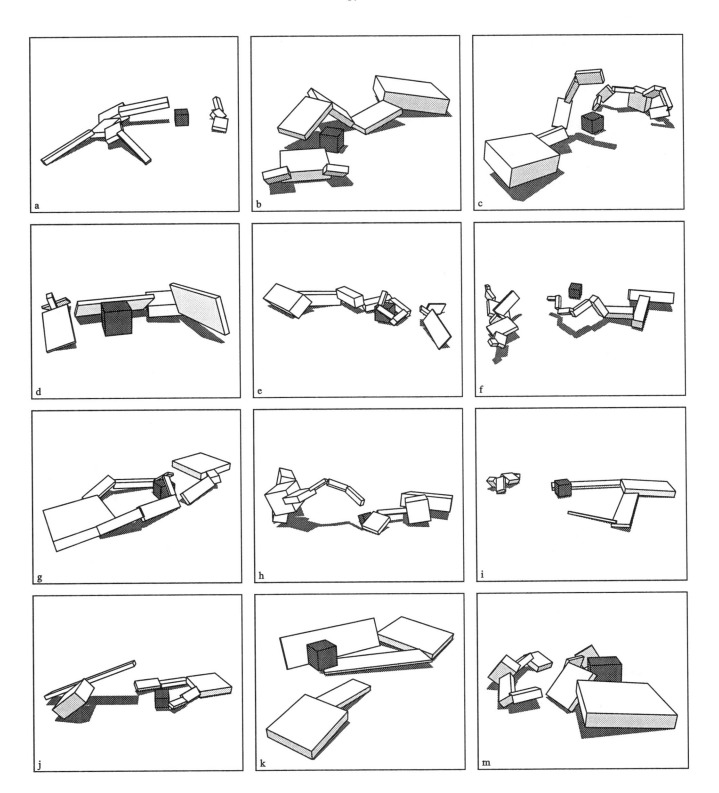

Figure 9: Evolved competing creatures.

In some evolutions the winners alternated between species many times with new strategies and counter-strategies. In other runs one species kept a consistent lead with the other species only providing temporary challenges.

After the results from many simulations were observed, the best were collected and then played against each other in additional competitions. The different strategies were compared, and the behavior and adaptability of creatures were observed as they faced new types of opponents that were not encountered during their evolutions. A few evolutions were also performed starting with an existing creature as a seed genotype for each species so they could further evolve to compete against a new type of opponent.

Figure 9 shows some examples of evolved competing creatures and demonstrates the diversity of the different strategies that emerged. Some of the behaviors and interactions of these specific creatures are described briefly here. The larger creature in figure 9b nudges the cube aside and then pins down his smaller opponent. The crab-like creature in 9c can successfully walk forward, but then continues blindly past the cube and over the opponent. Figure 9d shows a creature that has just pushed its opponent away from the cube, and the arm-like creature in 9e also jabs at its opponent before curling around the cube.

Most creatures perform similar behavior independently of the opponent's actions, but a few are adaptive in that they can reach towards the cube wherever it moves. For example the arm-like creature in figure 9f pushes the cube aside and then uses photosensors to adaptively follow it. If its opponent moves the cube in a different direction it will successfully grope towards the new location.

The two-armed creature in figure 9g blocks access to the cube by covering it up. Several other two-armed creatures in 9i, 9j, and 9k use the strategy of batting the cube to the side with one arm and catching it with the other arm. This seemed to be the most successful strategy of the creatures in this group, and the one in 9k was actually the overall winner because it could whisk the cube aside very quickly. However, it was a near tie between this and the photosensitive arm in 9f. The larger creature in 9m wins by a large margin against some opponents because it can literally walk away with the cube, but it does not initially reach the cube very quickly and tends to loose against faster opponents.

It is possible that adaptation on an evolutionary scale occurred more easily than the evolution of individuals that were themselves adaptive. Perhaps individuals with adaptive behavior would be significantly more rewarded if evolutions were performed with many species instead of just one or two. To be successful, a single individual would then need to defeat a larger number of different opposing strategies.

9 Future Work

Several variations on this system could be worth further experimentation. Other types of contests could be defined in which creatures compete in different environments and different rules determine the winners. Creatures might also be rewarded for cooperative behavior somehow as well as competitive, and teams of interacting creatures could be simulated.

Evolutions containing larger numbers of species should certainly be performed, with the hope of increasing the chances for emergence of more adaptive individuals as hypothesized above.

An additional extension to this work would be to simulate a more complex but more realistic environment in which many creatures simultaneously compete and/or cooperate with each another, instead of pairing off in one-on-one contests. Speciation, mating patterns, competing patterns, and even offspring production could all be determined by one long ecological simulation. Experiments like this have been performed with simpler organisms and have produced interesting results including specialization and various social interactions [18,24].

Perhaps the techniques presented here should be considered as an approach toward creating artificial intelligence. When a genetic language allows virtual entities to evolve with increasing complexity, it is common for the resulting system to be difficult to understand in detail. In many cases it would also be difficult to design a similar system using traditional methods. Techniques such as these have the potential of surpassing those limits that are often imposed when human understanding and design is required. The examples presented here suggest that it might be easier to evolve virtual entities exhibiting intelligent behavior than it would be for humans to design and build them.

10 Conclusion

In summary, a system has been described that can automatically generate autonomous three-dimensional virtual creatures that exhibit diverse competitive strategies in physically simulated worlds. A genetic language that uses directed graphs to describe both morphology and behavior defines an unlimited hyperspace of possible results, and a variety of interesting virtual creatures have been shown to emerge when this hyperspace is explored by populations of evolving and competing individuals.

Acknowledgments

Thanks to Gary Oberbrunner and Matt Fitzgibbon for Connection Machine and software support. Thanks to Thinking Machines Corporation and Lew Tucker for supporting this research. Thanks to Bruce Blumberg and Peter Schröder for dynamic simulation help and suggestions. And special thanks to Pattie Maes.

References

1. Angeline, P.J., and Pollack, J.B., "Competitive Environments Evolve Better Solutions for Complex Tasks," in *Proceedings of the 5th International Conference on Genetic Algorithms*, ed. by S. Forrest, Morgan Kaufmann 1993, pp.264-270.

2. Axelrod, R., "Evolution of Strategies in the Iterated Prisoner's Dilemma", in *Genetic Algorithms and Simulated Annealing*, ed. by L. Davis, Morgan Kaufmann, 1989.

3. Cramer, N.L., "A Representation for the Adaptive Generation of Simple Sequential Programs," *Proceedings of the First International Conference on Genetic Algorithms*, ed. by J. Grefenstette, 1985, pp.183-187.

4. Dawkins, R., *The Blind Watchmaker*, Harlow Longman, 1986.

5. Featherstone, R., *Robot Dynamics Algorithms*, Kluwer Academic Publishers, Norwell, MA, 1987.

6. de Garis, H., "Genetic Programming: Building Artificial Nervous Systems Using Genetically Programmed Neural Network Modules," *Proceedings of the 7th International Conference on Machine Learning*, 1990, pp.132-139.

7. Goldberg, D.E., *Genetic Algorithms in Search, Optimization, and Machine Learning*, Addison-Wesley, 1989.

8. Hart, J., "The Object Instancing Paradigm for Linear Fractal Modeling," *Graphics Interface*, 1992, pp.224-231.

9. Hillis, W.D., "Co-evolving parasites improve simulated evolution as an optimization procedure," *Artificial Life II*, ed. by Langton, Taylor, Farmer, & Rasmussen, Addison-Wesley, 1991, pp313-324.

10. Holland, J.H., *Adaptation in Natural and Artificial Systems*, Ann Arbor, University of Michigan Press, 1975.

11. Kitano, H., "Designing neural networks using genetic algorithms with graph generation system," *Complex Systems*, Vol.4, pp.461-476, 1990.

12. Koza, J., *Genetic Programming: on the Programming of Computers by Means of Natural Selection*, MIT Press, 1992.

13. Lindenmayer, A., "Mathematical Models for Cellular Interactions in Development, Parts I and II," *Journal of Theoretical Biology*, Vol.18, 1968, pp.280-315.

14. Lindgren, K., "Evolutionary Phenomena in Simple Dynamics," in *Artificial Life II*, ed. by Langton, Taylor, Farmer, & Rasmussen, Addison-Wesley, 1991, pp.295-312.

15. Mjolsness, E., Sharp, D., and Alpert, B., "Scaling, Machine Learning, and Genetic Neural Nets," *Advances in Applied Mathematics*, Vol.10, 1989, pp.137-163.

16. Ngo, J.T., and Marks, J., "Spacetime Constraints Revisited," *Computer Graphics*, Annual Conference Series, 1993, pp.343-350.

17. van de Panne, M., and Fiume, E., "Sensor-Actuator Networks," *Computer Graphics*, Annual Conference Series, 1993, pp.335-342.

18. Ray, T., "An Approach to the Synthesis of Life," *Artificial Life II*, ed. by Langton, Taylor, Farmer, & Rasmussen, Addison-Wesley, 1991, pp.371-408.

19. Reynolds, C., "Competition, Coevolution and the Game of Tag," to be published in: *Artificial Life IV Proceedings*, ed. by R. Brooks & P. Maes, MIT Press, 1994.

20. Sims, K., "Artificial Evolution for Computer Graphics," *Computer Graphics*, Vol.25, No.4, July 1991, pp.319-328.

21. Sims, K., "Interactive Evolution of Dynamical Systems," *Toward a Practice of Autonomous Systems: Proceedings of the First European Conference on Artificial Life*, ed. by Varela, Francisco, & Bourgine, MIT Press, 1992, pp.171-178.

22. Sims, K., "Evolving Virtual Creatures," *Computer Graphics*, Annual Conference Series, July 1994, pp.43-50.

23. Smith, A.R., "Plants, Fractals, and Formal Languages," *Computer Graphics*, Vol.18, No.3, July 1984, pp.1-10.

24. Yaeger, L., "Computational Genetics, Physiology, Metabolism, Neural Systems, Learning, Vision, and Behavior or PolyWorld: Life in a New Context," *Artificial Life III*, ed. by C. Langton, Santa Fe Institute Studies in the Sciences of Complexity, Proceedings Vol. XVII, Addison-Wesley, 1994, pp.263-298.

Altruism in the Evolution of Communication

David H. Ackley
Bellcore
445 South St
Morristown, NJ 07960
ackley@bellcore.com

Michael L. Littman
Brown University
Department of Computer Science
Providence, RI 02912
mlittman@cs.brown.edu

Abstract

Computer models of evolutionary phenomena often assume that the fitness of an individual can be evaluated in isolation, but effective communication requires that individuals interact. Existing models directly reward speakers for improved behavior on the part of the listeners so that, essentially, effective communication *is* fitness. We present new models in which, even though "speaking truthfully" provides no tangible benefit to the speaker, effective communication nonetheless evolves. A large population is spatially distributed so that "communication range" approximately correlates with "breeding range," so that most of the time "you'll be talking to family," allowing *kin selection* to encourage the emergence of communication. However, the emergence of altruistic communication also creates niches that can be exploited by "information parasites." The new models display complex and subtle long-term dynamics as the global implications of such social dilemmas are played out.

1 Models of the Evolution of Communication

Although schoolbook treatments often leave the impression that Darwinian evolution is about never-ending competition, a ceaseless struggle for survival by individual creatures in a nasty environment, it doesn't take much wide-eyed observation of the natural world to see that cooperation among individuals also plays a huge role. Using communication to accomplish tasks is a central example, as many other forms of cooperation presuppose a means of communication.

"Communication" is a very broad concept; here we focus only on evolutionary issues related to sending and receiving *initially arbitrary signals* about a shared environment. Focusing on signals avoids the complexities of syntax and compositional semantics, while initial arbitrariness excludes certain degenerate cases — a sudden change in direction, for example, might be an "incidental communication" to others that a predator is approaching. Though this may actually be the most prevalent means by which information moves between individuals, for present purposes it is uninteresting since such behavior presumably will be selected for even if the individual is alone.

We also wish to eliminate the possibility of "mimetic semantics," wherein an emitted signal somehow imitates the stimulus it represents, as in the use of a hissing sound to denote a snake. Such signals may be non-trivial from an evolutionary point of view, if they provide no direct value to the signaller. The question of why such signalling would evolve thus arises, and that is indeed the central question we wish to explore. However, although the *motivation for sending* a mimetic signal is non-trivial, the *mechanism for understanding* it is degenerate. Initial arbitrariness forces us to consider the harder problem of evolving both "speaking" and "understanding" abilities *simultaneously*.

There are, principally, two recent simulation models of the evolution of communication (MacLennan, 1991; Werner & Dyer, 1991; see also Hutchins, 1991, for an interesting but less closely related model), and it is useful to compare them briefly with the approach we are taking. MacLennan (1991) considers a population of simple machines, represented genetically (and phenotypically) by truth tables, and creates a shared environment through which the machines can pass initially arbitrary signals. A mostly conventional genetic algorithm (Holland, 1975; Goldberg, 1989) is used to evolve the population, based on a scoring function that measures how effectively communication is being used by the machines. Crucially, for our present purposes, the scoring function is such that the *speaker*, as well as the listener, is rewarded whenever a "match" occurs, meaning that the listener performed an action appropriate to the stimulus the speaker saw. One could imagine a circumstance where "truthful speech" by a speaker and "right action" by a listener causes food to rain down on both. MacLennan observes effective communication evolving in his machines, but other phenomena typically associated with communication do not occur. Lying, for example, is utterly pointless under such conditions: either both parties benefit or neither does.

The model of Werner & Dyer (1991) is different in a number of respects. Unlike MacLennan's model, there is no explicit scoring function; instead, effective communication allows "males" to find "females" more rapidly and thus increases the reproductive rate of individuals that communicate compared to those that do not. One could imagine this as a sort of "firefly" model, in which females in the grass below signal to males flying above, using basically arbitrary signals for more efficient mate-finding. In principle at least, one could imagine more subtle communication phenomena emerging in

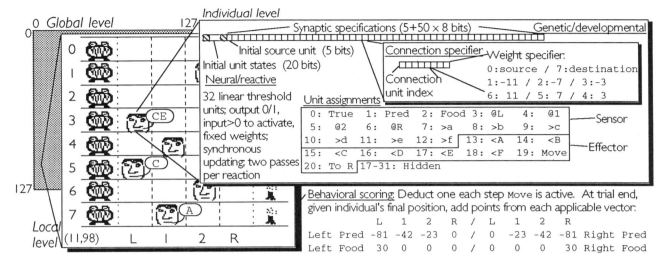

FIGURE 1. Model summary

this model — if, for example, during evolution the initial population divided itself into multiple species, a female of one species might send a misleading signal to a male of another species to impede his search for a suitable mate. Their model does postulate a spatially-distributed population, so one could imagine speciation along "territorial" lines occurring, and they discuss such possibilities. However — perhaps because of the globally-mixing reproduction strategy they employed — neither speciation nor any more complex communication phenomena were observed, so for present purposes the Werner/Dyer model is akin to the MacLennan model, in practice if not in principle, in that communication always turns out to be a win-win proposition for those involved.

2 Model Description: Worlds Within Worlds

Figure 1 sketches the models we explore. There are three organizational levels, which we refer to as *individual*, *local*, and *global*. The individual level captures the representation of genetic information in a *genotype*, the behavioral characteristics of a *body* or *phenotype*, and how a body is developed from a genotype. The local level captures the environment within which a group or *subpopulation* of individuals lives, the mechanism for allowing communication among the individuals of a subpopulation, the process of assigning a behavioral score to each individual in a subpopulation, and a within-subpopulation strategy for birth and death. The global level captures principally migration between subpopulations, although in two of the variations we present below, reproduction and migration are at least partially coupled and therefore reproduction is a global issue as well.

The level are interdependent in various ways; here we start at the local level, then go down to the individual level, and then up to the global level. We try to be as complete about details of the model as we can, not because we believe every detail to be critical — some of them certainly *seem* minor — but to be clear about what we have and have not built into the model "up front", and because, as experience with computer models of complex systems shows over and over again, it is nearly impossible to be certain *which* details don't matter.

2.1 Local level: A day in the life

First, we consider the behavior of individuals in a subpopulation, without worrying about how an individual implements any particular behavioral pattern. A subpopulation consists of eight individuals in a partially shared environment. The environment can be envisioned as eight horizontal *tracks* separated by walls that block *vision* but allow (non-localized) *sounds* to pass over. Each track is divided into four locations, labelled L(eft), 1, 2, and R(ight), and one individual lives in each track. An individual may move left or right along its track, one location per step, and may also *speak* on any or all of six independent audio channels (labelled A–F).

Nasty, flame-throwing predators (Pred) and delicious foods (Food) can appear at either end of the tracks, or both. When an individual is at L or R it will *see* either Pred, Food or nothing depending on what stimulus happens to be at that end; individuals at 1 or 2 see nothing. Also, each individual may *hear* (labelled a–f) the sum of the speech emitted by the subpopulation on the previous step.

A round of behavioral scoring, a *day*, begins by setting the individuals' scores to zero, and is then followed of 36 independent *trials*. The L and R stimuli are constant and identical across tracks during a trial. The eight individuals are initialized and placed at various starting locations in their tracks, and they then *react* for three *steps*. After the trial, each individual's behavioral score is adjusted as a function of the L and R stimuli, its final location, and how much it attempted to move, as shown in Figure 1.

The 36 trials consist of four repetitions of the nine possible combinations of L and R stimuli. The individuals' starting locations are randomized except that each individual begins in a different location on each repetition of a stimulus-pair, and exactly two individuals start in each location. This behavioral scoring procedure, though costly in simulation time, has the analytically useful property that unless the individuals of a subpopulation actually are communicating, the resulting individual behavioral scores are completely deterministic. We can compute, for example, that the best score possible in absence of communication is -12, and therefore any behavioral score greater than -12 is *proof* that that individual, on that day, definitely benefitted from signals emitted by the rest of its subpopulation.

At the end of the day, depending on the model variation involved, a *local reproduction* may occur. When it does, the following occurs: The eight individuals are ranked based on their behavioral scores. Two *parents* are chosen uniformly at random from the top half of the subpopulation, and an *offspring* is created (details in the next section). Then, one of the eight existing individuals is chosen uniformly at random and *killed*, and the offspring replaces it in the population.

2.2 Individual level: Genes and neurons

The previous section described the sensations and actions that an individual *may* experience and perform; which of them are *actually* used, and in what manner, is determined by the *brain* of the particular individual, which, in turn, is determined by the individual's genes. In the models we present in this paper, no "plasticity" or "adaptation" is involved — the genes completely and permanently determine the "wiring" and cognitive function of the individual.

An individual's brain is a synchronously updated neural network containing a total of 32 linear threshold units (Rumelhart, et al, 1986) assigned as follows: 12 *sensor units*, providing information about food, predators, location, and sound; 8 *effector units* controlling whether to move, which direction to go, and what to say; a *true unit* that always has value 1; and 11 *hidden units* that have no prespecified function.

The genome of an individual defines a wiring diagram and initial conditions for this architecture. A total of 448 bits of genetic information is divided up into three groups: 19 bits for the initial states of the effector and hidden units at the beginning of each trial; 405 bits of *synaptic specification genes* describing up to fifty connections between units; and 24 leftover bits, *pseudo genes* that are never decoded.

The *developmental process* grows a network from a genome, sequentially interpreting the synaptic specifications, operating in either *source mode* or *destination mode*. It begins in source mode. The 405 synaptic specification bits are: a five bit *initial source* group, and fifty *connection specifier* groups of eight bits each. The initial source yields an integer from 0 to 31, denoting a unit to use as the *current unit*. In source mode, new connections are created from the current unit to a unit determined by the next connection specifier; in destination mode, new connections are created *to* the current unit *from* a unit determined by the next connection specifier. A connection specifier is a five bit *connection unit* index and a three bit *weight specifier*. Two of the eight values of a weight specifier are *mode shift codes*: value zero means "set the current unit to the connection unit and enter source mode", and value seven means "if the connection unit is a sensor unit, do nothing, otherwise set the current unit to the connection unit and enter destination mode." The other six values of the weight specifier indicate that a specifically-weighted connection is to be grown between the current unit and the connection unit, as indicated in Figure 1.

This scheme for neural development, though devised mostly to conserve computer memory, has the effect that small changes in the genome can lead to large changes in the resulting phenotype. A single bit flip in the connection unit group of a mode-shifting connection specifier, in particular, can transfer many synapses from one source or destination to another.

At the beginning of each trial, the effector and hidden units are initialized from the initial states genes. On each step, the network *reacts*: first, the sensor unit values are determined from the environment: zero or one for the Pred, Food, L, 1, 2, and R units, as appropriate, and zero to eight for each of the six hearing ears — the sum of all speech from the previous time step (all zeros are always heard on the first step). Then, an *update pass* over the effector and hidden units is done: For each unit, the weighted sum of its inputs is computed, and if that sum is greater than zero, the unit adopts value one, otherwise it adopts value zero. Two passes over the network are performed each step, making it possible, for example, for a hidden unit to contribute to the decision about what to do now based on the latest environmental input. After both passes are concluded, the effector unit states are read off and executed — attempting to move left or right, or not, as directed, and contributing speech on channels A..F, or not.

The only individual level aspect we have not discussed is the way an offspring genotype is derived from two parent genotypes. There is one almost certainly inconsequential detail, and one very likely consequential detail, about how we do this. The basic mechanism we employ is *genetic recombination*, or *crossover*; the specific variation we use is, in the genetic algorithms literature, called *parameterized uniform crossover* (Spears & De Jong, 1991; Ackley, 1987) with a crossing probability of 0.05 per *byte* — and that is the almost certainly inconsequential detail: It is impossible, in this scheme, to cross more than once in a single byte of genotype.

The likely consequential detail is this: In these studies, we use *no mutation at all*. Recombination is the only *genetic operator* we employ. This is out of step with most genetic algorithms research — if anything, the trend in the last decade has been towards higher mutation rates, to combat the "premature convergence" problem that genetic algorithms often display on optimization tasks. One of us has actively championed that trend, in fact, in previous writing (Ackley, 1987). Why do we reverse direction in this case?

As we explored early versions of this model, hoping to see effective communication without "paying off" the speakers, we found ourselves turning the mutation rate lower, and seeing the results becoming more promising. Even a low mutation rate, it seemed, destroyed communicating subpopulations, eating them up from the inside out. With some trepidation, we decided to turn mutation off entirely... and the model behavior became much more interesting. We will return to this issue in Section 4.

2.3 Global level: Reproductive and migratory

Now we have discussed how an individual works, from its genes up to its sensorimotor interface and "cognitive function", and how individuals in subpopulations are evaluated, bred, and slaughtered. Just that much is a complete algorithm, in the sense that one could simulate one such group of eight individuals, and watch what happens. We did just that, during exploratory simulations — and what happens is not very interesting. The tiny group converges, genetically, in a flash, usually with very poor behavioral scores for the individuals. To make the model really cook, we need to have many such subpopulations evolving at once, and then consider what happens when individuals move between subpopulations. That is what the global level of the model does.

As in some prior work (Ackley & Littman, 1993), we designed the global structure of our model to match the overall characteristics of the fairly large parallel computer we were fortunate to have available. Overall, the world is a square *array* of 128x128 *cells*, each of which contains a subpopulation of eight individuals, for an aggregate population size of 131,072. The *neighborhood* of each cell are the eight closest cells on the array; the array overall is configured as a torus. A *quad* of cells is any 2x2 group of adjacent cells.

In the case studies described in the next section, we employ two global level mechanisms, alone and in combination. The first mechanism, *wind*, implements simple migration. *Windy days* occur at regular intervals, with a frequency that varies from case to case. On a windy day, first, all the normal daily events occur. A global wind direction is then chosen uniformly at random from the eight compass directions. In each subpopulation, an individual is then selected uniformly at random, then all the selected individuals migrate to the "downwind" neighbor cell. The immigrant immediately takes up residence in the track vacated by the emigrant.

The second mechanism, *festival*, implements a combination of reproduction and migration. *Festival days*, like windy days, occur at regular intervals with a case-specific frequency. On a festival day, first, the normal behavioral evaluation occurs. Then, instead of performing a local reproduction in each cell, festivals are held in each of 4,096 quads. The 32 individuals from the four cells of a quad are ranked in a single list based on their behavioral scores for that day, and a *quad reproduction* is performed, selecting two parents uniformly at random from the top quarter of the ranking, crossing them as in a local reproduction, killing an individual chosen uniformly at random from all 32, and placing the offspring in the subpopulation the dead individual was part of.

Two details conclude the description of the global level, and of the model overall: First, on successive festival days, the *phase* of the assembled quads is shifted, so that after four festivals, any given cell will have interacted with all eight of its neighbors. Second, in the case where it is possible to have *windy festival days*, the festival occurs first, and then the wind blows.

3 Case studies: Results and Observations

As the previous section should have made clear, although our model is but a pale shadow of even a small natural world ecology, it is a significant computational challenge. Despite efficient programming, the running time for each simulation reported here is measured in multiples of weeks. As a result — like many natural world experimenters, but unlike most artificial life researchers — we have not had the luxury of running many repetitions of each model variation to assess noise sensitivities. This is significantly less worrisome then it would be in a smaller model — for example, since we have over sixteen thousand subpopulations being simulated in parallel, random variations in the initial subpopulations tend to average out *spatially* in the early stages of a single run — but nonetheless, fair warning should be given.

We have focused on longitudinal empirical explorations, performing an extended simulation of each of three global strategies: wind-only, wind+festival, and festival-only. The festival-only strategy was so successful that we invested in a second run, varying only the pseudo-random number seed. Although the runs were unique in many details, the qualitative phenomena we discuss below appeared both times.

3.1 Case 1: Wind-only

In the first variation we used wind migration, with every fifth day being windy. We ended up letting it run for 13,110 simulated days before deciding we had seen most of the phenomena it was going to display. Some overall statistics are summarized in Figure 2. The upper curve represents the highest average behavioral score for a subpopulation and the lower, dotted curve displays the average of all behavioral scores over the entire array. The individual dots are average behavioral scores for randomly-selected subpopulations to help show the spread in the population. The average steadily grows and ends up just shy of the −12 mark. The mode subpopulation is −56 from around day 1000 to almost day 3000. From then on, the −12's dominate the array — the sample dots fuse into a line.

As we watched that data coming in, at first we were excited, when the maximum behavioral scores jumped to 42 — communication was definitely beginning to happen! — and then

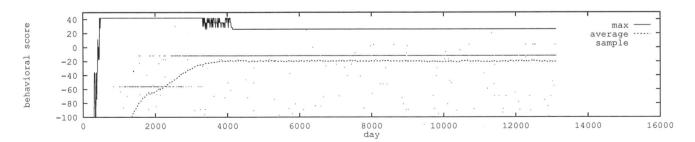

FIGURE 2. Behavioral scores verus time: Case 1 — Wind-only

we were disappointed when they fell back to 26, since it was "heading the wrong way" and we knew from first principles there were communication strategies that could score far higher than 42.

Our understanding was aided by watching "movies" of the spatial dynamics, based on data files written by the simulator. Plate 2 consists of six *frames* taken from the wind-only simulation. Each frame contains 128x128 pixels, one for each subpopulation, each assigned a color on the basis of the average of the behavioral scores in the corresponding subpopulation on a given day. The mapping from behavioral score to solid and horizontally striped colors, shown at the bottom of each plate, was chosen only to highlight score differences, and it is repeated beneath each of the frames along with a compressed histogram of the number of subpopulations versus subpopulation-average behavioral score.

We made movies for each of the model variations, and all shared distinctive features, supporting the idea that many of the effects are repeatable and depend more on the overall structure of the model and less on noise effects specific to a particular case. The following discussion of the initial part of the wind run can serve as an introduction to early features shared by all the cases.

We refer to an organism type by the score it obtains in a subpopulation of "clones," since that is deterministic. A creature described as, say, a "26," may receive a completely different score in a mixed population.

The nature of the neural architecture and the developmental process is such that, for a randomly generated genotype, by the far the most probable resulting behavior pattern is never to move at all. Such individuals receive behavioral scores of −696 and are so common at day 0 (Plate 2a) that subpopulations made up entirely of such stiffs are the most common shade of grey in the first frame. These sessiles are soon rooted out by more active strategies, which have spread and taken over small patches of the array by day 430 (Plate 2b).

By day 1620 (Plate 2c), more improvements have been discovered and spread to the majority of the array. The most popular organisms at this time score −56 and appear dark green. The −56's combine a default strategy of running to one end of the track with an overriding strategy of running the other way if a predator is seen. The medium green subpopulations visible in the same picture consist of −12's, which implement

the same strategy as the −56's except that they default to whichever end is nearer at the start of each trial. As mentioned in Figure 2.1, this is actually the best a non-communicating individual can do.

By day 3800 (Plate 2d), these self-reliant creatures dominate the array, forming the global "cellular" structures typical of these simulations. The borders between regions usually have lower behavioral scores resulting from crosses between incompatible organisms. (Some properties of such *mixing zones* are discussed in Ackley & Littman, 1994).

In all three cases we've explored, patches of communicating individuals do arise and expand. Some examples are visible in Plates 2c – 2f as yellow/orange (26's) and light green (4's) patches. Under the wind model, however, these communicating subpopulations are ephemeral. Before growing large, they are squeezed out by −12's that have discovered ways to mislead and deceive them. As we observed the wind model dynamics, we realized we "could have predicted it": After each windy day, at least potentially, "you'll be talking with strangers", so kin altruism has a hard time stabilizing, and cheaters have an easy time invading.

From that perspective we were impressed that, from day 4150 until we killed the run, the 26's manage to stave off extinction without ever controlling more than a handful of subpopulations at a time. They seem to survive because they "trust their ears" only in limited circumstances: Most of the time they follow the optimal non-communicating strategy of the −12's, and rely on signalling only to avoid fleeing from one predator to the other in the dangerous Left `Pred`/Right `Pred` trials. Given the "anything goes" nature of wind, the emergence of such "cautious communicators" was a satisfying result.

3.2   Case 2: Festivals and wind

Despite the persistence of the cautious communicators in the wind run, they never manage to hold a significant portion of the array for very long. We devised festival reproduction to increase the cohesion of groups in the hope that cautious communicators might be able to stabilize and more trusting communicators might appear.

Festival reproduction is in a sense a score-sensitive migration mechanism since only individuals with behavioral scores in the top quarter of their quad are given opportunities to reproduce into neighboring subpopulations. We expected that us-

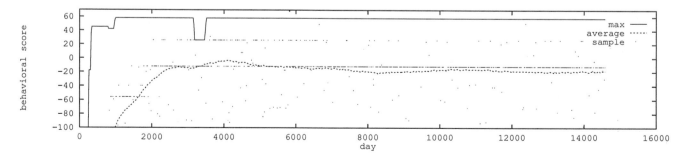

FIGURE 3.   Behavioral summaries versus time: Case 2—Wind+festival

ing more festivals than wind would allow for greater stability of high scoring subpopulations, and greater resistance to intruders from other subpopulations.

We ran the simulator with festival reproduction every other day and wind migration every 10th day for a span of 14,580 days. Our predications were only partially borne out, as can be seen in Figure 3 and Plate 3. In the figure, notice that the cautiously communicating 26's are populous enough to be clearly visible in the random samples. Furthermore, for a brief time between day 3000 and day 5000, the population average actually exceeds −12 : The average subpopulation was communicating.

Plate 3 tells the story of ongoing battles between cooperating communicators and competing individualists. By day 2000 (Plate 3a) a large cast of characters is onstage: dark green −56's, a rogue's gallery of green −12's, a vigorous species of yellow/orange 26's, and a colony of orange/red 58's. Over the next several hundred days, the 26's spread over the −12 species east and west of them, even as embedded species of −12 eats them up from the inside. By day 3440 (Plate 3b), the 26's hold substantial territory, but have been chopped completely in two by the (no-longer-embedded) −12's. Also, a group of light green 4's can be seen overrunning a patch of −12's (right of center). By day 4020 (Plate 3c) those 4's have been taken over by orange/red 58's, while marauding −12's continue to pressure the (now-reconnected) 26's.

The balance of power shifts over a few thousand days (Plates 3d and 3e) and by day 8400, few communicating species remain. A reemergence of 58's around day 11,000 (Plate 3f) is soon quashed by disruptive breeds of −12's. Communicators spread farther and faster in the wind+festival model than in wind-only but they are still unstable.

3.3   Case 3: Strictly festival

Wind migration tends to favor disruptive species, allowing them to infect and parasitize communicating species. By permitting migration by festival reproduction only, we hoped to

shift the balance still further towards group-cohesion. We ran the simulator using festival reproduction every two days for 99,980 days. A second run of 26,400 days yielded similar results. Only the longer run is reported here.

Figure 4 shows the behavioral summaries for the run. The global population average behavioral score exceeds −12 around day 5000, continues to rise through most of the run and reaches almost 100 — though that sharp decline in the average behavioral scores in the mid-20000's demands investigation.

Plate 4 illustrates some of the history of this run. Early on (day 2000, Plate 4a), the array is dominated by growing −12's (green) and dwindling −56's (dark green) but two large communicating populations (both orange/red) have appeared. On the left is a species of 59's, on the right, 58's.

By day 5600 (Plate 4b), species of −12's have taken over the entire array except for the two orange/red areas and a third light red patch of 62's that is just beginning to be conquered by a species of 142's (pink) — then the most successful communicators discovered.

As time passes, the 58's and 59's expand, and cautiously communicating 4's (light green) slowly replace the background species of −12's. At day 20,700 (Plate 4c), a group of striped orange 42's is growing from the base of the 142 patch, and the 58's and 59's have just met up. In direct competition, the 59's dominate the 58's and overrun their positions in the array. The leading edge of the 59 invasion wave is visible in the frame.

The increased genetic diversity resulting from the flood of 59's mingling with a new set of genes results in several breakthroughs which hone the 59's into individuals scoring 71 (dark red). At day 24,340 (Plate 4d), the new species can be seen sweeping back across the territory of the former 59's. At the same time, however, another new breed has started to spread. Individuals of this breed score an abysmal −175 when cultured in isolation yet they displace the new 71's very

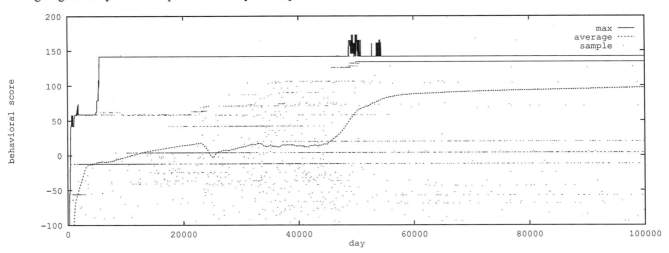

FIGURE 4.   Behavioral summaries versus time: Case 3—Strictly festival

rapidly. The −175 's decimate the 59/71 's only to be eliminated themselves by opportunistic −56 's and then −12 's. By day 48,780, Plate 4e, only traces remain of the orange/red populations.

That is the sequence of events that caused the temporary crash in the global average, and it highlights a basic difference between this world and the function optimizations performed by conventional genetic algorithms. The fitness scores in a GA form a total order — any pair of individuals always stand in the same relationship to each other: more, less, or equally "fit." With subpopulation-dependent behavioral scoring, there is no such simple dominance. In this particular simulation, for example, the 71 's are "more fit" than the surrounding −12 's since they slowly displace them. The −175 's are "more fit" than the 71 's since they destroy the 71 's in direct contact (see the next section) but, completing the circle, the −12 's are "more fit" than the −175 's and tend to replace them in the population.

By day 48,780 (Plate 4e), the after-effects of the 58-59 collision have mostly died down. The pink 142 's still hold nearly the same area of the array and a sequence of species has arisen from the ashes of the earlier conflagrations. These consist of 126 's, 128 's, 132 's and 134 's, and appears as various shades of pink spreading up towards the 142 's. Over the next 50,000 days (Plate 4f), the dark pink 134 's conquer most of the world. Successful communication dominates.

3.4 Analysis: War and Pestilence

The stories and pictures in the previous sections suggest the complexity of the evolution of communication in our simulations, but such global stories only describe *what* happened. To understand *why* things happened we must delve into the local and individual levels of the model.

Similar to Ray (1991), we use *culturing* to analyze interaction effects between various species at the local level. We take a sample of two evolved genotypes, call them A and B, from the simulator's data files. Using the local level of the simulator, we find behavioral scores for a subpopulation of 8 A's. Because of the way the evaluation procedure is designed, an individual cultured with clones will always receive the same behavioral score.

The score tells us how a completely converged subpopulation of A's would do "in the wild." We also evaluate a population of 8 B's, and then do 50 evaluations for each of the 7 possible combinations of *i* A's and 8-*i* B's. The results show how the scores of A's and B's change when in contact with members of the other species. (Note that they tell us nothing about the result of *mating* A's and B's.) We plot the results for both species as a function of the number of A's, with 95% confidence intervals if there is any variation.

As an example, we used this tool to examine the evolution of the −12 's. Figure 5 shows the result of culturing two different species of −12 's, an "archaic" breed from near the beginning of the festival run and a more "modern" one from close to the end. Though both species score −12 when cultured with clones, and also in silence (determined in a separate experiment), the archaic −12 's obtain terrible behavioral scores in the presence of even one modern.

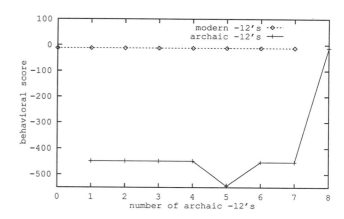

FIGURE 5. Archaic −12 's versus modern −12 's

The keys to success for the modern −12 's are two abilities: On the one hand they are broad-spectrum noise generators, and on the other hand they are themselves resistant to ambient noise. They are "ether warriors": They jam the signals of communicating groups, and confuse organisms that are sensitive to the ether, even if they don't signal themselves. An examination of the neural networks of these two species shows this clearly. The archaic −12 's are sensitive to sounds on several channels and speak on F. The modern −12 's have no "ears" at all and they are "wired for speech" on all channels (although only CDEF seem to occur in practice for this breed). They pollute the ether — tending to lower the score of communicators in the subpopulation — without affecting their own scores at all. Such "scorched ether" strategies are very common in the wind and wind+festival studies.

Under the festival-only model, it is generally not possible for one disruptive individual to enter and exterminate a high-scoring subpopulation without first understanding its signals. Thus, instabilities in this variation are of more subtle types. The fall of the 59/71 's to the −175 's is an excellent example of the takeovers that occur. The −175 's are perfectly adapted not only to parasitize the 71 's, but to wipe them out entirely. As seen in Figure 6, a single −175 in a population of 71 's ex-

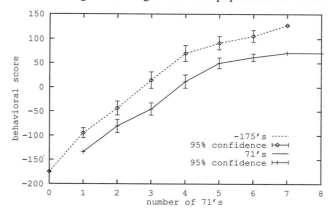

FIGURE 6. Altruistic 71's versus parasitic −175's

ploits the signals generated by the 71 's to obtain a superior behavioral score. The −175 's are then preferentially reproduced — there tend to be more −175 's and fewer 71 's. Although both species then score worse than they did when there were more 71's, in every possible combination, the −175's score higher. The 71's in the subpopulation are soon exterminated.

There is a second critical factor in the −175 epidemic discernible in Figure 6. Recall that in this simulation, the only way to migrate is to win at festivals. For taking over a single cell, all that is necessary is that the attacker score better than the defender in the cell: An invader to a group of 71's, for example, could take it over by scoring 50 if it could confuse the 71's into scoring 40's. In such a case, though, the invader would never win in festivals with adjacent uninfected groups of 71's. It would be stuck. In Figure 6, by contrast, we see that the canny −175's obtain scores that win festivals against neighboring 71's even when they've taken over fully half a cell, providing plenty of time to infect neighboring subpopulations before the inevitable crash.

An examination of the genomes reveals that the −175's are closely related to the 71's they wipe out, sharing 97.8% of genes. A naive notion of kin selection might predict increasing altruism with increasing levels of genetic overlap, but clearly this need not be. On the contrary, by sharing genes with the 71's, the −175's are in a good position to understand their signals and thereby exploit them.

4   Discussion: Mutation and Migration

We have demonstrated that effective communication based on the exchange of initially arbitrary signals can evolve and stabilize even when it provides no benefit to the individual speaker. We believe that the principal feature of our model that makes this possible is the *partial* alignment of communication and reproduction domains. To the degree they are the same, the potential beneficiaries of a speech act will tend to be close genetic kin to the speaker, and then the "kin selection" or "kin altruism" arguments from evolutionary theory (Hamilton, 1964, and many since) apply and one may hope to see the emergence of altruism. On the other hand, unless communication and reproduction domains are at least sometimes different, subpopulations will tend to converge, and the search for improved forms will be slowed. The case studies presented in the previous section only hint at the scope of the phenomena we have observed; here, we close with brief discussions of two issues that the models raise.

From an evolutionary computation point of view, perhaps the most striking aspect of these studies is that they make no use of mutation. As mentioned earlier, we were pushed to this design decision, against our own preconceptions, simply because it worked better — but what makes this case different?

In the majority of the evolutionary computation models in the literature, individuals are evaluated in isolation from each other. Consequently, the effects of a mutation are, initially, limited to only one individual. In our models — and in any co-evolutionary system in which relatively small groups of individuals can interact during the process of selection — the picture is quite different. Individuals in such systems will evolve towards *dependence* on the typical behavior patterns of the group they are part of whenever it provides an advantage. A mutation in one individual can have a much more dis-

ruptive effect, impacting other individuals immediately, in the same generation.

In his paradigm-creating book on genetic algorithms, Holland (1976) argued that crossover is more important than mutation as a genetic operator, and offered mathematical arguments for the hypothesis that the *reason* it is so important is because it improves the efficiency of an evolutionary search for highly-fit individuals. Although his arguments were insightful and elegant, in the nearly two decades of ensuing research, convincing empirical demonstrations in which crossover is more important than mutation have remained elusive; so much so that Holland has recently returned to this question, collaborating on a paper entitled "When will a genetic algorithm outperform hill-climbing?" (Mitchell, Forrest, & Holland, 1994, to appear).

Studying the evolution of communication, we are lead to propose a slightly different hypothesis: Crossover *is* a critical genetic operator, but not necessarily or even principally because it improves search in highly diverse populations. We suggest that the reason crossover is so important is because it amounts to a *convergence-sensitive variable mutation rate*. With a single mechanism, one can on the one hand generate radical new combinations when no particular strategy is dominant, as evidenced by a diverse genetic pool, while on the other hand preserve all the genetic structures that are "proven winners," as evidenced by their fixation in the population.

Note that to the degree that a system can be viewed as an isolated-individual optimization problem, with a "fitness function" that is stable over evolutionary time, Holland's account and ours are largely complementary takes on the same idea — "effective generation of new structures when diverse" versus "effective preservation of old structures when converged." In such systems his account is preferable because it makes non-trivial claims about why crossover's particular approach to new structure generation is desirable, whereas on the flip side there's basically only one possible strategy for old structure preservation.

However, to the degree that a system is best viewed as a co-evolving organization of relatively converged populations, where the behavioral score of assigned to a genome can change even if the genome remains constant, the situation is different. Clever search techniques that generate novel candidates by extracting information from behavioral scores efficiently are of less value, because the information necessary for success changes. On the other hand, the high-fidelity reproduction of individuals in converged groups has much higher value, because the group members can work together to improve all their behavioral scores if everybody cooperates. As we saw in Section 3, cheaters need not be strangers, so even "rare" changes due to mutation can have disastrous large-scale effects. We suspect many natural systems — and many human and artificial systems as well — are less like diverse population, static optimization problems, and more like co-evolutionary processes involving relatively converged populations. In such cases, crossover may outrank mutation, not because it searches effectively, but because it preserves selectively.

Still, even if this account of the roles of mutation and crossover is accurate, that wouldn't imply that the mutation rate should optimally be reduced all the way to zero, as we did in the studies reported here. The uniformity of color over large areas in the Plates, for example, suggests that there are many converged subpopulations in which no search at all is taking place (although, as the lone pixel visible in the middle of the pink patch in Plate 4d indicates, even deep in apparently converged regions there are some usually-silent variant alleles that occasionally are revealed via crossover.) The preliminary studies that prompted us to turn mutation off were all based on wind migration. In the festival model, by contrast, perhaps a small amount of mutation could enhance the search process without disastrous global consequences, and this is a possibility we are pursuing.

More generally, these studies suggest that intra-population genetic operators may be less important than inter-population migratory mechanisms in determining the qualitative behavior of distributed population models with subpopulation-dependent behavioral scoring. As a comparison between the wind+festival and strictly-festival studies shows, simple score-insensitive migration — even if relatively infrequent — can have a dramatic impact on the dynamics of the system.

There is a sense in which score-sensitive migration mechanisms such as the festival are fundamentally questionable. Consider ourselves as humans: As members of a communicating species we are products both of our genetic heritage and the communities within which we reside. Though we carry our genotypes with us when we (or our offspring) migrate, the communities we depended upon are left behind. For the artificial creatures in the festival model, although the participating individuals compete on the basis of their behavioral scores, those scores were obtained back in the individuals' "home" cells, not in the context of the whole quad. Such a score may be wholly unrepresentative of how that genotype would perform in another cell (even in absence of any genetic modifications). Using individual behavioral scores to determine who shall cross a score-dependent border is, in a significant way, an unprincipled use of the data.

To the degree that neighboring subpopulations differ from each other, a festival is comparing apples and oranges. As the genocide of the alruistic 75's at the hands of the grossly incompetent −175's showed, sometimes it can go wrong with disastrous consequences for certain species. But — barring degenerate "group selection" mechanisms, wherein entire subpopulations compete, reproduce, and displace each other as units — what choice is there? We didn't even bother with parallel simulations of the no-migration case, since the sequential simulator showed us that completely isolated subpopulations converge rapidly to usually awful scores. Using wind amounts to refusing even to attempt to compare apples and oranges. Despite its perplexing theoretical motivation, the success of the festival mechanism suggests that it is a game worth playing.

References

Ackley, D. H. (1987). *A connectionist machine for genetic hillclimbing*. Kluwer Academic Press: Boston.

Ackley, D. H. & Littman, M. L. (1994). A case for Lamarckian Evolution. In *Artificial Life III, SFI Studies in the Sciences of Complexity*, vol. **XVII**, edited by C. G. Langton, Addison-Wesley, 3–10.

Goldberg, D. (1989). *Genetic algorithms in search, optimization, and machine learning*. Addison-Wesley: Reading, MA.

Hamilton, W. D. (1964). The genetical theory of social behavior. *Journal of Theoretical Biology*, 1–32.

Holland, J. H. (1975). *Adaptation in Natural and Artificial Systems*. University of Michigan Press: Ann Arbor, MI.

Hutchins, E. & Hazlehurst, B. (1991) Learning in the Cultural Process. In *Artificial Life II, SFI Studies in the Sciences of Complexity*, vol. **X**, edited by C. G. Langton, C. Taylor, J. D. Farmer, & S. Rasmussen, Addison-Wesley, 689–706.

MacLennan, B. (1991). Synthetic Ethology: An approach to the study of communication. In *Artificial Life II, SFI Studies in the Sciences of Complexity*, vol. **X**, edited by C. G. Langton, C. Taylor, J. D. Farmer, & S. Rasmussen, Addison-Wesley, 631–655.

Mitchell, M., Forrest, S., & Holland, (1994, to appear). When will a genetic algorithm outperform hill-climbing? In *Advances in Neural Information Processing Systems 6*, edited by J.D. Cowan, G Tesauro, & J. Alspector. San Mateo, CA: Morgan Kaufmann.

Ray, T. S. (1991). An approach to the synthesis of life. In *Artificial Life II, SFI Studies in the Sciences of Complexity*, vol. **X**, edited by C. G. Langton, C. Taylor, J. D. Farmer, & S. Rasmussen, Addison-Wesley, 371–408.

Rumelhart, D. E., & McClelland, J. L. (1986) *Parallel Distributed Processing: Explorations in the microstructures of cognition. Volume 1: Foundations*. Cambridge, MA: The MIT Press (A Bradford Book), 63–64.

Spears, W. M. & De Jong, K. A. (1991). On the virtues of parameterized crossover. In *Proceedings of the Fourth International Conference on Genetic Algorithms*, edited by R. K. Belew & L. B. Booker. Morgan-Kaufman, 230–236.

Werner, G. M. & Dyer, M. G. (1991). Evolution of Communication in Artificial Organisms. In *Artificial Life II, SFI Studies in the Sciences of Complexity*, vol. **X**, edited by C. G. Langton, C. Taylor, J. D. Farmer, & S. Rasmussen, Addison-Wesley, 659–687.

Evolution of Metabolism for Morphogenesis

Hiroaki Kitano

Sony Computer Science Laboratory

3-14-13 Higashi-Gotanda, Shinagawa

Tokyo 141, Japan

kitano@csl.sony.co.jp

Abstract

This paper presents a simple computational model of development in multicellular organisms. Unlike previous models of the acquisition of development rules (such as L-system) through an evolutionary process, the new model acquires genetic codes for defining the metabolism within a cell. Morphogenesis is regarded as an emergent phenomena in this model. A genetic algorithm was used to evolve the genetic rules that determine the metabolism of cells. The model also captures intercellular communication, diffusion, and the active transport of chemicals. A new coupled map model, *super coupled map*, is proposed as a mathematical foundation for the dynamic behavior of the model. Experimental results demonstrate that the acquisition of genetic codes for simple metabolic reactions is possible, and a simple process of morphogenesis was observed.

Introduction

This paper describes a simple development model based on the evolution of metabolism. The ultimate goal of our research is to establish a model of the development of multicellular organisms, with particular emphasis on modeling neurogenesis and a *Perfect C. elegans*.

Understanding and simulating neurogenesis directly provides us with an insight into the emergence and evolution of intelligence. In addition, we expect novel computing principles and architectures to be derived from this research. *Perfect C. elegans* is an attempt to create a complete replica of *Nematode Caenorhabditis elegans (C. elegans)* in an electronic medium. We chose *C. elegans* because it is the simplest multicellular organism that is well understood from both its genetic and neural aspects [Sulston et al., 1983]. While being the largest organism for which we can obtain a detailed understanding in the foreseeable future.

Accomplishment of these tasks will require the fine-grain modeling of physics, chemical reactions, metabolism, and other phenomena. Although this research will take years of effort, this paper provides the first reports on our progress towards this goal, focusing on the preliminary results of modeling the development of a multicellular organism. The process of development from a single cell to a large number of differentiated cells is a genetically guided and environmentally affected process. The goal of this paper is to present a model of the development of a multicellular organism, integrating simple models of evolution, metabolism, chaotic behavior in cell activity, etc. Experiments demonstrate that the genetic codes for appropriate metabolism can be acquired through simulated evolution.

Previous Research on Development

Several development models has been proposed to date. For example, L-system, proposed by Lindenmayer [Lindenmayer, 1968; Lindenmayer, 1971] is a descriptive and syntactic approach to capturing an early stage of cell division. Several variations of L-system have been proposed, all being able to describe the early stage cell division of *Aster novae-angliae, Anabaena catenula, Callithamnion roseum,* and others. Kitano used a graph L-system to generate the structure of neural networks [Kitano, 1990; Kitano, 1994]. In Kitano's method (called *Neurogenetic Learning (NGL)*), a genetic algorithm is applied to the acquisition of graph rewriting rules, which are used to rewrite a graph representing the network structure. This approach exhibits better scaling properties and convergence speed, and gives us a more realistic picture of the development of neural networks than is possible with other methods (such as [Miller et. al., 1989]). Since this approach was first proposed in 1990, several research projects have been started to further augment the basic ideas behind the model.

However, NGL suffers from several drawbacks:

Central control: In NGL, the network configuration during development is described as a single connectivity matrix, resulting in central control of the process. There are two major problems: (1) there is no biological correspondence, and (2) central control is computationally inefficient for large-scale parallel implementations.

No physical constraints: Any two cells are allowed to establish a connection in NGL, due to a lack of physical constraints. This the lack of physical constraints again results in two major problems — no biological reality, and inefficient at parallel implementation.

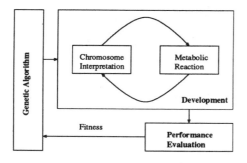

Figure 1: Overview of the model

Limited to neural networks: NGL is designed specifically to model the development of neural networks. It can not be applied to general morphogenesis modeling.

Deterministic model: L-system used in NGL is a deterministic and highly descriptive model. Thus, dynamic interaction between cells and the metabolism can not be captured.

Apart from models using the L-system, Kaneko and Yomo proposed an interesting model of cell differentiation [Kaneko and Yomo, 1994]. Their dynamic clustering model is based on global-coupled map, and simulates cell differentiation without being affected by genetic or positional information. Their model is quite interesting since they start from a model of metabolism within a cell. However, their model does not incorporate any evolutionary process. Fleischer and Barr proposed a model of multicellular development [Fleischer and Barr, 1994]. However, neither metabolism nor evolution were modeled.

In this paper, we present a new model that overcomes these problems. The model simulates simple metabolism within a cell. The development process is viewed as an emergent property of the cell activity.

The Model

Overview

In contrast to NGL, which acquires rewriting rules through evolutionary computing, our new model acquires metabolic reactions within each cell. Fig. 1 shows the overall organization of the model. A genetic algorithm is used to evolve a population of individuals. The chromosomes in each cell determine the metabolic reactions in that cell. Metabolic reactions are computed using a continuous value discrete time simulation. The production of certain chemicals is related to the synthesis of DNA. When the synthesis of DNA reaches a certain level, cell division takes place. After a certain number of simulation cycles, each individual will consist of a number of cells, some which are already differentiated. When this model is applied to neurogenesis, the model should be extended to simulate the formation of networks.

Super Coupled Map

Mathematically, the model uses a new type of coupled map. Coupled maps are mathematical formalization of a large-scale chaos, where a system consists of large number of elements each of which exhibits chaotic behavior. In a coupled map lattice (CML), a new state for the cell is determined from the state of the cell and its neighbours. Equation 1 formalizes this model in a case of one-dimension cell array.

$$x_i(t+1) = [1-e]f(x_i(t)) + \frac{e}{2}[f(x_{i-1}(t)) + f(x_{i+1}(t))] \qquad (1)$$

The other extreme is the global coupled map (GCM), described by equation 2.

$$x_i(t+1) = [1-e]f(x_i(t)) + \frac{e}{N}\sum_{j=1}^{N}f(x_j(t)) \qquad (2)$$

In this case, a new state for the cell is determined from the state of the cell and a global variable (e.g. the average energy level of the system). Both CML and GCM were proposed by Kaneko [Kaneko, 1992].

In this paper, we introduces a new map, *super coupled map (SCM)* whose one-dimension cell array formalization is described in equation 3.

$$x_i(t+1) = [1-e-g]f_i(x_i(t)) + \frac{e}{2}[f_{i-1}(x_{i-1}(t)) + f_{i+1}(x_{i+1}(t))] + \frac{g}{N}\sum_{j=1}^{N}f_j(x_j(t)) \qquad (3)$$

Fig. 2 graphically describes three types of coupled map. SCM differs from CML and GCM in two points:

1. SCM involves both local and global coupling, where CML and GCM formalizes either one of local or global coupling.

2. SCM allows function in each cell to be different, where CML and GCM assumes all cell emcompasses same function. In the above equations, CML and GCM only allow $f(x)$ to be identical in every cells in the system. SCM allows $f_i(x)$ to be different in each cell.

These differences are essential for modeling of biological systems, because, in the actual biological systems, (1) each cell is influenced by both local and global chemical concentration, and (2) each cell may take different state of gene regulation and expression.

Metabolic Reactions

One of the central issues related to this model is how to simulate the metabolic reactions in each cell. Chemicals and enzymes are represented by bit strings of a given length. The bit string of length n, representing chemicals and enzymes, means there are 2^n possible chemicals and enzymes in the model. The chromosome represents a set of rules governing the metabolic reactions. As shown in Fig. 3, a chromosome is a set of fragments, each of which represents a metabolic rule.

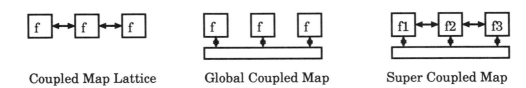

Figure 2: Three coupled maps

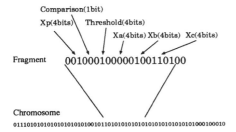

Figure 3: Chromosome and Fragment

For example, the fragment 000000010000100010010 is interpreted as;

$$x^0 < 0.5 \quad \Rightarrow \quad x^1 \rightarrow E^{1,2}(x^1 \rightarrow x^2) \qquad (4)$$

This reads as; if the concentration of x^0 within a cell is less than 0.5, then activate a metabolic reaction which creates enzyme $E^{1,2}$ from x^1. Enzyme $E^{1,2}$ acts as a catalyst for the reaction $x^1 \rightarrow x^2$.

At each cycle, these rules are applied to test whether the cell's chemical concentration triggers the activation or suppression of metabolic reactions. Thus, the metabolic reactions change dynamically during the cell's life span. In an actual implementation, a chromosome is decoded into a look-up table, with metabolic reactions being represented as a matrix, enabling matching to be performed efficiently.

Once metabolic reactions are determined by testing a rule set derived from the chromosome, the amount of each chemical in each cell can be determined from the following equations:

$$x_i^m(t+1) = x_i^m(t) + \frac{dx_i^m(t)}{dt} \qquad (5)$$

$$\frac{dx_i^m(t)}{dt} = \sum_{k=1}^{M} m(E^{k,m}x_i^k) - \sum_{k=1}^{M} m(E^{m,k}x_i^m)$$
$$+ Transp_i^m(t) + Diff_i^m(t)$$
$$+ LocTransp_i^m(t) + LocDiff_i^m(t) \quad (6)$$

$$m(S) = \frac{V_{max}S}{K_m + S} \qquad (7)$$

where x_i^m is a chemical concentration of chemical m in the i-th cell. The first term is a production of x_i^m from other chemicals, and the second term is production of other chemicals using x_i^m. When there is no enzyme (e.g. $E^{k,m} = 0$), no reaction occurs. This modeling is inspired from Kaneko and Yomo's model [Kaneko

and Yomo, 1994]. $m(S)$ is Michaelis-Menten equation, which provides reaction speed of enzyme-substrate complex. V_{max} and K_m are constant.

Active Transport and Diffusion

Cells take in chemicals from their environment by active transport and diffusion through their membranes.

$$Transp_i^m(t) = P\, t(\sum_{k=1}^{M} x_i^k(t), X^m(t)) \qquad (8)$$

$$Diff_i^m(t) = D[X^m(t) - x_i^m] \qquad (9)$$

$$t(a,x) = \frac{a \times x}{0.5x + a} \qquad (10)$$

where X^m is the chemical concentration of chemical m in the medium, and P and D are constants. In this equation, active transport corresponds to the amount of chemicals in the cell. Diffusion correlates to the difference in the chemical concentration inside the cell and that of the medium. The speed of active transport is limited according to the density of chemicals in the medium and activation level of each cell.

Cell Division

Cell division takes place when a DNA synthesis process reaches a certain level R. In this paper, we assume that DNA synthesis correlates with chemical x^0 by a factor γ. This is represented by the following equation:

$$\int_{t_{0(i)}}^{T} \gamma x_i^0(t)dt > R \qquad (11)$$

When a cell division takes place, chemicals in a cell is divided into two cells with certain level of fractuation.

Cell Death

Cell death takes place when the metabolism of the cell diminishes under a certain level S.

$$\sum_{k=1}^{M} x_i^{(k)}(t) < S \qquad (12)$$

Heuristics for modeling cell division and cell death were taken from a basic equations proposed in [Kaneko and Yomo, 1994].

Interaction with Neighbor Cells

Interaction between neighbouring cells is achieved through active transport and diffusion limited to adjacent, or near-by, cells. We have formalized the local interaction in two ways. The first method is to completely describe cell-to-cell interaction. The diffusion and active transport equation will be:

$$LocDiff_i^m(t) = DL\sum_{j}^{N} C(i,j)[X_{i,j}^m - x_i^m] \quad (13)$$

$$LocTransp_i^m(t) = P\sum_{j=0}^{N} t(\sum_{k=1}^{M} x_i^k(t), X_{i,j}^m) \quad (14)$$

where $X_{i,j}^m$ is the medium between cell i and j, and DL is a constant. $C(i,j)$ returns 1 when cell i and j are in contact. Otherwise, it returns 0.

The second method is to partition cell clusters into certain regional groups. For example, by assuming 16 regions, local interactions amoung cell in the same region can be simulated with far less computing.

$$LocDiff_i^m(t) = DL[X_r^m - x_i^m] \quad (15)$$

$$LocTransp_i^m(t) = Pt(\sum_{k=1}^{M} x_i^k(t), X_r^m) \quad (16)$$

$$(17)$$

In the simulation follows, we used the second method and 8 regions are assumed.

Paramters

Unless otherwise stated, following paramters are used in experiments in this paper; $P = 0.1$, $D = 0.1$, $PL = 0.1$, $DL = 0.1$, $R = 10.0$, $F = 0.1$, $S = 3.0$, $\bar{X} = 15.0$, $\delta = 0.1$, $V_{max} = 10.0$, $K_m = 5.0$, and Time Step $= 0.1$.

Results of Experiments

Cell Behavior

First, we chose to examine how this metabolic reaction model behaves using a hand-coded chromosome. The chromosome defines how metabolic reactions should be performed in each cell. Fig. 4 shows hand-coded metabolic reactions. Initially, X^1 exists in the medium and one type of enzyme, which instigates $x^1 \rightarrow x^2$, exists in the cell, such that the metabolic circuits that lead to DNA synthesis can be guaranteed.

Fig. 5 shows the results for a number of cells up to 600 cycles, together with the activity levels in each cell. Fig. 6 shows chemical concentration of x_0, x_1, and x_2 in cell-0 and in the medium. Due to the simplicity of the reaction circuit, change in chemicals in each cell shows a simple near-linear increase. Although this is a very simple and crude experiment, it illustrates how cell metabolism progresses and how cell division takes place.

Since this metabolic reaction is a slightly modified version of Kaneko and Yomo's experiment, it should generate similar, but not identical results to [Kaneko and

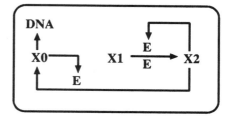

Figure 4: Metabolism within a cell

Yomo, 1994]. Since the figures show only an initial 600 cycles, it shows rather simple and linear changes in chemicals. With proper paramter settings and sufficiently longer cycles in experiments, more complex dynamics will emerge. In fact, Kaneko and Yomo discovered dynamic cell clustering, which small differences of distribution of chemicals at each cell division are magnified so that clusters of cells are formed. We have discovered, however, there are cases where activities of cells converged into a single coherent behavior. Fig. 7(Left) shows a case of coherent behavior. Global coupling was dominant since parameters were $D = 0.1$ and $DL = 0.001$. For this experiment, a large fractuation (± 0.5) was imposed for each chemical concentration at cell division. However, these differences diminished to form a single coherent behavior. One of the reason for this phenomena is that chemicals in each cells are balaced through active transport and diffusion. Active cells create more chemicals, which leak out to the medium, so that other cells which are in short of that types of chemicals take in these chemicals from the medium. Fig. 7(Right) is when the global diffusion constant (D) as well as local diffusion constant (DL) are kept low ($D = 0.001$, $DL = 0.0001$). Differences in activity level of each cell are amplified as the process continues.

Fig. 8 is when $D = 0.001$ and $DL = 0.005$. Activation level of cells are converge into several groups, but these are eventually converged into one coherent behavior. The dominance of local interactions force regional cells to converge into a coherent behavior within the region. Gradually, however, the effect of global coupling forces an entire cell group to act coherently.

Evolutionary Behavior

We introduced an evolutionary process to examine whether appropriate genetic codes for metabolic reactions can be acquired. An evolutionary process was added by using a genetic algorithm. cycle. The length of the chromosome is 200 and 1,000 bits, thus contains 10 and 100 fragments, respectively. We also run simulation using chromosome length of 40 and 100, but these run did not show cell division. The population size is 10. Elitist reproduction, combined with a proportional reproduction scheme, was used. Crossing over was 2-point crossover. Fitness is evaluated according to the amount of synthesized DNA existing at the 200th Of course, this is a great simplification, and does not necessarily reflect actual biological fitness in the real world. We chose to

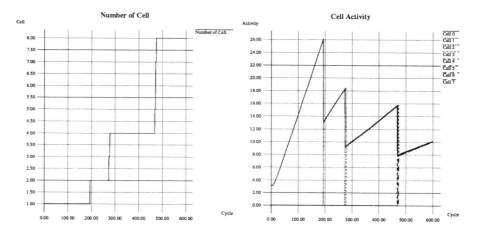

Figure 5: Number of Cells and Cell Activity

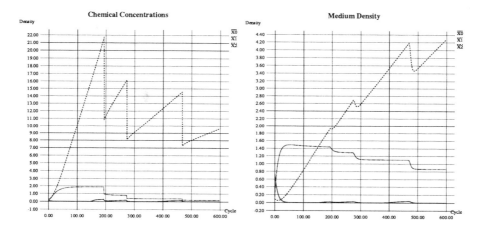

Figure 6: Chemical Density in Cell-0 and in Medium

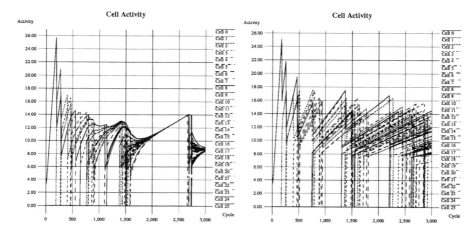

Figure 7: Activity Levels of Cells

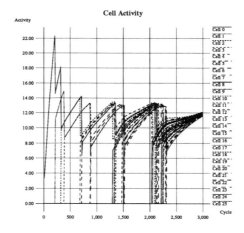

Figure 8: Activity Levels of Cells

use this fitness in this experiment because of its simplicity and it suffices as a means of examining whether this method is capable of acquiring the minimum required metabolism to cause cell division. In addition, a simple resource limitation is imposed, such that when the number of cells exceeds about 200, some cells suffer from lack of chemical supply from the medium.

Fig. 9 shows the number of cells in each generation, and the amount of synthesized DNA. Up to the 70th generation, the genetic code that defines the metabolic reactions is not sufficiently optimized to cause active cell division. Number of cells after 200 cycle were under 30. After the 70th generation, however, some individuals have acquired the genetic codes needed for active cell division. Once the genetic code that enables efficient metabolism has been acquired, cell division takes place explosively unless restricted by resource limitation. This is particularly seen in case of experiment with chromosome length 1,000.

Fig. 10 is a matrix representation of the metabolic reactions acquired by the 100th generation. It should be noted that this is a snapshot of the metabolic reaction matrix. Metabolic reactions could change at each cycle. In this particular example, an equation directly related to x^0 (excluding active transport and diffusion) is:

$$\frac{dx^0}{dt} = m(x^1 x^2) + m(x^5 x^{11}) + m(x^1 x^{10}) \\ -m(x^5 x^8) - m(x^8 x^{11}) - m(x^7 x^{15}) \quad (18)$$

Fig. 11 is a part of a rule set for metabolic reactions. For example, a rule-0 should be interpreted as; if the density of x^2 in the cell is over 3.25, then activate reaction which create enzyme $E^{15,11}$ from x^{14}. In effect, this reaction should be mapped to the matrix so that x^{11} is created from x^{14} and x^{15}. This is decoded from the same chromosome to matrix shown in Fig. 10. It should be noted that the rule-0 is not activated in the matrix since x^2 was less than 3.25 at this cell cycle. On the other hand, the rule-1 is activated because x^{11} is less than 3.00.

R.#	Xp		Thr		Xa		Xb	Xc
0 :	2	>	3.25	-->	14	->	(15	11)
1 :	11	<	3.00	-->	0	->	(7	15)
2 :	3	>	0.75	-->	8	->	(5	2)
3 :	1	<	3.00	-->	3	->	(12	13)
4 :	15	<	2.25	-->	1	->	(12	12)
5 :	11	>	1.25	-->	8	->	(5	10)
6 :	15	>	3.50	-->	10	->	(6	6)
7 :	1	<	1.75	-->	6	->	(8	2)
8 :	6	<	2.75	-->	13	->	(12	3)
9 :	11	<	0.25	-->	14	->	(15	1)
10 :	7	<	2.25	-->	10	->	(9	9)

Figure 11: A Part of Reaction Rules in DNA Sequence (Decoded)

Fig. 12 shows the number of cells in each cycle, and amount of DNA synthesis. Chemical density in the 0-th cell and in the medium are shown in Fig. 13. Although simulation uses 16 chemicals, density of only three chemicals are shown in the figure. These date show high complex and nonlinear dynamics of cell clusters. For example, in Fig. 13(Left), the concentration of x^2 in Cell-0 shows sharp drop before cell division take place (around 270th cycle, and other places). This is due to a metabolic rule which produces using x^2, activated around 270th cycle and regulated around 280th cycle. In addition, a quick recovery of x^2 implies that other rules which produce x^2 might have been activated for a short period of time around 280th cycle.

Morphogenesis

A simple process of morphogenesis was observed as a result of cell divisions. Fig. 14 shows an example of morphogenesis of an individual. It starts from a single cell and divide into two cells, and these cells cause cell divisions. This process continues. A simple kinematics has been imposed to determine how a group of cells are configured. An interesting phenomena is that

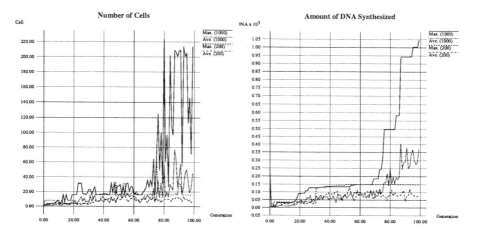

Figure 9: Number of cells and DNA synthesized in each generation

	0	1	2	3	4	5	6	7	8	9	10	11	12	13	14	15
0		2	1			11					1	5				
1									9	8		13		11	15	14
2		16	4		12		7	6	6				4			
3			6		13	2								13	5	
4																
5				10						3				15		13
6			15		12		6			13			4	9		2
7			9		4				2							
8	5					0		8	7							
9		11				7		5	7	10	9	1		14	13	
10		9							14	1					8	
11	8								0	13	9			9		
12		15	4		2			1					1			1
13				12					11			8	15			12
14		14				7		5							1	
15	7		13				13	2						6		

Figure 10: An Example of Acquired metabolic reactions (represented as a matrix)

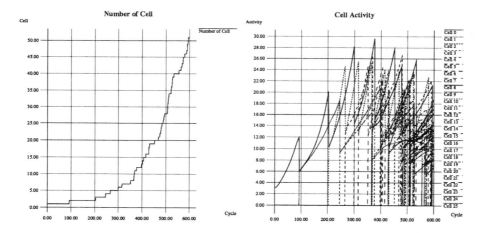

Figure 12: Number of Cells and Cell Activity

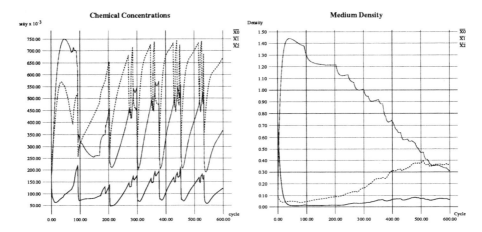

Figure 13: Chemical Density in Cell-0 and in Medium

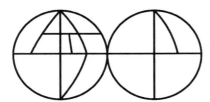

Figure 15: External View at the 16 cells state

frequency of cell division is different in the region, so that at the 16 cell state, it appears to have a primitive form of *vegitable pole* and *animal pole* (Fig. 15). A careful analysis should enable us to define L-system that describe this cell division process, but it would require an automated algorithm since the complexity of the phenonema is equivallent to an early stage of morphogenesis in actual living organisms. For this issue, we take a position that L-system is a syntactic description of the observed phenomena which emerged from the dynamics of the metabolism and a simple kinematics. Thus, a symbol dynamics interpretation of this simulation should leads to L-system grammar.

Discussions

Acquisition of Metabolism

Although our experiment are only preliminary, several interesting properties can be observed. For example, we discovered that chromosome length greatly affects the time needed to acquire metabolism. When the chromosome length is too short (e.g. 200 bits), only a few reactions can be established, thus reducing the probability of acquiring reaction pathways to cause cell division is very small. Hence, it takes longer to acquire such a pathway. Only 40 bits are used to define the reaction pathways used in the hand-coded chromosome. However, it is too inefficient to evolutionarily acquire the reaction pathways that lead to cell division with such a short chromosome. In the second experiment, we used a chromosome that was 40, 100, 1,000, and 2,000 bits

long. Pathways were acquired very quickly with longer chromosome. The implication of this is that the simulation should incorporate a mechanism for automatically adjusting the chromosome length.

Another issue is on how to constrain explosive cell division. Without resource constraints, cells will divide infinitely. To replicate a logistic curve as seen in actual development, reasonable resource limitations must be imposed, and the ability to autonomously adjust cell division must be reflected into comprehensive fitness.

Dynamics

Dynamics exhibited in this system was exremely interesting. In SCM model, both global and local interaction influence cell behavior. In either case, however, there are dynamics which force all cell to behave coherently. This is because chemicals in active cells are leaked into the medium so that less active cell can in take these chemicals, leading to balancing of all cell state. By supressing diffusion, this mechanism disappear and cells behave very differently. However, if there is no or very low interaction through the medium, the system is merely a set of independent dynamic systems. Complex and "meaningful" dyanmics appears when these parameters are set just right — perhaps at the edge of chaos. Results obtained by Keneko and Yomo [Kaneko and Yomo, 1994] are examples of dynamics in this region.

Use of two interaction channels in SCM enables simulation of realistic dynamics. Assuming that there is high local diffusion and and low global diffusion constant, such a system is largely affected by local interaction, with some global interaction. As seen in Fig. 8, cells within a region converged to exhibit coherent behavior, but, each region may behave very differently. Translating in biological terms, this is analogous to the situation when a group of cell is differenciated into a certain type of cells and other groups of cell are differenciated into different types of cells, but cells within each group are differenciated into the same type.

Dynamic of SCM itself is interesting. Due to the use of both global and local interaction channels, and differ-

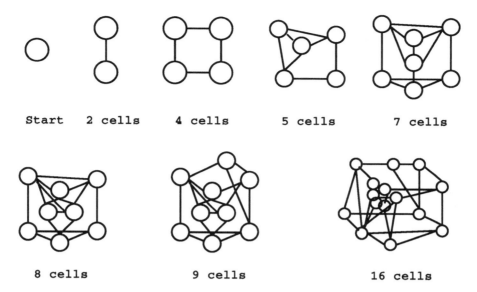

Start   2 cells   4 cells   5 cells   7 cells

8 cells   9 cells   16 cells

Figure 14: A Process of Morphogenesis

ent functions for each cell, SCM enables realistic simulation of dynamics of biological phenomena. Although this paper focus on evolution of metabolism, SCM can be applied to model dynamics of brain and other biological phenomena.

Unresolved Issues

First, it should be remembered that these results are only preliminary. Thus, many issues remain to be resolved. Some of these are:

- A more realistic model of gene expression must be incorporated. For example, the current model only incorporate epistasis indirectly through chemical concentration of the cell. Dependencies of gene expression must be attained through varying the chemical concentrations. The model should be extended to permit more direct epistasis as well as other important genetic phenonema.

- The growth of neural circuits was not tested in this paper. Growth of axons, selective stabilization, and other phenomena need to be modeled [Cowan et al., 1981; Changeux and Danchin, 1976; Edelman, 1984]. These are indispensable to the modeling of neurogenesis. We are currently extending our model to incorporate a simple model of axon growth and circuit formation.

- Kinematics must be introduced to simulate the physical constraints associated with cell division. Simulation of this aspect is particularly important to the *Perfect C. elegans* project which heavily involves morphogenesis.

- Information processing in each cell and between cells needs to be modeled. Intercellular interactions are important not only for the formation of appropriate body plans, but also for the emergence of behavior in the environment.

- Evolution of parameters need to be incorporated. Experiments in this paper assumed parameters are fixed. If these paramters are also evolved, we will be able to see whether the entire system evolve toward the edge of chaos.

Conclusion

In this paper, we proposed a new model of evolution of metabolism in multicellular organisms. Metabolic reactions within a cell were acquired through an evolutionary process simulated by a genetic algorithm. Interactions between cells are modeled using the super coupled map, a newly proposed model based on large-scale chaos.

Experimental results confirmed the possibility of acquiring a simple metabolic reaction rule, which is sufficient to cause growth of the organism. In addition, we discovered that (1) the dominant interaction patterns change as the development progresses, and (2) there are coherent phase, where all cells behave almost identical, and dynamic clustering phase, where cell groups are formed each of which exhibits different dynamics. Introduction of SCM was a key factor to identify various dynamics entailed in this complex system. Finally, we observed a possible analogue to the formation of the animal pole and the vegitable pole. It is still premature to relate our findings to actual biology, but we believe it worth further investigation.

Although the model and experiments reported in this paper are only preliminary, and much remains to be done to accurately simulate actual biological development and evolution, our results clearly indicate that the computer simulation of such phenomena can provide us with a great deal of insight into understanding complex biological phenomena.

References

[Changeux and Danchin, 1976] Changeux, J. and Danchin, A., "Selective stabilization of developing synapses as a mechanism for the specification of neuronal networks," *Nature,* 264, 705-712, 1976.

[Cowan et al., 1981] Cowan, W., Fawcett, J., O'Leary, D. and Stanfield, B., "Regressive Events in Neurogenesis," *Science,* 225, 1258-1265, 1981.

[Edelman, 1984] Edelman, G., "Modulation of cell adhesion during induction, histogenesis, and perinatal development of the nervous system," *Annu. Rev. Neurosci,* 7:339-377, 1984.

[Fleischer and Barr, 1994] Fleischer, K. and Barr, A., "A Simulation Testbed for the Study of Multicellular Development: The Multiple Mechanism of Morphogenesis," *Artificial Life III,* Addison Wesley, 1994.

[Kaneko and Yomo, 1994] Kaneko, K. and Yomo, T., "Cell Division, Differentiation and Dynamic Clustering," *Physica D,* 1994. (to appear)

[Kaneko, 1992] Kaneko, K., "Overview of coupled map lattices," *Chaos,* 2 (3), 1992.

[Kitano, 1994] Kitano, H., "Neurogenetic Learning: An Intergrated Model of Designing and Training Neural Networks using Genetic Algorithms," *Physica D,* 1994. (to appear)

[Kitano, 1990] Kitano, H., "Designing Neural Networks Using Genetic Algorithms with Graph Generation System," *Complex Systems,* Vol. 4, Num. 4, 1990.

[Lindenmayer, 1968] Lindenmayer, A., "Mathematical Models for Cellular Interactions in Development," *J. theor. Biol.,* 18, 280-299, 1968.

[Lindenmayer, 1971] Lindenmayer, A., "Developmental Systems without Cellular interactions, their Languages and Grammars," *J. theor. Biol.,* 30, 455-484, 1971.

[Miller et. al., 1989] Miller, G., Todd, P., and Hedge, S., "Designing Neural Networks using Genetic Algorithms," *Proceedings of the International Conference on Genetic Algorithms,* 1989.

[Sulston et al., 1983] Sulston, J., et al., "The Embryonic Cell Lineage of the Nematode Caenorhabditis elegans," *Developmental Biology,* 100, 64-119, 1983.

Competition, Coevolution and the Game of Tag

Craig W. Reynolds

Electronic Arts
1450 Fashion Island Boulevard
San Mateo, CA 94404 USA
telephone: 415-513-7442, fax: 415-571-1893
creynolds@ea.com
cwr@red.com

Abstract

Tag is a children's game based on symmetrical pursuit and evasion. In the experiments described here, control programs for mobile agents (simulated vehicles) are evolved based on their skill at the game of tag. A player's fitness is determined by how well it performs when placed in competition with several opponents chosen randomly from the coevolving population of players. In the beginning, the quality of play is very poor. Then slightly better strategies begin to exploit the weaknesses of others. Through evolution, guided by competitive fitness, increasingly better strategies emerge over time.

1. Introduction

Many of us remember playing the *game of tag* as children. Tag is played by two or more, one of whom is designated as *it*. The *it* player chases the others, who all try to escape. Tag is a simple contest of pursuit and evasion. These activities are common in the natural world, most predator-prey interactions involve pursuit and evasion. Tag also includes an aspect of role-reversal, both pursuit and evasion skills are required. *It*'s goal is to catch up with another player, to get close enough to reach out and touch the other player. At this point the pursuer shouts "Tag! You're *it*!" and the former evader becomes the new pursuer.

The game of tag serves here as a toy example, to study the use of competitive fitness in the evolution of agent behavior. Tag is intended as a simple model of behavior based on control of locomotion direction, or *steering*. By evolving a vehicular steering controller for the game of tag we have a test case to learn about evolving controllers for related, but more complex tasks.

We seek to automatically discover a controller through evolution based solely on competition between controllers. This approach stands in contrast to evolving controllers by pitting them against a static, predetermined expert strategy.

The use of *competitive fitness* has the significant advantage of avoiding the paradoxical need for an expert controller as a prerequisite for evolving an expert controller. Another advantage of competitive fitness is that since each fitness test is unique, there is no danger of overfitting a static expert.

2. Related Work

This research was originally inspired by Pete Angeline's elegant work on coevolution of players for the game of Tic Tac Toe, using competitive fitness [Angeline 1993]. That fitness could be tested by competition alone, even in the absence of an expert player, is a key insight in applying evolutionary techniques to the discovery of complex goal-directed behaviors. The work reported here extends Angeline's paradigm from games of pure strategy to those involving geometric motion. The competition in [Angeline 1993] was in the form of a single elimination tournament tree. The players in each new generation were paired up in competition. The winners of each pairing went on to a second round, and so on for several rounds until only one champion remained. A player's fitness was determined by the number of matches won before a defeat. See Table 1 for a comparison of several competitive architectures described in this section.

A similar approach has recently been used to coevolve strategies for the game of Othello [Smith 1994]. A genetic algorithm was used with competitive fitness to determine time-varying weightings for the static evaluator used in an alpha-beta search of the Othello game tree. In this work, two new players are placed in competition with each other, the score of their game determines the fitness of both.

Another paper on competitive evolution of behavior appears elsewhere in this volume [Sims 1994]. It describes experiments where the morphology and behavior of artificial creatures evolve through competition for control of an inanimate object. The creatures each try to get closest to the object, and if possible to surround it. Each new creature

Table 1:

competitive architecture	matches per generation of n	opponents per individual	reference
new versus all	$(n^2-n)/2$	n-1	[Koza 1992]
new versus several	nk	k	this paper
single elimination tournament tree	n-1	$\log_2 n$	[Angeline 1993]
new versus previous best	n	1	[Sims 1994]
new versus new	n/2	1	[Smith 1994]

is placed in competition with the best creature from the previous generation.

John Koza evolved pursuer-evader systems closely related to those reported here (see pages 423-428 of [Koza 1992]). But in that work the pursuers and evaders existed in separate populations. Their fitness was determined by comparison with a pre-existing optimal player. In the same book, Koza first discusses coevolution in Genetic Programming (pages 429-437) in the context of a discrete strategy game.

In the work reported here, the vehicle model, and the noise-tolerant Steady-State Genetic Programming system was taken from [Reynolds 1994a] and [Reynolds 1994c]. The vehicle model draws heavily from [Braitenberg 1984] and is equivalent to the *turtle* of the LOGO programming language.

Competitive fitness and coevolution were first explored in evolutionary computation in the context of the Iterated Prisoner's Dilemma in [Axelrod 1984], [Axelrod 1989], [Miller 1989], [Lindgren 1992] and [Lindgren 1994]. A restricted form of competitive fitness was used in [Hillis 1992] to drive the evolution of sorting networks: antagonistic test cases were coevolved to force generality. Coevolution and competitive fitness are fundamental to a wide class of ecological simulations studied in Artificial Life such as: ECHO [Holland 1992], Tierra [Ray 1992], BioLand [Werner 1993], PolyWorld [Yeager 1994], LEE [Menczer 1994], and the work reported in [D'haeseleer 1994]. Chapter six of [Kauffman 1993] is an authoritative analysis of "The Dynamics of Coevolving Systems."

The reader should understand that this work is *not* about optimal control theory, despite the fact that frequent mention is made of optimal strategies, and that this type of pursuer-evader system has a long history in the optimal control literature [Isaacs 1965]. The work reported here benefits from (but does not depend upon) the ability to compare evolved behavior with optimal behavior. It should be noted that this is practical only because of the extremely simple vehicle model used in these studies. Once slightly more complicated models are used, optimal behavior becomes a much more complicated issue, see for example [Merz 1971].

3. Experimental Design

In this work, Genetic Programming is used to evolve control programs for simulated vehicles, based on their ability to play the game of tag. Each new player's fitness is determined through competition with existing players. The game of tag provides a framework wherein the relative fitness of two players is judged through direct competition.

The vehicles are abstract autonomous agents, moving at constant speed on a two dimensional surface. Each vehicle's evolved control program is executed once per simulation step. Its job is to inspect the environment and to compute a steering angle for the vehicle. Angles are specified in units of *revolutions*, a normalized angle measure: 1 revolution equals 360 degrees.

For each player, at each simulation step: (1) its control program is run to determine a steering angle, (2) the

vehicle's heading is altered by this angle, (3) the vehicle is moved a fixed distance along its new heading, and (4) tags are detected and handled. The forward step length is typically 125% longer for *it*. The per-step steering angle is unconstrained: these vehicle can instantaneously turn by any amount. This is an abstract *kinematic* vehicle model, as opposed to a *physically realistic* model. In this work there is no simulation of force, mass, acceleration, or momentum. In contrast, a physically realistic model would serve to limit the turning radius.

In these experiments there are always two players in a tag game. The playing field is featureless. Each vehicle's environment consists of just one object: the opponent vehicle. The opponent's current heading is not relevant. In the absence of momentum, there is no correlation between the vehicle's current heading and its position in the next time step. For a given controller the entire state of the world consists of: a flag indicating who is *it*, and the relative position of the opponent's vehicle. The position information is presented to the control program in terms of x and y coordinates of the opponent, expressed in the local coordinate system of the controller's own vehicle. A vehicle's local coordinate system is defined with the positive y axis aligned with the vehicle's heading and the positive x axis extending to the vehicle's right. See Figure 1.

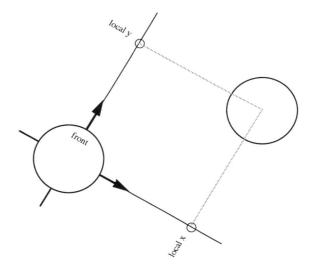

Figure 1: the input to a vehicle's controller is the position of the opponent vehicle, expressed in terms of local x and y coordinates.

While the traditional game of tag has no formal method of scoring, the intuitive goal is to avoid being *it*. Therefore, fitness is defined here to be the portion of time (simulation steps) spent not being *it*. Note that this is a normalized, bigger-is-better fitness metric: zero is worst, one is best. To determine a player's fitness it competes with other players.

To compare two players, a series of four games is played. Before each game the players are given random initial headings and are randomly positioned within a starting box measuring about 3.5 vehicle-body-lengths on a side. The two players alternate starting as *it* for each game of the series. If *it* tags the opponent, by getting to within one vehicle length, the *it* and non-*it* roles are reversed. Each game consisted of 25 simulation steps. A player's score for a game is the number-of-non-*it*-steps divided by 25. See Figure 2.

To determine a player's fitness, it is pitted against 6 other players. These opponents are chosen by uniform random selection from the existing population. Scores from these 24 games (a series of 4 games against each of 6 opponents) are averaged together to obtain the final fitness value. A value of 50% indicates that the player has an ability comparable to the population average. Fitness values above 50% indicate increasingly better players. Because opponents are randomly selected, and because initial conditions are randomized, these fitness values have significant variance and provide only a rough estimate of a player's actual fitness.

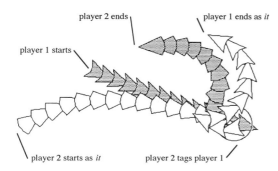

Figure 2: a game of tag between player 1: an evolved controller from run G, and player 2: the optimal strategy. As in all such diagrams in this paper, the pursuer (*it*) is shown in white, the evader is gray, and the site of a tag is indicated by a circle. Player 1 is shown here with a concave back edge and player 2 has a convex back edge. In this game player 1 begins by fleeing in a naive direction. Player 2 chases down and tags player 1. Player 2 then escapes because player 1 is too slow turning around. Player 1 was not it for 14 of the 25 steps so earned a score of 56%. Player 2 was not *it* for 11 steps so scored 44%.

In the discussion of experimental results below, reference will be made to an *optimal player*. Because of the unrealistically simple vehicle model used here, particularly the absence of momentum, the optimal strategy is quite straightforward. For the pursuer, the optimal strategy is to turn toward the evader (so the evader lies on the pursuer's positive y axis). Similarly the optimal strategy for the evader is to turn directly away from the pursuer (so that the pursuer lies on the negative y axis).

The "four quadrant arctangent" implements this optimal strategy. However the fitness criterion in these experiments is performance in tag competition. The optimality of a given controller is irrelevant to its fitness, all that matters is its performance in tag competition with the coevolving population. As will be seen below, controllers which are clearly non-optimal can still produce behavior which is asymptotically close to optimal performance.

4. Genetic Programming and Tag

The technique used to evolve computer programs in this work is known as Genetic Programming ("GP"), invented by John Koza. The best reference on this technique and its application is [Koza 1992]. In its original formulation, GP operated on a population of individuals generation by generation. Alternatively, GP can be combined with Steady State Genetic Algorithms [Syswerda 1991] as described in [Reynolds 1993a].

A very brief description of Steady State Genetic Programming (SSGP) follows. First a population of random programs is created and tested for fitness. In these experiments the population consisted of 1000 to 5000 programs. Thereafter SSGP proceeds by: (1) choosing two *parent* programs from the population, (2) creating a new *offspring* program from them, (3) testing the fitness of the new program as described in the previous section, (4) choosing a program to remove from the population to make room, and (5) adding the new program into the population. The parent programs are chosen in a way that favors the more fit while not totally ignoring the less fit, thus balancing *exploration* of the whole gene pool with *exploitation* of the current champions. In these experiments this choosing is done using *tournament selection*: seven individuals are chosen from the population at random, the most fit of those is selected as the winner. The recombination of two parents to form a new offspring is accomplished by the Genetic Programming *crossover* operator. GP crossover is a bit like "random cut and paste" done on balanced parenthetical expressions to guarantee the new program's syntactic correctness. After crossover, a program might be *mutated* by substituting one function for another, one terminal for another, or by crossover with a new random program fragment. .

Selecting a program to remove from the population could be done by using inverse tournament selection: removing

Table 2

function	usage	description	source
+	(+ a b)	a plus b	Common Lisp
–	(- a b)	a minus b	Common Lisp
*	(* a b)	a times b	Common Lisp
%	(% a b)	if b=0 then 1 else a divided by b	[Koza 1992]
min	(min a b)	if a<b then a else b	Common Lisp
max	(max a b)	if a>b then a else b	Common Lisp
abs	(abs a)	absolute value of a	Common Lisp
iflte	(iflte a b c d)	if a ≤ b then c else d	[Koza 1992]
if-it	(if-it a b)	if this player is *it* then a else b	this paper

the least fit of seven randomly chosen programs. However the greedy nature of SSGP, combined with the variance of fitness measures used in these experiments, leads to the possibility of a mediocre-but-lucky program receiving an undeservedly high fitness and going on to dominate the population. To combat this possibility, a modified removal policy was used in these experiments: half the time inverse tournament selection was used, the other half of the time an individual was selected for removal at random (without regard to fitness). All programs, even the best one, has a certain small but non-zero probability of being removed at each SSGP step. This approach reduces the possibility that the population could stagnate with a collection of mediocre-but-lucky programs. To survive, winning strategies must continue to perform well.

Another issue is that competitive fitness values are measured relative to the population at a certain point in time. These fitness value becomes less and less relevant as time passes. The hybrid SSGP removal policy ensures that the population is continually being recycled.

Because steady state genetic computation proceeds individual by individual, there is no demarcation of generations. However it is often convenient to describe the progress or length of a SSGP run in terms of "generation equivalents:" processing as many new individuals as there are programs in the population.

In order to evolve tag-playing control programs with Genetic Programming, we first choose a set of functions and terminals to form the language in which they will be expressed. Certain choices are dictated by the application itself, and some are intended to provide the logical and arithmetic "glue" with which GP will construct control programs. Three functions used here are specific to the underlying application, they provide the three parameters for the control programs. local-x and local-y are functions of zero arguments which return the x and y coordinates of the opponent player. The if-it macro provides conditional execution of its two subexpressions depending on whether this player is currently *it*. In addition to these functions, several others are provided to help form control programs. The choice is rather arbitrary, it was influenced by the author's domain knowledge and by preliminary attempts to write a tag-playing control program by hand. This function set provides for arithmetic, thresholding, sign manipulation, and conditional computation, see Table 2.

In addition to (local-x) and (local-y) mentioned above, the terminal set includes :random-0-1, an ephemeral random constant. After a new program is formed, any occurrences of this terminal are replaced by a pseudo-random floating point number between 0.0 and 1.0.

Evolved programs are subject to a size limitation which is measured in term of the total number of functions and/or terminals. When crossover produces a program whose size exceeds this limit, the *hoist* genetic operator [Kinnear 1994] is used to find a smaller (but hopefully still fit) subexpression. In these experiments the size limit was either 50 or 100.

These experiments used the if-it conditional in two different ways. In the early runs, a *syntactic constraint* was used to ensure that evolved programs always contained exactly one occurrence of if-it and it was always at the top (outermost) level of the program. That is, all programs in those runs had this structure:

(if-it *<pursuer-branch>* *<evader-branch>*)

The GP crossover operation was modified to exchange code fragments only within branches of the same type. With this constrained structure and crossover, the effect is to create two distinct gene pools, one for pursuit behavior and one for evasion behavior. Thus the frequency of code fragments are allowed to evolve differently in each population. This approach is similar to the Automatically Defined Functions described in [Koza 1992].

In later experiments this syntactic constraint was removed and the if-it conditional was treated as just another non-terminal. As a result there could be any number of if-it forms in an evolved program. There was no segregation of the gene pool, code fragments could cross over from the pursuit branch to the evasion branch, and vice-versa.

The Genetic Programming substrate used here was originally developed by the author on Symbolics Lisp Machines and subsequently ported to Macintosh Common Lisp (version 2.0p2). These experiments were run on Macintosh Quadra 950 workstations. In this implementation a fitness test consisting of 24 tag games takes 7 to 12 seconds to run, depending on program size.

5. Results

Seven evolution runs were performed using the basic experimental design described above. The runs varied in terms of population, the relative speed of the two players, the program size limit, whether mutation was used, whether the *it*/not-*it* branches were segregated, and the method used for selecting opponents. See Table 3. Runs A, C and G will be discussed in more detail in the following sections.

5.1 Run A

Run A had a population of 5000 individuals. Both players moved at the same speed. This puts *it* at a fundamental

Table 3:

run name	population size	*it* speed ratio	program size limit	mutation used	segregated branches	opponent selection
A	5000	1.00	50	no	yes	uniform
B	2000	1.25	50	no	yes	uniform
C	1000	1.25	50	yes	yes	uniform
D	1000	1.25	50	yes	yes	tournament
E	1000	1.25	50	yes	no	uniform
F	1000	1.25	50	no	no	uniform
G	1000	1.25	100	no	no	uniform

disadvantage: as long as the other player moves away, the best *it* can do is to follow at a constant distance, unable to close the gap.

Most of the initial, random programs in run A used ineffective strategies. A typical behavior was to bumble around aimlessly. Some of the early programs effectively had no steering behavior. These control program always returned zero (or some very small value) and the vehicle would simply travel in a straight line. While this is not a very effective pursuit strategy (for *it*), it is occasionally a good evasion strategy. If *it* happened to start off positioned behind you, running straight ahead is essentially optimal evasion. Since initial position and orientation is random, this non-turning evasion behavior would be effective between one-half and one-quarter of the time, depending on how good *its* pursuit strategy is. As a result, early in the run these non-turners began to proliferate through the population of evasion strategies.

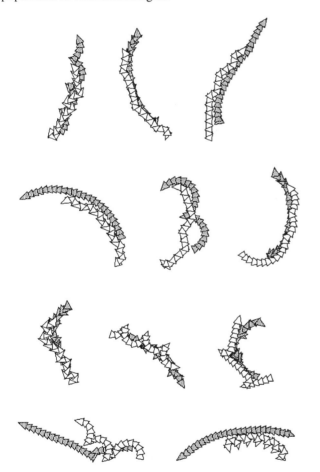

Figure 3: pursuit and evasion from run A in which both players moved at the same speed.

At this early stage, the most effective pursuit strategies appear to have been looping (constant steering angle) and "stumblers" that seemed to move erratically, but managed to creep slowly towards their target. The looping behavior was successful because it allowed the pursuer to cover more

ground, increasing its chance of blindly tagging the evader. The stumblers, while quite inefficient when compared to optimal pursuit behavior, were able to survive by *preying* on incompetent, aimless evaders. The large number of bad strategies in the population provided fertile ground for innovation: new strategies, even inefficient ones, could prosper by exploiting the weaknesses of others.

Later in run A, an improved evasion strategy appeared. The gist of it was: if the pursuer is behind you, go straight ahead, otherwise turn randomly. The effect is that the evader would twitch randomly for a few steps, then it would find itself pointing away from the pursuer and head straight away. Not only is this a fairly robust strategy, but it was easily implemented (hence easily evolved) in the programming language used in these experiments:

```
(if-it <pursuer-branch> (max 0 (local-y)))
```

This code fragment, and many variations of it, appeared in the population. Most of the variations were larger expressions which were functionally equivalent. Once this simple but effective evader appeared, the pursuers were in trouble. Because of the advantage mentioned earlier, that the evader need only move away from the pursuer in order to win, the pursuers could only achieve mediocre performance. In general the quality of pursuit in run A improved. The pursuers got to be very good at picking off the easy targets, the inefficient evaders. But this only served to improve the gene pool of the evaders. Soon only fairly good evaders remained, and they could easily escape from the best of the pursuers.

This sounds like it might have been the end of the story, but something interesting happened. In the interval between individuals 57000 and 76000, the pursuers developed several different techniques for following their targets, each with a characteristic and visually distinct pattern of motion (see Figure 3). Each of these patterns of motion had to have a certain measure of success to survive, but none could overcome the built-in advantage of the evaders in run A. There is a temptation (unsupported by any data) to see these distinct patterns of motion as something akin to evolutionary stable strategies (or perhaps species in environmental niches). It is also possible that these classes of motion arose solely as artifacts of the function set used in the GP representation.

One pursuit strategy seen in this run (and others) used a competent but inefficient "three phase" technique. These players would always do one of three things: turn left, turn right, or turn around. It is inefficient because turning left or turning right is a good idea only if the quarry is off to the side. If your opponent is directly ahead, a zig-zag path will slow your rate of progress. The code below is the pursuit branch of individual number 179120, unedited except to add comments:

```
(iflte 0.029628        ; if y > 0  (quarry ahead)
       (local-y)       ;
       (iflte 0.212021 ;    if x > 0.2  (quarry on right)
              (local-x) ;
              0.862946 ;       turn 49 degrees to right
              0.134561) ;      turn 48 degrees to left
       0.541760)        ;    turn 195 degrees (about face)
```

As evolution in run A proceeded, the evaders developed an unexpected behavior. For the majority of them life was easy, they could usually escape the pursuers. (Those who could not were quickly killed off.) One might guess that the evader population would settle down into using a simple, efficient strategy. Instead it seems that they had an overabundance of genetic material and were determined to use it! They adopted a complicated, elaborate strategy. Because both players move at the same speed, the pursuer is not a threat until it moves close to the evader. This defines a certain "threat radius" around the evader. If the pursuer is outside this radius it doesn't matter what the evader does, it could just as well bumble around aimlessly. Once the pursuer crosses the threat radius the evader must snap to attention and run away efficiently. These elaborate evaders are very large programs and hard to analyze in detail. One possible explanation is that an effective evasion strategy was discovered, but its implementation depended on the values of local-x and local-y being limited to certain small values. The result was an evader that would flee very efficiently when closely pursued, but otherwise would appear to randomly flip around. Subjectively this behavior looked for all the world like the evader was "saving its strength" -- moving slowly until the pursuer got close enough to be a threat. Yet this could not have been what actually happened, since there was no modeling of expended energy, nor did the fitness function reward conservation of energy.

A note about the selection of games shown in, for example, Figure 3. The choice of images for this paper was a biased, subjective process. An unedited random sampling of images from the run might have been more representative, but would have been much less visually interesting. These "photogenic" images were chosen because of their clarity, simplicity, and visual appeal. Several were typical, a few were unusual. Collecting these images was a process somewhat like nature photography. Lots of pictures were taken, many were discarded, and a few were selected that captured the spirit of the behavior.

5.2 Run C

Evolution of pursuit in run A was stymied by the evader's advantage. Run C sought to even out the competition by giving *it* a 25% speed advantage. Now a good pursuer could close in on and tag a good evader. This served to encourage efficient pursuit, which in turn encouraged improved evasion.

Run C used a population of 1000. Mutation was added in an attempt to help prevent the loss of diversity observed in earlier runs. Run C was evolved for 215 generation equivalents. The quality of pursuit and evasion improved steadily, becoming skillful and well matched. Many games consisted of a chase featuring near-optimal pursuit and evasion followed by a series of rapid tags with *it* alternating back and forth each step. After each tag the new *it* would just turn around, take one big step, and tag the other player. This was reminiscent of the games seen when two optimal players were pitted against each other.

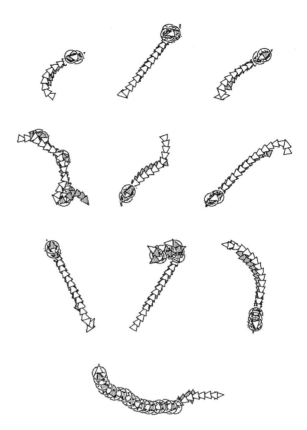

Figure 4: tag games from run C. Most consist of a successful chase followed by a series of tags.

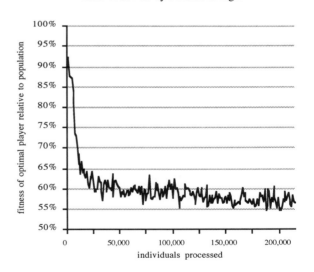

Figure 5: plot of the fitness of the optimal player placed in competition with the evolving population of run C. Since fitness is measured relative to the existing population, the decline in fitness shown here is indicative of the increasing average fitness of the population. Initially the optimal player avoided being *it* 93% of the time. After 22000 individuals (22 generation equivalents), the optimal player is *it* about 60% of the time. Thereafter the value drifts slowly down into the range between 55 and 60%.

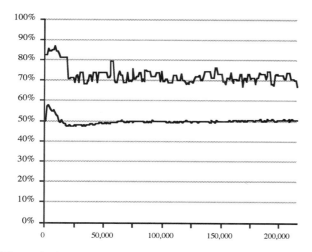

Figure 6: plots of best-of-population fitness (top) and average-of-population fitness (bottom) for 215 generations of run C.

After 215415 individuals were processed in run C there were four individuals with the same best fitness value. One of those was selected for further analysis. It was compared to the optimal player in a series of 100 games. The evolved player got a score of 49.3% indicating that was holding its own against the optimal player, being not-*it* just less than half of the time. The evolved program's size was 48. A few simplifications were made by hand to produce this size 38 version:

```
(if-it (max (% (local-y)
               (+ (min (local-y)
                       (min (abs (local-y))
                            (iflte 0.25235322
                                   (local-y)
                                   (* 0.09556236
                                      (local-x))
                                   (local-x))))
                  (local-y)))
            0.25235322)
       (iflte (local-y)
              0.0055459295
              (iflte (local-y)
                     (* (local-y)
                        (min (* (local-x) (local-x))
                             0.47677472))
                     (* 0.051421385 (local-x))
                     (* 0.107852586 (local-x)))
              0.47677472))
```

To visualize the overall behavior of a player's evolved control program, the two inputs can be used to define a perceptual field embedded in the local space surrounding the vehicle. Mapping the control program over a mesh of points in this field yields an array of steering angles. This map describes the stimulus-response relationship implemented by the control program as it maps an opponent's position into a steering angle.

Figure 7 shows plots of this relationship for the optimal player. Each arrow in this diagram indicates the player's steering angle response to an opponent in the vicinity of the arrow's tail. Figure 9 summarizes the behavioral mapping

for an evolved player from run C. This useful visualization tool has long been applied to the study of reactive robotic control system in the work of Ronald Arkin, see for example [Arkin 1987].

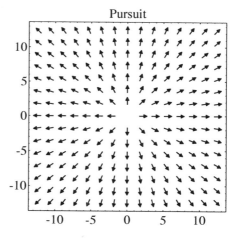

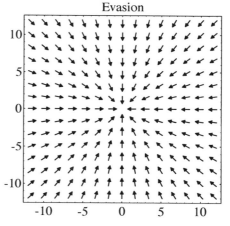

Figure 7: Plots of steering angle versus opponent position for the optimal player. These stimulus-response diagrams summarize the player's behavior.

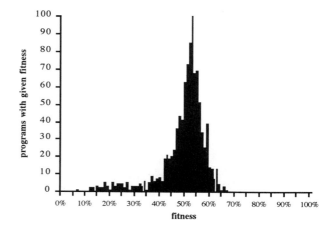

Figure 8: histogram of fitness distribution in population of run C after 215 generation equivalents.

An examination of Figure 9 suggests that a "foveal region" has developed in the perceptual field of both pursuer and evader. In the pursuer the forward facing quadrant has a near-optimal structure. In the evader the backward facing quadrant (particularly in the region close to the player and along its negative y axis) has a near-optimal structure.

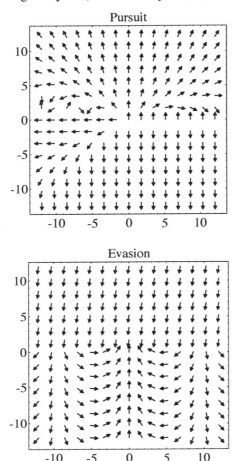

Figure 9: Plots of steering angle versus opponent position for the best-of-population player from run C after 215 generation equivalents

Figure 9 also indicates that in the "anti-foveal" direction the response is simply to turn halfway around. The pursuer will turn around if the evader is behind it, and the evader will turn around if the pursuer is in front of it. The general strategy appears to be to turn "quickly" toward (or away from) the opponent, to place them in your foveal region, wherein your response is close to optimal. The behavior is less organized in the direction perpendicular to the foveal axis, but the effect seems to be the same: after a step or two the vehicle is heading in the right direction.

Since the vast majority of the player's time is spent in this nearly optimal foveal region, there is correspondingly less selection pressure to "optimize" the behavior in other regions. This is particularly true given the relative small number of steps executed in these simulations. For example, sometimes an evader escapes because its behavior is more efficient, but sometimes it escapes because the 25 steps of the simulation have expired. There are very few tag games where non-optimal behavior in the off-foveal regions makes a measurable difference in fitness.

5.3 Run G

Run G did not segregate the pursuer and evader code and used a larger limit on program size. One of the changes seemed to make the problem harder to solve. The syntactic constraint used in earlier runs is a user-provided "hint" that there are two distinct cases to solve. This may lower the difficulty of the problem. Further comparisons of more runs would provide more reliable information about difficulty. Figure 10 shows the performance of the G population against the optimal player. Comparing this to Figure 5 shows that run C generally did better sooner.

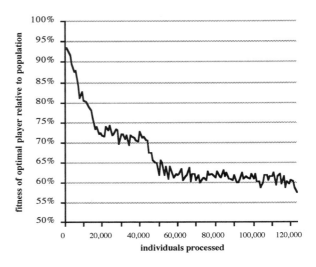

Figure 10: plot of the fitness of the optimal player placed in competition with the evolving population of run G over 123 generation equivalents.

Individual 113520 of run G was the best of population when it was created. The size 98 program is listed below. Its stimulus-response map is shown in Figure 11. This individual has many strange behavioral traits, for example pursuit behavior has a reasonable two phase strategy for opponents up to 5 units ahead but is very inept for opponents further away. The evasion behavior is strongly asymmetrical.

```
(% (% (if-it (abs (local-x)) (iflte (iflte (local-x)
0.57168305 (local-x) (+ (iflte (local-y) (iflte (local-y)
(if-it (local-x) (abs (local-x))) (iflte 0.40530929
0.26004231 (abs (local-x)) (local-y)) (if-it 0.40530929
0.57168305)) (min (abs (local-x)) (+ (local-x) (local-
x))) (local-x))) (local-x))) 0.57168305 (local-x) (+
(iflte (local-y) (iflte (local-y) (if-it (local-x)
(local-x)) (iflte 0.40530929 (local-x) (abs (iflte
(local-x) 0.37254661 0.32281655 (local-x))) (local-x))
(if-it 0.40530929 (abs (local-x)))) (min 0.1637349 (iflte
(local-x) (local-y) (abs (iflte (abs (local-x)) (max (if-
it (local-y) (abs 0.53183758)) (local-x)) 0.32281655
```

```
(local-x))) 0.53183758)) (local-x)) (local-x)))) (+
(local-x) (local-x)) (iflte (- (abs 0.53183758) (if-it
(% 0.57168305 (local-y)) (- 0.1637349 (local-y))))
0.40530929 (abs 0.53183758) 0.83426005))
```

6. Conclusions

Using the game of tag to test relative fitness, artificial evolution was able to discover skillful tag players in the form of vehicle-steering control programs.

These near-optimal players arose solely from direct competition with each other. Good results were obtained despite the considerable variance of fitness values inherent in the staging of individual games and the random selection of opponents. Fitness was based only on relative performance and did not require knowledge of the optimal strategy.

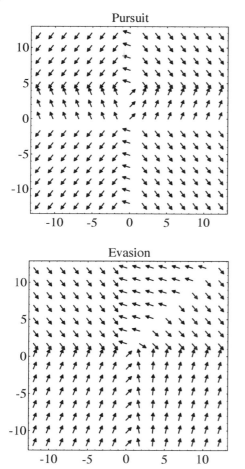

Figure 11: Plots of steering angle versus opponent position for the best-of-population player from run g after 113 generation equivalents

Since the optimal strategy *was* known for the problem studied here, it could be used as a benchmark. The player population evolved according to relative competitive fitness. Its progress in absolute terms could be measured by placing the evolved players in competition with the optimal player.

Figure 12: tag games from run G. A common flaw with all of these players is that they tend to have only two forward steering directions. Efficient pursuit and evasion require the ability to modulate turning angle. Some of these strategies look nothing like optimal behavior. They survive because they have adapted to their environment. For example the pursuer (in white) in the lower right image could never catch a good evader, but it is well suited to mowing down incompetent evaders.

Attempts to characterize the quality of evolved players are subject to differences of interpretation, but in at least one experimental configuration (run C) the population's average performance was within 10% of the optimal player, and the best of population individual performed within a few percentage points of optimal. Note that these quality metrics are only as meaningful as the somewhat *ad hoc* fitness measure used in these experiments.

In other runs described here, the quality of evolved players approached, but did not reach, that of the optimal player. It is not clear if this is because of a fundamental limitation of competitive fitness, a flaw in the experimental design, or simply because of limitations of genetic population size and length of runs.

7. Future Work

These experiments used a simple competitive behavior to test the concept of self organizing development of goal-directed behavioral control programs. The generally positive results found here encourage plans for more ambitious future projects.

Eventually the goal is to apply this technique to more complex games. But first, a more elaborate version of the game of tag is an obvious next step. A more complicated control problem, and a more visually interesting class of motion, would result from adding momentum to create a physically realistic model of vehicle motion in place of the current kinematic model. In a physically based model, the angle returned from evolved control programs would be interpreted as the direction of *steering force* to be applied to the vehicle's current momentum. If tag playing vehicles have momentum, their velocity and acceleration become relevant to the control problem. In the real world, the game of tag is usually played with multiple players and in the presence of obstacles. Adding these features, and sensory facilities for dealing with them, would create a rich environment for future studies. In place of the isolated, episodic, pair-wise tag games used in this work, it would be interesting to investigate an ongoing ecological simulation where a large population of agents interact with each other through the game of tag.

In [Angeline 1993] a individual's competitive fitness is determined in a single elimination tournament involving a generation's entire population. In the current work competitive fitness is tested against a randomly selected subset of the existing steady-state population. Several arbitrary choices where made when this selection mechanism was designed. Six opponents are selected with uniform random distribution. Is six too many or too few? Should the selection be skewed toward higher fitness opponents (say by tournament selection) as when parents are selected for crossover? Would judging fitness against the current leaders tend to improve evolvability or stifle it? This approach was attempted in run D but the results were inconclusive. On the other hand this approach is similar to that used in [Sims 1994] which produced good results.

Another approach to steady-state competitive fitness would be to use Angeline's technique of determining fitness based on standings in a single elimination tournament tree, but to do it in small batches of newly created individuals. A batch of 16 or 32 new individuals would be created, then pitted against each other in a single elimination tournament tree. Their standings in competition with their own litter-mates would determine their fitness. This approach is similar to that used in [Smith 1994] but would allow fitness values to reflect competition with a wider range of opponents.

The delightful variety of strategies that appeared in these experiments demonstrate that there are many, many way to approach a given task. The road to successful behavior may lie along any of those variations. The most robust evolutionary systems are those that can pursue many approaches at once. Geographically distributed, deme-based systems promote diversity and parallelism. They seem well-suited to competitive evolution of behavior [Ngo 1993] because the quality of competition is better when diversity is preserved.

Acknowledgments

This work is supported by Electronic Arts. The author wishes to thank Luc Barthelet, Vice President of Technology, for allowing research to coexist with product development. Additional thanks to Luc for help preparing the stimulus-response arrow diagrams. Thanks to John Koza and Pete Angeline for inspiration, encouragement, and critique. Thanks to James Rice for a detailed review. Special thanks to my wife Lisa and to our first child Eric, who was born at just about the same time as individual 15653 of run C.

References

[Angeline 1993] Angeline, P. J. (1993) and Pollack, J. B., Competitive Environments Evolve Better Solutions for Complex Tasks, in *Proceedings of the Fifth International Conference on Genetic Algorithms*, S. Forrest, Ed San Mateo, California: Morgan Kaufmann, pages 264-270.

[Arkin 1987] Arkin, R. C. (1987) Motor Schema Based Navigation for a Mobile Robot: An Approach to Programming by Behavior", *Proceedings of the 1987 IEEE Conference on Robotics and Automation*, Raleigh, North Carolina, pages 264-271.

[Axelrod 1984] Axelrod, R. (1984) *The Evolution of Cooperation*, Basic Books, New York.

[Axelrod 1989] Axelrod, R. (1989) Evolution of Strategies in the Iterated Prisoner's Dilemma, in *Genetic Algorithms and Simulated Annealing*, L. Davis editor, Morgan Kaufmann.

[Braitenberg 1984] Braitenberg, V. (1984) *Vehicles: Experiments in Synthetic Psychology*, MIT Press, Cambridge, Massachusetts.

[D'haeseleer 1994] D'haeseleer, P (1994) and Bluming, J., Effects of Locality in Individual and Population Evolution, in *Advances in Genetic Programming*, K. E. Kinnear, Jr., Ed. Cambridge, Massachusetts: MIT Press.

[Hillis 1992] Hillis, W. D. (1992) Co-Evolving Parasites Improve Simulated Evolution as an Optimization Procedure, in *Artificial Life II*, Santa Fe Institute Studies in the Sciences of Complexity, Proceedings Volume X, C. Langton, Ed., Addison-Wesley, Redwood City, California, pages 313-324.

[Holland 1992] Holland, J. H. (1992) *Adaptation in Natural and Artificial Systems*, second edition, MIT Press, Cambridge, Massachusetts.

[Isaacs 1965] Isaacs, R. (1965) Differential Games, John Wiley and Sons, New York.

[Kauffman 1993] Kauffman, S. A. (1993) *The Origins of Order*, Oxford University Press, New York.

[Kinnear 1994] Kinnear, K. E., Jr. (1994) Alternatives in Automatic Function Definition: A Comparison of Performance, in *Advances in Genetic Programming*, K. E. Kinnear, Jr., Ed. Cambridge, Massachusetts: MIT Press.

[Koza 1992] Koza, J. R. (1992) *Genetic Programming: on the Programming of Computers by Means of Natural Selection*, ISBN 0-262-11170-5, MIT Press, Cambridge, Massachusetts.

[Koza 1994] Koza, J. R. (1994) *Genetic Programming II: Automatic Discovery of Reusable Programs*, ISBN 0-262-11189-6, MIT Press, Cambridge, Massachusetts.

[Lindgren 1992] Lindgren, K. (1992) Evolutionary Phenomena in Simple Dynamics, in *Artificial Life II*, Santa Fe Institute Studies in the Sciences of Complexity, Proceedings Volume X, C. Langton, Ed., Addison-Wesley, Redwood City, California, pages 295-312.

[Lindgren 1994] Lindgren, K. (1994) and Nordahl, M., Artificial Food Webs, in *Artificial Life III*, Santa Fe Institute Studies in the Sciences of Complexity, Proceedings Volume XVI, C. Langton, Ed., ISBN 0-201-62494-X, Addison-Wesley, Redwood City, California, pages 73-103.

[Merz 1971] Merz, A. W. (1971) *The Homicidal Chauffeur: A Differential Game*, (PhD Thesis) Stanford University Center for Systems Research, Report SUDAAR 418.

[Menczer 1994] Menczer F (1994?) and Belew R. K., Latent Energy Environments, to appear in *Plastic Individuals in Evolving Populations*, Santa Fe Institute Studies in the Sciences of Complexity, Addison-Wesley (in press).

[Miller 1989] Miller, J. H. (1989) The Co-evolution of Automata in the Repeated Prisoner's Dilemma, Santa Fe Institute Report 89-003.

[Ngo 1993] Ngo, J. T. (1993) and Marks, J., Spacetime Constraints Revisited, Proceedings of SIGGRAPH 93 (Anaheim, California, August 1-6, 1993), in *Computer Graphics Proceedings*, Annual Conference Series, 1993, ACM SIGGRAPH, New York, pages 343-350.

[Ray 1992] Ray, T. S. (1992) An Approach to the Synthesis of Life, in *Artificial Life II*, Santa Fe Institute Studies in the Sciences of Complexity, Proceedings Volume X, C. Langton, Ed., Addison-Wesley, Redwood City, California, pages 371-408.

[Reynolds 1994a] Reynolds, C. W. (1994) An Evolved, Vision-Based Model of Obstacle Avoidance Behavior, in *Artificial Life III*, Santa Fe Institute Studies in the Sciences of Complexity, Proceedings Volume XVI, C. Langton, Ed., ISBN 0-201-62494-X, Addison-Wesley, Redwood City, California, pages 327-346.

[Reynolds 1994b] Reynolds, C. W. (1994) Evolution of Obstacle Avoidance Behavior: Using Noise to Promote Robust Solutions, in *Advances in Genetic Programming*, K. E. Kinnear, Jr., Ed. Cambridge, Massachusetts: MIT Press.

[Reynolds 1994c] Reynolds, C. W. (1994) Evolution of Corridor Following Behavior in a Noisy World, to appear in *From Animals to Animats 3: Proceedings of the Third International Conference on Simulation of Adaptive Behavior* (SAB94), in press.

[Sims 1994] Sims, K. (1994) *Evolving 3D Morphology and Behavior by Competition*, Artificial Life IV (in this volume), R. Brooks and Pattie Maes, Eds., MIT Press, Cambridge, Massachusetts.

[Smith 1994] Smith, R. E. (1994) and Gray, B., Co-Adaptive Genetic Algorithms: An Example in Othello Strategy, to appear in *The Proceedings of The Florida Artificial Intelligence Research Symposium 1994*, and available as *TCGA Report Number 94002*, The University of Alabama, Tuscaloosa.

[Syswerda 1991] Syswerda, G. (1991) A Study of Reproduction in Generational and Steady-State Genetic Algorithms, in *Foundations of Genetic Algorithms*, G. J. E. Rawlins, Ed. San Mateo, California: Morgan Kaufmann, pages 94-101.

[Werner 1993] Werner, G. M. (1993) and Dyer, M. G., Evolution of Herding Behavior in Artificial Animals, in *From Animals to Animats 2: Proceedings of the Second International Conference on Simulation of Adaptive Behavior* (SAB92), Meyer, Roitblat and Wilson editors, ISBN 0-262-63149-0, MIT Press, Cambridge, Massachusetts, pages 393-399.

[Yeager 1994] Yeager, L. (1994) Computational Genetics, Physiology, Metabolism, Neural Systems, Learning, Vision, and Behavior or PolyWorld: Life in a New Context, in *Artificial Life III*, Santa Fe Institute Studies in the Sciences of Complexity, Proceedings Volume XVI, C. Langton, Ed., ISBN 0-201-62494-X, Addison-Wesley, Redwood City, California, pages 263-298.

In Praise of Interactive Emergence, or Why Explanations Don't Have to Wait for Implementations

Horst Hendriks-Jansen
School of Cognitive and Computing Sciences
University of Sussex
Brighton BN1 9QH, UK
Email: horsth@cogs.susx.ac.uk

Abstract

In this paper I argue that Alife can and should make an immediate contribution to scientific explanations of human behavior, even though the development or evolution of robots with full human capabilities is likely to take many years. In support of my argument, I draw a clear distinction between the explanatory principles of the two main strands of Alife. My own interest is in situated robotics, not in the simulation of cost-benefit models inspired by behavioral ecology. Situated robotics provides existence proofs and explanatory principles for the interactive emergence of structured behavior. It points to a historical rather than a deductive, or probabilistic, or intentional explanation of behavior. I argue that the 'orienting attitudes' of ethology, in conjunction with autonomous models of this type, can provide principled explanations for the emergence of human thought, although the details of such explanations may take many years to work out. I present evidence from developmental psychology to suggest that, rather than possessing inborn knowledge about the world, human infants display species-typical activity patterns, selected for their power to attract the attention of adults and deceive them into intentional interpretations of the infants' behavior. Such activity patterns serve to establish interactive patterns of turn-taking, which are essential to the emergence of meaning. They can be said to perform a 'boot-strapping' role to launch the human infant into an environment of adults who think in intentional terms, communicate through language, and manipulate tools and artifacts.

1 Introduction

My interest is in explanation, not in design or engineering. The focus of this paper will therefore be on life as we know it rather than life as it could be or life as we might be able to evolve it artificially. What is more, I shall be concentrating on explanatory strategies which have a bearing on our own peculiar form of intelligent life. I believe that one of the major benefits of the branch of Alife in which I am interested (situated robotics) is the contribution it can make to a coherent naturalistic explanation of intentional behavior and conceptual thought as displayed by human beings.

This has become something of a heresy in Alife circles. The anti-human bias was spelled out some years ago by Brooks (1991b):

> I, and others, believe that human level intelligence is too complex and little understood to be correctly decomposed into the right subspecies at the moment and that even if we knew the subspecies we still wouldn't know the right interfaces between them. Furthermore, we will never understand how to decompose human level intelligence until we've had a lot of practice with a simpler level intelligence. (Brooks, 1991b, p140)

We may call this the no-explanation-without-implementation school of thought. I shall argue in this paper that, while it provided a necessary corrective to some of the wilder claims of classical and connectionist AI, it has now become counter-productive. Classical and connectionist models continue to dominate cognitive philosophy and are still making headway in developmental psychology. New discoveries concerning the early behavior of human infants, made possible by the increasing precision of experimental techniques, are being interpreted in accordance with such models (see Baillargeon, 1987, Diamond, 1991, and Spelke 1991). The experiments themselves are devised from a classical perspective, which assumes that the natural kinds for a scientific explanation of human thought have to be internal representations corresponding to entities in an objective world which is somehow pre-registered into objects, properties and relations. Development of sensori-motor skills is then conceptualized as a process of representational redescription rather than a complex process of interactive emergence involving a large number of situated activity patterns (Karmiloff-Smith, 1987).

While classical AI and connectionism thus invade the human sciences, there is a danger that Artificial Life is turning into a computational discipline specializing in slime-moulds and six-legged insects. I agree that we shall have a long time to wait for a genuine implementation of human intelligence (as opposed to systems which mimic rational choice and logical inference within some formally specified task-domain), but I believe that the new paradigm has much to tell us right now about the way in which the explanation of human behavior should be approached. We must not allow classical AI and connectionism to have it all their own way while we spend the next few hundred years pushing our situated implementations up their evolutionary hills. The paradigm based on decomposition by activity and interactive emergence provides valuable explanatory principles as well as solidly-based methods of implementation, and these principles can be applied before the implementations are achieved

2 Clarification of Terms

The notion of 'emergence' may be used in different senses. Nagel (1961) distinguishes two frequently encountered meanings in scientific explanation. The first implies that properties at higher levels are not necessarily predictable from properties at lower levels. As Nagel points out, that observation can be said to hold even for phenomena whose explanations are generally considered to be reductive in the 'hard' sense, such as the atomic theory of matter. The gross properties of water, like 'wetness', cannot be deduced from the properties of molecules, for the simple reason that molecular theory does not include statements about such properties.

The second meaning picked out by Nagel (1961) is a historical one. This draws attention to the fact that simple traits and forms of organization give rise, in the course of evolution, to more complex and irreducibly novel traits and structures. As Nagel points out:

> ...the question whether a property, process or mode of behavior is a case of emergent evolution is a straightforward empirical problem and can be resolved at least in principle by recourse to historical inquiry. (Nagel, 1961, p375)

I shall discuss some of the difficulties of resolving that question in section 4. In order to do so, I shall introduce a third type of emergence, which I have called 'interactive emergence', and which is defined more fully in that section.

The term 'merkwelt' derives from von Uexkull (1934) and has been used by Tinbergen (1951), and more recently, by Brooks (1986). It is used to emphasize the fact that a creature's perceptual world will be species-specific and depends on the creature's sensory modalities and resolving powers, its morphology, and its patterns of movement relative to the world.

'Affordance' is a term coined by Gibson (1979). It assumes an ecological approach to the explanation of perception and behavior, rather than the ethological approach I recommend in this paper. I use it in the next section to bring out the fact that the features relevant to an explanation of behavior cannot be conceptualized as abstract properties of the world. They are dynamic structures which emerge when a creature performs its species-typical activity patterns within a specific ecological niche.

'Natural kinds' may be seen as the scaffolding for scientific theory- building. Hypothetical groupings or explanatory entities, suggested by a growing theory, lead to fruitful inductions and become established as part of a new scientific paradigm. It has been suggested by Wilkes (1989) that one reason why the behavioral sciences are underdeveloped is their inability to arrive at a consensus about natural kinds. No acceptable, shared taxonomy exists in psychology. Traditional cognitive science can be seen as a (failed) attempt to establish such a consensus, proposing internal representations and computational formalisms as the natural kinds of psychology.

As Wilkes (1989) points out, the notion of natural kinds, and a recognition of their importance in scientific advance, are quite compatible with scepticism as to whether such hypothetical groupings are in any sense 'real' constituents of the world. 'Naturalness' may be glossed in terms of the likelihood that the chosen kinds can generate fruitful laws and generalizations.

3 Six Explanatory Principles Derived from Situated Robotics

Most of the key issues have been identified before by other authors (see Brooks, 1986, 1991a, 1991b, 1992, Chapman & Agre, 1987, Horswill, 1992, Kirsh, 1991, Smithers, 1992, Steels, 1991). I shall try to recapitulate the main points in a condensed form, using the engineering principles derived from situated robotics to define what I believe to be the central principles for explanations of human behavior and thought.

1. Understanding of autonomous behavior requires decomposition by activity rather than by function. Each component or activity module identified in this type of explanation interacts independently with the environment, and the interaction is continuous, dynamic and unplanned. This eliminates the need for internally stored information to facilitate planning, and for the transmission of information between modules. The underlying mechanism does not depend for its operation on the manipulation of semantic tokens representing states of the world or the agent's relation to the world.

2. Behavioral emergence therefore replaces the traditional functional hierarchy. Since the activity modules do not perform tasks that are logically defined subtasks of the overall system, their contribution to its performance can no longer be characterized from the design perspective favored by classical AI. An understanding of how a complex autonomous system is put together requires consideration of historical contingencies involving sensors, effectors and neural mechanisms that evolved in response to changing environmental conditions.

3. Much of the 'knowledge' that is used by an autonomous system is 'stored' in its environment. It does not need to be retrieved from memory, or recovered from sparse perceptual data, because it is by definition available as and when it is needed, and in the form in which it is needed. The individual activity patterns work by simple indicators which cause the

activities to be switched on and off and correctly oriented in the appropriate situations. In ethology such indicators are called sign stimuli and orienting stimuli. Most ethologists think of them as objective features, but interactive emergence demands a Gibsonian or Uexkullian point of view (see point 5).

4. The 'control mechanism' which makes the 'decisions' or 'choices' is thus inextricably embedded in the world. This does not mean that the system as a whole can be conceptualized as a behaviorist stimulus-response engine. Its perceptual and behavioral history will affect its responses to current stimuli, and the same stimulus may thus elicit different responses on different occasions. In the terminology of dynamical systems theory, perturbation by the environment alters the system's parameters and affects its evolution equation.

5. The creature's morphology, its sensori-motor characteristics, and the activity patterns it performs, all affect its 'merkwelt' (von Uexkull, 1934) and 'affordances' (Gibson, 1979), and consequently, the 'problems' it has to 'solve'. Tasks, problems, and solutions to problems cannot be characterized independently. This puts the dream of an 'implements relation' (Horswill, 1992) out of reach. A three-part equation relating classes of autonomous systems to task-descriptions and environmental categories would require an objective taxonomy for each. Interactive emergence implies that no such objective classification can be made that would be relevant to the generation or explanation of autonomous behavior.

6. The evolutionary and developmental history of the creature is important to an understanding of how it works. Each 'layer' on the way to a complex intelligent system has to be 'debugged' in the real world, either by natural selection or by the programmer. New *merkwelten* emerge from each successive layer. In Hendriks-Jansen (1993a) I argued that, in a naturally evolved creature, successive layers can confer the status of natural kind on the layers which preceded them, if it can be shown that the *merkwelt* resulting from the earlier layer was required for the evolution of the more recent one. This gives substance to the notion of a 'historical' (Sloman, 1978), or 'genetic' (Nagel, 1961) explanation of autonomous behavior resulting from interactive emergence.

4 Four Types of Scientific Explanation

Nagel (1961) distinguishes four types of explanation that are commonly used in science:

1. Deductive explanation, in which the *explicandum* is a logically necessary consequence of the explanatory premises, which must contain a set of initial conditions and a general law.

2. Probabilistic explanation, in which the explanatory premises do not formally imply the *explicandum* but make it probable.

3. Functional or teleological explanations, in which a thing or event is explained in terms of the function it performs in some larger whole, or the role it plays in bringing something about.

4. Genetic explanations. These are different from deductive, probabilistic, or teleological explanations because the explanatory premises take the form of a historical sequence of events. An important element in this notion is that explanation depends crucially on actual occurrences. There is no central law which constrains these occurrences. It would be impossible to predict an event at time t given the events at time 0. It is not even possible to explain, by some general principle, how the particular form taken by E_t followed from the structure of E_0. However, given the occurrence of all stages between E_0 and E_t, it becomes clear how the latter could have issued from the former.

The classic example of this type of explanation is, of course, evolutionary biology. Like all historical explanations, it lacks a deductive or probabilistic law. Natural selection does not confer predictive power. I argue this point in detail in Hendriks-Jansen (1993b), and also show that functional decomposition, and hence teleological explanations by Nagel's definition of the term, do not sit well with the theory of natural selection. This has led some authors (notably Popper, 1963) to question the scientific credentials of Darwin's theory, and it has led others (e.g. Gould & Lewontin, 1978) to point out that a plausible adaptive story can always be found to fit any explanatory hypothesis. However, the theory has stood the test of time and proved highly successful in providing a diversity of historical explanations, not only of morphological traits, but of behavioral entities as well.

The explanatory entities or natural kinds of a genetic or historical explanation are the successive stages in the emergence of a phenomenon, trait, or event. As Nagel (1961) pointed out, one of the difficulties with this type of explanation is the problem of deciding which specific occurrences, out of a more or less continuous historical line, deserve to be singled out as particularly significant. In Hendriks-Jansen (1993a) I argued that a principled choice can in fact be made, based on a slightly modified version of Millikan's (1984) notion of 'history of use'. As an existence proof to support my contention, I referred to an autonomous, wall-following and navigating robot developed by Mataric (Mataric, 1991, 1992, Mataric & Brooks, 1990).

The independent operation of four low-level reflexes in Mataric's robot, debugged in a specific office environment, results in the emergent activity of wall-following. There are no explicit instructions inside the robot which tell it to follow walls; no formal definition of walls is required to produce its behavior. The robot is therefore unlikely to follow a particular wall in precisely the same way on different occasions. Its route will depend on its approach in a particular instance, on noise in its sensor readings, and on the numerous unspecified contingencies of a dynamic environment. But its can be relied on to spend a considerable proportion of its time following walls, as long as it remains in an environment which is roughly similar to the one by which its wall-following behavior was 'selected'. This emergent behavior of wall-following produces a high-level *merkwelt* consisting of dynamic features with temporal extent which correspond only indirectly to features in the world. It allows the robot to identify 'landmarks' as reliable correlations between sonar readings averaged over time and temporal regularities in its own movements. Such a configuration, if found in a naturally occurring creature, would confer the status of natural kind on the lower level of emergent activity, since it would clearly establish a history of use for that activity in the creature's evolutionary past.

It is important to stress that the natural kinds of this type of explanation only come into existence as the result of the creature's situated activity. They are not entities which can be reduced to events in the creature's head, and neither can they be defined in terms of inputs and outputs, or by classifications of the environment. Interactive, situated behavior cannot be explained in terms of a deductive or generative law. It requires a historical explanation because there can be no rules to predict the sorts of behavior which might emerge. There are no short-cuts for deriving the sequence of events in a historical process.

5   Behavioral Ecology, Cost-Benefit Models and Natural Selection

It may seem that what I have claimed in the last two sections is flatly contradicted by important work in behavioral ecology, as well as by many of the simulations done in Alife. I am referring to 'economic' or cost-benefit studies based on such well-established notions as inclusive fitness and evolutionarily stable strategies. Clearly these theories make reference to natural selection, and clearly they also deal with classes of individuals defined in terms of abstract behavioral categories, such as 'hawks' and 'doves'. It would appear that there are obvious similarities between theories of this kind and what Nagel (1961) calls probabilistic theories, and that the effect of natural selection on behavior may thus be characterized in those terms.

The application of cost-benefit models in behavioral ecology results in explanations of behavior which account for observed correlations between certain environmental factors and formally defined classes of behavior by applying a logical calculus like game-theory. Such explanations are undoubtedly of great value in their own right. Computational models may be built which incorporate and modify this logic as well as the variables which the theory has identified. But these are not explanations in the sense that they can tell us anything about the underlying mechanisms; they remain economic or actuarial models which reproduce statistical correlations. It is a category mistake to treat the programs used in such models as explanations of the mechanisms which subtend the behavior of individual creatures.

Natural selection, writes McFarland (1985), has seen to it that there is an optimum balance between the amount of time spent by a brooding bird on incubation, and the amount of time she spends on foraging to keep up her strength. There will be a clear relation between different strategies (in this case the varying proportions of time spent on the two activities) and the bird's reproductive success. Since the two behaviors are both important to the production of offspring and compete for the bird's time during the period of incubation, there must be a measurable correlation between behavioral strategies and genetic survival.

The crucial question remains, however - namely what are the underlying mechanisms which might explain such a correlation? It will not do to say that they consist in a 'gene for foraging' and a 'gene for incubation', or that we are dealing with a mechanism that controls the amount of time spent on the two activities. As McFarland himself admits:

> ...such considerations are purely functional. They specify what animals ought to do to make the best decisions under particular circumstances, but they do not say anything about the mechanisms that animals might employ to attain these objectives. (McFarland, 1985, p456)

Foraging and incubation are functional categories which would need to be grounded in specific activity patterns to be linked to particular genes. The fact that cost-benefit bird (an actuarial abstraction like 'economic man') necessarily optimizes or maximizes its utility (because utility is by definition that which it optimizes) does not imply that individual birds contain choice mechanisms which work by optimization on entities like foraging and incubation. But the question which interests us as cognitive scientists is what sort of mechanism does subtend the bird's behavior. We would like natural selection to help us find clues to those underlying mechanisms, and given sufficient patience and care, it can in fact be made to do so, but economic models are not the appropriate tool for the job.

6   The Orienting Attitudes and Explanatory Principles of Ethology

The classic texts of ethology are often quoted in the literature on situated robotics. There is a feeling that the basic interests of the two disciplines coincide, even though their aims might be different.

Ethologists hold that behavior should be studied in the creature's natural environment, and not under laboratory conditions that have been specifically designed to elicit certain responses. In situated robotics, there is a corresponding emphasis on producing robots which will operate in 'natural' environments, meaning, in this case, that the environments should not be altered to suit the robot's limitations. Both disciplines share an awareness of the importance of context, and of the way an autonomous creature becomes embedded in its environment, so that its activities are to a very great extent shaped by it. There is, in both disciplines, an understanding that the global behavior of a creature may be the result of an interplay of diverse activities, each operating independently in response to action-specific stimuli, rather than a purposive succession of acts directed by a central controlling mechanism towards an internally represented goal. This leads, in autonomous agent research, to the principle of decomposition by activity, and in ethology, to an emphasis on deriving explanatory concepts from the study of simple behavior patterns and their interdependencies, rather than imposing some overall theoretical framework on behavioral diversity.

Hinde (1982) sums up what he calls the 'orienting attitudes' of ethology as follows:

1. Start all analysis and theorizing from a secure and extensive descriptive base.

2. Study behavior in the context of the environment by which it was selected.

3. Analyze in detail one specific type of behavior as it is displayed by one particular species in its natural environment; then compare with other, closely related behaviors, species and environments.

As will become clear in the next two sections, I believe it is these orienting attitudes which constitute the most important legacy of the early work in ethology, rather than the much-debated theoretical notions of fixed action pattern, sign stimulus, action specific energy, and hierarchical organization introduced by Lorenz (1937, 1939, 1952) and Tinbergen (1951). Ethology's orienting attitudes impose their own distinctive logic on explanation. Its emphasis on detailed analysis of specific activity patterns performed by particular species in their natural environment shifts attention from generative rules and formal task-descriptions to questions of taxonomy and history. The variety of naturally occurring behavior, and its diverse relations to the environment, rule out a single explanatory principle like the reflex arc or the negative feedback loop. Before anything like a mechanistic causal explanation can be attempted, the precise structure of each pattern of activity, and its relation to the structure of other activities and the environment, as well as its historical emergence, need to be understood.

Lorenz had an unfortunate obsession with 'innateness', which was probably a reaction against the behaviorist orthodoxy of his time. A somewhat simplistic view of ontogenesis and genetic transmission caused him to equate 'innate' with 'fixed' (both in the sense of 'totally determined by a genetic blueprint' and 'immutable in its structure'), and he was therefore forced to maintain that fixed action patterns were 'endogenously generated'. This led to a number of false starts. It tempted some ethologists to search for neurophysiological correlates of fixed action patterns and 'behavior centers', and it tempted others to think in terms of 'software' explanations involving 'programs' or 'systems' (see Tinbergen, 1951, Barlow, 1968, Baerends, 1976).

It can be argued (Hendriks-Jansen, in preparation) that all these attempts are denials of ethology's most important contribution, which was to draw attention to the situated, interactive and emergent character of natu-rally occurring behavior. The best corrective to such tendencies was provided by comparative psychologists like Schneirla (1966) and Lehrman (1970), who argued that no behavior is genetically fixed in the sense that Lorenz had decreed it to be. Species-typical activity patterns are emergent phenomena in three different senses of the word. They emerge in the species as a result of natural selection. They emerge in a maturing individual as the result of ontogenesis and learning. And they emerge every time they occur within the life of that individual as the result of interactions between the creature's low-level activities and its species-typical environment.

This means that there can be no straightforward neurophysiological correlates of complex activity patterns, just as there is no genetic blueprint for their structure. Nor can one think of units of behavior as being 'represented' in any way inside the creature. The explanation for the recognizable structure of species-typical activity patterns must not be sought in generative mechanisms; it is a consequence of the fact that the creature evolved, matured and acts within a particular ecological niche, ensuring that its emergent behavior displays the structure we observe.

Lehrman's and Schneirla's critiques brought to the surface certain fundamental implications of the ethological approach. If behavior is an emergent phenomenon in the three senses given above, it is unlikely that any universal laws can be discovered to describe and explain it. The problem is not just one of finding the right classifications which will point to causal generalities; it is the fact that the causal explanation is likely to be unique in each case.

There is nothing new about such a situation in science. It characterizes areas which are not susceptible to description in terms of formal rules, but only to what Marr (1982) called 'Type 2' explanations. As Marr himself explained it, in such cases the 'problem' is 'solved' by a large number of processes whose complex interaction is the simplest description of the processes involved. Marr believed that this made explanation impossibly difficult. However, that is so only if one adopts his position that behavior is a solution to some problem set by the environment. As I have argued in this paper, the natural environment does not set problems, and evolution does not solve any. Explanation of behavior should not be seen as a matter of characterizing the appropriate problem space, but as a matter of describing the space of emergent phenomena by classifying and relating the actual instances produced by genetic recombination and mutation acted on by natural selection.

Computational theory does not provide an appropriate framework for this type of explanation (Hendriks-Jansen, forthcoming), but situated robotics is its natural ally. Artificially built models capable of diverse levels of situated activity within a specific environment may serve to confirm hypotheses about the interactive emergence of naturally occurring behavior, and bring out 'family resemblances' between different kinds of behavioral dependency. Such models are models of behavior, not attempts to model the mind or brain. There is no presumption that the computational devices used to implement the behavior bear any resemblance to mechanisms inside the creature whose behavior is being modeled. However, a deeper understanding of behavioral dependencies may point to fruitful investigative strategies for the neural sciences.

7   The Emergence of Meaning in Human Infants

How does all this apply to an explanation of intentional behavior as manifested by human beings, whose activities seem so clearly governed by goals, plans and internal representations corresponding to the objects, properties and relations we all take to be the basic constituents of our world? An explanation in terms of activity modules and interactive emergence might be adequate to the activities of an ant or a spider, but surely we possess a central, directive intelligence which traffics in concepts? Surely those concepts play a crucial role in the generation of behavior, and must therefore figure in any explanation as causally effective entities?

Human beings do use concepts, but the lack of progress made by classical AI over the last thirty-five years has begun to raise some doubts as to whether symbolic manipulation, high-level planning by a central executive, and rules of inference working over semantic tokens are the most appropriate natural kinds for an explanation of human behavior and thought. Top-down analysis, both in the sense of deriving explanatory entities from folk-psychology, and in the sense of functional decomposition inspired by a design approach based on classical engineering, does not seem to have got us very far. I do not intend to rehearse all the well-known problems encountered by these strategies. Instead, I would like to offer some evidence from recent work in developmental psychology which suggests that a historical explanation in terms of interactive emergence might do better.

In the early 1970s, investigators started to report that very young infants, observed in their natural environment, perform activity patterns that are far more complex and far more varied than had previously been recognized. These investigators tended to be influenced by the orienting attitudes of ethology, as summed up in the last section. They tried to observe infants' behavior in the context in which it naturally occurred, deferred analysis and theoretical speculation until they had built up a solid descriptive base, and examined in great detail one particular type of behavior, rather than searching for evidence of a central law that might unify diverse behavioral phenomena.

Bateson (1975) provides a good example of how such a change in emphasis can render previously invisible (or in this case, inaudible) data perceptible to an observer. Her research centered on the acoustic analysis of infants' vocalizations during mother-infant interactions. It uncovered a wealth of coos and murmurs which were quite evidently picked up and interpreted by the mother, but which previous investigators appeared to have overlooked.

The difference seems to arise from a difference in sampling techniques, where [previous investigators] focused on describing an acoustic phenomenon and we are concerned with describing the acoustic aspects of an interpersonal process. (Bateson, 1975, p109)

The same point is made by Trevarthen (1977) about the rich variety of facial expressions and expressive gestures displayed by very young infants, by Brazelton (1979) about the infant's species-typical responses to adult handling, and by Schaffer (1977), Kaye (1979), Tronick et al (1979) and Newson (1979) about the cues and temporal patterning involved in mother-infant turn-taking.

All of these investigators made intensive use of film or video, and of the opportunities for micro-analysis which those media afford. They frequently remark that, prior to the development of such techniques, detailed investigation of activity was virtually impossible. Trevarthen (1977) writes that the actual movements of human beings - as opposed to the intentional acts in terms of which we normally describe and perceive them - were as difficult to observe before the invention of cine-photography as were the planets before the invention of the telescope. He speculates that this might be one of the reasons why psychology became a science of perception and cognition, rather than a science of activity patterns, gestures and facial expressions.

The explanatory entities of psychology have tended to be static entities like beliefs, desires, memories and mental states, whereas these new discoveries suggested that meaning emerges in dynamic interactions. More sophisticated recording and sampling techniques, allied to the orienting attitudes of ethology, revealed the importance of temporal patterning in the establishment and maintenance of mother-infant exchanges. It began to be realised that a large number of species-typical activity patterns are either present at birth, or else mature during the first few months of life, and that the main 'function' of many of these patterns is the establishment of an interactively emergent and typically human *merkwelt* of mother-infant 'dialogue'. Examples of such patterns are primitive reach-and-grasp (Trevarthen, 1977, von Hofsten, 1984, Fischer & Bidell, 1991), the early lip and tongue movements which Trevarthen called 'pre-speech' (Trevarthen, 1977, Papousek & Papousek, 1977), rhythmical stereotypies like supine kicking and hand-waving (Thelen, 1981), and burst-pause-burst in suckling (Kay, 1982).

The existence of such well-coordinated patterns of activity at a very early age, and the way they either disappeared at a later date, or were modified into more complex types of behavior, or served as the 'context' for the emergence of such later behavior, prompted a reappraisal of many of the accepted views about inborn abilities and learning.

It was realised that a rigid, stage-bound picture of development, such as had been proposed by Piaget (1954), could not adequately describe the processes that were being observed. Individual activity patterns had their own developmental profiles, and though, in a particular culture, their onsets and peaks might tend to occur at roughly the same ages for all children, this could not be ascribed to a succession of rigid, unifying internal structures. Some children follow idiosyncratic paths, and each mother-infant pair builds up its own repertoire of interactive patterns, substantially different from those of other mothers and infants (Schaffer, 1977, Bullowa, 1979, Kaye, 1982, Fogel & Thelen, 1987).

Adult skills and concepts emerge over time through a sequence of increasingly complex activity patterns. An ability like reaching-and-grasping, as 'implemented' in human beings, cannot be explained in terms of a formally defined competence or a specific set of internal commands contingent on appropriate feedback signals. Only a historical explanation can make sense of the relation between analogous but not necessarily homologous behaviors like the primitive reach-and-grasp of a two-month old infant and the visually guided reaching of an adult human being. These behaviors are not manifestations of a single competence, but they can be related in a situated context through history of use (von Hofsten, 1984, Fischer & Bidell, 1991).

Early forms of behavior can and usually do have their own adaptive advantages, which may have little to do with the 'function' of the adult form. The rhythmical stereotypy of supine kicking, which is almost certainly an early manifestation of the swing-stance cycle in walking, and may serve as a preparation in the sense of strengthening the muscles, also has expressive and communicative value. The mother tends to interpret variations in her infant's kicking as clues to the infant's moods and desires, and as responses to her own attempts at communication. Evolution seems to have hijacked an early form of adult walking for other purposes.

In many cases, it can be shown to be biologically advantageous to have an immature form of behavior at the early stages of development, rather than the fully developed form. The infant's inability to distinguish separate words in his mother's vocalizations probably allows him to treat her clauses as unitary utterances, equivalent to his own coos and murmurs, and thereby promotes the process of turn-taking in early 'dialogues', and provides a 'scaffolding' for the later parsing of clauses into meaningful sub-units (Bateson, 1979, Fernald, 1989, Hirsh-Pasek et al, 1987). Limited depth of field during the first months of life restricts the infant's resolution to objects at approximately twenty centimeters distance, and this, coupled with early fixation patterns, produces an *merkwelt* consisting predominantly of his mother's face, since she appears to be biologically primed to put herself into the optimum position (Turkewitz & Kenny, 1993, Johnson, 1993b).

Many of these early activity patterns appear to be typical to human beings. Burst-pause-burst in feeding does not occur even in the young of other phylogenetically advanced primates (Kaye, 1982). Human infants display a large number of facial expressions (Charlesworth & Kreutzer, 1973, Trevarthen, 1977) and rhythmical stereotypies (Thelen, 1981) which are absent in goril-

las and chimpanzees. Face-to-face exchanges between mothers and infants, and the complex interaction of activity patterns from which these emerge, seem also to be unique to the human species.

On the other hand, many of these species-typical activity patterns disappear, or become submerged, or disintegrate before the end of the first year. They are not functional components of adult behavior. Often their only reason for existence seems to be that they enable other, more advanced forms of behavior to develop within the situated context which they provide. This realization led the early researchers to formulate the notion of 'scaffolding' (Bruner, 1982, Newson, 1979, Kaye, 1982, Fischer & Bidell, 1991).

There are a number of different ways in which one can think of scaffolding. One can apply it to the supportive framework, usually provided by an adult, which enables the child to perform activities of which he may not be capable on his own until a somewhat later date. Thus, infants will demonstrate the ability to 'walk' if they are supported in the right way long before their leg muscles have developed sufficient strength to hold them up. This view of scaffolding stresses the intentional contribution of an adult. It sees the mother-infant dyad as two distinct causal entities, with the mother providing conscious support and guidance to enable her infant to learn new skills. The scaffolding continuously pushes the infant a little beyond his current capabilities, and it pushes him in the direction in which his mother wishes him to go. Scaffolding is then a pedagogical device, defined in terms of capabilities and tasks, and its nature and effect are presumed to be under the control of the adult. This is the concept of scaffolding as it was first introduced by Bruner.

A different, and for our purposes more interesting, view of scaffolding starts from the activity patterns themselves. It conceives of the mother-infant dyad as two tightly-coupled dynamic systems (Fogel, 1993). Recognizable patterns of behavior are seen, not as the manifestation of beliefs and desires or internally represented rules, but as emergent from continuous, low-level mutual adjustments between the two systems, much as wall-following emerges from the continuous adjustments of Mataric's robot to its particular environment. The activity patterns, rather than the presumed intentions of the mother and the level of skill of the infant, become the natural kinds for this type of explanation. Scaffolding then takes on a meaning similar to the dynamic *merkwelt* provided by wall-following for the emergence of landmarks and landmark navigation in Mataric's robot. Thus, the 'dialogue' between a suckling infant and the mother who jiggles him whenever he pauses in feeding (Kaye, 1979, 1982) constitutes a recognizable interactive pattern which emerges from low-level reflexes and interactively generated rhythms, and which establishes a habit of turn-taking on which later, face-to-face exchanges will be built. The 'pragmatics' of meaningful communication, in this view, precede any explicit 'content', and they are best conceptualized in terms of patterns of situated activity.

This notion of scaffolding stresses the importance of species-typical activity patterns, selected for their power to attract the attention of adults, deceive them into intentional interpretations, and establish the habit of turn-taking which is essential to human learning, rather than of inborn knowledge or innate cognitive structures. Movement subtended by a central pattern generator and/or simple reflexes requires no knowledge of the world and no *a priori* concepts of causation, space, time, objects, or their properties. Some of the species-typical activity patterns which newborn or very young infants are capable of performing may have no place in adult behavior. They may simply serve a 'boot-strapping' role to launch the infant into an environment of adults who think in intentional terms, communicate through language and manipulate tools and artifacts.

8 Conclusion

I have argued in this paper that an adequate understanding of intentional behavior and conceptual thought as manifested by mature human beings will only be achieved by adopting a historical perspective which starts from early, species-typical activity patterns and conceptualizes maturation and learning as interactive emergence. Such an explanation requires an ethological approach to the collection and interpretation of behavioral data. It calls for the identification of species-typical activity patterns through intra-species and cross-species comparisons, and for the formulation of hypotheses concerning the interdependence of such patterns. Situated robotics is the only computational discipline which is currently capable of testing such hypotheses.

I am not suggesting that we will be able to construct infant robots in the near future which can enter into meaningful dialogue with human mothers. What I have argued in this paper is that the explanatory framework of situated robotics, by providing simple existence proofs and a clear conceptual alternative to traditional explanations, can and should play a role in theories of behavior as well as its implementation. We must not allow classical AI and connectionism to have it all their own way in cognitive science, just because we have realised that their explanations leave out all the really difficult bits, such as dynamic, context-and activity-related perception and robust situated behavior.

Acknowledgements

I would like to thank Margaret Boden, Dave Cliff, Inman Harvey, Maja Mataric and Geoffrey Miller for comments related to the topics discussed in this paper.

Reference

Baerends, G. (1976). The functional organisation of behaviour. *Animal Behaviour, 24*, 726–738.

Baillargeon, R. (1987). Object permanence in $3\frac{1}{2}$ and $4\frac{1}{2}$ old infants. *Developmental Psychology, 23*(5), 655–664.

Barlow, G. (1968). Ethological units of behavior. In D.Ingle (Ed.), *The Central Nervous System and Fish Behavior*, pp. 217–232. University of Chicago Press, Chicago.

Bateson, M. (1975). Mother-infant exchanges: The epigenesis of conversational interaction. In D.Aaronson, & R.W.Rieber (Eds.), *Developmental Psycholinguistics and Communication Disorders*, pp. 101–113. New York Academy of Sciences, New York.

Bateson, M. (1979). The epigenesis of conversational interaction: A personal account of research development. In M.Bullowa (Ed.), *Before Speech: The Beginning of Interpersonal Communication*, pp. 63–78. Cambridge University Press, Cambridge, UK.

Brazelton, T. (1979). Evidence of communication during neonatal behavioral assessment. In M.Bullowa (Ed.), *Before Speech: The Beginning of Interpersonal Communication*, pp. 79–88. Cambridge University Press, Cambridge, UK.

Brooks, R. A. (1992). Artificial life and real robots. In Varela, F., & P.Bourgine (Eds.), *Proceedings of the First European Conference on Artificial Life*, pp. 3–10. MIT Press/Bradford Books, Cambridge, MA.

Brooks, R. (1986). Achieving artificial intellligence through building robots. Tech. rep. A.I. Memo 899, MIT A.I.Lab.

Brooks, R. (1991a). Challenges for complete creature architectures. In Meyer, J.-A., & Wilson, S. (Eds.), *From Animals to Animats: Proceedings of The First International Conference on Simulation of Adaptive Behavior*, pp. 434–443. MIT Press/Bradford Books, Cambridge, MA.

Brooks, R. (1991b). Intelligence without representation. *Artificial Intelligence, 47*, 139–159.

Bruner, J. (1982). The organisation of action and the nature of adult-infant transaction. In Cranach, M., & R.Harre (Eds.), *The Analysis of Action*, pp. 313–328. Cambridge University Press, Cambridge, UK.

Bullowa, M. (Ed.). (1979). *Before Speech: The Beginning of Interpersonal Communication.* Cambridge University Press, Cambridge, UK.

Carey, S., & Gelman, R. (Eds.). (1991). *The Epigenesis of Mind: Essays on Biology and Cognition.* Lawrence Erlbaum, Hillsdale, NJ.

Chapman, D., & Agre, P. (1987). Abstract reasoning as emergent from concrete ativity. In M.P.Georgeff, & A.L.Lansky (Eds.), *Reasoning About Action and Plans*, pp. 411–424. Morgan-Kauffman, Los Angeles, CA.

Charlesworth, W., & Kreutzer, M. (1973). Facial expressions of infants and children. In P.Ekman (Ed.), *Darwin and Facial Expression: A Century of Research in Review*, pp. 91–168. Academic Press, New York, NY.

Collis, G. (1979). Describing the structure of social interaction in infancy. In M.Bullowa (Ed.), *Before Speech: The Beginning of Interpersonal Communication*, pp. 111–130. Cambridge University Press, Cambridge, UK.

Diamond, A. (1991). Neurophysiological insights into the meaning of object concept development. In Carey, S., & Gelman, R. (Eds.), *The Epigenesis of Mind: Essays on Biology and Cognition*, pp. 67–110. Lawrence Erlbaum, Hillsdale, NJ.

Fernald, A. (1989). Intonation and communicative intent in mothers' speech to infants: Is the melody the message?. *Child Development, 60*, 1497–1510.

Fischer, K., & Bidell, T. (1991). Constraining nativist inferences about cognitive capacities. In Carey, S., & Gelman, R. (Eds.), *The Epigenesis of Mind: Essays on Biology and Cognition*, pp. 199–235. Lawrence Erlbaum, Hillsdale, NJ.

Fogel, A. (1985). Coordinative structures in the development of expressive behaviour in early infancy. In G.Zivin (Ed.), *The Development of Expressive Behavior: Biology-Environment Interactions*. Academic Press, Orlando.

Fogel, A. (1993). Two principles of communication: Co-regulation and framing. In J.Nadel, & L.Camaioni (Eds.), *New Perspectives in Early Communicative Development*, pp. 9–22. Routledge, London.

Fogel, A., & Thelen, E. (1987). Development of early expressive and communicative action: reinterpreting the evidence from a dynamic systems perspective. *Developmental Psychology, 23*(6), 747–761.

Gibson, J. (1979). *The Ecological Approaach to Visual Perception*. Houghton Mifflin, Boston.

Gould, J., & Lewontin, R. (1978). The spandrels of San Marco and the Panglossian paradigm: A critique of the adaptionist programme. *Proc. Royal Soc. London, 205*, 581–598.

Hendriks-Jansen, H. Brain-models, mind-models and models of situated behaviour. In D.Cliff (Ed.), *Evolutionary Robotics and Artificial Life*. Forthcoming.

Hendriks-Jansen, H. *Situated Activity, Interactive Emergence and Human Thought*. Ph.D. thesis, School of Cognitive and Computing Sciences, University of Sussex. In preparation.

Hendriks-Jansen, H. (1993a). Natural kinds, autonomous robots and history of use. In *Proceedings of 1993 European Conference on Artificial Life*, pp. 440–450.

Hendriks-Jansen, H. (1993b). Scientific explanations of behaviour: The logic of evolution and learning. Tech. rep. CSRP 298, University of Sussex, School of Cognitive and Computing Sciences.

Hinde, R. (1982). *Ethology: Its Nature and Relations with Other Sciences*. Oxford University Press, Oxford.

Hirsh-Pasek, K., Jusczyk, P., Wright-Cassidy, K., Druss, B., & Kennedy, C. (1987). Clauses are perceptual units of young infants. *Cognition, 26*, 269–286.

Horswill, I. (1992). Characterising adaption by constraint. In Varela, F., & P.Bourgine (Eds.), *Proceedings of the First European Conference on Artificial Life*, pp. 58–63. MIT Press/Bradford Books, Cambridge, MA.

Johnson, R. (Ed.). (1993a). *Brain Development and Cognition: A Reader*. Blackwell, Oxford.

Johnson, R. (1993b). Constraints on cortical plasticity. In Johnson, R. (Ed.), *Brain Development and Cognition: A Reader*, pp. 703–721. Blackwell, Oxford.

Karmiloff-Smith, A. (1987). Constraints on representational change: Evidence from children's drawing. *Cognition, 34*, 57–83.

Kaye, K. (1979). Thickening thin data: The maternal role in developing communication and language. In M.Bullowa (Ed.), *Before Speech: The Beginning of Interpersonal Communication*, pp. 191–206. Cambridge University Press, Cambridge, UK.

Kaye, K. (1982). Organism, apprentice and person. In E.Z.Tronick (Ed.), *Social Interchange in Infancy*, pp. 183–196. University Park Press, Baltimore.

Kirsh, D. (1991). Foundations of AI: The big issues. *Artificial Intelligence, 47*, 3–30.

Lehrman, D. (1970). Semantic and conceptual issues in the nature-nurture problem. In Aronson, L. (Ed.), *Development and Evolution of Behavior*, pp. 17–52. W.H.Freeman and Co, San Francisco.

Lorenz, K. (1937). The nature of instinct: The conception of instinctive behavior. In P.H.Schiller, & K.S.Lashley (Eds.), *Instinctive Behavior: The Development of a Modern Concept*, pp. 129–175. International University Press, New York.

Lorenz, K. (1939). Comparative study of behavior. In P.H.Schiller, & K.S.Lashley (Eds.), *Instinctive Behavior: The Development of a Modern Concept*, pp. 239–263. International University Press, New York.

Lorenz, K. (1952). The past twelve years in the comparative study of behavior. In P.H.Schiller, &

K.S.Lashley (Eds.), *Instinctive Behavior: The Development of a Modern Concept*, pp. 288–310. International University Press, New York.

Lorenz, K., & Tinbergen, N. (1938). Taxis and instinct: Taxis and instinctive action in the egg-retrieving behavior of the greylag goose. In P.H.Schiller, & K.S.Lashley (Eds.), *Instinctive Behavior: The Development of a Modern Concept*, pp. 176–208. International University Press, New York.

Marr, D. (1982). *Vision*. Freeman and Co, New York.

Mataric, M. (1991). Navigating with a rat brain: A neurobiologically inspired model for robot spatial representation. In Meyer, J.-A., & Wilson, S. (Eds.), *From Animals to Animats: Proceedings of The First International Conference on Simulation of Adaptive Behavior*, pp. 169–175. MIT Press/Bradford Books, Cambridge, MA.

Mataric, M. (1992). Integration of representation into goal-driven behavior-based robots. *IEEE Transactions on Robotics and Automation, 8*(3), 304–312.

Mataric, M., & Brooks, R. (1990). Learning a distributed map representation based on navigation behaviors. In *Proceeding of 1990 USA-Japan Symposium on Flexible Automation, Kyoto, Japan*, pp. 499–506.

McFarland, D. (1985). *Animal Behaviour: Psychology, Ethology and Evolution*. Pitman Publishing, London.

Millikan, R. (1984). *Language, Thought and Other Biological Categories: New Foundations for Realism*. MIT Press, Cambridge, MA.

Nagel, E. (1961). *The Structure of Science: Problems in the Logic of Scientific Explanation*. Routledge and Kegan Paul, London.

Newson, J. (1979). The growth of shared understandings between infant and caregiver. In M.Bullowa (Ed.), *Before Speech: The Beginning of Interpersonal Communication*, pp. 207–222. Cambridge University Press, Cambridge, UK.

Papousek, H., & Papousek, M. (1977). Mothering and the cognitive head-start: Psychobiological considerations. In H.R.Schaffer (Ed.), *Studies in Mother-Infant Interaction*, pp. 63–85. Academic Press, New York.

Piaget, J. (1954). *The Construction of Reality in the Child*. Basic Books, New York.

Popper, K. (1963). *Conjectures and Refutations: The Growth of Scientific Knowledge*. Routledge and Kegan Paul, London.

Schaffer, H. (1977). *Studies in Mother-Infant Interaction*. Academic Press, New York.

Schneirla, T. (1966). Behavioral development and comparative psychology. *The Quarterly Review of Biology, 41*, 283–302.

Sloman, A. (1978). *The Computer Revolution in Philosophy*. Harvester Press, Hassocks, Sussex.

Smithers, T. (1992). Taking eliminative materialism seriously: A methodology for autonomous systems research. In Varela, F., & P.Bourgine (Eds.), *Proceedings of the First European Conference on Artificial Life*, pp. 31–40. MIT Press/Bradford Books, Cambridge, MA.

Spelke, E. (1991). Physical knowledge in infancy: Reflections on Piaget's theory. In Carey, S., & Gelman, R. (Eds.), *The Epigenesis of Mind: Essays on Biology and Cognition*, pp. 133–169. Lawrence Erlbaum, Hillsdale, NJ.

Steels, L. (1991). Towards a theory of emergent functionality. In Meyer, J.-A., & Wilson, S. (Eds.), *From Animals to Animats: Proceedings of The First International Conference on Simulation of Adaptive Behavior*, pp. 451–461. MIT Press/Bradford Books, Cambridge, MA.

Thelen, E. (1981). Rhythmical behavior in infancy: An ethological perspective. *Developmental Psychology, 17*(3), 237–257.

Tinbergen, N. (1951). *The Study of Instinct*. Clarendon Press, Oxford.

Trevarthen, C. (1977). Descriptive analyses of of infant communicative behaviour. In H.R.Schaffer (Ed.), *Studies in Mother-Infant Interaction*, pp. 227–270. Academic Press, New York.

Tronick, E., Als, E., & Adamson, L. (1979). Structure of early face-to-face communicative interactions. In M.Bullowa (Ed.), *Before Speech: The Beginning of Interpersonal Communication*, pp. 349–372. Cambridge University Press, Cambridge, UK.

Turkewitz, G., & Kenny, P. (1993). Limitations on input as a basis for neural organisation and perceptual development: A preliminary theoretical statement. In Johnson, R. (Ed.), *Brain Development and Cognition: A Reader*, pp. 510–522. Blackwell, Oxford.

von Hoftsen, C. (1984). Developmental changes in the organisation of prereaching movements. *Developmental Psychology, 20*(3), 378–388.

von Uexkull, J. (1934). A stroll through the worlds of animals and men. In P.H.Schiller, & K.S.Lashley (Eds.), *Instinctive Behavior: The Development of a Modern Concept*, pp. 5–82. International University Press, New York.

Wilkes, K. (1989). Explanation — how not to miss the point. In Montefiore, A., & D.Noble (Eds.), *Goals, No Goals and Own Goals*, pp. 194–210. Unwin Hyman, London.

Spencer and Dewey on Life and Mind

Peter Godfrey-Smith

Department of Philosophy
Stanford University
Stanford, CA, 94305-2155
Email: pgsmith@csli.stanford.edu

Abstract

The theories of life held by Herbert Spencer and John Dewey are outlined and compared. Some different approaches within artificial life research are examined from these two perspectives. The relation between theories of life and theories of mind is also discussed.

1. Introduction

If it was possible to bring a collection of dead, famous philosophers into the present, and ask them what they thought of artificial life, I think few of them would have as much to say as Herbert Spencer and John Dewey. This paper is an attempt to work out some of what they might say. It is also an attempt to show how some of the differences between research programs and explanatory styles which exist in and around Alife are manifestations of some old and basic oppositions within science and philosophy. These oppositions have to do with the general nature of causal and explanatory relations between organic systems and their environments. Thirdly, it is about the relations between life and mind, and hence the relations between artificial life and artificial intelligence.

2. Spencer

Herbert Spencer (1820-1903) was a wide-ranging, speculative thinker who had a great influence on the intellectual scene in Victorian Britain. He wrote large-scale works in philosophy, psychology, biology and sociology. He also supported strongly *laissez-faire* views in economics and politics, and is often associated with the label "social Darwinism" (but see Bowler, 1989). Spencer published his evolutionary approach to psychology (1855) several years before Darwin's *Origin of Species* (1859). In his day he commanded the attention and respect of people like Darwin, T.H. Huxley and J. S. Mill, but

soon after the turn of the century his reputation fell like a stone and has barely shifted since. (See Richards, 1987, for a detailed account.) In fact, one of the few places in which his name has tended to come up in recent years is in some discussions of Alife and complexity (Farmer and Belin, 1992; McShea, 1991; Lewin, 1992; Levy, 1992).

Spencer's work is highly relevant to Alife, but this relevance is complicated. In this section I will briefly outline Spencer's overall view of the world, and then look at some of his ideas on the origins of complexity.

Spencer claimed to have found a general "law of evolution" which applies to the evolution of solar systems, planets, species, individuals, cultural artifacts and human social organizations. According to this law there is a universal trend of change from a state of "indefinite, incoherent homogeneity" towards a state of "definite, coherent, heterogeneity" (1872, p.396). That is, every system tends to change from a state in which the system has little differentiation of parts, little concentration of matter, and everything is much the same in structure, towards a state in which there is a variety of clearly distinguishable parts, where the individual parts differ from each other and are densely structured.

This basic trend appears to be "negentropic"; it tends towards increased differentiation and organization. Spencer struggled with the consequences of the second law of thermodynamics, which was being formulated and investigated around the same time as his work (Kennedy, 1978, p.43). He accepted that the universe as a whole must run down. But though the eventual fate of the universe is "omnipresent death," as long as the processes of organic and social evolution have the required resources there will be a growth in organization and differentiation. This process "can end only in the establishment of the greatest perfection and the most complete happiness" (1872, p.517). And once the universe has run down to death, it might start up again, produce life, and cycle this way indefinitely.

Spencer thought that a system like a galaxy will become more organized as a consequence of the fundamental properties of its constituents. However, once we reach the realms of biology and psychology, increases in complexity are the result of *external* factors. For Spencer, complexity in organic systems is explained in terms of complexity in the systems' environments.

Spencer has been labelled an "internalist" with respect to the explanation of biological complexity (McShea, 1991; Lewin, 1992). This is misleading, in my view. Spencer certainly thought that on the global scale complexity will inevitably develop by itself, but the manifestation of this trend in organic development is not "internalist." When Spencer is concerned with the properties of organic systems, most of the explanatory weight is borne by the environments of these systems. Internal properties of organic systems explain why these systems are the *types* of things which respond to their environments so sensitively, but the *particular* changes that any system undergoes are explained in terms of the specificities of its environment.

In fact, even when Spencer is discussing physical systems, and is not making use of biological mechanisms, his explanations for the trend towards complexity often have a roughly externalist character. For example, Spencer thought that any homogeneous system is unstable, as any new influence on the system will affect different parts of it differently -- parts on the inside of the system will experience the force differently from parts on the outside. The parts will hence respond differently and the whole system will become more heterogeneously structured.

In the remainder of this paper, the term "externalist" will be used specifically for explanations of internal properties of organic systems in terms of properties of their environments. Explanations of organic properties in terms of other internal or intrinsic properties of the organic system will be called "internalist." Spencer's externalism is seen clearly in the specific mechanisms he used to explain biological and psychological properties. Firstly, Spencer's biology was strongly adaptationist. He made use of both the inheritance of acquired characteristics, along the lines of Lamarck, and also Darwinian evolutionary mechanisms -- the "survival of the fittest." He viewed these two processes as different specific ways in which organisms respond adaptively to conditions in their environments (1866, part 3, chapters 9-13). His psychology was associationist, in the English tradition (1855). This is also an externalist program of explanation; complexity in the mind is explained in terms of

complexity in sensory experience. In fact, Spencer saw evolution and individual learning as basically the same type of thing; they are both modes of "equilibration" between organism and environment .

It is this recognition of an underlying similarity between learning and adaptive evolution that made it possible for Spencer to make an unusual move in his theory of mind, given what had gone before him. During the 17th and 18th centuries epistemological discussion was deeply concerned with the issue of whether the mind has an intrinsic, innate structure or whether whatever is in the mind has come in through sensory experience. "Rationalists" such as Leibniz took the former view and "empiricists" such as Locke took the latter. Spencer was basically an empiricist, but he was the first empiricist thinker I know of who did not care *at all* whether the mind has rich innate structure, as rationalists claimed it did. He in fact embraced this idea. Spencer did not regard this as a concession so long as the mind's innate structure has an adaptationist evolutionary explanation. Evolution is like a population-level learning process, so the basic empiricist pattern of explanation still applies (1855, part 4, chapter 7).

Spencer is often compared to Lamarck (1809/1984), and it is sometimes said that Spencer's evolutionary views are unoriginal because they are largely derived from Lamarck. It is true that Spencer learned a lot from Lamarck's ideas (which he first encountered second-hand though the criticisms of Lyell). But the relation between Spencer's ideas and Lamarck's is more interesting than mere imitation. Lamarck is remembered now for his claim that there can be evolution by the inheritance of characteristics which individuals acquire during their lifetimes in response to their environments. But this is a secondary mechanism in Lamarck's account. The basic mechanism for evolution, and the thing which explains complexity in particular, is an internal tendency in all living things which generates increases in the complexity of organization. This tendency is a consequence of the action of invisible inner fluids.

The phenomenon Lamarck explains in terms of the environment and the inheritance of acquired characteristics is the fact that the pattern in nature which shows the inevitable increase in complexity is an *imperfect* pattern. Adaptation to different environments is used by Lamarck to explain departures from an orderly progression of increases in complexity. In particular, in a perfectly static environment Lamarck thought there would be a clear linear scale with respect to complexity (1809/1984, p.69). Spencer would predict the exact opposite: in a perfectly constant and simple environment there would be nothing that could generate any increase in complexity (1866, p.83). Spencer would probably have added that if there ever was an environment like this, it would not last long. Spencer thought that environments inevitably tend to become more complex.

So Lamarck explained things other than progressive increases in complexity in terms of the environment, and had an internalist view of organic complexity. Spencer had an intrinsic mechanism for directional changes in the environment, and an externalist account of almost all organic properties, especially complexity.

3. Spencer on Life and Mind

The key to Spencer's conception of the organic world is his definition of life and mind. A single definition applies to both. For Spencer, life and mind are distinguished mainly by matters of degree and detail. Having a mind is an advanced mode of living. In a sense, for Spencer people are not just smarter than prawns, but also more alive than prawns.

Spencer thought that living systems are distinguished from inanimate ones by the existence of certain complex processes inside the system, and (more importantly) a special set of relations between internal processes and conditions in the systems' environments. The simplest formula he gave as an account of life was: "the continuous adjustment of internal relations to external relations" (1855, p.374, 1866, p.80; 1872, p.84).

Spencer was heading, I think, towards a view of life that we now might call "systems-theoretic" or cybernetic (Ashby, 1956). Living systems are self-preserving. They maintain their organization, and the discontinuity between system and environment. They do this by responding in particular ways to environmental events which, left to themselves, would tend to disrupt the organization of the living system. Living systems actively resist disruption and decay.

All vital actions.... have for their final purpose the balancing of certain outer processes by certain inner processes. There are unceasing external forces which tend to bring the matter of which organic bodies consist, into that state of stable equilibrium displayed by inorganic bodies; there are internal forces by which this tendency is constantly antagonized; and the perpetual changes which constitute Life, may be regarded as incidental to the maintenance of the antagonism. (1872, p.82)

Here Spencer uses the term "purpose" which is of course a suspicious one. But in my view the term "purpose" can simply be dropped from the passage above, and replaced with something as simple as "tendency," and Spencer's general intent is retained.

So Spencer anticipated a conception of life which has become quite popular during this century, a conception based upon the idea of self-preserving functional organization as characteristic of life.

In fact this general idea comes in several forms. It can appear in a relatively externalist form, in which there is a focus on self-maintaining *responses to the environment*. And it can also appear in a more internalist form, as exemplified by the "autopoietic" conception of life given by Maturana and Varela (1980). In this second form the focus is on self-*production* as characteristic of life, and the role of the environment is reduced. Spencer, with his perpetual focus on the environment, exemplifies the externalist form of this systems-theoretic view. He also anticipated some other central ideas of this account, such as the important roles played by (what we would call) negative feedback and homeostasis.

[T]o keep up the temperature at a particular point, the external process of radiation and absorption of heat by the surrounding medium, must be met by a corresponding internal process of combination, whereby more heat may be evolved; to which add, that if from atmospheric changes the loss becomes greater or less, the production must become greater or less. And similarly throughout the organic actions in general. (1872, pp.82-83)

In stressing this systems-oriented or cybernetic side of Spencer, I am downplaying another vocabulary he uses to talk about the basic properties of life. Spencer also says that life and mind are characterized by relations of *correspondence* between internal and external. He sometimes defined life in terms of a correspondence between internal relations and external relations.

The talk of correspondence in Spencer is much more problematic than the discussions of (what we would call) feedback and self-preserving responses to the environment. It was also controversial in his own day, and played an interesting role in the development of some aspects of the philosophical movement known as *pragmatism*. William James (the most famous pragmatist) published his first essay in philosophy on Spencer's view of mind (1878). John Dewey also used Spencer's talk of correspondence as an illustration of the type of position he was opposed to. We will look at Dewey's position later.

In my view, though Spencer talked a lot about correspondence as the mark of life and mind, this was actually less central to his picture than the idea that living systems are self-preserving. For Spencer, the *ways* in which living systems succeed in preserving the discontinuities between organism and environment involve relations of correspondence and "concord" between inner and outer. That is how they succeed in responding to environmental events in a way which prevents their dissolution and disruption.

How close is Spencer's account of life to contemporary views? There is no consensus view on what makes something alive. Many writers are very skeptical about the possibility of giving a definition of life or anything even close to a definition. Perhaps the closest thing to a consensus is the view that life is what philosophers call a "cluster concept." There is a list of properties that are associated with life, but to be alive a system does not have to have *all* of them. It only has to have some reasonable number of them -- this is deliberately supposed to be vague, and there will be a "grey area." That is certainly the impression one gets from the opening pages of many biology textbooks (see for example Curtis and Barnes, 1989, which lists 7 distinct basic properties. See also Mayr, 1982 chapter 2; Farmer and Belin, 1992).

For many modern writers though, there are at least two *types* of properties which are important in understanding what life is. Living systems firstly have a set of broadly "metabolic" properties, which involve the organization of the individual living systems and its relations to the environment. Homeostasis is an example. Secondly, they have a set of properties involving reproduction, and the relations between individuals. Some take this second family of properties to be more fundamental than the first, in fact. The strongest versions of this idea claim that life can be understood in terms of the capacity to evolve (Maynard Smith, 1993), or that life is a property of a population rather than an individual (Bedau and Packard, 1992).

Spencer's view of life is based more on individually self-maintaining properties, but he did not neglect reproduction. His view of reproduction and how it fits into life-history properties of organisms is interesting. Spencer thought that organisms come to a "moving equilibrium" with their environments -- all evolution and development has this character. But there is only so far an individual organism can go in this process, before it runs out of the internal properties of plasticity that are needed for further development and adaptation. Reproduction acts to jolt the organic system out of its temporary and imperfect equilibrium. It frees up each component of the system for further evolution by recombining these

components into a new individual (1866, part 2, chapter 10). Sexual reproduction is the most effective means for this, as organic material from two dissimilar individuals is united and this union will be much more plastic and less static in its properties than the two adults are individually.

In some ways this is quite a modern idea. Reproduction plays a role similar to the role played by the injection of noise into a hill-climbing system to prevent it sticking too readily on a local maximum. (Spencer even mentions physical processes of annealing when developing this view: 1866, p.274.) On this account there are not two distinct types of properties involved in life, individual-level and population-level properties. Rather, a single set of organism/environment relations are attained and maintained as a consequence of *both* individual-level and population-level activities.

Spencer, as I said, had a single definition of life and mind. He wanted to see even the highest cognitive capacities of humans as continuous with the most basic forms of organic action. Spencer and Dewey both saw cognition as something which emerges out of simpler modes of interaction with the world. Dewey sometimes called this an assumption of "continuity" (1938). In my view it is important to distinguish several different possible claims of "continuity" between life and mind.

Weak Continuity: Anything that has a mind is alive, although not everything that is alive has a mind. Cognition is an activity of living systems.

Strong Continuity: Life and mind have a common abstract pattern or set of basic organizational properties. The functional properties characteristic of mind are an enriched version of the functional properties which are fundamental to life in general. Mind is literally life-*like*.

These are both constitutive or ontological principles, principles about what life and mind *are*. There is also a continuity principle which has a purely methodological character.

Methodological Continuity: Understanding mind requires understanding the role it plays within entire living systems. Cognition should be investigated in this "whole organism" context.

Strong continuity implies weak continuity. If the pattern of organization characteristic of mind includes the pattern characteristic of life, then anything which thinks must have a lot of what it takes to be alive. The principle of methodological continuity is supported by both weak and strong continuity, though it is not strictly implied by either of them. It is also important that methodological continuity does not imply either of the constitutive principles.

It is a consequence of weak continuity that *artificial life must precede artificial intelligence* (or in the limit, be simultaneous with it). The same is true under strong continuity, and in addition to this, the strong continuity principle claims that once we have artificial life we also have the raw material, or an unrefined form, of artificial intelligence. We just need to get more of the same sort of properties.

Spencer is a clear example of someone who held the strong continuity thesis, and the methodological continuity principle as well. Spencer would have said that to build a living system you need to build a system that maintains itself in its environment, in the face of possible decay, by adjusting its internal processes and actions to deal with external events and relations. He also would have said that once you have a system that does a lot of this, and does it in a particular way, you have an intelligent system.

What is the "particular way" characteristic of intelligence? For Spencer, the most distinctive property of intelligent internal processing is (what we would call) its *serial* nature. A transition from parallel processing to serial processing, a transition which is never complete, marks the transition from merely living activity to real intelligence (1855, part 4, chapter 1). The contents of thought make up a single complex series. That is not to say that intelligent systems stop dealing with problems in parallel. The point is that only the serially structured part of the system's activities is the intelligent part. In addition, adaptation to conditions in the environment that are (i) highly changeable, (ii) spatially or temporally distal, (ii) compound, such as conjunctive or disjunctive, (iii) hard to discriminate, and (iv) abstract, involving superficially heterogeneous classes of events, all tend to demand the complex types of organic response characteristic of cognition.

So then: Spencer would have given a fairly simple recipe for artificial life, and one which is essentially the same as his recipe for artificial intelligence. I conjecture that Spencer would have had no qualms about the idea that systems which satisfy his criteria could be realized in software. For Spencer, life and mind are patterns of interaction which systems have to their environments. He might insist that an artificial life creature live in an environment which has the capacity to lead to the disruption of the living system. But I do not think this would preclude environments realized inside computers.

If Spencer was to look at some current work in and around artificial life, the work which would best exemplify his conception of biology would probably be work on classifier systems, "animats," systems constructed using genetic algorithms, and other *environment*-oriented work (Booker et al., 1989; Wilson, 1991; Todd and Miller, 1992). This is work which is directed at the chief focus of Spencer's biological thought -- the relation between internal complexity and environmental complexity. This family of ideas can be regarded as the externalist side of artificial life research. It seems to me to be a guiding idea in classifier work, animats, and genetic algorithm work more broadly, to think that the *fuel* for the development of organic complexity is environmental complexity: complex systems arise as solutions to complex environmental problems. Spencer would nod vigorously.

The work in Alife which I regard as furthest from Spencer's approach is work on cellular automata (Langton, 1992). This is because, as I understand it, this work is virtually environment-free. The "environment" for a cellular automaton is just the space it is in, a lattice of cells which can be in various states. The system changes via the local interactions of cells. The system can display complex dynamical behavior in which specific patterns or structures are preserved, but it does not generate these patterns as a response to a structured environment. Neither does it respond to potentially disruptive environmental events. The environment does not contain a set of intrinsic patterns which the organic system must adapt to or contend with. Work on cellular automata is one of the more internalist domains within Alife.

It is important to recognize that some research on cellular automata is not directly aimed at producing artificial life by producing a complicated cellular automaton. The stated aim in Langton 1992, for example, to investigation of the conditions under which "a dynamics of information" will emerge and "dominate the behavior of a physical system" (1992 p.42). So we need not regard all these systems as artificial life systems in their own right. On the other hand, cellular automata are very often intended to at least model or cast light on basic properties of the living. Langton does regard the existence of a "dynamics of information" as a very basic property of life. For Spencer, however, the basic properties of the living are a set of *organism/environment relations*, not a set of complicated internal properties or processes.

So in my view Spencer would not regard cellular automata as on the road to artificial life. But he might regard them as important models for a different reason. He might see them as illustrations of some parts of his view of the *inorganic* world. Spencer held that organic complexity is a response to environmental complexity. But he also thought that environments can get more complex under their own steam. The most basic laws of matter have the consequence that homogeneous physical systems are not stable, but will constantly generate new complexities of pattern. So cellular automata specifically, and the environment-free side of Alife work generally, might be illustrations of the possibility of heterogeneity of structure and activity arising as a consequence of local and intrinsic properties of basic elements of a physical system.

Lastly, what might Spencer think of systems like Tom Ray's "Tierra"? Tierra is a simulated ecology in which individual organisms do contend with an "environment," but the chief contents of this environment are other competing individuals, both of the same type or lineage and of other types. The "abiotic" aspects of the environment are just the fixed constraints exerted by the properties of the CPU, the operating system and the memory of the computer (Ray, 1992, p.374). Most environmental features are "biotic." Spencer would perhaps regard this as an unusual balance of biotic and abiotic, but as I understand Spencer's understanding of the concept of "environment" this would qualify as a simple but genuine case of life-like interaction between organic system and environment (1891, p.416). Or at least, it would be life-like to the extent that the "organisms" successfully interacted with these external features in a way which maintained their organization.

4. Dewey

John Dewey (1859-1952) is, on the face of it, a philosopher with little in common with Spencer. Spencer was a science-worshipper who nonetheless speculated from the armchair about factual matters, and an advocate of *laissez faire* economic libertarianism. Spencer believed in timeless laws of nature and the inevitability of universal progress, and built one of the more elaborate systems in English-speaking philosophy.

Dewey, on the other hand, was one of the great American liberal thinkers of the early and mid 20th century, especially in the domains of education and the theory of democracy. He knew a lot of science but also kept science in its place. He thought that metaphysical system-building is generally a sort of pseudo-inquiry which diverts people from addressing the real, concrete problems that beset us. He rejected guarantees of universal progress as attempts to falsely promise in advance the goods that can only be gained by hard work and practical problem-solving,

and which can never be guaranteed. So Spencer and Dewey differed about a lot.

It is not my goal here to say that deep down they were in agreement. But I do think the differences between them are often misunderstood, and it is possible to sharpen our understanding of the relations between them by focusing, as I am focusing here, on what they say about life, and the relation between life and mind.

Dewey is, beside Spencer, one of the very few major philosophers in the recent English-speaking tradition to think that a theory of life is of general philosophical importance -- to think that our theories of knowledge and inquiry, for example, should be linked to a general theory of living organization. For Dewey, making this link between inquiry and life was a way to overcome a "dualistic" view of the relations between mind and nature. Dewey established a picture of the relations between organism and environment in his theory of life, and was then able to make use of these relations in his theory of thought and inquiry. He sought to use his general position on organism/environment relations to avoid the artificial separations between mind and world which so often arise in epistemology.

Thus Dewey, like Spencer, conceived of life not as an intrinsic property of a system, but as something that involves certain relations to an environment. Life is a "transaction extending beyond the spatial limits of the organism" (1938, p.25). This transaction involves an exchange of energies between the system and the environment, in which states of disturbed equilibrium in the organism are changed back into states of equilibrium. Dewey admits that there are inanimate systems in which an external change causes disequilibrium followed by a restoration to equilibrium. The distinctive properties of living systems are the *ways* in which equilibrium is restored, and the consequences of this restoration (1929a, pp.253-254). Living systems reach this equilibrium in a way which tends to maintain the organization of the system. The living system acts to preserve its organization and integrity in the face of disturbances.

Iron as such displays exhibits characteristics of bias or selective reactions, but it shows no bias in favor of remaining simple iron; it had just as soon, so to speak, become iron-oxide. It shows no tendency in its interactions with water to modify the interaction so that consequences will perpetuate the characteristics of pure iron. If it did, it would have the marks of a living body, and would be called an organism. (1929a, p.254, see also 1929b, p.179)

So far this is not such a long way from Spencer's view, as I interpret it. It is a view of life based upon active self-maintenance. One difference, which is especially visible in his 1938 presentation, is that Dewey is more inclined to regard the organism plus environment as constituting a single system. Organic activities tend to preserve the *pattern of interaction* between organism and environment, rather than just preserving the organism itself (1938, pp.26-28).

Dewey also asserted "continuity" between life and mind, but earlier I distinguished several different continuity theses, and it is not easy to work out which ones Dewey held.

Dewey is certainly committed at least to the weak continuity thesis. He says: "The distinction between physical, psycho-physical [living], and mental is thus one of increasing complexity and intimacy of interaction among natural events" (1929a, p.261). It is harder to work out where he stands on strong continuity. He says that the general pattern of inquiry is "foreshadowed" by the general pattern of life (1938, p.34). This suggests strong continuity. All living systems respond to environmental dangers by acting on the world; intelligent inquiry is a specific *way* of approaching this basic aspect of life. However, Dewey also says that mind has a special relationship to language and communication. Only a communicating system in a social environment can literally think, because thinking is symbolic and symbolism is social (1929a, pp.211, 230; 1938, pp.43-44). I take all views on which having a language is necessary for thought to be views which deny strong continuity between life and mind (unless a very unusual view of the nature of life is taken). According to the strong continuity thesis, life is "proto-cognitive" or "proto-mental." But life is *not* proto-linguistic.

It is probably fair to say that Dewey did not accept strong continuity in a wholesale way, as Spencer did, but that he did think that *some* basic properties of cognition are formally similar to the basic properties of life. Living activity in general can be viewed as formally similar to problem-solving, although there is also more to genuine cognition than this.

In the discussion of Spencer I said that as a consequence of his holding the strong continuity thesis, Spencer would give basically the same recipe for artificial life and for artificial intelligence. To make a system which thinks you do not need to add something wholly new to a living system; you just increase the magnitude of certain properties it already has. If Dewey's view is that a complex living system becomes a thinking system only in virtue of its relations to other systems in a social context, the way to create real artificial intelligence is via creating a *society* of artificial life creatures, and having them deal with problems interactively.

I said that Dewey has a view of life which is not too far from Spencer's, and I supported this claim with the quote above about iron. But Dewey in fact thought there was a great difference between his view and Spencer's. He thought that Spencer was part of the problem. Dewey thought this in part because he focused on a side of Spencer which I downplayed in my discussion -- the side of Spencer in which he says that life involves a relation of "correspondence" between the internal and the external. Dewey thought that "corresponding" to the external was not a good way of staying alive.

> If the organism merely repeats in the series of its own self-enclosed acts the order already given from without, death speedily closes its career. Fire for instance consumes tissue; that is the sequence in the external order. Being burned to death is the order of "inner" events which corresponds to this "outer" order.
> [A]ll theories of psycho-physical parallelism, traditional theories of truth as correspondence, etc., are really elaborations of the same sort of assumptions as those made by Spencer: assumptions which first make a division [between organism and environment] where none exists, then resort to an artifice to restore the connection which has been willfully destroyed. (1929a, p.283)

Dewey's point here is that Spencer's view of life enforces a false separation between organism and environment, and then invents a magical new relation of "correspondence" to overcome this artificial problem. Dewey holds this view about the role of concepts of "correspondence" in theories of life in general and also in theories of mind. The idea that the purpose of thought is to "correspond" to an independent external realm is, for Dewey, a way to get around a problem which never existed. The false problem is the idea that mind and nature are completely different from one another and hence that there cannot be any straightforward, natural interactions between them.

I think Spencer is not *as* guilty of this charge as Dewey thinks. It is true that Spencer's official theory of knowledge did have a gulf between mind and nature of the sort that Dewey despised (see early chapters of Spencer, 1872). It is also true that Spencer said that "correspondence" was basic to life and mind, and was not clear about what correspondence is. But in my view, the central idea for Spencer was the idea that living systems act to preserve their organization in the face of environmental threats. The only "division" between the living system and the world on this account is the set of physical discontinuities which mark the distinction between inside and outside, and which the organism's actions maintain. Spencer's talk of correspondence is supposed to be part of an account of how this physical relation is maintained.

So what *is* the basic difference between Spencer's and Dewey's views of life?

5. Asymmetric Externalism

The difference which I think is most fundamental, the one on which many of their detailed disagreements depend, concerns their opinions of an attitude which I will call "asymmetric externalism." Spencer exhibits this attitude, and Dewey was against it.

As outlined earlier, I understand an *externalist* explanation of some property of an organic system as an explanation in terms of properties of the environment of the system. Adaptationist explanations of biological traits are often externalist, as are classical empiricist explanations of thought and knowledge. Empiricism explains what is believed or known in terms of what is experienced, in terms of what comes into the mind from outside.

An *asymmetrically* externalist view is something stronger than this. It is a program of explanation which explains internal properties in terms of external, and *also* explicitly or implicitly denies that these external properties are to be explained in terms of internal properties of the organic system. So what is denied is any significant level of feedback from the organic system on its environment. The organic system has its nature or trajectory determined by the environment, but the environment goes its own way. It is dynamically self-contained, rather than "coupled" to the organic system.

I said that classical empiricism is an externalist picture of thought. Is this also an asymmetrically externalist view? It is hard to say. Most of the famous empiricists like Locke and Hume did not *deny* that the thinking agent can act on the world and hence affect the future course of experience. They just did not discuss this very much. The important point is that Dewey *did* think, in effect, that orthodox epistemological views are asymmetrically externalist. He saw these views as holding that the business of thought is *conforming* to the world but leaving it untouched (1929b, p.110). This tradition views the ideal knower as a spectator who does not interfere with the course of the game. Dewey, on the

other hand, thought that effective inquiry and problem-solving do involve interfering with the world, transforming and reconstructing it. He was still basically an empiricist thinker, accepting that thought is a response to experience. To that extent Dewey accepts an externalist conception of mind, as all empiricism is externalist in this sense. But Dewey held that the agent's response to experience typically involves making changes to the world, and hence changing the future course of experience.

I also said that adaptationism in biology is externalist. Is it asymmetrically externalist? Again, it is hard to say in many cases, but one of the most important recent critiques of the adaptationist program can be understood as claiming that adaptationism *is* asymmetrically externalist. This critique is due to Lewontin (1983, 1991). Lewontin claims that orthodox adaptationist thought views organisms as the passive "objects" of evolution, when in fact they are subjects as well as objects. Organisms impact upon their environments, and hence alter the future course of the selection pressures to which they will have to respond. I understand Lewontin's attack on adaptationism as closely analogous to Dewey's attack on orthodox empiricist epistemology. Both argue that asymmetrically externalist views have to be replaced by views that recognize two-way interactions between organisms and environments.

Dewey prepares the way for his view of thought in his general view of life. And though his biological discussion is not as sophisticated as Lewontin's, the view of organism/environment relations which he develops in his view of life is designed to set up basically the same two-way or feedback-oriented picture that Lewontin supports. "Adjustment to the environment means not passive acceptance of the latter, but acting so that the environing changes take a certain turn" (Dewey, 1917, p.62). Dewey's view of the "transactions" between organism and environment which constitute life is a view based upon organic intervention in the world, as well as organic reaction to what the world does. Dewey sets up this "interactive" picture in his view of life, and his view of inquiry follows the same pattern. Thought and inquiry are responses to environmental problems, but their goal is not to generate internal states which merely correspond or conform to external things, but rather to make a change to environmental conditions; to adapt *them* to the goals of the organism.

This is the most fundamental difference between Spencer's and Dewey's views of life, and one which extends into their views of mind also. Dewey perceives Spencer as having what I call an asymmetrically externalist view of life and mind, and Dewey wants to replace this with a view based on two-way interaction. The difference between Dewey and Spencer about the role of the concept of "correspondence" as a feature of life and mind is a consequence of this more basic difference. Dewey saw the idea of correspondence as a relation between inner and outer as a typical philosophical product of the perspective he opposed.

Is it true that Spencer had an asymmetrically externalist picture? Is this charge justified? That is not a simple question. Spencer, like the other empiricists mentioned earlier and like many adaptationists, is more guilty of *neglecting* the phenomena of organic action on the world than he is of denying them. His picture of organic action is largely one of organic *re*-action. His discussions of life focus on organisms taking heed of environmental facts rather than making changes to them. On the other hand, Spencer also had a very holistic view of ecological systems. He saw ecological systems as tangled webs of relationships, and his general picture of the physical world was one in which changes in one place tend to ramify through to distant places. So changes to the composition or behavior of one species result in "waves of influence which spread and reverberate and re-reverberate" through the flora and fauna of the area, and this will influence in turn the future of the system behind the change. Spencer also thought that organisms also play an individual role in determining how complex their experience of their environment is (1866, pp.417-418). So there is a holistic side to Spencer, which would make it possible for him to accept some of Dewey's picture. But in general I would agree with Dewey that Spencer's picture is one in which organisms are far more the objects of external forces than they are subjects creating them. In particular, Spencer conceives of progress in the biological world as an inevitable consequence of basic laws governing all matter. Organic progress is achieved because all life is shackled to the inevitable advance of complexity in every environment. Dewey, on the other hand, views any progress which can be attained as something which we (and other living systems) must create and bring about ourselves. The universe is not going to do the work for us.

So with respect to specific projects in Alife, I conceive of Dewey as finding an area of real agreement with Spencer, but also disagreeing with him on other points. Dewey would agree with Spencer in seeking to place organism/environment relations as the central focus of research. Consequently, I think he would share with Spencer the view that cellular automata and other "environment-free" systems are not models of basic properties of living organization. Dewey, like Spencer, opposes the idea that basic properties of organic self-maintenance can be understood in an internalist way, "as a sort of unrolling push from within" (1917, p.62).

On the other hand, while I think Spencer would regard classifier systems, animats, and the like as on just the right track, Dewey might think of some systems of this type as suspiciously low in two-way interactions or feedback relations between organism and environment. Even within the environment-oriented conception of life, which is common to Spencer and Dewey, there is a major difference between views which are for the most part asymmetrically externalist and views which are based on two-way interactions between organism and environment. Alife organisms can be placed in environments which are fixed, or have their own autonomous principles of change, or alternatively they can be placed in their environments in such a way that their own actions determine the future structure of their world. For Dewey, some classifying might be too close to mere "spectating." The form of intervention in environments that Dewey has in mind is something more than just eating when there is food and not when there is poison. A thoroughly Dewey-oriented Alife system would feature rich connections in both directions between organism and environment; the environment would pose problems for the organism, and the organism would not just adjust and adapt itself to environmental events, but would intervene in the environment's own course and alter its trajectory. This alteration would in turn bring about new problems and new possibilities for organic action and control. Co-evolutionary models have some of these properties (Kauffman and Johnsen, 1992).

If Dewey did express a view such as this, some might reply to him: "one step at a time!" In these early days of research it might be reasonable to idealize towards a more Spencerian picture, and neglect, for a period, the complexities of two-way interaction between organism and environment. Even Dewey admits that the simplest forms of life tend to accommodate their environments rather than intervene in them -- he regards intervention in the environment as a sign of "'higher'" life (1917, p.62, Dewey's own scare-quotes). Perhaps there is no need to build all these properties into simple systems of the type we have now. On this view, the first problems to deal with really *are* problems like when to eat, how to keep the system intact, and how to avoid the poison and predators.

Others may think that research should follow Dewey's lead from the start, and self-consciously avoid generating a picture of life in which the environment calls the shots and the organism just responds. Those who are impressed with Dewey's arguments about the conceptual quagmires and dead-ends that result from accepting an asymmetrically externalist perspective should perhaps prefer this latter view.

Acknowledgement:

I have benefitted from discussions of these matters with Tom Burke, Richard Francis, Yair Guttmann, Richard Lewontin, Greg O'Hair, and Peter Todd. This work has been supported with a grant from the Office of Technology Licensing, Stanford University.

References

Ashby, W.R. 1956. *An Introduction to Cybernetics*. Reprinted New York: Wiley, 1963.

Bedau, M.A. and N.H. Packard. 1992. Measurement of evolutionary activity, teleology, and life. In Langton et al. 1992

Booker, L.B., D.E. Goldberg and J.H. Holland. 1989. Classifier systems and genetic algorithms. *Artificial Intelligence* 40: 235-282.

Bowler, P. 1989. *Evolution: The History of an Idea*. Revised edition. Berkeley: University of California Press.

Curtis, H. and N.S. Barnes. 1989. *Biology*. 5th edition. New York: Worth.

Dewey, J. 1917. The need for a recovery of philosophy. Reprinted in *The Philosophy of John Dewey*, edited by J.J. McDermott. Chicago: Chicago University Press, 1981.

Dewey, J. 1929a. *Experience and Nature*. Revised edition. Reprinted New York: Dover, 1958.

Dewey, J. 1929b. *The Quest for Certainty*. Reprinted in *John Dewey: The Later Works, 1925-1953. Volume 4: 1929*, edited by J.A. Boydston. Carbondale: Southern Illinois University Press, 1988.

Dewey, J. 1938. *Logic: The Theory of Inquiry*. New York: Henry Holt.

Farmer, J.D. and A.d'A. Belin. 1992. Artificial life: The coming evolution. In Langton et al. 1992.

James, W. 1878. Remarks on Spencer's definition of mind as correspondence. *Journal of Speculative Philosophy* 12: 1-18. Reprinted in *William James: The Essential Writings*, edited by B. Wilshire. New York: Harper and Row, 1971.

Kauffman, S.A and S. Johnsen. 1992. Co-evolution to the edge of chaos: Coupled fitness landscapes, poised states and co-evolutionary avalanches. In Langton etal. 1992.

Kennedy, J.G. 1978. *Herbert Spencer*. Boston: Twayne.

Lamarck, J.B. 1809/1984. *Zoological Philosophy*. Translated by Hugh Elliot. Chicago: University of Chicago Press, 1984.

Langton, C. ed. 1989. *Artificial Life.SFI Studies in the Sciences of Complexity* Reading, MA: Addison-Wesley.

Langton, C.G., C. Taylor, J.D. Farmer, S. Rasmussen, eds. 1992. *Artificial Life II. SFI Studies in the Sciences of Complexity, Vol.X*. Reading, MA: Addison-Wesley.

Langton, C.G. 1992. Life at the edge of chaos. In Langton et al. 1992.

Levins, R. and R.C. Lewontin. 1985. *The Dialectical Biologist*. Cambridge MA: Harvard University Press.

Levy, S. 1992. *Artificial Life: The Quest for New Creation*. New York: Pantheon.

Lewin, R. 1992. *Complexity: Life at the Edge of Chaos*. New York: Macmillan.

Lewontin, R. C. 1983. The Organism as the Subject and Object of Evolution. Reprinted in Levins and Lewontin 1985.

Lewontin, R.C. 1991. *Biology as Ideology: The Doctrine of DNA*. New York: Harper.

Maturara, H. and F.J. Varela. 1980. *Autopoiesis and Cognition: The Realization of the Living*. Dordrecht: Reidel.

Maynard Smith, J. 1993. *The Theory of Evolution*, 3rd edition. Cambridge: Cambridge University Press.

Mayr, E. 1982. *The Growth of Biological Thought*. Cambridge MA: Harvard University Press.

McShea, D. 1991. Complexity and evolution: what everybody knows. *Biology and Philosophy* 6: 303-24.

Meyer, J-A, and S.W. Wilson. 1991. *From Animals to Animats: Proceedings of the First International Conference on the Simulation of Adaptive Behavior*. Cambridge MA: MIT Press.

Richards, R. 1987. *Darwin and the Emergence of Evolutionary Theories of Mind and Behavior*. Chicago: University of Chicago Press.

Spencer, H. 1855. *Principles of Psychology*. London: Longman, Brown and Green.

Spencer, H. 1872. *First Principles of a New System of Philosophy* 2nd edition. New York: Appelton.

Spencer, H. 1866. *Principles of Biology*. Volume 1. New York: Appelton.

Todd, P.M. and G.F. Miller. 1991. Exploring adaptive agency II: Simulating the evolution of associative learning. In Meyer and Wilson 1991.

Wilson, S.W. 1991. The animat path to AI. In Meyer and Wilson 1991.

CROSSOVERS GENERATE NON-RANDOM RECOMBINANTS UNDER DARWINIAN SELECTION

Gene Levinson

Genetics & IVF Institute
3020 Javier Road, Fairfax, VA 22031

Department of Human Genetics
Medical College of Virginia
Richmond, VA

Tel. (703) 698-3902; FAX (703) 849-1792
email p01314 @ psilink.com

Abstract

Genetic diversification is not the only consequence of recombination. When recombination occurs in a gene pool that is under selective pressure, the result is non-random linkage of selected sequence elements with high potential fitness. Computer simulations based on the genetic algorithm show that populations of DNA sequences evolve more rapidly *via* homologous recombination than by point mutations alone, especially under mild selective pressure. More dramatic effects are observed in models based on illegitimate recombination and the exon theory of genes. In the context of widely observed features of genome organization and evolution, results suggest that a synergism between recombination and selection may have played a major role in Darwinian molecular evolution, a role that is obscured by the semantics of the Modern Synthesis.

1. Introduction

A recent review of the genetic algorithm (GA; Forrest, 1993) posed an intriguing question: can data generated by an idealized computer model provide useful insights about natural evolutionary systems? Data presented here suggest that simulations based on the GA can demonstrate important features of Darwinian DNA sequence evolution that are obscured by the prevailing neo-Darwinian paradigm, also known as the Modern Synthesis (MOS).

The MOS (*ca.* 1937-1953) provided an important conceptual link between Darwin's theory of natural selection and classical Mendelian genetics (Dobzhansky et al., 1977; Futuyma, 1986). Experiments with model organisms such as *Drosophila* focused attention on mutations as a primary source of heritable variation. Apparently as a defense against alternative theories that invoked teleological or Lamarckian mechanisms for evolution, the MOS emphasized the random nature of such genetic variation: "Mutation merely provides the raw material of evolution", wrote Huxley (1953), who popularized the MOS; "it is a random affair, and takes place in all directions...[The effects of mutations] are not related to the needs of the organism, or the conditions in which it is placed". In the MOS paradigm, most mutations are deemed harmful because they produce random changes in genes, which are likely to be destructive. However, some may by chance prove useful, conferring an advantage on the survival or reproduction of individual organisms. This would lead to the preferential replication and expansion of such useful mutations in a natural population. Beneficial mutations would therefore be expected to accumulate in the genes, and over evolutionary time, adapt each species to the changing requirements of its ecological milieu.

This elegant conceptual model was only strengthened by subsequent discoveries of molecular genetics that revealed the structure of DNA, its replication, and the nature of the genetic code. Consequently, to this day, the MOS paradigm (updated with molecular genetics) is the predominant evolutionary theory in textbooks and popular accounts. It provides a common conceptual framework for a vast array of empirical observations concerning DNA sequences, and has spawned an extensive literature on the subject of molecular evolution.

In accord with the MOS paradigm, recombination has been viewed as a shuffling process that generates random constellations of alleles, which may by chance prove useful.

Here, recombination is defined as it is in classical genetics, as exchanges between homologous chromosomes[1] that come together as a result of sexual outcrossing: "All that recombination can do is to produce a more random distribution from a less random one; technically, it produces 'linkage equilibrium[2]' from 'linkage disequilibrium' " (Maynard Smith, 1988a). A logical consequence of this widely accepted view is the following dilemma: why, despite the selective disadvantages both of males (who do not bear young) and of lost energy spent on mate acquisition, has sexual reproduction been maintained in most taxonomic lineages? A host of plausible explanations have been advanced; many of these arguments have been summarized elsewhere (Bell, 1982; Michod and Levin, 1988).

In view of this history, it may at first seem paradoxical that the genetic algorithm (GA)[3] a software technique based on Darwinian evolution by natural selection, has little use for mutations, while recombination is a crucial component. This distinction was clearly pointed out by Holland (1992) in his classical mathematical treatment of the subject, and is also discussed in more recent publications from the computer science community (Forrest, 1993; Koza, 1992).

In most GA applications, random strings are used to represent potential solutions to computational problems. Subjection of these strings to repeated cycles of selection, expansion, and recombination can result in the production of non-random sequences that represent remarkably effective solutions to otherwise intractable problems. Fitness enhancement by recombination depends on linkage of potentially useful elements from previously selected string populations. Here, recombination is simply defined as reciprocal breaking and joining events between two strings. After each successive selection cycle, the probability for generating recombinant strings with higher fitnesses rapidly increases, since strings with partial fitness have already been selectively expanded in the population.

Two crucial distinctions between the GA and the MOS paradigm are that (1) *recombination represents a form of mutation whose outcome is not random, but is sequence-dependent*, and (2) *in the context of selection, non-random sequence motifs with proven fitness potential are available for assembly from previous selection cycles*. The effectiveness of the GA, and its Darwinian substance, raises an interesting question: might recombination play a similar role in *biological* evolution to that seen in GA applications? In the present study, computer simulations are used to examine and compare the evolution of DNA

sequences that takes place in models based on the MOS paradigm vs. the GA paradigm.

2. Materials & Methods

2.1. Hardware & Software

Simulations and related tools were run on an IBM-compatible 486 DX2 local bus personal computer with 340 MB hard drive and 16MB RAM, running at 66 MHz (Gateway 2000, North Sioux City, SD). Programs were written by the author in C++ and compiled with Borland C++ software, v. 3.1 (Borland International, Scotts Valley, CA). Data analysis and graphing were facilitated by Borland Quattro Pro software, v. 5.0.

2.2. Simulated DNA Sequence Evolution

In each initial sequence population, 100 ASCII character strings of 36 nucleotides (A, C, G and T) represented populations of idealized genes. Each triplet of 3 nucleotides (read left to right) coded for one amino acid, as defined by the genetic code. In each generation, sequences were subjected to variation, selection and expansion.

Selection was arbitrarily for sequences containing lysine triplets[4] (AAA or AAG), which have a 2/64 probability of occurrence in a random sequence. Fitness was defined by the number of lysine triplets in the population. Each lysine triplet was assigned a fitness score of 1 point, such that 12 X 100 points represented a perfect (maximal) fitness score. In data shown, each type of allowed variation was set at 1 event per string per generation; reciprocal exchanges counted as two events. (Exploration of alternative frequencies of variation frequencies in data not shown failed to reveal additional relevant trends).

Variation types included: (a) point mutation (defined as random replacement of a single nucleotide of a random string by A, C, G or T); (b) reciprocal crossover (defined as linkage of nucleotides 1 to n from random string *a* to nucleotides n+1 to 36 from random string *b*, and visa versa, where n is a random number between 1 and 36; or (c) gene conversion (defined as replacement of a random segment of random string *a* by a corresponding segment from random string *b*, without altering string *b*.

GA simulations of homologous recombination[5] (Figure 1) began with various input populations, each containing 100 identical copies of a   lysine depleted, pseudo-random 36mer.

---

[1] homologous chromosomes: paired chromosomes, one from each parent, each containing the same set of genes.
[2] Linkage equilibrium: random assortment of DNA sequences between homologous chromosomes.
[3] See Goldberg et al. for a GA bibliography.

[4] "Lysine triplet" refers to the DNA sequence that specifies the amino acid lysine in a protein chain; for simplicity, messenger RNA intermediates are ignored.
[5] Homologous recombination takes place between homologous chromosomes (note 1).

TABLE 1

Figure	Type	Pmut?	Xovr?	Gcnv?	Heter?	Lys?	Student's Paired t Tests under 10% attrition, against:	
1a	MOS	yes	no	no	no	no	1b: p = .003	1c: p = .0001
1b	GA	yes	yes	no	no	no	1c: p = .097	
1c	GA	yes	no	yes	no	no		
4a	MOS	yes	no	no	yes	yes	4b: p = .013	
4b	GA	yes	yes	no	yes	yes		
5a	no var.	no	no	no	yes	yes	5b: p < .0001	
5b	GA	no	yes	no	yes	yes		
5c	GA	no	yes	no	yes	no		

Table 1. **Distinctive features of evolutionary simulations. Abbreviations: MOS: modern synthesis model; GA: genetic algorithm model; no var: no variation; Pmut: point mutation; Xovr: reciprocal crossovers; Gcnv: gene conversions; Heter: heterogeneous input sequences; Lys: lysine triplets in input populations. Paired t tests were performed for mean fitnesses at 10% attrition after 300 generations; p values below 0.05 are significant.**

GA simulations of illegitimate recombination (Figures 4 & 5) began with initial populations each containing 100 *diverse* pseudo-random 36mers, containing lysine triplets at the expected frequency of 1/32 unless otherwise noted. Illegitimate events were restricted to exchanges between (and not within) the 12 triplets, to simulate exchanges within introns of split genes[6]. After the variation and expansion phases of each cycle, strings were ranked according to the number of lysine triplets they contained.

Ten selection stringencies were examined; these determined the attrition rate, defined as the fraction of lower-ranking strings to be eliminated in each cycle. Surviving strings were randomly chosen for expansion, restoring each population to 100 strings at the end of each cycle. Parameter selection was controlled by options entered on the command line and in an auxiliary file. Numeric data and verification of parameter settings were automatically written to auxiliary ASCII comma files, which were then imported into Quattro Pro notebooks for numeric and statistical analysis and graphing. For graphs of evolutionary time-courses, each parameter set was repeated in 10 separate simulations, using separate pseudo-random input sequence populations. Population fitness means from these 10 separate runs were then plotted for each of 300 (or 40) generations.

3. Results & Discussion

In the data to follow, the MOS and GA models were compared in computer simulations of evolving DNA sequences. Sequence evolution was documented by graphing the mean fitness of DNA sequence populations over 300 (or 40) cycles, under various degrees of selective pressure (selection stringency[7]). Comparisons thus reveal differences in rates of fitness increase, maximal fitness attained, and dependence on the stringency of selection, under the MOS and GA models. MOS models were based on point mutations, while two types of GA models were based on either reciprocal crossovers or gene conversion events (see **Materials & Methods** for definitions). Fitness was arbitrarily defined as the number of lysine triplets [AAA or AAG] in sequences of 36 nucleotides that code for 12 amino acids according to the biological genetic code. In all of the graphs to follow, each plotted point represents mean fitness values obtained from 10 independent simulations with distinct input populations. Results shown are reproducible in separate collections of runs. Distinctive features of each of the simulations are summarized in **Table 1**.

3.1 GA Models of Homologous Recombination

The first set of models compared the time course for appearance of new lysine triplets in homogeneous populations that were initially devoid of them. Each input population of 100 identical pseudo-random DNA sequences was first depleted of lysine triplets. By definition, these populations had zero initial fitness, and any lysine triplets that appeared in the populations would have to be created by mutation and/or recombination events.

---

[6] Genes of higher organisms are split into coding segments (exons), which specify protein sequence, flanked by non-coding segments (introns), which do not code for protein but may contain other useful information or properties.

---

[7] Selection stringency is defined by the attrition rate: the fraction of strings eliminated by selection in each cycle.

1a: Mutation Only
Initial Pop: Ident. Lys-Deplet.

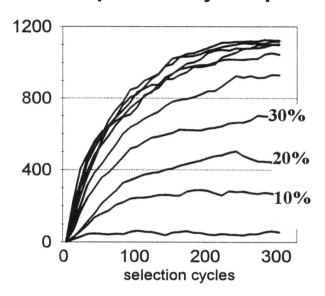

1c: Gene Conv. + Mutation
Initial Pop: Ident. Lys-Deplet.

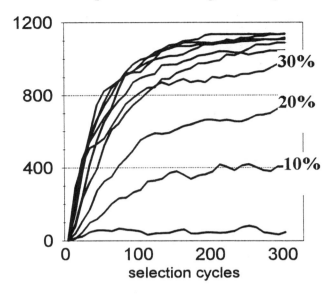

Figure 1: Computer simulations comparing evolution of DNA sequence populations selected for lysine codons, under MOS and GA models. Fitness scores (total number of lysines in population) are indicated on the Y axis. Each curve represents mean scores from 10 independent simulations with distinct input populations. Each initial populations contained 100 identical lysine-depleted sequences. Each curve represents a specific selection stringency ranging from 0% to 90% attrition; for clarity, only 10-30% curves are labeled (on the right). **a:** point mutations only. **b:** crossovers + point mutations. **c:** gene conversion + point mutations.

1b: Crossover + Mutation
Initial Pop: Ident. Lys-Deplet.

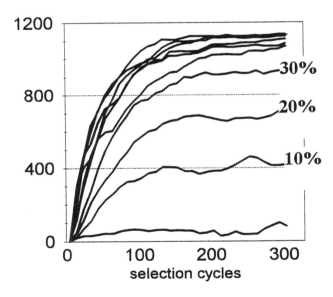

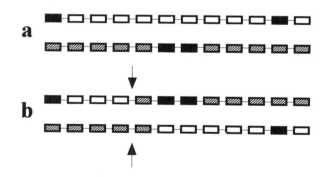

Figure 2: Computer Model for Reciprocal Homologous Recombination. Each box represents a triplet that codes for an amino acid in a protein; black boxes represent lysine-coding triplets. A reciprocal exchange (at arrows) between upper (light) and lower (shaded) sequences results in the transfer of two lysine triplets to the upper strand, which now has three lysine triplets. **a:** original sequences. **b:** sequences after reciprocal crossover.

Results obtained under the MOS model are summarized in **Figure 1a.** Graphs of these simulations are gently curved and asymptotic, reflecting gradual, cumulative increases in population fitness[8]. Rates of increase and maximal fitnesses obtained were dependent on selection stringency. For the highest selection stringencies, total population fitnesses approached the theoretical maximum of 1200 over hundreds of generations. For the lowest selection stringencies, an apparent fitness plateau was reached at sub-optimal fitnesses, over the 300 generations studied.

Under the GA model, reciprocal crossovers (defined in **Materials & Methods**) were also allowed in addition to point mutations: see **Figure 2** for a diagram of reciprocal recombination between homologous chromosomes.

Simulations involving both mutation and reciprocal recombination are summarized in **Figure 1b.** Comparison with Figure 1a reveals an accelerated time course for fitness increases in the presence of recombination. Moreover, higher fitnesses were ultimately attained under the GA model. These trends are especially apparent under mild selective pressure (low attrition rate). Final fitness means attained under 10% attrition (the lowest selective pressure examined) are significantly different by the paired t test (**Table 1**).

A second type of GA model was also examined, based on gene conversion events instead of reciprocal crossovers. Results of these simulations are shown in **Figure 1c.** Gene conversion events result in the conversion of one DNA sequence to that of another, while leaving the latter unchanged, as diagrammed in **Figure 3**. An important consequence for molecular evolution is that multi-gene families can evolve as a group, transcending individual selective forces (Li and Grauer, 1991). Both gene conversion and concerted evolution have been widely reported in the evolutionary literature (Li and Grauer, 1991; Lewin, 1994; Reynaud et al., 1989). Results graphed in **Figure 1c** closely resemble those seen with reciprocal recombination (Figure 1b). Final fitness means under 10% attrition did not differ significantly between these two types of recombination, while both were significantly higher than those generated by point mutation alone (**Table 1**). These data demonstrate that both types of recombination (Figures 1b & 1c) generate similar fitness enhancements, as compared to the MOS model (Figure 1a).

3.2. Summary: GA Models of Homologous Recombination

In the simulations above, homologous recombination acted as a creative force that generated novel sequences by linking (or possibly creating) lysine triplets from previously selected parent sequences. When exchanges took place under selective pressure, assembly of numerous lysine triplets was likely to occur. Consequently, strings with high fitness were rapidly produced by a *synergism* between recombination and selection over several generations. Even mild selective pressure combined with recombination resulted in rapid fitness increases. These effects, ignored by population genetic models based on the MOS, are revealed by these simulations with idealized DNA sequences. Presumably, the mechanism for these changes is no different than that previously described for computational applications of the GA (Forrest, 1993; Holland, 1992; Koza, 1992). These data underscore the fact that *recombination can represent a form of non-random, sequence-dependent mutation, particularly in a selected population.* This fits with a molecular definition of recombination (Maynard Smith, 1988b), as breaking and joining events between DNA sequences. Despite their simplicity, the models demonstrate the dramatic influence that recombination could have on molecular evolution.

This molecular view is obscured by a classical genetic definition of recombination as exchanges between homologous chromosomes derived from two parents (Maynard Smith, 1988b). The classical definition usually[9] focuses on outcrossing-- production of new combinations of genes *via* sexual reproduction-- rather than *assembly* of selected motifs to generate *new genes*. The semantics of the classical definition tends to obscure the observation, demonstrated in the GA simulations, that prior selection would have already reproduced "building blocks" with useful properties in previous generations. Others have previously described similar effects in terms of "schema" theory and the building-blocks model (Forrest, 1993; Holland, 1992; Koza, 1992).

The homogeneous, lysine-depleted populations used to obtain results in Figures 1b and 1c may have unfairly limited potential fitness enhancements, because production of new lysines by mutation was rate limiting in these models. Although these conservative GA models do show statistically significant differences over the MOS model, the magnitude of these effects is not large. It is important to note that in natural populations, there is more sequence diversity (polymorphism) than in the idealized models, and

---

[8] Fitness is defined here as the number of lysine codons in a sequence, which determines its rank prior to selection. This definition should not be confused with population genetics terminology.

[9] There are exceptions: geneticists have known for many years that recombination *within* a gene can also have profound consequences.

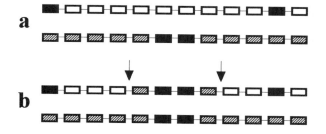

Figure 3: Computer Models for Gene Conversion. Event (arrows) replaces a segment of the upper sequence with that of the lower sequence, without altering the latter. Other symbols as in Figure 2.

4a: Mutation Only
Initial Pop: Diverse Random

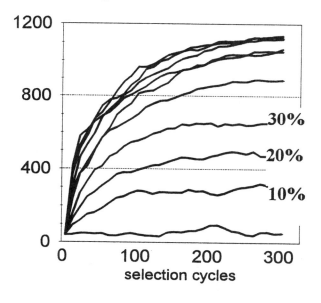

4b: Crossover + Mutation
Initial Pop: Diverse Random

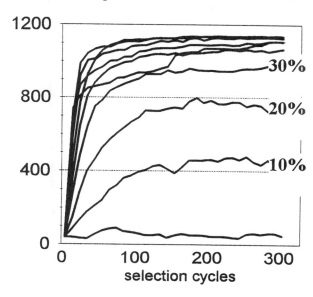

Figure 4. Computer simulations comparing evolution of DNA sequence populations selected for lysine codons, under MOS and GA models. Fitness scores (total number of lysines in population) are indicated on the Y axis. Each curve represents mean scores from 10 independent simulations with distinct input populations. Each initial population contained 100 diverse random sequences (*not* depleted of lysines). Each curve represents a specific selection stringency ranging from 0% to 90% attrition; for clarity, only 10-30% curves are labeled (on the right). **a:** point mutations only. **b:** crossovers + point mutations.

effects may be more rapid. This may also have important theoretical ramifications[10].

In addition, an even greater source of sequence diversity must also be considered, since recombination in real genomes[11] is not limited to exchanges between homologous sequences. This brings us to the subject of illegitimate recombination.

3.3. Illegitimate Recombination & the Exon Theory of Genes

The molecular definition of recombination-- breaking and joining events between DNA molecules-- includes not only homologous crossovers, but also exchanges between DNA sequences bearing only limited sequence similarity, which need not have common ancestry. Such so-called illegitimate recombination events have been studied for some time, both in the laboratory and by inference from molecular sequence data (Li and Grauer, 1991; Radding, 1978). Numerous examples of sequences that were presumably derived from illegitimate recombination have been reported in a variety of taxonomic groups. When viewed in the context of the complex genome, illegitimate recombination takes on special significance as an evolutionary mechanism. Several broadly accepted features of genome organization and evolution, including split genes, repetitive non-coding sequences, and gene duplication, provide an array of potential target sequences for illegitimate recombination events. The exon theory of genes (Gilbert, 1987) hypothesized that genes[12] are assembled from shorter fragments that are linked by illegitimate recombination events. Introns, the non-protein-coding sections of genes, provide a vast domain in which illegitimate recombination events could take place without compromising the integrity of (protein-coding) exons. Several examples supporting this model have been published, including cases of "exon shuffling", where entire exons have been swapped by illegitimate exchanges between introns of two or more alleles (reviewed in Gilbert, 1987). In addition, simple repetitive motifs, ubiquitous in introns and other non-coding domains, have been implicated as hot-spots for illegitimate recombination events (reviewed in Levinson and Gutman, 1987a and Gaillard and Strauss, 1994). In view of these kinds of evidence, the exon theory of genes appears sufficiently well-grounded to provide a platform for a second type of

GA model based on illegitimate reciprocal crossovers or illegitimate gene conversion events.

3.4. GA Models of Illegitimate Recombination

In GA models based on illegitimate reciprocal recombination, twelve exons, each consisting of only a single triplet, are flanked by introns, where recombination events take place. To simplify computation, introns are not represented in strings representing the DNA sequences, but are virtual. The model is idealized such that all introns are assumed to have sufficient sequence similarity (in simple repeats or other repetitive elements) to permit illegitimate exchanges between any pair of strings. Therefore, the recombination component in these simulations is formally equivalent to those that occur in the gene conversion model (Figure 1c), except that exchanges can only take place *between* triplets, and not within them.

An important difference from the homologous recombination models above (section 3.1) is that the input sequences are made diverse. The rationale for this feature is that since exchanges are not limited to homologous sequences, linkages between a variety of distant or unrelated sequences should be permitted; this is consistent with naturally occurring events. A second important difference in these models is that lysine triplets were *not* removed from the input populations, unless otherwise specified. This is consistent with the exon theory of genes, where pre-existing coding elements are subject to linkage by exchanges within introns. For comparative purposes, simulations were first performed in MOS counterparts, where only point mutations were allowed. These MOS simulations with diverse, lysine-containing input populations are summarized in **Figure 4a.**

Figure 4b displays results obtained with a GA model of illegitimate recombination, similar to the MOS model above but with reciprocal crossovers allowed (between introns). Here, fitness increases were greatly accelerated over the MOS counterpart (compare with Figure 4a). Differences in maximal attained fitnesses before reaching fitness plateaus were especially pronounced under less stringent selective pressure. Fitness means attained after 300 cycles of 10% attrition were significantly higher for the GA model than its MOS counterpart. (Table 1). In general, effects on fitness in this model were more striking than those observed with homologous recombination and homogeneous, lysine-depleted sequences (compare to Figure 1b).

Rapid fitness increases under the illegitimate model (Figure 4b) are not unexpected. The populations of 100 pseudo-random sequences already contained approximately 37 lysine triplets by chance alone, which were randomly

---

[10] Indeed, sequence diversity *per se* might be subject to positive selection for this very reason: this would represent a novel hypothesis for the observed excess polymorphism seen in natural populations (see Lewontin, 1974).

[11] Genome: the entire DNA complement of an individual, which is characteristic of its species.

[12] Gene: typically refers to a genetic unit that codes for a single protein chain.

5a: Selection Only
Initial Pop: Diverse Random

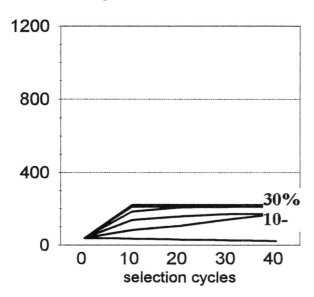

5c: Crossover Only
Initial Pop: Diverse Lys-Deplet.

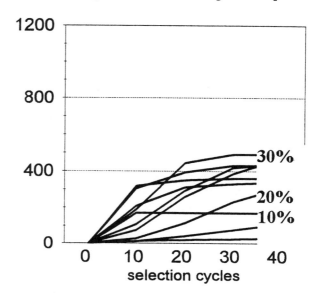

5b: Crossover Only
Initial Pop: Diverse Random

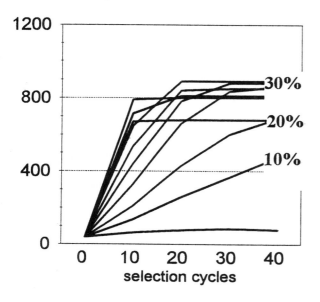

Figure 5. **Computer simulations of the evolution of DNA sequence populations selected for lysine codons, in the absence of point mutations.** Fitness scores (total number of lysines in population) are indicated on the Y axis. Each curve represents mean scores from 10 independent simulations with distinct input populations. Each initial population contained 100 diverse random sequences. Other features as in Figure 4. **a:** selection only. **b:** crossovers only. **c:** as in b, but lysine-depleted.

distributed among the 12 possible amino acid positions. Such randomly dispersed lysine triplets were available for immediate linkage by illegitimate exchanges. In contrast, in the homologous GA model, lysine triplets had to first be created before they could be linked, which imposed limits on fitness increases.

3.5. Illegitimate Recombination in the Absence of Point Mutations

It was therefore of interest to determine whether fitness increases would also occur in the *absence* of point mutations. For comparison, simulations were first performed in a model with selection only. These results are summarized in **Figure 5a**. In the absence of point mutations or recombination, only unmodified sequences could be selected, such that low fitness plateaus were rapidly established.

In dramatic contrast to these results, when illegitimate recombination was permitted (in the *absence* of point mutations), rapid fitness increases were seen, as shown in **Figure 5b**. In the first few generations, the time course for these rapid effects resembled those seen in Figure 4b. Not surprisingly, final mean fitnesses attained after 300 generations of 10% attrition differed significantly between Figures 5a and 5b (Table 1). These comparisons demonstrate that even in the absence of point mutations, recombination within selected sequence populations can generate sequences with high fitness, by linking existing elements into new combinations.

Recombination could produce strings containing large numbers of lysine triplets by two kinds of mechanisms. First, reciprocal recombination or gene conversion could splice segments such that existing lysine triplets are linked together in a recombinant string. In addition, recombination could also *create* new recombinant lysine triplets, by splitting and fusing triplets during exchanges. Such effects would be expected to be smaller than those resulting from triplet linkage, since the probability of generating new lysines by fusion is small. To examine this more subtle effect, diverse input sequences were depleted of lysine triplets prior to the simulation. Under those conditions, all lysine triplets would have to be created by fusion between existing triplets *via* crossover events. As shown in **Figure 5c**, rapid fitness increases were seen in early generations, even under non-stringent selection. Maximal fitnesses attained under these conditions were smaller, as expected, than those resulting from triplet linkage. These results demonstrate that the creative potential of recombination extends even to the level of the triplets themselves.

Taken together, results show that illegitimate recombination in selected sequence populations is a potent creative force, even in the absence of point mutations.

Simulations also demonstrate that classical point mutations and molecular recombinant mutations can act in a complementary fashion during DNA sequence evolution. Both types of variation-- point mutations and recombination-- may contribute to Darwinian sequence evolution. *Recombination, however, appears to generate more rapid effects that are less dependent on the stringency of selection.*

Rapid time-course and low dependence on selection stringency for fitness increases would have important consequences for molecular evolution. It would mean that even slight fitness advantages, that accrue to recombinant sequences by linkage of useful sequence motifs, could result in their rapid fixation in natural populations. Recombination may therefore represent an important mechanism for molecular sequence evolution that has been heretofore neglected by evolutionary biologists.

3.6. Applicability to Real Organismal Evolution

Returning to the question (Forrest, 1993) raised in the introductory paragraph, what insights about natural evolutionary processes can these idealized GA models provide? In particular, are trends observed in these simulations applicable to the natural history of genomic evolution in real organisms? An intriguing recent observation in bacterial cells is that specifically under selective pressure, a class of "adaptive mutations" is observed that takes place only in strains that are capable of recombination (Harris et al., 1994). These types of mutations, which have in previous work been controversial because they appear specifically in response to selective pressure (reviewed in Thaler, 1994), might be caused by an adaptive mechanism that promotes recombination when the bacterial cells are stressed.

More generally, in natural populations, selection would presumably be multi-factorial and dependent on changing ecological circumstances. The simple additive fitness models above cannot capture the richness of these natural processes. Genome organization, however, does provide abundant examples of conserved sequence motifs that are reused again and again. This can be seen both in comparisons between evolutionary lineages, and in comparisons between members of multi-gene families. As conserved sequences evolve, they should constitute a pool of potentially useful elements from which new sequences could arise, and DNA sequence data from a variety of organisms are certainly consistent with this expectation (Li and Grauer, 1991; Lewin, 1994; Genetics Computer Group, 1994).

The assertion that frequent recombination events could generate new functional sequences is consistent with a host of observations too numerous for exhaustive review. Following are a few examples (see Li and Grauer, 1991; Levinson and Gutman, 1987a; Lewin, 1994): first, many genes are members of multi-gene families (Ohno, 1970) that have exchanged segments following gene duplication. Second, we find that conserved sequences and motifs, such as zinc fingers and homeoboxes and response elements, are reused repeatedly among divergent multi-gene family members. Most genes, and most functional motifs, come from other genes. Third, the literature is rich with examples of concerted evolution, where gene conversion events homogenize widely dispersed repetitive sequences. Fourth, simple repetitive sequences are widely dispersed throughout the genome, and are known hotspots for illegitimate recombination and unequal crossing over (Radding, 1978; see also Gaillard and Strauss, 1994). The same is true of satellite DNA in general (Levinson and Gutman, 1987a). Fifth, there are many documented cases of exon shuffling (Gilbert, 1987). Sixth, recombination within a gene can generate a recombinant enzyme with a new substrate specificity (Hall and Zuzel, 1980). Seventh, in the chicken, gene conversion (rather than somatic mutation) is responsible for broad functional diversity in antibody V regions (Reynaud et al., 1989). Eighth, genome mapping finds evidence for site-specific recombination rates and linkage heterogeneity (Buetow et al., 1991). To summarize, the concept that functional motifs-- segments that have specific receptor or DNA-binding potential, catalytic potential, folding potential, secondary structure, etc. (Lewin, 1994; Gillespie, 1991, Alberts et al., 1989)-- should be reused in evolution-- is in accord with several observed features of genomic evolution.

Another significant fact is that the generation of novel, functional recombinants, *without disruption of existing functions*, is made possible by gene conversion events. Gene conversion events, by definition, transfer short sections of functional genes in a one-way fashion, such that parent sequences need not be altered in any way. Conserved parent sequences are likely to contain potentially useful elements. The sequences therefore represent a reservoir of previously selected, potentially functional motifs, that could be assembled by fortuitous recombination events without disrupting existing functions. Examples where functional domains of genes have been spliced together to generate new motifs have been cited by others as evidence for the exon theory of genes (reviewed in Gilbert, 1987). As noted above, there is also considerable evidence that simple repetitive motifs, which are abundant within introns and other non-coding segments, can be hotspots for recombination, and that concerted evolution *via* gene conversion is widespread. Thus, two essential elements that are required for a GA model-- an enriched pool of potentially useful sequence elements, and opportunities for recombination-- are well supported by empirical data.

The term "exaptation" has been used (Brosius and Gould, 1992) to distinguish adaptations that, having evolved under one set of selective pressures, are subsequently deployed for other purposes. The hypothesis presented here--that conserved sequence motifs can be reused in the form of recombinant genes-- postulates exaptation at the molecular level.

Production of new functional elements from a pool of existing sequences should also have important consequences for the evolution of non-coding DNA, since there should be selective pressure for expansion of repetitive sequences that *promote* recombination. This might help explain the ubiquitous presence of simple sequences such as satellites, minisatellites, and microsatellites, that constitute a large percentage of the genomes of higher eukaryotes (reviewed in Lewin, 1994; Levinson and Gutman, 1987a; Gaillard and Strauss, 1994). Genes that specifically control the rates at which simple sequence elements are generated-- such as the mutator gene -- have also been described (see Levinson and Gutman, 1987b, Parsons et al., 1993), and could be subject to positive selection (Maynard Smith, 1988c). Illegitimate recombination events may have been more frequent in ancient evolving genomes, if repair genes were less efficient (or less active) than their modern counterparts.

Non-random linkage of elements with high potential fitness requires some degree of prior selection, to enrich the population for such elements. The MOS paradigm fails to encompass this synergistic interaction between recombination and selection. Erroneous views concerning the role of recombination in the evolutionary process ensue from this conceptual limitation. This is evident in discussions concerning the evolution of sex, which assume that recombination functions only as a randomizing process that promotes linkage equilibrium (Maynard Smith, 1988a; Michod and Levin, 1988; Bell, 1982). This view is inconsistent with results presented here, which demonstrate linkage of pre-selected elements *via* recombination-- resulting in the production of non-random sequences with high fitness potential. In view of these findings, the synergism between recombination and selection can be added to the list of proposed reasons (Michod and Levin, 1988; Bell, 1982) why, despite its intrinsic limitations, sexual reproduction has been maintained in most taxonomic lineages.

3.7. Darwinian Evolution Revisited

A synergism between selection and recombination would have profound implications for Darwinian evolutionary theory (Darwin, 1966), since it would represent a type of mutation that is decidedly non-random, in the sense described by Huxley (see Introduction; Huxley, 1953). While Huxley's statement certainly applies to point mutations, it cannot strictly apply to recombination events that take place within a selected sequence population. In this latter case, the elements that are candidates for linkage have been specified by prior selection. Recombinant mutations do *not* take place in all directions: the range of possibilities is constrained by the population of parent sequences. Moreover, the consequences of each crossover event are dependent on the content of the parent sequences. The effects of a crossover event *can be related* to the needs of the organism, or to possible biological uses, if that event combines two previously selected, conserved elements with proven functional potential: this is exaptation at the molecular level. No teleological, Lamarckian, or anti-Darwinian mechanisms need be invoked to support this assertion.

The fitness criterion used in these studies-- incremental increase in the number of triplets for lysine-- is an obvious oversimplification of the kinds of selective pressures that are apt to operate during natural history. In real ecologies, the DNA sequences of a taxonomic lineage are, by definition, exhibiting sufficient fitness in each generation to allow survival and reproduction of the species. Although the lysine fitness model is adequate for the MOS vs. GA comparisons in the present study, future studies may allow a more realistic fitness criterion-- that of survival and reproduction *per se*-- to be modeled in a meaningful way. That goal is congruent with new approaches in the emerging field of Artificial Life (Langton, 1989; Langton, Langton et al., 1992; Langton, 1994).

"Perhaps the most obvious area for extending the GA is to the study of evolution itself", wrote Mitchell and Forrest in a recent paper (1993). "Although ideas from evolution have provided inspiration for developing interesting computational techniques, there have been few attempts to better understand the evolutionary systems which inspired them." Data presented here suggest that such efforts may bear many fruits.

ACKNOWLEDGMENT

I thank M.Z. Radic for stimulating discussions and helpful suggestions related to the manuscript.

REFERENCES AND NOTES

Alberts, B., Bray, D., Lewis, J., Raff, M., Roberts, K., and Watson, J.D. 1989. *Molecular biology of the cell*, 2nd ed. New York: Garland Publishing.

Bell, G. 1982. *The masterpiece of nature: the evolution and genetics of sexuality*. Berkeley: University of California press.

Brosius , J. and Gould, S.J. 1992. On genomenclature": a comprehensive (and respectful) taxonomy for pseudogenes and other "junk DNA". *Proc Natl Acad Sci USA* 89:10706-10710.

Buetow, K.H., Shiang ,R., Yang, P., Nakamura, Y., Lathrop, G.M., White, R., Wasmuth, J.J., Wood, S., Berdahl, L.D., Leysens, N.J., Ritty, T.M., Wise, M.E., and Murray, J.C. 1991. A detailed multipoint map of human chromosome 4 provides evidence for linkage heterogeneity and position-specific recombination rates. *Am. J. Hum. Genet.* 48: 911-925.

Darwin, C. 1966. (reprinted from 1859 edition) *On the origin of species by means of natural selection*. Facsimile edition. Cambridge: Harvard University Press.

Dobzhansky, T., Ayala, F.J., Stebbins, G.L., and Valentine, J.W. 1977. *Evolution*. San Francisco: WH Freeman, pg. 17.

Forrest, S. 1993. Genetic algorithms: principles of natural selection applied to computation. *Science* 261: 872-878.

Futuyma D. 1986. Evolutionary Biology, 2nd ed. Sunderland, MA: Sinauer Associates.

Gaillard, C. and Strauss, F. 1994. Association of poly(CA)·poly(TG) DNA fragments into four-stranded complexes bound by HMG1 and 2. *Science* 264: 433-436.

Genetics Computer Group. April 1994. *Program manual for the GCG package*, version 8, 575 Science Drive, Madison, WI 53711.

Gilbert, W. 1987. The exon theory of genes. *Cold Spring Harbor. Symp. Quant. Biol.* LII: 901-905.

Gillespie, J.H. 1991. *The causes of molecular evolution*. New York: Oxford University Press.

Goldberg, D.E., Milman, K., and Tidd, C. 1992. *Genetic algorithms: a bibliography*. illiGAL library report No. 92008, Dept. of Genl. Engineering, University of Illinois, Urbana, IL 61801; library@gal1.ge.uiuc.edu.

Hall, B.G. and Zuzel, T. 1980. Evolution of a new enzymatic function by recombination within a gene. *Proc. Natl. Acad. Sci. USA* 77: 3529-3533.

Harris, R.S., Longerich, S., and Rosenberg, S.M. 1994. Recombination in adaptive mutation. *Science* 264: 258-260.

Holland, J.H. 1992. (1st ed. 1975) *Adaptation in natural and artificial systems*. Cambridge: MIT Press.

Huxley, J. 1953. *Evolution in action.* New York: Harper & Row, pg. 36.

Koza, J. 1992. *Genetic programming: on the programming of computers by means of natural selection.* Cambridge: MIT Press.

Langton, C.G., ed. 1989. Artificial Life. Proceedings of an interdisciplinary workshop on the synthesis and simulation of living systems held September, 1987 in Los Alamos, New Mexico. Redwood City, CA: Addison-Wesley.

Langton, C.G., Taylor, C., Farmer, J. D., and Rasmussen, S., eds. Artificial Life II. Proceedings of the workshop on artificial life held February, 1990 in Santa Fe, New Mexico. Redwood City, CA: Addison-Wesley.

Langton, C.G., ed. 1994. Artificial Life III. Proceedings of the workshop on artificial life held June, 1992 in Santa Fe, New Mexico. Redwood City, CA: Addison-Wesley.

Levinson, G. and Gutman, G.A. 1987a. Slipped-strand mispairing: a major mechanism for DNA sequence evolution. *Mol. Biol. Evol.* 4: 203-221.

Levinson, G. and Gutman, G.A. 1987b. High frequencies of short frameshifts in poly-CA/TG tandem repeats borne by bacteriophage M13 in *Escherichia coli* K-12. *Nucleic Acids Res* 15: 5323-38.

Lewin, B. 1994. *Genes V.* New York: Oxford University Press.

Lewontin, R.C. 1974. *The genetic basis of evolutionary change.* New York: Columbia U. Press.

Li, W.H. and Grauer, D. 1991. *Fundamentals of molecular evolution.* Sunderland, MA: Sinauer Associates.

Maynard Smith, J. 1988a. In Michod and Levin, pg. 113.

Maynard Smith, J. 1988b. In Michod and Levin, pp. 106-107.

Maynard Smith, J. 1988c. Selection for recombination in a polygenic model: the mechanism. *Genet. Res.* 51: 59-63.

Michod RE and Levin BR, eds.1988. *The evolution of sex: an examination of current ideas.* Sunderland, MA:Sinauer Associates.

Mitchell, M. and Forrest, S. 1993.Genetic algorithms and artificial life. *Artificial life,* in press. Pre-printed as Santa Fe Institute working paper #93-11-072, Santa Fe, New Mexico.

Ohno, S. 1970. *Evolution by gene duplication.* Berlin: Springer Verlag.

Parsons, R., Li ,G.M., Longley, M.J., Fang, W.H., Papadopoulos, N., Jen, J., de la Chapelle, A., Kinzler, K.W., Vogelstein, B., and Modrich, P. 1993. Hypermutability and mismatch repair deficiency in RER+ tumor cells. *Cell* 75: 1227-1236.

Radding, C.M. 1978. Genetic recombination: strand transfer and mismatch repair. *Ann. Rev. Biochem.* 47: 847-880.

Reynaud, C.-A., Dahan, A., Anquez, V., Weill, J.-C. 1989. Somatic hyperconversion diversifies the single V_H gene of the chicken with a high incidence in the D region. *Cell* 59: 171-183.

Thaler, D.S. 1994. The evolution of genetic intelligence. *Science* 264: 224-225.

STEPS TOWARDS CO-EVOLUTIONARY CLASSIFICATION NEURAL NETWORKS.

Jan Paredis

RIKS

Postbus 463, NL-6200 AL Maastricht

The Netherlands

(jan@riks.nl)

Abstract.

This paper proposes two improvements to the genetic evolution of neural networks (NNs): life-time fitness evaluation and co-evolution. A classification task is used to demonstrate the potential of these methods and to compare them with state-of-the-art evolutionary NN approaches. Furthermore, both methods are complementary: co-evolution can be used in combination with life-time fitness evaluation.

Moreover, the continuous feedback associated with life-time evaluation paves the way for the incorporation of life-time learning. This may lead to hybrid approaches which involve genetic as well as, for example, back-propagation learning. In addition to this, life-time fitness evaluation allows an apt response to noise and changes in the problem to be solved.

1. Introduction.

Over the last years NNs have proven to be robust classifiers. Particularly, the use of back-propagation for supervised learning has received much attention. Current research is continuously pushing the performance of these algorithms.

An alternative approach uses genetic search to find neural nets that correctly classify a set of given pre-classified training examples. This way, the advantages of both NNs and genetic algorithms (GAs) are exploited: NNs are known to be good pattern associators whereas GAs have proven to be good function optimizers. Or as Whitley (1989b) puts it: "when optimizing the weights of a NN a GA must find a set of weighed connections that allow the NN to solve a given problem".

This technique has multiple advantages: 1) it can be applied to any NN architecture (i.e. there is no restriction to multi-layered perceptrons), 2) it can deal with sparse and delayed feedback, and 3) it does not need gradient information. Despite remaining open research issues, genetic learning is becoming roughly competitive with back-propagation (Whitley 93). In this paper we improve the genetic search along a new dimension: through the use of life-time fitness evaluation.

The structure of this paper is as follows: Section 2 introduces the classification problem on which our methods are tested and evaluated. Section 3 describes a state-of-the-art genetic algorithm as point of comparison. Sections 4 and 5 successively add life-time fitness evaluation (LTFE) and co-evolution to this standard algorithm. Section 6 discusses how LTFE allows steady-state reproduction systems to deal with noisy fitness evaluations. Section 7 describes a few other application areas and possible extensions of the approach proposed. Finally, conclusions are given.

2. Experimental Set-Up

Figure 1 depicts the classification task used in our experiments: a neural network has to represent a correct mapping from $[-1,1] \times [-1,1]$ to the set of classes {A, B, C, D} given 200 pre-classified, randomly selected, examples. This problem is designed such that the regions in the instance space ($[-1,1] \times [-1,1]$) corresponding to the different classes have different degrees of complexity: class A consists of a circular convex region, class B is the union of two disjoint convex regions, C consists of one non-convex region, and D is the union of two non-convex regions.

Figure 2 depicts the NN architecture used: a standard multi-layer perceptron with one hidden layer consisting of 12 hidden nodes. The coordinates of the point to be classified are clamped on the input nodes of this network. Each output node is associated with one class. Hence, our NN has four output nodes.

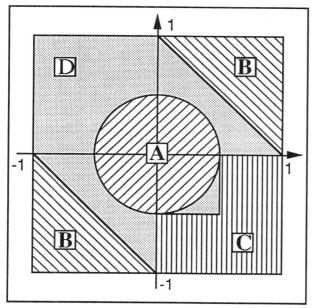

Figure 1: Test classification problem

As can be seen from figure 1, a correct neural net should classify, for example, the point (0.2,0.1) as belonging to class A. This classification proceeds as follows: 1) the coordinates 0.2 and 0.1 are clamped on the associated input nodes of the NN, 2) feed-forward propagation is executed - here the standard sigmoidal activation function is used, 3) the result of the classification is the class corresponding to the output node with the highest activity level. The NN is said to classify the example above correctly when the output node associated with class A is the most active one.

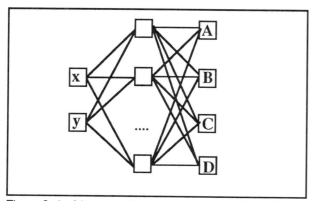

Figure 2: Architecture of the neural net classifier

So far the description of the test problem and the NN architecture to be used. Now we proceed to the evolutionary search for appropriate connection weights such that a NN can solve the test problem. The algorithm is introduced in three steps each time adding one feature. This is done not only for reasons of clarity but also to allow empirical testing of the role of these features.

The experiments described below were repeated with various crossover operators: one-point, two-point, four-point, six-point, twelve-point, and uniform crossover. These operators control the way weights are inherited from parents by the children. A description of these operators can be found in (Goldberg 1989), and (Eshelman, Caruana, and Schaffer 1989), amongst others. In addition, the experiments were executed with population sizes of 100 and 200. In both cases, the same trend appeared clearly. Hence, only the experiments with a population size of 100 are described here. The extensive testing - using different crossover operators and population sizes - confirms the same trends in all experiments. This strongly supports the validity of the approach presented.

3. Approach 1: The "Traditional" Approach

3.1 Description of the Algorithm

The traditional approach operates on a population of NNs. Figure 3 depicts this population data structure. The algorithm used here is based on GENITOR (Whitley 1989b), a well known genetic algorithm specifically designed for evolving NNs and further improved over the years (Whitley 93).

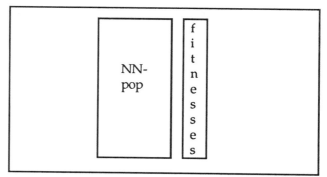

Figure 3: Traditional approach: population set-up

The algorithm can be described as follows. First, the initial population of NNs is created. The weights of these NNs are randomly drawn from the closed interval [-100,100]. Furthermore, each NN has an associated fitness indicating its classification accuracy. Here, the fitness of a NN is the number of examples - from the set of 200 training examples - it classifies correctly. Next, GENITOR repeats the following steps: 1) SELECT two parents (NNs) from the population. This selection is biased towards highly ranked individuals in the population which is sorted on fitness; 2) a new individual is generated from these parents through the application of MUTATION and CROSSOVER operators (see below); 3) the child is tested on the set of pre-classified examples in order to calculate its FITNESS; 4) if this fitness is higher than the minimum fitness in the population then the child is INSERTed into the appropriate rank location in the NN population. At the same time, the individual with the minimum fitness is removed. The pseudo-code below describes the basic cycle of the algorithm:

```
nn1:= SELECT(nn-pop)    ; parent1
nn2:= SELECT(nn-pop)    ; parent2
child:= MUTATE-CROSSOVER(nn1,nn2)
f:= FITNESS(child)
INSERT(child,f,nn-pop)
```

We use a simple genetic representation of a NN: a linear string of its weights, with weights belonging to links feeding into the same node located next to each other. In accordance with (Paredis 91) and (Whitley 93) the weights are encoded directly as real numbers. Adaptive mutation is used: the smaller the Euclidean distance between the weight vectors of the parents the higher the probability of mutation. Mutation replaces a weight by a new randomly generated weight.

3.2 Empirical Results

Table 1 summarizes the results of the traditional algorithm. It shows the number of offspring (in thousands) generated before a NN with a given level of classification accuracy (i.e.: 70%, 80%, 85%, 90%, and 95% of the 200 examples are correctly classified) is obtained. The table shows the results for all 6 crossover operators mentioned above. For each such crossover 5 runs were executed. The best-so-far was averaged over these 5 runs. Each run used a population size of 100 and generated a total of 50000 offspring. The table shows, for example, that - on average - one-point crossover reaches a solution which correctly classifies 85% of the examples after generating 10500 offspring. Note also that none of the crossover operators reaches an average best-so-far of 95% of correct classifications within the allowed 50000 offspring.

The results of this algorithm are extremely poor. The computational effort needed to obtain good solutions is unexpectedly large. This is because most time is spent on fitness evaluation: each child is tested on all 200 examples. Moreover, most of these tests are uninformative: in the beginning of a run the majority of examples is wrongly classified, at later stages most of them are correctly classified. Hence, the individual tests have a relatively low discriminative power: only a few of the 200 examples are relevant.

	>70	>80	>85	>90	>95
1pt	2.5	8	10.5	17	-
2pt	2.5	5	7.5	13.5	-
4pt	3	8	10	-	-
6pt	3.5	5	11	21.5	-
12pt	4	7	10	22	-
unif	6	10.5	14	20	-

Table 1: Approach 1: Empirical Results

Nevertheless one can observe a typical route followed by the algorithm. In the initial population, most of the NNs - which all consist of randomly assigned weights - assign one class to all examples. Or, in other words, for a given network of the initial population there is virtually always only one output node which is most active and this independent of the example clamped on. Hence, those networks that always assign the class which has most members in the training set are among the fittest. During later phases, more and more regions belonging to other classes are correctly classified by the NNs until finally a network implements boundaries that (almost) precisely match those in the problem statement.

Initially, the weak performance of this algorithm came as a surprise. All the more since a closely related algorithm developed within the context of artificial life (Paredis 91) had no such heavy computational demands. In that paper a NN maps environmental situations to motor actions such that an animal could survive in a simulated world. The same type of classification NN as discussed above was used to implement this mapping.

A closer inspection of the differences between both algorithms reveals what is going on: the animal NN receives *continuous feedback*. Each animal has an internal food store that we refer to as its energy. This energy is lowered when it moves or gets injured. Eating food, on the other hand, increases an animal's energy. An animal's probability of reproduction is proportional to its energy. In this way, the energy plays the role of an approximate, continuously updated, fitness value. The traditional classification approach discussed in this section, on the other hand, calculates the fitness in one go. It looks as if such complete, costly evaluations involving all examples constitute considerable overkill. This difference between both types of fitness evaluation suggests the use of life-time fitness evaluation for our classification task as well.

4. Approach 2: Adding Life-time Fitness Evaluation

4.1 Description of the Algorithm

In this approach the set of all pre-classified examples plays a much more prominent role (figure 4 depicts the data structures used) than in the previous approach. The fitness of a NN is now calculated on the basis of the examples it "encounters" (see below) instead of on all 200 examples.

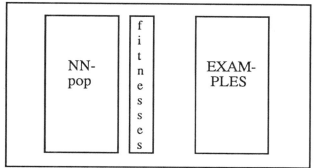

Figure 4: Life-time fitness evaluation: population set-up

Here, the fitness of a NN is defined as the number of successful classifications of the last 20 examples it encountered. Hence, each NN has a history which contains the results of its 20 last classifications. First, the initial NN population is created. The fitness of these initial NNs is defined as the number of correct classifications of 20 randomly chosen examples. Next, the following cycle is repeated. First, twenty - this number is arbitrarily chosen - NNs and examples are paired. During such an ENCOUNTER the NN classifies the example. The result of such an encounter is 1 if this classification is correct or 0 if incorrect. This result is pushed on the history of encounters the NN was involved in. At the same time, the result of the least recent encounter is removed from the NN's history. Now the fitness can be UPDATEd as well: it is set equal to the number of successful classifications in its history. After the execution of the DO-loop, the standard approach is followed: two parents are SELECTed, a child is created and INSERTed into the population.

One more point should be made here. The selection of NNs involved in an encounter is biased in the same way as the selection of NNs for reproduction: the fittest NN is 1.5 times more likely to be selected

than the NN with median fitness. Hence, the fitter NNs are tested more often: they have to prove themselves. The DO-loop at the beginning of the basic cycle described above takes care of this. The examples, on the other hand, are randomly chosen with uniform distribution. Hence, the name U-SELECT in the pseudo-code. This same distribution is used to select the 20 examples which are used to calculate the fitness of newly generated NNs. The pseudo-code below describes the algorithm's basic cycle:

```
DO 20 TIMES
    nn:= SELECT(nn-pop)
    ex:= U-SELECT(examples)
    res:=ENCOUNTER(nn,ex)
    UPDATE-HISTORY-AND-FITNESS(nn,res)
nn1:= SELECT(nn-pop)     ; parent1
nn2:= SELECT(nn-pop)     ; parent2
child:= MUTATE-CROSSOVER(nn1,nn2)
f:= FITNESS(child)
INSERT(child,f,nn-pop)
```

By now, the difference with the previous approach should be clear. The crucial point is the use of continuous partial fitness feedback. This life-time fitness evaluation allows for early detection of good and bad solutions. Once a NN is considered potentially good it is put under closer scrutiny.

4.2 Empirical Results

In order to allow for comparison between both approaches the accuracy of an NN is still defined as the number of correct classifications of *all* 200 examples. Hence, in this approach the accuracy is not equal to the fitness. That is because the fitness only takes into account a subset of the examples. Of course, reproduction is based on fitness and not on accuracy. Despite this difference, the term "average best-so-far" will still be used instead of the more precise description "average best-accuracy-so-far".

When comparing this approach with the previous one, one should also acknowledge the improved computational efficiency resulting from the smaller number of classifications to be done during fitness evaluation. The execution of the basic cycle is actually more than four times quicker than in the first approach. We allow the runs to create 100000 offspring, which is only double the amount allowed before. This means that the total computational effort spent on a run is still half of that spent in the first approach.

Now we can compare both approaches. The large majority of entries in table 2 reach their accuracy level within triple the number of offspring needed in the traditional approach[1]. Hence, these results clearly show that the use of life-time evaluation considerably reduces CPU time. Moreover, only three entries in table 2 require more than 50000 offspring to reach their corresponding accuracy. These three only need marginally more offspring: they all reach their accuracy level after the generation of less than 60000 offspring.

	>70	>80	>85	>90	>95
1pt	3.5	8.5	17.5	35	-
2pt	3.5	11.5	24.5	34	-
4pt	6	17	26.5	51.5	-
6pt	11	25.5	31.5	59.5	-
12pt	7	15	25	46	54.5
unif	13.5	25	41.5	-	-

Table 2: Adding Life-time Fitness Evaluation: Empirical Results

So far the increased execution speed of the algorithm. The increase in solution quality obtained is fairly small and not very significant. This can be seen from table 2 as well. Now one crossover - twelve-point - reaches an average best-so-far classification accuracy of more than 95%. The first approach, on the other hand, never yields such a high quality average best-so-far.

The next section describes how the concept of co-evolution, as introduced by Hillis (1992), can be integrated with life-time fitness evaluation. In this way, co-evolution can be used to even further improve the performance on our classification task.

5. Approach 3: Co-evolution of NNs and Examples

5.1 Description of the Algorithm

In this last step the examples have a fitness as well. Figure 5 depicts this. Just as in Hillis (1992) the two co-evolutionary species - here examples and NNs - interact through their fitness. The fitness of a NN is still defined as the number of successful classifications of the twenty most recently encountered examples. The fitness of an example, on the other hand, is the number of times it was *incorrectly* classified by the NNs it encountered most recently.

In the strictest sense, the examples do not form a real population because they are never replaced: the population consists all the time of the same 200 pre-classified examples. There is one more important

difference with the previous approach. The fitness of examples now allows for a biased selection of examples as well. Similar to fit NNs, fit examples are more often involved in an encounter. Or, in other words, the more difficult an example is, the more often it will be presented to the NNs.

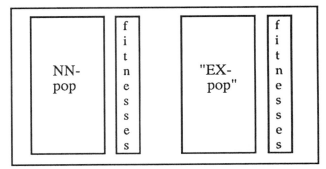

Figure 5: Co-Evolution of NNs and examples

The do-loop in the code below now ensures that the fitnesses keep up with the changing populations - or the ranking within them. The function TOGGLE implements the inverse fitness interaction between examples and NNs. It changes a 1 into a 0 and vice versa.

```
DO 20 TIMES
    nn:= SELECT(nn-pop)
    ex:= SELECT(ex-pop)
    res:= ENCOUNTER(nn,ex)
    UPDATE-HISTORY-AND-FITNESS(nn,res)
    UPDATE-HISTORY-AND FITNESS(ex,TOGGLE(res))
nn1:= SELECT(nn-pop)    ; parent1
nn2:= SELECT(nn-pop)    ; parent2
child:= MUTATE-CROSSOVER(nn1,nn2)
f:= FITNESS(child)
INSERT(child,f,nn-pop)
```

5.2. Empirical Results

The fitness interaction between examples and NNs gives rise to an interesting dynamic behavior in both populations. As soon as the NNs are becoming successful on a certain set of examples then these examples get a lower fitness and other, not yet solved examples will be selected more often. This process forces the NNs to concentrate now on these more difficult examples (because of the fitness proportional selection of the individuals involved in an encounter).

This considerably focuses the genetic search, as one can see in table 3. Again, comparisons with the previous approaches can be made along two dimensions: computational demand and solution quality.

The decrease in computation demand is substantial. The basic cycle of the algorithm takes

---

[1] Only three entries -- all belonging to the six-point crossover -- require more than triple the offspring generated in approach 1.

marginally more time than in the previous approach. This is due to the additional bookkeeping of examples (fitness calculation and sorting). Nonetheless, the execution of a basic cycle is still 3.75 times faster than in the traditional approach. The small increase in computation per cycle, pushes down significantly the number of cycles to be executed. In fact all entries - except for one-point crossover - require less offspring than the previous approach (table 2) to reach their accuracy level. This difference is outspoken for the higher accuracy levels in particular. In comparison with the traditional approach the improvement is even more spectacular. Although the basic cycle is 3.75 times quicker, half of the entries in table 3 need less cycles than the first approach to reach their accuracy level!

	>70	>80	>85	>90	>95
1pt	4	9	20	60	-
2pt	3.5	9.5	12	17	42.5
4pt	4	6.5	11	26	74
6pt	4	7.5	11	21.5	58
12pt	4	5.5	8	12.5	36.5
unif	4.5	7	10	16	47.5

Table 3: Adding Co-evolution: Empirical Results

As Hillis (1992) argues a co-evolutionary approach has two advantages over the traditional one. Primo, it helps prevent large portions of the NN population from becoming stuck in local optima. Secundo, fitness testing becomes much more efficient: one mainly focuses on examples not yet solved. This last point can be easily observed in our classification task: after a while only the examples which are situated near the boundaries between different classes have a high fitness value and reside at the top of the population. This way, the pitfalls of the traditional approach described in section 3 are avoided.

The use of LTFE within co-evolution has several advantages. One is the fine-grained nature of the algorithm. This allows for easy parallel implementation. A massively parallel architecture with each processor containing one NN and a (copy of) an example can be used. Efficient machine-level shift operations can change the partners involved in the encounters. Local selection and reproduction can then be implemented using the fine-grained parallel genetic algorithm described in (Spiessens and Manderick 1991).

In addition to the potential for parallel implementation, the use of LTFE has another advantage: it makes the algorithm robust under noisy fitness evaluation. The next section discusses this in more detail.

6. Steady-State Reproduction, Noise, and LTFE.

Syswerda (1989) introduced the term steady-state reproduction for the one-at-a-time reproduction used in systems like GENITOR. However, these systems are not very robust. They perform best on problems with a deterministic fitness evaluation function. Noisy fitness functions might pose serious problems to steady-state reproduction systems. Consider, for example, an individual which, due to noise, is assigned too high a fitness. In steady-state systems this fitness remains associated with that individual as long as it stays in the population. During all this time its overestimated genetic material may spread throughout the population. This problem does not occur when generational replacement is used. In that case, each generation creates an entirely new population and calculates the fitness of these members anew at each generation. Hence, an individual is re-evaluated each time it survives a generation. This way noisy evaluations do not substantially decrease the reliability of the sampling.

Life-time fitness evaluation does provide a good way to deal with noise within steady-state reproduction. This due to the continuous fitness feedback which averages out the noise in the fitness evaluation. Hence, the research described here takes up the challenge described in Davis 1991:

```
Although no one has yet published a paper
about steady-state reproduction with noisy
evaluation functions, it seems possible that
modifications to the steady-state algorithm
can be devised that will outperform genera-
tional replacement on such problems. This is
a fertile topic for future research.
```

7. Current Status and Outlook

The integrated use of LTFE and co-evolution has been shown effective on a classification task. A companion paper (Paredis, under submission) applies this same technique on the completely unrelated task of constraint satisfaction. In this case, the two competing populations contain potential solutions and constraints, respectively. Path finding is another application area currently under investigation.

In addition to the application to other tasks the current framework can be extended in yet another way. All examples discussed above have used prey-predator relations to improve the power of artificial search. Obviously, many other mechanisms - not necessarily based on inverse fitness interaction - exist in nature. Symbiosis is such an important and widely oc-curring counter example. It consists of a positive fit-ness feedback in which a success on one side improves the chances of survival of the other.

As far as we know, the use of symbiosis to en-hance the power of evolutionary search is as yet un-explored. The author has just finished a first investi-gation into the symbiotic evolution of solutions and their genetic representations (i.e. the ordering of the genes). A representation adapted to the solutions cur-rently in the population speeds up the search for even better solutions which in their turn might progress optimally when yet another representation is used.

The research mentioned above supports the generality and effectiveness of the combined use of LTFE and co-evolution.

8. Conclusion.

The integrated use of life-time fitness evaluation and co-evolution clearly boosts the performance of genetic neural net search. This in terms of solution quality as well as computation demand. This performance in-crease can be attributed to two factors. Primo, the partial nature of life-time fitness evaluation allows for early detection of particularly good or bad NNs. Secundo, co-evolution concentrates its effort on highly relevant training examples. The continuous updating of life-time fitness evaluation allows to track changes in the environment or in the problem to be solved. Moreover, this feedback makes the algo-rithm less vulnerable to noisy fitness evaluations. Hence, the approach presented here is robust as well as adaptive.

This paper describes the use of Artificial Life techniques - such as life-time fitness evaluation and co-evolution - on a standard Artificial Intelligence task such as classification. Further research is needed to investigate the use of these techniques on other tasks. Another avenue for future research is the integration of life-time fitness evaluation and life-time learning. The experiments described in this pa-per show the strength of our algorithm over a large spectrum of conditions. At the same time many inter-esting research issues are still open for further explo-ration.

Acknowledgments.

The author is indebted to Steven Wellink for inter-face design and to Desiree Baaten for proofreading this paper.

References.

Davis, L., (1991), (ed.), *Handbook of Genetic Algorithms*, Van Nostrand Reinhold, New York.

Eshelman, L.J., Caruana, R.A., Schaffer, J.D., (1989), *Biases in the Crossover Landscape*, Proc. Third Int. Conf. on Genetic Algorithms, Morgan Kaufmann.

Goldberg, D.E., (1989), *Genetic Algorithms in Search, Optimization & Machine Learning*, Addison-Wesley, Reading, Mass.

Hillis, W.D., (1992), *Co-evolving Parasites Improve Simulated Evolution as an Optimization Procedure*, in Artificial Life II, Langton, C.G.; Taylor, C.; Farmer, J.D., and Rasmussen, S., (eds), Addison-Wesley, California.

Spiessens, P., and Manderick, (1991), A Massively parallel Genetic Algorithm: Implementation and First Analysis, Proc. Fourth Int. Conf. on Genetic Algorithms, Morgan Kaufmann.

Paredis, J., (1991), *The Evolution of Behavior: Some Experiments*, in Proceedings of Simulation of Adaptive Behavior: From Animals to Animats, Meyer, and Wilson (eds.), MIT Press/Bradford Books.

Paredis J., (under submission), Co-evolutionary Constraint Satisfaction.

Syswerda, G., (1989), *Uniform Crossover in Genetic Algorithms*, Proc. Third Int. Conf. on Genetic Algorithms, Morgan Kaufmann.

Whitley, D., (1989a), *The Genitor Algorithm and Selection Pressure: Why Rank-Based Allocation of Reproductive Trails is Best.* Proc. Third Int. Conf. on Genetic Algorithms, Morgan Kaufmann.

Whitley, D., (1989b), *Optimizing Neural Networks using Faster, more Accurate Genetic Search.* Proc. Third Int. Conf. on Genetic Algorithms, Morgan Kaufmann.

Whitley, D., (1993), *Genetic Reinforcement Learning for Neurocontrol Problems*, Machine Learning, 13: p. 259-284, Kluwer Academic Publishers.

Self-organisation in a system of binary strings

Wolfgang Banzhaf

Department of Computer Science, Dortmund University

Baroper Str. 301, 44221 Dortmund, GERMANY

banzhaf@tarantoga.informatik.uni-dortmund.de

Abstract

We discuss a system of autocatalytic sequences of binary numbers. Sequences come in two forms, a 1-dimensional form (operands) and a 2-dimensional form (operators) that are able to react with each other. The resulting reaction network shows signs of emerging metabolisms. We discuss the general framework and examine specific interactions for a system with strings of length 4 bits. A self-maintaining network of string types and parasitic interactions are shown to exist.

Introduction

Sequences of binary numbers are the most primitive form of information storage we know today. They are able to code any kind of man-made information, be it still or moving images, sound waves and other sensory stimulations, be it written language or the rules of mathematics, just to name a few. As the success of von-Neumann computers has shown over the last 50 years, binary sequences are also sufficient to store the commands that drive the execution of computer programs. In fact, part of the success of the digital computer was due to the universality of bits and their interchangeability between data and programs.

It is not far-fetched to expect that the physical identity between operators (programs) and operands (data) may also play an essential role in self-organisation. We have proposed to consider a simple self-organising system [1], in which sequences of binary numbers are able to react with each other and sometimes even to replicate themselves. This ability of binary strings was a result of the proposition to consider binary strings similar to sequences of nucleotides in RNA. RNA sequences which presumably stood at the cradle of life [2, 3], seem capable of self-organisation and come in at least two alternative forms, a one-dimensional genotypic form and a two or three-dimensional phenotypic form. We proposed to consider binary strings in analogy and to provide for a second, folded and operative form of strings. Technically, we considered as this alternative a two-dimensional

matrix form that is able to perform operations on other one-dimensional binary strings.

Reactions between binary strings

The fundamental ideas of this model have been outlined elsewhere (see ref. [1],[4],[7] for details). Here we only give a brief overview of what has been learned so far.

Let us consider sequences

$$\vec{s} = (s_1, s_2, ..., s_i, ..., s_N). \qquad (1)$$

of binary symbols $s_i \in \{0, 1\}, i = 1, ..., N$ organised in 1-dimensional strings.

Then we ask the question: Does there exist an alternative form of these strings, that is (i) reversibly transformable into the form (1), and is (ii) operative on form (1)? The answer is surprisingly simple and well known from mathematics: Yes, there are operators with the above capabilities, known as matrices.

Thus, we require the existence of a mapping $\mathcal{M}$

$$\mathcal{M} : \vec{s} \to \mathcal{P}_{\vec{s}} \qquad (2)$$

which transforms $\vec{s}$ into a corresponding 2-dimensional matrix form $\mathcal{P}_{\vec{s}}$ of the sequence which should be unique and reversible. This mapping is simply a spatial reorganisation of the information contained in a sequence and may be termed a *folding*, in close analogy to the notion used in molecular biology.

The most compact realization of such a 2-dimensional form would be a quadratic matrix. For a string with a quadratic number of components $N, N \in \mathcal{N}_{sq}$ with $\mathcal{N}_{sq} = \{1, 4, 9, 16, 25, ...\}$, the procedure is straightforward: Any systematic folding (examples are shown in Figure 1) would do. Since folding is not yet very sophisticated, and different configurations may be obtained by a renumbering of string components, we shall consider here the topological folding of Figure 1 (b) only.

In the more general case of N being a non-quadratic number, different generalizations are reasonable. Here

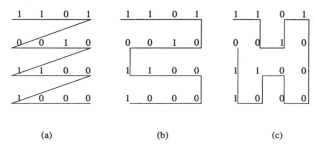

Figure 1: Some two-dimensional compactly folded forms of a string in an example with $N = 16$ binary numbers: $\vec{s} = $ (1101001011001000).
(a): non-topological folding, (b) and (c): topological foldings.

we shall only discuss a compact folding [1] in non-square matrices, where

$$N_1 \times N_2 = N. \qquad (3)$$

In order to treat non-quadratic cases similar to the quadratic case, a bias should be used in the direction of the most compact solution, i.e.

$$N_i = \sqrt{N} + \varepsilon_i, \qquad i = 1, 2 \qquad (4)$$

with $|\varepsilon_i|$ as small as possible.

Table 1 gives the resulting 2-dimensional form for strings up to $N = 10$. One can see that strings with a length corresponding to a prime number are somewhat special as they do not allow any compactification in the 2-dimensional form.

The interaction between a 2-dimensional form of a string and a 1-dimensional form can be considered a reaction between the two strings. As an example, let us assume an operator $\mathcal{P}_{\vec{s}}$ was formed from string $\vec{s}$. This operator might now "react" with another string, $\vec{s}\,'$, producing thereby a new string $\vec{s}\,''$:

$$\mathcal{P}_{\vec{s}}\, \vec{s}\,' \Rightarrow \vec{s}\,'' \qquad (5)$$

The notion here is that some sort of raw material (analogous to energy-rich monomers in Nature) is continuously supplied to allow the ongoing production of new strings based on the information provided by the cooperation of $\mathcal{P}_{\vec{s}}$ and $\vec{s}\,'$.

A typical example of an interaction is given in Figure 2 for the simple case of strings of the same quadratic length N. $\vec{s}\,'$ might be considered as concatenated from $\sqrt{N}$ segments with length $\sqrt{N}$ each. The operator $\mathcal{P}_{\vec{s}}$ acts on each of these segments sequentially, and performs semilocal operations. In this way, it moves down the string in steps of size $\sqrt{N}$ until it has finally completed the production of a new string $\vec{s}\,''$.

The particular algorithm for assembling new components "0" and "1" into strings that we have examined in

---

[1] Compact foldings do not have any spacing between adjacent string elements

Length	Compact folded form
1	$(\, s_1 \,)$
2	$(\, s_1 \quad s_2 \,)$
3	$(\, s_1 \quad s_2 \quad s_3 \,)$
4	$\begin{pmatrix} s_1 & s_2 \\ s_4 & s_3 \end{pmatrix}$
5	$(\, s_1 \quad s_2 \quad s_3 \quad s_4 \quad s_5 \,)$
6	$\begin{pmatrix} s_1 & s_2 & s_3 \\ s_6 & s_5 & s_4 \end{pmatrix}$
7	$(\, s_1 \quad s_2 \quad s_3 \quad s_4 \quad s_5 \quad s_6 \quad s_7 \,)$
8	$\begin{pmatrix} s_1 & s_2 & s_3 & s_4 \\ s_8 & s_7 & s_6 & s_5 \end{pmatrix}$
9	$\begin{pmatrix} s_1 & s_2 & s_3 \\ s_6 & s_5 & s_4 \\ s_7 & s_8 & s_9 \end{pmatrix}$
10	$\begin{pmatrix} s_1 & s_2 & s_3 & s_4 & s_5 \\ s_{10} & s_9 & s_8 & s_7 & s_6 \end{pmatrix}$

Table 1: Compact topological string folding with length up to $N = 10$. Each folding comes also with the transposed matrix.

more detail, is:

$$s'_{i+k\sqrt{N}} = \sigma \left[\sum_{j=1}^{j=\sqrt{N}} P_{ij} s_{j+k\sqrt{N}} - \Theta \right] \qquad (6)$$

$$i = 1, ..., \sqrt{N} \qquad k = 0, ..., \sqrt{N} - 1$$

with $\sigma[\;]$ being the squashing function

$$\sigma[x] = \begin{cases} 1 & \text{for } x \geq 0 \\ 0 & \text{for } x < 0 \end{cases} \qquad (7)$$

and Θ used as an adjustable threshold. Eq. (6) may be interpreted as a combination of Boolean operations, applied separately in each segment k of the string if $\Theta = 1$.

The consistent generalization of eq. (6) for interaction of non-quadratic strings and for strings of different length is straightforward: Suppose a matrix of size $N_1 \times N_2$ is interacting with a string of length N_3. The operator locally interacts with N_1 elements of the second string in order to generate one component of the new string. This operation will be repeated N_2 times, then the operator moves on to interact with the next N_1 elements of the second string. The newly produced string will thus consist of N_4 elements with

$$N_4 = \lceil \frac{N_3}{N_1} \rceil \times N_2. \qquad (8)$$

where $\lceil x \rceil$ are Gaussian parentheses giving the next larger integer to x.

In mathematical terms, the interaction reads:

$$s''_{i+kN_2} = \sigma \left[\sum_{j=1}^{j=N_1} P_{ij} s'_{j+kN_1} - \Theta \right] \qquad (9)$$

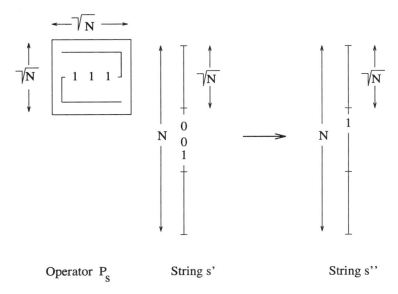

Operator $\mathcal{P}_s$ String s'   String s''

Figure 2: An operator $\mathcal{P}_{\vec{s}}$ of matrix dimension $\sqrt{N} \times \sqrt{N}$ (derived from string $\vec{s}$) acts upon a string $\vec{s}'$ consisting of $\sqrt{N}$ segments of length $\sqrt{N}$ each to produce a new string $\vec{s}''$.

with

$$i = 1, ..., N_2 \qquad k = 0, ..., \lceil \frac{N_3}{N_1} \rceil - 1.$$

This interaction is generally length-changing — either resulting in a shorter or a longer product strings. The particular direction of this length-change depends on the relation of N_1 to N_2: If $N_1 > N_2$ then $N_4 < N_3$, and the new string is shorter. If, however, $N_1 < N_2$ then $N_4 > N_3$, and the new string is longer.

The different types of possible reactions between strings are listed in Table 2. For a given N, say $N = 4$, the reactions form a reaction network, and we shall observe in the next Section the behavior of such a network.

Reactants	Product	Description
$s + s'$	s''	Heterogeneous reaction
$s + s'$	s	Replication
$s + s'$	s'	Replication
$s + s$	s'	Heterogeneous self-reaction
$s + s$	s	Self-replication

Table 2: Characterization of different polymerization reactions.

Dynamics of a sample system

Every reaction vessel is only able to keep a finite number of strings, say M. The reactions discussed in Section 2, however, continuously produce new strings. Therefore, a competitive dynamics has to be implemented by providing for an overflow mechanism for the reaction vessel. Since at present we would like a well-stirred reaction vessel without any spatial structure for reactions, the removal of strings will be a random process, hitting each sort of strings with a probability proportional to its concentration. For each newly produced string, one string is removed from the vessel. Whereas this random process does not influence in any way the constitution of the vessel content, due to different reaction channels producing new strings, a change in the composition of the content will happen over time.

There are, however, some potentially "lethal" strings in such systems. A string is said to be lethal if it is able to replicate in an unproportionally large number in almost any ensemble configuration. For eq. (6), this happens to be the case for two self-replicating string types [2], $s^{(0)} = (0, 0, ..., 0, 0)$ and $s^{(2^{N-1})} = (1, 1, ..., 1, 1)$. The former is able to replicate with every other string, the latter with most of the other strings.

In order to balance this tendency of the system we prohibit production of $s^{(0)}$ and discourage production of $s^{(2^{N-1})}$. In other words, $s^{(0)}$ will not be added to the vessel, if the reaction product should be $s^{(0)}$. Instead, a randomly selected string will be copied. We deal with $s^{(2^{N-1})}$ in a more gentle way by providing a means of non-deterministic string removal due to decay processes. The fewer the number of "1"'s a string contains, the more stable it becomes. The chance to decay therefore depends on the string feature

$$I^{(k)} = \sum_{i=1}^{N} s_i^{(k)}, \qquad k = 1, ..., M. \qquad (10)$$

---

[2] We shall name strings with decimal numbers corresponding to their binary sequence

$I^{(k)}$ measures the number of "1"'s in string k and determines a probability

$$p^{(k)} = (I^{(k)}/N)^n \qquad (11)$$

which determines whether a string should be removed. Usually, we set the parameter n to $n = 1$. In any case, the decay probability of $s^{(2^{N-1})}$ is 1. Once a string decays, its place might be filled
(i) with a later reaction product or
(ii) with a copy of a randomly selected string in the vessel. The latter method has the advantage of allowing a constant string number M in the vessel and is adopted here.

One sweep through the algorithms hence consists of the following steps:

STEP 1:

Generate M random binary strings of length N each

STEP 2:

Select a string and fold it into an operator by forming a compact matrix

STEP 3:

Select another string and apply the operator generated in STEP 2

STEP 4:

Release the new string, the old string and the operator (as string) into the reaction vessel, provided it is not an $s^{(0)}$. Otherwise go to STEP 2.

STEP 5:

Remove one randomly chosen string in order to compensate for the addition of a string in STEP 4

STEP 6:

Select one string and substitute it according to the probability of (11) with the copy of a randomly selected string

STEP 7:

Go to STEP 2

M sweeps through this algorithm are called a generation.

For a discussion of the system's dynamic behaviour we use as observables the concentrations $x_i(t)$ of all the different string types $s^{(i)}$ with:

$$x_i(t) = m_i(t)/M \qquad (12)$$

where $m_i(t)$ is the number of actual appearances of string type $s^{(i)}$ in the vessel at time t.

If we run a system by seeding it with an initial composition of M random strings, we regularly observe a transition into a (mostly fixed point) attractor. Due to different rates of production of different sorts, an initial composition will change until an equilibrium is reached. During the transition, new sorts are produced, already present sorts disappear, and every now and then a coexistence between sorts is reached for some time. As long as new sorts are created by interactions between already present sorts, the network has to reorganise itself in order to incorporate the newly emerging reaction channels between the different sorts. After some time, however, no new string sorts arrive, and the system reaches a steady state. Thus, the system behaves as one of the metabolic networks that are discussed in Bagley et. al. [5, 6]. As long as we have a small number of sorts, we can easily describe the system by a set of deterministic differential equations for the time development of string sort concentrations.

Deterministic rate equations were derived in [1] and are given here as a summary: We assume continuous non-random concentration functions $y_i(t)$ of the different string types $i, 1 \leq i \leq n_S$, which are considered to approximate the time averaged concentrations $<x_i>_t$:

$$y_i(t) \cong <x_i>_t, \qquad 0 \leq y_i(t) \leq 1 \qquad (13)$$

The deterministic rate equations in $y_i(t)$ read:

$$\dot{y}_i(t) = A(t)y_i(t) + \left[B_i y_i(t) + \sum_{k \neq i}^{n_S} C_{ik} y_k(t) - D_i \right] y_i(t) +$$

$$\sum_{j,k \neq i}^{n_S} W_{ijk} y_j(t) y_k(t) - \frac{y_i(t)}{\sum_k y_k(t)} \Phi(t) \qquad (14)$$

where B_i, C_{ik}, W_{ijk} are coupling constants derived from a reaction table containing all sorts $1...n_S$. D_i determines a selection term

$$D_i = p^{(i)} \qquad (15)$$

and $A(t)$ reflects the addition of strings due to random copies

$$A(t) = \sum_{i,j}^{n_S} a_{ij} y_i(t) y_j(t) + \sum_i^{n_S} D_i y_i(t) \qquad (16)$$

where

$$a_{ij} = \begin{cases} 1 & \text{if the reaction of } s^{(i)} \text{ and } s^{(j)} \text{ produces } s^{(0)} \\ 0 & \text{otherwise .} \end{cases}$$
$$(17)$$

Finally, $\Phi(t)$ is a flow term that enacts competition between the various string sorts $s^{(i)}$ by enforcing constancy of the overall sum of concentrations.

The reaction table listing the interactions between string types (cf. Table 3) can be used to derive interaction graphs for various situations. In Figure 3 we have depicted all interaction graphs that can be generated from Table 3 if we start the reaction vessel with one out of 2^{N-1} string types (here $N = 4$). Functionally identical graphs are not depicted. Figure 3 illustrates the variety of interactions emerging from a start with different string types. It ranges from self-replication over parasitic interaction to entire metabolisms. From an interaction graph it is evident, what kind of attractor may be approached.

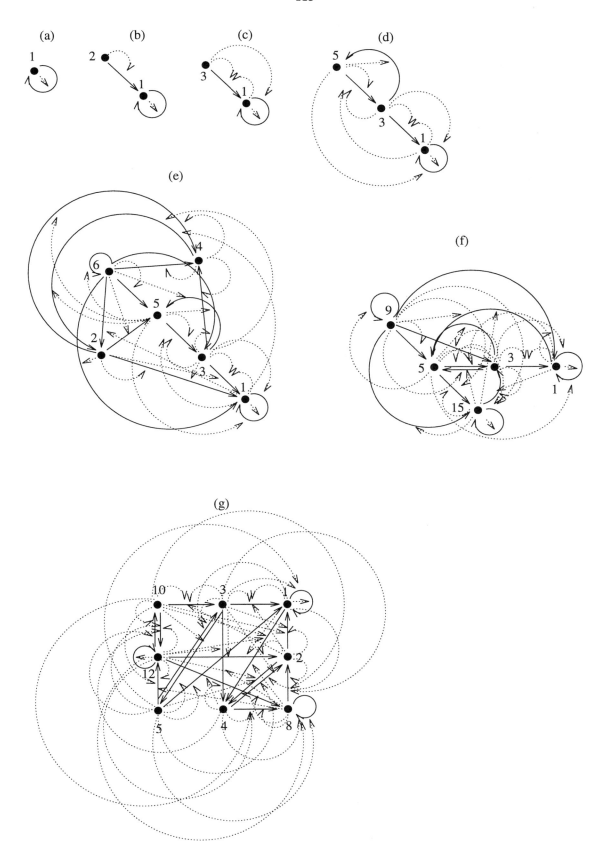

Figure 3: Interaction graphs of the system with $N = 4$. These graphs include all string sorts that are produced if the upper left string sort is used as the one and only initial sort in the reaction vessel. Solid lines connect the two string sorts participating in a reaction. Dashed lines indicate operator sort. (a) Self-replicator, also realized by sorts $s^{(7)}, s^{(12)}$ and $s^{(15)}$. (b) - (f): Simple and complicated parasitic interactions, (b) also realized by pairs $(s^{(11)}, s^{(15)}), (s^{(13)}, s^{(15)})$ and $(s^{(14)}, s^{(15)})$.

Operator	String														
	1	2	3	4	5	6	7	8	9	10	11	12	13	14	15
1	1	0	1	2	3	2	3	0	1	0	1	2	3	2	3
2	0	1	1	0	0	1	1	2	2	3	3	2	2	3	3
3	1	1	1	2	3	3	3	2	3	3	3	2	3	3	3
4	0	4	4	0	0	4	4	8	8	12	12	8	8	12	12
5	1	4	5	2	3	6	7	8	9	12	13	10	11	14	15
6	0	5	5	0	0	5	5	10	10	15	15	10	10	15	15
7	1	5	5	2	3	7	7	10	11	15	15	10	11	15	15
8	4	0	4	8	12	8	12	0	4	0	4	8	12	8	12
9	5	0	5	10	15	10	15	0	5	0	5	10	15	10	15
10	4	1	5	8	12	9	13	2	6	3	7	10	14	11	15
11	5	1	5	10	15	11	15	2	7	3	7	10	15	11	15
12	4	4	4	8	12	12	12	8	12	12	12	8	12	12	12
13	5	4	5	10	15	14	15	8	13	12	13	10	15	14	15
14	4	5	5	8	12	13	13	10	14	15	15	10	14	15	15
15	5	5	5	10	15	15	15	10	15	15	15	10	15	15	15

Table 3: Reactions table for the simulations of a $N = 4$ system. It was generated using a variant of (6) with topological folding.

The dynamics of the parasitic interactions of Figure 3 is examined by integrating eq. (14). Figure 4 - 6 show the results of a simulation. The transition of the string composition is clearly visible. In [1, 4] we have shown that simulations on the reaction level agree completely with the integration of rate equations used here.

A simple metabolism emerges if we do not start with one sort only, but with two or more from the outset. Figure 7 shows the interaction graph of this self-maintaining network of reactions. This graph is somewhat special as each reaction channel is of nearly equal strength. A search through the space of all combinations of 2 initial sorts uncovers that the self-replicator $s^{(12)}$ plays some special role. Usually, as soon as even a spurious concentration of $s^{(12)}$ is present, together with one other sort (except $s^{(1)}$), the metabolic attractor emerges. Figure 8 gives two examples.

It is interesting to note that there are many closed subsets of elements within even a simple $N = 4$ system. In Tables 4, 5 we give a complete list of them, ordered according to their complexity in terms of participating string sorts. Following [8], a closed subset is defined as the set $\mathcal{A}^*$ of elements from the ensemble of string types $\mathcal{N}_S = \{s^{(1)}, s^{(2)}, ..., s^{(2^{N-1})}\}$,

$$\mathcal{A}^* \subseteq \mathcal{N}_S \tag{18}$$

that might be produced by all different sequences of n reactions,

$$R_n(\mathcal{A}) = \cup_{i=0}^{n} r_n(\mathcal{A}) \tag{19}$$

starting from an initial set $\mathcal{A} \subseteq \mathcal{N}_S$, for $n \to \infty$, with

$$r_0(\mathcal{A}) = \mathcal{A} \tag{20}$$

$$r_n(\mathcal{A}) = \cup_{i=0}^{n-1} r_i(\mathcal{A}) \circ r_{n-i-1}(\mathcal{A}) : \tag{21}$$

$$\mathcal{A}^* = \lim_{n \to \infty} R_n(\mathcal{A}). \tag{22}$$

Closed subsets are important organisational structures, especially in the light of the fact, that we can only populate part of sequence space, once the component number strings increases.

We should keep in mind, that we have dealt here with a system consisting of 4-bit components. The complexity of interactions in such a simple system as Figure 3 demonstrates, is astonishing. We expect the two basic behavioral classes, parasitic interaction and metabolism, to emerge in a variety of forms in systems with longer strings.

Evolution

As we have seen, the dynamics in this small system quickly settles into one of its attractor states. The question, however, arises, whether there is a perspective for evolution, that is, for a sequential exploration of possibilities. For evolution to happen, an occasional mutation of one string into another should lead to a cascade of newly

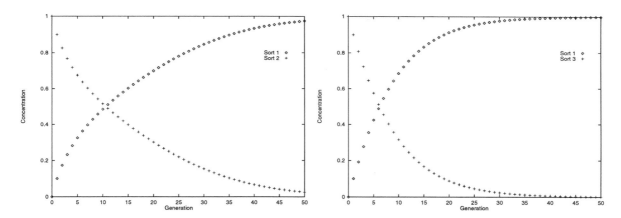

Figure 4: Dynamics of the interaction graph Figure 3(b) and (c).

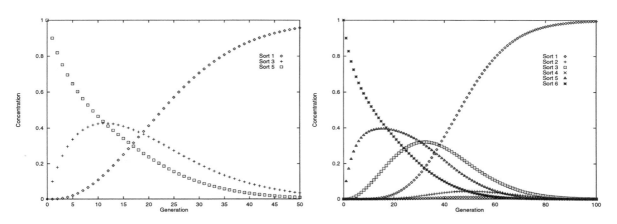

Figure 5: Dynamics of the interaction graph Figure 3(d) and (e).

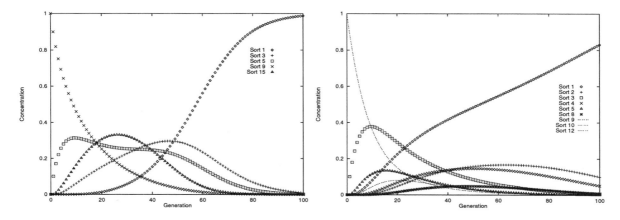

Figure 6: Dynamics of the interaction graph Figure 3(f) and (g).

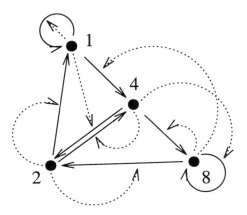

Figure 7: Reaction graph of the metabolism of $N = 4$.

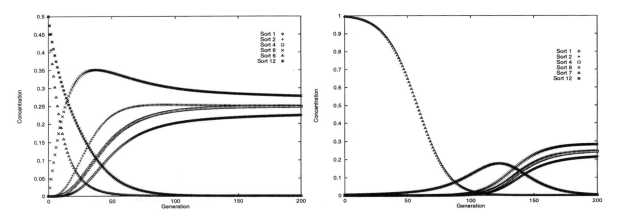

Figure 8: Dynamics of the interaction graph Figure 7. Left: Equal concentration of $s^{(6)}$ and $s^{(12)}$, at the outset; Right: High concentration of $s^{(7)}$, low concentration of $s^{(12)}$ at the outset.

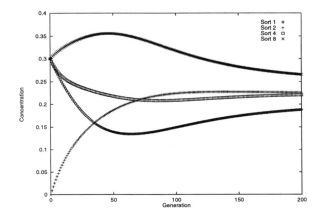

Figure 9: Dynamics of the interaction graph Figure 7. Starting with three different string sorts.

1	2	3	4	5	6	7	8
1	(1,2)	(1,2,3)	(1,2,3,4)	(1,2,3,4,5)	(1,2,3,4,5,6)	(1,2,3,4,5,6,7)	(1,2,3,4,5,8,10,12)
(4)	1,3	(1,2,4)	(1,2,4,8)	(1,2,3,4,8)	(1,2,3,4,5,7)	1,3,5,7,11,13,15	1,3,5,7,9,11,13,15
7	(4,8)	1,3,5	1,3,5,7	(1,2,4,8,12)	(1,2,3,4,8,12)		
(8)	7,15	(4,8,12)	1,3,5,15	1,3,5,7,15			
15	(8,12)	7,11,15	7,11,13,15	1,3,5,9,15			
	13,15	13,14,15		7,11,13,14,15			
	14,15						

Table 4: Closed subsets of elements with up to 8 members. First column: Self-replicators. In parenthesis: Subsets which occasionally produce the destructor.

9	11	13	15
(1,2,3,4,5,8,10,12,15)	(1,2,3,4,5,6,8,9,10,12,15)	(1,2,3,4,5,7,8,10,11,12,13,14,15)	(1,2,3,4,5,6,7,8,9,10,11,12,13,14,15)

Table 5: Closed subsets with more than 8 members. All subsets occasionally produce the destructor.

produced string types, that lead to a new equilibrium. We have shown this to happen in a $N = 9$ system [7], and will adopt the results learned there.

We have been using a mutation as a motor for occasional change. A mutation hits each string with a probability depending on its size. We define q to be the probability that one element of a randomly selected string changes to another symbol, here "0" to "1" and vice versa. Since each element may be hit, this is a length dependent change and the probability that at least one error occurs in a string is $Q(1) = Nq$, with the provision that $q << \frac{1}{N}$. Evidently, this mutation probability depends linearly on the concentration of string sorts in the reaction vessel. That is to say, a more successful string sort will spawn more variations. Two-bit mutations are then $Q(2) = (Nq)^2$ where we neglect the fact that sometimes back-mutation may happen. In Nature, at least on instance of this type of mutations occurs in mutations caused by cosmic radiation.

Mutation does open up new transformation pathways between string sort, something Bagley et al. term a stochastic metadynamics [6].

Suppose we start our system by sort $s^{(7)}$. Since this is a self-replicating string sort, nothing interesting will happen, unless the mutation process introduces one of its nearest neighbors $s^{(3)}, s^{(5)}, s^{(6)}, s^{(15)}$. The reaction table shows, that the appearance of $s^{(6)}$ will have no consequence, whereas the appearance of $s^{(3)}, s^{(5)}, s^{(15)}$ allows the system to switch to another attractor. Figure 10, left, shows the effect of introducing $s^{(5)}$. As a result, the interaction graph of Figure 3 (d) comes into play, and $s^{(1)}$ dominates. Figure 10,right, is the evolution from selfreplicator $s^{(15)}$ to the metabolism consisting of $s^{(1)}, s^{(2)}, s^{(4)}, s^{(8)}$. This has been achieved by

introducing $s^{(12)}$, a two-bit mutation from $s^{(15)}$, in spurious concentration.

Conclusion

We have examined a very simple self-organising system. The main idea was to introduce a second form of the information carriers of our system, the sequences of binary numbers. This has been accomplished by using an operative matrix form for the strings. We then have defined a particular interaction between matrices and strings and considered the interaction itself as some sort of a reaction with input and output. The low-level ("atomic") computations in the system have thus been likened to chemical reactions in the real world.

It has been shown that closed subsets of strings exist which can be considered as organisations. Under the assumption of one particular folding, these subsets of strings might be studied in their 2-dimensional matrix form alone, effectively yielding an interesting class of mathematical objects that are closed under the proposed non-linear interaction.

We also dealt with the dynamics of the competitive system naturally emerging, with reactions going on between different species of strings. As in other artificial systems [5, 6, 9, 10, 11] an attractor state was reached relatively quickly, beyond which nothing interesting happened any more. However, we already demonstrated powerful evolutionary effects brought about by the inclusion of a mutation or the potential of length changing interactions. Systems with longer strings will certainly possess different metabolic networks, and it is clear that the behavioral flexibility in such systems will be enormous.

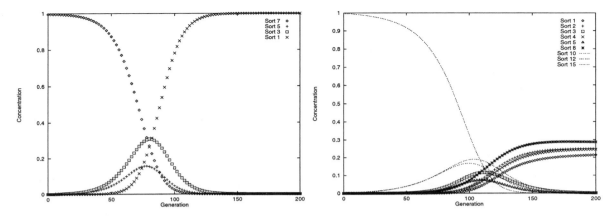

Figure 10: Evolutionary dynamics. Left: A 1-bit mutation causes $s^{(5)}$ to appear. This leads to $s^{(1)}$ as the dominant string sort after the transition. Right: A 2-bit mutation causes $s^{(12)}$ to appear. The result is the emergence of the metabolism of Figure 7.

Acknowledgement

I have enjoyed discussions with Dr. Walter Fontana. Mr. Helge Baier has provided me with the data of Table 4 and 5.

References

[1] W. Banzhaf, *Self-replicating sequences of binary numbers*, Computers and Mathematics, **26** (1993) 1

[2] M. Eigen, *Steps toward Life: a perspective on evolution*, Oxford University Press, 1992

[3] M. Eigen, P. Schuster, *The Hypercycle - A principle of natural self-organization, Part A-C*, Naturwissenschaften **64** (1977) 541 and **65** (1978) 7, 341

[4] W. Banzhaf, *Self-replicating sequences of binary numbers — Foundations I and II*, Biological Cybernetics, **69** (1993) 269 and 275

[5] R.J. Bagley, J.D. Farmer, *Spontaneous Emergence of a Metabolism*, in: C.G. Langton, C. Taylor, J.D.Farmer, S. Rasmussen (Eds.), *Artificial Life II.* Addison-Wesley, Redwood City, CA, 1991, 93

[6] R.J. Bagley, J.D. Farmer, W. Fontana, *Evolution of a Metabolism*, in: C.G. Langton, C. Taylor, J.D.Farmer, S. Rasmussen (Eds.), *Artificial Life II.* Addison-Wesley, Redwood City, CA, 1991, 141

[7] W. Banzhaf, *Self-replicating sequences of binary numbers — Foundations III*, to be published

[8] W. Fontana, L. Buss, *"The arrival of the Fittest": Toward a Theory of Biological Organization*, Bulletin of Mathematical Biology, submitted

[9] W. Fontana, *Algorithmic Chemistry*, in: C.G. Langton, C. Taylor, J.D.Farmer, S. Rasmussen (Eds.), *Artificial Life II.* Addison-Wesley, Redwood City, CA, 1991, 159

[10] J.D. Farmer, S.A. Kauffman, N.H. Packard, *Autocatalytic Replication of Polymers*, Physica **D22** (1986) 50

[11] S. Rasmussen, C. Knudsen, R. Feldberg, *Dynamics of Programmable Matter*, in: C.G. Langton, C. Taylor, J.D.Farmer, S. Rasmussen (Eds.), *Artificial Life II.* Addison-Wesley, Redwood City, CA, 1991, 211

Effects of Tree Size on Travelband Formation in Orang-Utans: Data Analysis Suggested by a Model Study

Irenaeus J. A. te Boekhorst[1]
Ethology and Socio-Ecology Group
Padualaan 14, 3508 TB Utrecht,
the Netherlands

Pauline Hogeweg
Bioinformatica
Padualaan 8, 3584 TB LA Utrecht
the Netherlands
ph@marvel.bio.ruu.nl

Abstract

To investigate sociality in orang-utans (Pongo pygmaeus), we designed an artificial (MIRROR-) world where individual entities (ORANGs) search for food (FIGs and FRUIT). ORANGs are steered by simple foraging rules that are activated only by the local information they obtain from the environment in which they dwell. Concerning the distribution of food sources and the population composition of ORANGs, this environment resembles the orang-utan habitat at Ketambe (Sumatera, Indonesia). In the MIRROR world, ORANGs are observed to form temporary aggregations in the huge but uncommon FIG trees. This is because the enormous crops of these trees permit each ORANG to feed until satiated and hence each one departs separately. However, when feeding in the much smaller FRUIT trees with others, ORANGs jointly deplete a resource before they are satiated and subsequently travel to another food source; since the next nearest food tree is the same for all co-feeding group-members, they do so together. Thus, travel bands are formed especially in periods of FRUIT abundance and emerge as a self-organizing property of basic foraging rules in interaction with the structure of the habitat. The ORANG-world was used as a "search image" in the analysis of field data collected in Ketambe. This analysis provided indications for a similar mechanism in the observed orang-utans population.

1 Introduction

Orang-utans are relatively solitary animals (Rodman, 1973), but occasionally they form parties (MacKinnon, 1974; Rijksen, 1978; Galdikas, 1978, 1979). These take two forms: first, as "travel bands" when orang-utans jointly leave a food tree and travel together for a certain amount of time (cf "social groups" in Rijksen, 1978 and Galdikas, 1978, 1979) and second, as "temporary aggregations". In the latter,

animals part after having fed together (Rijksen, 1978).

In the Ketambe area (Sumatera, Indonesia), orang-utans aggregate less in times of fruit scarcity (Sugardjito et al., 1987). Since orang-utans are mainly frugivorous, this correlation was interpreted as reflecting the costs of feeding competition and active mutual avoidance by the apes. Apart from sexual consorts (Schürmann, 1981, 1982; Schürmann & van Hooff, 1987), the benefits of grouping for orang-utans, however, remain unclear: because of their large body size and arboreal mode of life, protection against predators is an unlikely explanation.

Alternatively, parties could emerge as a side effect of behavior that operates independent of any advantage of grouping itself and in that case considerations about benefits of being in groups are void. Along these lines, Sugardjito et al. (1987) interpreted formation of parties in strangling fig trees. These huge trees are relatively uncommon, but produce enormous crops (Rijksen, 1978) throughout the year and are regularly revisited by orang-utans (Sugardjito et al., 1987). The fact that strangling fig trees are important "meeting places" of orang-utans might simply be caused by the apes' preference for figs (Sugardjito et al., 1987). In other words, a simple feeding rule steered by direct information obtained from the local environment (presence of fig trees in the neighbourhood), might be all that is needed for the formation of parties in fig trees. For travel bands and parties in the smaller fruiting (non-fig) trees, however, Sugardjito et al. (1987) were not able to give a similar explanation and therefore suggest that "in this case orang-utans actively join each others company and stay together for no other reason than the benefits of grouping itself." (pg. 36). The purpose of this study is to assess whether this is necessarily true.

The paradigm that simple behavioural rules operating on the level of individual entities may suffice to explain structures on a higher level of complexity (in this case parties) has been elaborated in the so-called "MIRROR modelling strategy" by Hogeweg & Hesper (1979, 1981, 1988, 1989). MIRROR modelling has been applied to study various socioinformatic (Hogeweg & Hesper, 1983, 1985) and etho-ecological processes (Hogeweg & Hesper, 1990). In this fashion, Hogeweg was able to construct an artificial "MIRROR world" inhabited by simulated CHIMPs whose community paralleled many aspects of that of real chimpanzees (Hogeweg & Hesper, 1991; te Boekhorst & Hogeweg, 1994).

---

1. Current affiliation: AI Lab, Department of Computing Science, University Zürich-Irchel,
Winterthürerstrasse 190, CH 8057 Zürich,Switzerland
boekhors@ifi.unizh.ch

By uncovering the self-structuring impact of simple rules, a MIRROR model helps to arrive at a minimal description of a (social) system, but at the same time generates insight into its dynamics. For example, analysis of the CHIMP world revealed that CHIMPs often travel together, especially when the habitat contained many small-sized food trees. This was due to the fact that a small tree is depleted by co-feeding CHIMPS before they are satiated. Consequently, they synchroneously leave a tree and because foraging CHIMPs always go to the nearest next food tree in their moving direction, a travel band originates. This clearly offers an alternative explanation to the notion that orang-utans travel together because of the advantages of grouping.

In this study, we investigate whether a similar mechanism as in CHIMPs indeed leads to the formation of travel bands in fruit trees by orang-utans. To this end, we study a MIRROR world of ORANGs and corollaries of the suggested mechanism are tested with field data from Ketambe. The derived hypotheses are:

1. Travel bands originate especially in (small) fruiting trees, but temporary aggregations most often in (large) fig trees.

2. The chance of orang-utans still being together in a next tree is larger when they previously fed together in a fruit tree than in a fig tree.

3. The formation of travel bands is associated with an increased frequency of visiting fruit trees and therefore takes place especially during a season of increased fruit availability. Temporary aggregations, in as far as they originate in fig trees, should however occur with about equal frequencies throughout the year.

4. Visiting more fruit trees means more travelling. As in CHIMPs, day range distance should therefore be positively correlated with being more often in travel bands.

2 Material and Methods

2.1 Fieldwork

The study site (4.5 km^2) is bordered by two rivers (Ketambe and Alas) and grades in the slopes of the Gunung Leuser mountains. The largest (=lower) part of the area lies about 350 above sea level and is relatively flat. The vegetation is a mixed tropical rainforest (Richards, 1952) composed of small-to-medium sized trees (up to about 30 m) and only a few emergents (mainly strangling fig trees, genus Ficus, Moraceae) (Rijksen, 1978; Sugardjito et al., 1987). Monthly counts of the number of fruiting (non-fig) trees along a standard trail by van Schaik (1986) display a clear seasonality with a peak between May and September and a fruit-scarce period from January up to and including March. The other months show an overlap in fruit availability with one or both seasons and form a transition period (te Boekhorst et al., 1990).

The orang-utan population of Ketambe consists of about ten residents (one adult male, three subadult males, three adult females with dependent offspring and about two to four adolescents of both sexes). Besides these, about an equal number of non-residents (with a comparable sex/age distribution) visit the area during the fruiting season (te Boekhorst et al., 1990).

The orang-utan data stem from so-called focal animal samples (Altmann, 1974) and have been collected by Schürmann (from 1975 to 1980, see Schürmann, 1982) and Sugardjito (from 1980 to 1983; te Boekhorst joined him in 1982). A focal animal sample consists of one or more consecutive days during which a target animal was followed from dawn to dusk. A continuous written protocol of the focal animals' activities was kept and a map of his travel route was made. If possible, food trees visited by the focal animal were identified at species level or otherwise their vernacular name was noted. The identity of other orang-utans with whom a focal animal was seen in a party was recorded. A party was defined as an assemblage of individuals within eyesight of the observer (about 30 m.). In addition, Sugardjito noted if a target animal left a food (fig or fruit) tree with others and stayed with them for at least half an hour. If so, these associations are called travel bands and are distinguished from temporary aggregations, in which animals parted after having fed together (cf. Rijksen 1978).

From the data we calculated day range distances and scored the identity of the food source (fig, fruit, leaves, insects) in which a focal animal encountered other orang-utans and whether or not the target was still in (partly) the same company the next tree it was recorded in. These data allowed us to draw up so-called "food tree chains" (= uninterrupted series of visits to subsequent food trees by a party). Following a Markovian approach, we estimated from the chains the probability of orang-utans still being together in the next food tree, depending on the type of food exploited in the previous tree. Transition rates between food tree types and fission probabilities are summarized in diagrams, whose construction is explained in Appendix 1.

In total, 217 focal animal days from 8 target animals are analysed in this paper.

2.2 Processing of the field data

Unless mentioned otherwise, each focal animal sample was considered as one data point; daily visiting frequency of food trees (figs and fruits) and day range distance were averaged over the number of days of a focal sample. Formation of parties was expressed as the percentage of days of a focal animal sample a target was seen at least once in a party of a given type.

Tests for (Kendall and Spearman rank) correlations between the above mentioned variables were carried out for each focal animal separately. Only when the correlation coefficients of all the animals tested were in the same direction, their probabilities were combined by means of Fisher's Combination Test (FCT, see Sokal & Rohlf, 1981). Animals whose correlation coefficients equaled zero were excluded from this test. Significance level is set at $\alpha = 0.05$ two tailed, but because of the small sample size we consider cases in which $p < 0.10$ or all correlations are of the same sign

- even though their combined probability exceeds α - as trends worth of mentioning.

2.3 The Model

In the model, we created an artificial habitat that resembles Ketambe in size (4 km²), number of ripe orang-utan food trees (FIGs and FRUITs) and composition of the orang-utan (ORANG) population in two seasons (FRUITing and NONFRUITing), which differ in food availability. Based on figures from Fernhout (in Rijksen, 1978) and van Schaik (unpubl.), we have set the maximum number of simultaneously fruiting non-fig (=FRUIT) trees at 2000 for the FRUIT season and at 1000 for the NOFRUIT season. For both seasons, the maximum number of ripe FIG trees was put at 50. FRUIT trees were dispersed at random, but FIG trees were given a degree of clumping (cf. Rijksen, 1978; van Schaik & Mirmanto, 1985): half of the FIG trees grow in one quarter of the area.

The MIRROR world is inhabited by eight resident ORANGs (one adult male, two adult females, two young males and two young females) and a similar number of visiting ORANGs (of the same sex/age composition). Visitors occur only in the FRUIT season.

The movements of ORANGs are dictated by a few foraging decisions and only two "social" rules (Figure 1; parameter values are presented in table 3 and more details are given in Appendix 2.)

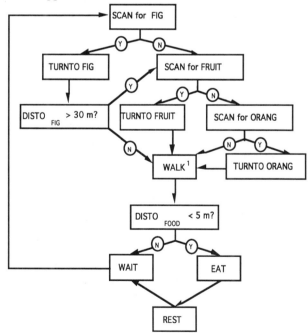

Figure 1. Flow chart of the ORANG algorithm. [1] If males encounter a higher ranking male, they flee in opposite direction. For further explanation, see text.

The first foraging rule instructs an ORANG to scan the environment for a large food (=FIG) tree. Although an experienced (= adult resident) ORANG "knows" to find such

a tree even when it is out of sight (but not further than 100 meter away from the ORANG), it only goes straightforward to a FIG tree when one is well within range of vision (< 30 m). Otherwise, the ORANG moves "slowly" in the direction of the distant FIG tree. With "slow", we mean that after every step the ORANG looks for other food (FRUIT) trees (which can be spotted from 30 m.) in the direction of the distant FIG tree and eventually enters them. Upon entering a food source, an ORANG feeds in it until satiated or the tree is emptied. Next, the ORANG leaves and rests in an adjacent tree after which the procedure starts again with the first foraging rule. When neither FIG nor FRUIT trees are noted, the first "social" rule becomes active and the ORANG checks if a congener is in the vicinity. A congener is noticed from a considerable distance (100 m) (this is a reasonable assumption, because orang-utans make their presence known by their noisy movements such as branch breaking etc.), after which the ORANG moves "slowly" (as defined above) towards it. If the approach brings it within 5 meter of any food tree, the ORANG enters it and starts eating. The idea behind this is that the presence of others might indicate the presence of a food tree. In case neither food trees nor other ORANGs are found, the ORANG moves on searching for food (by the first foraging rule).

Because of the observed hostility among adult orang-utan males (Galdikas, 1985; Mitani, 1985, 1989), a second social rule was specified: any male avoids the adult resident (=dominant) male and subadult males move away from any adult male.

Grouping by ORANGs is studied in principle in the same fashion and by the same variables as in orang-utans (see table 1). In the MIRROR-world, an encounter between ORANGS is scored when they come within 10 meter of each other. If they are separated by a distance of 10 meter or more for longer than 20 minutes, they are considered to have parted unless a new encounter between them occurs within this period (in which case the 20 minute timer is reset). If the tree of departure is not the same as the one in which the ORANGs met, they have travelled together to the next tree and a travel band is recorded. Otherwise the party is scored as a temporary aggregation.

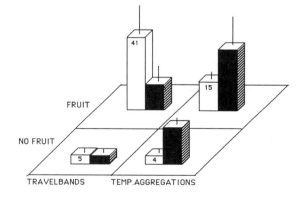

Figure 2. Mean number of times ORANGs occurred in parties in two seasons. Data from eight targets, each followed for 33 days. Black bars: parties formed in FIG trees; white bars: parties in FRUIT trees.

	Ketambe						Model*			
	fruit	N	no fruit	N	T	p	FRUIT	NO FRUIT	T	p
Daily Visiting Frequency of										
FIG/fig trees	1.85 ± 0.29	8	1.87 ± 0.25	7	6	NS	2.52 ± 0.12	2.74 ± 0.17	13	NS
FRUIT/fruit trees	8.14 ± 1.35	8	3.23 ± 0.69	7	0	0.03	6.44 ± 0.56	3.67 ± 0.55	0	0.007
Day Range Distance	953 ± 88.49	8	858 ± 64.18	7	7	NS	933 ± 56.52	985 ± 101.93	11	NS
% Solitary Days	14.61 ± 4.60	8	30.18 ± 10.88	7	6	NS	26.75 ± 7.11	47.25 ± 2.27	1	0.03
Party Size	2.13 ± 0.24	7	1.50 ± 0.13	6	0	0.03	1.84 ± 0.13	1.21 ± 0.03	0	0.02

* all based on 8 focal animals

Table 1. Comparison of basic variables between field data (Ketambe) and model output.

To investigate the origin of grouping in the ORANG world, we counted for each season how often travel bands and temporary aggregations emerged in FRUIT- and FIG trees. The frequencies of the various categories of parties, averaged over the total scores of eight individual ORANGs, are presented in figure 2. As can be seen from this figure, parties occur mostly in the FRUIT period. Travel bands are formed especially in this season in FRUIT trees. In contrast, temporary aggregations are formed mostly in FIG trees. The occurrence of temporary aggregations in FIG trees is less seasonally biased than the other categories.

3 Results

3.1 Comparisons between MIRROR-world and Ketambe

Results from the MIRROR world agree fairly well with field data from Ketambe for a number of basic variables (Table 1): orang-utans and ORANGs range about 900 meter a day and more often visit fruit/FRUIT trees than fig/FIG trees, especially in the fruiting season. Fig/FIG trees are visited about two - three times a day and this frequency is not much affected by seasonality. Being in (any type of) party, however, is: both in the ORANG world and in Ketambe, party size is significantly larger in the fruiting season than in the fruit-scarce period. Accordingly, the percentage of days a target animal is solitairy is higher for the fruit-scarce period than for the fruit-rich period in both data sets (although the difference is not significant for Ketambe) (See also Fig. 5).

3.2 Tests of Model Corollaries by Field Data

Travelbands originate especially in fruiting trees

As in the MIRROR-world, the formation of parties in Ketambe is associated with food tree type: temporary aggregations originate more often in fig trees and travel bands more frequently in fruit trees than expected by chance alone ($\chi^2 = 18.86$, df = 1, p < 0.005). In figure 3, this association is depicted in relation to party size. Temporary aggregations are on average larger than travel bands. For each size class,

except the largest travel bands, the proportion of parties formed in fig trees is higher for temporary aggregations than for travel bands.

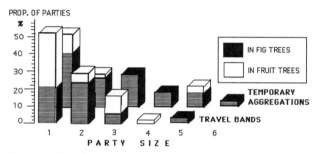

Figure 3. Size distributions of temporary aggregations (mean size: 2.34, N = 38) and travel bands (mean size: 1.74, N = 63).

The chance of staying together is higher after feeding fruits than figs

To investigate what happens after a focal animal had been feeding together with others in the same food tree, we casted Schürmann's data on food tree chains into diagrams as explained in Appendix 1. These diagrams show how being together in a given type of food tree influences the probability for a party to be in a next resource. We made diagrams for three periods (fruiting season, non-fruiting season and a transition period between these two seasons) by averaging the diagrams of individual focal animals (fig.4). In all three periods, the fission rate after a party had been in a fig tree is larger than when in a fruit tree. Conversely, the combined transition rates to other food types is higher for fruit trees than for fig trees. This was verified by analysing a 2x2 contingency table of "food type of previous tree" (fig / fruit) versus "next state" (alone / still in party in next tree). Apart from one adolescent male target, contingency tables of all (six) focal animals showed the same trend and the combined results of the χ^2 tests was significant (FCT: $\chi^2 = 22.11$, df = 12, 0.025 < p< 0.05). Another consistent feature is the high return rate in fruit trees, meaning that once in a fruit tree the next tree visited by a party is often again a fruit tree. Further, note the

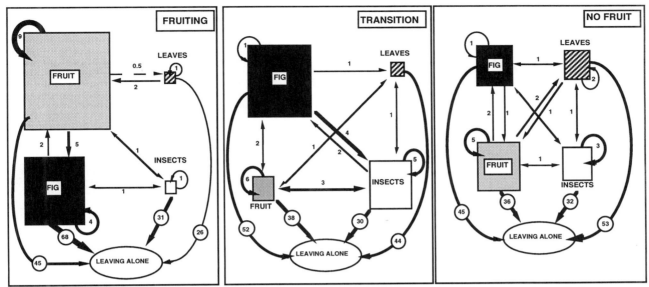

Figure 4. Transition rates and fission probabilities of parties in various types of food trees. For explanation, see Appendix 1.

high fission rate after a party had been feeding on leaves (in transition and no-fruiting season) and the outspoken seasonal differences: the diagram for the fruiting season is dominated by parties in fig and fruit trees, but the other two periods show a more diverse picture due to the increased importance of insects and leaves.

Formation of travelbands is associated with visiting more fruiting trees

During the transition period between the two seasons, some extraordinary large fig trees were revisited many times. Correspondingly, the presence of parties in fig trees was very high , while at the same time the exploitation of fruit trees was low in that period (Fig.4) (in general, orang-utans visit fig trees at the expense of fruit trees: correlation between average daily frequency of visiting fig- and fruit trees, FCT: χ^2 = 18.78, df = 8, p < 0.005).

As expected, we found a general trend for orang-utans to be

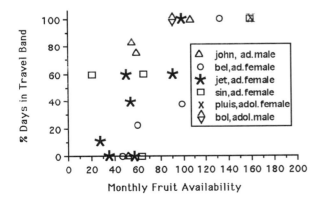

Figure 5. Correlation between fruit availabilty and presence in travel bands. FCT on Kendall Rank Correlation Tests for each of the focal animals: χ^2 = 15,42; df = 8; 0.05 < p < 0.10.

more often in travel bands (formed in fruit trees) when visiting more fruit trees (FCT: χ^2 = 14.41, df = 8, 0.05 < p < 0.10); the low exploitation rate of fruit trees in the transition period may thus have caused the striking reduction of parties in fruit trees in that time of the year (fig.4).

Because fruit trees were especially visited in the fruiting season (table 1), the association between being in party and visiting fruit trees implies that - just as in the MIRROR-world - parties should be formed more often when more fruit is available. This appears to be especially true for travel bands (fig.5), but does not hold for temporary aggregations (the correlations are even negative for all animals tested, but the combined probability is not significant. FCT: χ^2 = 9.69, df = 6, 0.10 < p < 0.25).

Being more often in travelbands increases day range distance

Not surprisingly, orang-utans travel further when they visit more fruit trees. This is evident from correlations between day range distance and visiting frequency of fruit trees (FCT: χ^2 = 20.52, df = 8, p < 0.005). Consequently, correlations between day range distance and the formation of travel bands in fruit trees are positive for all focal animals tested (although their combined probability is not significant, FCT: χ^2 = 10.77, df = 6, 0.10 < p < 0.25).

To tease out the mutual effects of visiting fruit trees, being in a group, day ranging and fruit availability, we performed the following series of tests. First, the effect of group size was eliminated by establishing the correlation between daily frequency of visiting fruit trees and day range distance for solitary days only. These correlations are positive and significant for all focal animals. Second, for solitary days the relationship between visiting fruit trees and ranging is measured for the separate seasons (Fig. 6) to control for the impact of fruit availability. As can be seen from figure 6, also when orang-utans are alone fruit trees are visited much more often in the fruiting season than in the other two periods. Day

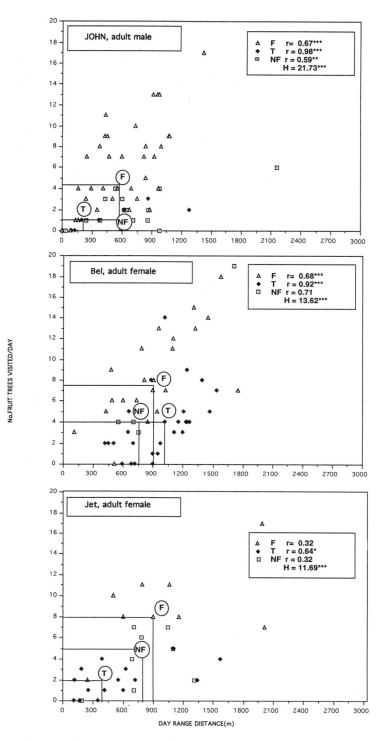

Figure 6. Relationship between day range distance and daily visiting frequency of fruit trees. Data come from the solitary days of three focal animals. Data and Spearman Rank correlation coefficients are presented for three seasons: F = fruiting season; T = transition period; NF = non-fruiting season. H is the test statistic for differences in visiting frequency among seasons. Lines connect the medians of visiting frequency with those of day range distance. Kruskal-Wallis tests for differences in day range distance among seasons are not significant.
* 0.025 < p < 0.05; ** 0.01 < p < 0.025; *** p < 0.001.

A. DATA

Individual (I)	focal sample [f(I)]	Social Condition (S) Solitary	In Group
JOHN, adult male	1	97	521
	2	288	100
BEL, adult female	3	771	970
	4	650	625
	5	700	900
JET, adult female	6	698	221
	7	293	455
	8	750	798
	9	328	340
	10	375	678
SIN, adult female	11	750	828
	12	125	400

B. ANOVA Table

Source	SS	df	MS	F	p
Between Individuals					
I	667108.48	3	222369.49	4.54	0.09
Error = f(I)	588828.98	8	73603.62		
Within Individuals					
S	124043.84	1	124043.84	4.63	0.06
S*I	68174.28	3	22724.46	0.88	0.51
Error = S*f(I)	214211.18	8	26776.4		

Table 2. Average day range distance for days on which the target animal was seen solitary or in a group. Comparisons are made within the same sample.

range distance, however, does not differ between the periods. Finally, the influence of being in groups was assessed. Hereto, we compared mean visiting frequency and day range distance between solitary and non-solitary days within the same focal animal study (Table 2). Possible individual bias was accounted for by applying a two factor (Individuals, Social Conditions) ANOVA with focal samples nested within Individuals. This is in effect a two-way design with repeated measurements on one factor (Cohen & Holiday, 1982; Damon & Harvey, 1983) and was analysed as such with the GLM procedure in SAS.

Day range distances tended to be larger on days when the target was seen "in group" than when it was alone ($F_{[1,8]}$ = 4.63, p = 0.06 in table 2 B). This was found for being in travel bands (ANOVA for the comparison between solitary days and travel band days: $F_{[1,4]}$ = 5.41, p = 0.08) rather than for being in temporary aggregations ($F_{[1,6]}$ = 2.73, p = 0.15). Accordingly, on days when focal animals where in travel bands, they tended to visit more often fruiting trees than on solitary days ($F_{[1,4]}$ = 5.77, p = 0.07), but this was not true for being in temporary aggregations ($F_{[1,6]}$ = 0.24, p = 0.64).

4 Discussion

4.1 Costs versus Environmental Structure as Determinants of Sociality

Two of our findings have been interpreted by others as reflecting costs of sociality in orang-utans. First, when fruit is scarce, orang-utans are more often alone than otherwise (see also Sugardjito et al, 1987). Second, when in groups,

orang-utans travel further than when alone (see also Galdikas, 1978, 1979, 1979); presumably, food trees are emptied sooner if more than one orang-utan exploits the same tree and the animals therefore have to visit more resources to meet their nutritional demands (Galdikas, 1978, 1979; Rodman, 1984; Wrangham, 1986 a,b). But if these costs are the reason for the solitariness of orang-utans, as maintained by the above mentioned authors, why then should they aggregate and travel together at all ? Note that in the MIRROR-world, where cost-benefit considerations do not apply, such a dilemma does not exist, because the very mechanism that forces ORANGs to go to the next tree is the same one that keeps them together. In the MIRROR world this is accomplished by a rule that directs still hungry ORANGs, after having jointly emptied a FRUIT tree, to the next nearest resource. In the NOFRUIT season this rule is infrequently activated because of the reduced number of FRUIT trees: the resulting low degree of sociality in food scarce periods is largely due to reduced environmental structure. In other words, with respect to staying together or being alone neither active maintenance of party cohesion nor active avoidance of feeding competition but the fabric of the forest is essential. Also, the formation of travel bands by ORANGs when FRUIT is abundant has nothing to do with relaxed competition for food allowing them to do so (as has been suggested for orang-utans by Sugardjito et al., 1987).

4.2 Sufficiency of TODO rules

Another important feature of this model (and of MIRROR models in general) is the simple "TODO" (=do what there is to do) form of the rules: the behaviour of an individual is triggered by nothing more than its direct experience and local information of its environment (Hogeweg & Hesper 1985).

Is it meaningful to postulate such simple constructs for "real" orang-utans ? One might object that the model assumptions are too simplistic and the rules are bad, because they are not "optimal" and too non-commitant.

First of all, we want to stress that ORANGs are not orang-utans. Some properties of ORANGs are at most hypothetical for the real apes and the MIRROR world they inhabit is not meant to resemble a particular world closely. Instead, a MIRROR world is a paradigm system rather than a (simulation) model sensu stricto and created to study which complicated interaction patterns emerge from certain uncomplicated precepts.

Indeed, ORANGs are extremely simple entities, whereas orang-utans possess highly developed information processing abilities (see for example Lethmate, 1977). It is not our aim to argue against these differences, but for a species to be blessed with impressive cognitive capacities does not imply that sophisticated decisions permeate all aspects of its social life. Also, we do not claim a complete lack of socialness in orang-utans. Young orang-utans are very playful and often team up with one another. Furthermore, orang-utans form consort pairs that can stay together for weeks and attract other individuals (Rijksen, 1978). These are certainly important additional mechanisms for grouping, but they cannot explain why travel bands form mainly in small trees during a fruiting season (actually, most consorts in

Ketambe seem to occur during transition or no-fruiting periods).

The foraging decisions of ORANGs obviously do not take into account that it may benefit them to avoid competition by feeding alone. Instead of pondering about how good it might be for an animal if it should do something else than observed, we feel it is more fruitful to stick with the empirical fact that orang-utans after all do travel together. Furthermore, our model (as MIRROR models in general) is not concerned with effects on fitness, but with the self-structuring properties of behavioural rules, which themselves may or may not be the result of natural selection. Of course, one may argue that implicit costs and benefits are included in the model: if orang-utans do not search for food they die and ORANGs that react on the noise of others are better able to obtain food than those who don't. However, we consider these effects of such a basic nature that labeling them explicitly as costs does not deeply increase our insight in the behavior of the ORANGs. Indeed, showing that the "costs" are of such a trivial kind that one can safely ignore the term was one of the starting points of this study.

How can one be sure that the chosen model and not any other set of rules applies to real orang-utans ? This is a general problem shared by all model systems and not just of MIRROR models.

Although it is impossible to prove that the same process taking place in ORANGs also occurs in orang-utans, our inability to falsify a number of corollaries lend support to the existence of a comparable mechanism. The probability that a party endures and travels to a next tree is higher when its members feed together in a quickly depleted food source (fruiting tree) than in one that provides enough items to satisfy several individuals (figs, leaves). Therefore, temporary aggregations originate mostly in fig trees, whereas formation of travel bands particularly takes place in fruiting trees and is associated with an increased visiting frequency of these much smaller resources. This also explains why travel bands emerge especially during the fruiting season but temporary aggregations not. Another check of the model would be to see whether the composition of parties (in terms of the sex/age classes of its members) agrees with that of Ketambe, but with this topic we will deal in a later paper.

In the MIRROR-world, the enlarged population in the FRUIT season obviously enhances encounter rate and travel band formation by chance. Because ORANGs are speed up when in travel bands, this encounter rate - and hence the formation of new travel bands - is even further increased. It is this positive feedback, brought about by extremely simple mechanisms, that is responsible for the structure observed in the MIRROR world. We think it also is in the real world: in Ketambe, orang-utans visit more fruiting trees as a direct response to the increased supply of fruit, but also as an effect of grouping. Our interpretation of the dynamics of travel band formation in orang-utans is summarized in figure 7. It is based both on the dynamics in the ORANG world and the statistical analysis of real data. The hypotheses underlying the latter came from the interaction between model building and field work.

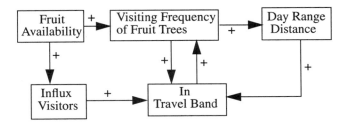

Figure 7. Proposed mechanism for travel band formation in orang-utans.

As a final remark, we stress that although the basic idea behind the ORANG model has a general purport, the results described in this paper apply solely to the Ketambe situation. Because the arboreal mode of life of this large ape requires a continuous intake of high-calory diet, even small deviations in the density and distributions of fruit might bring about drastic differences in social structure between habitats.

Appendix 1

Transition Rates and Fission Probabilities

As an example, the data of a subadult male orang-utan gathered during a non-fruiting season are worked out in figure 8. First, the "food tree chains" are summarized in a transition matrix. Here, rows represent the types of initial food trees (in which a party was formed and a chain started) and columns those of the next tree a party was seen in. Cells contain the transition frequencies (f_{ij}) that denote how often tree type j was visited next after tree type i by a party. Counts in the column END_i signify how often tree type i was the last tree of a chain (i.e. the focal animal was alone in the next tree). Row totals (R_i) are the total number of times a party visited food source i, column totals (C_j) give the total frequency tree type j was a "next tree". Substracting column totals from the corresponding row totals yields the number of times food tree type i was the starting point of a chain. Dividing f_{ij} by the corresponding row total R_i gives an estimate of the transition rate a_{ij}, which is defined as

$$\alpha_{ij} = Pr\{transition\ from\ tree\ type\ i\ to\ j \mid party\ is\ in\ i\}$$

To facilitate interpretation, all values are scaled to the total number of transitions ($\Sigma\ R_i + \Sigma\ C_j$), multiplied by 100 and brought together in diagrams such as the one illustrated in figure 8. In these diagrams, food tree types are ordered anti clockwise according their importance as starting points. The size of the squares representing the food types is proportional to the relative frequency with which they were visited by a party ($R_i/\Sigma R_i$, lower left rectangles in the squares). Width and direction of the arrows symbolize transition rates. The arrows descending from a food type and pointing to "leaving alone" stand for the splitting up of a party after entering that food

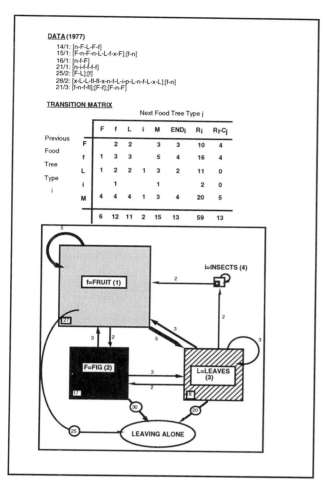

DATA (1977)
14/1: [n-F-L-F-f]
15/1: [F-n-F-n-L-L-f-x-F];[f-n]
16/1: [n-f-F]
21/1: [n-i-f-f-f-f]
25/2: [F-L];[f]
28/2: [x-L-L-fl-fl-x-n-f-L-i-p-L-n-f-L-x-L];[f-n]
21/3: [f-n-f-fl];[F-f];[F-n-F]

TRANSITION MATRIX

Next Food Tree Type j

Previous Food Tree Type i	F	f	L	i	M	END$_i$	R$_i$	R$_f$C$_j$
F		2	2		3	3	10	4
f	1	3	3		5	4	16	4
L	1	2	2	1	3	2	11	0
i		1			1		2	0
M	4	4	4	1	3	4	20	5
	6	12	11	2	15	13	59	13

Figure 8. Food tree chains and derived representation of transition rates and fission probabilities. Between brackets: uninterrupted series of visits to subsequent resources by a party; n = tree in which target made a nest; F = figs; f = fruits; L = leaves; i = insects. P(hloeem), fl(owers) and trees in which a party only went through (X) are taken together as M(iscellaneous) in the transition matrix and are not included in the diagram. For further explanation, see text.

type; here, the encircled numbers of the arrows express the fission rate relative to the visiting frequency of the corresponding food type (i.e. as $100*(END_i/R_i)$).

During his field work, Schürmann marked about 3500 food trees in which orang-utans were observed feeding with coded, metal tags. Unfortunately, Schürmann did not record non-tagged trees in which the target did not feed, leading to an underestimation of the length of the food tree chains. However, since we have no reason to suspect that such non-tagged trees occurred more around some food tree types than around others, this cannot have biased our principal aim (the comparison of transition - and fission rates between food tree types).

Appendix 2

Additional Information on the ORANG Model

Both FIG- and FRUIT trees are productive for on average seven consecutive days. When emptied by ORANGs within this period, FIG trees regenerate FIGs the next day. An empty FIG tree is substituted by a productive one elsewhere. For FRUIT trees this replacement is carried out only in case they are depleted by ORANGs. This operation saves the trouble of generating non-used FRUIT trees every time and ensures that the number of trees is not too strongly reduced by the feeding activities of the ORANGs. Also, it prevents ORANGs to stay in the same area for unnatural lengths of time.

ORANGs are restrained from disproportionate feeding on FRUITs by rejecting too small FRUIT trees (< 0.5) in favor of larger ones nearby (size is expressed as the number of hours a single ORANG needs to empty a tree). Excessive returning to a FIG tree is anticipated by letting an ORANG scan for FRUIT trees instead of FIG trees if it became satiated with FIGs (that is after two hours of continuous feeding) during its last meal. However, a certain degree of revisiting FIG trees is realistic and promoted by their large size (see table 3), the regeneration of FIGs on the next day and the ORANG's preference for FIGs. The latter is implemented in the model by the high priority of scanning for FIGs and a rule that tells an ORANG to turn in the direction of the last visited FIG tree if six hours later still no new FIG tree is encountered. We like to emphasize that these adaptations of the original model reflect the field impressions of the first author.

A distinction is made between "experienced" (= adult residents) and "unexperienced" (visitors and young residents) ORANGs. The latter category do not "remember" the location of distant FIG trees, but can only spot them within 60 m. Furthermore, young ORANGs travel faster and rest less than adult ones (young orang-utans, probably because they are less hampered by bodyweight, appear to range further per day than adult ones. Data not shown). In the model this is realized by multiplying walking speed and resting time by an age factor. The values of this factor and other parameters are summarized in table 3.

PARAMETER			VALUE
Angle of Vision			
	Scanning for:	FIG trees/other ORANGs	150 degrees
		FRUIT trees	60 degrees
Travelling			
	STEPSIZE		10-14 m
	AGEFACTOR		
		Adult ORANGs	1
		Young ORANGs	0.6
	WAIT		0.03-0.05 hr
	SPEED:= (STEPSIZE*AGEFACTOR)/WAIT		
		Adult ORANGs	0.20-0.47 km/hr
		Young ORANGs	0.33-0.78 km/hr
Feeding/Food tree size			
	Size FIG trees (*)		4.0-10.0 hr
	Size FRUIT trees (*)		0.2-0.8 hr
Resting			
	Duration resting bout at time t:		
	REST(t):=AGEFACTOR*EAT(t-1)*0.3		

* expressed as ORANG hours feeding

Table 3. Model parameters and their values.

Acknowledgments

This work formed part of I.J.A.t.B.'s Ph.D. thesis, supervised by Jan van Hooff who is acknowledged for his continuous support. The first author was financially supported by the Dobberke Foundation and the Lucy Burgers Foundation and receives a grant from the Swiss National Science Foundation. He thanks Lottie Hemelrijk for her encouraging discussions. Many thanks also to Chris Schürmann and Jito Sugardjito for data, discussion and companionship. Last but not least, I.J.A is grateful to the guys from the AI Lab for the team spirit, especially Dimitrios Lambrinos who helped me a lot with the preparation of this paper.

References

Altmann, J. 1974. Observational study of behaviour: samplingmethods. *Behaviour* 49: 227-267.

Boekhorst, I. J. A. te., C. L Schürmann, and J. Sugardjito. 1990. Residential status and seasonal movements of wild orang-utans in the Gunung Leuser Reserve (Sumatera, Indonesia). *Anim.Behav.* 39: 1098-1109.

Boekhorst, I. J. A. te and P. Hogeweg. Selfstructuring in artificial "CHIMPS" offers new hypotheses for male grouping in chimpanzees. *Behaviour* (in press).

Cohen, L. and M.Holliday. 1982. *Statistics for Social Scientists*. New York: Harper & Row.

Damon, R. A. and W. R. Harvey. 1987. *Experimental Design, ANOVA and Regression*. New York: Harper & Row.

Galdikas, B. F. 1978. Orangutan adaptation at Tanjung Puting Reserve, Central Borneo. Ph. D. Thesis, University of California, Los Angeles.

Galdikas, B. F. 1979. Orangutan adaptation at Tanjung Puting Reserve: mating and ecology. In *The Great Apes*, edited by D. A. Hamburg and E. R. McCown. Menlo Park, California: Benjamin/Cummings.

Galdikas, B. F. 1985. Adult male sociality and reproductive tactics among orangutans at Tanjung Puting. *Folia Primatol.* 45: 9-24.

Goodall, J. 1986. *The Chimpansees of Gombe*. Cambridge: Belknap.

Hogeweg, P. 1988. MIRROR beyond MIRROR, puddles of life. In *Artificial Life I* edited by C. G. Langton. SFI Studies in the Sciences of Complexity, Proceedings Vol. VI. Redwood City, CA: Addison-Wesley.

Hogeweg, P. 1989. Simplicity and complexity in MIRROR universes. *Biosystems* 23: 231-246.

Hogeweg, P. and B. Hesper. 1979. Heterarchical self structuring systems: concepts and applications in biology. In *Methodologyof System Modelling and Simulation* edited by B.P. Zeigler etal. Amsterdam: North-Holland.

Hogeweg, P. and B. Hesper. 1981. Two predators and one prey in a patchy environment: an application of MICMAC modelling. *J.theor.Biol.* 93: 411-432.

Hogeweg, P .and B. Hesper.. 1983. The ontogeny of the interaction structure in bumble bee colonies: a MIRROR model. *Behav. Ecol.Sociobiol* 12: 271-283.

Hogeweg, P. and B. Hesper. 1985. Socioinformatic Processes: MIRROR modelling methodology. *J. theor. Biol.* 113: 311-330.

Hogeweg, P. and B. Hesper. 1990. Individual-oriented modelling in ecology. *Mathl.Comput.Modelling* 13: 83-90.

Hogeweg, P. and B. Hesper. 1991. Evolution as pattern processing. TODO as substrate for evolution. In *From Animals to Animats* edited by J.A. Meyer and S.W. Wilson. Cambridge, MA: MIT Press.

Lethmate, J. 1977. Problemlöseverhalten von orang-utans (Pongo pygmaeus). *Fortschritte der Verhaltungsforschung, beihefte zur Zeitschr. für Tierpsychol.* 19. Berlin: Paul Parey.

MacKinnon, J. R. 1974. The ecology and behaviour of wild orangutans (Pongo pygmaeus). *Anim. Behav.* 22: 3-74.

Mitani, J.C. 1985. Mating behaviour of male orangutans in the Kutai Reserve, East Kalimantan,Indonesia. *Anim Behav.* 33: 392-402.

Mitani, J.C. 1990. Experimental field studies of asian ape social systems. *Int. J. Primatol.* 11, 2: 103-126.

Richards, P. W. 1952. *The tropical rainforest; an ecological study*. New York: Cambridge University Press.

Rijksen, H. D. 1978. *A Field Study on Sumateran Orang-utans (Pongo pygmaeus abelii, Lesson1827): Ecology, behaviour, and conservation*. Wageningen, the Netherlands: H. Veenman & Zonen.

Rodman, P. S. 1973. Population composition and adaptive organization among orangutans of the Kutai Reserve. In *Comparative Ecology and Behaviour of Primates* edited by R. P. Michael and J. H. Crook. London: Academic Press.

Rodman, P.S. 1984. Foraging and social systems of orangutans and chimpanzees. In *Adaptations for foraging in nonhuman primates: Contributions to an organismal biology of prosimians, monkeys, and apes* edited by P.S. Rodman and J.G.H. Cant. New York: Columbia University Press.

Schaik, C. P. van. 1986. Phenological changes in a Sumateran rain forest. *J. Tropical Ecol.* 2: 327-347.

Schaik, C. P. van and E. Mirmanto. 1985. Spatial variation in the structure and litterfall of a Sumatran rain forest. *Biotropica* 17, 196-205.

Schürmann , C. L. 1981. Courtship and mating behavior of wild orangutans in Sumatra. In *Primate Behavior and Socio biology* edited by A. B. Chiarelli and R. S. Corruccini. Berlin: Springer.

Schürmann, C. L. 1982. Mating behaviour of wild orang utans. In *Biology and Conservation of the Orang Utan* edited by L. E. M. de Boer.. Den Haag: Junk.

Schürmann, C. L. & van Hooff, J. A. R. A. M. 1986. Reproductive strategies of the orang-utan: new data and a reconsideration of existing socio-sexual models. *Int. J. Primatol.* 7: 265-287.

Siegel, S. 1956. *Nonparametric Statistics for the Behavioral Sciences*. New York: McGraw-Hill.

Sokal, R. & Rohlf, F. 1981. *Biometry*. San Fransisco: W. H. Freeman.

Sugardjito, J., te Boekhorst, I. J. A. & van Hooff, J. A. R. A. M. 1987. Ecological constraints on the grouping of wild orang-utans (Pongo pygmaeus) in the Gunung Leuser   National Park, Sumatera, Indonesia. *Int. J. Primatol.* 8: 17-41.

Wrangham, R.W. 1986a. Evolution of social structure. In *Primate Societies* edited by B. Smuts, D. L. Cheney, R. M. Seyfarth, R. W. Wrangham, and T. T. Strushaker. Chicago: University of Chicago Press.

Wrangham, R. W. 1986b. Ecology and social relationships in two species of chimpanzee. In *Ecological Aspects of Social Evolution: Birds and Mammals* edited by D. I. Rubinstein and R. W. Wrangham. Princeton: Princeton University Press.

A Biologically Inspired Immune System for Computers

Jeffrey O. Kephart

High Integrity Computing Laboratory
IBM Thomas J. Watson Research Center
P.O. Box 704, Yorktown Heights, NY 10598

Abstract

Computer viruses are the first and only form of artificial life to have had a measurable impact on society. Currently, they are a relatively manageable nuisance. However, two alarming trends are likely to make computer viruses a much greater threat. First, the rate at which new viruses are being written is high, and accelerating. Second, the trend towards increasing interconnectivity and interoperability among computers will enable computer viruses and worms to spread much more rapidly than they do today.

To address these problems, we have designed an immune system for computers and computer networks that takes much of its inspiration from nature. Like the vertebrate immune system, our system develops antibodies to previously unencountered computer viruses or worms and remembers them so as to recognize and respond to them more quickly in the future. We are careful to minimize the risk of an auto-immune response, in which the immune system mistakenly identifies legitimate software as being undesirable. We also employ nature's technique of fighting self-replication with self-replication, which our theoretical studies have shown to be highly effective.

Many components of the proposed immune system are already being used to automate computer virus analysis in our laboratory, and we anticipate that this technology will gradually be incorporated into IBM's commercial anti-virus product during the next year or two.

1 Introduction

Unique among all forms of artificial life, computer viruses have escaped their playpens and established themselves pervasively throughout the world's computing environment. Of the roughly 100 to 200 million PC and Macintosh users in the world, at least several hundred thousand, and perhaps over a million, have been afflicted at one time or another. Computer viruses have found a niche on all of the world's continents, including Antarctica [1][1], and most of its countries.

---

[1]The "Barrote" virus was discovered at Spanish and Argentinian scientific bases in Antarctica when it triggered on January 5th, 1994. Machines booted on or after that date displayed a pattern of jail-like bars with the legend "Virus Barrote" (Spanish for "Virus Jail"), and would halt the PC (and thus any scientific experiments that were being conducted).

A sufficiently amoral artificial life enthusiast might view the success of these artificial creatures in the real world as amazing, amusing, and admirable, but most responsible citizens regard computer viruses (and those who write them) with abhorrence. Even though just a small minority of viruses are intentionally harmful, the vast majority of them are poorly-written, poorly-tested, buggy pieces of software that create problems that are often time-consuming to diagnose. According to a Dataquest survey [2] and spokesmen for several different insurance companies [3], a virus spreading among several PC's in a company costs (on average) several thousands of dollars in down-time and data lossage; one company interviewed by Dataquest reported a $2 million dollar loss due to a single incident. At least one insurer offers a $100,000/year policy for damage due to computer virus infection [3].

Computer viruses are serious business. They have engendered an entire anti-virus industry, consisting of hundreds of researchers and developers who are employed by dozens of companies around the world. At least one such company, devoted almost exclusively to anti-virus software, is traded on the Nasdaq stock exchange.

Currently, the arms race between virus authors and anti-virus developers is roughly even. During any particular moment, it is typical for a few viruses to be increasing in prevalence, and other formerly prevalent ones to be on the decline [4]. However, two alarming trends threaten to turn the balance in favor of virus authors:

1. The rate at which new viruses are being written is quite high, and appears to be accelerating. Human experts who analyze and find cures for viruses are already swamped, and their ability to keep pace with the large influx of new viruses is being questioned.

2. The continuing increase in interconnectivity and interoperability among the world's computers enhances the ability of any particular virus to spread, and the rapidity with which it does so. The current strategy of periodically distributing updates to anti-virus software from a central source will be orders of magnitude too slow to keep up with the spread of a new virus.

In the near future, computers will somehow need to automatically recognize and remove previously unknown viruses on the spot soon after they are discovered. Fortunately for us, Nature has already invented a remarkably effective mechanism for recognizing and responding rapidly to viruses and other undesired intruders, even in cases where the intruder has never been seen before: the vertebrate immune system. The success of the vertebrate immune system in protecting its host from a wide array of viruses and other undesirables that are continually mutating and evolving has inspired us to design and implement an immune system for computers that is founded on similar principles. Various components of the immune system are already being used to automate the task of computer virus analysis in the laboratory. Over the next year or two, the immune system will be phased gradually into IBM's anti-virus software.

This paper is organized as follows. Section 2 briefly discusses the two trends mentioned above, and why they threaten to overwhelm current anti-virus technology. Appealing to biological analogy, section 3 motivates and presents a biologically inspired design for an immune system for computers and computer networks. Section 4 concludes with a brief discussion of important issues that remain to be resolved.

2 Why current anti-virus techniques are doomed

There are a variety of complementary anti-virus techniques in common usage [5, 6]. *Activity monitors* alert users to system activity that is commonly associated with viruses, but only rarely associated with the behavior of normal, legitimate programs. *Integrity management systems* warn the user of suspicious changes that have been made to files. These two methods are quite generic, and can be used to detect the presence of hitherto unknown viruses in the system. However, they are not often able to pinpoint the nature or even the location of the infecting agent, and they often flag or prevent legitimate activity, and so can disrupt normal work or lead the user to ignore their warnings altogether.

Virus scanners search files, boot records, memory, and other locations where executable code can be stored for characteristic byte patterns that occur in one or more known viruses. They tend to be substantially less prone to false positives than activity monitors and integrity management systems. Scanners are essential for establishing the identity and location of a virus. Armed with this very specific knowledge, *repairers*, which restore infected programs to their original uninfected state, can be brought into play. The drawback of scanning and repair mechanisms is that they can only be applied to known viruses, or variants of them; this requires that scanners and repairers be updated frequently.

Debates over the relative merits of the various anti-virus techniques have largely subsided, and many of the major anti-virus vendors now offer packages that usefully integrate scanners and repairers with activity monitors and integrity management systems.

In the remainder of this section, I shall describe the typical method by which scanners and repairers are updated, and demonstrate why it can be expected to become untenable in the near future, given projected trends in viral influx and increased interconnectivity among computers.

2.1 Virus scan/repair updates

Whenever a new virus is discovered, it is very quickly distributed among an informal, international group of virus collectors who exchange samples among themselves. Many such collectors are in the anti-virus software business, and they set out to obtain information about the virus which enables:

1. detection of the virus whenever it is present in a host program, and

2. restoration of an infected host program to its original uninfected state (which is usually possible.)

Typically, a human expert obtains this information by disassembling the virus and then analyzing the assembler code to determine the virus's behavior and the method that it uses to attach itself to host programs. Then, the expert selects a "signature" (a sequence of perhaps 16 to 32 bytes) that represents a sequence of instructions that is guaranteed to be found in each instance of the virus, and which (in the expert's estimation) is unlikely to be found in legitimate programs. This "signature" can then be encoded into the scanner, and the knowledge of the attachment method can be encoded into the repairer.

Such an analysis is tedious and time-consuming, sometimes taking several hours or days, and even the best experts have been known to select poor signatures — ones that cause the scanner to report false positives on legitimate programs.

2.2 Viral influx and its consequences

One reason why current anti-virus techniques can be expected to fail within the next few years is the rapid, accelerating influx of new computer viruses. The number of *different* known DOS viruses over the last several years can be fit remarkably well by an exponential curve. [2] Currently, it is approximately 2000, with two or three new ones appearing each day — a rate which already taxes to the limit the ability of anti-virus vendors to develop detectors and cures for them. Were this trend to hold up (Fig. 1), there would be approximately 10 million different DOS viruses by January, 2000 — about 100,000 new ones per day! Of course, curve extrapolation of a phenomenon that depends largely on human sociology and psychology should be regarded very skeptically, but it is not impossible that virus writers could be so prolific. To do so, they would have to automate both the writing and the distribution of viruses. Already, the beginnings of a trend towards automated virus-writing is evinced by the Virus Creation Laboratory, a menu-driven virus toolkit circulating among virus writers' bulletin boards. Even if the rate at which new viruses appear were to suddenly plateau at a level not much higher than what it is

---

[2]Note that is *not* the same as the growth in prevalence of any particular viral strain. Even for the minority of viruses that are successful in any degree, the growth in prevalence is strongly sub-exponential, perhaps even roughly linear.

today, the number of different DOS viruses could easily reach the tens of thousands by the year 2000, and the burden on current anti-virus techniques to detect and eradicate so many viruses would be severe.

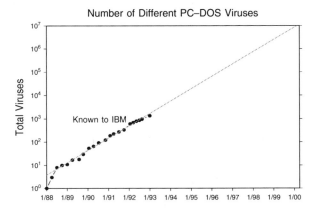

Figure 1: Number of different known DOS viruses vs. time (logarithmic scale). Straight line is the best exponential fit of the data through mid-1993. **Warning:** Extrapolation of the exponential trend beyond 1993 should be regarded *very* skeptically.

2.3 Interconnectivity and its consequences

It is unfortunate, but hardly surprising, that increased interconnectivity and interoperability among computers — designed to facilitate the flow of desirable information — also facilitates the flow of computer viruses. Biological diseases have always taken advantage of technological advances which enhance man's mobility [8]; it is natural that computer diseases should make opportunistic use of advances in the mobility of information.

One can expect increased networking to be reflected in increases in two important epidemiological parameters: the overall rate at which a given infected individual computer spreads a virus and the number of partners with which that individual has potentially infectious contacts. The first factor is related to one of the most fundamental results of classical mathematical epidemiology [10]. If the average rate at which infection can spread from one individual to another is sufficiently low, widespread infection is impossible. Above a well-defined critical threshold, however, epidemics can occur. As a simple way of explaining the existence of a sharp threshold, imagine that an individual has the flu. If, during that individual's period of contagion, he or she can be expected to infect 0.9 other people, the strain of flu will sooner or later die out. However, if that individual can be expected to infect 1.1 other people, there is likely to be a flu epidemic. The second factor, increased promiscuity, has apparently been given little attention by theoretical epidemiologists until our own study of it [7, 11]. We have found that a topology in which each individual has several "neighbors" to which it can spread infection is more conducive to epidemics than one which is sparsely connected — even when the infection rate along each link is adjusted so as to keep the total the same in the two cases.

Thus, to the extent that technological advances will increase the contact rate and promiscuity among computers, we can expect computer virus epidemics to become more likely, to spread faster, and to affect more computers. Experience with the Internet worm, which spread to hundreds or perhaps thousands of machines across the world in less than one day in 1988 [9], shows that even today's computing environment is vulnerable to a spread rate that is about two orders of magnitude faster than the typical timescale of monthly updates. While it is true that updates might be made somewhat more frequently, this would not solve the problem. The updates must be distributed to customers, and the customers must install them. Given the time, money, and effort involved, it is not surprising that many customers blissfully continue to use anti-virus software that is more than a year out of date.

3 An immune system for computers

Imagine that, every time a new strain of the common cold began to make its rounds, researchers at the Center for Disease Control had to race to find a cure for it. They would have to make sure that the cure worked properly for all sorts of people, and did not cause any allergic or other adverse reactions. The problem of distributing the cure to billions of people worldwide would be overwhelming.

This scenario is clearly ludicrous — we could not have survived as a species if we relied on a central agency to defend us against every disease. Yet this is precisely how we defend ourselves against computer viruses today! Time is running out on this approach, and a different alternative is sought.

Rather than relying on a central authority to protect them from all ills, humans and other vertebrates carry around their own individual immune systems. The vertebrate immune system exhibits some remarkable properties, including [12]:

1. Recognition of known intruders.

2. Elimination/neutralization of intruders.

3. Ability to learn about previously unknown intruders.

   - Determine that the intruder doesn't belong.
   - Figure out how to recognize it.
   - Remember how to recognize it.

4. Use of selective proliferation and self-replication for quick recognition and response.

Phrased in this way, it is evident that these fundamental properties are desirable for computers as well. The remainder of this section describes how each of these functions are being implemented in our design of the computer immune system, and compares our implementation with Nature's implementation of the vertebrate immune system. At the end of the section, the various elements will be assembled into a complete sketch of the proposed computer immune system.

3.1 Recognizing Known Intruders

The vertebrate immune system recognizes particular antigens (viruses and other undesirable foreign substances) by means of antibodies and immune cell receptors which bind to epitopes (small portions of the antigen, consisting of at least 4 to 6 amino acids).

It is interesting to note that an *exact* match to the entire antigen is not attempted; in fact, it is almost certainly a physical impossibility. No antibody molecule or immune-cell receptor could be perfectly specific to a given antigen because matching occurs at surfaces, not throughout volumes. T cell receptors can see the inner portions of antigen, but only after the antigen has been consumed by a macrophage or other cell, which then presents pieces of the antigen on it surface, where they can be seen by other cells.

Similarly, in the computer immune system, a particular virus is not recognized via an exact match; rather, it is recognized via an exact or fuzzy match to a relatively short sequence of bytes occurring in the virus (a "signature", as described in section 2). Although matching to a small portion of the virus is not necessitated in this case by the laws of chemistry, it has some important advantages. In particular,

1. it is more efficient in time and memory, and

2. it enables the system to recognize variants.

The issues of efficiency and variant recognition are relevant for biology as well.

For both biological and computer immune systems, an ability to recognize variants is essential because viruses tend to mutate frequently. If an exact match were required, immunity to one variant of a virus would confer no protection against a slightly different variant. Similarly, vaccines would not work, because they rely on the biological immune system's ability to synthesize antibodies to tamed or killed viruses that are similar in form to the more virulent one that the individual is being immunized against.

3.2 Eliminating Intruders

In the biological immune system, if an antibody meets up with an antigen, the two bind together, and the antigen is effectively neutralized. Thus recognition and neutralization of the intruder occur simultaneously. Alternatively, a killer T cell may encounter a cell that exhibits signs of being infected with a particular infecting agent, whereupon it kills the host cell. This is a perfectly sensible course of action. A biological virus co-opts its host cell's machinery, matter and energy into synthesizing viral proteins that are assembled into copies of the virus. Eventually, the host's cell wall is ruptured, resulting in the death of the host and the release of hundreds or thousands of viruses into the intercellular medium. By killing an infected host cell, a killer T cell is merely hastening the execution of a cell that was slated to die anyway , and it prevents the virus from completing the replication process.

If the computer immune system were to find an exact or fuzzy match to a signature for a known virus, it could take the analogous step of erasing or otherwise inactivating the executable file containing the virus. This is a valid approach. However, an important difference between computer viruses and biological viruses raises the possibility of a much gentler alternative.

From the body's point of view, cells are an easily-replenished resource. Even if biological viruses didn't destroy infected cells, an infected host cell would hardly be worth the trouble of saving; there are plenty of other cells around that can serve the same function. In contrast, each of the applications run by a typical computer user are unique in function and irreplaceable (unless backups have been kept, of course). A user would be likely to notice any malfunction. Consequently, it would be suicidal for a computer virus to destroy its host program, because the ensuing investigation would surely lead to its discovery and eradication. For this reason, all but the most ill-conceived computer viruses attach themselves to their host in such a way that they do not destroy its function. The fact that host information is merely rearranged, not destroyed, allows one to construct repair algorithms for a large class of non-destructive viruses for which one has a precise knowledge of the attachment method.

3.3 Learning to Recognize Unknown Intruders

When the biological immune system encounters an intruder that it has never seen before, it can immediately recognize the intruder as non-self, and attack it on that basis. Over the course of days or weeks, through a process of mutation and selective proliferation (see the next subsection), it "learns" to fabricate antibodies and B- and T cell receptors capable of recognizing that particular intruder very efficiently. By some unknown means, the immune system is able to "remember" the antigen (*i. e.* it retains immune cells with the proper receptors for recognizing that antigen) for decades after the initial encounter, and thus it is ready to respond much more quickly the next time that antigen is encountered.

To be effective, an antibody or receptor for a particular antigen must bind to that antigen (or close variants of that antigen) with high efficiency, and it must *not* bind to self proteins — otherwise, the host would be likely to suffer from an auto-immune disease. The biological immune system reduces the chances of recognizing self by subjecting immature immune cells to a training period in the thymus, during which those possessing self-recognizing receptors are eliminated.

Unfortunately, the notion of "self" in computers is somewhat problematic. We can not simply regard the "self" as the set of software that was pre-loaded when the computer was first purchased. Computer users are continually updating and adding new software. It would be unacceptable if the computer immune system were to reject all such modifications and additions out of hand on the basis that they were different from anything else that happened to be on the system already. While the biological immune system can usually get away with presuming the guilt of anything unfamiliar, the computer immune system must presume that new software is innocent until it can prove that it is guilty of containing a virus.

The thorny issue of what constitutes "self" for computer software, interesting as it is, can be regarded as a side-issue. The actual problem that both the vertebrate and the computer immune system must solve is to distinguish between harmful and benign entities. Due to the high degree of stability of body chemistry in individual vertebrates during their lifespans, their immune systems can replace the difficult problem of distinguishing between benign and harmful entities by the much simpler one of distinguishing self from non-self. This is a nice hack, because "self" is much easier to define and recognize than "benign". The immune system can simply implement the strategy "know thyself" (and reject all else). Although this errs on the side of false positives (*i.e.* falsely rejecting benign entities), rejection of foreign benign entities is generally not harmful (except in cases of blood transfusion or organ transplantation, which have been introduced much too recently to have affected the course of evolution).

By contrast, false rejection of legitimate software is extremely harmful. It worries users unnecessarily, and can cause them to erase perfectly legitimate programs — leading to hours or days of lost productivity. After such an experience, users are often tempted to stop using anti-virus software, leaving themselves completely unprotected. Thus a false positive indentification of a virus may be much more harmful than the virus itself. For this reason, self/non-self discrimination is not by itself an adequate means for distinguishing between harmful and unharmful software.

The process by which the proposed computer immune system establishes whether new software contains a virus has several stages. Integrity monitors, which use checksums to check for any changes to programs and data files, have a notion of "self" that is as restrictive as that of the vertebrate immune system: any differences between the original and current versions of any file are flagged, as are any new programs.[3] However, evidence of a non-self entity is not by itself enough to trigger an immune response. Mechanisms that employ the complementary strategy of "know thine enemy" are also brought into play. Among these are activity monitors, which have a sense of what dynamic behaviors are typical of viruses, and various heuristics, which examine the static nature of any modifications that have occurred to see if they have a viral flavor.

In the computer immune system, integrity monitors and generic know-thine-enemy heuristics are periodically or continually on the lookout for any indications that a virus is present in the system. If one of the virus-detection heuristics is triggered, the immune system runs the scanner to determine whether the anomaly can be attributed to a known virus. If so, the virus is located

and removed in the usual way. If the anomaly can *not* be attributed to a known virus, either the generic virus-detection heuristics yielded a false alarm, or a previously unknown virus is at large in the system.

At this point, the computer immune system tries to lure any virus that might be present in the system to infect a diverse suite of "decoy" programs. A decoy program's sole purpose in life is to become infected. To increase the chances of success in this noble, selfless endeavor, decoys are designed to be as attractive as possible to those types of viruses that spread most successfully. A good strategy for a virus to follow is to infect programs that are touched by the operating system in some way. Such programs are most likely to be executed by the user, and thus serve as the most successful vehicle for further spread. Therefore, the immune system entices a putative virus to infect the decoy programs by executing, reading, writing to, copying, or otherwise manipulating each of them. Such activity tends to attract the attention of many viruses that remain active in memory even after they have returned control to their host. To catch viruses that do not remain active in memory, the decoys are placed in places where the most commonly used programs in the system are typically located, such as the root directory, the current directory, and other directories in the path. The next time the infected file is run, it is very likely to select one of the decoys as its victim. From time to time, each of the decoy programs is examined to see if it has been modified. If one or more have been modified, it is almost certain that an unknown virus is loose in the system, and each of the modified decoys contains a sample of that virus. These virus samples are stored in such a way that they will not be executed accidentally.

The capture of a virus sample by the decoy programs is somewhat analogous to the ingestion of antigen by macrophages or B cells [12]. It allows the intruder to be processed into a standard format that can be parsed by some other component of the immune system, and provides a standard location where information on the intruder can be found. In the biological immune system, the T cells that recognize the antigen are selected according to their ability to bind to fragments of the antigen that are presented on the surface of cells that have ingested (or been infected by) the antigen. Likewise, in the computer immune system, the infected decoys are then processed by another component of the immune system — the signature extractor — so as to develop a recognizer for the virus. The computer immune system has an additional task that is not shared by its biological analog: it must attempt to extract from the decoys information about how the virus attaches to its host, so that infected hosts can be repaired (if possible).

Unfortunately, the proprietary nature of our methods for deriving a virus's means of attachment to its host forbid any discussion of them here. Briefly, the algorithms extract from a set of infected decoys information on the attachment pattern of the virus, along with byte sequences that remain constant across all of the captured samples of the virus.

---

[3]An interesting alternative to traditional integrity monitoring via checksums, inspired by the detailed mechanisms by which the vertebrate immune system learns to recognize "self", has been studied recently by Forrest, Perelson, Allen, and Cherukuri [13].

Next, the signature extractor must select a virus signature from among the byte sequences produced by the attachment derivation step. The signature must be well-chosen, such that it avoids both false negatives and false positives. In other words, the signature must be found in each instance of the virus, and it must be very unlikely to be found in uninfected programs.

First, consider the false negative problem. The samples captured by the decoys may not represent the full range of variable appearance of which the virus is capable. As a general rule, non-executable "data" portions of programs, which can include representations of numerical constants, character strings, work areas for computations, *etc.* are inherently more likely to vary from one instance of the virus to another than are "code" portions, which represent machine instructions. The origin of the variation may be internal to the virus (*e.g.* it could depend on a date). Alternatively, a virus hacker might deliberately change a few data bytes in an effort to elude virus scanners. To be conservative, "data" areas are excluded from consideration as possible signatures. Although the task of separating code from data is in principle somewhat ill-defined, there are a variety of methods, such as running the virus through a debugger or virtual interpreter, which perform reasonably well.

The false positive problem is more interesting. In the biological immune system, false positives that accidentally recognize self cause auto-immune diseases. In both traditional anti-virus software and the proposed computer immune system, false positives are particularly annoying to customers, and so infuriating to vendors of falsely-accused software that it has led to at least one lawsuit against a major anti-virus software vendor. (So one could say that health is also an issue in this case!)

Briefly, the automatic signature extractor examines each sequence of S contiguous bytes (referred to as "candidate signatures") in the set of invariant-code byte sequences that have presented to it, and for each it estimates the probability for that S-byte sequence to be found in the collection of normal, uninfected "self" programs. Typically, S is chosen to be 16 or 24. The probability estimate is made by

1. forming a list of all n-grams (sequences of n bytes; $1 \leq n \leq n_{max}$) contained in the input data (n_{max} is typically 5 or 8),

2. calculating the frequency of each such n-gram in the "self" collection (in the case of signatures that are to be distributed worldwide, we use a half-gigabyte corpus of ordinary, uninfected programs),

3. using a simple formula to combine the n-gram frequencies into a probability estimate for each candidate signature to be found in a set of programs similar in size and statistical character to the corpus, and

4. selecting the signature with the lowest estimated false-positive probability.

Characterizations of this method show that the probability estimates are poor on an absolute scale, due to the fact that code tends to be correlated on a longer scale than 5 or 8 bytes. However, the relative ordering of candidate signatures is rather good, so the method generally selects one of the best possible signatures. In fact, judging from the relatively low false-positive rate of the IBM AntiVirus signatures (compared with that of other anti-virus vendors), the algorithm's ability to select good signatures is *better* than can be achieved by typical human experts.

Having automatically developed both a recognizer and a repair algorithm appropriate to the virus, the information can be added to the corresponding databases. If the virus is ever encountered again, the immune system will recognize it immediately as a known virus. A computer with an immune system could be thought of as "ill" during its first encounter with a virus, since a considerable amount of time and energy (or CPU cycles) would be expended to analyze the virus. However, on subsequent encounters, detection and elimination of the virus would occur much more quickly: the computer could be thought of as "immune" to the virus.

3.4 Self Replication and Selective Proliferation

In the biological immune system, immune cells with receptors that happen to match a given antigen reasonably well are stimulated to reproduce themselves. This provides a very strong selective pressure for good recognizers, and by bringing a degree of mutation into play, the immune cell is generally able to come up with immune cells that are extremely well-matched to the antigen in question.

One can view this as a case in which self-replication is being used to fight a self-replicator (the virus) in a very effective manner. One can cite a number of other examples in nature and medical history in which the same principle has been used very successfully. The self-replicator need not itself be a virus. In the case of the worldwide campaign against smallpox, those who were in close contact with an infected individual were all immunized against the disease. Thus immunization spread as a sort of anti-disease among smallpox victims [10].

We propose to use a similar mechanism, which we call the "kill signal", to quell viral spread in computer networks. When a computer discovers that it is infected, it can send a signal to neighboring machines. The signal conveys to the recipient the fact that the transmitter was infected, plus any signature or repair information that might be of use in detecting and eradicating the virus. If the recipient finds that it is infected, it sends the signal to *its* neighbors, and so on. If the recipient is not infected, it does not pass along the signal, but at least it has received the database updates — effectively immunizing it against that virus (see Fig. 2).

Theoretical modeling has shown the kill signal to be extremely effective, particularly in topologies that are highly localized or sparsely connected [4, 11].

Kill Signals

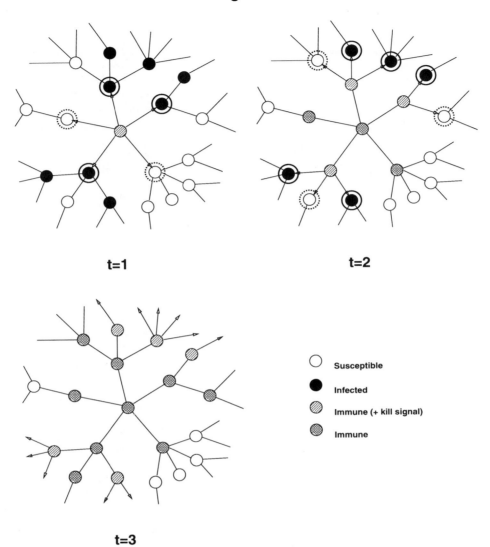

Figure 2: Fighting self-replication with self-replication. When a computer detects a virus, it eliminates the infection, immunizes itself against future infection, and sends a "kill signal" to its neighbors. Receipt of the kill signal results in the immunization of uninfected neighbors; infected neighbors are both immunized and prompted to send kill signals to their neighbors. Thus detection of a virus by a single computer can trigger a wave of kill signals that propagates along the path taken by the virus, destroying the virus in its wake.

Immune System Overview

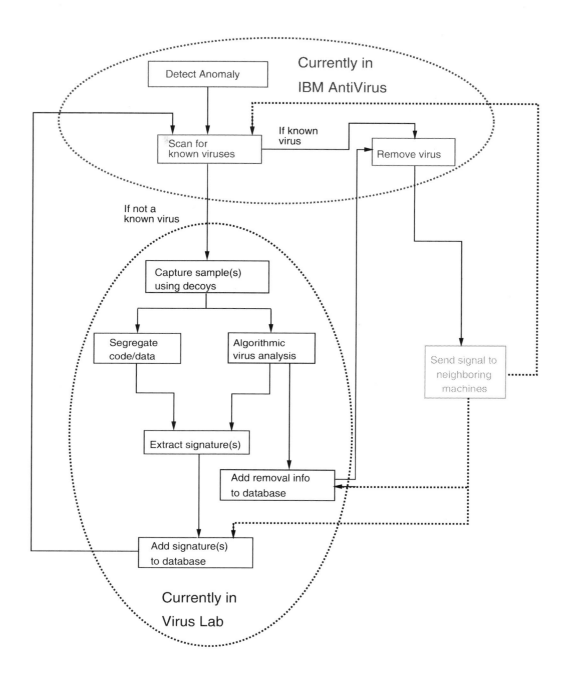

Figure 3: The main components of the proposed immune system for computers and their relationship to one another.

3.5 Computer Immune System: Schematic and Implementation

Fig. 3 sketches the relationship among various components of the proposed computer immune system. Some are already integrated into the current version of IBM AntiVirus. The components of the immune system that deal with unknown viruses are currently being used in a slightly different capacity: to extract signatures and repair information automatically from newly-discovered viruses. This enables us to keep pace with the influx of new viruses with just one human virus expert who analyzes viruses half-time, as opposed to the dozen or more virus analysts employed by some other anti-virus software vendors.

When a raft of new viruses is received, it is presented to an automatic "triage" machine situated in IBM's virus isolation laboratory. First, the triager scans the putative viruses using the current version of IBM AntiVirus. Any samples infected with a virus that is already detected by IBM AntiVirus are immediately dismissed from further consideration. The triager then executes each of the remaining infected samples one or more times, and (for each infected sample) exercises a set of six decoy programs so as to entice the virus to infect them. Each of the decoy programs is examined from time to time to see if it has been modified. Any decoys that *have* been modified are stored away in a form such that they cannot be executed, and the triage machine is automatically rebooted to eliminate the virus from memory. The triage script goes through the same routine for the next putatively infected sample, and so on until all the original samples have been given a reasonable chance to infect the decoys.

Putatively infected samples which successfully infect decoys are placed in the archive of confirmed viruses, and the infected decoys are placed in special directories for further processing. Putatively infected samples which fail to infect any decoys may contain recalcitrant viruses that for some reason were not in the infecting mood, or they may not contain a virus at all. On a rainy day some weeks hence, another attempt will be made to coax them into infecting the decoys.

Typically, a given virus sample will infect two or three of the six decoys. During the last three years, the triager has been used to capture samples of over 2000 different PC/DOS viruses.

The infected decoys are then processed by the algorithmic virus analyzer, which extracts information that is useful in repairing viruses. Still in early prototype, the analyzer is able to supply useful information for about 90% of the viruses that it has seen. A debugger is used to execute each infected decoy; any executed instructions are obviously code (rather than unexecutable data), and as such are eligible for consideration as part of a signature for the virus.

Next, the automatic signature extractor takes as input all byte sequences which appear in each infected decoy and which have been established as code, selects a signature, and provides an estimate of the maximum number of mismatches between scanned data and the signature

that can be considered a match. During three years of constant improvements, the automatic signature extractor has been used to extract signatures for roughly 1500 different PC/DOS viruses. In addition, it has been used to evaluate several hundred signatures that had been extracted by expert humans.

The automatically-extracted signatures and repair information are then subjected to a variety of independent tests. The signatures are run against a half-gigabyte corpus of legitimate programs to make sure that they do not cause false positives, and the repair information is checked out by testing on samples of the virus, and further checked by a human expert. Finally, the detection and repair databases used by IBM AntiVirus are updated, and the new version is distributed to customers worldwide.

The remaining component of the immune system, the kill signal, is the only one that has not yet been implemented; it is currently being evaluated via theoretical modeling.

4 Conclusion

An immune system for computers is desirable and feasible. As suggested in Fig. 3, most of the necessary components are already in use in one form or another. Some already exist in IBM AntiVirus itself. Others are presently in use in the virus laboratory, for the purpose of updating the databases employed by IBM AntiVirus to recognize viruses and repair infected files.

One of the technical issues that remains to be explored further is the kill signal. Further simulation will help to establish the exact circumstances under which a node should send signals to its neighbors, and for what length of time these signals should be sent. Further analysis and simulation must be conducted to assess the effectiveness of various fail-safe mechanisms that have been proposed to deal with the propagation of erroneous kill signals, which could result from false positives, software bugs, or intentional subversion by malicious users. The biological immune system has invented various inhibitory mechanisms which may turn out to be of use to us.

We anticipate that, as the design for our computer immune system evolves, it will be influenced, not just by what Mother Nature has invented, but also by theories invented by immunologists to explain the observed function of the immune system [14, 15, 16]. In fact, we may offer new employment opportunities for theoretical immunologists, because our criteria for success are different: a proposed mechanism need not be a correct description of biology; it only has to work!

Acknowledgments

I would like to thank all the members of IBM's High Integrity Computing Laboratory, particularly Steve White, Bill Arnold, Dave Chess, and Greg Sorkin, for numerous discussions about automatic virus analysis and the design of the immune system. Bill Arnold and Dave Chess invented the idea of using decoy programs. Bill Arnold implemented the decoy-infection routines and the viral-code identifier on DOS machines in the virus isolation laboratory, and invented "kill signals" (and coined their name). Greg Sorkin invented the algorithms that produce a concise description of a virus's method of attachment to its host.

References

[1] Amodio, Jorge. Submission to Virus-L digest 7:4. January 18, 1994.

[2] Dataquest. 1991. *Computer Virus Market Survey for National Computer Security Association.* San Jose.

[3] Sulit, Beth K. Nothing to sneeze at. *Risk & Insurance,* August 1992, 1.

[4] Kephart, Jeffrey O. and Steve R. White. Measuring and modeling computer virus prevalence. *Proceedings of the 1993 IEEE Computer Society Symposium on Research in Security and Privacy.* Oakland, California, May 24–26, 1993, 2–15.

[5] Spafford, E. H. 1991. Computer viruses: A form of artificial life? In D. Farmer, C. Langton, S. Rasmussen, and C. Taylor, eds., *Artificial Life II. Studies in the Sciences of Complexity,* 727–747. Redwood City: Addison-Wesley.

[6] Kephart, Jeffrey O., Steve R. White, and David M. Chess. *Computers and epidemiology.* IEEE Spectrum, May 1993, 20–26.

[7] Kephart, Jeffrey O. and Steve R. White. "Directed-graph epidemiological models of computer viruses," *Proceedings of the 1991 IEEE Computer Society Symposium on Research in Security and Privacy,* Oakland, California, May 20–22, 1991, 343–359.

[8] McNeill, William H. 1977. *Plagues and Peoples.* New York: Doubleday.

[9] Spafford, E. H. 1989. The Internet worm program: an analysis. *Computer Comm. Review* 19, 1.

[10] Bailey, Norman T. J. 1975. *The mathematical theory of infectious diseases and its applications,* second edition. New York: Oxford University Press.

[11] Kephart, Jeffrey O. 1994. How topology affects population dynamics, in C. Langton, ed., *Artificial Life III. Studies in the Sciences of Complexity,* 447–463. Redwood City: Addison-Wesley.

[12] Paul, William E., ed. 1991. *Immunology: Recognition and Response ... Readings from Scientific American.* New York: W. H. Freeman and Company.

[13] Forrest, Stephanie, Alan S. Perelson, Lawrence Allen, and Rajesh Cherukuri. Self-nonself discrimination in a computer. *Proceedings of the 1994 IEEE Computer Society Symposium on Research in Security and Privacy,* Oakland, California, May 16–18, 1994.

[14] Perelson, Alan S., ed. 1988. *Theoretical Immunology, Parts One and Two.* Redwood City: Addison-Wesley.

[15] Forrest, Stephanie and Alan S. Perelson. 1990. Genetic algorithms and the immune system. *Proceedings of Workshop on Parallel Problem Solving from Nature.* New York: Springer-Verlag.

[16] Seiden, P. E. and F. Celada. 1992. A model for simulating cognate recognition and response in the immune system. *J. Theor. Biology* 158, 329.

Egrets of a Feather Flock Together

Yukihiko Toquenaga

Institute of Biological Sciences,University of Tsukuba

toquenag@kz.tsukuba.ac.jp

Isamu Kajitani

Institute of Engineering Mechanics,University of Tsukuba

kajitani@kz.tsukuba.ac.jp

Tsutomu Hoshino

Institute of Engineering Mechanics,University of Tsukuba

hoshino@kz.tsukuba.ac.jp

Abstract

Group foraging and colonial formation in avian species are examined with an ALife model using Genetic Algorithm and Neural Network. Horn's classical model predicts that colonial birds are more beneficial than territorial ones in a patchy resource environment, and the reverse is true when resources are evenly distributed. The weak point of the model is confusion between colony formation and flock foraging and implicitly assumes perfect knowledge of resource distribution by foraging birds. The authors made an ALife model that realized both flock foraging and colonial formation simultaneously during evolution in a patchy environment. The reference organisms were egrets which make colonial nests in the breeding season and colonial roosts in the non-breeding season. In the computer model, artificial egrets used the presence of other egrets as an indicator to locate resource-rich patches in a resource-clumped environment. On the contrary, egrets were less reliant on other individuals as a source of foraging information in evenly distributed resource conditions. Colonial nesting was also induced only in a patchy environment where the foraging efficiency was always higher than the evenly distributed condition. Local enhancement played an important role in achieving colonial and flock foraging.

INTRODUCTION

Group living is one of the most exciting themes in behavioral and evolutionary ecology (Rubenstein 1978, Clark and Mangel 1986, Pulliam and Caraco 1984). The way of group living ranges from a simple gathering of conspecies to a systematic social assemblage shown in several social animals (Wilson 1975). Territoriality is the alternative strategy that is also observed in a wide range of animal taxa. It also ranges from the individual to groups, and further social territoriality (Bacon et al. 1991). Contrasts between group and solitary living suggests some underlying mechanisms to diverge the two strategies in the course of evolution.

There are two axes to consider in the observation of group living in nature. The first is coloniality, or group formation for nesting or roosting. This axis is closely related to the reproductive and mating systems of organisms. The second axis is flock formation during foraging. The two axes may be further divided into two important aspects when we discuss the evolution of group living: predation and efficient foraging.

An anti-predator hypothesis can be applied to both coloniality and flock foraging. Group living provides selfish benefits to group members, such as "dilution effect" and "confusion effect" (Krebs 1981). It also provides the "many eyes" effect (Powell 1974, Pulliam 1973) or defensive activity against predators (e.g., mobbing). On the contrary, group living may make groups to be so conspicuous that predators can easily detect them. Tradeoffs between benefits and disadvantages of group living against predation have been discussed in many theoretical works (e.g., Packer and Abrams 1990, Pulliam 1984).

The other important aspect of group living is efficient foraging by coloniality and flock formation. Coloniality may provide an information center where roosting individuals exchange information about beneficial patches (Clode 1933, Ward and Zahavi 1973). On the other hand, flock formation provides "many eyes" for detecting prey and may bring a "beating effect" as observed in Cattle Egrets (Scott 1984). In some avian species, it has further evolved to systematic cooperative foraging as discovered in Cormorants (Morrison and Slack 1977).

In this article, the authors discuss the evolution of group living with a focus on efficient foraging. Our reference organisms are egrets and herons (Ciconiformes: Ardeidae: *Egretta* spp.) that form heronries in breeding seasons and colonial roosts in non-breeding seasons. In the following discussion, the authors first review two conventional foraging models of group-living organisms in the light of resource distribution. Then the authors introduce an ALife model that succeeded to lead to the simultaneous evolution of colonial and flock foraging by acquiring a simple behavioral rule, named "local enhancement." Finally the results of the simulation are compared with those of several other ALife models.

CONVENTIONAL MODELS FOR GROUP FORAGING

Horn's Model

Horn's model (Horn 1968) has been repeatedly cited to demonstrate the relationship between the resource distribution and the roosting strategy in avian species (cf. Brown et al. 1992). The model assumes a square universe of 4 x 4, and each node is a candidate of a resource patch (Fig. 1). Birds inhabiting the universe are assumed to perform central foraging from their nests or roosts. They have two choices: colonial or territorial roosting.

If resources are evenly distributed (upper panels in Fig. 1), it is optimal for birds to make territorial roosts (small triangles) from which the individuals only visit four neighboring patches to minimize traveling costs (distance) for foraging. On the contrary, if the same amount of resources are concentrated in a single patch that moves around the universe (lower panels in Fig. 1), the birds should form a single colonial roost (the large triangle) at the center of the universe. In general, the model predicts that birds are territorial if resources are evenly distributed. When resources are clumped and highly unpredictable, birds should form colonial roosts.

One weak point of Horn's model is that both central and group foraging were simultaneously implemented at the beginning. In other words, birds in this model know exactly where resource patches would appear. Moreover, they know how to return to their roosts without wandering. The territorial foragers never explore resource piles in other territories.

The similar misleading simplification may also be seen in famous models of ant trail-making behaviors, which are frequently used as examples of "emergence" in ALife studies (e.g., Collins and Jefferson 1991). In those models, ants first start to forage randomly and then form pheromone trails after detecting resource piles. On the contrary, the ants can detect their colony perfectly and, thus, return to their colony with relative ease. This efficient colony-orientation is fairly unbalanced with their poor ability to find their resources at the beginning of the simulation.

Group foraging can be divided into two phases. At the first phase individuals form the center of their activities, such as a breeding colony and a communal roost. The formation of a colony or a roost ensures that these individuals execute central foraging. The second phase is the formation of flocks at foraging sites. Note that the scale is different between the two phases. The colonial formation is an event at a larger scale or a higher level phenomenon than the flock foraging. Sometimes these two phases are independently evolved, as in the case of the Diurnal Activity Center (DAC) which was observed in European starlings (Caccamize and Morrison 1986). However, in herons and egrets, it is likely that these two phases have evolved simultaneously.

Flock Foraging and Local Enhancement

Flock formation causes foraging efficiency in several ways, such as the beating effect, the many-eye effect, cooperative systematic hunting, and the copying of efficient foragers. The target statistics are the mean and the variance of an individual's intake within flock foragers. Flock foraging causes not only high mean values but also less variation in an individual's intake among flock mem-

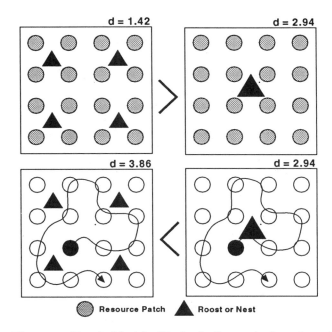

Figure 1: Horn's Model. Circles indicate the location of foraging sites. Triangles indicate the location of nests or roosts. The metric d is the mean travel distance for foraging from roosts.

bers (Chantrey 1982). This low variability among flock foragers reduces the risk of a no-capture experience for each individual (Thompson et al. 1974). Several field studies showed that high mean and low variance of intake were achieved in avian species (e.g., Brown 1988a 1988b, Morrison and Slack 1977).

A key aspect is information parasitism among flock members. There are several ways to accomplish this parasitism, such as the information center strategy (Brown 1988a, 1988b), social learning, or the copying the foraging tactics of skillful flock mates (Thompson et al. 1974). Among them, local enhancement is the simplest way in which foragers use the existence of other individuals as a key for finding beneficial foraging sites (Thorpe 1963). This simple rule is adopted in herons and egrets in nature (Armstrong 1971, Wait 1981, Weatherhead and Greenwood 1981).

Local enhancement can be expressed using the Polya-Eggenberger process (Arthur 1987, Wade 1990). The formulation is as follows:

$$Pr[\bigcap_{j=0}^{k}(X_j = x_j)]$$

$$= \binom{n}{x_0, x_1, \ldots, x_k} \frac{\prod_{i=0}^{k} c_i(c_i+1)\ldots(c_i+x_i-1)}{c(c+1)\ldots(c+n-1)}$$

$$c = \sum_{j=0}^{k} c_j, n = \sum_{j=0}^{k} xj$$

where k is the number of patches, n is the total number of foragers arriving in the patches, c_j is the initial number of foragers in each patch, and x_j is the number of foragers added in each patch.

An assumption is made that there are only two patches that egrets choose for foraging sites. If there is no resource depletion in each patch, the change in the proportion of egrets at one site corresponds to a stochastic process described in the above formula. This process consists of two phases. The first phase resembles an ordinal random walk, but it soon changes over time to a specific canalizing phase (Fig. 2).

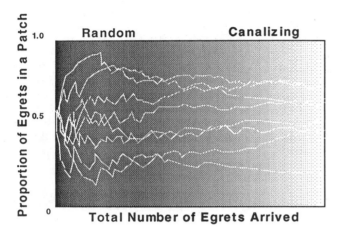

Figure 2: Polya Eggenberger Process

If resource are clumped, the canalizing phase of the Polya-Eggenberger process continues to gather foragers until resources are completely depleted (the left panel of Fig. 3). Poorer patches should be refused quickly, and egrets should be attracted more and more by those patches that contain other egrets. This process guarantees a high mean and low variance of an individual's intake within flock members. On the contrary, if the resource is evenly distributed, the canalizing phase will seldom be achieved due to the frequent occurrence of resource depletion in each patch. Thus, the dynamics of foragers cannot be distinguished from a simple random walk (the right panel of Fig. 3).

This mathematical formulation unfortunately becomes intractable in an analytical sense when resource depletion is incorporated. If a numerical calculation is the only way to solve this formulation, there is no advantage in pursuing this top-down approach. It cannot create alternative foraging behaviors because they have to be pre specified at the beginning of the simulation. Our interest is not only to compare the benefits of alternative strategies, but also to foster those strategies in given environments.

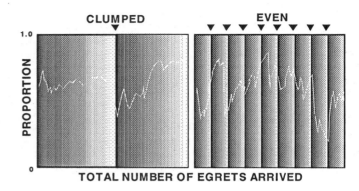

Figure 3: Polya-Eggenberger Process with Resource Depletion

MODEL: WANDERING EGRET BRAINS

The common problem of the above two conventional models is that they can only compare the evolutionary strength (e.g., in terms of E.S.S) of contradicting foraging strategies that are previously prepared or specified by the modelers. In real nature, foraging strategies should have emerged during evolution under given environments. Moreover, the above two models do not explain emergence of central foraging per se. To achieve the simultaneous emergence of central and group foraging, the authors constructed an ALife model to simulate foraging strategies of egrets equipped with self-reference neural networks.

Foraging Arena and Egret's View

Artificial egrets forage in a habitat of 30 x 30 cells. Each cell is assigned one of seven types of land use. They are paddy land (PL), shallow water (SW), PL & SW, Deep Water (DW), Dry Land (DL), Lotus Field (LF), and LF & SW. These 7 types of land use are commonly observed in the habitat of natural egrets and herons around Tsukuba, Ibaraki prefecture, in the middle part of Japan. The amount of prey in each cell is assigned by a random variable obeying β-distribution under a constraint that the total amount of the resource is 3,000. Each cell is then assigned an appropriate land type depending on the relative abundance of prey density (Table 1).

Land Type	PL	SW	PL SW	DW	DL	LF	LF SW
Value	3	2	3	0	0	4	3

Table 1: Relative Prey Abundance for Land types

The prey amount in a cell decreases by a unit when the cell is foraged by an egret. If at least one egret is foraging in a cell, the cell is labeled as White Egret

143

(WE) instead of the initial cell type. Thus, a foraging egret may choose from a total of eight cell types. Each egret can view 9 x 9 cells around itself (Fig. 4). The 81 cells are further divided into nine blocks, and each block consists of nine cells. A foraging egret is assumed to be standing or flying at the fourth cell in the fourth block (the shaded cell in the upper panel of Fig. 4). There is no reflection at the edge of the arena, and an egret disappears from the arena if its 9 x 9 view protrudes the edge.

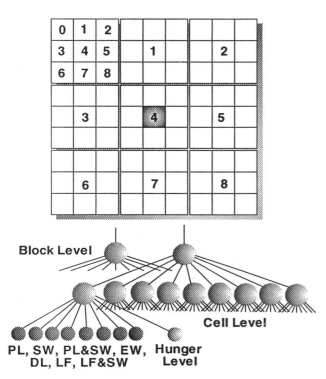

Figure 4: Egret's View and Action Network

Action Network

An individual egret has a 3-layer-action-network for its foraging behavior (Fig. 4). At the lowest level, there are nine nodes. Eight of them correspond to eight cell types (PL, SW, PL&SW, DW, DL, LF, LF&SW, and WE). Each node returns 1 or 0 depending on whether it matches the type of the patch at which the egret is looking. The ninth node receives the egret's hunger level as the input value. The hunger level is the amount of prey that the egret has obtained so far. The second layer of the neural network corresponds to each cell in a block of an egret's view. Every node has an identical 9-footed network. Each node at the top layer corresponds to each block of an egret's view.

Note that this network is quite different from ordinal multi-layer neural networks. The important point is that the network is a simple assemblage of a 9-footed basal node (Fig. 5). An intuitive image of this network is as follows. First an single-eyed egret views 81 cells with its

eye that consists of nine nodes. A total 81 inputs are further subtotaled into nine blocks. Thus, the layers of this network simply corresponds to a hierarchical view of egrets rather than the non-linear propagation of input signals in ordinal neural networks.

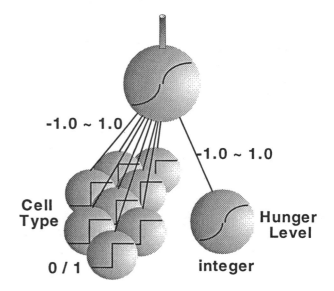

Figure 5: Structure of a Basal Node

Foraging and Reinforcement Learning

The weight of each path of the neural network is an important aid in the evolution of foraging behavior. Each egret has a set of 9 bit strings, each of which encodes the weight value of each path in the 9-footed network. The weights are real values and are initially specified randomly with a range from -1 to 1. These values are later modified by GA operations in the course of evolution.

Ackley and Littman's approach (Ackley and Littman 1991) is applied to accomplish reinforcement learning of the foraging behavior. Other than the big action network consisting of 81 nodes, each individual has a small evaluation network that determines whether reinforcement learning is executed in each foraging bout. The structure of the evaluation net is the same 9-footed network as the lowest nods in the action network (Fig. 5).

Learning by egrets are performed as follows. An egret first looks around and puts down its action network on the 81 cells within its view. The action net quickly subtotals return values from basal nodes. Then the egret chooses the block that returns the maximum output value. Next it chooses the cell that returns the maximum output within the chosen block and puts down its evaluation net on the cell. If the evaluation network is fired, the return value from the basal node of the action net is evaluated as the right answer. If the return value is positive, the desired target is set to 1. If the return value is negative, the desired target is set to -1. Then the weights of the 9-footed basal node of the action net are rearranged to achieve the desired target as its return value by the back propagation method (Fig. 6).

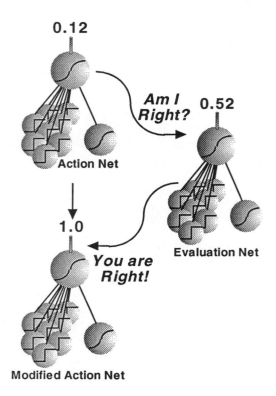

0.12

Am I Right? **0.52**

Action Net

1.0

You are Right!

Evaluation Net

Modified Action Net

Figure 6: Reinforcement Learning

There are two important remarks. The first one is that all 9-footed-basal nodes of the big action network simultaneously change their weights to 9 paths. It means that after the reinforcement learning, an egret uses a modified eye for the following foraging. The second point is that the weight change of the paths is not inherited by the next generation. The offspring network weights are the same as those of parents without reinforcement learning if there is no mutation and crossing-over. The chosen cell by the action net is a candidate of landing. If the action net chooses the same cell in the next trial, then the egret lands on the cell and starts foraging. The egret gains a unit of reward if the patch contains some prey. In some cases, the egret does not land on the first chosen cell because the following reinforcement learning changes its preference and the action net does not choose the cell again.

Nesting

Beside the action and evaluation networks for foraging behavior, each egret has a third neural network for nest formation. The network structure is the same as the 9-footed-basal node of the action net or the evaluation node. Egrets always have a positive motivation to make nests. Egrets try to seek nest sites every step after choosing candidate cells. Each egret puts its nest formation network on the selected cell. The decision to form a

nest is executed when the sum of the output from the 9-footed-node of the nesting net exceeds 0.5.

If the nesting network is fired, the egret puts the net on 81 cells in its view. and starts the hierarchical selection of a nest site as the same way that it picks up a candidate cell with its action network. No reinforcement learning was applied for the nest formation. The weights of the nine paths of the nest formation net are also modified by GA operations.

Reproduction

Hunger level is an integer value. The initial hunger level for each egret is five units and it decreases by one unit as time elapses. If the egret hits a prey, he gains five units. Its hunger level is multiplied by five if it returns to its own nest. Individuals that have a hunger level less than unity are eliminated from the foraging arena.

Each individual is constituted of three sets of bit-strings. Each set corresponds the action, evaluation, and nest-making network, respectively. Each set is further constituted of nine sets of bit-strings that code the weight of paths in the 9-footed networks.

Egrets that make nest(s) are targets of GA operations. Each breeding individual is assumed to be hermaphrodites. Although no egrets and herons are hermaphrodites in nature, this simplification is valid because no sex related specificity is implemented in this model. Crossing over is applied among the breeding individuals and they exchange a part of bit-strings that codes the weights of the three networks.

The number of offspring for each individual depends on the hunger level of the mothers. Mutation occurs in the bit-strings of offspring at birth. Offspring start to forage at the nest where they were born.

Initial Settings

The initial population size was 400 individuals who were randomly distributed in the test fields. The crossing over rate and mutation rate were always set at 0.9 and 0.5, respectively.

The clumped and even distributed resource patches were generated with β-random procedure to compare with Horn's model. A static but highly clumped distribution was adopted for the patchy environment instead of an irregularly moving resource-rich patch. The two parameters for β function were both set to 0.001 in the resource-clumped condition. In the even resource distribution, the two parameters were set to two. Thus, the clumped and even distributions corresponded to the bath-tab and unimodal β functions, respectively.

The realized frequency of each land type is shown in Table 2. Note that there were only three cell types (DW, DL, LF) in the clumped resource distribution. It was due to the random generator not producing values that corresponded with the other land types. On the contrary, all land types appeared in the even resource environment.

Four replicates were conducted changing seeds for the random generator for each resource distribution. All other parameters, such as mutation and crossing over

Lande Type	PL	SW	PL SW	DW	DL	LF	LF SW
Clumped	0	0	0	410	465	25	0
Even	91	397	83	61	54	118	96

Table 2: Realized Frequency of Land Types

rates, were common in each replicate. Each run was performed for 10,000 generations.

SIMULATION RESULTS

There was little variation in the results among the four replicates for all aspects of the simulation. Thus, the results are credible despite the small sample size.

Nest Formation

Figure 7 gives a typical example of the change of nest distribution through generations in the clumped and evenly distributed resource arenas. Back ground patterns show resource abundance. Lighter colors indicate higher resource patches. The maximum and the minimum values are 300 and 0, and they are assigned pure white and pure black, respectively. When the resources were clumped (the left hand side panels in Fig. 7), egrets formed several large colonies as time elapsed. Note that the colonies were not always located at the resource abundant cells. Moreover, the largest three colonies seldom changed their location throughout generations.

If the resources were evenly distributed (the right hand side panels in Fig. 7), egrets never formed a large colony during 10,000 generations. Nests were dispersed all over the arena and there was no tendency to locate the nests at resource rich cells.

In the following two sections, all figures have the same implementation. For Generations less than 1,000, the resolution of the graphs is set to 10 generations. For 1,000 or more generations, the resolution was set to 100 generations. Thus, the graphs look busy during the first 1,000 generations and become calm in the following generations. All statistical analyses were conducted with the resolution of 100 generations.

Population Dynamics

Figure 8 depicts a typical example in the change of the total population size of foraging egrets in the two test arenas. The mean population size in the clumped resource condition almost always exceeded that in the even resource condition. The mean population size through generations was significantly larger in the clumped resource condition than in the even resource environment for all replicates (mean for clumped = 316.419 mean for even = 228.004, nested ANOVA result was F = 325.351, d.f.=7, P <<.01). Irregular fluctuations were observed in the population dynamics with the clumped resource field.

A typical example of the change of m*/m of foraging and nesting egrets in the test field is observed in Fig. 9.

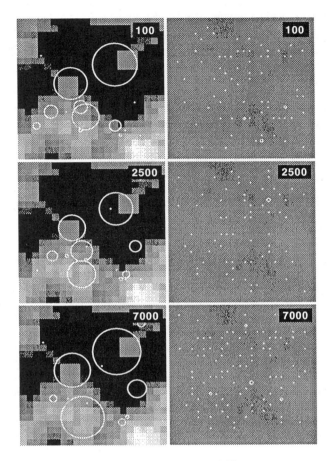

Figure 7: Distribution of Nests

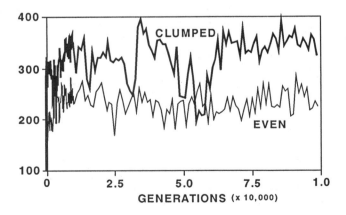

Figure 8: Total Number of Egrets

The m*/m value in the clumped resource field was much higher than that in the even resource field (mean for clumped = 60.112, mean for even = 5.176, F = 3317.156, d.f.=7, P << .01). However, the tendency of foraging egrets to form flocks was not constant and highly variable throughout generations under the clumped resource condition.

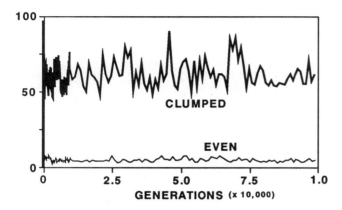

Figure 9: m*/m of Foraging Egrets

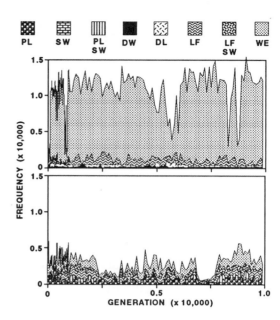

Figure 10: Number of Egrets Attracted by Each Cell Type during Foraging

Figure 10 shows the number of landings by foraging egrets on each patch type. In the resource clumped field (upper panel of Fig. 10), most egrets landed on the patch where other egrets were foraging (labeled as WE). The magnitude of the number of egrets landing on the WE cells was much higher than in the other cell types. The second best is LF which was assigned the highest reward among the cell types.

The attractiveness of egrets was also prominent in resources that were evenly distributed (lower panel of Fig. 10). However, the relative abundance of attractiveness was quite small compared with that observed in the resource clumped field. The second and following best cell types (LF, PL, LF&SW, and PL&SW) shared almost the same amount of relative abundance.

Egrets themselves were attractive while they made nests for breeding in the resource clumped patches (upper panel of Fig. 11). However, the dominance of WE was not as strong as the attractiveness in foraging individuals. The second best was LF that frequently predominated during 10,000 generations.

Where resources were evenly distributed, the attractiveness of egrets decreased and the other patch types also attracted nesting egrets (lower panel of Fig. 11). The attractiveness of each cell type changed frequently throughout generations.

Analysis of Action Network

Among the three networks implemented the action net was used in every bout of foraging and nesting, and, thus, should have recorded the most important foot prints in the evolution of group foraging behavior.

Figure 12 shows the changes in the mean weight for each path of the 9-footed-basal nodes in the action net-

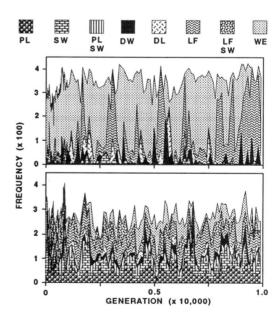

Figure 11: Number of Egrets Attracted by Each Cell Type during Nest Formation

work. In the resource clumped distribution (Fig. 12), the weights for DW and DL were significantly negative throughout 10,000 generations (sign test, P << .01) for three replicates. On the contrary, weights for LF and WE were significantly positive (sign test, P << .01). Weights for the other five land types showed neither positive nor negative tendencies. These weights had no correlation between each other.

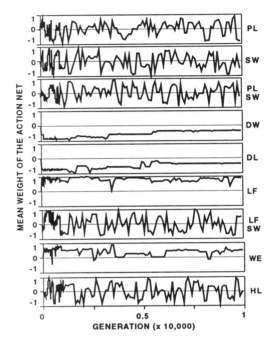

Figure 12: Mean weights for Action Net in the Resource Clumped Condition

When resources were evenly distributed (Fig. 13), the weights for DW and DL were significantly negative and LF was significantly positive (sign test, P << .01), as in the clumped distribution. However, the weight for WE did not significantly deviate from zero. Instead the weights for SW and PL&SW were significantly negative for three replicates (sign test, P << .01). Correlations between these weights were very low and most of them were not significant.

Figures 14 and 15 show the change in the variance of weights for each path of the action network. When resources were clumped (Fig. 14), variances were very low for those weights that showed significant deviations from zero (DW, DL, LP, and WE). On the contrary, when resources were evely distributed (Fig. 15), variances were relatively high even for the weights that showed significant deviation from zero (see DW, DL, and LF).

DISCUSSION

Our ALife simulation demonstrates that group foraging and coloniality emerged simultaneously in the patchy environment. The modeled egrets did not have prior knowledge as to which land type had rich resources nor

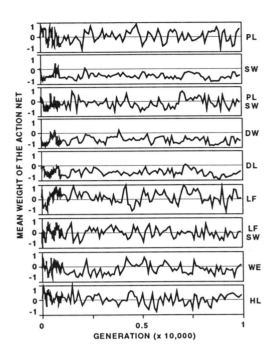

Figure 13: Mean weights for Action Net in the Even Resource Condition

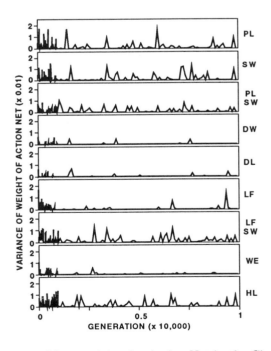

Figure 14: Mean weights for Action Net in the Clumped Resource Condition

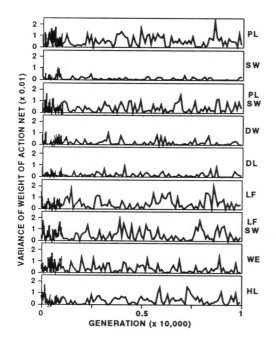

Figure 15: Mean weights for Action Net in the Even Resource Condition

how to change foraging strategies. All what the authors previously implemented were simple stimuli from the land types, hunger levels, and random weighting values of neural networks.

Foraging Behavior

The most interesting and important aspect is the emergence of local enhancement. This simple and local rule was applied to forage in the resource clumped environment (Fig. 10). Even within an even resource environment, local enhancement was used to some extent (Fig. 10).

In the clumped resource environment there were only four cell types. Among them, LF (Lotus Field) contained the maximum amount of prey . Figure 12 and 14 show that the weight for LF was nearly unity with small variation during evolution. The mean weight for WE (White Egrets) had a greater degree of variance although this was still relatively small and remained positeve.

On the contrary, the weights for DW (Deep Water) and DL (Dry Land) never exceeded zero. Thus, the artificial egrets have learnt to discriminate good and bad paths using the proximate cue of existence of other egrets in their view. Note that the values for DW and DL came close to zero as generations passed. This may correspond to the sealding effect as discussed in Ackley and Littman (1991).

Where resources were evenly distributed, the artificial egrets seemed to be confused by variable information on land types. They were thoughtful and landed much less frequently in any kinds of land types than in the clumped resource arena (Fig. 10). However, they evaluated cor-

rectly that SW, DW, and DL were resource poor types. The relative resource values for LF, PL, SW, PL&SW, and LF&SW were not so different from each other in the even resource condition. Even in this difficult situation, the egrets seemed to succeed in evaluating LF as the most favorable patch type (Fig. 11).

One confusing result was that, even though the weight value for WE showed a negative tendency during the 10,000 generations, the egrets seemed to use the existence of others as a cue for landing during foraging bouts (Fig. 10). This may be a result of complicated interactions with other stimuli, however this consideration is beyond the scope of the crude analysis of the action networks employed in this study.

Nest Formation

Local enhancement also played an important role in nest formation in the patchy environment. The cue of WE was most frequently used as the nesting stimulus (Fig. 11). The results from nest distribution suggest that egrets did not always make colonies at the resource abundant patches (Fig. 7).

Moreover, the egrets were conservative about the location of large colonies. This may be simply because all the individuals that had evolved were those which survived and bred in the consecutive generations. Figure 8 demonstrates that the population size dramatically decreased and then rapidly increased at the beginning of simulations. Offspring starting their foraging game at their birth place were the survivors of bottleneck(s) at the early stage of evolution. In this sense, historicity might have played an important role in this sense. The relationship between bottlenecks and philopatry is also discussed in several avian species (e.g., Ovenden et al. 1991).

In an even resource environment, the existence of conspecies did not work as a cue for nest formation (Fig. 11). The number of nests was fewer than in the clumped resource condition, but the relative nest abundance against that in the clumped distribution was much higher than the landing frequency of foraging individuals (Fig. 10). Figure 7 also suggests that there was no philopatry in the even resource environment.

Information Transfer

Evolution of coloniality is a controversial topic in behavioral ecology. The information centre theory (ICT) is one of the candidates to explain the colony formation (Brown 1988a, 1988b, Clode 1993, Galef 1991), which is also rejected in some studies (e.g., Mock et al. 1988, Richner and Marclay 1991). In the present ALife model, the colonies in the patchy environment were not information centers at all. Information was rather transferred by simple local enhancement among wandering egrets at foraging sites. Of more interest is that artificial egrets sharing the same colony were often highly related at the "genetic" level. The preference of patch types coded in genetics might generate the leaving patterns that would support the ICT.

Local enhancement has both positive and negative aspects. If the target individual is foraging on a resource rich patch, it guarantees efficient foraging for the followers. However, an excess of attractiveness accelerates the rate of resource depletion in the patch. Egrets moderately relied on the existence of other individuals where resources were evenly distributed. It is possible that the trade-off between the benefits and costs of social information had been resolved (Valon 1989, 1993).

Flexibility and Robustness of Adaptive Behaviors

Several key aspects require consideration in order to achieve the simultaneour realization of colonial and flock foraging, as developed in the present study. The first is the masking of the land value of patch types. Foraging egrets did not have prior knowledge of the actual resource abundance of each patch. In this sense, the present situation resembles the language acquiring game simulated by Maclennan (1991). The arbitrary matching of actual resource amounts and land types might cause highly fluctuating dynamics in local enhancement (Fig. 10 and 11).

The second aspect is an egret's view point. The model adopted a hierarchical view consisted of 81 cells for each egret. The view enabled the efficient collection and abstraction of information at multiple scales of patches (Kotliar and Wiens 1990). The number of view layers was fixed in the present model, but it may have been improved if the number of layers was adjustable to resource patchiness.

In order to force ALives to find their own answers to problems, it is necessary for models to be equipped with highly flexible meta-models, such as GA, CA, and Neural Network. Combining these meta-models makes ALife simulations a more powerful tool to investigate evolutionary processes (e.g., Lindgren's IPD model, Ray's Tierra, and Kauffman's NK-model). Those meta-models are often used for asking open-ended simulations. The concept of open-ended implicitly means that the systems are robust against changes in the environment. Robustness is achieved by obtaining a high variation in strategies and behaviors to respond to variable situations. Pleiotropy and epistasis in genetic codes are important to realize such robustness in evolving systems.

Models incorporating learning may behave differently from that predicted by optimal strategy models (Ollason 1987). In the present model, neural networks equipped with GA played an important role to find robust answers from a huge and complicated domain spanned by parameters.

There was a slight discrepancy between the prediction of Horn's model and our simulation results. In Horn's model, the territorial birds gain more in an even resource environment than in a clumped one. In the present model, the total gain was lower in even resource distribution. The answers obtained by iterated learning may not be the optimal answers for our foraging arena. However, the biological environment around artificial egrets changed rapidly and irregularly by interactions among patch qualities and egrets. To live on in such a situation, egrets should behave robustly rather than optimally.

Territoriality, the Next Target

Many theoretical works and models have asked where birds make colonies and what size the colonies should be (e.g., Brown et al. 1990). Those discussing the relationship of resource distribution are modified version of Ideal Free Distribution models (Cairns 1989, Gibbs 1991, Gibbs et al. 1987). Birds in the models are fairly omniscient and know exactly where the resource patches are distributed.

Individual Based Models (Judson 1994) are applied to construct general views of the evolution of coloniality (e.g., Wolff 1994). With the help of Object Oriented Programming and Artificial Intelligence, the models create hypothetical individuals that behave like birds in the computer. The problem is the implementation of each individual. In those models, rules of the birds are all specified by the modeller and there is no emergence of colonial and flock foraging.

The present study succeeded in emerging colonial and flock foraging in a patchy environment. However, the artificial egrets could not emerge territoriality. The territoriality and coloniality sometimes coexist in natural systems (Marion 1989). Several avian studies have showed that young birds prefer flock foraging but the adults prefer solitary or territorial foraging (Henderson and Hart 1991, van Vessem et al. 1984).

Among egrets and herons, the foraging behavior varies from species to species. Gray herons (*Ardea sinerea*) and great egrets (*Egretta alba*) prefer to forage in solitary whereas cattle egrets (*Bubulcus ibis*) often forage in groups. Little egrets (*Egretta garzetta*) often forage in solitary and occasionally form flocks. In the middle part of Japan, these egrets form heterospecific colonies.

What is needed to implement such complex territoriality in the present model? Should putting another basal network and letting it control the repelling behavior against other egrets be considered? How many new neurons are necessary to achieve age related and species specific territoriality? In what way can we analyze results of interactions between multiple networks?

In making weak ALife simulations, the most important thing that we have to remind ourselves is the elimination of extra information from the target objects. For example, several types of information, such as the absolute coordinate axis and the criterion of good patch, were excluded or masked in the present model. That information might be useful if we try to simulate animal behaviors in the light of the optimal theories or conventional kinetic models for prediction. However, in ALife models, not the programmer but the ALives themselves have to seek and obtain what is good behavior in a given artificial world.

ACKNOWLEDGEMENTS

We thank Mr. Tomomi Anzai for valuable discussion on the earlier version of our model. We also deeply thank Ms. Annette Raper for helping to polish up English.

This study is partially supported by the University of Tsukuba Research Projects.

References

Ackley, D. and M. Littman. 1991. Interactions between learning and evolution. Artificial Life II, edited by Langton Christpher G. , C. Taylor , J. D. Farmer and S. Rasmussen. Addison-Wesley:New York,487-510

Armstrong, E. A. 1971. Social signalling and white plumage. Ibis 113:534

Arthur, W. B. , Y. M. Ermoliev and Y. M. Kaniovski. 1987. Path-dependent processes and the emergence of macro-structure. European J. Operational Research 30:294-303

Bacon, P. J. , F. Ball and P. Blackwell. 1991. A model for territory and group formation in a heterogeneous habitat. J. Theor. Biol.:148,445-468

Brown, C. R. 1988. Enhanced foraging efficiency through information centers: A benefit of coloniality in cliff swallows. Ecology 69:602-613

Brown, C. R. 1988. Social foraging in cliff swallows: Local enhancement, risk sensitiveity, competition, and teh avoidance of predators. Anim. Behav. 36:780-792

Brown, C. R. , B. J. Stutchbury and P. D. Walsh. 1990. Coice of colony size in birds. TREE 5:398-403

Caccamise, D. F. and D. W. Morrison. Avian communal roosting; Implications of diurnal activity centers. Am. Nat. 128:191-198

Cairns, D. K. 1989. The regulation of seabird colony size: A hinterland model. Am. Nat. 134:141-146

Clark, C. W. and M. Mangel. 1986. The evolutionary advantages of group foraging. Theor. Popul. Biol. 30:45-75

Clode, D. 1993. Colonially breeding seabirds: predators or prey ? Trends Ecol. Evol. 8:336-338

Collins, R. J. and D. R. Jefferson. 1991. AntFarm: Towards simulated evolution. Artificial Life II, edited by Langton Christpher G. , C. Taylor , J. D. Farmer and S. Rasmussen. Addison-Wesley:New York,579-601

Galef, Jr. B. G. 1991. Information centre of Norway rats: sites for information exchange and information parasitism. Anim. Behav. 41:295-301

Gibbs, J. P. 1991. Spatial relationships between nesting colonies and foraging areas of great blue herons. Auk 108:764-779

Gibbs, J. P. , S. Woodward M. L. Hunter and A. E. Hutchinson. 1987. Determinants of great blue heron colony distribution in coastal marine. Auk 104:38-47

Henderson, I. G. and P. J. B. Hart. 1991. Age-specific differences in the winter foraging stragegies of rooks Corvus frugilegus. Oecologia 85:492-497

Horn, H. S. 1968. The adaptive significance of colonial nesting in the brewer's blackbird Euphagus cyanocephalus. Ecology 49:682-694

Judson, O. P. 1994. The rise of the individual-based model in ecology. TREE 9:9-14

Kotliar, N. B. and J. A. Wiens. 1990. Multiple scales of patchiness and patch structure: a hierarchical framework for the study of heterogeneity. OIKOS 59:253-260

Krebs, J. R. and N. B. Davies. 1881. An introduction to behavioural ecology. 1st ed. Blackwell:Oxford

MacLennan, B. 1991. Synthetic ethology: an approach to the study of communication. Artificial Life II, edited by Langton Christpher G. , C. Taylor , J. D. Farmer and S. Rasmussen. Addison Wesley:New York,631-658

Marion, L. 1989. Territorial feeding and colonial breeding are not mutually exclusive: the case of the grey heron.(Ardea cinerea) J. Anim. Ecol. 58:693-710

Mock, D. W. , T. C. Lamey and B. A. Thompson. 1988. Falsifiability and the information centre hypothesis. Ornis Scand. 19:231-248

Morrison, M. L. and R. D. Slack. 1977. Role of flock feeding in Olivaceous cormorants. Bird Band 48:277-279

Ollason, J. G. 1987. Learning to forage in a regeneraing patchy environment: Can it fail to be optimal. Theor. Popul. Biol. 31:13-32

Wust-saucy, J. R. R. A. , N. R. Bywater and R. W. G. White. 1991. Genetic evidence for philopatry in a colonially nesting seabird, the fairy prion. (Pachyptila turtur) Auk 108:688-694

Packer, C. and P. Abrams. 1990. Should co-operative groups be more vigilant than selfish groups? J. Theor. Biol. 142:341-357

Powell, G. V. N. 1974. Experimental analysis of the social value of flocking by starlings (Sturnus vulgaris) in relation to predation and foraging. Anim. Behav. 22:501-505

Pulliam, H. R. On the advantages of flocking. J. Theor. Biol. 38:419-422

Pulliam, H. R. and T. Caraco. 1984. Living in groups: Is there an optimal group size? 2nd ed. Behavioural ecology, edited by Krebs J. R. and N. B. Davis. Brackwell:Oxford,122-147

Richner, H. and C. Marclay. 1991. Evolution of avian roosting behavior: A test of the information centre hypothesis and of a critical assumption. Anim. Behav. 41:433-438

Rubenstein, D. I. 1978. On predation, competition, and the advantages of group living. Perspectives in ethology, edited by Bateson P. P. G. and P. H. Klopfer. Plenum Press:New York

Scott, D. 1984. The feeding success of cattle egrets in flock. Anim. Behav. 23:1089-1100

Thompson, W. A. , J. Vertinsky and J. R. Krebs. 1974. The survival value of flocking in birds: a simulation model. J. Anim. Ecol. 43:785-820

Thorpe, W. H. 1963. Learnign adn instinct in animals. Methuen:London

Valone, T. J. 1989. Group foraging, public information, and patch estimation. Oikos 56:357-363

Valone, T. J. 1993. Patch information and estimation: A cost of group foraging. Oikos 68:258-266

van Vessem, J. , J. D. Draulans and A. F. de Bont. Movements of radio-tagged Grey Herons *Ardea cinerea* during the breeding season in a large pond area. Ibis 126:576-587.

Wade, M. J. and S. G. Pruett-Jones. 1990. Female copying increases the variance in male mating success. Proc. Natl. Acad. Sci. USA 87:5749-5753

Wait, R. K. 1981. Local enhancement for food finding by rooks (*Corvus frugilegus*) foraging on grassland. Z. Tierpsychol. 57:14-36

Ward, P. and A. Zahavi. 1973. The importance of certain assemblages of birds as 'information centres' for food finding. Ibis 115:517-534

Weatherhead, P. J. and H. Greenwood. 1981. Age and condition bias of decoy-trapped birds. J. Field Ornithol. 52:10-15

Wilson, E. O. 1975. Sociobiology. Harvard University Press:Cambridge

Wolff, W. F. 1994. An individual-oriented model of a wading bird nesting colony. Ecol. Model. 72:75-114

A Model of the Effects of Dispersal Distance on the Evolution of Virulence in Parasites

C. C. Maley

The Artificial Intelligence Laboratory *

NE43-803

Massachusetts Institute of Technology

cmaley@ai.mit.edu

617-253-6567

Abstract

The effects of differing dispersal distances on the evolution of virulence in parasites was explored through the use of a configuration model. Both hosts and parasites were individually represented on a two-dimensional surface. Each parasite had a "gene" for virulence that determined the amount of energy the parasite removed from its host in a single time step. When a parasite reproduced, its offspring was placed near to, or far away from the parent depending on a dispersal distance parameter. The offspring's virulence was also mutated slightly at reproduction. Distance of dispersal had a small but statistically significant, positive effect on the evolution of the parasite's virulence. That is, those parasites that were forced to disperse their offspring further away, evolved a higher level of virulence than the parasites that dispersed their offspring locally.

Introduction

There is a growing interest in the application of evolutionary theory to disease control and treatment [Ewald, 1983, Williams and Nesse, 1991]. At the same time, computer modeling is opening up new areas of evolutionary theory by allowing us to ask questions that were previously extremely difficult to test in the lab [Koza, 1994, Lindgren and Nordahl, 1994, for example]. This paper represents the cross fertilization of these two research projects in a model of the evolution of virulence in parasites.

The Evolution of Virulence

Why is the parasite that causes the common cold less deadly than the protozoa that causes malaria (genus *Plasmodium*) or the bacteria that cause cholera (*Vibrio cholerae*)? Are these parasites becoming more or less virulent as they coevolve with humans, and why? These questions are neither academic nor intractable.

*This research was made possible through the generosity of the MIT AI Lab. It was supported by NSF grant 9217041-ASC and ARPA under the HPCC program. I would like to thank W.D. Hamilton, L.D. Hurst, S. Rich, R.C. Berwick, E.S. Johnson, and M.S. Golden for their helpful comments and support.

The Theory I am using the term parasite in the broadest possible understanding. A parasite is any organism that gains its nutrients from other living organisms and usually has a deleterious effect on those organisms [Begon *et al.*, 1990]. It was long thought that a parasite's relationship to its host would evolve to become less deleterious, or virulent, over time. This was based on an essentially group selectionist, or "good of the species" argument. Parasites that did less damage to their hosts would thereby preserve their hosts for longer and thus live longer themselves as compared to parasites that were more damaging to their hosts. This kind of argument still persists today despite advances in the theory. Anderson and May have published extensive analyses of the coevolution of parasites and hosts.

> ...formal studies make it clear that the coevolutionary trajectory followed by any particular host-parasite association will ultimately depend on the way the virulence and the projection of transmission stages of the parasite are linked together: depending on the specifics of this linkage, the coevolutionary course can be toward essentially zero virulence, or to very high virulence, or to some intermediate grade. [Anderson and May, 1982]

In other words, there are significant pressures on parasites that might select for varying levels of virulence, not just for avirulence or commensalism. This can be seen formally in a simple equation for the reproductive rate given by May and Anderson (1990).

$$R_0 = \frac{\lambda(N)}{\alpha + b + v} \tag{1}$$

Here N is the current population size of the hosts and $\lambda(N)$ is the rate of successful transmission of the parasite offspring to other hosts as a function of the population density of the hosts. The denominator contains the variables that might reduce the pool of hosts available for infection. Here, α is the disease induced death rate of the hosts, b is the death rate for the hosts from causes other than the disease, and v is the recovery rate of the hosts. After a host has recovered, it is considered to be immune to the disease.

Given this relatively simple equation and the assumption that selection puts pressure on the parasites to maximize R_0, May and Anderson show that the virulence of the parasite, α, may be expected to evolve toward any number of results. For example, if transmission, λ, and recovery, v, are independent of virulence, α, then R_0 is maximized by avirulence, $\alpha \to 0$, a commensal relationship. However, such independence is not the norm. If we assume that the likelihood of transmission, λ, increases as a linear function of the virulence of the parasite, α, then R_0 is maximized by maximum virulence, $\alpha \to \infty$. This would evolve regardless of the extreme likelihood that the parasites would thereby drive their hosts to extinction and thus perish themselves. May and Anderson (1990) conclude, "What are needed are factual data about the functional relationships among λ, v and α, derived from the life history details of specific host-parasite associations. In no single instance do we have enough information to determine the likely evolutionary trajectory dictated by maximizing R_0 in [equation 1]."

The Observations In fact, parasite reproduction is generally correlated with disease severity [Ewald, 1983]. As Ewald puts it, the question then becomes, "What environmental conditions select for different levels of parasite reproduction and, hence, for different levels of severity?" Two examples help to illustrate the point. Various species of protozoa in the genus *Plasmodium* cause the disease malaria in human hosts. These parasites are transmitted to new hosts through the biting of mosquitoes that transfer small amounts of blood from human to human. The high levels of virulence in these parasites is hypothesized to be an adaptation to their mode of dispersal. By reproducing in great numbers the parasite achieves two effects. First, any small amount of blood taken from the host is likely to contain a few protozoans. Second, by causing a sever reaction, the host is immobilized and thus more easily bitten by mosquitoes. In contrast, the common cold is transmitted through social contacts and so a parasite that immobilized its host would thus be reducing its likelihood of transmission to new hosts. This may help to explain why the cold is a relatively mild affliction.

Ewald (1983) collected data on a large number of human diseases and found a correlation between disease severity and parasites transmitted through biting insects. The cholera epidemic in India in the 1950's and 1960's provides a tantalizing evidence of the relationship between mode of parasite transmission and virulence. The bacteria that cause cholera (*Vibrio cholerae*) are generally transmitted through drinking-water supplies. The bacteria cause diarrhea in their hosts. The diarrhea may contaminate the water supplies either through poor sewage disposal or through the washing of contaminated sheets. In any case, the parasite's chances of being transmitted are not decreased by immobilizing its host. In fact, its chances of getting into the water supply may even increase if care-takers wash contaminated bed-clothes in the water supply of a down-stream community.

As water supplies were purified in cholera-endemic regions in India during the 1950's and 1960's, the milder agent of cholera, the El Tor type of *Vibrio*

cholerae, displaced the more dangerous form, classical *V. cholerae*. [Ewald, 1993]

Unfortunately, the evidence for the evolution of virulence in relation to the parasite's mode of dispersal remains largely observational, and thus only correlational [Ewald, 1983]. Ewald (1983) goes on to predict that parasites that are highly mobile should cause severe diseases.

The correlation between virulence and parasite mobility fits with a body of evolutionary theory that analyzes the continuum from purely vertical transmission, from parent to offspring, to purely horizontal transmission, i.e., transmission between unrelated hosts. When a parasite is transmitted vertically, it maximizes its reproductive success by maximizing its host's reproductive success. Thus, the parasite should evolve to a commensal, if not symbiotic relationship with the host species. However, a parasite that is transmitted horizontally to unrelated hosts has no particular interest in preserving its current host. Thus horizontally transmitted parasites are expected to evolve high levels of virulence [Williams and Nesse, 1991, Herre, 1993, Nowak, 1991, Hurst, 1991].[1]

Configuration Models of Parasites

Differential and difference equations have been used for many years to model population dynamics [May, 1973]. These models lump individuals together into homogenous groups that can be represented by a single number, the number of individuals in that group. They may break a population down into a number groups depending on any number of characteristics, like age, sex, weight, etc. [Caswell, 1989]. However, in each subgroup, the individuals within the groups are still treated as identical. In this way, these models may describe the distribution of individuals in the population. Caswell and John (1992) label these distribution models. In contrast, a relatively new form of computer model, often called an individual-based model, explicitly represents each individual in the population as a separate data structure. The model keeps track of the configuration of all of the individuals in the population. Since the distribution model is also based on the individuals in the population, this new form of model may be called a configuration model [Caswell and John, 1992].

Although configuration models are relatively new to ecological modeling [Huston *et al.*, 1988], they have been essential to artificial life. Emergent dynamics are often observed through the local interaction of individuals. This is the basis of cellular automata models as well as many of the early models in artificial life [Langton, 1992, Fontana, 1992, Koza, 1992, Taylor *et al.*, 1989, Travers, 1989, Reynolds, 1987, for example]. Recently, attention has been focused on the theory behind the construction of configuration models [Maley and Caswell, 1993,

---

[1] It should be noted that definitions of virulence vary amongst these authors. For example, Herre (1993) operationalizes virulence as a reduction in lifetime reproductive success of the host, rather than disease induced death rate as per Ewald (1983) and May and Anderson (1990).

Baveco and Lingeman, 1992]. configuration models open up the possibility of explicitly representing the interactions between individuals. On occasion this has produced evidence that contradicts or extends the analysis of distribution models [Boerlijst and Hogeweg, 1992, Collins and Jefferson, 1992].

In the case of a host-parasitoid interaction, a configuration model can explicitly model the reproduction and transmission of parasites. This allows an intuitive and direct analogy between the dynamics of the model and the dynamics of the biological system. Baveco and Lingeman (1992) have constructed such a model to examine the population dynamics of a host- parasitoid system in a patchy environment.

Baveco and Lingeman locate the hosts and parasites in a 5 × 5 grid of patches. Both hosts and parasites have a three stage life history; eggs, larvae, and adults. Hosts can only be parasitized in their larval stage. Parasite transmission in their system is density-independent. Similarly, host reproduction is density-independent. Furthermore, the life span of an individual is predetermined at birth. The individual will die after the determined number of time steps unless it is a larval host that becomes infected by the parasites and so dies before it can reproduce. Parasites disperse from a patch to a neighboring patch when no hosts remain in their patch.

Baveco and Lingeman found that if the reproduction rate of the parasites was greater than the reproduction rate of the hosts, the parasites quickly drove the hosts to extinction. On the other hand, if the reproduction rate of the parasites was less than the reproduction rate of the host, the parasite population never caught up to the host population. This later case is probably due to the fact that host reproduction was assumed to be density-independent. However, in the former case, the extinction of the hosts was found to be a robust effect for most of their parameter settings. Baveco and Lingeman tried increasing the dimensions of the environment from 5 × 5 up to 10 × 10. They tentatively conclude that an environment of 10 × 10 patches is about the minimum size necessary for the persistence of the host population.

Baveco and Lingeman were not trying to model the evolution of virulence. In fact, they were not trying to model evolution at all. They took a classical distribution model [Nicholson and Bailey, 1935] of an ecological interaction between hosts and parasites and examined the dynamics of the system when all the individuals were explicitly represented. However, a similar model might be constructed where the parasites might have a "gene" for virulence. Given the possibility of mutation at the time of reproduction, the parasites might evolve varying levels of virulence over time in response to the selective pressures in the model.

A Model of the Evolution of Virulence

Ewald (1983), May and Anderson (1982 and 1990) all predict that virulence will evolve in relation to the mode of transmission of the parasite. However, mode of transmission is not an easily quantifiable characteristic. The

problem can be simplified by focusing on the distance over which the parasites disperses. Whether it is transmission through biting insects, contaminated water supplies, or through sneezing, the dispersal of the parasites can be quantified by the distance of dispersal from the infected host. The question can now be asked, what is the effect of the distance of dispersal on the evolution of virulence in parasites?

The Hypothesis

Hamilton (1992) suggested that we should expect parasites that disperse over short distances to evolve to be less virulent than parasites that disperse over long distances. This follows from the continuum from vertical to horizontal transmission. Parasites that reproduce locally will be likely to share their host with their offspring. Thus they damage their own reproductive fitness by damaging their host. On the other hand, a parasite that disperses over long distances is less likely to share its host with its offspring, and so can exploit its host with near impunity.

General Structure

The hypothesis as it stands requires a system of only four significant components: host organisms, parasite organisms, spatial structure in the environment so that the parasites can disperse to new hosts, and a parasite gene for level of virulence that can evolve over time. It is necessary to use a configuration model because the spatial relationships between organisms directly affects the dynamics of the model. Because the hypothesis does not involve the immune response of the host, host immune systems were excluded from the model.

I chose to use an energy flow model [Dewdney, 1989, Rizki and Conrad, 1986, Taylor et al., 1989]. Each discrete time step, a quantified amount of energy flows from the environment into the host organisms. At the same time, parasites are removing energy from the hosts to support the loss of energy in the parasites due to their metabolism. Additional energy is used by both hosts and parasites when they reproduce and that energy is given to the newborn organisms.

Hosts and parasites interact in a two-dimensional environment, or grid. The edges of the grid are connected to form a toroidal surface. Each patch in the grid can hold a single host. Each host can hold a virtually unlimited number of parasites while it still lives.[2] Death occurs when an organism's energy level drops to zero.

Hosts are relatively uninteresting. They only have one attribute: their energy level. Hosts cannot move. They can reproduce when their energy reaches a certain threshold. However, the newborn only survives if there is an empty patch in the grid adjacent[3] to the parent.

---

[2] The maximum number of parasites allowed to infest a single host was set to 500. Since the actual number of parasites infesting a host tended to stay below 10, this should not have influenced the dynamics of the model.

[3] Every patch has eight adjacent patches corresponding to the eight points of the compass. Since the edges of the grid wrap around, the surface actually forms a torus with no edge

Since the hosts have no "genes" they cannot evolve.

Parasites have two attributes: energy and virulence. The virulence attribute is the number of energy units the parasite takes from its host each time step. Like hosts, parasites reproduce when their energy level reaches a certain threshold. However, they have genes for virulence, so the virulence attribute may mutate (e.g., ±6) in the newborn parasite. Furthermore, there is a parameter in the model, dispersal distance, that restricts where the newborn parasite may be randomly placed in the environment.

For each time step, the model executes four steps. 1. The energy levels in the hosts are updated in parallel. 2. The energy levels in the parasites are updated in parallel. 3. Newborn parasites are placed in the environment. 4. Newborn hosts are placed in the environment.

Energy flows in the following way: First, each host gains a fixed amount of energy, *food*, each time step. Second, each host loses energy equal to the sum of the virulences of the parasites infesting it. Third, each parasite gains its *virulence* worth of energy units each time step, as long as its host is still alive, *host energy* > 0. Fourth, each parasite loses a fixed amount of energy, *cost of living* each time step. In addition to this basic energy flow, an organism loses a fixed amount of energy, *cost of reproduction*, each time it reproduces The newborn organism starts off with that amount of energy.

Parameters

Although the basic structure of the model has only four components, there are a host of details that must be specified in order to model the life histories and interactions of the organisms. These details can be divided into three categories: fundamental parameters, parameters that can be derived from previously defined parameters, and the initial conditions in the model at the beginning of a simulation. [In all cases the default settings will be put in brackets.]

Fundamental Parameters The fundamental parameters in the model are the bottom line. They are used to derive reasonable quantities for many other parameters in the model, and so define basic choices about the scale of the model. There are only four such fundamental parameters: *dispersal, reproduction rate, host cost of reproduction* , and *size* .

Dispersal distance is the independent variable in the experiment. The *dispersal distance* represents the maximum distance between a parasite and the placement of its offspring. *Dispersal distance* defines the size of the square, with the parental parasite's patch in the center, in which the offspring parasite is randomly placed. The probability of placement for a parasite was the same for every patch within that square.

Reproduction rate is the number of time steps it would take an uninfested host to build up enough energy to reproduce. In other words, it is a sort of generation time. *Reproduction rate* determines the granularity of time in the model. [2]

effects.

Host cost of reproduction is the number of energy units transferred from a host parent to its offspring at birth. This determines the basic energy scale or granularity in the model. The larger the number the finer the grain. [1000]

Finally, *size* determines the relative scale between the hosts and parasites. It is roughly the number of parasites that can fit in a host. The larger the number, the smaller the parasites. [5]

Derived Parameters Most of the parameters in the model are derived from *size, host cost of reproduction*, and *reproduction rate*. They generally determine the specifics of energy flow through the model.

Energy enters the cycle as *food* . *Food* is the amount of energy given to a host each time step. *food = host cost of reproduction / reproduction rate*. [500]

Similarly, energy leaves the cycle in the *cost of living for parasites*. This is the amount of energy removed from every parasite each time step. *cost of living for parasites = food / size*. [100]

Hosts automatically reproduce when their energy level reaches the *host reproduction threshold* . At that point, the host loses the *host cost of reproduction* amount of energy. *Host reproduction threshold = 2× host cost of reproduction*. [2000]

Parasites also lose energy when they reproduce equal to the *parasite cost of reproduction = host cost of reproduction / size*. [200]

Like hosts, parasites automatically reproduce when their energy level rises above the *parasite reproduction threshold = host reproduction threshold / size* . However, unlike hosts, parasites might be so virulent that their energy level jumps far above the *parasite reproduction threshold* . Parasites are forced to reproduce multiple times until the repeated deductions of the *parasite cost of reproduction* lowers the parasite's energy level below the *parasite reproduction threshold*. In this way, highly virulent parasites reproduce more quickly than less virulent parasites. [400]

The *mutation factor* scales the range of mutation of the virulence gene in the parasites. [8] It is not strictly a derived parameter since it does not formally depend on other parameters. However, it has been set such that the actual range of mutation ($\frac{food}{2 \times mutation\,factor \times size}$ $[= 6]$) in the virulence gene is small compared to the amount of energy entering a host, *food*.

Finally, for ease of programming, there is a *carrying capacity of the hosts* parameter that puts a strict limit on the number of parasites that may infest the host. This was set high enough that in actuality it never affected the dynamics of the model. [500]

Initial Conditions The only pieces that remain to describe are the initial settings of the characteristics in the organisms at the start of a simulation. Hosts only possess a single characteristic, energy, while parasites possess both energy and virulence characteristics. Hosts start with an energy that is determined randomly in the range of 0 to *host reproduction threshold*. However, any newborn host starts out life with *host cost of reproduc-*

Parameter	Description	Default
energy	Energy units accumulated by an organism	variable
virulence	Amount of energy a parasite removes from its host	variable
dispersal distance	Radius of dispersal for parasite offspring	1-15
reproduction rate	Uninfested host generation time	2
host cost of reproduction	Energy penalty for host reproduction	1000
size	Size scale between hosts and parasites	5
food	Energy given to a host each time step	500
cost of living for parasites	Energy taken from a parasite each time step	100
host reproduction threshold	Amount of energy triggering host reproduction	2000
parasite cost of reproduction	Energy penalty for parasite reproduction	200
parasite reproduction threshold	Amount of energy triggering parasite reproduction	400
mutation factor	Scales the range of mutation in virulence	8
carrying capacity of the hosts	Maximum number of parasites in a host	500

Table 1: The variables, fundamental parameters, and derived parameters that define the model.

tion worth of energy units. Similarly, the parasite's initial energy was set randomly between 0 and the *parasite reproduction threshold,* and a newborn parasite starts life with *parasite cost of reproduction* energy units. The parasites initially placed in the model have virulences that are determined randomly in the range from the *cost of living for parasites* to $3 \times$ *cost of living for parasites.* [100 - 300] [4]

Quirks and Caveats

This model of a host-parasitoid system is extremely simplified. The principle of parsimony suggests that the bare minimum of detail should be included so as to test the hypothesis. Beyond this, there were a number of decisions that had to be made more or less arbitrarily. Usually ease of model construction prevailed over the choices.

Major Dynamics Left Out The most glaring exemption is the lack of resistance genes in the hosts. This removes the entire dynamic of coevolution. It also removes the possibility for sexual reproduction in hosts. Hamilton, Axelrod, and Tanese (1990) argue that sexual reproduction may have evolved as an adaptive response to parasitism.

We lack data on the relationship between parasite reproduction and virulence in general [May and Anderson, 1990, Ewald, 1983], or λ and α in equation 1. Therefore I have let the relationship be linear as a first approximation. Specifically, a parasite will produce $\lceil \frac{energy - parasite\ reproduction\ threshold}{parasite\ cost\ of\ reproduction} \rceil$ number of offspring in one time step. Furthermore, the model does not contain any sense of host recovery or immunity (v in equation 1).

On a more physical level, there is no dissipation of energy during transfer from environment to host, from host to parasite, and from parent to offspring. Specifically, parasites gain exactly the same amount of energy as their host loses in a single time step.

No distinction was made in the model between microparasites and macroparasites. Nor is there any representation of life histories of the parasites and host beyond the dispersal distance. Finally, there is no interspecific competition between different parasites species within the hosts. The model only simulates intraspecific competition amongst the parasites.

Additional Details is in the Model When a host dies all of the parasites infesting it die. However, when a host reproduces, half of the parasites infesting the parent are transmitted to its offspring. This allows parasites to be transmitted vertically.

There are a number of quirks surrounding reproduction. If a host tries to reproduce when there are no empty adjacent patches, the host will lose the *host cost of reproduction* amount of energy, but no new host will be created, and so the parental host will retain all of its parasites. Similarly, there are a few rare cases where two parasites will try to place offspring in the same position of a single host, thus causing a "collision" since the model operates in parallel. In this case, the more virulent offspring will win out and the other parasite offspring is killed. In an even more unlikely case, a parasite will try to place an offspring into a host that already has the *carrying capacity of the hosts* number of parasites infesting that host. Again, the parasite's offspring is killed under these conditions.

The Experiment

The model was constructed in the language C^* and was run on a Thinking Machines CM5 parallel computer. Once the model had been constructed, the experiment was straight-forward. What is the relationship between distance of dispersal and virulence in parasites? The independent variable was *dispersal distance* and the dependent variable was the *virulence* in the population of parasites after a specified amount of time.

Method

Each run of the model began with a host in every patch of the 30×30 environment. One of the hosts was also infected by a single parasite. The model began by simulating the invasion of the host population by the parasite. Initial runs indicated that the oscillations of the host and parasite population sizes were so violent as to crash in

---

[4]If a parasite has a *virulence* less than the *cost of living for parasites* then it will inevitably die before reproducing.

many cases when the *size* parameter was too large, so *size* was set at five for all of the runs.

The virulences of the parasites were sampled after a specified number of time steps. In order to determine an appropriate number of time steps, the model was run a few times under different dispersal conditions for extremely long periods of time (e.g., 10,000 time steps) and the average virulence was observed over time. Virulence seemed to level out after about 4,000 time steps in most cases, so the sampling time was set to be 5,000 time steps. At the end of this period of time, the virulence for each parasite in the population was saved in a file for later analysis.

The independent variable, *dispersal distance*, ranged from 1 to 15. If the *dispersal distance* was 1, parasites could only transmit their offspring to the nearest neighbor hosts. In contrast, with a *dispersal distance* of 15, a parasite could transmit its offspring anywhere in the environment. The model was run under a *dispersal distance* of 1, 2, 4, 8, and 15. For each of these possible settings the model was run five times using different random number generator seeds. In total, data was collected for 25 runs of the model.

Results

All of the runs of the models produced oscillations in the host and parasite populations that were stable enough that neither population crashed in the sampled interval. However, the resulting sizes of the parasite populations varied from about 125 to 250 at the 5,000 time step mark.

An analysis of the variance in the resulting populations indicated that dispersal had a statistically significant effect on the evolved virulence of the parasites ($p < 0.0001$), as did the seed for the random number generator ($p < 0.05$). The line of best fit for the $virulence = 594.23 + 0.91 dispersal - 0.11 seed$.

Discussion

Why is the parasite that causes the common cold less deadly than the protozoa that causes malaria (genus *Plasmodium*) or the bacteria that cause cholera (*Vibrio cholerae*)? Perhaps the differences in virulence depend in part on the differences in the dispersal mechanisms of the parasites. The model gives some evidence that the distance over which a parasite transmits its offspring has a significant effect on the parasite's evolution of virulence. However, according to the model, this effect is relatively small. This is probably due to the fact that selection pressure against a high virulence is relatively small compared to the advantage of rapid reproduction in the short dispersal distance conditions. High virulence boosts a parasite's reproductive success primarily by allowing the parasite to reproduce quickly. Only on a secondary level does it reduces the resources in the area upon which the parasite's offspring must depend. Even if a parasite kills off its host, the parasite's offspring could have been transmitted to any of that host's eight neighbors. Thus, a high virulence only removes a fraction of the resources available to its offspring.

It is likely that the details of the life history of a real biological parasite would have a much stronger effect on its evolution of virulence than just its dispersal distance. For example, many parasites of humans are transmitted through the close physical contact of medical workers with their patients. The medical workers carry the parasites from patient to patient. Even though dispersal distance is relatively short in this case, it is to the parasite's reproductive advantage to have a severe effect on its human host, so that the host is taken to the hospital.

One oddity of the model ought to be addressed. It was observed that a high value for *size* (e.g., 100) caused the populations to crash for most settings of the *dispersal distance* parameter. That is, when many parasites could infect a host, the populations tended to crash. As *size* was adjusted downwards, the populations began to enter stable oscillations. This is probably due to the resultant infection rate in the population of hosts. Once a host has been infected, its days are numbered. The parasites begin draining off energy making it less likely that the host will be able to reproduce. Even if it does manage to reproduce, half of its parasites are transmitted to the offspring. Thus, if there are many parasites in a host, the only introduction of uninfected hosts must come from the reproduction of other uninfected hosts. Meanwhile, parasite virulence levels evolve to the point of draining all of the host's food in one time step. A large *size* parameter means that parasites can reproduce relatively easily. Thus a single infected host can be the staging ground for many new infections in a single time step. In this way, a parasite population can sweep through the host population until all the hosts are infected. Given even moderate levels of virulence, this spells disaster for both species.

In contrast, when the *size* parameter is small, only a few parasites can infect a single host. The result is that there is some chance that no parasite will be transmitted to an infected host's offspring. More importantly, a single infected host can only be used as a staging ground for the transmission of a few new parasites each time step. Thus, infection rates in the host population are lower. This maintains a resource of uninfected hosts producing uninfected offspring. The high levels if infection in the host population is probably the effect of two dynamics. First, it is relatively easy for a parasite to transmit its young to a new host. In fact, the survival of a parasite during transmission is almost 100 percent. Second, the hosts lack an immune system which might also serve to reduce infection rates in the host population.

The model is highly simplified. In the biological world, a parasite's mode of transmission is not fixed, and may evolve along with virulence. It is also clear that there are many complex interactions between real parasites and the host's immune system, as well as between different parasite species, not to mention other factors [Begon *et al.*, 1990]. These effects are likely to be dramatic compared to the effect of dispersal distance on virulence. However, it was the hope that by excluding these factors, and thus considering all things to be equal, it may be possible to begin to pick apart the complexities of the evolution of virulence in parasites.

References

[Anderson and May, 1982] R. M. Anderson and R. M. May. Coevolution of hosts and parasites. *Parasitology*, 85:411–426, 1982.

[Baveco and Lingeman, 1992] J. M. Baveco and R. Lingeman. An object-oriented tool for individual-oriented simulation: Host-parasitoid system application. *Ecological Modelling*, 61:267–286, 1992.

[Begon et al., 1990] M. Begon, J. L. Harper, and C. R. Townsend. *Ecology*. Blackwell Scientific Publications, Oxford, UK, 1990.

[Boerlijst and Hogeweg, 1992] M. Boerlijst and P. Hogeweg. Self-structuring and selection: Spiral waves as a substrate for prebiotic evolution. In Christopher G. Langton et al., editor, *Artificial Life II*, pages 255–276, Redwood, CA, 1992. Addison-Wesley.

[Caswell and John, 1992] Hal Caswell and A. M. John. From the individual to the population in demographic models. In D. DeAngelis and L. Gross, editors, *Individual-Based Models and Approaches in Ecology*, pages 36–61, New York, 1992. Chapman and Hill.

[Caswell, 1989] Hal Caswell. *Matrix Population Models: Construction, Analysis, and Interpretation*. Sinauer Associates, Sunderland, MA, 1989.

[Collins and Jefferson, 1992] Robert J. Collins and David R. Jefferson. The evolution of sexual selection and female choice. In F. J. Varela and P. Bourgine, editors, *Toward a Practice of Autonomous System*, pages 327–336, Cambridge, MA, 1992. MIT Press.

[Dewdney, 1989] A. K. Dewdney. Simulated evolution: Wherein bugs learn to hunt bacteria. *Scientific American*, pages 138–141, May 1989.

[Ewald, 1983] Paul W. Ewald. Host-parasite relations, vectors, and the evolution of disease severity. *Annual Reviews of Ecological Systems*, 14:465–485, 1983.

[Ewald, 1993] Paul W. Ewald. The evolution of virulence. *Scientific American*, 268:56–62, April 1993.

[Fontana, 1992] Walter Fontana. Algorithmic chemistry. In Christopher G. Langton et al., editor, *Artificial Life II*, pages 159–211, Redwood, CA, 1992. Addison-Wesley.

[Hamilton et al., 1990] William D. Hamilton, R. Axelrod, and R. Tanese. Sexual reproduction as an adaptation to resist parasites (a review). *Proceedings of the National Acadamy of Science, USA*, 87:3566–3573, 1990.

[Hamilton, 1992] William D. Hamilton, 1992. Personal communication. Zoology Dept., South Park Rd., Oxford, OX1 3PS, UK.

[Herre, 1993] Edward Allen Herre. Population structure and the evolution of virulence in nematode parasites of fig wasps. *Science*, 259:1442–1445, March 5 1993.

[Hurst, 1991] Laurence D. Hurst. The incidences and evolution of cytoplasmic male killers. *Proceedings of the Royal Society of London B*, 244:91–99, 1991.

[Huston et al., 1988] M. Huston, D. DeAngelis, and W. Post. New computer models unify ecological theory. *Bioscience*, 38(10):682–691, 1988.

[Koza, 1992] John R. Koza. Genetic evolution and co-evolution of computer programs. In Christopher G. Langton et al., editor, *Artificial Life II*, pages 603–630, Redwood, CA, 1992. Addison-Wesley.

[Koza, 1994] John R. Koza. Artificial life: Spontaneous emergence of self-replicating and evolutionary self-improving computer programs. In Christopher G. Langton, editor, *Artificial Life III*, pages 225–262, Redwood, CA, 1994. Addison-Wesley.

[Langton, 1992] Christopher G. Langton. Life at the edge of chaos. In Christopher G. Langton et al., editor, *Artificial Life II*, pages 41–93, Redwood, CA, 1992. Addison-Wesley.

[Lindgren and Nordahl, 1994] K. Lindgren and M. G. Nordahl. Artificial food webs. In Christopher G. Langton, editor, *Artificial Life III*, pages 73–104, Redwood, CA, 1994. Addison-Wesley.

[Maley and Caswell, 1993] Carlo C. Maley and Hal Caswell. Implementing i-state configuration models for population dynamics: An object-oriented programming approach. *Ecological Modelling*, 68:75–89, 1993.

[May and Anderson, 1990] R. M. May and R. M. Anderson. Parasite-host coevolution. *Parasitology*, 100:S89–S101, 1990.

[May, 1973] R. M. May. *Model Ecosystems*. Princeton University Press, Princeton, NJ, 1973.

[Nicholson and Bailey, 1935] A. J. Nicholson and V. A. Bailey. The balance of animal populations. part i. *Proceedings of the Zoological Society of London*, 3:551–598, 1935. Referenced in Baveco and Lingeman, 1992.

[Nowak, 1991] Martin Nowak. The evolution of viruses. competition between horizontal and vertical transmission genes. *Journal of Theoretical Biology*, 150:339–347, 1991.

[Reynolds, 1987] C. W. Reynolds. Flocks, herds, and schools: A distributed behavioral model. *Computer Graphics*, 21(4):25–34, 1987.

[Rizki and Conrad, 1986] M. M. Rizki and M. Conrad. Computing the theory of evolution. *Physica D*, 22:83–99, 1986.

[Taylor et al., 1989] C. E. Taylor, D. R. Jefferson, S. R. Turner, and S. R. Goldman. Ram: Artificial life for the exploration of complex biological systems. In Christopher G. Langton, editor, *Artificial Life*, pages 275–296, Redwood, CA, 1989. Addison-Wesley.

[Travers, 1989] M. Travers. Animal construction kits. In Christopher G. Langton, editor, *Artificial Life*, pages 421–442, Redwood, CA, 1989. Addison-Wesley.

[Williams and Nesse, 1991] George C. Williams and Randolph M. Nesse. The dawn of darwinian medicine. *The Quarterly Review of Biology*, 66(1):1–22, 1991.

Innate Biases and Critical Periods:
Combining Evolution and Learning
in the Acquisition of Syntax

John Batali
Department of Cognitive Science
University of California at San Diego
La Jolla, CA 92093-0515
batali@cogsci.ucsd.edu

Abstract

Recurrent neural networks can be trained to recognize strings generated by context-free grammars, but the ability of the networks to do so depends on their having an appropriate set of initial connection weights. Simulations of evolution were performed on populations of simple recurrent networks where the selection criterion was the ability of the networks to recognize strings generated by grammars. The networks evolved sets of initial weights from which they could reliably learn to recognize the strings.

In order to recognize if a string was generated by a given context-free grammar, it is necessary to use a stack or counter to keep track of the depth of embedding in the string. The networks that evolved in our simulations are able to use the values passed along their recurrent connections for this purpose. Furthermore, populations of networks can evolve a bias towards learning the underlying regularities in a class of related languages.

These results suggest a new explanation for the "critical period" effects observed in the acquisition of language and other cognitive faculties. Instead of being the result of an exogenous maturational process, the degraded acquisition ability may be the result of the values of innately specified initial weights diverging in response to training on spurious input.

1 Introduction

One of the most popular, and controversial, explanations of the ability of children to learn their native languages, is that humans are born with innate biases to be able to recognize and produce certain kinds of linguistic structures (Chomsky, 1987). Human languages are apparently too complex, and the evidence available to the language-learner is too sparse, for general-purpose learning mechanisms to accurately acquire linguistic competence (Gold, 1967; Wexler & Culicover, 1980).

It has been suggested that first language acquisition is the result of the maturation of some kind of language-specific acquisition device which subsequently disappears when the child gets older (Lennenberg, 1967). One line of evidence for this hypothesis is the apparent existence of a "critical period" for language acquisition (Newport, 1990). Children not exposed to a language before the age of five or so are often not able to learn language as well as those exposed to language while younger.

We are motivated by such issues (and the attendant controversies) to explore the interactions between evolution and learning in the acquisition of syntax. The proposals for "innate biases" and a "language acquisition device" are very vague, and, at best, promissory notes to be cashed out when and if we understand better how language is processed in the brain, and therefore what innate biases could be like, and what sorts of acquisition devices might be used. And of course if a bias or an acquisition device is innate, it must have evolved somehow.

Artificial neural connectionist networks constitute a concrete (and somewhat biologically plausible) model of computation and learning. However the precise limits of their computational power have yet to be determined. Fodor & Pylyshyn (1988) argue that neural networks cannot represent recursively structured representations in a way that allows for the sorts of mental inferences that people can perform. One response to this argument is that neural networks have been shown to be capable, in principle, of performing any computation that a universal Turing machine can perform (Pollack, 1987; Siegelmann, 1993). However the practical question remains, for any given task, whether a network can be *trained* perform the task upon exposure to examples.

General purpose neural network training algorithms, for example weight updating by backpropagation of error (Rumelhart, Hinton and Williams, 1986), are useful for a very wide range of problems, but whether or not a given network and learning algorithm will converge on a solution for a given episode of training is crucially dependent on the initial connection weights of the network. Small differences in the initial weights can determine whether or not the training algorithm will converge on a solution, how quickly convergence occurs, and the specific properties of the solutions that are found (Kolen & Pollack, 1990).

Such considerations suggest a specific version of an innateness hypothesis: The initial values of the connection weights of a neural network are innate, the result of evolutionary processes where the selection criteria are based

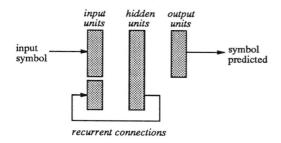

recurrent connections

Figure 1: A simple recurrent network for recognizing strings generated by a grammar. Symbols from the string are presented sequentially to the input units. The network is trained to predict the next symbol in the string. Each unit in a layer has connections to each of the units in the next layer; these connections are not shown. Recurrent connections from hidden units feed back to the input layer. The specific numbers of input units, hidden units, output units, and recurrent connections varied in the different experiments.

on the ultimate performance of the networks after their connection weights are modified by training. The innate initial weights assure that the networks will efficiently and reliably find good solutions to the tasks they face.

Since the innate biases of the networks are realized as a set of specific values of initial connection weights, those biases are degraded if the weights of the networks diverge from those that evolved. Such divergence can be the result of training on spurious input. This suggests an explanation of some critical period effects that does not require any appeal to maturational processes or special-purpose language acquisition devices.

2 Networks & Grammars

Recurrent neural networks (as shown in figure 1) have been shown to be capable of recognizing and generating strings satisfying simple grammars (Jordan, 1986; Servan-Schreiber, Cleeremans and McClelland, 1988; Elman, 1990a). A typical training regimen (and the one used in the experiments reported here) consists of presenting the symbols from a string generated by the grammar sequentially to the input units of the network. Activation is then fed forward through the network. The activations of the output units are taken as a prediction of the next symbol in the string. The prediction is compared with the actual symbol that appears next, and the error between the predicted and correct value is backpropagated through the network to adjust the connection weights between units.

One way of characterizing the complexity of formal languages is in terms of the kinds of automata that can recognize strings from the language (Hopcroft & Ullman, 1979). The "regular languages," for example, can be recognized by finite-state machines. Recurrent neural networks are quite capable of being trained to implement finite state machines; they can learn to use the values passed along their recurrent connections to rep-

resent the state (Cleeremans, Servan-Schreiber and McClelland, 1989). Hence recurrent neural networks can be trained to recognize strings from regular languages.

Human languages are more complex than regular languages (Chomsky, 1957). In particular, all human languages allow for recursive nesting of phrase structure. The simplest class of formal languages which can generate strings exhibiting recursive phrase structure is the class of "context-free" languages. Recognition of strings in a context-free language requires, in addition to storing the state the recognizer is in, the use of a "pushdown stack" to keep track of the depth of nesting in the string.

An example of a syntactic pattern which requires a context-free grammar to express is that of the "center-embedded" constructions in natural languages. In such a construction some constituent, for example a noun phrase, must be matched with another constituent, for example a verb phrase, across some intervening material, for example a relative clause, which might exhibit the same structure.

Recall Fodor & Pylyshyn's (1988) arguments, mentioned above, that networks are incapable of manipulating recursive constituent structure. Christiansen (1992) wonders, based on evidence from natural language corpora, and psycholinguistic experiments, whether it is really sensible to presume that human languages can exhibit very deep nested structures. He suggests, in fact, that the limitations of networks in processing embedded structures match closely with those observed in humans, thus the network's limitations *support* their appeal as models of human language processing. While this response is intriguing, if we choose to accept Fodor & Pylyshyn's challenge, and take the ability to recognize whether strings were generated by a context-free grammar as a minimal criterion for the ability to handle recursive constituent structure, then we can demonstrate that ability by showing that part of the network is implementing a pushdown stack or some computationally equivalent device.

The results of Gold (1967) show that if a general-purpose learning algorithm is shown strings from a formal language, only regular languages can be learned. Gold lists several ways out of this predicament, including the one we are exploring: that the learner possesses innate biases to learn specific classes of languages. With respect to learning context-free languages, this means that the network must possess an innate bias to learn to implement and manipulate something equivalent to a stack.

Investigations of the ability of neural networks to learn to recognize strings exhibiting recursive phrase structure have begun to show how networks can represent that structure. For example in (Elman, 1990b) and (Weckerly & Elman, 1992) networks trained to recognize strings containing nested relative clauses and center-embedded constructions, respectively, use different trajectories of the activation values of their hidden units to represent the depth of embedding in the strings. Pollack (1990, 1991) shows how neural networks can implement dynamical systems and suggests that some of the attractors of such systems can perform the functions of a pushdown stack. Giles, et al. (1990) and Sun, et al. (1990) describe networks equipped with an external stack, which can learn to manipulate the stack and recognize context-free languages.

Elman (1993) addresses the difficulty of training networks to recognize recursive phrase structure. His solution involves beginning training with short input strings and gradually increasing the lengths of the strings. This training schedule is successful because the embedded material (e.g., relative clauses), exhibits the same sequential pattern as that of the structure it is embedded within. Even with the use of special training routines, learning to recognize strings generated by a grammar that allows recursively nested structure is difficult, time-consuming, and not guaranteed to succeed in any particular training trial.

3 Simulated Evolution

A compelling intuition, first presented by Baldwin (1896), is that evolution and learning can work synergistically. Suppose we are considering a species of animals whose fitness depends on their behavior. If the behavior is encoded entirely genetically, evolution will have difficulty locating a solution of relatively high fitness if such solutions are very small regions in the space of possible solutions. An animal whose behavior is modifiable by learning, on the other hand, can locate a highly adaptive solution if evolution places it near, rather than precisely on, the solution. The combination of evolution and learning thus broadens the optima in the solution space and gives evolution more of a gradient to work with. Hinton & Nowlan (1987) and Belew (1989), have explored this idea in computational simulations.

Another potential interaction between evolution and learning is illustrated by Kolen & Pollack's (1990) demonstration of the extreme sensitivity of the backpropagation network training algorithm to the values of networks' initial connection weights. Belew, McInerney, and Schraudolph (1991) suggest that the initial weights for networks be located by a genetic algorithm, where the fitness measure is based on the network's ultimate performance on a task after training. Nolfi, Elman and Parisi (1990) describe a population of simple creatures, controlled by networks whose initial weights are selected by a genetic algorithm, that evolve the ability to learn to behave appropriately for their environment.

In our experiments, the networks initial weights were specified directly by a sequence of real numbers. The initial weights of the first generation were uniformly distributed between -1 and $+1$. After each of the networks in the population was trained, its fitness was assessed. The top third of the networks survived unchanged into the next generation, with their connection weights reset to the initial values they had before training. Each of these networks was also used to create two offspring. For each offspring, a copy of the parent's initial weights was copied to the offspring. The offspring's initial weights were then modified by adding a random vector to the initial weight vector. The values at each position of the random mutation vector were normally distributed with a standard deviation of 0.05. The new offspring networks and the best networks from the previous generation were then trained, and the cycle repeated each generation.

There are a number of different approaches to the simulation of evolution (Goldberg, 1989) and the specific details described above were chosen mostly on the basis of trial-and-error. The direct connection (indeed identity) between the genotype and initial phenotype of these networks is obviously not biologically plausible, and we are exploring more realistic models of the interactions among genetics, development, and learning. On the other hand the genotype must be able to *somehow* affect the initial weights. Rather than try to duplicate the details of mutation and other genetic events, especially given the complexity that the development process introduces to the modeling of gene expression, we treated their cumulative effects as a random vector affecting the whole genome. Subsequent experiments suggest that simulating sexual reproduction by creating new initial weight sequences by combining those of two parents would speed up the evolutionary search, but no such crossover operation was used in the simulations reported here.

4 Learning a Simple Context-Free Language

A very simple context-free language consists of strings of the form $a^n b^n$, that is: some number of tokens of the symbol a, followed by the same number of tokens of the symbol b; thus: ab, aaabbb, aaaaaabbbbbb, etc.

Any machine that can recognize whether a string is in this language or not must somehow count up the number of a's it sees, and count down each time it sees a b. While in general, context-free languages require a pushdown stack that can store arbitrary symbols at each level of the stack, for this language all that is required is a counter which can be incremented, decremented, and compared with zero.

A recognizer for this language must also keep track of which of two states it is in: an initial state corresponding to the sequence of a's, and a second state corresponding to the sequence of b's. During the period the machine is in the initial state, it may see either an a or a b; if it sees an a, it remains in the initial state and increments the counter; when it sees a b, it enters the second state and decrements the counter. While in the second state, it must only see b's, and must decrement the counter for each one.

The networks used for learning this language had 3 input units, 10 hidden units, 3 output units and 7 recurrent connections from hidden units to the input layer. Training consisted of presenting the networks strings from the $a^n b^n$ language preceded and terminated by 'space' character. Prediction error was backpropagated through the network to update the connection weights. (The backpropagation learning rate for all of these experiments was 0.1. No momentum term was used.) Each network was trained for a total of 500,000 characters, which works out to about 33,000 strings. (The strings ranged in length

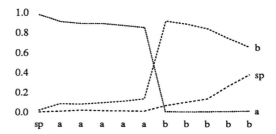

Figure 2: Performance of a randomly initialized network after training on strings in the $a^n b^n$ language. Values plotted are the activation values of output units, after the characters from the given string are presented to the network. The symbol 'sp' stands for the string-delimiting 'space' character. This network had an average prediction error of 0.251 for the set of test strings.

from 4 to 26 characters.) Each network's average per-character prediction error for a set of 12 test strings from the language was then measured.

To assess the ability of networks to find solutions to this problem, 513 randomly initialized networks were trained on strings from the language. The mean value of the networks' per-character prediction error was 0.244, with a standard deviation of 0.031. The best network in this set achieved an average prediction error of 0.168.

The performance of a randomly initialized network after training on strings from the $a^n b^n$ language is shown in figure 2. The values plotted are the activation values of the network's output units, after each of the characters from the string is shown to the network. The network correctly predicts the appearances of the a character for the first half of the string, and once the first b is seen, the prediction for a drops to zero. In the second half of the string, the prediction for b, initially high, begins to drop off, and the the prediction for the 'space' character begins to rise. Although these predictions accord with the statistics of the training strings, the network has hardly learned the language very well. It should be predicting only b until as many b's as a's have been seen, at which point it should predict only the 'space' character. The behavior of the network shown in figure 2 is representative of virtually all of the randomly initialized networks.

A population of 24 of these networks was used as the initial generation of an evolutionary simulation. The fitness value of a network was proportional to the inverse of the average per-character prediction error it achieved after training on 500,000 characters from strings from the language.

By the 155th generation of this simulation, the best network had an error of 0.151 and the average error for all of the networks in the population was 0.179. This is 2.2 standard deviations better than the randomly initialized networks achieved.

Figure 3 shows the activation values of some of the units of a network that evolved in this experiment, as it is

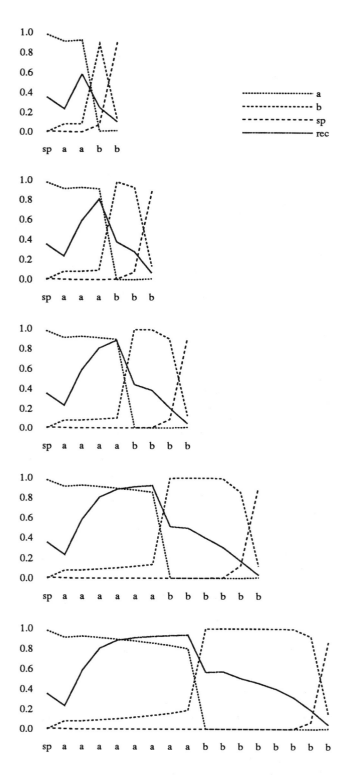

Figure 3: Activation values of units of an evolved network after training, while recognizing strings of various lengths. Plots labeled a, b, and sp show activation values of output units. The plots labeled rec show the activation values of a unit feeding one of the network's recurrent connections.

shown a string of characters. Note that the network predicts the a character very strongly after seeing the initial 'space' character, and then predicts a until it sees a b. From then on it predicts b until it has seen the same number of a's as b's, at which point it predicts the 'space' character.

The strong prediction of a for the first half of the string is interesting because this prediction does not match the statistics of the strings the network was trained on. Since a b can follow an a at any point in the string, the statistics would predict a and b at roughly equal levels throughout the first half of the string (or perhaps an initially higher value for a, decreasing uniformly throughout the first half of the string). However if the string is long enough, the average error the network receives over the length of the string is smaller if it receives a one-time error of 1.0 at the point where the first b appears, rather than an error of about .5 for each character in the first half of the string. The network has found a solution that reflects not the statistical regularities in its input, but the correct underlying rule.

The plots labeled **rec** show the activation value of one of the units feeding a recurrent connection in the network as it is presented the characters from the string. The activation value of this unit illustrates how the network manages to recognize the strings correctly. The unit is behaving as a counter: each a increases its activation value until b is seen, after which the value decreases. When the activation value of this unit approaches zero, the network predicts the 'space' character. As described above, the behavior of this unit is precisely what is required for the network to recognize the language $a^n b^n$.

The networks that emerged in the final generation of the simulation were able to learn to recognize strings whose level of embedding was 12, which was the level of embedding in the longest string the networks were trained on. If the networks were shown longer (and therefore more deeply embedded) strings, they would usually still be predicting b strongly at the end of the string.

5   Learning from a Class of Languages

The population of networks in the first experiment was able to combine evolutionary search with backpropagation learning to learn a particular language. An attractive solution would have been for a network to need no training at all — it would be "born" with the connections already correct to recognize the target language. This solution is not implemented in humans. Children are able to learn whatever language they are exposed to, no matter what language their parents spoke. While there are no innate biases to learn *particular* languages, it is possible that all human languages share certain abstract structural features, and there are innate biases to learn languages with those structural features.

We explored this idea by simulating the evolution of a population of networks trained on languages generated by a class of related grammars. In each generation, each of the members of a population of networks was trained on a particular language. A network's ability to

learn the language it was trained on was used to compute the network's fitness value, and hence whether it survived into the next generation and reproduced. But in the next generation the network's offspring would, in general, face a different language. To flourish over the generations, networks had to develop biases not for a specific language, but for those properties shared by all of the languages in the class.

Each of the languages in the class used the symbols a, b, c, and d. In a given language, each of these four symbols is assigned to one of three categories: the 'push' symbols, the 'pop' symbols and the 'idle' symbols. In a grammatical string from any language in the class, each 'push' symbol must be matched on its right by a 'pop' symbol. Any number of 'idle' symbols can appear anywhere in a string, except that the first symbol in the string must be a 'push' symbol, and the last symbol in the string must be its matching 'pop' symbol.

Each language had at least one 'push', 'pop', and 'idle' symbol, and one of the three categories had two symbols. The 36 possible languages defined this way fall into three subclasses, corresponding to the languages with two 'push' symbols, two 'pop' symbols and two 'idle' symbols. There are 12 languages in each subclass, differing only in which symbols are assigned to which category.

For example one language in the class has a and b as the 'push' symbols, d as the 'pop' symbol and c is the 'idle' symbol. The string 'baadcadcdd' is in this language.

The intuition being explored here is that the specific lexical items used in languages are their most arbitrary features. Underneath lexical differences, languages may share aspects of linguistic structure (word order, case systems, phonological or morphological processes, etc.), and there may be underlying regularities common to all languages. Thus a language with c and d as its 'push' symbols, a as its 'pop' symbol, and b as its 'idle' symbol, would be structurally very similar to the language in the previous paragraph, although its strings would look quite different.

To recognize a string from one of these languages, an automaton must keep track of the value of a counter. Whenever it sees a 'push' symbol, it should increment the counter; when it sees a 'pop' symbol, it should decrement the counter; and when it sees an 'idle' symbol, it should keep the counter at the same value. When the counter reaches zero, the end of the string has been found. This computation is involved in recognizing all languages in the class, and hence the ability to quickly learn to perform it would be an ideal innate bias for the networks to acquire. Then, when exposed to a specific language during training, the network would only have to learn the specific mapping of the symbols to their categories.

In this experiment, networks with 5 inputs, 10 hidden layers, 5 outputs and 1 recurrent connection were used. The single recurrent connection was used in order to make the language-learning task as difficult as possible

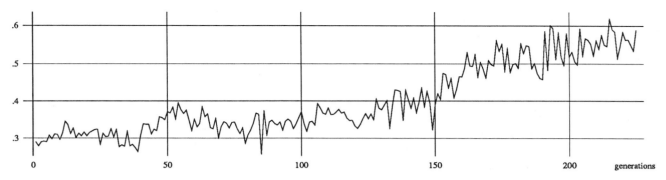

Figure 4: The evolution of networks trained on languages from the class of context-free languages. The value plotted is the average of the pred scores of the networks in each generation.

for the networks, given the relative simplicity of the languages compared to the $a^n b^n$ language.

The networks were trained on strings of characters as in the first experiment. However in this simulation, the performance of a network after training was assessed by computing the average value of its 'space' output unit at the end of each of a set of test strings. This value will be referred to as the "pred" value of the network. Ideally, it ought to be 1.0 — indicating that the network has correctly predicted the end of each of the strings. This value is more informative than the average per-character prediction error for these languages, as the the specific character that can follow another is much less constrained than in the earlier language. As before, the character prediction error seen was used for backpropagation training of the networks.

The difficulty of this task was assessed by training 436 randomized networks with varying numbers of hidden units and recurrent connections for 500,000 characters on languages from the class. The average pred value after training was 0.375 with a standard deviation of 0.117. The best network had a pred value of 0.827. Of 103 networks with a single recurrent connection, the average pred value was 0.280 with a standard deviation of 0.057; the best network had a pred value of 0.457.

The evolutionary simulation was organized in a similar fashion as the first experiment, with the pred value of each network used to compute its fitness. Figure 4 shows a record of the simulation. The value plotted at each generation is the average pred value for the networks in the population. After an initial period of relatively aimless search, the members of the population steadily improve their aptitude at learning the languages. (This run represents about one week's computation on a Sun Sparc Station 10.)

Figure 5 illustrates the performance of one of the networks that evolved. The activation of the 'space' output unit and the unit feeding the recurrent connection are plotted against the symbols of a string. The recurrent connection is used as a counter: the 'push' symbols **a** and **b** increment its value; the 'pop' symbol **d** decrements it, and the 'idle' symbol **c** modifies the value only slightly. When the recurrent value decreases far enough, the network signals the end of the string.

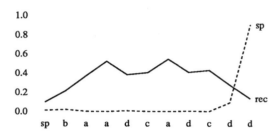

Figure 5: Activation values of units in one of the networks that evolved to learn languages from the class of context-free languages, after being trained on a specific language from the class. For this language, the symbols **a**, and **b** are 'push' symbols, **d** is the 'pop' symbol, and **c**, is the 'idle' symbol. The solid line plots the activation value of the unit feeding into the single recurrent connection of the network after the given character is seen. The dotted line plots the network's prediction of the end of string 'space' character.

After 150 generations, the average performance of the entire population is better than the best performance for the randomly initialized networks, and the performance of the population steadily increases after that. The networks are indeed developing an innate bias towards learning the languages in the class.

Some aspects of the way this innate bias works were investigated by looking at the performance of a network from a late generation of the simulation before it had been trained on any language. Its connection weights were therefore those it had inherited. When tested on a string from a language, its 'space' output unit is strongly activated: it is almost 1.0 for each character of the string; and the recurrent activation is essentially 0.0 throughout the string. Thus the "newborn" network is apparently hypothesizing the minimal language consistent with the data it has seen: namely the empty set. (This is a rather simple and extreme version of the sorts of strategies studied in formal learning theory (Wexler & Culicover, 1980).) Furthermore, this prediction of the 'space' symbol is innately associated with a zero recurrent input. This association will be reinforced later when the recurrent connection begins to be used as a counter, and its being zero signals the end of the string. Further investigation of the innate biases of networks to learn related languages is discussed below.

6   Critical Periods for Learning Languages

The development of a number of animal faculties shows a so-called "critical period" effect: Unless the animal is exposed to the appropriate environmental input at a certain period of its development, it may never acquire the full faculty. For example Marler (1991) describes how song sparrows deprived of the opportunity to hear the songs of other sparrows between the ages of 20 and 50 days will ultimately learn very reduced versions of the normal song of their species.

There is evidence for the existence of a critical period in the acquisition of language by humans. Newport (1990) describes how the ultimate ability of deaf children to master sign language decreases markedly with the age of first exposure, with significant degradation noticeable if the children were older than around five years when they first encountered sign language. Similar results are obtained in studies of the ability to learn a second language.

Lennenberg (1967), proposes that the critical period is the result of the maturation, and subsequent disappearance, of a specific language-learning mechanism. Newport suggests an explanation in which the child's initially limited cognitive abilities assist in the ability to learn language, for example by limiting the size of utterances that the child can store in memory and analyze. A related idea was explored by Elman (1993) in training recurrent neural networks to recognize recursively embedded strings by initially limiting the lengths of the strings, corresponding to either limited cognitive resources, or externally restricted input.

The experiments described here suggest a new explanation of critical period effects. A consequence of a network's innate biases being realized as specific values of initial connection weights, is that this bias can be degraded substantially if the weights of the network diverge from those that evolved. The backpropagation learning algorithm changes the weights of the network in response to an error signal based on a comparison between an expected output and the actual output of the network. If the network is trained on spurious input — strings not from the language it will eventually learn — its connection weights will be adjusted to values that differ from those that evolved. With enough such training, the weights of the network will diverge so far from the innate values that the network will have lost its innate ability to learn a language.

We presume that the network is trained on input patterns available in its environment. Ordinarily, these will be strings from the language that is ultimately to be learned. Deprived of exposure to a language, the network will be trained on whatever unstructured input exists; we model such unstructured input as random strings presented to the network.

Marchman (1993) proposes an explanation of the critical period effect that also does not invoke an exogenous maturational process. Her explanation, based on a neural network learning model, is that the network becomes "entrenched" in a particular solution after some training, and cannot find its way to new solutions. The explanation being presented here is compatible with Marchman's account, and indeed entrenchment effects will be described below. The difference is that in our account the effect is the result of the network's weights being removed from some ideal zone for learning, as opposed to being stuck in some other solution, as in Marchman's proposal.

One of the networks that evolved to learn languages from the class was trained for varying periods with random strings. After some number of characters of such training, the network was then trained on one of the languages in the class, for 500,000 characters. As illustrated in the left hand plot of figure 6 the ultimate performance of the network on learning a language decreases with the length of exposure to random input.

More dramatic degradation in the performance of the network is observed if the evolved network is trained on one language for a while, and then switched to another, as is illustrated in the right hand plot of figure 6. The network was first trained for some number of characters on languages from one of the three language subclasses. Language class 1 has two 'push' characters; language class 2 has two 'pop' characters, and language class 3 has two 'idle' characters. After the network was trained for the indicated number of characters on the other language, it was then trained for 500,000 characters on a language from class 1. (If the initial training was on a class 1 language, the subsequent training was on a different class 1 language.) The final average pred values for a set of test strings from the language is plotted. Note that there is little interference from a language of the same class, but that a language from one class can decrease the effectiveness of the network at learning a language from another class, sometime to *less* than the average performance of randomized networks at learning the same language.

Apparently what is happening here is that the innate biases the networks have evolved tend to place the networks' initial weights in zones from which training will guide the networks to good solutions. A period of random training will jiggle the weights around, but won't move them out of the zone right away. On the other hand, exposure to structured input will move the weights out of the initial bias zone towards a particular solution. Once out of that initial zone, a network will have a hard time finding its way back into it. In fact, enough training on structured input of the wrong kind (here, a different language class) will actually put the weights of the network into a zone from which learning cannot reach the optima available to most randomly initialized networks. This illustrates Marchman's notion of "entrenchment."

The behavior of the network illustrated by the right-

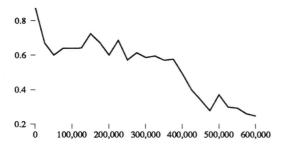
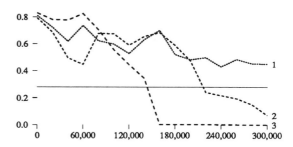

Figure 6: Performance of a network in learning a language after exposure to spurious input. One of the networks that evolved to learn from the class of related languages was chosen. The plot on the left illustrates the final pred value for the network after first being trained on the indicated number of characters from random sequences, and then trained on 500,000 characters from a language in the class. The plot on the right shows the final pred value for the network after first being trained on strings from a language from each of the three subclasses of languages for the indicated number of characters, and then trained on 500,000 characters from a language of class 1. The solid line indicates the average performance of randomly initialized networks. Both plots display average values taken over 20 training runs each. Note that the scales of the abscissas are different in the two plots.

hand plots in figure 6 also shed light on how the networks' innate biases work. The fact that there is less interference between members of the same language class suggests that the initial weights of the networks are such that they can quickly locate solutions that exploit the commonalities of the languages of the same class. Once such a solution is in place, it is possible for a network to learn a new mapping from symbols to their categories. On the other hand, once a network begins to develop a solution for a language of one class, that solution strongly interferes with the network's ability to learn both a new language's structure and its mapping from symbols to categories.

7 Learning Temporal Boolean Functions

Some of these issues were explored by investigating the evolution of networks on a simpler (and therefore more quickly trainable) recognition task. Each of the networks in a generation was trained on one of four temporal versions of two-input boolean functions. The four functions are:

xor   exclusive or of current and previous inputs
-xor   negation of xor
prev   value of previous input
-prev   negation of previous input

These functions were chosen because they were found by experiment to be the most difficult of the 16 temporal boolean functions for randomly initialized networks to learn.

An evolutionary simulation of 200 generations was performed with a population of 24 networks. Each network had 1 input unit, 3 hidden units, 1 output unit, and 1 recurrent connection. In a generation, each network was trained with 100,000 characters from one of the temporal boolean functions, and the network's final average prediction error was used to determine its fitness. As

with the networks evolving to learn languages from the class of context-free languages, the offspring of networks which did well in a particular generation at learning the function they were trained on would, in general, face a different function than their parent did. Hence the selective pressure was toward acquiring whatever innate bias would help learn all of the functions, not just a specific one.

The following table shows the performance of networks after being trained on 100,000 characters from the temporal boolean functions. The column headed "Randomized" shows the final average prediction error of networks with random initial weights; the column headed "Evolved" shows the final average prediction error of networks whose initial weights were found by the evolutionary simulation. The results in this table are averaged over 400 training runs each.

| boolean | Randomized | | Evolved | |
function	mean error	standard deviation	mean error	standard deviation
xor	.178	.088	.0151	.036
-xor	.182	.088	.0006	5.00×10^{-6}
prev	.205	.077	.0001	7.35×10^{-7}
-prev	.202	.080	.0001	1.02×10^{-6}

The networks have apparently evolved some sort of bias towards learning the boolean functions. At least some of the nature of this bias is illustrated in the table below. A network from the last generation of the simulation run was given a sequence of inputs before it was trained on any of the functions. The responses of its output unit and the unit feeding its recurrent connection are shown in the center columns. Note the activation value of the recurrent unit is very close to the input to the network. This value will be available to the network as input when the next input is seen. This innate response of the network means that when trained on the temporal boolean function, the network just has to learn

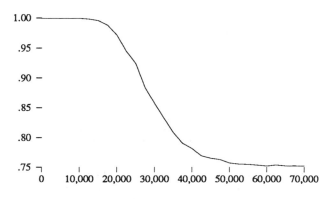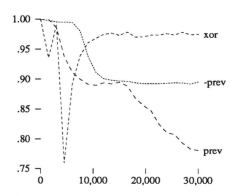

Figure 7: Performance of one of the networks that evolved to learn temporal boolean functions after exposure to spurious input. The value plotted is 1.0 minus the average error of the network for a set of test strings after training for 100,000 characters, after being exposed to spurious input for the indicated number of characters. The plot on the left shows the result for a network exposed to random sequences. The plot on the right shows the results for the network exposed first to the indicated temporal boolean function, and then to the **-xor** function. The values on both graphs are averaged over 400 training runs.

the non-temporal version of that boolean function. The right hand columns in the table show the performance of the network after training with 100,000 characters from the **xor** function. The behavior of the recurrent values haven't changed at all, and the network is correctly computing the temporal **xor** function.

input	correct output	Before Training actual output	Before Training rec value	After Training actual output	After Training rec value
1	1	.19	.96	.97	.96
0	1	.10	.04	.98	.04
0	0	.16	.04	.02	.05
1	1	.19	.96	.97	.96
1	0	.12	.96	.03	.96
0	1	.10	.04	.98	.04

Figure 7 shows the performance of the network after exposure to spurious input. The left hand plot shows how the network's ultimate prediction error after training is affected by initial training on random sequences; it exhibits the classic shape of a critical period effect[1] — for a short period (in this case amounting around 20,000 characters) the spurious input has no effect at all on the ultimate performance of the network. The ultimate performance then drops rapidly, until after enough time (around 50,000) characters, the behavior of the network is the same as that of randomly initialized networks.

The plot on the right of figure 7 displays the final performance of the network after first being trained on one temporal boolean function for the indicated number of characters, and then trained on the **-xor** function.

The different functions have different effects on the ultimate ability of the network to learn **-xor**. Learning **xor** first has very little effect on the ultimate performance of the network (except for a very strong effect at 4500 characters, which will be discussed below).

Learning **-prev** has almost no effect until around 7000 characters and then begins to have a slight, but significant effect. Learning **prev** has a rather dramatic effect on the ultimate performance of the network, decreasing it to that of randomly initialized networks after 30,000 characters.

The glitch at 4500 characters in the plot labeled **xor** provides a clue to how learning another function interferes with the subsequent learning of **-xor**. As was shown above, the networks have evolved a set of initial weights that result in their recurrent connections passing the input value back into the network with the next character. When the network has been trained on 4500 characters of the **xor** function, that behavior still occurs, and the network is also producing output values that are just beginning to approximate **xor**: the output values when a 1 is expected are slightly more than .5 and the output values when a 0 is expected are slightly less than .5.

If the function being learned is then switched to **-xor**, the network can respond to the new training in one of two ways: either by changing the values of weights computing the boolean function from the character input and the recurrent input, or by changing the values of the weights that result in the recurrent connection feeding the current input back into the network. The latter choice is made most of the time, and the recurrent connections begin to carry intermediate values. Since the recurrent pattern was a major source of the networks innate bias towards learning the functions, they then behave essentially as randomly initialized networks do.

After as few as 2000 more characters of initial training on **xor** however, this effect disappears and the network is able to learn **-xor** correctly even after 30,000 characters of exposure to **xor** first. Apparently after solidly learning **xor**, the network is able to locate a solution that involves negating an output unit without modifying the weights that determine the values passed along its recurrent connection.

---

[1]See, for example, figure 2 of Newport (1990), or figures 2.18 and 2.19 of Marler (1991).

In these experiments, the innate biases of the networks implement a partial solution to the problem that the networks face when learning each of the temporal boolean functions. Spurious input can degrade the network's ultimate performance in two ways: either by drifting the weights of the network away from the values implementing the innate partial solution; or by interrupting the standard development of the innate partial solution toward its being a complete solution for some particular function.

8 Discussion

Figures 3 and 5 demonstrate that recurrent neural networks can learn to implement a counter in the course of being trained to recognize strings from context-free languages, and therefore satisfy the requirement posed by the theory of formal automata. The networks are limited to a finite value for the maximum depth of recursive nesting in the strings, but so would any actual hardware implementation of an abstract automaton.

The languages our networks recognize are members of a subclass of the full class of context-free languages: they are members of the class of languages recognizable by *deterministic* automata. It is difficult to imagine how a network could efficiently implement a nondeterministic automaton — the network would somehow have to encode a complex representation of a search state in its activation values. On the other hand, it has been argued that deterministic automata are sufficient for processing natural languages (Marcus, 1980).

In order for the networks to be able to learn to recognize context-free languages, they had to start learning with proper sets of initial connection weights. The values of those initial weights were found by a simulation of evolution. The enormous investment in computational power illustrated by evolutionary searches like that shown in figure 4 make it questionable whether such searches really are a good way to obtain networks which can recognize strings from grammars. However the point is not to construct individual networks capable of recognizing specific languages. Instead, these results demonstrate that the members of a species can evolve biases to learn different, but related, languages reliably and efficiently. These biases would be crucial for the members of a species whose subgroups face different, but related, communicative situations.

The innate biases the networks acquire seem to confer sensitivity to the underlying regularities of a class of languages. In the case of learning the temporal boolean functions, the networks inherited a partial solution to the task they had to learn. Such sensitivity to underlying regularities, and partial solutions, makes the learning task faced by the network easier. Only the more superficial aspects of the language need to be learned upon exposure to examples. This was the initial motivation for the proposal of innateness in the first place: a difficult (or impossible) learning task is made tractable by innate mechanisms or biases. However no domain-specific innate acquisition mechanism needs to be introduced.

Instead all that is required is a general-purpose learning mechanism in an appropriate initial configuration.

The sensitivity of the networks to the underlying regularities of the class of languages could make one suspect that the network has an innate set of formal constraints characterizing the abstract form of a grammatical string. Language acquisition on this model consists in discovering specific rules consistent with this set of general constraints (Pinker, 1989). But any constraints the network possesses, and any rules it learns, are represented only implicitly, in the connection weights of the network. Indeed it might be best not to think of these initial connection weights as a specific set of formal constraints and rules, but as the states of a dynamical system interacting with an environment (the set of training examples). It seems likely that the behavior of such systems would submit to quite a different sort of analysis than that of formal learning theory (Wexler & Culicover, 1980), leading to a different characterization of the process of human language acquisition than that of acquiring rules subject to innate constraints.

Though these investigations were motivated partly by Fodor & Pylyshyn's (1989) critical analysis of the adequacy of neural network models of cognition, the results presented here are a long way from a refutation of their argument. The networks have not been shown to manipulate abstract recursive structures to perform valid inference, or any other of the sort of semantically sound "structure sensitive" operations that Fodor & Pylyshyn take to be criterial of cognition.

It is important, however, to note that their arguments for the existence of recursively structured mental representations rest on observed human abilities to perform certain tasks: to find two sentences to be synonymous, for example; or to draw a valid inference from a set of premises. It is a *hypothesis* that recursively structured representations are needed to account for these abilities. The networks described here, for example, are able to recognize recursively characterizable structures with no recursively structured representations. If it turns out that more and more cognitively relevant abilities can be achieved without recursively structured representations, than the fact that neural networks are, or are not, capable of such representations becomes less relevant.

It is likely that there are a number of interacting explanations for the critical period effects observed in the development of some cognitive abilities. For one thing, it is possible there are adaptive benefits for animals to acquire cognitive abilities in a certain sequence or at certain times (it is probably beneficial for children, for example, to learn language while rather young). Also there might be a value towards focusing cognitive effort on one problem at time, perhaps under the control of general maturational processes.

The explanation of the critical period effect proposed here is a direct consequence of innate biases being realized as the initial state of a general-purpose learning mechanism. Such effects would be observed whether or not any other maturational process or mechanism exists.

We are currently exploring better ways to analyze the initial weights of the evolved networks, to better understand exactly what the innate biases are encoding. We are also attempting to evolve networks that can learn more complex and more realistic grammars. All of our evolutionary simulations so far have involved formal languages that we imposed on the networks to learn. We intend to explore the evolution of populations of networks whose fitness is based on their performance on a communication task, to see what kinds of languages arise.

9   Conclusion

The proposal of "innateness" as a solution to puzzles in cognitive development has been around since ancient times. With the advent of computational modeling of evolutionary processes comes the possibility of evaluating concrete proposals for the mechanisms of innateness. The experiments reported herein are an exploration in that direction.

The ability to recognize and generate complex temporal patterns, necessary for language as well as other activities, is important enough to expect that some innate ability to learn to do so would be beneficial. Recurrent neural networks could implement that ability, with the innate biases realized as appropriate values for the networks' initial connection weights.

Acknowledgments

I would like to thank Rik Belew and Jeff Elman for valuable advice and encouragement.

References

Baldwin, J. M. (1896). A new factor in evolution. *American Naturalist*, 30, pp. 441–451.

Belew, R. K. (1989). When both individuals and populations search: Adding simple learning to the genetic algorithm. *Proceedings of the 3rd International Conference on Genetic Algorithms*, J. D. Schaefer, (ed.), Morgan Kaufman.

Belew, R. K., McInerney, J., and Schraudolph, N. N. (1991). Evolving networks: Using the genetic algorithm with connectionist learning. In *Artificial Life II, SFI Studies in the Sciences of Complexity, Volume X*. C. G. Langton, C. Taylor, J. D. Farmer, and S. Rasmussen, (eds.), Addison-Wesley, pp. 511–547.

Chomsky, N. (1957). *Syntactic Structures*. The Hague: Mouton.

Chomsky, N. (1987). *Knowledge of Language: Its Nature, Origin, and Use*. New York: Praeger.

Christiansen, M. (1992). The (non)necessity of recursion in natural language processing. *Proceedings of the Fourteenth Annual Meeting of the Cognitive Science Society*, Indiana University, Bloomington.

Cleeremans, A., Servan-Schreiber, D. and McClelland, J. L. (1989). Finite state automata and simple recurrent networks. *Neural Computation*, 1, pp. 372–381.

Elman, J. L. (1990a). Finding structure in time. *Cognitive Science*, 14, pp. 179–211.

Elman, J. L. (1990b). Distributed representation, simple recurrent networks, and grammatical structure. *Machine Learning*, 7, pp. 195–225.

Elman, J.L. (1993). Learning and development in neural networks: The importance of starting small, *Cognition*, 48, pp. 71–99.

Fodor, J., & Pylyshyn, Z. (1988). Connectionism and cognitive architecture: A critical analysis. *Cognition*, 28, pp. 3–71.

Giles, G. L., Sun, G. Z., Chen, H. H., Lee, Y. C., and Chen, D. (1990). "Higher order recurrent networks & grammatical inference." In *Advances in Neural Information Processing Systems*, 2, D. S. Touretzky (ed.), San Mateo CA: Morgan Kaufman.

Gold, E. M. (1967). Language identification in the limit. *Information and Control*, 16, pp. 447–475.

Goldberg, D. E. (1989). *Genetic Algorithms in Search, Optimization, and Machine Learning*. Reading, MA: Addison-Wesley.

Hinton, G., & Nowlan, S. J. (1987). How learning can guide evolution. *Complex Systems*, 1, pp. 495–502.

Hopcroft, J. E. & Ullman, J. D., (1979). *Introduction to Automata Theory, Languages, and Computation*. Addison-Wesley.

Jordan, M. (1986). *Serial Order: a Parallel Distributed Processing Approach*. ICS Report No. 8604. Institute for Cognitive Science; University of California at San Diego.

Kolen, J. F., & Pollack, J. B. (1990). Back propagation is sensitive to initial conditions. *Complex Systems 4*, pp. 269–280.

Lennenberg, E. H. (1967). *Biological Foundations of Language*. New York: John Wiley & Sons, Inc.

Marchman, V. A. (1993). Constraints on plasticity in a connectionist model of the English past tense. *Journal of Cognitive Neuroscience*, 5, pp. 215–234.

Marcus, M. (1980). *A Theory of Syntactic Recognition for Natural Language*. Cambridge, MA: MIT Press.

Marler, P. (1987). The instict to learn. In *The Epigenesis of Mind: Essays on Biology and Cognition*. Susan Carey & Rochel Gelman (eds.), Hillsdale, New Jersey: Lawrence Erlbaum Associates. pp. 37–66.

Newport, E. (1990). Maturational constraints on language learning. *Cognitive Science*, 14, pp. 11–28.

Nolfi, S., Elman, J. L., and Parisi, D. (1990). *Learning and Evolution in Neural Networks.* CRL Technical Report 9019. Center for Research on Language. University of California at San Diego.

Pinker, S. (1989). *Learnability and Cognition.* Cambridge, MA: MIT Press.

Pollack, J. B. (1987). *On Connectionist Models of Natural Language Processing.* PhD Thesis, Computer Science Department, University of Illinois, Urbana. Available as Memorandum MCCS-87-100, Computing Research Laboratory, New Mexico State University.

Pollack, J. B. (1990). Language acquisition via strange automata. In *Proceedings of the Twelfth Annual Conference of the Cognitive Science Society.*

Pollack, J. B. (1991). The induction of dynamical recognizers. *Machine Learning,* 7, pp. 227–252.

Rumelhart, D. E., Hinton, G. E., and Williams, R. J. (1986). Learning internal representations by error propagation. In *Parallel Distributed Processing: Explorations in the Microstructure of Cognition, Volume 1: Foundations.* David E. Rumelhart, James L. McClelland and the PDP Research Group. Cambridge, MA: MIT Press, pp. 318–362.

Servan-Schreiber, D., Cleeremans, A., and McClelland, J. L. (1988). *Encoding sequential structure in simple recurrent networks.* CMU Technical Report CMU-CS-335-87). Carnegie-Mellon University, Department of Computer Science.

Siegelmann, H. T. (1993) *Foundations of Recurrent Neural Networks.* PhD Dissertation. Rutgers University, Graduate Program in Computer Science.

Sun, G. Z., Chen, H. H., Giles, G. L., Lee, Y. C., and Chen, D. (1990). Connectionist pushdown automata that learn context-free grammars. *Proceedings of the International Joint Conference on Neural Networks, Vol. 1.* M. Caudill (ed.) Hillsdale, NJ: Lawrence Erlbaum.

Weckerly, J., & Elman, J.L. (1992). A PDP approach to processing center-embedded sentences. In *Proceedings of the Fourteenth Annual Conference of the Cognitive Science Society.* Hillsdale, NJ: Erlbaum.

Wexler, K., & Culicover, P. (1980). *Formal Principles of Language Acquisition.* Cambridge, MA: MIT Press.

DYNAMICS OF SELF-ASSEMBLING SYSTEMS
—ANALOGY WITH CHEMICAL KINETICS—

Kazuo Hosokawa, Isao Shimoyama, and Hirofumi Miura

Department of Mechano-Informatics
Faculty of Engineering, The University of Tokyo
7-3-1 Hongo, Bunkyo-ku, Tokyo, 113 Japan
e-mail: hosokawa@leopard.t.u-tokyo.ac.jp
isao@leopard.t.u-tokyo.ac.jp
miura@leopard.t.u-tokyo.ac.jp

Abstract

In this paper, we propose a new analyzing method for self-assembling systems. Its initial purpose was to predict the yield—the final amount of desired product—of our original self-assembling mechanical model. Moreover, the method clarifies the dynamical evolution of the system. In the method, the quantity of each intermediate product is adopted as state variables, and the dynamics which dominates the state variables is derived. The behavior of the system is reduced to a set of difference equations with a small degree of freedom. The concept is the same as in chemical kinetics or in population dynamics. However, it was never applied to self-assembling systems. The mathematical model is highly abstracted so that it is applicable to other self-assembling systems with only small modifications.

1. Introduction

One goal of this research is making self-reproducing machines. There was some pioneering work on machine self-reproduction in the 1950s (von Neumann 1966, Penrose 1959). Recently, Ichikawa (1992) demonstrated a one-dimensional self-reproducing machine using microprocessors. The authors aim at self-reproduction such as found in living things, where the difference of complexity between basic unit and complete body is large. Such systems will be realized using nanotechnology, i.e. a technique by which things are handled on a molecular scale, in the near future (Drexler 1986).

Making self-reproducing machines is a difficult problem. The problem should be divided into subproblems. Self-assembly, which we discuss in this paper, is one of the subproblems. Self-assembly is a phenomenon in which basic units come together and form a structure spontaneously. Such processes are widely observed at protein supermolecules, for example viruses, flagella of bacteria, and ribosomes (Lehninger 1975). The authors think that self-assembly is available as a process in self-reproduction.

In the biological world, self-assembly has been investigated in detail for a long time. For example, Casper (1980) studied viruses, and Jones and Aizawa (1991) studied flagella of bacteria.

Whereas self-assembly is spontaneous aggregation of substance, self-organization has a more general meaning. It is the appearance of global patterns from uniform space, caused by local interactions. The patterns consist not only of substance but also of energy field or velocity field, and so on. Self-organization has been studied in various fields: physics, chemistry, biology, and economics. Synergetics (Haken 1978) or Dissipative Structure (Nicolis and Prigogine 1977) are trials of mathematical systematization of self-organization. They have succeeded in explaining simple phenomena such as Bérnard convection or Belousov-Zhabotinsky reaction. The methods described in the present paper are similar to these frameworks.

There have been several physical experiments on self-assembly in real world: Fukuda and Nakagawa (1989) and Murata, Kurokawa, and Kokaji (1993) in robotics, Cohn, Kim, and Pisano (1991) and Yeh and Smith (1994) in micro electro mechanical systems. The physical approach is important as a complement of computer simulation in Artificial Life. Ichikawa (1992), Fukuda and Nakagawa (1989), and Murata, Kurokawa, and Kokaji (1993) aim at immediate applications; they made complicated systems using microprocessor for each unit.

The authors believe that there are unknown general principles in self-assembly or self-organization and consider that more basic research is necessary to discover them. For this purpose, we started with most primitive mechanical models. In this paper, we discuss a important problem common to self-assembling systems: the *yield problem*. It can be solved using a new analyzing method, which provides besides final yield knowledge such as time variation of initial, intermediate, and final products. The theory is verified by experimental results.

2. Mechanical Model

Figure 1 and fig. 2 show a basic unit of the mechanical model used for our experiments. The triangular body is made of polyurethane (density: 1.07). The permanent magnets made of neodymium have a surface magnetic flux density of 2.8 kG.

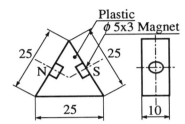

Figure 1. Basic unit of the mechanical model (Dimension: mm).

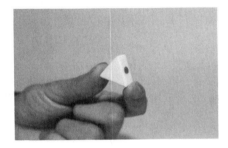

Figure 2. Photograph of the basic unit.

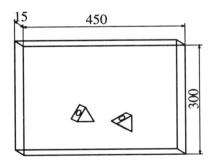

Figure 3. The case for shaking the units (Dimension: mm).

Many units are put into a flat box illustrated in fig. 3. Their movement is restricted to a plane. An electric motor rotates the box at a rate of 4.0 r.p.m and shakes the units thoroughly (fig. 4). The units are bonded to each other by magnetic force. Six units form a complete body (fig. 5) —a regular hexagon. We used 20 or 100 basic units for the experiments.

Figure 4. The experimental apparatus.

3. Theory and Experiment

3.1 Yield Problem and Its Solution

Testing the mechanical model described above, the authors encountered a problem which has never been studied: the *yield problem* of self-assembling systems. For example, if the process starts with 12 basic units, the ideal goal is to form two complete hexagons as in fig. 5(a). But actually the system usually reaches a final state like fig. 5(b). Increasing the quantity of starting units cannot be a solution. As is generally known, yield is a chemical term which stands for the final amount of desired product. Yield is two in fig. 5(a), and is zero in fig. 5(b). We define the yield problem as "problem of predicting the final amount of complete bodies in an arbitrarily given self-assembling system."

This problem can be solved completely using the following method. Moreover, this method clarifies the dynamical evolution of the system. The method will be a useful tool when we design a self-assembling system.

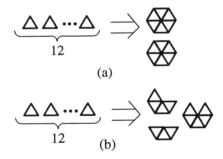

Figure 5. Explanation of the yield problem. (a)The ideal case. Yield is two. (b)An actual case. Yield is zero.

The procedure is summarized as follows:

1. Let the quantity of every intermediate product be a state variable. This greatly reduces the degree of freedom of the system.

2. Derive the dynamics governing the state variables from the information of elementary processes. This phase is similar to chemical kinetics (Moore 1983) or population dynamics (Haken 1978).

3. Calculate the variation of the state variables from given initial conditions according to the dynamics.

In the case of our mechanical model, there are one initial, four intermediate, and one final product shown in fig. 6. They are represented by symbol $X_1 \sim X_6$ according to the number of constituent units. However, X_1 is abbreviated to X. Symbol x_i represents quantity of X_i ($i = 1 \sim 6$). These are state variables of the system.

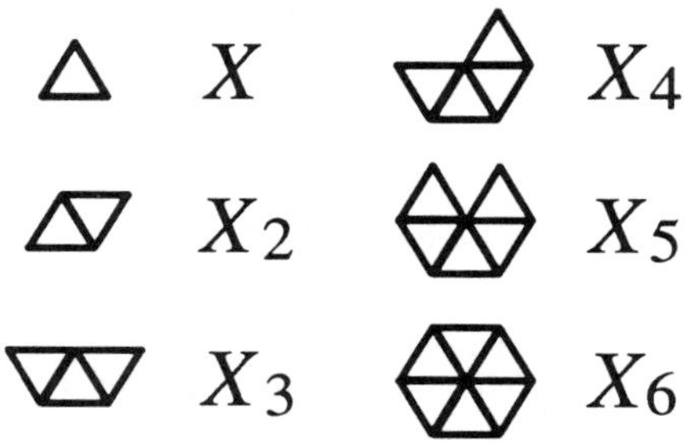

Figure 6. Initial, intermediate, and final products of the system.

Elementary processes are described similarly to chemical reactions as follows:

$$\begin{array}{ll} 2X \rightarrow X_2, & X + X_2 \rightarrow X_3, \\ X + X_3 \rightarrow X_4, & X + X_4 \rightarrow X_5, \\ X + X_5 \rightarrow X_6, & 2X_2 \rightarrow X_4, \\ X_2 + X_3 \rightarrow X_5, & X_2 + X_4 \rightarrow X_6, \\ 2X_3 \rightarrow X_6. \end{array} \qquad (1)$$

The elementary processes are often called "reactions" hereafter. We assumed that all reactions are bimolecular reactions. In other words, we neglected reactions in which more than two clusters are bonded simultaneously such as $3X \rightarrow X_3$.

If every x_i are large enough, the state vector $\boldsymbol{x} = (x_1, \cdots, x_6)$ obeys the difference equation

$$\boldsymbol{x}(t+1) = \boldsymbol{x}(t) + \boldsymbol{F}(\boldsymbol{x}(t)), \qquad (2)$$

where t is the parameter corresponding to time. Speaking more strictly, t represents the number of collision between clusters. And $\boldsymbol{F} = (F_1, \cdots, F_6)$ is a mapping from 6-dimensional space to itself, where F_i signifies the expected value of increment of x_i per one step.

$$F_i = \sum_j \nu_{ij} P_j, \qquad (3)$$

where ν_{ij}, corresponds to the stoichiometric number, is coefficient of X_i in the j-th reaction formula in (1) (The order can be decided arbitrarily). The coefficient ν_{ij} has a positive sign if X_i is product (in the right side of the formula), and has a negative sign if X_i is reactant (in the left side of the formula). For example, if the reaction $2X_2 \rightarrow X_4$ is decided to be the first reaction, $\nu_{21} = -2$, $\nu_{41} = 1$, $\nu_{i1} = 0$ ($i \neq 2$, 4). Transposing the terms formally, the reaction formulae in (1) can be rewritten as follows:

$$0 \rightarrow \sum_{i=1}^{6} \nu_{ij} X_i, \qquad (j = 1, 2, \cdots, 9). \qquad (4)$$

In eq. (3), P_j is the probability of the j-th reaction occurring at the present step. If the reaction is caused by X_l and X_m, P_j consist of two factors:

$$P_j = P_{lm}^{\mathrm{b}} P_{lm}^{\mathrm{c}}, \qquad (5)$$

where P_{lm}^{c}, the collision probability, is the probability of X_l and X_m colliding, and P_{lm}^{b}, the bonding probability, represents the conditional probability of bonding on the condition that they collide. Since the definition of t, one collision between two clusters occurs in one step of t. P_{lm}^{c}—the probability that the two clusters are X_l and X_m—is assumed to be equal to the probability that X_l and X_m are chosen when arbitrary two clusters are picked up from $S = x_1 + x_2 + \cdots + x_6$ clusters. Therefore, assuming that $x_i \gg 1$, we obtain

$$P_{lm}^{\mathrm{c}} = \begin{cases} 2x_l x_m / S^2, & l \neq m, \\ x_l^2 / S^2, & l = m. \end{cases} \qquad (6)$$

Thus, F_i of this system can be expressed explicitly as

$$\begin{aligned} F_1(\boldsymbol{x}) =& (-2P_{11}^{\mathrm{b}} x_1^2 - 2P_{12}^{\mathrm{b}} x_1 x_2 - 2P_{13}^{\mathrm{b}} x_1 x_3 \\ & - 2P_{14}^{\mathrm{b}} x_1 x_4 - 2P_{15}^{\mathrm{b}} x_1 x_5)/S^2, \\ F_2(\boldsymbol{x}) =& (P_{11}^{\mathrm{b}} x_1^2 - 2P_{12}^{\mathrm{b}} x_1 x_2 - 2P_{22}^{\mathrm{b}} x_2^2 - 2P_{23}^{\mathrm{b}} x_2 x_3 \\ & - 2P_{24}^{\mathrm{b}} x_2 x_4)/S^2, \\ F_3(\boldsymbol{x}) =& (2P_{12}^{\mathrm{b}} x_1 x_2 - 2P_{13}^{\mathrm{b}} x_1 x_3 - 2P_{23}^{\mathrm{b}} x_2 x_3 \\ & - 2P_{33}^{\mathrm{b}} x_3^2)/S^2, \\ F_4(\boldsymbol{x}) =& (2P_{13}^{\mathrm{b}} x_1 x_3 + P_{22}^{\mathrm{b}} x_2^2 - 2P_{14}^{\mathrm{b}} x_1 x_4 \\ & - 2P_{24}^{\mathrm{b}} x_2 x_4)/S^2, \\ F_5(\boldsymbol{x}) =& (2P_{14}^{\mathrm{b}} x_1 x_4 + 2P_{23}^{\mathrm{b}} x_2 x_3 - 2P_{15}^{\mathrm{b}} x_1 x_5)/S^2, \\ F_6(\boldsymbol{x}) =& (2P_{15}^{\mathrm{b}} x_1 x_5 + 2P_{24}^{\mathrm{b}} x_2 x_4 + P_{33}^{\mathrm{b}} x_3^2)/S^2. \end{aligned} \qquad (7)$$

The value of P_{lm}^{b} depends on the definition of "collision." This problem will be discussed more deeply in section 6. We calculated P_{lm}^{b} only from the clusters' geometrical properties on the following assumption. "The bonding probability P_{lm}^{b} is equal to the probability that there is a pair of bonding faces of X_l and X_m such that each face can be seen from the other face completely when the clusters are placed on the plane at random." We will explain this calculation in detail giving an example in appendix. The result of this calculation is shown in table 1.

The solution of eq. (2) with the initial condition of $\boldsymbol{x}(0) = (100, 0, \cdots, 0)$ is shown in fig. 7.

Table 1. The bonding probability $P_{l;m}^{b}$.

m \ l	1	2	3	4	5	6
1	0.472	0.444	0.417	0.278	0.139	0
2		0.389	0.333	0.222	0	0
3			0.250	0	0	0
4				0	0	0
5	Symmetric				0	0
6						0

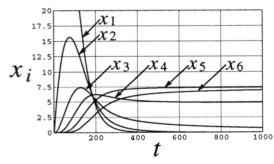

Figure 7. The solution of eq. (2) with the explicit $\boldsymbol{F}$ of eq. (7) and the initial condition of $\boldsymbol{x}(0) = (100, 0, \cdots, 0)$.

Since the correspondence between t and real time is vague, we compare the experimental results with calculated results in $\boldsymbol{x}$ space. The solution generates a trajectory in the 6-dimensional space spanned by $x_1 \sim x_6$. Figure 8 shows the projections of the trajectory on some coordinate planes. Plotted points represent experimental data. They include halfway states as well as final states. Experimental results are rather scattered. The number of starting units $x_1(0) = 100$ was not large enough to verify the theory precisely. But we consider that the theory was confirmed roughly.

3.2 Master Equation

The method described in section 3.1 is not so efficient in case of small x_i. In this case, it is advantageous to consider not the mean value of x_i but the probability distribution. The following method is known as *master equation* (Haken 1978).

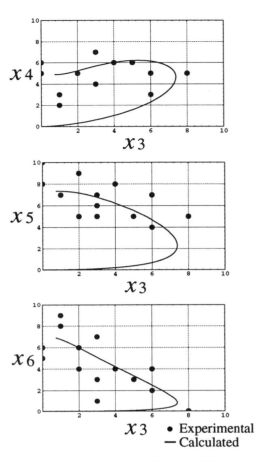

Figure 8. Projections of the trajectory of the state variables.

In this section, the state vector $\boldsymbol{x}$ is treated as discrete, whereas it is treated as continuous in the other sections of this paper. Let $p(\boldsymbol{x}; t)$ denote the probability of the system being in the state $\boldsymbol{x}$ at step t. The master equation describes the dynamics which dominates the probability distribution $p(\boldsymbol{x}; t)$.

$$p(\boldsymbol{x}; t+1) - p(\boldsymbol{x}; t)$$
$$= \sum_{\boldsymbol{x}'} w(\boldsymbol{x}, \boldsymbol{x}')p(\boldsymbol{x}'; t) - p(\boldsymbol{x}; t)\sum_{\boldsymbol{x}'} w(\boldsymbol{x}', \boldsymbol{x}), \quad (8)$$

where $w(\boldsymbol{x}, \boldsymbol{x}')$ denotes the probability of a transition from $\boldsymbol{x}'$ to $\boldsymbol{x}$. It is independent of t in this system. The first term in the right side of eq. (8) represents the probability that the system comes to the state $\boldsymbol{x}$ from other states at the step $t \to t+1$. The second term represents the probability of the system changing to other states from the present state $\boldsymbol{x}$.

The transition probability $w(\boldsymbol{x}, \boldsymbol{x}')$ can be derived in a similar way as in section 3.1. Where $\boldsymbol{x} = (x_1, x_2, x_3, x_4, x_5, x_6)$, there are nine $\boldsymbol{x}'$ which give nonzero $w(\boldsymbol{x}, \boldsymbol{x}')$ corresponding to nine kind of reactions of (1). For example, where $x' = x + (2, -1, 0, 0, 0, 0)$ corresponding to the reaction

$$2X \to X_2, \qquad (9)$$

$w(x, x')$ equals the probability of reaction (9) occurring at state x'. Therefore,

$$w(x, x') = P_{11}^{b} \cdot \frac{x_1 + 2}{S + 1} \cdot \frac{x_1 + 1}{S}. \qquad (10)$$

This is approximated by $P_{11}^{b} x_1^2 / S^2$ in section 3.1 on the assumption of $x_i \gg 1$.

The explicit form of eq. (8) is

$$p(x; t+1) - p(x; t) = \frac{1}{S(S+1)} \times$$

$$[P_{11}^{b}(x_1 + 2)(x_1 + 1)p(x + (2, -1, 0, 0, 0, 0); t)$$
$$+ 2P_{12}^{b}(x_1 + 1)(x_2 + 1)p(x + (1, 1, -1, 0, 0, 0); t)$$
$$+ 2P_{13}^{b}(x_1 + 1)(x_3 + 1)p(x + (1, 0, 1, -1, 0, 0); t)$$
$$+ 2P_{14}^{b}(x_1 + 1)(x_4 + 1)p(x + (1, 0, 0, 1, -1, 0); t)$$
$$+ 2P_{15}^{b}(x_1 + 1)(x_5 + 1)p(x + (1, 0, 0, 0, 1, -1); t)$$
$$+ P_{22}^{b}(x_2 + 2)(x_2 + 1)p(x + (0, 2, 0, -1, 0, 0); t) \qquad (11)$$
$$+ 2P_{23}^{b}(x_2 + 1)(x_3 + 1)p(x + (0, 1, 1, 0, -1, 0); t)$$
$$+ 2P_{24}^{b}(x_2 + 1)(x_4 + 1)p(x + (0, 1, 0, 1, 0, -1); t)$$
$$+ P_{33}^{b}(x_3 + 2)(x_3 + 1)p(x + (0, 0, 2, 0, 0, -1); t)]$$
$$- \frac{1}{S(S-1)}p(x; t) \times$$
$$[P_{11}^{b}x_1(x_1 - 1) + 2P_{12}^{b}x_1 x_2 + 2P_{13}^{b}x_1 x_3$$
$$+ 2P_{14}^{b}x_1 x_4 + 2P_{15}^{b}x_1 x_5 + P_{22}^{b}x_2(x_2 - 1)$$
$$+ 2P_{23}^{b}x_2 x_3 + 2P_{24}^{b}x_2 x_4 + P_{33}^{b}x_3(x_3 - 1)].$$

Generally, the master equation gives more information but requires more computation than the dynamics of the mean value described in section 3.1. In the case of this mechanical model, the method in section 3.1 needs 6 degrees of freedom (DOF). On the other hand, the master equation has a DOF which equals the number of possible x. The number depends on the initial condition. If the initial condition is $x(0) = (N, 0, \cdots, 0)$ (i.e. $p((N, 0, \cdots, 0); 0) = 1$, other $p(x; 0) = 0$), possible x are the sets of nonnegative integers which satisfy

$$x_1 + 2x_2 + 3x_3 + 4x_4 + 5x_5 + 6x_6 = N. \qquad (12)$$

The numbers of possible x calculated by computer are summarized in table 2. It increases with N. Moreover, the number of steps until convergence becomes larger. Therefore, the limitation to calculate the master equation numerically within reasonable cost is approximately

$N = 20$. The master equation is not suitable if N is greater than this order.

Table 2. The number of possible different states x with respect to the number of initial basic units N.

N	Num. of x
10	35
20	282
100	189,507

In the case of $N = 20$, the possible final states are listed in table 3. The same table also contains each final probability obtained by calculation and experiment. In calculation, the solution is regarded as converging when $t = 200$. Each $p(x; 200) < 10^{-2}$ in state x which is not listed in table 3. The experimental results are obtained by 100 trials.

Table 3. Final states and their probabilities obtained by calculation and experiment.

Final State x	Probability	
	Calc.	Exp.
(0, 0, 0, 1, 2, 1)	0.4587	0.38
(0, 0, 1, 0, 1, 2)	0.1965	0.24
(0, 0, 0, 2, 0, 2)	0.1582	0.20
(0, 0, 1, 3, 1, 0)	0.0811	0.11
(0, 0, 0, 0, 4, 0)	0.0631	0.03
(0, 1, 0, 0, 0, 3)	0.0333	0.04
(0, 0, 0, 5, 0, 0)	0.0056	0

4. Activating Mechanism

In the case of self-assembly of viruses or bacterial flagella, the final state never becomes like fig. 5(b). This is because they are assembled in a more orderly manner than our mechanical model described above. Casper (1980) wrote that their assembly can be controlled by switching the subunits from an inactive, unsociable form to an active, associable form. Penrose (1959) also designed an activating mechanism for his model.

Figure 9 shows our activating mechanism. The unit has two magnets, the same kind as described in section 2, but they slide in cylinders. The dimensions of the unit were designed carefully. In the inactive state, the two magnets of a unit attract each other and are both pulled towards the center of the triangular body (fig. 9(a)). The attraction of two inactive units is too weak to form a stable combination (fig. 9(c)). An inactive unit is activated by pulling out one magnet from the center to the outside (fig. 9(b)). In this case, since the other magnet becomes almost free, the unit can activate another inactive unit and can be linked together with it (fig. 9(d)). In addition, two active units can be linked together.

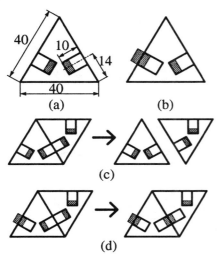

Figure 9. The activating mechanism. (a)The inactive state (Dimension: mm). (b)The active state. (c)Two inactive units cannot be linked together. (d)An active unit can activate another inactive unit and can be linked with it.

This system can be analyzed using the method in section 3.1. Let I and A symbolize inactive and active units, respectively. There are 13 permitted reactions:

$$
\begin{aligned}
&I + A_2 \rightarrow A_3, \quad I + A_3 \rightarrow A_4, \\
&I + A_4 \rightarrow A_5, \quad I + A_5 \rightarrow A_6, \\
&2A \rightarrow A_2, \quad\quad A + A_2 \rightarrow A_3, \\
&A + A_3 \rightarrow A_4, \quad A + A_4 \rightarrow A_5, \\
&A + A_5 \rightarrow A_6, \quad 2A_2 \rightarrow A_4, \\
&A_2 + A_3 \rightarrow A_5, \quad A_2 + A_4 \rightarrow A_6, \\
&2A_3 \rightarrow A_6.
\end{aligned}
\tag{13}
$$

Assuming that the bonding probability P^{b} depends only on the geometry of the units, we use the same P^{b} as in section 3.1. Therefore, I, A, and X as well as A_i and X_i are regarded as equivalent concerning P^{b}.

Where x_0 and x_i ($i = 1 \sim 6$) denote the quantities of I and A_i, respectively, $\boldsymbol{x} = (x_0, x_1, \cdots, x_6)$ obeys a difference equation written in the form

$$
\begin{aligned}
&\boldsymbol{x}(t+1) = \boldsymbol{x}(t) + \boldsymbol{F}(\boldsymbol{x}(t)), \\
&\boldsymbol{F}(\boldsymbol{x}) = (F_0(\boldsymbol{x}), F_1(\boldsymbol{x}), \cdots, F_6(\boldsymbol{x})).
\end{aligned}
\tag{14}
$$

We calculated the solution with the initial condition of $\boldsymbol{x}(0) = (100, 10, 0, \cdots, 0)$. The solution is regarded as converging at $t_f = 5000$. The final state vector is

$$
\boldsymbol{x}(t_f) = (56.1, 0, 0, 0, 0, 0, 9.0).
\tag{15}
$$

The terms written as zero in eq. (15) are smaller than 10^{-6}. Ten A_6 were expected to be formed since the system started with ten active seeds (A). However, it was found that only nine A_6 were formed because of combination of two active intermediate parts A_i and A_j occurring once on the average during the process.

5. Application to Penrose's Work

The method in this paper can be applied to Penrose's work (1959). At first, we will explain his self-reproducing machine briefly. Figure 10 illustrates the basic "double-hook units." Two units form a complete body. A complete body grows combining with neighbor units. If it grows to the length of four units, it divides at the center into two complete bodies.

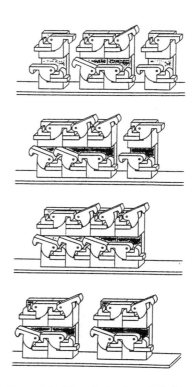

Figure 10. Penrose's "double-hook units" (from Penrose, 1959 with permission).

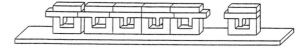

Figure 11. Penrose's "counting device" (form Penrose, 1959 with permission).

Where X represents a basic unit, the elementary processes can be expressed as follows:

$$2X \to X_2, \tag{16}$$

$$X + X_2 \to X_3, \tag{17}$$

$$X + X_3 \to 2X_2, \tag{18}$$

$$X_2 + X_3 \to \begin{cases} X_2 + X_3, \\ X + 2X_2, \end{cases} \tag{19}$$

$$2X_3 \to \begin{cases} 2X_3, \\ 3X_2, \\ 2X + 2X_2, \\ X + X_2 + X_3. \end{cases} \tag{20}$$

In (19) or (20), there are several possible results. We assume that each result occurs with the same probability.

In the same manner as in the previous sections, x_i represents the quantity of X_i. Assuming all the bonding probabilities P^b equal to 1, the dynamics which dominates $x = (x_1, x_2, x_3)$ is

$$x(t+1) = x(t) + F(x(t)),$$
$$F(x) = (F_1(x), F_2(x), F_3(x)),$$
$$F_1(x)$$
$$= (-2x_1^2 - 2x_1 x_2 - 2x_1 x_3 + x_2 x_3 + \frac{3}{4}x_3^2)/S^2,$$
$$F_2(x) \tag{21}$$
$$= (x_1^2 - 2x_1 x_2 + 4x_1 x_3 + x_2 x_3 + \frac{3}{2}x_3^2)/S^2,$$
$$F_3(x) = (2x_1 x_2 - 2x_1 x_3 - x_2 x_3 - \frac{5}{4}x_3^2)/S^2,$$
$$S = x_1 + x_2 + x_3.$$

Penrose designed two attachments. Figure 11 shows the "counting device" (CD), which prevents more than four units from coming close together. If this mechanism is attached to the double-hook unit, the reactions (19)(20) are prohibited. In this case, F in eq. (21) is rewritten as

$$F_1(x) = (-2x_1^2 - 2x_1 x_2 - 2x_1 x_3)/S^2,$$
$$F_2(x) = (x_1^2 - 2x_1 x_2 + 4x_1 x_3)/S^2, \tag{22}$$
$$F_3(x) = (2x_1 x_2 - 2x_1 x_3)/S^2.$$

The second attachment is the "activating cam-lever" (ACL). Figure 12 shows double-hook units with both attachments. The ACL are attached above the CD. The unit with tilted lever is active. Two inactive units which have horizontal levers cannot come close together. The ACL prohibits reaction (16). Therefore, in the dynamics of the system shown in fig. 12, F is rewritten as

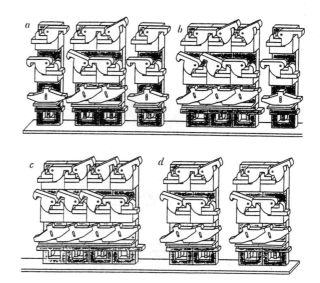

Figure 12. Penrose's "complete self-reproducing machine" (form Penrose, 1959 with permission).

$$F_1(x) = (-2x_1 x_2 - 2x_1 x_3)/S^2,$$
$$F_2(x) = (-2x_1 x_2 + 4x_1 x_3)/S^2, \tag{23}$$
$$F_3(x) = (2x_1 x_2 - 2x_1 x_3)/S^2.$$

There are four cases according to whether each attachments exists or not. We calculated the evolution of the system for these four cases with the initial condition of $x = (100, 1, 0)$ and summarized the results in fig. 13.

First, in the presence of CD, the solution converges more rapidly, however, the yield is low. This is because the reactions of the systems with CD stop when x_1 becomes zero, whereas X_3 can be used as material to continue the processes in the absence of CD.

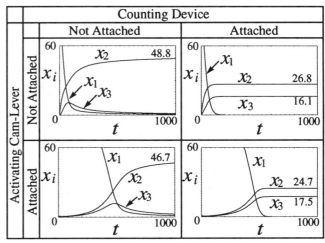

Figure 13. The solutions of Penrose's systems.

Second, the reaction proceeds more rapidly in the absence of ACL. It is natural that the reaction velocities greatly differ at the beginning when many X exist, since the reactions among X can occur in the absence of ACL. From the standpoint of velocity, systems without ACL are better, but those with ACL have a large advantage of controlling the process with active seeds.

We conclude that the system with ACL and no CD is the best of the four cases.

6. Discussion

The bonding probability P^b depends on the definition of "collision" which is difficult to define clearly. If collision is defined as any contact, however slightly it is, it is inconvenient that the sizes of clusters must be considered in calculating the collision probability P^c. We have chosen another way, adding a condition to the relative velocity of clusters having contact. For example, fig. 14 shows a collision between X and X_3. The arrow represents the velocity of the center of gravity of X in a coordinate system located on X_3. This can be defined as a collision when the length d is less than a defined length d_0. In this case, d_0 remains arbitrary. If a small d_0 is adopted, which means strong definition of collision, P^b increases. In reverse, a large d_0 decreases P^b. However, it is considered that their ratio (i.e. $P_{11}^b : P_{12}^b : P_{13}^b : \cdots$) does not vary so much with d_0.

Then, let us investigate how the solutions in this paper (except those in section 3.2) are affected when we change P^b keeping their ratio constant. The difference equations are written in the form

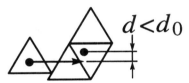

Figure 14. Example of collision between X and X_3.

$$x(t+1) = x(t) + F(x(t)). \qquad (24)$$

When bonding probabilities P^b vary by a factor k simultaneously, then eq. (24) changes into

$$x(t+1) = x + kF(x(t)). \qquad (25)$$

On the other hand, F has a remarkable property such that

$$F(ax) = F(x), \quad (a\text{: arbitrary constant}). \qquad (26)$$

Therefore, performing the scale transformation

$$kx = y, \qquad (27)$$

eq. (24) is rewritten as

$$y(t+1) = y(t) + kF(y(t)). \qquad (28)$$

In other words, changing P^b by a factor k is equivalent to a scale transformation (27). In this case, the solution becomes similar to the solution of the original equation with the initial condition which is $1/k$ times as large as original. When k has the order of $1/5 \sim 5$, it is difficult to consider that the solutions are qualitatively different from each other. In fact, there were no such cases as far as the authors calculated. For example, fig. 15 shows the solution of eq. (2) with the explicit form of F of eq. (7), and the initial condition of $x = (10, 0, \cdots, 0)$. It is almost similar to fig. 7. Thus, as regards the bonding probabilities, only their ratio is important for the present rough discussion. It seems that a precise definition of collision is not necessary yet.

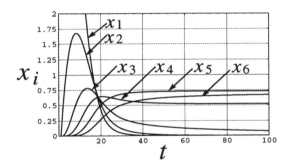

Figure 15. The solution of eq. (2) with the explicit F of eq. (7) and the initial condition of $x(0) = (10, 0, \cdots, 0)$.

7. Conclusion

1. Quantities of every intermediate products can be state variables which represent the state of a self-assembling system.
2. The dynamics, which rules the time variation of the state variables, can be derived from elementary processes using an analogy with chemical kinetics or population dynamics.
3. In the case of the number of units being small, it is efficient to calculate the time variation of the probability distribution using the master equation.
4. This method is a useful tool to obtain quantitative knowledge when we design a self-assembling system.

The methods described in this paper succeed in rather simple self-assembling systems, however, there remains the problem whether the methods are still efficient in more complex systems with a large number of states.

Appendix

We will explain the calculation of the bonding probability P^{b} in detail using the example of P_{11}^{b}. In section 3.1, P_{lm}^{b} was assumed to be equal the probability that

> "there is a pair of bonding faces of X_l
> and X_m such that each face can be
> seen from the other face completely" (A.1)

when the clusters are placed the plane at random. Let the two units be a and b. The plane can be divided into four regions using a unit for the basis as shown in fig. 16.

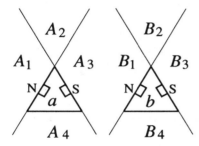

Figure 16. Divisions of the plane using the units for the basis.

When b is in the region A_1 from the standpoint of a, if b directs its S-pole-face to a, (A.1) is true. In other words, a should be in the region B_2 or B_3. The probability of that is

$$P(b \text{ in } A_1)P(a \text{ in } B_2 \cup B_3) = \frac{1}{3} \times \frac{1}{2} = \frac{1}{6}. \quad (A.2)$$

In the case of b being in the region A_3, the probability of (A.1) is 1/6 as well. And in the case of b being in the region A_2, a should be in the region B_1, B_2, or B_3. So, the probability is

$$P(b \text{ in } A_2)P(a \text{ in } B_1 \cup B_2 \cup B_3) = \frac{1}{6} \times \frac{5}{6} = \frac{5}{36}. \quad (A.3)$$

Thus, P_{11}^{b} is obtained as the sum of these probabilities:

$$P_{11}^{\mathrm{b}} = \frac{1}{6} + \frac{1}{6} + \frac{5}{36} = \frac{17}{36} \simeq 0.472. \quad (A.4)$$

References

Casper, D. L. D. 1980. Movement and Self-Control in Protein Assemblies—Quasi-Equivalence Revised. *Biophysics Journal* 32: 103-138.

Cohn, M. B., Kim, C. J., and Pisano, A. P. 1991. Self-Assembling Electrical Networks: An Application of Micromachining Technology. *Transducers '91* : 490-493.

Drexler, K. E. 1986. *Engines of Creation—The Coming Era of Nanotechnology*: Anchor Books.

Fukuda, T. and Nakagawa, S. 1989. A Study on Dynamically Reconfigurable Robotic Systems. *Transactions of Japan Society of Mechanical Engineers* 55 C: 114-118.(in Japanese).

Haken, H. 1978. *Synergetics — An Introduction*: Springer-Verlag.

Ichikawa, Y. 1992. One-Dimensional Self-Reproducing Machine. *Journal of the Robotics Society of Japan* 10: 266-272. (in Japanese).

Jones, C. J. and Aizawa, S. 1991. The Bacterial Flagellum and Flagellar Motor—Structure, Assembly and Function. *Advances in Microbial Physiology* 32: 109-172.

Lehninger, A. L. 1975. *Biochemistry*: Worth Publishers, Inc.

Moore, W. J. 1983. *Basic Physical Chemistry*: Prentice-Hall, Inc.

Murata, S., Kurokawa, H., and Kokaji, S. 1993. Self-Repairable Mechanical System. *The Workshop of the Robotics Society of Japan* 11: 149-152.(in Japanese).

Nicolis, G. and Prigogine, I. 1977. *Self-Organization in Nonequilibrium Systems—From Dissipative Structures to Order through Fluctuations*: John Wiley & Sons, Inc.

Penrose, L. S. 1959. Self-Reproducing Machines. *Scientific American* June: 105-114.

von Neumann, J. 1966. *Theory of Self-Reproducing Automata*: University of Illinois Press.

Yeh, H. J. and Smith, J. S. 1994. Fluidic Self-Assembly of Microstructures and its Application to Integration of GaAs on Si. *IEEE Micro Electro Mechanical Systems* : 279-284.

FROM LOCAL ACTIONS TO GLOBAL TASKS:
STIGMERGY AND COLLECTIVE ROBOTICS

R. Beckers [1,2], O.E. Holland [1,3] and J.L. Deneubourg [2]

[1] ZiF-Universität Bielefeld, Wellenberg 1, D-33615 Bielefeld
[2] Centre for non-linear phenomena and complex systems, Université Libre de Bruxelles, CP 231, B-1050 Brussels
[3] Faculty of Engineering, University of the West of England, Coldharbour Lane, Frenchay, Bristol BS16 1QY
R.Beckers: beckers@NOV.ZIF. uni-bielefeld.de
O.Holland: o_hollan@csd.uwe-bristol.ac.uk

Abstract

This paper presents a series of experiments where a group of mobile robots gather 81 randomly distributed objects and cluster them into one pile. Coordination of the agents' movements is achieved through stigmergy. This principle, initially developped for the description of termite building behaviour, allows indirect communication between agents through sensing and modification of the local environment which determines the agent's behaviour. The efficiency of the work was mesured for groups of one to five robots working together. Group size is a critical factor. The mean time to accomplish the task decreases for one, two and three robots respectively, then increases again for groups of four and five agents, due to an exponential increase in the number of interactions between robots which are time consuming and may eventually result in the destruction of existing clusters. We compare our results with those reported by Deneubourg *et al.* (1990) where similar clusters are observed in ant colonies, generated by the probabilistic behaviour of workers.

1. Introduction

There is a class of natural systems in which large numbers of simple agents collectively achieve remarkable feats through exploiting a single principle. They offer a spectacular existence proof of the possibility of using many simple agents rather than one or a few complex agents to perform complex tasks quickly and reliably. It is therefore surprising that the systematic exploitation of this principle has been neglected within the field of robotics. The natural systems we refer to are social insects - ants, termites, wasps, and bees. The principle is that of stigmergy, recognised and named by the French biologist P.P. Grassé

(1959) during his studies of nest building in termites. Stigmergy is derived from the roots 'stigma' (outstanding sign) and 'ergon' (work), thus giving the sense of 'incitement to work by the products of work'. It is essentially the production of a certain behaviour in agents as a consequence of the effects produced in the local environment by previous behaviour.

When they start to build a nest, termites modify their local environment by making little mud balls and placing them on the substrate; each mud ball is impregnated with a minute quantity of a particular pheromone. Termites deposit their mud balls probabilistically, initially at random. However, the probability of depositing a mud ball at a given location increases with the sensed presence of other mud balls and the sensed concentration of pheromone. The first few random placements increase the other termites' probability of putting their loads at the same place. By this blind and random game little columns are formed; the pheromone drifting across from neighbouring columns causes the tops of the columns to be built with a bias towards the neighbouring columns, and eventually the tops meet to form arches, the basic building units. Finally, as the influence of other stigmergic processes comes into play (e.g. processes involving water vapour and carbon dioxide concentrations, and modulated by the presence of the queen), the whole complex and highly differentiated nest structure is produced, with the royal cell, brood nurseries, food stores, air circulation, communication and foraging tunnels, and other areas all contained within one of the largest non-excavated structures built by any creature except man.

The use of stigmergy is not confined to building structures. It also occurs in cooperative foraging strategies such as trail recruitment in ants, where the interactions between foragers are mediated by pheromones put on the ground in quantities determined by the local conditions of the environment. For example, trail recruiting ant species are able to select and preferentially exploit the richest food source in the neighbourhood (Pasteels, Deneubourg and Goss 1987; Beckers *et al.* 1990) or the shortest path between the nest and a food source (Beckers, Deneubourg and Goss 1992). This strategy takes advantage of the characteristics of the trail-laying and trail-following mechanisms of the ants in combination with their essentially probabilistic behaviour: the probability that an ant follows a trail is a non-linear function of the trail's pheromone concentration, and the probability that an ant lays a pheromone spot depends on the characteristics of the recently-encountered food source and the environment (Beckers, Deneubourg and Goss 1994). When a trail between a single food source and the nest is first established, its pheromone concentration is low, and a high proportion of ants lose the path before reaching the food or the nest. As more and more journeys are made along the trail, the pheromone concentration increases progressively and so does the accuracy of trail following; finally the majority of the foragers will successfully use that trail. Where there are multiple food sources, or multiple trails of different lengths to the same food source, the non-linear dependence of the probability of successful trail-following on pheromone concentration sharply favours the rate of increase in strength of trails which are already strong, or short, or lead to rich food sources; as a strong trail recruits and retains ants, it reduces the number of ants available to strengthen other trails; evaporation and breakdown of the pheromone continually reduce the strength of all trails; the net result is that a single trail becomes dominant, and it is usually the 'best' choice from the point of view of length and richness of food source. The important factor is that very small fluctuations in the pheromone concentrations of different trails, occurring at the beginning of the recruitment, are amplified and determine the eventual outcome of the collective decision making process (Beckers *et al.* 1990; Deneubourg and Goss, 1989).

The stigmergic principle also appears to organise corpse-gathering behaviour in ant colonies. Observations show that these insects tend to put the corpses of dead nestmates together in cemeteries which occur in certain places far from the nest and which grow in size with time. If a large number of ant corpses are scattered outside a nest, the ants from the nest will pick them up, carry them about for a while, and drop them; within a short time it will be seen that the corpses are being put into small clusters, and as time goes on the number of clusters will decrease and their size will grow until eventually all the corpses will be in one or two large clusters. The emergence of these clusters has been studied by Deneubourg *et al.* (1990), who

showed that a simple mechanism involving the modulation of the probability of dropping corpses as a function of the local density of corpses was sufficient to generate the observed sequence of the clustering of corpses.

These examples from social insects show how global problems can be solved by exploiting the interactions between workers, and between workers and the environment. These processes give rise to self-organized structures which are not represented explicitly in each or any agent, but which guide and influence the actions of individual agents. The work described in this paper explores the possibility of extending these principles to robotics.

How would it be best to put stigmergy to work in robots? The traditional computational paradigm of robotics typically involves sensing the environment, then detecting features, then constructing or modifying a world model, then reasoning about the task and the world model in order to find some sequence of actions which might lead to success, then executing that action sequence one step at a time while updating the world model and replanning if necessary at any stage. Doing any of these is intractable in at least some domains; doing all of them in an unstructured dynamic environment fast enough to survive in that environment has turned out to be a practical impossibility regardless of the hardware resources available. Behaviour-based architectures, inspired by biology and epitomised by Brooks' subsumption architecture, have changed all that (Brooks 1986). A behaviour-based robot essentially consists of a small number of simple modules, each of which is capable on its own of sensing some limited aspect of the environment, and of controlling part or all of the robot's effector system to achieve some limited task; these modules are embedded in a simple architecture which uses low bandwidth communication between the modules to select which module or modules actually has access to the effectors at any time. The overall simplicity means that such systems have excellent real time performance even with modest resources. The subsumption architecture uses a hard-wired priority scheme for the selection process; the highest priority behaviour active at any time gains control of the output of the whole robot. By a careful choice of modules, and ingenious exploitation of the interactions between behaviours, environment, and tasks, Brooks and others have shown that robots can be constructed which can carry out sophisticated and complex tasks reliably in unstructured dynamic environments (Flynn and Brooks 1989; Connell 1990).

The fit between stigmergy and behaviour-based robotics is excellent. It is the essence of stigmergy that the consequences of behaviour affect subsequent behaviour. Behaviour-based systems deal directly in behaviour. Conventional robots are too slow to cope with an environment containing other moving robots, and too expensive for anyone to be able to experiment with large numbers of them; behaviour-based robots cope well with unstructured dynamic environments and are cheap. We might expect the biological principle of stigmergy to fit better with the biologically inspired architectures of behaviour-based robots than with the alien computational paradigm of conventional robotics. Finally, in 'synthesising phenomena normally associated with natural living systems' and getting them to do something useful in the real world, combining stigmergy with behaviour-based robotics might help to make artificial life look a little less remote than is sometimes the case.

Behaviour-based robotics has given new force to the branch of AI concerned with situated agents and embedded systems. As well as effective slogans ('the world is its own best model' - Flynn and Brooks 1989) and important new ideas ('emergent functionality' - Steels 1991) the field has generated a deep conviction that systems for the real world must be developed in the real world, because the complexity of interactions available for exploitation in the real world cannot be matched by any practical simulation environment. It is for this reason that we have chosen to implement stigmergic mechanisms directly on behaviour-based robots rather than undertaking any preliminary simulation studies; we do however recognise that simulation may be a valid and useful method for investigating stigmergic phenomena in general.

2. Materials and Methods

We decided to develop a system using multiple robots to gather together a dispersed set of objects into a single cluster, much like the corpse-gathering behaviour of ants. As a first step towards achieving this task using stigmergy, a robot was designed which could move small numbers of objects and which was more likely to leave them in locations where other objects had previously been left. This was accomplished by effectively sensing a very local density via a simple threshold mechanism. The plan was to evaluate the performance of the robots with this mechanism and to develop the mechanism and the behaviours as necessary until the task could be performed reliably.

The battery-powered robots (Figure 1) are built on a 21x17cm platform. A 12v motor powered wheel is positioned at the mid-point of each long side, with a castor wheel at the mid-point of one of the shorter sides; this allows the robot to move forwards or backwards in a straight or curved trajectory, and to turn on the spot (see also: Beckers, Deneubourg and Goss 1993). Each robot carries a 17cm wide aluminium forward-facing C-shaped gripper with which it can push objects. The objects used are circular pucks, 4cm in diameter and 2.5cm in height. The robots are run in a square arena 250x250cm; before the

start of each run, 81 pucks are placed on a regular 25cm grid in the arena (Figure 2a).

Figure 1: Robot equipped with a gripper for object gathering. Experiments where carried out with one to five robots of the same type.

The robots are equipped with two IR sensors for obstacle avoidance, and a microswitch which is activated by the gripper when a certain number of pucks are pushed. For the experiments reported here, this number is set to three. The robots have only three behaviours, and only one is active at any time. When no sensor is activated, a robot executes the default behaviour of moving in a straight line until an obstacle is detected or until the microswitch is activated (pucks are not detected as obstacles). On detecting an obstacle, the robot executes the obstacle avoidance behaviour of turning on the spot away from the obstacle and through a random angle; the default behaviour then takes over again, and the robot moves in a straight line in the new direction. If the robot is pushing pucks when it encounters the obstacle, the pucks will be retained by the gripper throughout the turn. When the gripper pushes three or more pucks, the microswitch is activated; this triggers the puck-dropping behaviour, which consists of backing up by reversing both motors for 1 second (releasing the pucks from the gripper), and then executing a turn through a random angle, after which the robot returns to its default behaviour and moves forwards in a straight line. The obstacle avoidance behaviour has priority over the puck-dropping behaviour.

The robots operate completely autonomously and independently; all sensory, motor, and control circuitry is on board, and there is no explicit communication (IR or radio link) with other robots or with the experimenters. The robots only react to the local configuration of the environment.

At the start of each experiment, the robots are placed in the centre of the arena, each pointing in a different direction. Every 10 minutes of runtime, the robots are stopped manually, the sizes and positions of clusters of pucks are recorded, and the robots are restarted. A cluster is defined as a group of pucks separated by no more than

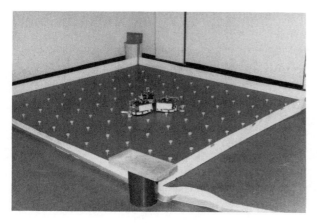

a.

b.

c.

d.

Figure 2: The initial setup (a) and time evolution of a typical experiment involving a group of three robots. Phase I (b), occuring after approximately 10 min, is charaterized by a large number of small clusters constituted by 1 to 10 pucks. In phase II (c), some clusters grow rapidly and the environment becomes more heterogeneous. Finally, phase III (d) is characterized by the competition between a small number (2-3) of large clusters and evolves towards the clustering of all objects in one pile.

one puck diameter. The experiment continues until all 81 pucks are in a single cluster. Experiments reported here have used one to five robots working simultaneously.

3. Results and analysis

From a qualitative point of view, experiments have three more or less distinct phases, regardless of the number of robots. At the start, the arena contains only single pucks (Figure 2a). In the first phase, a robot typically moves forwards scooping pucks into the gripper one at a time; when three have been gathered, the robot drops them,

leaving them as a cluster of three, and moves off in another direction. Within a short time, most pucks are in small clusters which cannot be pushed (Figure 2b). In the second phase, the robot removes one or two pucks from clusters by striking the clusters at an angle with the gripper; the pucks removed in this way are added to other clusters when the robot collides with them. Some clusters grow rapidly in this phase. After a time, there will be a small number of relatively large clusters (Figure 2c). The third and most protracted phase consists of the occasional removal of a puck or two from one of the large clusters, and the addition of these pucks to one of the clusters, often to the one they were taken from in the first place. To our initial surprise, the process eventually results in the formation of a single cluster (Figure 2d).

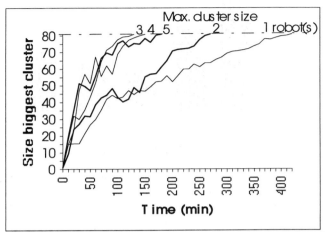

Figure 3: Time evolution of cluster formation until completion of the task (81 pucks in one cluster) for experiments involving one to five robots.

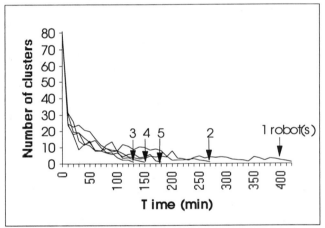

Figure 4: Time evolution of the number of clusters showing an exponential decrease from 81 to 1 cluster for experiments involving one to five robots.

If the experiment is allowed to run on, a puck or two will occasionally be removed from this single cluster, but they are inevitably returned to it as there is no other structure within the arena which can trigger the puck-dropping behaviour. As the number of robots increases, the number of pucks likely to be in transit from the cluster back to the cluster tends to increase, and the stable state is a dynamic equilibrium. Because the robots have no means of detecting that the task has been completed, they carry on working just the same (interestingly enough, so do ants.)

Figure 3 and 4 show the results, in terms of number of clusters and maximum cluster size respectively, from five representative experiments run under identical circumstances and using one, two, three, four and five robots. Phase 1 is clearly seen in all five experiments in the steep fall in the number of clusters by the time the first observations were taken after 10 minutes. Phase 2 is where cluster size and number of clusters are both most variable, because the largest cluster is still relatively small and vulnerable to being broken up, and there are still plenty of clusters of one and two pucks which can rapidly be added to any of the existing clusters. Phase 3 can be seen in the steady and surprisingly regular rise in the size of the largest cluster, which is always the 'same' cluster once its size rises above about 25.

Phases 2 and 3 require some explanation. The puck-dropping behaviour cannot differentiate between a local concentration of three pucks, and one of more than three pucks. What process is organising the net transfer of pucks from smaller to larger clusters when the robots are unable to discriminate between them with their sensors? And what role is played by stigmergy? The answer is surprisingly subtle.

Because the robots turn through random angles whenever they meet a wall, meet another robot, or drop pucks, they may be regarded as following a succession of random straight-line paths through the environment. For a given cluster in a given location, a straight-line path may or may not lead to a collision. The outcome of any collision in terms of whether any pucks are added to or taken away from the cluster depends of the number of pucks carried by the robot at the time of the collision, and on the relationship between the course of the robot and the point of contact with the cluster. It is only possible to remove pucks if the collision is almost tangential to the cluster; a more 'frontal' collision will trigger the puck-dropping behaviour.

The probability that a random path produces a frontal or tangential collision with a cluster is a function of the size, shape and position of the cluster. The stigmergic coupling operates as follows: if a robot adds pucks to a cluster, or removes pucks from it, the consequent change in size and shape alters the probability that a subsequent random path taken by that (or any other) robot will strike the cluster frontally or tangentially, thereby affecting the probability of adding or removing further pucks in the future.

We can now consider the dynamics of this process in a little more detail. Assume for convenience that all clusters are roughly circular, and that the spatial distribution of random paths in the arena is approximately uniform. Consider the five possible combinations of situation (number of pucks being carried by a robot on a random path) and outcomes affecting a given cluster (number of pucks added to or removed from the cluster; there is no need to consider outcomes that leave the cluster unchanged):

Situation A: the robot is not carrying a puck
    Outcome (i): 1 puck removed from the cluster
             (ii): 2 pucks removed from the cluster
Situation B: the robot is carrying 1 puck
    Oucome (i): 1 puck added to the cluster
             (ii): 1 puck removed from the cluster
Situation C: the robot is carrying 2 pucks
    Outcome (i): 2 pucks added to the cluster

In order to remove a single puck, a robot needs to strike a cluster almost tangentially, describing a chord only a small distance inside the circumference; to remove two pucks, it must describe a chord an additional distance inside the circumference. The probability that a random path will produce one of these outcomes will be simply proportional to the relevant distances. Since these distances will both reduce slightly with increasing cluster size, the probabilities of the associated outcomes will also reduce slightly with cluster size. In order to add a single puck, a robot carrying one puck must strike the cluster so that its original course describes a chord further in from the circumference than the distance for removing one puck; the probability of this outcome is proportional to the radius of the cluster minus the distance allowing the removal of one puck. This probability is therefore much greater than the probability of removing one puck, and increases with cluster size slightly faster than the radius increases. (The radius will of course increase as the square root of the number of pucks in the cluster.) A robot carrying two pucks will add them to a cluster wherever it strikes the cluster, and so the probability of this outcome is proportional to the radius of the cluster and increases as the square root of the number of pucks in the cluster.

We can now summarise the expected effects of each situation on a cluster as a function of the size of the cluster. Situation A can only remove pucks from the cluster, and the probability of doing so decreases with increasing cluster size. Situation B will tend to add pucks to the cluster because the probability of B(i) is greater than that of B(ii), and the probability of doing so increases with increasing cluster size. Situation C can only add pucks to the cluster, and the probability of doing so again increases with increasing cluster size. Whatever the situation, it will therefore always be the case that larger clusters will be more likely to gain pucks and less likely to lose pucks than smaller clusters. Stigmergy is therefore active in controlling both the rate of gaining and of losing pucks; either outcome (gaining or losing) alters the size of a cluster and therefore increases the probability of a robot producing the same outcome in that location in the future. Since the total number of pucks in the environment is constant, the inevitable result will be the eventual formation of a single cluster containing all the pucks.

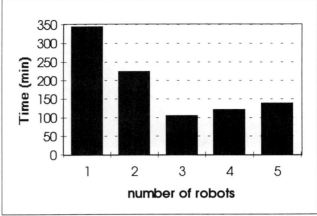

Figure 5: Mean time (averaged on 3 replications) for the clustering of 81 pucks in one pile for one to five robots working simultaneously. For these experiments the optimal group size appears to be three.

The stigmergic principle allows a single agent to interact with the effects of its own previous actions; this is how a single robot achieves the task. From the standpoint of conventional robotics, it is in many ways remarkable that adding one, two, three or four more identical agents still allows the task to be completed, especially since the agents cannot communicate with one another, have no information about position, and there is no explicit specification of where the single large cluster is to be built. It is even more remarkable that the time to completion of the task decreases progressively with the addition of one and two agents (Fig. 5). This may be understood as follows: for most of the time, the robots operate in parallel unaffected by direct interactions with the others, but their behaviour is influenced by the previous behaviour of the others via stigmergy, mediated through the configuration of pucks and clusters. When they do meet, they will lose some 'working' time in avoiding each other, but since they arrive on random courses and leave on random courses, the basis of the stigmergic action will not be disturbed; if they meet when carrying pucks, the interaction may result in pucks being abandoned or transferred; again, they will lose some working time but the stigmergic process will not be affected. Finally, due to the priority of the obstacle avoidance behaviour, two robots meeting near a cluster may destroy it while turning away from each other. If the frequency of interaction of n robots is sufficiently low, the task might be expected to be completed almost n times faster than with a single robot; on the other hand, if it is sufficiently high, clusters might be destroyed so often that the task duration is extended, possibly indefinitely. The results accord with this analysis. Figure 5 shows the mean time to completion for three replications of each condition. We felt that a strict interpretation of 'completion' was appropriate because the curves in Figure 4 all approach the state of completion reasonably smoothly, even though the stable end state is a dynamic equilibrium. The gains from parallel working appear to be maximised by three robots.

In order to evaluate the hypothesis that robot-robot interactions might be responsible for this degradation of performance, further experiments were carried out. Pucks were distributed in five equal clusters in the arena, and the interactions between robots were counted for a twenty minute period for each number of robots. The results are plotted in Figure 6, and show a positively accelerated increase with number of robots. A typical interaction between two robots lasts 4 seconds, so 100 interactions consume over 13 robot-minutes; since the difference in number of interactions between three and four robots is just over 100, the potential gain in total working time supplied by the fourth robot of 20 robot-minutes would be reduced to under 7 robot minutes by the increase in interactions.

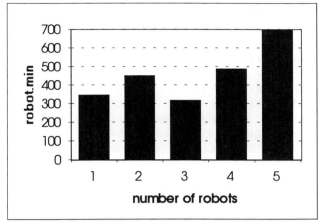

Figure 7: Efficiency (in robot.minutes) for one to five robots working simultaneously. The optimal group size is three, after which the efficiency, in terms of time and energy consumtion, decreases.

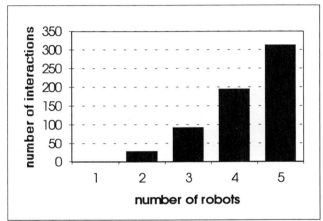

Figure 6: Number of interactions occuring during 20 min in a typical experiment involving 1 to five robots. As interactions are counted only avoiding behaviours between two robots.

Figure 7 shows the average effort required for completion in robot-minutes. The interest here is to look for signs of synergy, which would be indicated by a decrease in the total effort accompanying an increase in the number of robots. This appears to occur with three robots in relation to one and two, but because of the small difference and the small number of samples, further replications will be required to resolve this. It would not be surprising if synergy occurred; we have noticed that the spatial distributions look noisier with more robots, and a certain amount of noise may break up smaller clusters faster than large clusters (which are more robust) which could well speed the task. In the rapid loss of efficiency following the addition of a fourth and fifth robot it becomes obvious that there must be an optimal group size above which the gain of adding a supplementary robot is more than offset by the loss of time due to the increased number of interactions.

Our target was to get the robots to do something useful. Whilst it would certainly be useful to put a team of cleaning robots into a dirty environment on evening and to come back next morning to find all the dirt in a single pile, it would be even more useful to find the dirt in a pile in a designated place. We reasoned that the rapid positive feedback from a 'seed cluster' might induce the formation of the final pile in that location. However, in order to prevent the seed cluster being broken up by accident in the early stages, we used a large saucer instead of a cluster of pucks (anything low enough not to trigger obstacle avoidance would have done). Three robots formed the final cluster round the saucer in 126 minutes - slightly longer than the average finishing time without a seed cluster, but proof that the robots could be induced to form a single cluster in a designated place.

We then wondered what would happen if we provided two seed clusters of different sizes, and so a competition between a large plate and the saucer was organised, again with three robots. The results emphasised the importance of developing real-world systems in the real world. When the experiment was stopped after 300 minutes, the plate was on top of a cluster containing most of the pucks, and the saucer was being moved gradually towards the plate. Neither of these would have happened during a simulation; either might be exploitable by a subsequent development of the system to achieve some relevant task in the real world.

4. Discussion

In this instance, stigmergy has been shown to be able to control and coordinate a number of robots so that a potentially useful task can be performed. It is worth noting that it seems to be a robust technique, able to cope with occasional robot failure (a stopped robot simply becomes a static obstacle; other robots avoid it and any pucks it was carrying are soon scavenged). Another less frequently considered advantage of multiple robotics is that, if speed gains can be made by adding additional robots without reconfiguring any of the robots already working on the task, then the speed of the task can be controlled by changing the number of robots; with a single robot, the only way to speed up is to make the robot work faster.

Some studies of multiple robots attempt to achieve coordination by explicit and direct communication between robots (Arkin, Balch and Nitz 1993). It is possible to view stigmergy as an indirect method of communication - assuming that the object of direct communication is to affect the behaviour of the other robot, we could say that a robot which causes another to produce a certain behaviour through stigmergy has had an implicit communication with that robot through the environment (Mataric 1993). But stigmergy is by no means an inferior form of communication when the object of the communication is to cause a particular behaviour to be produced in a particular location. Consider what a direct communication requires: the sending robot must encode and transmit a message about what is to be done, and where it is to be done (implying a knowledge of location, unless this is coded by the site of origin of the message); this message is local in time and space, and so only those robots close enough and not otherwise engaged will be free to receive the message; they must then decode the message, and either remember it for long enough to get to the place and carry out the action, or remember it for even longer while they carry out some other more important task. A stigmergic communication requires no encoding or decoding, no knowledge of place, no memory, and it is not transient; all it requires is that a robot passes near enough to the location where the communication was placed to be affected by it. As we saw above, random wandering is an effective way of achieving this, though of course it may not be efficient. In fact, the environment can be regarded as a sort of external memory, accessible to all. Pursuing this analogy, the use of volatile pheromones in the environment may represent a type of short-term memory. Perhaps stigmergy is best regarded as the general exploitation of the environment as an external memory resource; it is certainly possible to investigate computational schemes which take this approach (Bull and Holland, 1994).

Acknowledgments

This work has been realised as a contribution to the research group on 'Prerational Intelligence' held in 1993-'94 at the 'Center for Interdisciplinary Research' , University of Bielefeld. We would like to thank the 'Center for Interdisciplinary Research for the financial support of this project. We also are grateful to Richard King and 'Lionheart Electronics' for technical help and discussions.

References

Arkin R.C., Balch T., and Nitz E. 1993. Communication of behavioural state in multi-agent retrieval tasks. *Proc. IEEE International Conference on Robotics and Automation*, Atlanta GA, Vol. 3, 588-594.

Beckers R., Deneubourg J.L., Goss S. 1992. Trails and U-turns in the selection of the shortest path by the ant Lasius niger. *J. Theor. Biol.* 159, 397-415

Beckers R., Deneubourg J.L., Goss S. 1993. Self organised groups of interactive robots. *Proc. ECAL 93*, in press

Beckers R. Deneubourg J.L. and Goss S. 1994. Modulation of trail-laying in the ant Lasius niger (Hymenoptera:Formicidae) and its role in the collective selection of a food source. J. Ins. Behav., ...

Beckers R., Deneubourg J.L., Goss S., Pasteels J.M. 1990. Collective decision making through food recruitment. *Ins. Soc.* 37, 258-267

Brooks R.A. 1992. A robust layered control system for a mobile robot. *IEEE Journal of Robotics and Automation*, RA-2, April, pp. 14-23.

Bull, L. and Holland, O.E. 1994. Internal and external representation: a comparison in evolving the ability to count. *AISB Workshop on Models or behaviours - which way forward for robotics?*, Leeds, April 1994

Connell J.H. 1990. Minimalist mobile robotics: a colony-style architecture for an artificial creature. *Academic Press*, San Diego, CA.

Deneubourg J.L.and Goss S. 1989. Collective patterns and decision making. *Ecology, Ethology, and Evolution*, 1, 259-311

Deneubourg J.L., Goss S., Franks N,R. Sendova-Franks A., Detrain C., Chretien L. 1990. The dynamics of collective sorting: robot-like ants and ant-like robots. In: *Simulation of Adaptive Behaviour: from animals to animats*. Meyer J-A and Wilson S. (eds), MIT Press 356-365

Flynn A.M. and Brooks R.A. 1989. Building robots: expectations and experiences. *IEEE / RSJ, International Workshop on Intelligent Robots and Systems, IROS '89*, Tsukuba, Japan, 236-243.

Grassé, P.P. 1959. La reconstruction du nid et les coordinations inter-individuelles chez *Bellicositermes natalensis* et *Cubitermes* sp. La theorie de la stigmergie: Essai d'interpretation des termites constructeurs. *Ins. Soc.*, 6, 41-83

Mataric M.J. 1993. Designing emergent behaviour: from local interactions to collective intelligence. in: *From animals to animats 2*, eds.: J.-A. Meyer, H.L. Roitblat and S.W. Wilson, MIT-Bradford.

Pasteels J.M., Deneubourg J.L., Goss S. 1987. Self organisation mechanisms in ant societies (i) Trail recruitment to newly discovered food sources. In: Pasteels J.M. and Deneubourg J.L. (eds) *From individual to collective behavior in social insects.* Experientia Supplementum 54 Birhauser Verlag Basel 155-175

Steels L. 1991 Towards a theory of emergent functionality. in: *From animals to animats*, eds.: J.-A. Meyer and S.W. Wilson, MIT-Bradford.

HOW TO EVOLVE AUTONOMOUS ROBOTS: DIFFERENT APPROACHES IN EVOLUTIONARY ROBOTICS

*Stefano Nolfi   ****Dario Floreano**   *****Orazio Miglino**   ******Francesco Mondada**

*Institute of Psychology, National Research Council
15, Viale Marx - 00187 - Rome - Italy
e-mail: stefano@kant.irmkant.rm.cnr.it
**Laboratory of Cognitive Technology,
AREA Science Park - Trieste, Italy
e-mail: dario@psicosun.univ.trieste.it
***Department of Psychology, University of Palermo
Viale delle Scienze, Palermo, Italy
e-mail: orazio@caio.irmkant.rm.cnr.it
****Laboratory of Microcomputing
Swiss Federal Institute of Technology, Lausanne, Switzerland
e-mail: mondada@di.epfl.ch

Abstract

A methodology for evolving the control systems of autonomous robots has not yet been well established. In this paper we will show different examples of applications of evolutionary robotics to real robots by describing three different approaches to develop neural controllers for mobile robots. In all the experiments described real robots are involved and are indeed the ultimate means of evaluating the success and the results of the procedures employed. Each approach will be compared with the others and the relative advantages and drawbacks will be discussed. Last, but not least, we will try to tackle a few important issues related to the design of the hardware and of the evolutionary conditions in which the control system of the autonomous agent should evolve.

1. Introduction

In the last few years new approaches that involve a form of simulated evolution have been proposed in order to build autonomous robots that can perform useful tasks in unstructured environments (Brooks, 1992; Cliff, Husband and Harvey, 1993). The great amount of interest in this new approach is due to dissatisfactions with traditional robotics and Artificial Intelligence and to the belief that interesting robots may be too difficult to design. There are two main reasons why strong difficulties arise in designing a control system for autonomous robots:

(a) it is extremely difficult to co-ordinate the parts of a robot, both at the level of mechanics and of the control system; it is also hard to predict the interaction between these two levels. As Cliff, Harvey, and Husband noted (1993) the complexity of the design scales faster than the number of parts or modules within the system; rather, it scales with the number of possible interactions between parts and modules.

(b) autonomous robots interact with an external environment and, therefore, the way in which they behave in the environment determines the stimuli they will receive in input (Parisi, Cecconi, and Nolfi, 1990). Each motor action has two different effects: (1) it determines how well the system performs with respect to the given task; (2) it determines the next input stimuli which will be perceived by the system (this last point strongly affects the success or the failure of a sequence of actions). Determining the correct motor action that the system should perform in order to experience good input stimuli, is thus extremely difficult because any motor action may have long term consequences. Also, the choice of a given motor action is often the result of the previous sequence of actions. A final source of uncertainty in the design of the system is the fact that often the interaction between the system and the environment is not perfectly known in advance.

Thus, it would appear reasonable to use an automatic procedure, such as a genetic algorithm, that gradually builds up the control system of an autonomous agent by exploiting the variations in the interactions between the environment and the agent itself. It remains to be determined if it is feasible. In particular we should answer the questions: What to evolve? And, how to evolve it?

The choice of what to evolve is controversial. Some authors have proposed to evolve controllers in the form of explicit programs in some high-level language. Brooks (1992) proposes to use an extension of Koza's genetic programming technique (Koza, 1990). Dorigo and Schnepf (1993) propose to use a form of classifier system. Others propose to evolve neural networks controllers (Cliff, Husband and Harvey, 1993; Floreano and Mondada, in press; Miglino, Nafasi, Taylor, 1994; Nolfi, Miglino and

Parisi, 1994). We think that evolving neural networks is the most promising way for a number of reasons:

(a) Neural networks can easily exploit various form of learning during life-time and this learning process may help and speed up the evolutionary process (Ackley and Litmann, 1991; Parisi and Nolfi; in press).

(b) Neural networks are resistant to noise that is massively present in robot/environment interactions. This fact also implies that the fitness landscape of neural networks is not very rugged because sharp changes of the network parameters do not normally imply big changes in the fitness level. On the contrary it has been shown that introducing noise in neural networks can have a beneficial effect on the course of the evolutionary process (Miglino, Pedone, and Parisi; 1993).

(c) We agree with Cliff, Harvey, and Husband (1993) that the primitives components manipulated by the evolutionary process should be at the lowest level possible in order to avoid undesiderable choices made by the human designer. Synaptic weights and nodes are low level primitive components.

The methodology used to evolve control systems for autonomous robots is not well established. The large population size and the number of generations required for the emerging of interesting form of behaviors with the evolutionary techniques implies that a large number of robots must be evaluated. This fact has often restricted most of the experiments to computer simulations and declaration of the intentions to move to physical robots. However, traditional wisdom tells us that computer simulations are of limited usefulness for predicting the behavior of real robots.

In this paper we want to show different examples of applications of evolutionary robotics to real robots by describing three different approaches to develop neural controllers for mobile robots. In all the experiments described below real robots are involved and are indeed the ultimate means of evaluating the success and the results of the procedures employed. Each approach will be compared with the others and the relative advantages and drawbacks will be discussed. Last, but not least, we will try to tackle a few important issues related to the hardware design of autonomous agents and to the new methodological issues in the analysis of the system.

2. The evolution of an ability to navigate by using the physical robot

Floreano and Mondada (in press) developed neural controllers for autonomus agents that should perform a navigation task by using an evolutionary approach. The robot used was Khepera, a miniature mobile robot (Mondada, Franzi and Ienne, 1993). Khepera has a circular shape with a diameter of 55 mm, a height of 30 mm and a weight of 70g; it is supported by two wheels and two small teflon balls. The wheels are controlled by two DC motors with an incremental encoder (10 pulses per mm of advancement of the robot), and can move in both

directions. The robot is provided with eight infra-red proximity sensors. Six sensors are positioned on the front of the robot, the remaining two on the back. A motorola 68331 controller with 256 Kbytes of RAM and 512 Kbytes ROM manages all the input-output routines and can communicate via a serial port with a host computer. Khepera was attached to the host by means of a lightweight aerial cable and specially designed rotating contacts. This configuration allowed a full track and record of all important variables by exploiting the storage capabilities of the host computer; at the same time it provides electrical power without using time-consuming homing algorithms or large heavy-duty batteries.

Figure 1. Khepera, the miniature mobile robot.

The robot was put in an environment consisting of an arena with internal walls of irregular shape. The external size of the arena was approx. 80x50 cm. The walls were made of light-blue polystyrene and the floor of a gray thick paper. Such environment was illuminated from above by a 60 watt light bulb.

The authors' goal was to develop a robot that could avoid obstacles while keeping the straightest possible trajectory at the fastest speed. The evolutionary training was a standard genetic algorithm as described by Goldberg (1989) with fitness scaling and roulette wheel selection, biased mutations (Montana and Davis, 1989), and one-point crossover. The population size was set to 80 and each individual performed 80 actions. The neural network architecture was fixed and consisted of a single layer of synaptic weights from eight input units (clamped to the sensors) to two output units (directly connected to the motors) with mobile thresholds, logistic activation functions, and recurrent connections. Synaptic connections and thresholds were coded as floating point numbers on the chromosomes. Each motor action lasted 300 ms. The fitness criterion (F) was a function of the average rotation speed of the two wheels (V), the algebraic difference between signed speed values of the wheels (DV), and the activation values of the proximity sensor with the highest activity (I):

$$F = V * (1 - sqrt(DV)) * (1 - I)$$

F has three components: the first one is maximized by speed, the second by movement in a straight direction, and the third by the avoidance of obstacles. What is important to notice is that the whole evolutionary process was carried out entirely on the robot. This means that each individual network of each generation was evaluated by letting the robot move in the real environment for 80 time steps (the "brain" of each individual being sequentially injected in the same physical robot).

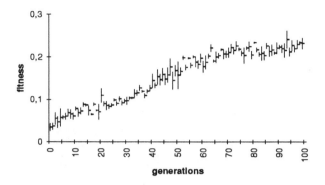

Figure 2. Best individual fitness throughout generations. Values are averaged over three runs (S.E. displayed).

Kepera genetically learned to navigate and avoid obstacles in less than 100 generations (figure 2). However, around the 50th generation the best individuals already exhibited a nearly optimum behavior. Their navigation was extremely smooth; they never bumped into walls and corners, and tried to keep a straight trajectory. This allowed them to perform complete laps of the corridor without turning back. The whole experiment lasted about 60 hours.

3. The evolution of an exploration ability by using a simulated approach.

Miglino, Nafasi, Taylor (1994) evolved recurrent neural networks to control a mobile Lego robot that should explore an open arena by using a rough simulator of the robot/environment interaction. The on-board computer used was a Miniboard 2.0 developed at the MIT Media Laboratory, Cambridge Massachusetts. It was a single board computer, optimized for controlling small DC motors and receiving data from various electronic sensors. The CPU was a Motorola 6811 micro-chip, an 8-bit microprocessor with 256 bytes of internal RAM and 12K bytes of electronically erasable programmable ROM. The robot relied on two wheels for its locomotion and had two optosensors located on the main frame. For each bout of sensory stimulus, the robot performed one of four fixed actions: a) go forward 10 cm; b) go backward 10 cm; c) turn left 45 degrees; or d) turn right 45 degrees. Because of

wear and other unpredictable causes the effective action could slightly vary.

The robot was expected to explore the greatest percentage of an open arena within an allotted number of steps. The arena was 2.6 x 2.6 meters. At its center was a white-colored square, 2.0 x 2.0 meters, marked into a 20x20 grid, 10 cm per side. Surrounding this was a black border area, so that the optosensors on the robot could detect whether it was on the white or on the black surface. The lighting was provided by ordinary fluorescent room lights. The number of steps allowed was 400 and lasted about 6 minutes depending on the condition of the battery, the sequence of steps, etc.

The robot was controlled by a recurrent neural network (Elman, 1990), with 2 sensory units, 2 output units, 2 hidden units, and 1 memory unit. A simulated robot was trained in a simulated environment that was represented in a simplified way with respect to the real environment. It contained 26 X 26 cells, each representing 10 cm square, with a central white grid of 20 X 20 cells. The robot was considered as always located above the center of a single 10 cm. square. The simulated optosensors always sensed one cell ahead and behind with respect to the robot's current location. Actions in the simulated environment were represented as jumps from cell to cell.

A simple genetic algorithm (Holland, 1975; Goldberg, 1989) was used to evolve the weights for the neural network connections. The genotype of each individual in the population was represented by a vector of 17 integer numbers. Each individual was randomly positioned in the simulated environment 10 times, and, at each new starting point, was let free to move for 400 steps. Networks were scored for the number of cells touched by the simulated robot and visited for the first time. Those networks with higher scores were selected for reproduction. A population size of 100 individuals per generation was used. The top 20 individuals were allowed to reproduce by generating 5 offspring each. Mutations were introduced by randomly modifying 10% of the offspring genes. The simulation lasted 600 generations (about 3 hours using a SUN SparkStation).

The simulations showed that an efficient explorative behavior emerged throughout generations. Three different individuals representative of different phases of a particular simulation were transferred into the physical robot and tested in the real environment. Despite the fact that the trajectories of the robots in the real environment significantly differed from the trajectories observed in the simulated environment, the authors showed that the correlation between the fitness values observed in the two conditions were fairly high (0.73).

4. The evolution of a navigation ability using a hybrid (simulated/physical) approach

Nolfi, Miglino and Parisi (1994) developed neural controllers for autonomus agents that should perform a navigation task by using a hybrid approach. The robot used was Khepera (see section 2). A simulator of the interaction between such a robot and an environment similar to that described in section 2. was built. The environment was a rectangular box 60x35 cm with an obstacle of 30x5 cm placed in the center. The walls and the obstacle were made of wood and had natural wood color.

In order to build the simulator the authors sampled the enviroment by letting Khepera turn 360° and by recording the sensory activations at different distances with respect to a wall (an automatic procedure that can be used to also sample other types of objects was developed). The activation level of each of the eight infra-red sensors was recorded for 180 different orientations and for 20 different distances. In addition, the authors sampled how and how much their own Khepera moved and turned for some of the 20x20 possible states of the two motors (the result of the other symmetrical states was computed without actually sampling them). These information was used by the simulator to set the activation level of the neural network inputs and to compute the displacements of the robot in the simulated environment during the first phase of the evolutionary process. The physical shape of Khepera, the environment structure, and the actual position of the robot were accurately reproduced in the simulator with floating point precision.

The neural network architecture was fixed and consisted of a feed-forward neural network with eight input units (coding the 8 infra-red sensors), 2 hidden units, and two output units (coding the state of the motors). Mobile thresholds and logistic activation functions were used. Synaptic connection and thresholds were coded as floating point numbers on the chromosomes. Each motor action lasted 100 ms. A simple genetic algorithm was used to evolve the weights for the neural networks. The genotype of each individual in the population was represented by a vector of 24 real numbers. Each individual was evaluated by randomly positioning it in the simulated environment 2 times and then leaving it free to move for 500 steps each time. The same fitness function described in section 2 was used. A population 100 in size was used and the top 20 individuals were allowed to reproduce by generating 5 offspring each. Mutations were introduced by randomly modifying 20% of the offspring genes. Noise was added to the sensory activation values.

We ran 3 experiments starting with different randomly assigned weights. The first part of the simulation, performed in the simulated environment, lasted 300 generations (about 1 hour using an IBM RISK/6000). Then the evolutionary process continued in the real environment for 30 generations.

Figure 3 shows the performances of the best network throughout generations in the simulated environment. Figure 4 shows the performances of the same networks tested in the real environment (performances of the 30 additional generations evolved in the real environment are also shewn). Performance of the evolved networks

significantly decreased if tested in the real environment. On the other hand, performances similar to that obtained in the simulated condition were obtained by continuing the evolutionary process in the real environment for only few generations. This means that the performance decrement in the transfer to the real robot is not due to a failure of the evolved behavioural strategies, but rather to a mismatch between the simulated and the real sensory-motor apparatus. The fast recovery rate documents that only few adjustments were needed in order to achieve a successful behaviour in the real environment.

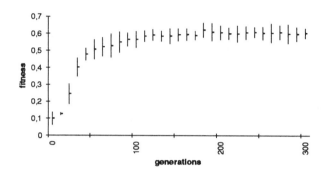

Figure 3. Performances of the best individuals throughout generations tested in the simulated environment. (S.E. displayed).

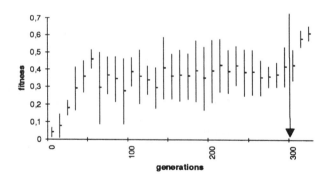

Figure 4. Performances of the best individuals throughout generations tested in the real environment. The first 300 generations have been evolved in simulated environment, the last 30 generations in the real environment (S.E. displayed).

5. Simulation versus physical approaches

The experiments described in sections 2 and 4 show that, at least in the case of simple (but not trivial) tasks, the evolutionary process can be carried out on real robots. Evolving a control system in a real environment is certainly time-consuming, but it appears to be feasible.

Nevertheless, as emerged from sections 3 and 4, computer simulations can also be useful. As several researchers pointed out (Brooks, 1992; Floreano and

Mondada, in press), there are several reasons why those who want to use simulative models to develop control systems for real robots may enconter problems:

(a) Sensor reading should not be confused, as happens in some simulative models, with the description of the world. Real sensors do not separate objects from the background, they do not operate in a stable coordinate system, and they do not give information regarding the absolute position of objects.

(b) Numerical simulations do not usually consider all the physical laws of the interaction of a real agent with its own environment, such as mass, weight, friction, inertia, etc.

(c) Physical sensors deliver uncertain values and commands to actuators have very uncertain effects, whereas often simulative models use grid-worlds and sensors which return perfect information.

(d) Different physical sensors and actuators perform differently because of slight differences in the electronics and mechanics or because of their different positions in the robot. This fact is usually ignored in simulative models.

Some of these problems may be easily eliminated by designing the simulative models carefully. For example, one should avoid using sensors or fitness functions that cannot be implemented on a real robot or that use information which is not available to the robot in the real environment (see also next section). Designing simulators based on samples of the real environment, as shown in section 4, can also avoid very difficult problems like the fact that identical sensors may respond differently. Finally, noise can be added to the simulated sensors (as shown in section 4) and introduced in the simulated actuators in order to take into account the fact that physical sensors and actuators do not perform accurately. The right amount of noise that should be introduced in order to emulate the inaccuracy of the physical sensors and actuators can even be measured by confronting the behavior of the robot in both the real and simulated environments by using different levels of noise (see Miglino, Nafasi, and Taylor, 1994). We believe that, as for the experiments described in sections 3 and 4, simulative models can be useful in developing control systems for real robots when these special solutions are taken into account.

However, one should not expect control systems which have evolved in a simulated environment to behave exactly the same as in the real environment. This is not necessary. We can be satisfied by an above-zero performance after the transfer to the real world. Once we have obtained this, the evolutionary process can be continued in the real environment for a few generations and produce perfectly adapted individuals (see section 4). From this point of view the evolution in the simulated environment can be interpreted as a selection for correlated characters and the change from the simulated environment to the real environment as a change in the environment (Prof. Charles Taylor, personal communication but see also Falconer, 1981). Different kinds of hybrid approaches may be also pursued: for example, the most promising individuals of a population evolving in a simulated environment may be tested in the real environment, or tests in the real environment can be made at given intervals during the evolutionary process.

Another important reason for using simulative models is that they can allow preliminary studies of the evolutionary process. It is well known that genetic algorithms are sensitive to the initial conditions. Different runs of the same simulation may produce solutions with different performances. Evolution in simulated environments, being usually less time-consuming than evolution in real worlds, may allow us to ascertain to what extent a specific evolutionary process is sensitive to the initial conditions and therefore what the probability would be that a limited number of simulations in real environments could produce desired performances. Similarly, simulative models can be used to set a number of important initial parameters such as selecting a good architecture for the neural network or finding the best organization of the environment to be used during the training process.

Even if it is certainly true that as the studied problems will become more and more complex it will be more and more difficult to build useful simulators, we think that at the moment, the use of simulation can still be helpful.

6. The automatic evaluation of the individuals

The development of a control system for an autonomous robot implies the evaluation of a very high number of different individuals. This fact forces to adopt an automatic way of evaluating individuals. This implies that the fitness function should compute only information that is available to the robot through its internal or external sensors. The fitness function used in the experiments described in sections 2 and 4 is a good example of that. It uses information that is available to the robot through its Infrared sensors and through its internal sensors of the state of the motors. On the contrary, the fitness function used in the simulations described in section 3 may make the automatic evaluation of the system in the real environment difficult (unless one has some sort of device to detect the robot's exact position). In simulations all information is available and therefore the fitness function can be freely designed. On the other hand one should consider that even if the entire evolutionary process is to be performed in the simulated environment, the fact that the fitness function cannot be implemented in the physical condition can create serious limitations in evaluating the real robot performance.

Additional sensors may be dedicated exclusively to the fitness evaluation. In some experiments that we are conducting we try to evolve a robot that should stop close to objects of small sizes and ignore objects of larger sizes. In order to automatically evaluate individuals we painted the floor black around small objects and used this information for rewarding individuals that went close to small-size objects. The fitness was computed by reading the value of an infra-red sensor positioned under the robot; this value is not provided to the robot neural network. Therefore once the evolutionary process ended, we could remove the black spots on the floor without affecting the robot behavior (i.e. the black spots and the sensor under the robot are used only in order to evaluate the individuals' fitness).

7. Hardware requirements

The evolutionary approach substantially affects also the physical characteristics of the robots employed. Within the classic approach, where the control system implementation is necessarily preceded by a modelization phase, the hardware designer is faced with many important constraints. During the modelling of the interaction between the robot and the environment, the model of the robot itself plays a crucial role. Thus, the engineer must design the robot in order to make its modeling process feasible and simple enough. This leads to the choice of sensors with very linear response, actuators with a limited number of degrees of freedom (but geometrically optimal), low-noise electronic devices that can take highly precise measures, etc. Unfortunately, these systems may result as being sub-optimal and not very efficient in the real world where intrinsic noise at all levels, non-linearities, and complex shapes are the basic characteristics.

The evolutionary approach does not require the choice of a specific control system. Thus, all it is needed is some general requirement concerning the proper functioning of the sensors and of the actuators provided. Obviously, this requirement is to be taken into consideration by the designer, but it can be greatly simplified by the adoption of a greater number of devices. Such an approach would hence result in robots which are provided with more (and possibly redundant) sensors than traditional robots. However, these sensors would be basically simpler, without corrections for intrinsic non-linearities, special protections from noise, and highly precise measuring devices. It is up to the evolutionary mechanism to exploit these non-linearities or somehow amend them, and properly combine various measures to extract the information necessary for a proper functioning of the system itself. The working tools would drastically change as well. As we have already pointed out, within the classic approach the modelling phase plays an important role, whereas the analysis of the robot behavior is reduced to a confirmation of the model. On the contrary, within the evolutionary approach the modeling process can be reduced or completely avoided in order to leave more space to behavior analysis which is indeed a key point of the procedure. It is thus necessary to devise a set of new tools and methodologies that could allow a better observation, measurement, and analysis of the robot behavior. This may also facilitate the automatic evaluation and final performance monitoring.

Khepera is an attempt in this direction. Its sensory devices are very simple, but still allow a wide range of interesting experiments. In the basic version, Khepera is provided with eight proximity sensors which are based on emission and reception of infrared light. These sensors afford both the measurement of ambient infrared light, and of obstacle proximity by detection of the reflected infrared light emitted by the robot itself. These measures are not very precise, do not have linear characteristics, and strongly depend upon external factors, such as the object materials, color, illumination conditions, etc. Additional sensors can be easily added thanks to a good hardware and software modularity of the system. The miniaturization of the system itself and the environment tools developed for this experimental platform are clearly designed in the direction of the analysis, rather than toward the classic design approach. Khepera has been conceived and designed in order to be easily used on the top of a desk, close to a workstation and connected to it via a serial line that also supplies the electricity to the robot. This configuration affords optimal experiment conditions by allowing easy monitoring of the robot, the real environment, and the computer. Such a configuration is particularly well-suited for experiments in evolutionary robotics where the robot may display "pathological" form of behaviors for long periods of time (initial generations), e.g. crashing into walls or simply pushing against obstacles. With some simple precautions (e.g., walls in polistyrene) and thanks to physical laws according to which homotetically reduced objects offer greater mechanical robustness, these experiments can be carried out without problems.

8. Conclusions

A methodology has not yet been well established in order to evolve control systems for autonomous robots. In this paper we showed three different examples of the application of evolutionary techniques to real robots. The results of our experiments showed that evolving control system in real environment is feasible even if it is certainly time-consuming (see section 2). We showed that also simulations can be useful in evolving control systems for real robots. In some cases even a rough model of the real robot and environment can be enough to evolve control systems that can then be transferred to the physical robot (see section 3). There are ways of designing simulative models that significantly reduce the discrepancies between the simulated and real conditions. In particular we showed how to design the simulative model by sampling the real environment through the real sensors and actuators of the physical robot may result in software models that approximate much better the real environment condition (see section 4). We believe that one should not expect the control system which evolved in the simulated environment to behave exactly the same in the real environment. We rather believe that a hybrid approach in which part of the evolution process is performed in the simulated environment and part in the real one would be more fruitful (see section 4).

Another important issues is whether or not these approaches can be applied to more complicated tasks and which tasks are particolarly suited to them. We are exploring different directions. In a current experiment Floreano and Mondada (1994) provide the environment with a zone where the robot battery gets automatically charged (a simulated charge based on a hardware prototype under test) and an oriented light source. The robot is also provided with a few more sensors (light sensors), but the fitness function is exactly the same as the one employed for obstacle avoidance. The difference with respect to the experiment described in section 2 is that the robot can recharge its battery and thus prolong its own life if it passes over the charging zone. By employing the same evolutionary procedure, the robot learns to keep itself "alive" by periodically returning to the charging zone. This emergent homing behavior is based on the autonomous development of an internal topographical map that allows the robot to choose the appropriate trajectory as function of its location and of its remaining energy. In another current experiment the authors try to address the issue of object recognition for grasping using the miniature robot Khepera with an added gripper module. The environment is a surface with a number of objects and obstacles. The objects, as in natural situations, feature different shapes and sizes. The robot is expected to autonomously learn to approach only those objects that can be physically grasped by its own gripper module.

Acknowledgments

We would like to thank Domenico Parisi and Charles Taylor who provided important comments. Dario Floreano has been partially supported by C.N.R grant n. 93.01065.PF67 (co-ordinator: Walter Gerbino), Francesco Mondada by the Swiss National Research Foundation (project n. PNR23), and Stefano Nolfi and Orazio Miglino by P.F. "ROBOTICA", C.N.R., Italy.

References

Ackley, D. H., M. L. Littman, 1991. Interactions between learning and evolution. In *Artificial Life II*, edited by C. G. Langton, J. D. Farmer, S. Rasmussen, C. E. Taylor. Addison-Wesley. Reading, Mass.

Brooks, R. A. 1991. New approaches to robotics. *Science* **253**:1227-1232.

Brooks, R. A. 1992. Artificial life and real robots. *In Toward a Practice of Autonomous Systems: Proceedings of the First European Conference on Artificial Life* edited by F. J. Varela, P. Bourgine. Cambridge, MA: MIT Press/Bradford Books.

Cliff D. T., I. Harvey, P. Husbands. 1993. Explorations in Evolutionary Robotics. *Adaptive Behavior* **2**: 73-110.

Collins R., D. Jefferson. 1991. Representations for artificial organisms. In *Proceedings of the Simulation of Adaptive Behavior*, edited by J. A. Meyer, S. Wilson. Cambridge, MA: MIT Press/Bradford Books.

Dorigo M., U. Schnepf. 1993. Genetis-based machine learning and behavior based robotics: a new syntesys. *IEEE Transaction on Systems, Man and Cybernetics*, **23**:141,153.

Elman J. L. 1990. Finding structure in time. *Cognitive Science, 2: 179-211.*

Falconer D. S. 1981. *Introduction to Quantitative Genetics*, Longman, London.

Floreano D., F. Mondada. 1994. Emergent homing behaviour in a mobile robot. *Technical Report* LAMI n. DF94.14I, Swiss Federal Institute of Technology, Lausanne.

Floreano D., F. Mondada. In press. Automatic Creation of an Autonomous Agent: Genetic Evolution of a Neural-Network Driven Robot. In *From Animals to Animats 3: Proceedings of Third Conference on Simulation of Adaptive Behavior,* edited by D. Cliff, P. Husbands, J. Meyer, S. W. Wilson. MIT Press, Bradford Books.

Goldberg D. E. 1989. *Genetic Algorithms in Search, Optimization and Machine Learning*. Reading, Mass.: Addison Wesley.

Holland J. H. 1975. *Adaptation in Natural and Artificial Systems*. University of Michigan Press, Ann Arbor.

Koza J. R. 1990. Genetic Programming. A Paradigm for Genetically Breeding Populations of Computer Programs to Solve Problems. *Technical Report* STAN-CS-90-1314. Stanford University Computer Science Dept.

Miglino O., R. Pedone, D. Parisi. 1993. A noise Gene for Econets. In *Proceedings of Genetic Algorithms and Neural Networks*, edited by M. Dorigo, Reading, Mass.: Addison Wesley.

Miglino O., K. Nafasi, C. Taylor. 1994. Selection for Wandering Behavior in a Small Robot. *Technical Report*. UCLA-CRSP-94-01. Department of Cognitive Science, University of California, Los Angeles.

Mondada F., E. Franzi, P. Ienne. 1993. Mobile Robot miniaturisation: A tool for investigation in control algorithms. In: *Proceedings of the third International Symposium on Experimental Robotics*, Kyoto, Japan.

Montana D., L. Davis. 1989. Training feed forward neural networks using genetic algorithms. In *Proceedings of the*

Eleventh International Joint Conference an Artificial Intelligence edited by N.S. Sridharan. San Mateo: Morgan Kaufmann.

Nolfi, S., O. Miglino, D. Parisi. 1994. Phenotypic plasticity in evolving neural networks: Evolving the control system for an autonomous agent. *Technical Report* n. PCIA-94-04, Institute of Psychology, C.N.R., Rome.

Parisi, D., F. Cecconi, S. Nolfi. 1990. Econets: Neural networks that learn in an environment. *Network* 1:149-168.

Parisi, D., S. Nolfi. In press. How learning can influence evolution within a non-Lamarckian framework. In *Plastic Individuals in Evolving Populations*, edited by R. K. Belew, M. Mitchell. SFI Series, Addison-Wesley

Evolving Visual Routines

Michael Patrick Johnson, Pattie Maes and **Trevor Darrell**
MIT Media Laboratory
Cambridge, Mass.
aries/pattie/trevor@media.mit.edu

Abstract

Traditional machine vision assumes that the vision system recovers a a complete, labeled description of the world [Marr, 1982]. Recently, several researchers have criticized this model and proposed an alternative model which considers perception as a distributed collection of task-specific, task-driven visual routines [Aloimonos, 1993, Ullman, 1987]. Some of these researchers have argued that in natural living systems these visual routines are the product of natural selection [Ramachandran, 1985]. So far, researchers have hand-coded task-specific visual routines for actual implementations (e.g. [Chapman, 1993]). In this paper we propose an alternative approach in which visual routines for simple tasks are evolved using an artificial evolution approach. We present results from a series of runs on actual camera images, in which simple routines were evolved using Genetic Programming techniques [Koza, 1992]. The results obtained are promising: the evolved routines are able to correctly classify up to 93% of the images, which is better than the best algorithm we were able to write by hand.

1 Introduction

One of the hardest problems to be solved when building a real robotic autonomous agent is the perception problem. Traditional machine vision assumes that the vision system produces a labeled, perfectly resolved model of the world, distinguishing and representing all objects [Marr, 1982]. This turns out to be an extremely hard, if not impossible, problem to solve.

Active vision is a new paradigm for doing machine vision that has received a lot of attention recently [Ballard, 1989, Aloimonos, 1993]. The insight that active vision brings to the problem is that almost all of the time an intelligent agent doesn't even *need* all that information, since it is involved in a particular task that only requires specific knowledge of certain objects. Several researchers have proposed ways of allowing the central control system to actively control what is processed by the vision system in a *task-dependent* manner (e.g. [Ballard and Whitehead, 1990]). For example, if I want to pick up the coffee cup on my desk, I don't need to process and represent all the paper clips, papers, every character on every paper, and all the rest of the clutter usually on a desk. I just want information on my hand, the cup and anything that might be in the way of my hand and the cup. Furthermore, that information will be different information than I might extract if I were intent on drawing the cup. Thus, active vision requires more work on the part of the central system.

To explain how the internal active processing could be implemented, Shimon Ullman proposes the *visual routines* model [Ullman, 1987], which describes a set of primitive routines that can be applied to an input image in order to find certain spatial relations between objects as well as other useful information. He suggests that given a specific task by the central system, the visual routines processor compiles an appropriate visual routine and applies it to a base representation, perhaps changing it in some way or returning a result. It may be asked to process something else in response to that answer, and so on. Ullman does not go into detail on how routines are developed in the first place, how they are stored or how they are chosen and applied. An obvious answer might be that the central system does some intelligent reasoning in order to determine their application.

Ramachandran, a human vision researcher, disagrees with this solution and suggests the "utilitarian theory" of perception [Ramachandran, 1985]. He argues that since many other systems in the human body are collections of *ad hoc* pieces that all function in their own way but tend to work together, perception is likely to be the same. He argues that this is the manner in which natural selection works — anything that works will be used whenever it works. Hence, he concludes, perception is like a "bag of tricks" selected by evolution to solve various perceptual subproblems. Whichever trick has proven to work for a particular problem will be the one used for that problem in the future.

In creating a computer vision system, we have at our disposal a powerful alternative — the programmer. This person can design by hand visual routines that solve a given task, test them and refine them, hopefully reaching an optimal solution. This may be a reasonable approach in limited cases, such as the Sonja system by David Chapman [Chapman, 1993], which has a set of routines designed by Chapman for solving different visual tasks involved in playing a particular video game. The set of routines is fairly small and fixed, allowing a programmer to work them out by hand.

Another similar example (and the one which we will focus on) is the simple vision system used by the ALIVE (Artificial Life Interactive Video Environment) virtual environment project [Maes *et al.*, 1994, Maes *et al.*, 1993]. In this system, a user can interact in real-time with a computer graphics creature using gestures which are interpreted by a vision system. This system employed a set of hand-coded low-level heuristics for solving specific simple visual tasks involved in efficiently processing camera input of a person interacting with the system. These tasks included perception problems such as "Find the hands" and "Is the person sitting?" and "Is the person pointing?".

The problem with this method is that it ties up a lot of programmer-hours fiddling with knobs and parameters and conditions, trying to get something that works well all the time. If a new problem is approached, a new set of routines has to be created. Also, it is difficult for the programmer to anticipate all cases in which the program should work.

This paper describes a way of automatically producing a visual routine appropriate for a given task by using simulated evolution of computer programs in the style of John Koza [Koza, 1992]. The paper is organized as follows. Section 2 discusses previous work on visual routines. Section 3 describes the particular perception task that we concentrated on, namely finding the left and right hand of a person in a black and white silhouette of the person, which is extracted from a real camera image. Section 4 discusses the actual Genetic Programming (GP) implementation used for the experiments. Section 5 discusses results of the GP runs. Section 6 lists possible future directions. Finally, section 7 gives a summary.

2 Visual Routines

Ullman [Ullman, 1987] breaks the visual system into three main areas: the base representation, the visual routines processor and the higher level components. The base representation is the result of initial, parallel processing of the retinal image. It is bottom-up and uniform in the sense that the same processing occurs across the whole image (which he calls spatial parallelism). This could involve resolution edge detection, color processing, etc. The visual routines processor performs the tasks that require a more directed, specific or inherently sequential set of processes. The higher level components consist of recognition memory and task formulation.

Ullman makes the distinction between universal routines and specific routines. Universal routines are routines that are processed automatically in order to form a basis for deciding which specific routines must be used. In the absence of a specific task, universal routines operate. They might, he suggests, isolate prominent areas of the image, do simple spatial relations tasks, and perform crude color and shape characterizations. An example might be the process that leads to a saccade toward a pop-out area of the visual field; that is, an area that is significantly different from its surroundings due to motion, color, orientation or depth.

Finally, Ullman proposes a set of specific routines which he admits is not comprehensive, but seem to be useful for certain problems.

- *Indexing* involves shifting the focus of attention to important areas of the visual field. For example, shifting focus to "pop-out" areas or in order to peform a visual search for specific objects.

- *Marking* allows the system to remember a location or describe areas of processing for other operations.

- *Ray intersection* involves tracing along a ray between two points (one may be "at infinity," or out of the critical processing area) and count or place markers at intersections with boundaries. This is useful for inside-outside decision tasks (for example, deciding if an indicated point lies inside or outside a closed contour).

- *Bounded Activation* or *coloring* fills in an area surrounding a marked point, stopping at a boundary. What constitutes a boundary is a function of the task, so must be a parameter to the operation. For example, consider again the inside-outside task. People can still tell if a point is inside a circle with a dashed line for a border (see [Ullman, 1987]). Thus, the boundary need not be solid.

- *Boundary tracing* is another sequential operation that traces a boundary (with the same parameter as above to define a boundary), searching for various facts, such as whether the curve is closed or not, or whether there are objects along it.

Other routines can be applied to the results of these computations. For example, a point could be marked and the area around it colored. Checking whether the color spread to some point known to be outside the curve (using a bounding box of the curve, say), will tell us whether the point was inside a closed contour or not.

David Chapman applied similar operations to the domain of visual attention in his Sonja video game playing system [Chapman, 1993]. Sonja can find gaps in a boundary (ostensibly doors), track moving pop-out objects (such as the player's icon) with a marker, color a region and cast rays, for instance[1]. In the game, characters move around a two-dimensional world full of impenetrable walls. To move to a certain marked point (say there is something desirable there) a ray is cast from the current location to the desired location to detect any intersections with walls. If an intersection point is found, a path around the obstacle is constructed. Chapman hand-coded his visual routines to solve these problems using a circuit specification language. A large amount of programming effort was required to get these to work

---

[1] However, Chapman does not provide a model of "low-level" vision for Sonja; his primitives are actually implemented by directly referencing the models which underly the screen representation. This is appropriate for a closed video game system, but ignores issues such as how the primitives will perform in noisy and/or unexpected situations that occur in real-world environments.

correctly. An automatic technique for finding visual routines is thus quite desireable.

This is exactly the problem which we focus on in this paper. Inspired by the way in which nature may have evolved a range of special-purpose visual routines [Ramachandran, 1985], we try to automatically develop visual routines for particular tasks that are relevant to a system using artificial evolution. This work is related to the research described in [Cliff D. T. and I., 1993], which also evolves visually-guided behaviors. However, in contrast to that research, the work presented here uses actual camera images, rather than simulated ones.

3   The Problem — Find the Hands

The simple image processing application we focus on for the preliminary study is that of finding the hands in the bitmap silhouette of a person. A solution to this problem was actually necessary for the ALIVE virtual environment project [Maes *et al.*, 1994] (see Section 3.1). For ALIVE, a visual routine was written by hand for this task, applied when hand location was desired. We restrict the problem slightly for this study, to see if genetic programming can solve subgoals of the overall task. Specifically, we ask that the genetic programming system evolve a program for finding the "left" or "right" hand in the silhouette when the hand is below or at the shoulder line and the person is facing mostly toward the camera. The "left" hand is defined to be the one on the left side of the bitmap, since we don't have enough information to be sure the person isn't facing the wrong way. We define the "right" hand in like manner. Furthermore, our evolved solution need not find both hands at once — we evolved programs for finding the left hand and right hand separately. It is probably equivalent to learning to find just one side and use a mirrored image to find the other, but by doing them separate we can try to get around problems such as noise caused by inconsistent lighting. The evolution for one side might find a different solution better for the hardware than using the mirrored version of the other.

3.1   Fitness Case Generation

The ALIVE system allows a user to interact via gestures with a computer graphics behavior-based agent. The user's video image (taken from a video camera) is *superimposed* on the real-time graphical world. The composite image is then displayed on a life-size projection screen in front of the user. The effect is as if the user were looking into a magical mirror — they see a mirror image of themselves (their live video image) superimposed on the graphics world. They interact with the creatures in the virtual world by gestures. For example, they can point and the creature will walk in the indicated direction. They can wave and it will return.

ALIVE uses single color background subtraction, a method used ubiquitously for special effects. The user stands in front of a monochromatic blue wall with a camera pointed at them. The algorithm then replaces anything of the shade of blue in the background with computer graphics output and leaves the rest as live video. The effect is that the user is superimposed on the graphics. Technically, it returns a binary image with bits on

where the video should go and bits off where the graphics should go. This output is used directly as input for the vision system and is the source of our fitness case silhouettes (see Figure 1). We used a total of 46 fitness cases in the experiments described below (and used 46 mirrored silhouettes as evaluation test cases).

The fitness cases were produced from the larger, noisier original images produced by the vision system by finding the largest connected component and returning its bounding box and centroid. For our purpose, we presume some other universal routine has already performed this task for us. Notice that some of the cases are extremely difficult as part of the hand is in front of the body or not easily locatable. A person labelled the "correct" hand locations. For the difficult cases, a best guess was given as a location near the waist. It is probably not proper to ask the program to find the hands in these cases; indeed, the original ALIVE system will only look for hands when it is likely that they are visible in the silhouette.

4   Genetic Programming Implementation

Based on Koza's original Genetic Programming (GP) algorithm, we implemented a typed GP system for solving the restricted hands problem. We refrain from a full discussion of GP here and give only salient points. Koza discusses this system and variations of it at great length in his book [Koza, 1992]. The system used here is similar to Koza's except for the addition of the type specification described below and several other small improvements.

The genetic operations used in this work are copying, crossover and mutation. Copying involves copying a certain percentage of individuals to the new generation roughly in proportion to their fitness (Koza calls this "fitness-proportionate reproduction"). Crossover chooses two individuals at random, again roughly proportional to fitness, and swaps randomly chosen subtrees. Mutation picks a random subtree and replaces it with a newly generated subtree. Better results were obtained using the tournament selection method described in [Koza, 1992], which picks k individuals at random and chooses the one with the best fitness (effectively a tournament between the individuals, hence tournament selection). This can be used for copying and for selecting the two individuals for crossover. Tournament selection seems to keep the system from converging prematurely due to one individual swamping the population.

4.1   The Typed Genetic Program

Our implementation is typed; the programmer must specify a type for each terminal and a *signature* for each function. A signature is comprised of a return type and a list of types for the arguments. Only forms consistent with this type specification will be allowed. This makes the space of unmeaningful programs smaller, as well as allowing the primitives to be more specialized by not having to deal with arbitrary typed input. This in turn speeds up fitness evaluation. Koza does have several examples of a typed system in his book, but they seem less general than the one used here.

Figure 1: Fitness Cases. X shows the desired location. + shows the best program's result.

The programmer specifies the top level return type, eliminating the problems of forms returning incorrect data types. For example, in our experiments the top level type is a point. The GP system chooses randomly among the primitives of the appropriate type to fill in the argument slots recursively, starting at the root node. The maximum depth of the original programs is a parameter.

Types make crossover a little harder. One of the two mating programs is picked at random. Then we pick a random subtree in it and find the return type. We then choose randomly among the subtrees with appropriate root type in the other mate. If there are none, no crossover occurs and the individuals are copied. Also, if crossover will produce a program deeper than the maximum allowable depth of a program (another parameter) the crossover does not occur and the parents are copied.

Mutation chooses a node at random and replaces it with a randomly generated subtree of the same type as the original node. The depth of the created subtree is again limited by a parameter.

One drawback of the typed GP used is that overloading is not allowed — programs can have one and only one signature (return type and arg list) to make the type of a given primitive static. If it could have multiple types, when it is picked to do crossover, the system would have to find out what type it is returning in the given program. To do this it has to traverse the entire subtree. Thus, if a function can take two types of arguments, the programmer must make a unique function for each case. This was not a problem in our experiment.

A big advantage of the typed GP is that we can constrain the system in many useful ways. By having a function that returns a special type that exists no where else in the terminal and function set and specifying it as the root type, we can force the system to put that function at the root. This can be used to optimize the arguments of a particular top-level function. Another advantage is that we can create special terminal types that only certain functions use. If we want a function to take integers in a certain range, we make a set of terminal variables of a special type, say "smallint" and set them to constant integers. Thus, if larger integers are used elsewhere, the protected function won't get them. This aspect of the typed system was invaluable in making the *percent* terminal, described below.

4.2 The Primitives

Two alternatives were available. One was to evolve a sequential set of actions that used side effects on an internal representation and left a marker on the answer. The other was to try and evolve a form that would evaluate and return the answer purely functionally. We decided it would be interesting to try the latter first, since this was essentially how the ALIVE solution worked and it would allow us to sanity check the GP system with a known hand-coded solution. Future research will experiment with the side-effecting primitives, which is closer in spirit to Ullman's work.

In a GP system with a limited number of fitness cases and sufficiently powerful primitives, it is likely that solutions will be too specific. For example, if there is some

way to use the primitives to recognize each fitness case and return the fixed answer for that case, GP will tend to exploit it. Such solutions are often fragile and will not generalize to new cases. Since we had a small number of fitness cases, we limited the power of our primitives. The primitives were carefully chosen to be "weak" in the sense that they had little observational power. For instance, there are no conditionals and no observers for the values of the terminals that could be used in conditionals. We discuss generalization results in Section 5.

Terminals In our implementation, the terminals actually hide a fair amount of preprocessing. We assume that the set of universal routines as proposed by Ullman has already operated and that we have found the moving object, subtracted the background, and found the bounding box and centroid of the silhouette. Our terminal set is:

- Centroid point of the silhouette (*centroid*)

- Bounding Box Top Left point (*bb-tl*)

- Bounding Box Bottom Right point (*bb-br*)

- Percentages (0%, 10%, 20%, ..., 100%)

We defined a *point* type as an ordered pair of integers. The first three terminals are of this type. Percentages are floating point values, 0.0, 0.1, 0.2, ..., 1.0.

Functions We chose a set of primitive functions that was similar to that used in the original ALIVE code. These were the following:

- Point Operators (*pt+, pt-, point-between*)

- Feature Detectors (*find-bottom-edge, find-top-edge*)

- Point List Filters (*leftmost-point, rightmost-point, average-point*)

The first two operators add and subtract points. The third, *point-between*, takes two points (which define a rectangle) and two percents and returns the point in the rectangle specified in normalized coordinates by the percents.

The edge detectors look for a fixed size edge in a rectangular window specified by two points. One detector finds a transition from dark to light as the y-value increases, the other a transition from light to dark. These are implemented as matched filters. A fixed-size edge template (five pixels wide by four high, with half the pixels on) is convolved around the image to produce the number of matching points at each location. The set of points with the maximum correlation is then calculated and returned. It is hoped that the system will discover a small enough window of the image to apply the filters on so that the result set is small and the points near the desired solution.

The point list filters do the obvious things — they return the point with the smallest x-coordinate, point with the largest x-coordinate, and average all the points in the list, respectively. These are necessary for dealing with arms held out to the sides, where each edge detector would return a long list of matches along the edge.

Since the routine supplies the window arguments to the edge detector, they can be invalid for several reasons. For example, they can be outside the area of the image. For these degenerate calls, we just return the top left window point arbitrarily. It might be more intelligent to only look at the part of the window within the image, but we prefer that a proper program exhibits a finer control on the windows chosen.

4.3 The Fitness Measure and Fitness Cases

Koza describes the concept of a "hit." When a program gets the answer on a fitness case (henceforth *cases*), it is considered to get a hit. The GP system stops running when it gets hits on all the cases or reaches a maximum generation. We define a hit to be getting within three pixels of the hand position. The hands are about five pixels wide, so this error margin is still usually within the area of the hand.

The fitness of a routine ρ where $\rho(i)$ is the point returned by the routine on fitness case i, is the number of missed cases plus the logarithm of the error (Euclidean distance) between $\rho(i)$ and the actual hand location summed over fitness cases, plus some other constraints:

$$Fitness(\rho) = \\ (N - Hits) + \gamma \, Error(\rho) + \alpha \, Size(\rho)$$

where

$$Error(\rho) = \\ \left(\sum_{i \in Cases} \log(\|\rho(i) - actual_i\| + 1.0) \right)$$

The length term was a parsimony constraint added to satisfy the Occam's Razor impulse for simple-looking solutions. Also, simple solutions are often more general, which we desire. The *size* of a form is the total number of nodes (internal and leaves) in it. We chose a coefficient that would only make length matter when the answer was "close" to a fitness of zero (specifically, $\alpha = 0.1$). Thus, of two programs with the same error, the shorter would be more fit.

We chose a γ of 1.0 in the trials reported below, making misses greater than around ten pixels be considered exponentially less important. Since we are trying to optimize hits, we also add the number of misses to the function.

Early runs used just a sum of absolute errors instead of the logarithmic sum. This tended to reduce overall fitness, but didn't optimize the number of hits — that is, it found only average results on all cases rather than trying to optimize accuracy. It seemed that the system was missing potentially very good solutions because it happened to be *way* off on just a few bad cases. In order to maximize hits, we first tried replacing the sum of absolute errors with the number of misses. This fared poorly, probably due to lack of gradient information on whether one solution was better than another, even if both had the same number of hits. A noticable improvement was discovered by adding the logarithmic sum, which effectively considered points way off or very far off to be about the same fitness. This has the effect of giving gradient information for near misses, but just some bad score if it wasn't close at all. Several suggestions for ways to avoid this "sensitivity to fitness function," which seems to be a problem in GP, are discussed in Section 6.

5 Results

First, we note that finding the "perfect" answer to every fitness case isn't necessary in the ALIVE environment, since it involves a time series of silhouettes. A post-processing part of the system can remove outliers, use Kalman filtering or other methods to clean up errors. Also, unless the hands are doing something interesting, like moving around, we don't care where they are, like just hanging to the side of the person. So we don't mind if it misses hard cases, but care a lot if it misses easy ones.

We did multiple runs with a population size of 500. Copy percent was zero, crossover was 85%, and mutation was 15%. The effective copy rate was still about 10% due to failed crossovers[2], however. The initial program maximum depth was five, the maximum program depth after crossover or mutation was eleven, and the maximum mutation subtree depth was four. A worst-case combinatorial analysis on the size of this search space with these depth restrictions is simple. Given a maximum depth of d and noticing that the branching factor is less than two (those functions with more arguments only have two that can recurse), we compute that there are a maximum of 2^{d-1} internal nodes (where functions can be added) and 2^{d-1} terminal nodes. Note that there can be more than 2^{d-1} terminal nodes due to the percent terminals for the point-between functions. If every internal node was a point-between, we'd have another 2^{d-1} terminals, for a total of 2^d. If there are α terminals and β functions to choose for each point, the total number of maximal trees of depth d is $\alpha^{2^d} \beta^{2^{d-1}}$. Since all size trees of depth $1, \ldots, d$ are allowable, we have a search space of size:

$$\sum_{i=1}^{d} \alpha^{2^i} \beta^{2^{d-1}}$$

For our parameters, our search space is approximately 10^{2166}. This is astronomically huge. Of course, not all of the programs in this space will be distinct, but this gives us an idea of the area we are searching for good solutions. Also, the parsimony constraint will tend to make us look at smaller programs, unless the really good ones are larger, in which case it will have to search larger programs.

Results for the left hand runs are summarized in Figures 2 and 3. These figures represent the best of generation individual's hits and fitness, averaged over twelve

---

[2]When crossover cannot be completed due to size restraints, it fails and the parents are copied.

separate *tabula rasa* runs. The error bars show a standard deviation from the mean. The best left-hand program found across all runs was:

```
(POINT-BETWEEN
  (LEFTMOST-POINT
    (FIND-BOTTOM-EDGE
      TL
      (LEFTMOST-POINT
        (FIND-BOTTOM-EDGE
          (POINT-BETWEEN TL BR 20% 70%)
          (LEFTMOST-POINT
            (FIND-BOTTOM-EDGE
              (POINT-BETWEEN TL CENTROID 0% 90%)
              (POINT-BETWEEN TL BR 0% 80%)))))))
    TL 30% 10%)
```

which received 43 hits out of the possible 46, for an accuracy of 93%. Its predictions are shown in Figure 1.

The first hand-coded solution we created received only 4 hits and a standardized fitness of 78.6. It was:

```
(LEFTMOST-POINT
  (FIND-BOTTOM-EDGE
    TL
    (POINT-BETWEEN TL BR 20% 90%)))
```

This just looks for a bottom edge in a window on the left side of the image. It returns the leftmost maximal correlation point. Since the correct locations are defined as the center of the hand, this doesn't often get close enough for a hit. The next hand-coded program was better:

```
(POINT-BETWEEN
  (LEFTMOST-POINT
    (FIND-TOP-EDGE TL (POINT-BETWEEN TL BR 20% 90%)))
  (LEFTMOST-POINT
    (FIND-BOTTOM-EDGE TL (POINT-BETWEEN TL BR 20% 90%)))
  50%
  50%)
```

This finds a top and bottom edge and averages them, hoping to get the center of the hand. This gets 12 hits and a fitness of 79.12. The best hand-coded left-hand program, which took about an hour of tweaking, was:

```
(PT- (LEFTMOST-POINT
      (FIND-BOTTOM-EDGE TL (POINT-BETWEEN TL BR 20% 90%)))
  (PT- (POINT-BETWEEN TL CENTROID 0% 10%) TL))
```

This gets 40 hits (87% accuracy) and a fitness of 26.64. It finds the bottom edge of the hand and moves the answer up a little. It is encouraging that we can find a solution as good or better than this one automatically in about the same amount of time.

Clearly most runs had converged by 100 generations. Since this is equivalent to evaluating 50,000 individuals, (500 per generation times 100 generations) we decided to compare results against 50,000 randomly created programs (with no evolution) to see if GP was more efficient. These results are summarized in Figures 4 and 5. Note that the y-axis is logarithmic scale. Most random programs get zero hits. The best are about 15 hits. This pales in comparison to the consistent good results using GP, with similar computation expenditure.

Results for the right-hand runs are summarized in Figures 6–9, similarly to the left hand runs. These results are less good than those obtained for the left hand problem. Most notably, the GP algorithm seems to converge more slowly and with a wider variance than the left-hand cases. This is probably due to the fact that the right hand cases are much harder or more unfair. For instance, in many of these cases the hands are somewhere in front of the body where our weak primitives will not work. The position given as the "answer" by a

human was a guess as to where the hand probably was, using knowledge of human anatomy, etc. Probably it is not right to ask that it find the hand in these cases since its primitives cannot locate it here. Even so, it manages to find the hands in almost all the cases where we would care where the hand is. One subset of cases it misses are when the hands are raised above the shoulder since this subset of problems requires a very different solution than the majority of the other cases. These results seem to suggest that it is harder to converge on a solution when there are fitness cases that are unsolvable.

The best right-hand program evolved across all runs was:

```
(RIGHTMOST-POINT
 (FIND-TOP-EDGE
  (POINT-BETWEEN
   (RIGHTMOST-POINT
    (FIND-TOP-EDGE
     (RIGHTMOST-POINT
      (FIND-BOTTOM-EDGE
       (POINT-BETWEEN BR CENTROID 0% 60%)
       (POINT-BETWEEN
        (POINT-BETWEEN BR CENTROID 0% 60%)
        CENTROID
        80% 90%)))
     (POINT-BETWEEN
      (LEFTMOST-POINT
       (FIND-BOTTOM-EDGE
        (POINT-BETWEEN TL CENTROID 80% 60%)
        (POINT-BETWEEN BR CENTROID 0% 60%)))
      (POINT-BETWEEN TL TL 80% 10%)
      0% 40%)))
   (RIGHTMOST-POINT
    (FIND-BOTTOM-EDGE
     (POINT-BETWEEN BR CENTROID 0% 60%)
     (POINT-BETWEEN TL CENTROID 80% 90%)))
   100% 10%)
  (POINT-BETWEEN
   CENTROID
   (RIGHTMOST-POINT
    (FIND-BOTTOM-EDGE
     (POINT-BETWEEN TL BR 80% 40%)
     (POINT-BETWEEN
      CENTROID
      (POINT-BETWEEN BR BR 60% 30%)
      100% 80%)))
   100% 70%)))
```

which received 32 hits, for an accuracy of 70%. Its predictions are shown in Figure 1. A surprise was that it managed to get case 34 correct, where both hands are on the same side of the body. It also failed to get seemingly easy ones, such as 6 and 25.

The best hand-written right-hand program was:

```
(PT-
 (RIGHTMOST-POINT
  (FIND-BOTTOM-EDGE (POINT-BETWEEN TL BR 80% 0%) BR))
 (PT- (POINT-BETWEEN TL CENTROID 0% 10%) TL))
```

which received 26 hits (57%) and a fitness of 54.67. This does the equivalent of the left-hand program, looking in the right 20% of the bounding box for the rightmost bottom edge and moving it up a little.

We also tested whether these solutions would work on other fitness cases than those in the training set. We created 46 "new" fitness cases by mirroring the existing cases through the vertical axis. In this manner, we could also compare how the programs evolved for that particular side compared to the one evolved for the other side. Testing the best left-hand program on these 46 flipped cases produced 26 hits (57% accuracy) and a fitness of 53.3. This is not as good as the evolved version for the right side; however, the best right-hand program gets 32 hits (70%), which is not far from that. Therefore, it does generalize to some degree.

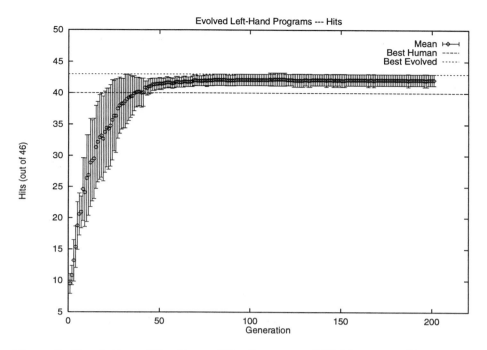

Figure 2: Population: 500. Copy: 0%. Crossover: 85% Mutate: 15%. 46 fitness cases. Mean and Std. Dev. over 12 trials.

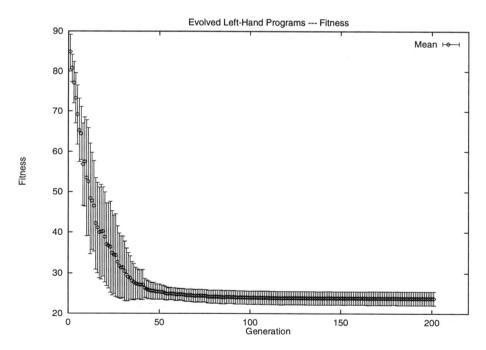

Figure 3: Population: 500. Copy: 0%. Crossover: 85% Mutate: 15%. 46 fitness cases. Mean and Std. Dev. over 12 trials.

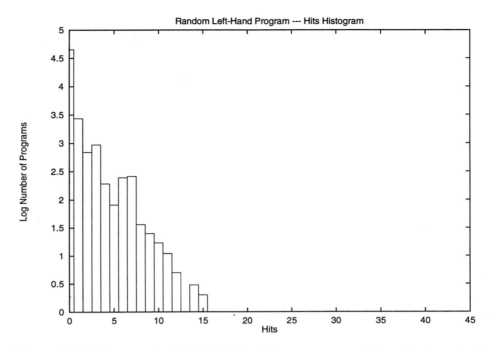

Figure 4: Hit histogram for 50,000 left-hand programs generated randomly and not evolved. Note: y-axis is log number of programs.

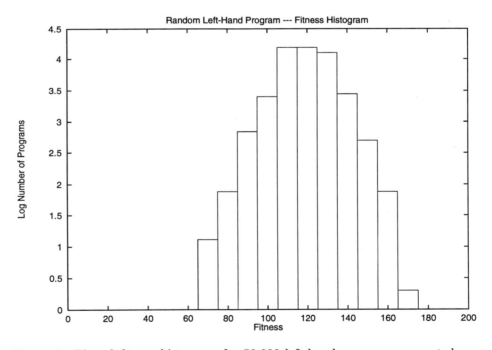

Figure 5: Binned fitness histogram for 50,000 left-hand programs generated randomly and not evolved. Note: y-axis is log number of programs.

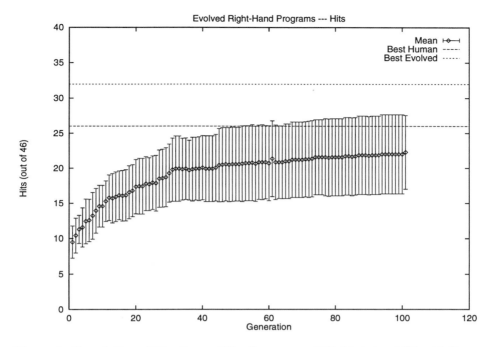

Figure 6: Population: 500. Copy: 0%. Crossover: 85% Mutate: 15%. 46 fitness cases. Mean and Std. Dev. over 15 trials.

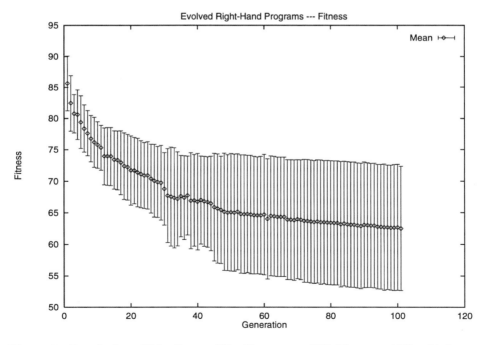

Figure 7: Population: 500. Copy: 0%. Crossover: 85% Mutate: 15%. 46 fitness cases. Mean and Std. Dev. over 15 trials.

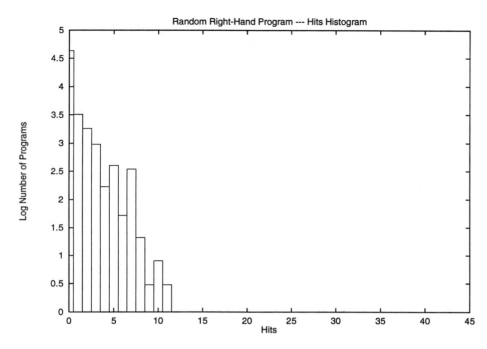

Figure 8: Hit histogram for 50,000 random right-hand programs. Note: y-axis is log number of programs.

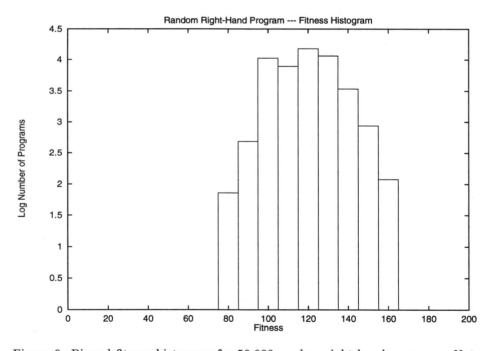

Figure 9: Binned fitness histogram for 50,000 random right-hand programs. Note: y-axis is log number of programs.

We also tested the best right-hand program on the flipped left-hand cases. It gets 27 hits (59%) and a fitness of 54.5. This is not nearly as good as the evolved programs for the left side, but it is still much better than random. Clearly some generalization is occurring. A probable explanation for the poor performance is that there were less "good" fitness cases in the right hand set from which to learn, hobbling the evolution. It also had to trade-off for solutions it could not solve.

Finally, we did 22 cross-validation runs for the left hand GP search. Eighty percent of the 46 fitness cases (30) were chosen randomly to serve as the fitness cases for each separate run. A best solution was evolved for each of these sets of fitness cases and tested against the remaining 10 cases. Results were promising: a mean accuracy of 77% ± 21% was attained.

6 Future Research

We plan to try the Ullman and Chapman style primitives in future experiments. By allowing the primitives to side-effect an incremental representation, we hope to make the search space simpler. For example, being able to store a value that is the result of a difficult computation in a register is useful when you need that result in several places. In a functional approach, you must evolve the hairy subroutine in multiple places.

We also want to use many more fitness cases to be sure the solutions are not overfitting the set. One way to get more fitness cases is to use a computer model of a human body and project it into a silhouette. In this way, we get the silhouette and know where the hand is in the image (from the projection). Thus, we'd have a continuum of fitness cases to sample from each time and could vary the person's height and size. This would get a human supervisor out of the loop, as well. Even more interesting would be to use a different Genetic Algorithm to coevolve configurations of this model with the solutions. Coevolution has been shown to be faster and more robust than a fixed fitness function in some cases [Koza, 1991, Hillis, 1991]. One might also want to add a weighting to the fitness cases based on some notion of their statistics; that is, more common postures should be focused on rather than contorted postures. Finally, giving the system a time series of silhouettes will add a level of power — it can use its previous estimates of hand position to figure out cases where the hand passes in front of the body, for example.

Another way in which we could get the human supervisor out of the loop would be to use a second sensory modality, e.g. sound or touch, to tell the creatures in ALIVE where the hand is at some point in time, so that it can learn about one sensory modality (vision) based on how it correlates to a second sensory modality.

7 Conclusions

We applied Genetic Programming to the problem of creating a simple visual routine, namely locating the left and right hands in a silhouette of a person. Results were promising. We hope to extend the system in terms of speed of computation and use of solutions in a real system. We also hope to try the system with more general primitives on a diverse set of problems, including template matching in spatio-temporal sequences. The current GP algorithm was sensitive to the choice of fitness function. It is hoped that some new methods, such as coevolution, can be used to solve this problem.

Acknowledgements

Dave Glowacki (dglo@cs.berkeley.edu) provided the public-domain C GP implementation used (and modified) in the research described in this paper.

References

[Aloimonos, 1993] Y. Aloimonos. *Active Perception*. Lawrence Erlbaum Associates, Inc., Hillsdale, 1993.

[Ballard and Whitehead, 1990] D. Ballard and S. Whitehead. Active perception and reinforcement learning. *Proceedings of the Seventh International Conference on Machine Learning*, 1990.

[Ballard, 1989] D. Ballard. Reference frames for animate vision. *Proceedings of IJCAI-89 Conference, Detroit*, 1989.

[Chapman, 1993] D. Chapman. *Vision, Instruction and Action*. MIT Press, Cambridge, Massachusetts, 1993.

[Cliff D. T. and I., 1993] Husbands P. Cliff D. T. and Harvey I. Evolving visually guided robots. *Proceedings of the second international conference on Simulation of Adaptive Behavior (SAB 92), MIT Press*, 1993. Ed. by J.-A. Meyer, H. Roitblat and S. Wilson.

[Hillis, 1991] D. Hillis. Co-evolving parasites improve simulated evolution as an optimization procedure. *Artifical Life II, SFI Studies in the Sciences of Complexity*, vol. X, 1991.

[Koza, 1991] J. R. Koza. Genetic evolution and co-evolution of computer programs. *Artificial Life II, SFI Studies in the Sciences of Complexity*, vol. X, 1991.

[Koza, 1992] J. R. Koza. *Genetic Programming*. MIT Press, Cambridge, Massachusetts, 1992.

[Maes *et al.*, 1993] P. Maes, B. Blumberg, T. Darrell, and S. Pentland. Alive: An artificial life interactive video environment. *Visual Proceedings of Siggraph '93, ACM Press*, 1993.

[Maes *et al.*, 1994] P. Maes, T. Darrell, B. Blumberg, and S. Pentland. The alive system: Full-body interaction with animated autonomous agents. *Submitted for publication*, 1994.

[Marr, 1982] D. Marr. *Vision*. W.H. Freeman, San Francisco, CA, 1982.

[Ramachandran, 1985] V. S. Ramachandran. Apparent motion of subjective surfaces. *Perception*, 14:127 – 134, 1985.

[Ullman, 1987] S. Ullman. Visual routines. *Readings in Computer Vision*, pages 298 – 327, 1987. Ed. by Martin A. Fischler and Oscar Firschein.

Evolving Sensors in Environments
of Controlled Complexity

Filippo Menczer and **Richard K. Belew**
Cognitive Computer Science Research Group
Computer Science and Engineering Department
University of California, San Diego
La Jolla, CA 92093–0114, USA
Email: {fil, rik}@cs.ucsd.edu

Abstract

Sensors represent a crucial link between the *evolutionary* forces shaping a species' relationship with its environment, and the individual's *cognitive* abilities to behave and learn. We report on experiments using a new class of "latent energy environments" (LEE) models to define environments of carefully controlled complexity which allow us to state bounds for random and optimal behaviors that are independent of strategies for achieving the behaviors. Using LEE's analytic basis for defining environments, we then use neural networks (NNets) to model individuals and a steady-state genetic algorithm to model an evolutionary process shaping the NNets, in particular their sensors. Our experiments consider two types of "contact" and "ambient" sensors, and variants where the NNets are not allowed to learn, learn via error correction from internal prediction, and via reinforcement learning. We find that predictive learning, even when using a larger repertoire of the more sophisticated ambient sensors, provides no advantage over NNets unable to learn. However, reinforcement learning using a small number of crude contact sensors does provide a significant advantage. Our analysis of these results points to a trade-off between the genetic "robustness" of sensors and their informativeness to a learning system.

1 Introduction

A defining characteristic of the artificial life (ALife) perspective, in common with earlier ethological research, is a holistic view of organisms and their environment, as part of a single system to be modeled. A critical aspect of all ALife models, then, is how the organism/environment "interface" is defined. This specification effectively cleaves the modeler's problem into two, a model of the internal "cognitive" system the organism uses to control its behavior, and a model of environmental change, due in part to these behaviors. In many ALife models, the complexity of the behaviors displayed by the organism is evaluated without taking into account the role played by the sensory system and motor system engineered by the modeler. Rather than assuming

any such *a priori* division, we propose to investigate the evolution of the organism/environment interface itself.

The sensory system in particular provides a strong coupling between environmental complexity and difficulty of the survival task. Intelligent behavior may result from complex sensors and trivial processing of this information, or simple sensors and clever processing (Menczer and Parisi 1992; Todd and Wilson 1993; Cliff, Husband, and Harvey 1993). In fact, experiments by Miglino and Parisi (1991) in which the arbitrary specification of the sensory interface appeared to have important consequences for learning helped to motivate our work.

Sensors — transducers from external environmental signals to the internal cognitive system — therefore represent a crucial link between two distinct adaptive forces. They are simultaneously phenotypic features shaped by evolutionary forces to define a species' relationship with its environment, and the "input" channels on which a cognitive system's abilities to act and learn are wholly dependent. Features of the sensory system such as determinism, reliability, information content, computational complexity, signal/noise level, locality, etc. determine the difficulty of the task to be performed by the organism. The same holds for the motor system, but in this paper we will limit our attention to the sensory system.

In section 2 we begin by sketching a new class of "latent energy environments," and an ALife simulator called LEE underlying our experiments. LEE allows environments of carefully controlled complexity to be defined in terms of the temporal and spatial distribution of a small number of "atoms" and simple "chemical reactions" between atoms that together define the "latent energy" available to support one or more species. This construction lets us make useful analytic statements (e.g., bounding the range of potential solutions), while still permitting the sort of "open-ended" evolutionary experimentation missing from genetic algorithm (GA) simulations using a pre-specified, exogenous fitness function. Section 3 uses LEE as the basis of an analysis of minimal and optimal behaviors possible in these environments. In section 4, we use LEE as the basis of experiments using neural networks (NNets) to model individuals and the GA to model an evolutionary process shaping the NNets, in particular their sensors. We report on experiments using two classes of sensors, "contact" (tactile) sensors which allow an organism to sense only directly adjacent space, and "ambient" (olfactory) sensors which provide information from a broader spatial range. We consider variants where the NNets are not allowed to learn, learn via error correction from internal prediction, and via reinforcement learning.

2 Latent Energy Environments

The same ALife coupling of organisms with environments mentioned above brings with it a major methodological problem within ALife research: results reporting behaviors of different organisms in different environments are incommensurate. It is therefore difficult to assess whether an apparently superior behavior is the consequence of more sophisticated adaptive techniques or due to relative *complexity* of the environments. For example, should we be more impressed by the abilities of "amoebas" that successfully evolve to forage for foods (as in (Nolfi, Elman, and Parisi 1990; Parisi, Cecconi, and Nolfi 1990)), by "ants" capable of following "trails" (as in (Collins and Jefferson 1992)), or by "agents" that avoid predators (as in (Ackley and Littman 1992))? It would be very desirable to be able to define artificial environments of controlled complexity, within which a wide range of ALife techniques might be directly compared.

On the other hand, traditional genetic algorithms can be viewed as having gone too far towards analytically tractable models of the environment. The typical GA models every environment as a "fitness" function returning a value $f(x)$ for each phenotype x, and the process of evolution as finding the value x^* optimizing f. GA theory has progressed considerably within this frame, for example stating conditions on f which make a search for the optimum "GA-hard" or "GA-easy," but casting evolution as a search for optimality with respect to a fixed, exogenously defined function leaves much to be desired in terms of biological veridicality. An ALife environment must also support the specification of strategies of arbitrary and unanticipated complexity, allowing creative, "open-ended" evolution.

Our work builds on several other attempts to define generic conditions on environmental complexity that are analytically tractable without constraining evolution's creative potential. The physicist Otto Rössler observed that varying food density could be used as a simple metric on the complexity of an environment facing an organism (Rossler 1974). Even a random walk is an adequate foraging technique if food is abundant, but more coherent movement is required if the foodstuffs are more widely spaced; in even more scarce environments, the foraging organism may need to depend on landmarks, cognitive maps, etc. Others have pursued similar analyses of environmental complexity as a function of spatial food density (Wilson 1991; Todd and Wilson 1993; Todd, Wilson, Somayaji, and Yanco 1994).

Food density is clearly a useful dimension for ALife simulation, but this single dimension of environmental variability must be extended to include other factors if we are to be able to test the full repertoire of ALife models. Holland's recent ECHO model describes an extraordinarily rich domain in which many intra- and inter-specific evolutionary questions might be explored (Holland 1992; Jones and Mitchell 1993). A small set of basic elements and rules by which these are combined, metabolized and replicated are defined, and conventional constructs such as "organism," "species," "predator," etc. emerge indirectly. In fact the range of possible experiments, or outcomes to a single experiment in ECHO is so broad that, like real biological systems, its behavior is often extremely difficult to analyze.

"Latent energy environments" (LEE) are a modest step from the single dimension of food density towards the richness of the ECHO model. The first step is to break up the conventional notion of "food." Food is typically modeled as a spatially localized element of the environment, consumption of which is directly correlated with an organism's fitness. Instead of food, LEE environments are filled with "atoms" (e.g., A, B and C) that must be combined by organisms in order to realize the "energy" given off by various "chemical reactions".[1] These reactions may be exothermic or endothermic (i.e., give off positive or negative energy) and may give off other atoms: a reaction can thus be represented as $A + B \rightarrow Energy + Byproducts$. A LEE environment is specified by defining the set of atoms, their spatial and temporal distribution, and the chemical reactions over these atoms. The set of possible reactions can be represented by the entries of a symmetric reaction table representing the "physics" of the environment, which typically remains constant throughout a simulation.

We can now define two critical features of a LEE environment. The *latent energy* of an environment is the total potential energy available from all reactions over the atoms in the environment.[2] By controlling the rate at which atoms of each type are introduced, we can regulate the amount of potential energy available in its latent form, while actual energy can ultimately be realized only upon dynamically choosing a reaction sequence. Second, the amount of *work* required to release the latent energy is defined in terms of the distance one of the atoms in a reaction must be moved in order to be brought into contact with the other. This is just the correct metric for evaluating organisms who must harvest energy to survive, and whose fundamental behavior, as will be shown in the next subsection, is movement. By controlling the spatial distributions of atom types in the world, we can dynamically regulate the amount of work required to combine them.

In short, we have quantitative knowledge of and control over the latent energy available in a world, how much work must be done in order to realize this energy, and how difficult it therefore is for an organism (or ecology of

---

[1] It is important to caution that our use of physical and chemical terminology ("atoms," "reactions," "energy," "work," etc.) is metaphoric. In particular, in the context of our investigation of sensors, we are not trying to model the low-level physiological physics of receptors.

[2] In fact, latent energy is not well-defined for arbitrarily specified reactions. If the reactions are carefully defined, however, the amount of energy latent in an environment can be derived directly from the atomic distributions.

organisms) to survive, without specifying just how the latent energy is to be realized by one or more species. The monitoring of population dynamics makes it possible to compare ecologies in different environments, in terms of how efficiently each exploits the energy latent in its own environment.

2.1 Modeling individuals' behavior

The defining feature of the LEE model is that energy in an environment can only be released through the *behaviors* of organisms. Behaviors induce reaction sequences, thus catalyzing the transformation of latent energy into actual, life-giving energy. The notion of behavior — actions taken by the organism that change the world and/or the organism's relationship to the world — immediately binds an organism to its environment in an intrinsic way. Thus we take *movement* to be our canonical behavior, in part because this overt action is most easily observed.

In the experiments reported here we model the mapping from *sensory* states to *motor* actions using well-studied feed-forward *neural networks*. An aspect of real organisms that is often forgotten by cognitive scientists is that these three components of an organism's cognitive apparatus are but a subset of a more inclusive *soma*, with adaptive consequences in its own right. We briefly describe each of these four components of LEE organisms below.

Soma There are two features of an organism's body that are relevant to the experiments reported here. First, each organism is assumed to have an internal cavity we call its *gut*. We will assume that every gut is capable of holding a single atom. An organism automatically ingests any atom over which it travels, causing this atom to be placed in its gut. If the gut already contains an atom and the new atom causes them to be involved in a reaction, the reaction is performed and the organism either accrues (exothermic reaction) or loses (endothermic reaction) energy. The second feature of a body is that it determines the placement and orientation of an organism's sensors and motors. While none of these somatic characteristics are varied or allowed to evolve in the experiments reported here, the gut will be shown to function as an important form of "memory" in section 3.2, and two different sensor configurations are considered below. More generally, LEE's somatic model has been designed to explore what often seems ill-founded discrimination into "cognitive" and "non-cognitive" features.

Sensors The sensory system receives information from both the world "outside" the organismic membrane and from "inside" its internal soma. Sensors define the mapping from these sources of information to the brain's input. In general, LEE sensors may differ in range, directionality, sensitivity, resolution, specificity, accuracy, etc. Here we will consider only two types of sensor, *contact* and *ambient*. Contact sensors provide a binary presence/absence indication of a single atom type in the space directly adjacent to the sensor. Con-

tact sensors are sufficient to support avoidance behaviors, but not approach. Ambient sensors provide an indication of a particular atom's presence, summed over cells in a neighborhood of the sensor and weighted inversely according to their distance.[3] Ambient sensors can underlie approaching behavior only if the brain possesses either a memory to compare *temporal* differences in a single sensor, or multiple ambient sensors placed and oriented differently to compare *spatial* differences (see also Littman 1994).

Motors Motors function as the output effectors of the organism, changing its location or orientation with respect to a fixed environment referential frame. In general, they may differ in range, energetic efficiency, accuracy, etc. In these experiments we use exclusively a very simple motor system used by others (Nolfi, Elman, and Parisi 1990; Parisi, Cecconi, and Nolfi 1990). Two binary output channels (left and right) control a single motor allowing one of four moves: stand still (00), turn 90 degrees right (10) or left (01), or go one cell forward (11).

Neural network The signals from each sensor are used as the inputs to a feed-forward neural network with a single hidden layer of units, and the NNet's output units are used to control the motors. We will consider two variants of this basic NNet, each using a different source of feedback for learning. In the experiments of section 4.3, we follow the work of Nolfi, Elman, and Parisi (1990). The NNet's output is extended to produce not only a motor action, but also a *prediction* of the sensory input it will receive after that action. This prediction is compared with the actual sensory information following the action and prediction error is minimized by a standard back-propagation technique. In the experiments of section 4.4, prediction-based error-correction learning is replaced by *reinforcement* learning. NNet weights are modified only immediately after actions catalyzing a non-zero energy reaction using a simplified version of associative reward-penalty.

Putting two organisms together Figures 1 and 2 show two examples of fully specified organisms. Both organisms use a pair of sensors for the gut and the motor system described above. The organism in figure 1 uses a pair of contact sensors, a hidden layer of two units, and reinforcement learning. The one in figure 2 uses three pairs of ambient sensors, eight hidden units, and prediction learning. The second organism may seem unnecessarily complicated, but recall that predictive learning, like approaching, demands a difference, in either space or time. To allow a spatial comparison of signals, we use two sensors directed towards opposite sides of the organism, each with a range of 2 moves; to break the lateral symmetry of these two sensors, we add a third sensor with range 3,

---

[3]In these experiments the metric is defined behaviorally, in terms of the number of moves required of the organism to reach this location. Given the simple motor system, this distance is well-specified.

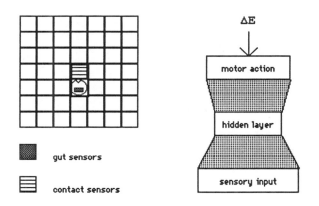

Figure 1: Sensory system of LEE organisms for the avoidance task (left); architecture of the corresponding NNets (right). All connections are modifiable by reinforcement learning (the arrow indicates the source of the reward-penalty signal).

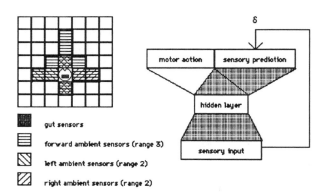

Figure 2: Sensory system of LEE organisms for the approaching task (left); architecture of the corresponding NNets (right). The connections shown in dark are modifiable by prediction learning (the arrow indicates the source of the teaching signal for back-propagation).

oriented frontally. Each sensor is replicated twice to allow full information (as will become clear in section 3.2), for a total of six ambient sensors and two gut sensors. Note also that the weights from the hidden layer to the motor units cannot be modified by learning. If these NNet characteristics are to change, it must be due to evolutionary forces.

2.2 Modeling populations' evolution

The same model of an organism can and must be viewed from two distinctly different levels of analysis. The description above was in terms of the *individual's* behaviors, the soma, sensors, etc. that generated them, and the relevant time scale was that of the behaviors. However, this same organism is simultaneously a member of an evolving *population,* with the relevant time scale being that of generations. These two perspectives on LEE are suggested by the two windows from the Macintosh interface to the LEE simulator shown in figure 3.[4] One shows a population of organisms scattered over an environment containing a uniform distribution of two atom types, and the other shows an expanded view of an individual's NNet.

In LEE the construct tying an individual's behaviors to its evolutionary fitness is the energy it accrues in its gut by "metabolizing" reactions. If ever the organism runs out of energy, it dies. If it collects more than a fixed threshold of energy in this way, it asexually reproduces. A critical observation is that this form of fitness is "endogenous" in LEE environments, resulting directly from competition for resources within the population rather than being externally computed as an evaluation of individuals (Mitchell and Forrest 1994).

After a LEE environment is specified, a simulation is begun by randomly placing an initial population of organisms within it. For each time step, every organism executes a basic life cycle:

```
for each cycle
  for each alive organism
    sense world
    feed forward activations
    move         /* automatically ingest */
    digest       /* catalyze reactions */
    learn        /* change NNet weights */
    if (energy > α)
      reproduce  /* copy genotype */
      mutate     /* new genotype */
    end
    else if (energy < ω) die
  end
  replenish world
end
```

This cycle is repeated indefinitely, with organisms foraging independently in a shared world as some die, others reproduce, and their offspring repeat the process.

[4]C source code and documentation for release 1.* of LEE (Menczer and Belew 1993) is available by anonymous ftp from cs.ucsd.edu (132.239.51.3) in the pub/LEE directory. From the World Wide Web, use URL http://www-cse.ucsd.edu/users/fil.

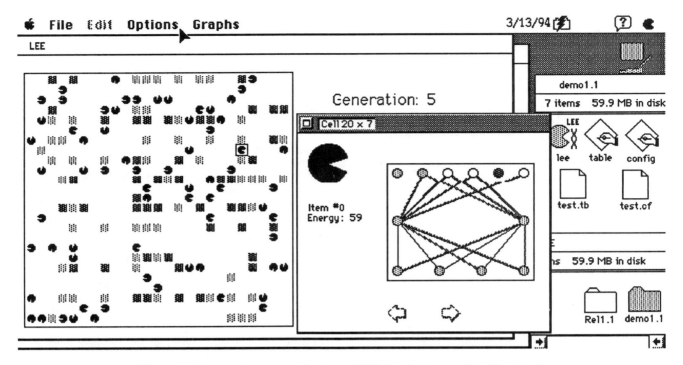

Figure 3: Interactive screen during a LEE simulation on the Macintosh.

Energy is always conserved: at reproduction, parents split energy with their offspring. An important consequence of the "steady-state" genetic algorithm (DeJong and Sarma 1993) is that the population size does *not* remain constant throughout an experiment. Analysis of this dynamic variable, and its connection to individual behaviors, is considered in the next section.

In addition to specification of selective pressures deriving from relative fitness, a Darwinian process must also have a mechanism for generating variation. In these experiments we will consider only random mutations (i.e., there is no sexual recombination/cross-over) of two classes of genotypic characters. The first are the types of atoms the sensors sense (A,B,C, etc.). With small probability (0.05) an offspring's sensor may signal a different atom type than the corresponding sensor in its parent. The second class of genotypic features are the NNet's weights. In all of the experiments reported here, weights are represented as floating-point numbers, and mutation performed by adding random, uniformly distributed noise.

3 Analysis

3.1 Fitness

Mitchell and Forrest (1994) point out the lack of theory regarding ALife environments with endogenous fitness. How can we measure how well behaviors are adapted in such environments? In LEE, the characteristics of an environment allow us to establish a connection between the size of an evolving population and the average adaptation of the observed behaviors in the population. The finiteness of environmental resources imposes a limit on

how many individuals can be sustained: we identify this maximum population size with the "carrying capacity" of the environment. For example, extinction indicates that the carrying capacity of the environment is insufficient to support a population large enough to withstand stochastic fluctuations.

To see how to quantitatively estimate optimality of behaviors by monitoring the population size throughout an experiment, consider the relationship between latent energy and population size in a particular, very simple and regular reaction table for three atoms, A, B and C:

$$
\begin{array}{c|ccc}
 & A & B & C \\
\hline
A & -\beta E & E & \# \\
B & E & -\beta E & \# \\
C & \# & \# & \#
\end{array}
\tag{1}
$$

where $E > 0$, $|\beta| < 1$, $\#$ indicates no reaction, and there are no byproducts. These conditions insure the non-zero-sum energetic content on the environment necessary to avoid extinction at the onset of the evolutionary process.

Notice that table 1 is symmetric, so order of ingested atoms does not matter. For $\beta > 0$ we can characterize two different strategies, one for combining A and B atoms catalyzing exothermic reactions, and one combining homogeneous atoms catalyzing endothermic reactions. In fact, a single parameter can be used to measure the relative frequency of these two strategy types (see below). Since we know C atoms to be causally irrelevant, we choose to leave them out of the environment entirely.

Table 1 needs to be augmented with knowledge about the spatiotemporal distributions of the atoms in order to determine the environmental complexity of the survival task. Let us assume the environmental resources are introduced with uniform distribution in space and time, at rate of replenishment $r(A) = r(B) = r$ (measured in atoms per unit time and area). It is important to note that these resources are shared by the population: these conditions lead to density-dependence of fitness (Stearns 1992). Finally let us define a small constant cost, ϵ, incurred every time that an organism makes a move. It is then easy to show (Menczer and Belew 1994) that *at equilibrium* (i.e., for constant population), the expected population size p is given by:

$$p = \frac{rE}{\epsilon}(\eta + \eta\beta - \beta) \qquad (2)$$

where η is the time-average of the fraction of catalyzed reactions yielding energy E. Equation 2 provides an important link between optimality of behavioral strategies, expressed through η, and population size. In particular, the case of optimal behavior ($\eta = 1$) corresponds to the maximum sustainable population, i.e., the carrying capacity:

$$p_{max} = \frac{rE}{\epsilon}. \qquad (3)$$

Conversely, to estimate the optimality of a population's behavior at equilibrium by measuring the population size, we can solve (2) for η:

$$\eta = \frac{\frac{p\epsilon}{rE} + \beta}{1 + \beta}. \qquad (4)$$

Another useful bound results from behaviors in which individuals are as likely to catalyze exothermic as endothermic reactions. This results in a "random choice" population level:

$$p_{\eta=1/2} = \frac{rE}{2\epsilon}(1 - \beta). \qquad (5)$$

Equations 3 and 5 give us an upper bound and a baseline, respectively, to measure the average fitness of the population. We have shown elsewhere (Menczer and Belew 1994) how to use these equations to make accurate predictions about the outcomes of simulations with simple environments, and to compare behaviors in different environments.

When evolved behaviors allow organisms to make a more efficient use of the latent energy, an increase in η follows and there is a transition to a new equilibrium. How can we characterize at the individual level the increase in fitness observed at population level? During these phases density dependence is no longer the sole fitness factor, and thus carrying capacity is neither a sufficient predictor of population size nor a satisfactory fitness measure. Population size and average age, on the other hand, remain very well correlated at all times (the correlation coefficient being $\mu \sim 0.8$). This suggests a connection between age and individual fitness. To clarify the connection, let us introduce a fitness measure that can be used without bias. *Expected reproductive success* is typically defined as:

$$R_0 = Exp\left[\frac{\#offsprings}{lifetime}\right]. \qquad (6)$$

To find the relationship between R_0 and population size, we can use two sensible assumptions. First, the number of offspring generated up to age a grows linearly with a:

$$n(a) \approx \rho a \qquad (7)$$

where ρ is a constant called *reproductive rate*. Second, in agreement with measured age distributions in LEE (Menczer and Belew 1994), the number of organisms having age a at any given instant follows the *Euler-Lotka* equation (in the continuous limit):

$$N(a) \approx \begin{cases} pe^{-a} & a \leq a_{max} \\ 0 & a > a_{max}. \end{cases} \qquad (8)$$

Using (7) we find

$$R_0 = \int_0^{a_{max}} n(a)da \approx \frac{\rho}{2}a_{max}^2. \qquad (9)$$

To find a_{max} notice that, from (8),

$$\begin{aligned} N(a_{max}) &\sim 1 \\ pe^{-a} &\sim 1 \\ a_{max} &\sim \ln p \end{aligned} \qquad (10)$$

and thus we conclude:

$$R_0 \propto (\ln p)^2. \qquad (11)$$

Equation 11 establishes a link between expected individual fitness and population size which remains valid even when the behavior of the organisms is improving and thus equation 4 is of no assistance. Furthermore, it provides a way to measure selective pressures between different behaviors from differences in population size.

3.2 A priori strategies

Using only contact sensors, it is not possible to approach atoms. Atoms must be encountered by some type of "blind" foraging pattern. Given the toroidal structure of the LEE world, always moving straight ahead is typical. With only blind foraging allowed, and for the environment described by table 1, the *optimal* behavior is:

```
if the gut is empty, go forward
else if the forward cell's content is
  the same as the content of the gut,
  turn
else go forward
```

To implement this optimal strategy, the sensory system should ignore C's and discriminate A from B with both gut and external sensors. Assuming the availability of gut and contact sensors giving accurate discriminating information, an (example of) optimal NNet function can be represented by the following table:

Gut	Envt	Left	Right
A	A	1	0
A	B	1	1
B	A	1	1
B	B	0	1

Note that this table simplifies the space of potential behaviors from the entire repertoire of available sensors in order to focus on the four consequential decisions. To make the discriminations necessary for this optimal strategy, one of the gut's two sensors must sense A and the other B, and the same must be true of the two environmental sensors.

If some sensors give information that is only partially discriminating, then only suboptimal strategies are possible. For example assume that the gut's sensors are as above but the environmental sensors both sense A. With this limited information, the best possible strategy is:

```
if the gut is empty, go forward
else if both the gut and the forward
  cell's contents are A,
  turn
else go forward
```

Critically, the NNet implementation of this *sub-optimal* strategy:

Gut	Envt	Left	Right
A	A	1	0
A	?	1	1
B	A	1	1
B	?	1	1

depends on learning a function very much like "inclusive OR," which is known to be much simpler than the "exclusive OR" function underlying the optimal strategy above.

4 Experiments

4.1 Controls

The first experiment is made of two control simulations aimed at measuring the baseline and best performance obtainable by evolving the network connections alone. There is no learning during life, and no evolution of sensors. Later experiments can be compared with these controls to see how much evolved sensors are better (worse) than the worst (best) possible sensors we can design. Unless otherwise stated, the experiments of this and the following subsections use organisms with the architecture shown in figure 1.

The worst possible sensors are those that leave the organism blind, i.e., without any information whatsoever about the gut or the external environment. This is done by setting all sensors to C, so that the inputs of the NNet are always zero: the adapted weights must implement constant actions, independent of the environment, i.e., a "random choice" strategy. The best possible sensors are those allowing organisms to discriminate between A and B atoms. This is done by setting both sensor pairs to A,B: the adapted weights may use the "full information" signals to implement the optimal strategy.

The results are shown in figure 4. Population averages and error bars (in this and the following experiments) are over different simulation runs. Random choice, as expected, cannot show any improvement past the very initial phase in which the strategy to always move sets in.

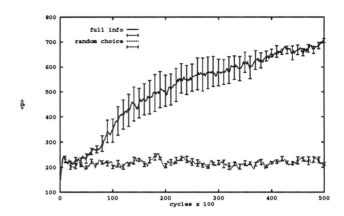

Figure 4: Population size in control simulations.

Full information sensors, on the other hand, allow near-optimal convergence. To see this, note that the population size at the end of the simulations corresponds to an average fraction $\eta \sim 0.87$ of positive energy reactions (from equation 4). Another way to assess the optimality of the evolved population is by the selective pressure in terms of reproductive success ratio: $R_0/R_0(p_{max}) \sim 0.93$ (from equation 11). These measures match the observation that the majority of organisms at the end of the simulations exhibit behaviors corresponding to the optimal strategy (cf. section 3.2).

4.2 Evolution of sensors without learning

We now want to test whether evolution alone can adapt sensory systems while coadapting connection weights with the information provided by the sensors in order to produce good behaviors. Thus no learning occurs during life in the experiments reported in this subsection, and the sensory system evolves as illustrated in section 2.2. The only gut and contact sensor pairs which afford the information necessary for the optimal strategy are A,B and B,A. In either case, we expect the population to converge to optimal percentages of 50% for A (and B) sensors, and 0% for C sensors.

Figure 5 reports the resulting evolution of sensors in this experiment by plotting percentages of A and C summed over all sensors. The B percentage is not shown for clarity (it can be obtained by subtracting those of A and C from 100%, and its expected optimum is identical to that of A). Although the tendency is as predicted, the large error bars indicate the importance of random genetic drift in these experiments. Upon repeating the simulation for different runs, sensor percentages converge to highly dispersed values.

Our interpretation of the large drift is that only about 5% of the possible sensor configurations (4 out of the total $3^4 = 81$) allow the optimal strategy to evolve. Furthermore, these configurations do not all share the same coadapted connection weights, so the resulting behavior is very fragile in the face of mutations on the sensory system. In contrast, a suboptimal strategy like the one shown in section 3.2 can be implemented with $28/81 \sim 35\%$ of the sensor configurations. While this strategy results in somewhat inferior performance, it is more robust because not only more configurations can achieve it, but more importantly, many of them function with the same network weights. This follows naturally from some of the sensors being redundant with respect to the suboptimal strategy: these sensors can be mutated without affecting the network behavior. Such redundance is also the source of the observed genetic drift: *robustness* and *drift* are two faces of the same selective pressure against optimality. To be sure, optimal sensors are generated from mutations at all times, but they are destroyed before the connections evolve to the coadapted weights. The fact that the suboptimal strategy is the winning one is also confirmed by looking at how organisms at the end of a run behave in the environment.

One way to increase the selective pressure toward the optimal strategy is to increase the environmental complexity. This is easily done with LEE by dynamic changes in the parameters describing the environment, such as the reaction table entries or the spatiotemporal distributions of resources. We have compared the experiment above with one in which the environment becomes increasingly hostile by simply increasing the β parameter linearly with time, at the rate $d\beta/dt = 1.1\times 10^{-4} cycle^{-1}$. To see that this corresponds to increasing complexity, notice that the random choice behavior ceases to be a viable strategy, resulting in inevitable extinction as soon as $\beta \geq 1$.

The population levels for constant $(d\beta/dt = 0)$ and variable environments are plotted in figure 6. Notice that the minimum for the changing environments is reached at the "zero-sum" value $\beta = 1$ (3000 cycles). After that, the population average performance is consistently better in the more complex environments. It is important that, although the population levels are about the same at the and of the simulation, they do not correspond to equal performances: for example, recall from (4) that $p \sim 700$ corresponds to $\eta \sim 0.87$ in a fixed environment, but to $\eta \sim 0.97$ in a changing environment. In fact, almost all the organisms evolved in complex environments exhibit the optimal behavior at the end of the simulation. However, the population reaches the same size in the two experiments: evolution has reached its reproductive success goal independently of environmental complexity (cf. equations 6 and 11).

4.3 Prediction learning

The so-called "Baldwin effect" describes an indirect mechanism by which learning can increase fitness of an individual, short of direct, Lamarckian inheritance of the knowledge acquired via learning (Baldwin 1896; Belew 1990; Ackley and Littman 1992). It is reasonable to hypothesize, then, that learning may increase selective pressure towards more informative sensors and consequently reduce the drift observed above.

A requirement for any learning mechanism in our model is the assumption that there be no additional "teaching" input available to the organism, beyond the

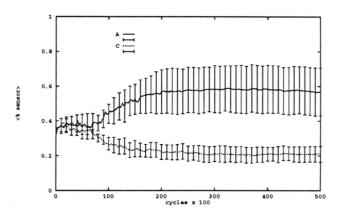

Figure 5: Sensor percentages in simulations with evolution alone (no learning).

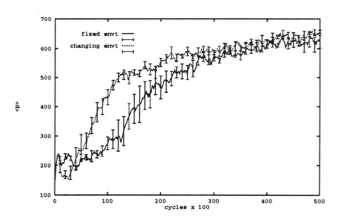

Figure 6: Population size in simulations with evolution alone (no learning) and different environmental complexity.

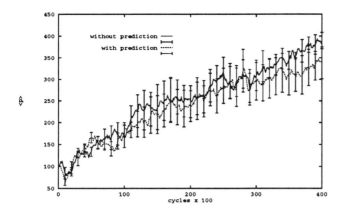

Figure 7: Population size in simulations with ambient sensors, with and without prediction learning.

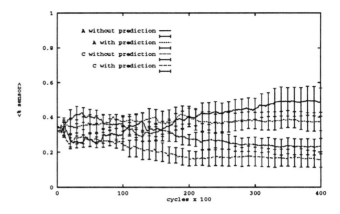

Figure 8: Evolved sensor percentages in simulations with ambient sensors, with and without prediction learning.

relatively neutral stimulus of the organism's environment. This would seem to preclude many successful "supervised" learning techniques, e.g., back-propagation (Rumelhart, Hinton, and Williams 1986). However, if the organism is simply forced to *predict* the expected outcome of its actions, differences between its expectations and the actual outcome can generate the same sort of error information, without any additional teacher (Nolfi, Elman, and Parisi 1990; Parisi, Cecconi, and Nolfi 1990). This motivates our first choice, prediction learning using a standard back- propagation algorithm.[5] If an organism's learning system can converge on a consistent explanation of the world that can be of aid in the survival task, we may expect that the Baldwin effect will facilitate the evolution of sensors.

The organisms of this experiment use the architecture shown in figure 2. To increase the selective pressure against the random choice strategy, we have modified the environments by clustering the A and B sources with gaussian distributions. This way, the random choice strategy costs more work. To avoid initial extinction, we have compensated this added complexity using a smaller β. Results are plotted in figures 7 and 8 for population size and sensor percentages, respectively. The observed performance is consistently generated by evolved behaviors which perform some approaching, but as far as combining atoms correspond to the suboptimal strategy discussed above. Good approaching strategies can improve fitness only to a limited extent in our environments, therefore these results basically show that prediction learning provides no significant advantage over evolution alone, neither for fitness improvement nor for drift reduction.

This apparently disappointing result is actually of assistance in understanding the nature of the information afforded by ambient sensors and their interaction with the prediction learning algorithm. One difficulty related to these sensors is the *ambiguity* of their signals, because the function of the environment state that they compute

<hr />

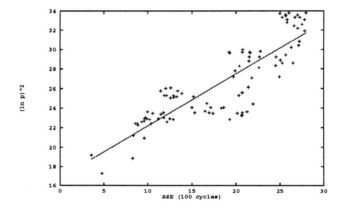

Figure 9: Scatter plot of prediction error and expected reproductive success (from (11)). Linear regression fit is also shown.

integrates signals over the neighborhood and thus is non-invertible (many-to-one). In this sense, the information they provide is harder to use than for contact sensors, even though theoretically it allows prediction because it is less local. Another difficulty related to prediction learning is that the prediction task is *not well correlated* with the survival/reproduction task. For example, one reliable prediction in these environments is the absence of C atoms, i.e., a zero signal from all inputs corresponding to C sensors. Therefore the learning process drives phenotypes toward behaviors that are not useful because they are based on blind (C) sensors. Figure 9 shows a scatter plot of R_0 fitness and sum-squared-error during a typical run with fixed, full information sensors. The prediction error is positively correlated with reproductive success ($\mu \sim 0.86$), confirming that evolution selects against good predictors and thus this type of learning is unlikely to facilitate evolution in this type of experiment.

4.4 Reinforcement learning

From the result of the previous experiment we certainly cannot draw the conclusion that learning in general cannot interact constructively with evolution, but rather the observation that prediction learning is not appropriate for our simulations. Therefore we need another unsupervised learning algorithm, with the same motivation as for prediction learning: to determine whether the Baldwin effect can facilitate the evolution of informative sensors. However, we seek a learning algorithm whose error criterion is better correlated with the fitness of the survival task, and for this we turn to reinforcement learning. A number of authors have explored the use of *associative reinforcement learning* in conjunction with NNets, GAs, and ALife (Ackley and Littman 1990; Ackley and Littman 1992; Barto and Anandan 1985; Whitley, Dominic, and Das 1991). Evolution itself can be viewed as a form of reinforcement search algorithm, but the reinforcement signal (death or reproduction) is heavily delayed and of relatively little use during life (Ackley and Littman 1992). If however a useful reinforcement signal is made available by an organism's interaction with its environment, then it can be used to determine an error information directly from the association between inputs and actions. Such a signal is indeed present during the life of organisms in LEEs: the changes in internal energy following their actions. An increase in energy corresponds to a reward, a decrease to a penalty, and we expect the corresponding error criterion to be more closely related to the survival task than prediction error.

To use reinforcement learning in experiments on the evolution of sensors, we go back to the simpler situation in which only contact sensors are used. So the only actions that determine the fitness are those choosing whether or not to move forward when a food is sensed by the contact sensor, based on the gut content. Notice that this setting avoids the problem of delayed reinforcement altogether. Therefore we use reinforcement to modify weights only immediately after actions catalyzing reactions. The implemented algorithm is a simplified version of associative reward-penalty (Barto and Anandan 1985). When a change in energy $|\Delta E| > \epsilon$ occurs, a two-valued reinforcement signal

$$s = sgn(\Delta E) \qquad (12)$$

is generated (this does not violate our assumption that there be no external teacher, since the reinforcement can be generated internally, without intervention of cognitive skills). This is used to produce a teaching action pattern o_i^t from the last output o_i:

$$o_i^t = \begin{cases} H\left(o_i - \frac{1}{2}\right) & s = +1 \\ 1 - H\left(o_i - \frac{1}{2}\right) & s = -1 \end{cases} \qquad (13)$$

where H is the Heaviside step function and the activation o_i has not yet been clipped to 0/1. The classic delta rule[6] with back-propagation is then used for weight correction.

Figures 10 and 11 show the results of adapting sensors with evolution alone and with reinforcement learning, respectively. A significant reduction in drift and

---

[6]Learning rate=0.1.

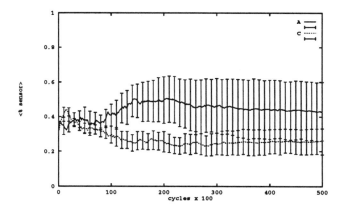

Figure 10: Evolved sensor percentages in simulations without reinforcement learning.

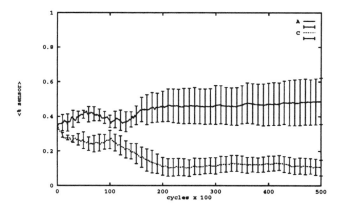

Figure 11: Evolved sensor percentages in simulations with reinforcement learning.

increased selective pressure toward informative sensors (e.g., against C sensors) is observed in the simulations with learning. This reflects the fact that in a greater fraction of the runs the population converges to full information sensors (approximately 50% A, 50% B and no C) and optimal behaviors as defined in section 3.2, and provides evidence for our hypothesis that the Baldwin effect can facilitate the evolution of sensors.

5 Conclusion

We believe that LEE represents both a rich theoretical framework and a useful simulation tool for the ALife community. In this paper we have demonstrated how it can be used to model biologically realistic selection pressures and thereby measure the complexity of a survival task. An important feature of the model is its emphasis toward behaviors which are *adapted to* as opposed to *optimal for* their environments. This adresses a common criticism of biologists toward adaptationist models of evolution as an optimization procedure (Gould and

Lewontin 1979).

We have reported on the use of LEE to investigate the evolution of sensors, something that has turned out to be an unexpectedly difficult problem due to the interdependence between the NNet implementing a behavior and the sensory interface producing the information upon which the behavior must be based. We have shown that, if the cognitive modifications provided by learning are well correlated with those required for survival in a particular environment, then phenotypic learning can assist adaptation in that environment by facilitating the evolution of appropriate genotypes.

The model has been successfully applied to other issues, such as assessing the adaptive advantages of random individual plasticity in non-stationary environments (Menczer 1994) and studying the evolution of age at maturity in the presence of cultural transmission by imitation (Cecconi, Menczer, and Belew 1994). In the near future we intend to use LEE for research on several other issues, including: development and morphology of both somatic and cognitive characteristics; sexual reproduction; emergence of speciation; multiple species ecologies; and predator-prey population dynamics. Further work is necessary to better understand the relationships between different types of learning and evolution in environments of controlled complexity.

Acknowledgments

This project originated from conversations with Stefano Nolfi and Domenico Parisi at the Institute of Psychology of the National Research Center in Rome, Italy. Stefano Nolfi, Jeff Elman, Greg Linden, and Federico Cecconi have contributed to the code. We also thank Peter Todd, Federico Cecconi, and everyone in the Cognitive Computer Science Research Group for helpful discussions and suggestions on many modeling and experimental issues.

References

Ackley, D. and M. Littman (1990). Generalization and scaling in reinforcement learning. In D. Touretzky (Ed.), *Advances in Neural Information Processing Systes (NIPS2)*, pp. 550–557. San Mateo, CA: Morgan Kaufmann.

Ackley, D. and M. Littman (1992). Interactions between learning and evolution. In C. Langton, C. Taylor, J. Farmer, and S. Rasmussen (Eds.), *Artificial Life II*, Volume X of *Santa Fe Institute Studies in the Sciences of Complexity*, Reading, MA, pp. 487–507. Addison Wesley.

Baldwin, J. (1896). A new factor in evolution. *American Naturalist 30*, 441–451.

Barto, A. and P. Anandan (1985). Pattern recognizing stochastic learning automata. *IEEE Transactions on Systems, Man, and Cybernetics 15*, 360–375.

Belew, R. (1990). Evolution, learning, and culture: Computational metaphors for adaptive algorithms. *Complex Systems 4*, 11–49.

Cecconi, F., F. Menczer, and R. Belew (1994). Learning and the evolution of age at maturity. (in preparation).

Cliff, D., P. Husband, and I. Harvey (1993). Analysis of evolved sensory-motor controller. In *Second European Conference on Artificial Life (ECAL93)*, Brussels.

Collins, R. and D. Jefferson (1992). Antfarm: Toward simulated evolution. In C. Langton, C. Taylor, J. Farmer, and S. Rasmussen (Eds.), *Artificial Life II*, Volume X of *Santa Fe Institute Studies in the Sciences of Complexity*, Reading, MA, pp. 579–601. Addison Wesley.

DeJong, K. and J. Sarma (1993). Generation gaps revisited. In L. Whitley (Ed.), *Foundations of Genetic Algorithms 2*, pp. 19–28. San Mateo, CA: Morgan Kaufmann.

Gould, S. and R. Lewontin (1979). The spandrels of san marco and the panglossian paradigm: A critique of the adaptionist programme. *Proceedings of the Royal Society of London B205*, 581–598.

Holland, J. (1992). *Adaptation in Natural and Artificial Systems* (2 ed.). Cambridge, MA: Bradford Books (MIT Press).

Jones, T. and M. Mitchell (1993). Introduction to the echo model. Working paper 93-12-074, Santa Fe Institute.

Littman, M. (1994). Memoryless policies: Theoretical limitations and practical results. In J.-A. Meyer, P. Husband, and S. Wilson (Eds.), *From Animals to Animats 3: Proc. Third Intl. Conf. on Simulation of Adaptive Behavior*. Camridge, MA: MIT Press/Bradford Books. (in press).

Menczer, F. (1994). Changing latent energy environments: A case for the evolution of plasticity. Technical Report CS94-336, University of California, San Diego.

Menczer, F. and R. Belew (1993). Lee: A tool for artificial life simulations. Technical Report CS93-301, University of California, San Diego.

Menczer, F. and R. Belew (1994). Latent energy environments. In R. Belew and M. Mitchell (Eds.), *Plastic Individuals in Evolving Populations*, Santa Fe Iinstitute Studies in the Sciences of Complexity. Reading, MA: Addison Wesley. (in press).

Menczer, F. and D. Parisi (1992). Recombination and unsupervised learning: effects of crossover in the genetic optimization of neural networks. *Network 3*, 423–442.

Miglino, O. and D. Parisi (1991). Evolutionary stable and unstable strategies in neural networks. In *IJCNN*, Volume 2, pp. 1448–1453. Piscataway, NJ: IEEE.

Mitchell, M. and S. Forrest (1994). Genetic algorithms and artificial life. *Artificial Life 1*(3).

Nolfi, S., J. Elman, and D. Parisi (1990). Learning and evolution in neural networks. Technical Report CRL-TR-9019, University of California, San Diego.

Parisi, D., F. Cecconi, and S. Nolfi (1990). Econets: Neural networks that learn in an environment. *Network 1*, 149–168.

Rossler, O. (1974). Adequate locomotion strategies for an abstract environment – a relational approach. In M. C. M, W. G. W, and M. Cin (Eds.), *Physics and Mathematics of the Nervous System*, pp. 399–418. New York, NY: Sringer-Verlag.

Rumelhart, D., G. Hinton, and R. Williams (1986). Learning internal representations by error propagation. In D. Rumelhart and J. McClelland (Eds.), *Parallel Distributed Processing: Explorations in the Microstructure of Cognition*, Volume 1. Cambridge, MA: Bradford Books (MIT Press).

Stearns, S. (1992). *The Evolution of Life Histories*. New York, NY: Oxford University Press.

Todd, P. and S. Wilson (1993). Environment structure and adaptive behavior from the ground up. In J.-A. Meyer, H. Roitblat, and S. Wilson (Eds.), *From Animals to Animats 2: Proc. Second Intl. Conf. on Simulation of Adaptive Behavior*, pp. 11–20. Cambridge, MA: MIT Press/Bradford Books.

Todd, P., S. Wilson, A. Somayaji, and Y. Yanco (1994). The blind breeding the blind: Adaptive behavior without looking. In J.-A. Meyer, P. Husband, and S. Wilson (Eds.), *From Animals to Animats 3: Proc. Third Intl. Conf. on Simulation of Adaptive Behavior*. Camridge, MA: MIT Press/Bradford Books. (in press).

Whitley, L., S. Dominic, and R. Das (1991). Genetic reinforcement learning with multilayer neural networks. In R. Belew and L. Booker (Eds.), *Fourth International Conference on Genetic Algorithms*, pp. 562–569. San Mateo, CA: Morgan Kaufmann.

Wilson, S. (1991). The animat path to ai. In J.-A. Meyer and S. Wilson (Eds.), *From Animals to Animats: First Intl. Conf. on Simulation of Adaptive Behavior*, pp. 15–21. Cambridge, MA: Bradford Books (MIT Press).

Traffic at the edge of chaos

Kai Nagel[a,c,d] and **Steen Rasmussen**[a,b,c]

[a]TSA-DO/SA, MS-M997 and [b]CNLS, MS-B258, Los Alamos National Laboratory,
Los Alamos, NM 87545, U.S.A.
[c]Santa Fe Institute, 1660 Old Pecos Trail, Santa Fe, NM 87505, U.S.A.
[d]Zentrum für Paralleles Rechnen ZPR, Universität zu Köln, 50923 Köln, Germany*
kai@zpr.uni-koeln.de, steen@lanl.gov

Abstract

We use a very simple description of human driving behavior to simulate traffic. The regime of maximum vehicle flow in a closed system shows near-critical behavior, and as a result a sharp decrease of the predictability of travel time. Since Advanced Traffic Management Systems (ATMSs) tend to drive larger parts of the transportation system towards this regime of maximum flow, we argue that in consequence the traffic system as a whole will be driven closer to criticality, thus making predictions much harder. A simulation of a simplified transportation network supports our argument.

1. Introduction

More and more metropolitan areas worldwide suffer from a transportation demand which largely exceeds capacity. In many cases, it is not possible or, even not desirable to extend capacity to meet the demand [1]. In consequence, a consistent management of the large, distributed, man-made transportation systems has become more and more important. Examples of such activities include the construction of fast mass transit systems, the introduction of local bus lines, design of traveler informational systems and car pooling to improve the use of current capacity, introduction of congestion pricing, and in the long term also guidance of the urban planning process towards an evolution of urban areas with lower transportation needs.

Unfortunately, the man-made transportation systems are highly complex, which makes them very difficult to manage. Due to the complexity of the dynamics of these systems, control decisions often lead to counter-intuitive results. In fact, management measures may even have consequences opposite to their intention. A clear example of how this can happen is the addition of a new street in a particular road network which leads to a *reduced* overall capacity [2]. The reason for this dynamical response is an extreme example of the general conflict between the individual traveler's optimal travel plans (Nash Equilibrium) and the travel plans that give overall maximal throughput; the System Optimum [3]. At the level of a metropolitan region, the transportation dynamics is the aggregated result of thousands or, in some cases, millions of individual trip-making decisions for the movement of people and goods between origins and destinations. And every decision is based on incomplete information of the state of the transportation system as a whole. Since complete global knowledge of the current (and future) state(s) of a transportation system seems very difficult to obtain, future informational based control strategies probably to a large extent should be based on self-organizing local strategies. However, that would still not take away the tension between global and local transportation optima which is one of the many reasons why predictability is very difficult in such systems.

There is another source of unpredictability which may very well become more dominating in a foreseeable future: Assume that all these management measures and modern information technology succeed in moving the transportation system closer towards higher efficiency. Then we face another problem. In road traffic systems, there is a critical regime around maximal capacity, as we shall see, which implies that transportation systems are very sensitive to small perturbations in this regime. Small perturbations will generate large fluctuations in congestion formation and thus travel times.

This is the topic for our paper.

One method of dealing with the inherent complexities of the large transportation systems is to represent the systems and generate their dynamics through simulation. The most straightforward way seems to be a bottom-up microsimulation of the dynamics of all travelers and loads at the level of where the transport decisions are made. Starting with a generation of travel demands and trip decisions, then routing, over traffic, eventually the consequences for congestion frequencies, travel time, air quality etc. are generated and can thus be analyzed. This is the approach used by the TRANSIMS project [4], which this work also is a part of. Note that all the performance properties that we may be interested in in a transportation system (in fact in any man-made system) are emergent properties from the interacting objects in the system. They are nowhere explicitly represented at the level of the interacting objects. They are generated through the dynamics.

The advantage of a microsimulation approach is that the system dynamics is being generated through the simulation with all its emergent properties without any explicit assumptions or aggregated models for these properties. The major disadvantages of a complete microsimulation are extremely high computational demands on one side and perhaps explanatory problems on the other. The inclusion of many details of reality may be excellent for generating a dynamics which is close to the system under investigation, but it does not necessarily lead to a better understanding of the basic (minimal) mechanisms that cause the dynamics. Therefore the TRANSIMS project also includes the investigation of much simpler and computationally less demanding models as the one we are going to discuss here.

In this paper, we concentrate on an extreme case of such a simplified transportation system. Out of the many modes of transportation (bus, train, ...) we only include vehicular traffic, and we assume that all vehicles as well as drivers are of the same type. Our model includes only single lane traffic, and the driving behavior is modeled by only a few very basic rules. The travelers may have individual routing plans so that they know the sequence of links and exits they want to use to go from their origin to their destination on a given transportation network. They can also re-plan depending on their earlier experiences of travel time. The approach is extendable to, e.g., multi-lane traffic and/or different vehicle types [5, 6].

We use numerical techniques from Computational Physics (cellular automata [7, 8]), and because of this similarity together with the resulting high computational speed [9], we are able to use methods of analysis originating in Statistical Physics (critical phenomena, scaling laws) [10]. We obtain results which are easy to interpret in the context of everyday experience, which is the more surprising as it is common belief that traffic is deeply coupled to the unpredictability of human behavior and cannot be modeled in terms of simple cellular automata rules.

Other large transportation microsimulation projects which are also dealing with different aspects of the dynamic complexities of large transportation systems are PARAMICS [11] and TRAFF/NETSIM [12].

The main part of this paper is divided into two parts. The first one deals with results for "traffic in a closed loop", i.e. without ramps or junctions. We review recent results about the connection between jams, maximum throughput, and critical behavior; and we present new results about the relation of these phenomena to travel times. In the second part of the paper, we turn to networks. We concentrate on a simple (minimal) example, which is nevertheless sufficient to discuss some of the issues we believe are important, especially our prediction that traffic systems become more variable when pushed (by traffic management) towards higher efficiency. Our simulation results support this prediction. We finish with a conclusion.

2. Single lane traffic in a closed loop

2.1 Single lane cellular automata model

Our freeway traffic model has been described in detail in Ref. [13]. Therefore, we only give a short account of the essentials.

The single lane version of the model is defined on a one-dimensional array of length L, representing a (single-lane) freeway. Each site of the array can only be in one of the following seven states: It may be empty, or it may be occupied by one car having an integer velocity between zero and five. This integer number for the velocity is the number of sites each vehicle moves during one iteration. Before the movement, rules for velocity adaption ensure "crash-free" traffic. The choice of five as maximum velocity is somewhat arbitrary, but it can be justified by comparison between model and real world measurements, combined with the aim for simplicity of the model. In any case, any value $v_{max} \geq 2$ seems to give qualitatively the same results (i.e. the emergence of branching jam waves). For every (arbitrary) configuration of the model, one iteration consists of the following steps, which are each performed simultaneously for all vehicles (gap := number of unoccupied sites in front of a vehicle):

- **Acceleration of free vehicles:** Each vehicle of speed $v < v_{max}$ with $gap \geq v + 1$ accelerates to $v + 1$: $v \rightarrow v + 1$.

- **Slowing down due to other cars:** Each vehicle (speed v) with $gap \leq v - 1$ reduces its speed to gap: $v \rightarrow gap$.

- **Randomization:** Each vehicle (speed v) reduces its speed by one with probability $1/2$: $v \rightarrow \max[v - 1, 0]$ (takes into consideration individual fluctuations).

- **Movement:** Each vehicle advances v sites.

The three first steps may be called the "velocity update". The randomization step condenses three different behavioral patterns into one single computational rule: (i) Fluctuations at free driving, when no other car is close; (ii) Non-deterministic acceleration; (iii) Overreactions when slowing down.

Already this simple model gives realistic backtraveling disturbances, as can be seen in the top two pictures of Fig. 1. In addition, one obtains a realistic fundamental diagram for, e.g., the flow q versus the density ρ. Fig. 2 gives simulation results for: (i) short time averages in a large system, (ii) long time averages in a large system, (iii) long time averages in a small system. These results are obtained for a closed system with periodic boundary conditions, i.e. "traffic in a closed loop". The small system means $L = 10^2$, a long system has $L \geq 10^4$.

Figure 1 *(next page)*. Space-time plots at different resolutions of traffic at different densities. *Left column:* Density $\rho = 0.07$, slightly below the regime of maximum flow. *Right column:* Density $\rho = 0.1$, slightly above the regime of maximum flow. Resolutions are from top to bottom 1:1, 1:4, and 1:16. In other words, in the top row, each pixel corresponds to one site (x, t), and one can follow the movement of individual cars from left to right. In the bottom row, 16×16 pixel of the space-time information are averaged to one pixel of the plot.

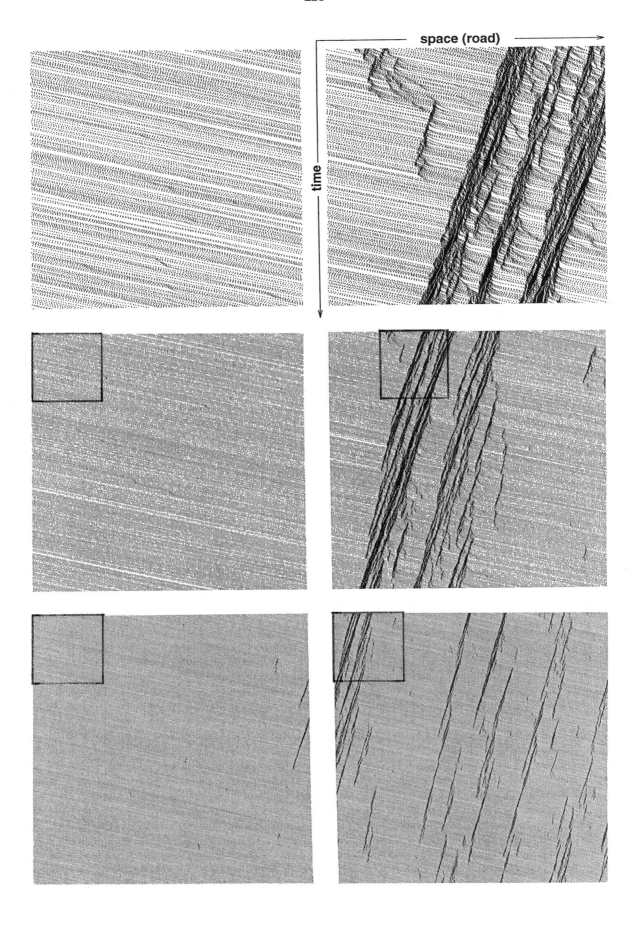

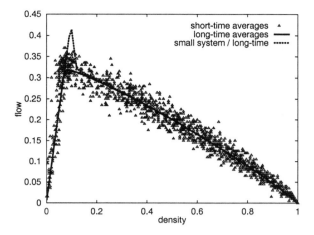

Figure 2. Fundamental diagram of the model (throughput versus density). Triangles: Averages over short times (200 iterations) in a sufficiently large system ($L = 10,000$). Solid line: Long time averages (10^6 iterations) in a large system ($L = 10,000$). Dashed line: Long time averages (10^6 iterations) for a small system $L = 100$.

Measurements are done at one fixed place in the system; technically, we measure

$$\rho = \frac{1}{T} \sum_{t=t_0}^{t_0+T-1} \frac{1}{v_{max}} \sum_{x=x_0}^{x_0+v_{max}-1} \delta(x,t)$$

and

$$q = \frac{1}{T} n_T .$$

$\delta(x,t)$ is 1 if x is occupied at time t and 0 elsewhen; the sum over v_{max} sites is necessary so that each bypassing vehicle is really "seen" by the algorithm. n_T simply is the number of vehicles which passed at x_0 during the measurement time T.

According to Fig. 2, our model reaches capacity ($=$ maximum throughput) $q_{max} = 0.318 \pm 0.001$ at a density of $\rho^* := \rho(q_{max}) = 0.086 \pm 0.002$ for large systems ($L \geq 10^4$). In addition, the figure shows that for a smaller system ($L = 10^2$) the maximum throughput is much higher. This means that short segments behave differently from long ones!

A comparison with real traffic measurements [13] indicates that it is reasonable to assume that, at least to the order of magnitude, one site occupies about $7.5\,m$ (which is the space one car occupies in a jam), one iteration is equivalent to about 1 second, and maximum velocity 5 corresponds to about 120 km/h.

It should be noted that the model so far can be treated analytically [14]. The analytical results, however, are more difficult to obtain; and the analytic methodology is not extendable to more complicated situations like multi-lane traffic, ramps, or networks ([5, 6], and see below).

Other work using (mostly even simpler) cellular automata for traffic flow on roads is, e.g., [15, 16].

2.2 Critical life-times of traffic jams

Looking closer at traffic pattern near capacity (as in Fig. 1), one makes at least two observations:

- Already at densities lower than ρ^* (left column of Fig. 1), the system displays spontaneous jams. They are sometimes very rare: In Fig. 1 their existence only shows up in the bottom picture of the left column, near the right of the picture.

- Space-time plots of systems near ρ^* have a remote resemblance to a directed percolation transition [17] and the emergence of the giant component in random graphs [18] in the way that jams at densities ρ^* have a finite life-time, whereas there seem to be jams of infinite life-times and spanning jam-clusters above ρ^* (right column in Fig. 1).

A jam-cluster is roughly defined in the following way: Spontaneous formation of a jam is caused by one car accidently coming too close to the one ahead of it, which leads to a lower speed than normal. Other cars which have to slow down because of this car are "in the same jam". The life-time T_{life} of this jam-cluster is the time until this structure is dissolved (i.e. no more cars with speed lower than normal).

In a more quantitative treatment, we measure the distribution of jam life-times using closed systems with different densities. We plot the number $N(\geq t)$ of jams with a life-time longer than t as a function of t in a doublelogarithmic plot [19]. (Technically,

$$N(\geq t) := \sum_{\tilde{t}=t}^{\infty} n(\tilde{t}) ,$$

where $n(\tilde{t})$ is the number of jams with a life-time exactly $= \tilde{t}$ in a given simulation run or number of runs.) For a true percolation-like transition [20] one would expect a behavior as depicted in Fig. 3a. Roughly speaking, the curves mean that, at low densities, long life-times are very improbable. However, at densities higher than critical, the system should be dominated by a few "very longlived" jams, which only leave room for shortlived jams between them. This leads to the $N(\geq t)$-curve becoming horizontal for large t. And in between one would expect a "critical" density, at which these curves converge towards a straight line ($N(\geq t) \propto t^{-\alpha}$) for $\rho \to \rho_c$ and $t \to \infty$.

In practice, we find a more complicated behavior for our system [19, 21]. The following is a short interpretation:

- The model has a certain probability p_{spont} of the spontaneous initiation of a new jam, which depends on the density of cars and on the amount of fluctuations which happen when cars move at full speed.

- This probability provides an upper cut-off on the length-scale and on the time-scale, up to which the model can display critical behavior.

"True" critical behavior can be recovered, when the model is redefined in a way that $p_{spont} = 0$. In terms of the model, this means that one has to reduce the fluctuations of free driving (i.e. undisturbed by other cars) to zero. Note that this leaves the fluctuations at accelerating and at slowing down unchanged. Once all cars

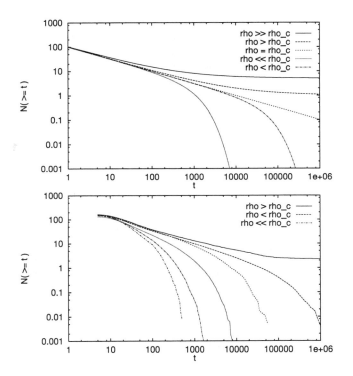

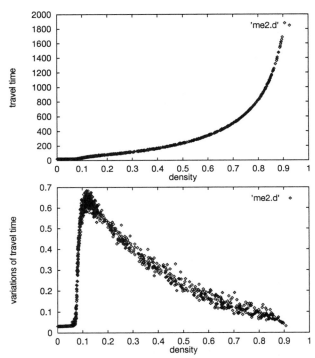

Figure 3. Theoretical (*top*) and simulated (*bottom*) distribution functions of life-times of traffic jams for a model where the spontaneous initiation of jams is impossible ($p_{spont} = 0$). The curves show, for different densities, the number of jams with a life-time larger or equal than t as a function of t. (y-axis arbitrary units)

Figure 4. Travel time and variations of travel time as a function of density. System size $L = 10^3$, length of traveled subsection $l = 10^2$, measured time $T = 10^5$ time-steps.

have reached maximum speed (if density allows that), no new jam may initiate itself spontaneously. One can then externally initiate one jam at a time (e.g. by picking on car randomly and reducing its speed by 1) and measure the properties of this jam. Doing this with different densities, we obtain the results of Fig. 3 bottom, which show that in this particular limit, the model corresponds indeed closely to the theoretical picture.

Please refer to the above-mentioned references for a more complete description.

2.3 Variability and predictability of travel times

Measuring the life-time distribution of traffic jams is convenient for a theoretical understanding, but it is not very useful for everyday traffic. The probably most important reason for this is that life-times of jam-clusters are practically not amenable to measurements.

A quantity which is much easier to measure and which is extremely relevant in the context of transportation management is the individual travel-time and its variation from vehicle to vehicle using the same route. For the following simulations, we still use a closed loop of size L. We define a subsegment of length $l < L$ and measure, for each car, the time t_l between entry and exit of this subsegment.

The relative variation of travel-times is defined as

$$\sigma(t_l) := \frac{\sqrt{\langle (t_l - \langle t_l \rangle)^2 \rangle}}{t_l}.$$

$\langle \ldots \rangle$ denotes the average over all cars during the simulation; $\langle t_l \rangle$ therefore is the average travel-time for all cars during the simulation.

Results of these measurements as a function of density are shown in Fig. 4. We use a system of length $L = 10^3$ and measure trips along a designated subsegment of $l = 100$. The simulation runs for 10^5 time steps, and every time a car finishes a complete travel along the measurement subsegment, its travel time is taken into account for the average.

One clearly sees that both the travel time and the vehicle-to-vehicle fluctuations are approximately constant up to a density around 0.09. There, the travel time starts to rise as a function of density, and the fluctuations go up very steeply and reach a maximum near $\rho = 0.11$. In other words, one can not only show that the region of maximum throughput shows near-critical behavior in a theoretical sense, but also that this behavior has practical consequences: It implies that, passing from slightly below to slightly above capacity, one comes from a regime where the travel time is predictable with an accuracy of approx. $\pm 3\%$ to a regime where the error climbs up to $\pm 65\%$ or more.

2.4 Traffic at the edge of chaos

Summing up our results, we obtain the following picture: Near maximum throughput, our model shows scaling of jam life-times and high variability of travel time, features which indicate a critical phase transition. But the lifetime scaling shows an upper cut-off; and the density of maximum throughput does not exactly coincide with the density of maximum fluctuations. Thus, the transition is not truly critical (although we use the word criticality throughout the text). However, it becomes exactly critical in the limit of zero fluctuations for free driving (i.e. not influenced by other cars).

A helpful concept for understanding critical phase transitions in discrete systems is the notion of "damage spreading" [22]: One simulates two identical copies of the system. At a certain point, a minimal change in one of the copies is made and then the time evolution of the *differences* between the systems is observed.

In our traffic system, "damaging" means to change the velocity of one randomly picked car by -1. This car then causes a jam of a certain life-time; and downstream of this jam, the traffic pattern will be different from the undisturbed model. After this jam has dissolved, the *spatial* amount of damage extends from the disturbed car to the last car involved in the jam, and this length is proportional to the life-time of the jam. For the limit $p_{spont} \rightarrow 0$ (but $p_{spont} \neq 0$), i.e. where spontaneous initiation of a jam becomes rare, one obtains the following picture:

- For low densities $\rho << \rho_c$, jams are usually short-lived (i.e. with an exponential cut-off in the life-time distribution). As a result, the average amount of spatial damage is small.

- When approaching the critical density ρ_c, jams become increasingly long-lived, with the result that the amount of spatial damage becomes larger and larger. Ultimately, exactly at the critical point, a damage of size infinity (in the thermodynamic limit) is possible.

- Above the critical point ($\rho > \rho_c$), the jam caused by the disturbance will (in the average) survive forever, thus (in the average) causing infinite damage.
  However, traffic for $\rho >> \rho_c$ is characterized by the existence of many jams quasi-randomly distributed over the system. So the additional jam caused by the disturbance will not change the *statistical* properties of the system.

All these observations are similar to conventional damage spreading observations in cellular automata (CA) [23]: The damage is limited for class I and class II CA, and it *can* be infinite for class VI CA. For class III CA, the damage is practically always infinite, but does not change the statistical properties of the system.

In summary, for the limit $p_{spont} \rightarrow 0$, $p_{spont} \neq 0$ we have in our probabilistic CA a phase transition of the traffic patterns similar to the one found in more conventional and determinstic CA. The control parameter in our case is the density, whereas in CA rule space it is still an open question how to derive an order parameter from the rules [23, 24], or if this is at all possible [25].

But, as stated initially in this section, a more realistic version of the model with p_{spont} significantly different from zero moves the point of maximum throughput away from the critical point. Therefore, we have, similar to other systems [26], the existence of traffic in the *vicinity* of "the edge of chaos".

The rest of this paper will be devoted to arguing why this regime is of special importance for transportation.

3. A simple transport network

We now move away from the single lane closed loop system to a single lane highway network with ramps connecting the different segments. The travelers on this network have route plans so that they know which ramps they need to exit to reach their individual destinations. We assume that each traveler always has the same origin/destination pair. Each traveler remembers the last travel time for each alternative route between his or her origin and destination. The network may have traffic density sensors at specified locations which can be used to identify congested areas and perhaps introduce toll for the use of such links. The travelers are able to re-plan depending on their aggregated transportation costs which is their remembered travel time plus eventual toll. Such a sensor setup is an example of a (Advanced) Transportation Informational System (ATIS), and the introduction of toll for the use of highly congested links is a simple example of an (Advanced) Transportation *Management* System (ATMS) [27, 28]. The rationale behind such a toll policy is to make the highway traffic more efficient by pushing a larger part of the system towards the density corresponding to maximum flow. Interestingly, this implies that more traffic intentionally will be moved into the critical regime as defined above which in turn will increase the fluctuations of the travel times as well as the non-predictability of transportation system dynamics. This effect is the topic of this section.

Some attempts have been made earlier to use CA techniques to simulate simple representations of network traffic. Most of the models map network traffic on particles hopping on a 2-D square grid [29, 30, 31]. These models are very useful to understand the transition from a free flowing to a jamming phase in urban traffic, but in these models the maximum flow of traffic is given entirely by the capacities of the intersections [30], which is not always realistic, e.g. for arterials. Closest to our approach is [32], which however was never used in order to do systematic studies like the one presented here.

3.1 Ramps

In order to simulate this simple network, we first need reasonable algorithms for transferring vehicles from one road to another at junctions. This involves two parts: Including the vehicle into the traffic stream on the target road; and then deleting it from the source road.

Unfortunately, introducing an additional car into a given traffic stream can cause some problems. Just adding the ramp-inflow to the traffic on the main road easily leads to disturbances which (i) block the traffic on the main road, and (ii) lead to an outflow, downstream

228

from the ramp, which is *below* capacity. For this reason, we chose an algorithm where access to the main road is only possibly when there is sufficient space between vehicles. We believe that this is realistic enough to represent metered ramps (i.e. ramps with regulated access), and since we are often concerned with the analysis of future traffic systems, it seems appropriate to model a technically advanced traffic control system here.

The algorithm works as follows. Imagine a ramp, as in common experience, as two parallel stretches of road; these parallel stretches have a length of 5 sites in the model. The target stretch is part of a longer road and therefore is connected at both ends, whereas the source stretch is only backwards connected. If there is a vehicle (velocity v) on the source stretch, then

- it looks, on the target stretch, for the next car ahead (which may be its neighbor; $\rightsquigarrow gap_{forward}$); and
- it looks for the next car behind on the target stretch ($\rightsquigarrow gap_{backwards}$).
- Then the following rules are applied:

$$IF \ (gap_{forward} > v \ .AND. \ gap_{backwards} > v_{max} \ )$$
$$change\_lane$$
$$v = \max(v_{max}, gap_{forward}) \ on \ new \ lane$$
$$ELSE$$
$$IF \ (\ v \geq 1 \ ) \ go\_one\_site\_backwards \qquad (*)$$
$$v = 0$$
$$ENDIF$$

One may imagine that this is emulating a ramp metering system, where a technical device upstream of the ramp determines where to fit in a car. The car then gets a green light and arrives at maximum speed, in between two other cars on the target road.

Line (*) is a technicality. It is necessary because the global velocity update may reaccelerate the speed of the vehicle to one and then move it one site ahead. If the car repeatedly fails to change to the target lane, then the car would slowly advance on the changing area and ultimately leave it.

The details of this algorithm will probably not matter for our results, as long as it allows maximum flow downstream from the ramp. That this indeed is the case is shown in Fig. 5, which may be compared to Fig. 1. It gives the fundamental diagram for a system with two road segments where one is a closed loop and the other one provides an alternative route for a certain length, connected to the main road by one exit and one entry ramp. Half of the vehicles use this alternate lane. Density and throughput are measured on the undivided part.

3.2 Nash Equilibrium versus System Optimum

An important issue in the context of a transportation network is the difference between Nash Equilibrium (NE; = User Equilibrium, UE) and System Optimum (SO); the optimum dynamics of an individual traveler versus the situation where the capacity of the transportation system is used in the most optimal manner. These two systems states are often in conflict.

This conflict can perhaps best be illustrated in terms of a simple transport network example (a variation

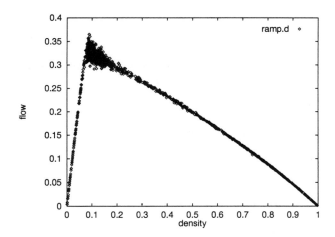

Figure 5. Fundamental diagram for ramp. A circular segment of length $L = 10^3$ may be partially bypassed by a second segment. 50% of the traffic uses the bypass; at the end of the bypass, it again merges with the main stream of the traffic. The measurements were taken at the part of the main segment where no bypass exists.

of [33]), especially as we will use the same example for simulation experiments later on.

Imagine a road from A to B with capacity q_{max}, with a bottleneck with capacity q_{bn} shortly before B. (Fig. 6 shows the same network, albeit different traffic patterns.) Further imagine that there exists an alternative, but longer route between A and B. On the direct route from A to B additional travelers from C have to go to destination D. First assume that there are no travelers with origin in C.

If many drivers are heading from A to B, they will, without knowing anything about the overall traffic situation, all enter the direct road. In consequence, a queue builds up from the bottleneck.

A Nash-Equilibrium is defined as a situation where no agent (= driver) can lower his or her cost (= decrease travel time) by unilaterally changing behavior. Assuming that the drivers have complete information, this implies that the waiting time in the queue exactly compensates for the additional driving time on the alternative route.

Now assume that there are additional travel demands from C to D (see Fig. 6), the exit for the latter lying shortly before the bottleneck. Obviously, this traffic is suffering from the bottleneck queue upstream (= left) of the bottleneck, and from these travelers point of view it would be much better if the queue were located to the left of the ramp that the travelers from C use to enter the link. Note that moving the queue further upstream does not make any difference for the drivers originating in A.

This example illustrates that one easily finds situations where there are better overall solutions than the NE. (Technically, a SO is reached when the sum of all individual costs (= travel times) is minimal.) Recent simulation results [34] indicate that the SO could give

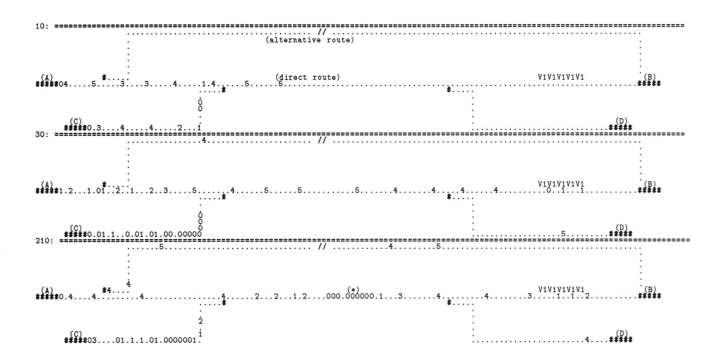

Figure 6. Schematic network representation with traffic, showing time-steps 10, 20, and 210 of a particular simulation. Traffic entering at (A) is bound for (B) and may use the "direct route" or the "alternative route". Traffic entering at (C) is bound for (D). The bottleneck is denoted by V1V1V1V1 (maximum speed 1). One observes that the traffic coming from (C) has difficulties entering into the main stream; and—in time-step 210—a disturbance denoted by (*) has traveled backwards from the bottleneck.

performance advantages of about 15% for realistic situations.

A way to push a traffic system from a NE towards a SO is to keep the density on each road at or below ρ^*, the density of maximum throughput. Then there would not exist queues anywhere in the system, thus ensuring that additional traffic could proceed undisturbed. Note that this could for instance imply (in the limit of a perfect implementation) that drivers have to wait to enter the road network until sufficient capacity is available for them.

3.3 Travel plans and individual decision logic

In our simple network, there are only two different types of travelers: Travelers from A to B, and travelers from C to D. Travelers from A to B can choose between the direct and the longer alternate route. In order to make decisions, each AB-driver remembers his or her last travel-time on each of the two routes.

A traveler calculates expected costs [3] according to

$$cost_{direct} = toll + \alpha \cdot t_{direct}$$

and

$$cost_{alt} = \alpha \cdot t_{alt}$$

where $cost_{direct}$ and $cost_{alt}$ are the expected costs for the two route choices, $toll$ is the toll for the current day

(see below), t_{direct} and t_{alt} are the remembered travel times for each route, and α is a conversion factor which reflects trade-off between time and money. α could be different for each driver, but is uniformly equal to one in this work. (α reflects "standard values of time", VOT, which can be looked up for traffic systems.)

Then, each driver chooses the cheaper route, except that there is a 5% probability of error (which gives each driver a chance from time to time to update her information about the other possibility).

As long as the traffic dynamics is deterministic and completely uniform, this scheme leads to a Nash equilibrium [3]. However, in our case of stochastic traffic dynamics, this is no longer true: There might well be a decision rule different from the one above where at least one traveler is better of, for example by triggering from some kind of day-to-day oscillation between the two routes and taking advantage of it. In other words, by dealing with stochastic traffic dynamics, the notions of economic equilibrium theory have to be used with care.

3.4 Space-time dynamics

Before we discuss how to determine the toll, we shortly turn to a space-time plot of the direct route from A to B (Fig. 7). As in each part of Fig. 2, vehicle movement is to the right and time is downward. The figure contains the first 300 time-steps, and then time-steps 2000 to 2950.

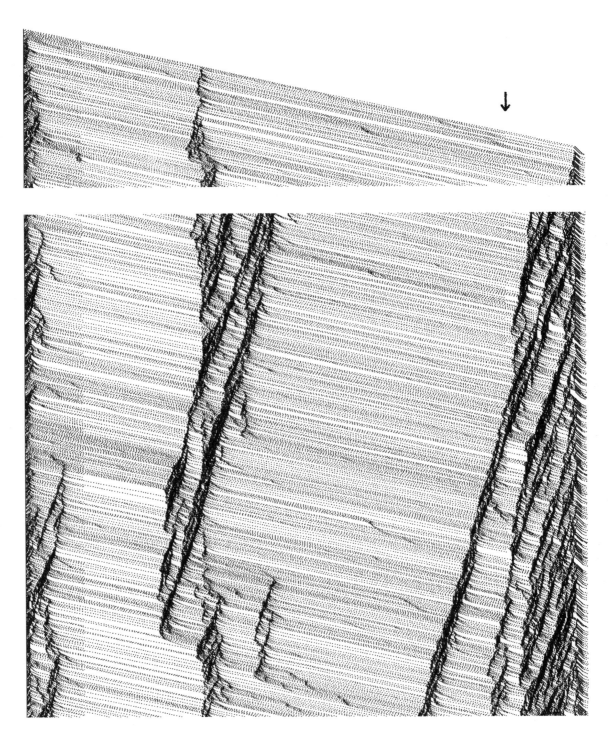

Figure 7. Space-time plot of the main segment (A-B) of the network. The cars are injected at the left. About one half inch to the right, a change in gray indicates the junction where vehicles to the alternative route leave. Another one and a half inches to the right, the jam structures indicate the on-ramp for travel from C. About one inch from the right, another change in gray indicates the off-ramp to D (see arrow on top). Very close to the right is the bottleneck, together with jams emerging from this region and traveling backwards into the system.

The major dimensions of the system are:

- direct route from A to B: 1021 sites (full size of plot)
- cars leave for the alternative route at position 111
- cars coming from C enter at position 322 and leave again at position 881
- the bottleneck ($v_{max} = 1$) extends from position 1001 to 1011.

20% of the A-B vehicles are preselected to leave at the junction for the longer route, as can be seen in the picture by a change of the gray shading. The entry-point of the C-D vehicles is marked by the permanent existence of a disturbance, which is very often connected to other disturbances which travel "backwards" through the system.

The point of exit for the C-D vehicles is covered by dense traffic most of the time, but it may be seen near the top right of the figure as a change in gray shading and as a sudden stop of some trajectories.

The bottleneck is visible at the very right edge of the figure, where the trajectories of the vehicles are diagonally pointing downwards to the right.

The striking feature of this picture is the graphic illustration of the highly dynamic and (seemingly) nonlinear structure of traffic patterns. Vehicles do *not* wait orderly in front of the bottleneck, but instead self-organize into backwards moving jam waves. If one of these waves reaches back into an area with higher density (in our case the junction where the C-D cars leave), then the survival probability of this jam wave suddenly becomes much larger, and it may move deeply back into the system. A single snap-shot of such a traffic situation could not uncover the origin of such a wave. The implications to traffic measurement and modeling are important.

Something similar is true for the region where the C-D traffic stream enters the main road. It is not a process where both traffic streams line up to wait until they can jointly proceed. Instead, it is often even possible that the additional traffic enters into the main stream without causing a major disturbance right away. But the locally enhanced densities is unstable and leads sooner or later to the initiation of a disturbance, which then travels backwards to the junction (and often beyond).

These results indicate that the methodology of queueing networks [35] has to be handled with care when applied to vehicular traffic.

3.5 Congestion detection, toll and travel pricing

For simplicity, assume that the current toll is based on some traffic observation on the last period (day). Let us further assume that each driver only drives this trip once in each period (day). (Note that this is oversimplistic, and further investigations are needed to make it work for, e.g., workdays versus weekend-days.)

Algorithmically, we can proceed as follows: (i) The traffic microsimulation is executed for one period. Each driver updates her travel time information just after arrival at the destination. (ii) After all cars have reached their destinations, the toll is adapted according to the average value provided by the sensor. (iii) Each A-B driver makes her route choice. (iv) The next microsimulation period starts. — This results in a day-to-day evolution of the decision pattern [36]. The procedure is actually very similar to standard game-theory [37], except that we obtain the pay-off from the microsimulation and not from a predefined matrix.

A critical question remains: Where should one place the traffic sensor(s) for the determination of the toll? Placing it *inside* the bottleneck is not very useful, because traffic there is always at or below the "efficient" density (i.e. at or below the density corresponding to maximum throughput).

Intuitively, it would make sense to measure the length of the queue in front of the bottleneck. However, as we showed in the last section, the dynamics of the traffic does not lead to the built-up of a regular queue but to a system of backtraveling jams instead, which makes this approach infeasible.

Therefore, we chose to measure the average density on the segment upstream of the bottleneck, i.e. between the exit to D and the bottleneck. Then, the next question is, which should be the target density for the control algorithm? When one is measuring traffic upstream of a bottleneck [38], then stationary traffic can never reach maximum throughput: Either traffic operates at densities corresponding to flow rates lower than the bottleneck capacity, or dense traffic builds up. Traffic can only "use" the part of the fundamental diagram which is below the capacity of the bottleneck; in consequence, densities are either far below or far above the ones corresponding to maximum throughput.

However, having some knowledge about the bottleneck is not really helpful: In a more complicated traffic network, it may be the case that further downstream from one bottleneck there is another one, which has even lower capacity. Or the bottleneck may be the on-ramp to a crowded major road: Here the performance of the bottleneck depends on the time-dependent and fluctuating load on the main road.

We therefore follow a simplistic and completely local approach here, which will nevertheless prove to be quite effective. Assume that the toll is operated by a local "toll agent", who does not have any global knowledge. However, she knows the fundamental diagram (flow as function of density) of *her* sensor area. If she wants to keep the traffic at maximum flow, she has to keep the density in the correct range, i.e. near $\rho = 0.08$. We implement this by the following rules:

> IF ($\rho < 0.06$) THEN
>   toll = toll - delta
> ELSE IF ($\rho > 0.10$) THEN
>   toll = toll + delta
> ENDIF ,

where *delta* is an external parameter.

According to our arguments above, is not obvious that this approach will produce meaningful decision behavior: The toll agent tries to keep the traffic at a density regime which is dynamically impossible because of the bottleneck downstream. It is not clear, a-priori, what effect this will have, and it was one of our main points of interest in how this control mechanism would work.

4. More simulation results: How to play traffic games

4.1 Technical Set-up

In the following, we describe one particular simulation run in more detail. We used a network of overall size 1962 sites, composed of the following parts:

- direct route from A to B: 241 sites (smaller than for Fig. 7 to reduce computational demand)
- alternate route from A to B: 1570 sites (*much* longer than direct)
- connection from C to main route: 103 sites
- connection from main route to D: 48 sites
- length of the section shared by A-B and C-D-travelers: 101 sites
- thus, overall length from C to D: 252 sites
- length of bottleneck (with maximum velocity reduced to one): 10 sites
- position of the bottleneck: starts 20 sites before reaching B

The density ρ_{toll} for the update of the toll is measured between the junction where the vehicles heading for D leave the main route, and the start of the bottleneck.

We have $N_{AB} = 16000$ vehicles which want to travel from A to B, and the same amount $N_{CD} = 16000$ which want to travel from C to D. At each "day" of the simulation, they are lined up outside the simulated system in the same sequence; and they enter the system at their respective entry points as soon as the simulated traffic allows it (cf. Fig. 6).

When the vehicles enter the system, they already have decided on their travel plans, so they just execute these plans. The simulation runs until all vehicles have reached their respective destinations. Then the toll is updated and drivers decide their route for the next day, as described above.

4.2 A simulation of 200 periods (days)

We describe 200 days of a simulation where the toll was kept at zero during the first 100 days, and in addition all A-B-travelers were forced to use the direct route during the first 50 days.

Fig. 8a shows results for the trip times and the adaptive toll, Fig. 8b the vehicle-to-vehicle variations of the trip time (as defined earlier), and Fig. 8c the day-averaged density, on selected road sections. These sections are: (i) the section where the density for the toll adaption is measured, (ii) the section of the main road between the on-ramp from C and the off-ramp from D, and (iii) the alternative route.

Even when allowed (i.e. after day 50), not many of the A-B drivers use the new option of the alternative route. This is to be expected, since it is more than six times longer than the direct one. In consequence, travel times and fluctuations do not change much.

After day 100, the adaptive tolling starts and fairly quickly reaches a stationary value around 260. As the

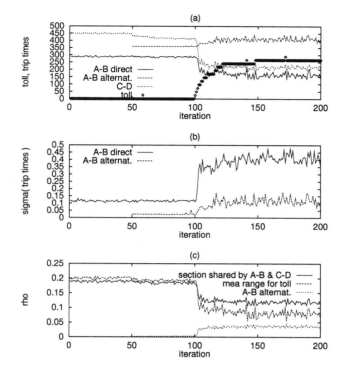

Figure 8. Simulation output for 200 iteration of the simple corridor network model. Time-steps 1-50: No adaption; 51-100: drivers can choose alternative route; 101-200: drivers can choose alternative route, and the toll adapts in order to keep the density at the specified level. *Top:* Average trip times for the direct and for the alternative route from A to B as well as for the route from C to D, and toll for the direct route from A to B. *Middle:* Vehicle-to-vehicle fluctuations of trip time for the direct and for the alternative route from A to B. *Bottom:* Densities on the segment shared by A-B-direct travelers and C-D-travelers, on the segment shortly before the bottleneck used for determination of the toll, and on the alternative route from A to B.

"toll" line in Fig. 7c indicates, this keeps indeed ρ_{toll} near the specified range between $\rho = 0.06$ and 0.10. In addition, the density on the main segment (used by both A-B and C-D travelers) drops to around 0.11, above, but close to the density of maximum throughput.

Travel times for C-D and for A-B-direct travelers go down (Fig. 8a); and the toll just offsets the time gain for use of the direct route: $time_{direct} + \alpha \cdot toll \approx time_{alternat.}$; remember that $\alpha = 1$.

Vehicle-to-vehicle fluctuations (Fig. 8b) for the use of the alternative road go up from ca. 2% to around 12%, and for the use of the direct road from ca. 11% to around 42%. Moreover, the *day-to-day* fluctuations also seem to go up in all measurements.

All this is in agreement with our intuition that traffic management can indeed make traffic more efficient, but may in addition lead to higher fluctuations and, as a consequence, lower predictability, since the system is driven closer to capacity and thus to the edge of chaos.

One should distinguish between two different kinds of fluctuations: Fluctuations due to the dynamics, and fluctuations due to the learning. The fluctuations in the latter might be due to the specifics of the chosen learning scheme, especially the lack of historic information beyond the last day. More realistic assumptions about the learning and en-route information are claimed to avoid that [39]. However, the results for the vehicle-to-vehicle fluctuations (i.e. the σ as defined in the text) only depend on the fact that the traffic density is driven towards the critical value. A less fluctuating learning scheme should therefore even *increase* our values for σ.

5. Conclusion

We started out establishing/reviewing some facts for a simple closed-loop single-lane system:

- Traffic at maximum capacity is in a regime which is critical up to an upper cut-off.
- This upper cut-off depends on the probability p_{spont} for the spontaneous initiation of a jam.
- The predictability of travel times sharply decreases when the density goes above this point.

This leads to the observation that advanced flow control will not only affect traffic flow, but will moreover drive large portions of the system towards the critical regime. The main reason for this is that the most efficient use of a traffic system takes place when all parts operate at densities at or below capacity. Systems designed for the management of traffic flows will reroute traffic from overcrowded roads to undercrowded ones, thus driving both closer to criticality. (We use criticality in this text even for the "not truly" critical situations, as discussed in the text.) Once traffic is near the critical region, further control inputs will have very unpredictable consequences.

More precisely, the following occurs. If one assumes complete information and rational decisions by everybody, the traffic will aim towards a Nash Equilibrium (NE). As nowadays drivers do not have complete information, we assume that they do something like bias their decisions towards "safe" routes, preferring e.g. shorter routes over longer ones even if both yield the same travel time.

Advanced Traveler Information Systems (ATIS) [27, 28] are developed to enhance the amount of information available. As explained above, this will push the traffic system closer to the NE and therefore—because it spreads traffic out over the network—closer to criticality.

Moreover, traffic management will aim beyond the NE towards a System Optimum (SO). A necessary condition for a SO is that no part of it is operating above the density ρ^* of maximum throughput, which will drive the system again closer to criticality.

This implies that the approximation of deterministic, predictable traffic patterns would be less and less correct the more one approaches high performance of the traffic system. In consequence, traffic assignment methods based on relaxation to equilibrium would no longer be meaningful: The changes in the traffic patterns due to one relaxation step would get lost in the changes due to the inherently fluctuating dynamics, and the algorithm would never converge. An open question is inhowfar one

can replace the equilibrium quantities by statistical averages (e.g. many Monte-Carlo runs); this is a topic of future research.

One envisaged way [34] of reaching the SO is to give each driver individual route guidance instead of complete traffic information. If one doubts that this will lead to high user acceptance, then congestion pricing seems to be the only alternative. Our simulation results support the idea that already locally operating agents can achieve this in an efficient manner.

In the text, we discuss the case of tolling on a specific road segment upstream of "the" bottleneck. This demands prior knowledge about the system.

However, one can imagine a completely local algorithm in the following way (see also [40]): Assume that *every* road segment in the system is operated by a simple economic agent. This agent wants to keep the operation of the segment as efficient as possible, and the only measure she has is to go up or down with the toll. The agent knows the performance characteristics (i.e. throughput q as a function of density ρ) of her segment, and from this she obtains the density which corresponds to maximum flow and therefore to maximum road performance. The agent then tries to keep the density on her segment at this particular density, increasing the toll when the density becomes too high, and else decreasing it. In a real network, we would expect that the toll for most segments turns out to be zero.

This tolling scheme gives the impression that every agent locally drives her segment towards criticality ($=$ maximum flow), but the situation is more complicated. In most cases, it is not the traffic *inside* bottlenecks which is tolled, but the overcrowded segments *upstream of* the bottlenecks. But because of the bottleneck, these upstream segments usually cannot operate at maximum throughput: As soon as the incoming flow is more than the bottleneck capacity, dense traffic builds up, and the segment switches from operation far below to far above the critical point (see text). Nevertheless, our results show that this still leads to having more parts of the network near criticality, as a result of collective effects.

In an economic context, we therefore have a local aiming for high performance, which happens to coincide with criticality. But even though the criticality very often cannot be reached locally by this mechanism, it drives the *global* system closer towards criticality: Local maximization of efficiency leads to global criticality [41].

Or in short: The fact that, in a complex system, high performance often has the downside of high variability seems also to be true in transportation systems.

Acknowledgments

KN is also a member of the "Graduiertenkolleg Scientific Computing Köln–St. Augustin". We thank S.R. Nagel, D. Stauffer, and others for making us aware that phase transitions which are not "truly" critical are still interesting and maybe even more relevant for reality. Nick Vriend, Kay Axhausen, and the anonymous referees provided helpful comments. We are especially grateful to Chris Barrett for carefully reading and commenting on the first draft. ZPR Köln and TSA-DO/SA (LANL) provided most of the computer time.

References

[1] Meyer, M.D., and E.J. Miller. 1984. *Urban Transportation Planning*. McGraw-Hill Series in Transportation.

[2] Cohen, J., and F. Kelly. 1990. A Paradox of Congestion in a Queuing Network. *J. Appl. Probablilty* 27:730-734.

[3] Arnott, R., A. De Palma, and R. Lindsay. 1993. A structural model of peak-period congestion: A traffic bottleneck with elastic demand. *The American Economic Review* 83(1):161-179.

[4] TRANSIMS, The TRansportation ANalysis and SIMulation System project at the Los Alamos National Laboratory.

[5] Rickert, M. *Outflow from 2-lane traffic simulations.* In preparation.

[6] Rickert, M. 1994. *Simulationen zweispurigen Autobahnverkehrs auf der Basis von Zellularautomaten.* Master Thesis, Univ. of Cologne.

[7] Wolfram, S. 1986. *Theory and Applications of Cellular Automata.* Singapore: World Scientific.

[8] Stauffer, D. 1991. Computer simulations of cellular automata. *J. Phys. A* 24:909-927.

[9] Nagel, K., and A. Schleicher. 1994. Microscopic traffic modeling on parallel high performance computers. *Parallel Comput.* 20:125-146.

[10] Landau, L.D., and E.M. Lifshitz. 1986. Statistical physics, Course in Theoretical Physics, Vol 5. Oxford: Pergamon Press.

[11] Wylie, B.J.N, D. McArthur, and M.D. Brown. 1992. PARAMICS parallelisation schemes. EPPC-PARAMICS-CT.10, Edinburgh.

[12] Rathi, A.K., and A.J. Santiago. 1990. The new NETSIM simulation program. *Traffic Engineering + Control* May 1990, 317-320.

[13] Nagel, K., and M. Schreckenberg. 1992. A cellular automaton model for freeway traffic. *J. Phys. I France* 2:2221.

[14] Schadschneider, A., and M. Schreckenberg. 1993. Cellular automaton models and traffic flow, *J. Phys. A* 1993, 26:L679.

[15] Nagatani, T. 1993. Self-organization and phase transition in traffic-flow model of a two-lane roadway. *J. Phys. A* 26:L781-L787.

[16] Vilar, L.C.Q., and A.M.C. de Souza. 1994. Cellular Automata Models for General Traffic Conditions on a Line. Preprint.

[17] Kinzel, W. 1983. Directed percolation. In *Percolation structures and processes*, edited by G. Deutscher, R. Zallen, and J. Adler. A. Hilger.

[18] Bollobas, B. 1985. *Random graphs.* London: Academic Press.

[19] Nagel, K. 1994. Life-times of simulated traffic jams, *Int. J. Mod. Physics C.* In press.

[20] Stauffer, D., and A. Aharony. 1992. *Introduction to percolation theory.* London: Taylor & Francis.

[21] Nagel, K., M. Paczuski, P. Bak. 1994. In preparation.

[22] Stanley, E.H., D. Stauffer, J. Kertész, and H.J. Herrmann. 1987. Dynamics of spreading phenomena in two-dimensional Ising models. *Phys. Rev. Lett.* 95(20):2326-2328.

[23] Langton, C. G. 1992. Life at the edge of chaos. In *Artificial Life II*, edited by C. Langton et al. Santa Fe Institute Studies in the Science of Complexity, Vol. 10. Redwood City, CA: Addison-Wesley.

[24] Bagnoli, F., R. Rechtman, and S. Ruffo. 1992. Damage spreading and Lyapunov exponents in cellular automata. *Phys. Lett. A* 172:34-38.

[25] Mitchell, M., J.P. Crutchfield, P.T. Hraber. 1994. Dynamics, computations, and the "Edge of Chaos": A re-examination. In *Integrative Themes*, edited by G. Cowan, D. Pines, and D. Melzner. Santa Fe Institute Studies in the Sciences of Complexity, Vol. 19, Reading, MA: Addison-Wesley.

[26] de Sousa Vieira, M., G.L. Vasconcelos, and S.R. Nagel. 1993. Dynamics of spring-block models: Tuning to criticality. *Phys. Rev. E* 47:R2221.

[27] Hall, R.W. 1993. Non-recurrent congestion: How big is the problem? Are traveler information systems the solution?. *Transpn. Res. C* 1(1):89.

[28] IVHS AMERICA. 1993. *Surface transportation: Mobility, technology, and society. Proceedings of the IVHS AMERICA 1993 annual meeting.* Washington, D.C.: IVHS AMERICA.

[29] Biham, O., A. Middleton, and D. Levine. 1992. Self-organization and a dynamical transition in traffic-flow models. *Phys. Rev. A* 46:R6124.

[30] Nagatani, T. 1993. Jamming transition in the traffic-flow model with two-level crossings. *Phys. Rev. E* 48(5):3290-3294.

[31] Cuesta, J.A., F.C. Martínez, J.M. Molera, and A. Sánchez. 1993. Phase transitions in two-dimensional traffic flow models. *Phys. Rev. E* 48(N6):R4175-R4178.

[32] Schütt, H. 1991. *Entwicklung und Erprobung eines sehr schnellen, bitorientierten Verkehrssimulationssystems für Straßennetze.* Schriftenreihe

der AG Automatisierungstechnik TU Hamburg-Harburg No. 6.

[33] Catoni, S., and S. Pallottino. 1991. Traffic Equilibrium Paradoxes. *Transp. Sc.* 25:240-244.

[34] Mahmassani, H.S., and S. Peeta. 1993. Network performance under system optimal and user equilibrium assignments: Implications for Advanced Traveler Information Systems. *Transportation Research Record* 1408:83-93.

[35] Simão, H.P., and W.B. Powell. 1992. Numerical methods for simulating transient, stochastic queueing networks. *Transpn. Sci.* 26:296.

[36] Mahmassani, H.S., G.-L. Chang, and R. Herman. 1986. Individual Decisions and Collective Effects in a Simulated Traffic System. *Transp. Sc.* 20(4):258-271.

[37] Axelrod, R. 1984. *The Evolution of Cooperation.* New York: Basic Books.

[38] Hall, F.L., B.L. Allen, and M.A. Gunter. 1986. Empirical analysis of freeway frow-density relationsships. *Transpn. Res. A* 20A(3):197–210.

[39] Axhausen, K. 1994. Personal communication.

[40] MacKie-Mason, J.K., and H.R. Varian, Pricing the internet, preprint 1994.

[41] Bak, P., C. Tang, and K. Wiesenfeld. 1988. Self-organized criticality. *Phys. Rev. A* 38:368.

A Phase Transition in Random Boolean Networks

James F. Lynch *
Department of Mathematics and Computer Science
Clarkson University
Potsdam, N. Y. 13699-5815
jlynch@sun.mcs.clarkson.edu

Abstract

A random boolean network consists of interconnected boolean gates where the inputs, the boolean function, and the initial state of each gate are chosen randomly. In this article, each gate has two inputs. It is shown that the behavior of these random networks depends on a simple parameter derived from the probability distribution used in assigning boolean functions to gates. When the parameter is greater than 0, with probability asymptotic to 1, the random networks exhibit very stable behavior: almost all of the gates stabilize, almost all of them are very weak, i.e., the effect of perturbing any one of them lasts only finitely many steps and thus does not change the state cycle that is entered, and the state cycle is bounded in size. When the parameter is less than 0, the random networks behave chaotically: the fractions of stable gates and very weak gates are strictly less than 1. Random networks on the border where the parameter is 0 show the onset of chaos: with probability asymptotic to 1, almost all of the gates are still stable and very weak, but the state cycle size is unbounded. In fact, the average state cycle size is superpolynomial in the number of gates.

1   Introduction

One of the oldest and best known models of self organization is the random boolean network of S. Kauffman (1970). It was originally developed as a model of the genome of a cell, but has become a paradigm for the emergence of order in complex systems. Virtually all of the work on random boolean networks has been computer simulations; only recently have mathematical results been proven about them. This article will summarize these results, present new unpublished ones, and describe the proof techniques where space permits. In a broad sense, the theorems support the conclusions drawn by many years of computer simulation. They show that many classes of random boolean networks exhibit certain kinds of stability. On the other hand, it will be shown that there are other classes of random boolean networks that behave chaotically; and, perhaps most significantly,

there is a sharp transition between the classes of stable networks and the chaotic ones. Also surprisingly, the class of networks originally investigated by Kauffman lies precisely on this threshold.

A simple parameter, the difference of two probabilities, governs the three modes of behavior. If the parameter is greater than 0, then the random networks are stable with high probability, and when it is less than 0, they are chaotic. Random networks on the border where the parameter is 0 display the onset of instability. Similar thresholds have been observed in other types of complex systems (Derrida and Pomeau, 1986, Derrida and Weisbuch, 1986, Langton, 1990, Schulman and Seiden, 1978), and this phenomenon may be widespread. The results described in this article appear to be the first actual proofs about this kind of transition in behavior for random boolean networks.

2   Discussion of Results

A boolean network is a collection of interconnected boolean gates that operate synchronously on a discrete time scale. That is, at each time step t, each gate is in a state 0 or 1. The state of the gate at the next step $t+1$ is determined by the states of the gates that have connections to the gate and by the boolean function that the gate computes. (All the connections are assumed to be one way, i.e., directed.) For example, suppose gate i receives connections from gates j_1, j_2, j_3, and the boolean function computed by i is the OR of its three inputs. Then i will be in state 1 at time $t + 1$ if and only if at least one of j_1, j_2, or j_3 is in state 1 at time t. The state of the network at any particular time is the list of the states of its gates at that time. The parameter n will always be the number of gates in such a network and k will be the maximum number of inputs to any gate. Of course, all gates can be forced to have exactly k inputs simply by adding dummy inputs to gates with fewer than k inputs.

Because the evolution of the genotype of a cell is subjected to unpredictable events, a model of the genome should try to incorporate this randomness. At present, it is not possible to describe the probability distributions that have led to actual genomes in nature. Since these distributions are not known, the computer models of a genome are usually based on the following kind of

---

*Research supported by NSF Grant CCR-9006303.

simple construction. A random boolean network results from three independent random processes. First, the connections are constructed by choosing, for each gate, its k inputs from the $\binom{n}{k}$ possibilities. Next, each gate is assigned a boolean function from the 2^{2^k} possibilities. Last, an initial state for the network is chosen from the 2^n possibilities. Most of the experiments involving random boolean networks have used uniform distributions in each of the three steps, i.e., all choices are equally likely. In particular, each boolean function has probability $1/2^{2^k}$ of being assigned to a particular gate.

If such a network is allowed to run, since there are only finitely many states, it will eventually repeat a state. From that point on, it will cycle through a sequence of states indefinitely. This is called a state cycle. Of course, the time at which the state cycle is entered and the size of the state cycle could be as large as 2^n or some other rapidly growing function of n, which is enormous even for rather modest values of n. It appears that k is very important in determining the behavior of a random boolean network. Specifically, the stability of a random network seems to depend critically on k. Three measures of stability that have been considered are:

1. The proportion of gates that stabilize, i.e. eventually stop changing.

2. The proportion of weak gates, i.e., gates that can be perturbed without affecting the state cycle that is entered.

3. The size of the state cycle that the system eventually enters.

The second and third of these measures are finite discrete analogues of criteria that are used to characterize chaos in dynamical systems. A small proportion of weak gates is similar to sensitivity to initial conditions, and a large state cycle is similar to nonperiodicity. A fourth notion that has been used implicitly by Luczak and Cohen, 1991, and Lynch, 1993a and 1993b, to analyze weak gates, will be referred to explicitly in this article. A gate is very weak if the the effect of perturbing it lasts only finitely many steps. Since very weak implies weak, a large proportion of very weak gates is a clear indication of stability, while a small proportion gives some evidence of instability.

When $k = n$, a random network will almost surely exhibit chaotic behavior. This is not surprising since such a network is equivalent to a random functional digraph on 2^n elements, where each element corresponds to a sequence of n 0's and 1's. It can be shown that as n increases, almost all such networks behave chaotically in all three senses. Elementary arguments show that most of the gates do not stabilize and are not weak. A well-known result from Harris, 1960 states that the average size of a cycle in a random functional digraph on m elements is asymptotic to $\sqrt{m/(8\pi)}$. Taking $m = 2^n$, it immediately follows that the average size of the state cycle in a $k = n$ random network is on the order of $2^{n/2}$. As k decreases, computer simulations indicate that the behavior of the random net remains chaotic until $k = 2$, when it changes dramatically. Most of the gates stabilize and are weak, and the state cycle size is much smaller than $2^{n/2}$. In fact, Kauffman has conjectured that the average state cycle size is on the order of $\sqrt{n}$. He has also investigated $k = 2$ networks without constant gates, where each of the remaining gates has probability $1/14$. The simulations show only a slight increase in state cycle size.

In spite of extensive experimental work on these networks, comparatively little has actually been proven about them. The first article containing formal proofs of stability in the $k = 2$ model is Łuczak and Cohen, 1991. They show that as $n \to \infty$, for almost all random boolean networks with n gates, the number of stable gates and the number of very weak gates is asymptotic to n. They also give a nontrivial upper bound on the state cycle size: with probability asymptotic to 1, it is less than $2^{n\omega(n)/\log n}$ for any increasing function $\omega(n)$. In Lynch, 1993a, it was shown that by giving a slight bias to the probability of certain of the boolean functions assigned to the gates (on the order of $\log \log n/\log n$), for almost all random boolean networks with n gates, the state cycle size can be bounded above by n^γ, for some γ. However, the proof failed when the bias was reduced to 0, i.e., for the random model originally studied by Kauffman. This suggested two lines of research. First, a more extensive analysis of random boolean networks with nonuniform probabilities of the boolean functions might be possible. This could be a step toward understanding more realistic models of the genome. Also, the breakdown of the proof at the uniform distribution hinted at a threshold phenomenon.

Treating all 16 of the two argument boolean functions individually seems to be a complex undertaking. A classification of the boolean functions due to Kauffman (1990) has proven useful. He referred to certain boolean functions as canalyzing. This will be defined precisely in the next section, but for now it suffices to note that among the canalyzing functions are the two constant functions; i.e., the function that outputs 0 regardless of its inputs and its negation that always outputs 1. Further, among the two argument boolean functions, there are only two noncanalyzing functions: the EQUIVALENCE function that outputs 1 if and only if both of its inputs have the same value, and its negation the EXCLUSIVE OR.

The remaining two argument boolean functions are nonconstant canalyzing functions. Typical examples are the OR and the AND functions. With the uniform distribution on assigning boolean functions to gates, i.e., each function has probability $1/16$, an average of $7/8$ of the gates of a random boolean network will be canalyzing. Kauffman (1990) suggested that the stability observed in his simulated networks was due to the presence of many canalyzing gates. He conjectured that other classes of random networks with a high probability of canalyzing gates would also be stable. However, the results in this article show that it is not the proportion of canalyzing gates that affect the stability of the network; rather it is the relative proportions of constant to noncanalyzing gates that is the critical factor.

Let a (respectively c) be the probability that the boolean function assigned to a gate is constant (respectively noncanalyzing). In Lynch, 1993b, it was shown that when $a > c$, with probability asymptotic to 1, the random boolean network is very stable in all three senses: almost all of the gates are stable and very weak, and the state cycle size is bounded. For completeness, these results will be summarized in §5.

Next, §6 explains why $a = c \neq 0$ is a border between stable and chaotic behavior. This includes Kauffman's well-studied $k = 2$ model as a special case $a = c = 1/8$. When $a = c$, the first two kinds of stability still hold (although the bounds here are not as tight), but the state cycle size is unbounded for almost all networks. In fact, the average state cycle size is greater than any polynomial in n. Thus, the network still appears to be stable when viewed locally, i.e. at the level of a typical gate, but large state cycles are a global symptom of the beginning of instability.

Finally, §7 shows that $a < c$ characterizes a region of unstable random networks. In this case, almost all of the networks have significant fractions of unstable gates and gates that are not very weak. Good bounds on the state cycle size are not known for this case.

It is worth comparing these results with Kauffman's computer simulations. When all 16 boolean functions are equally likely, the simulations indicate a rather slow growth rate in the average state cycle size. But as was pointed out, this is a special case of $a = c$, where the average state cycle size is superpolynomial in n. To reconcile this difference, it should be noted that $a = c$ is on the border of the region $a > c$, where state cycles are very small. Possibly networks of the size used in the simulations were just beginning to show the appearance of large state cycles. Another explanation may be that large state cycles are concentrated in a small fraction of the networks. This could still result in a large average size, even though most networks have small state cycles. The computer simulations also indicated that networks without constant gates, where the remaining 14 boolean functions were equally likely, also displayed stable behavior. Specifically, the average state cycle size was still small. This is a special case of $a < c$, but there is not such a clear distinction between the analytic results of this article and the simulations because there is no proof that the average state cycle size is large in this case.

Phase transitions have been studied in other models of complex systems, most notably cellular automata. These are lattices of gates in a Euclidean space. The inputs to a gate are usually in some small neighborhood of the gate, and each gate obeys the same transition rules. Langton (1990) discovered a remarkably simple parameter, λ, that characterizes stable and chaotic cellular automata. It is the ratio of transitions that enter the quiescent state to the total number of transitions. Furthermore, cellular automata that have the capability to perform complex computations appear to be near the border between stability and chaos. An example is Conway's game of Life, perhaps the most famous cellular automaton. Schulman

and Seiden (1978) have studied a stochastic version of this game. They introduced a stochastic component in the transition rules of the game, controlled by a parameter referred to as the temperature. Their experimental results indicate that there is a phase transition as the temperature passes through a certain value: the density of live cells changes from 0 to approximately 0.37. Other examples of cellular automata modeling a wide variety of physical processes may be found in Manneville et. al., 1989.

An interesting variant of boolean networks was introduced in Derrida and Pomeau, 1986. In their version, at each time step, the inputs and the boolean functions are randomly chosen before the system updates its state. Since the connections and functions in Kauffman's network do not change with time, they refer to his systems as quenched and theirs as annealed approximations. They consider the Hamming distance between iterations of two starting configurations. That is, given two starting configurations that differ in a certain number of gates, they study how this difference changes in successive iterations of those configurations. They show that k, the number of inputs to each gate, has a critical value at 2. When $k \leq 2$, the ratio of the distance between the configurations to the total number of gates converges to 0 with high probability, but it remains positive when $k > 2$. Proving similar facts for the quenched model seems quite difficult, but computer simulations of the quenched model show very close agreement to the theoretical results obtained for the annealed approximation. Thus, their results support the evidence that $k = 2$ is a critical value for stability in the quenched model. Of course, state cycles do not exist in the annealed approximation because the rules change at each step, so the annealed approximation cannot be used to analyze that kind of stability. A number of other questions were investigated by Derrida and Pomeau, and related results are contained in Derrida and Weisbuch, 1986.

Perhaps the most significant feature in the proofs of the results in this article is the use of combinatorial methods from random graph theory. This subject was invented by Erdős and Rényi (1959, 1960), and has become an active area of mathematics. The monograph of Bollobás (1985) is a standard reference. Random graph theory is concerned with the properties of large, randomly constructed finite structures. The most commonly studied type of random graph is defined in terms of its size (number of vertices) and a parameter known as the edge probability. This is a function $p(n)$ of the size n of the graph which takes values in the range 0 to 1. The random graph on n vertices is constructed by choosing independently for each pair of vertices, that there is an edge between them with probability $p(n)$. It is interesting to note that many important graph-theoretic properties have sharp thresholds. That is, changing the growth rate of $p(n)$ very slightly can change the asymptotic probability that a large random graph has the property from 0 to 1. For example, the probability that a random graph has at least five vertices that are connected to each other is asymptotic to 0 when $p(n) = n^{-\alpha}$ and $\alpha > 5/4$, but it is asymptotic to 1 when $\alpha < 5/4$. There is a similar threshold for the property that the graph is connected when $p(n) = \alpha \log n/n$. When $\alpha < 1$, almost surely the graph will not be connected, but when $\alpha > 1$, almost surely it will be connected. Thus it should not be surprising that the techniques of random graph theory are useful in studying complex systems consisting of many parts. In fact, this was predicted by Kauffman (1990): "the expected size and structure of the resulting forcing structures [connected structures of gates whose state cannot change after a certain number of steps] is a mathematical problem in random graph theory."

The proofs of this article involve forcing structures and the complementary notion of unforced structures. When $a > c$, the unforced structures must be small. This is the key fact used in proving that networks in this region are stable in all three senses. When $a = c$, the unforced structures become slightly larger. They are still small enough so that most gates are stable and very weak. However, this is the point at which many unstable structures start to appear. In particular, structures known as vortices appear, and there are enough of them to cause large state cycles.

3 Definitions

Precise definitions of the terms that were introduced in the previous section are now given. Let n be a natural number. A *boolean network* B with n gates is a triple $\langle D, F, x \rangle$ where D is a directed graph with vertices $1, \ldots, n$ (referred to as *gates*), $F = \langle f_1, \ldots, f_n \rangle$ is a sequence of boolean functions, and $x = (x_1, \ldots, x_n) \in \{0,1\}^n$ (the set of 0-1 sequences of length n). In this article, each gate will have indegree two, and each boolean function will have two arguments. Gate j is an *input* to gate i if (j, i) is an edge of D. B is a finite state automaton with state set $\{0,1\}^n$ and initial state x. The pair $\langle D, F \rangle$ defines the transition function of B in the following way. For each $i = 1, \ldots, n$ let $j_i < k_i$ be the inputs of i. Given $y = (y_1, \ldots, y_n) \in \{0,1\}^n$, $B(y) = (f_1(y_{j_1}, y_{k_1}), \ldots, f_n(y_{j_n}, y_{k_n}))$. That is, the state of B at time 0 is x, and if its state at time t is $y \in \{0,1\}^n$, then its state at time $t+1$ is $B(y)$.

The first set of definitions pertains to the aspects of stability that will be studied.

Definitions 3.1 *Let* $B = \langle D, F, x \rangle$ *be a boolean network.*

1. *Let* $B^t(x)$ *be the state of* B *at time* t, *and* $f_i^t(x)$ *the value of its* ith *component, or gate, at time* t.

2. *Since the number of states is finite, i.e.* 2^n, *there exist times* $t_0 < t_1$ *such that* $B^{t_0}(x) = B^{t_1}(x)$. *Let* t_1 *be the first time at which this occurs. Then* $B^{t+t_1-t_0}(x) = B^t(x)$ *for all* $t \geq t_0$. *The set of states* $\{B^t(x) : t \geq t_0\}$ *is the* state cycle *of* B, *to distinguish it from a cycle of* D *in the graph-theoretic sense.*

3. *Gate* i stabilizes *in* t *steps if for all* $t' \geq t$, $f_i^{t'}(x) = f_i^t(x)$.

4. *Gate* i *is* weak *if, letting* $\overline{x}^i$ *be identical to* x *except that its* ith *component is* $1 - x_i$,
$$\exists t_0 \exists d \forall t (t \geq t_0 \Rightarrow B^t(x) = B^{t+d}(\overline{x}^i)).$$

In other words, changing the state of i *does not affect the state cycle that is entered.*

5. *Gate* i *is* very weak *if*
$$\exists t (B^t(x) = B^t(\overline{x}^i)).$$

In other words, changing the state of i *causes only a temporary change in the state of the network.*

The next definitions describe a property of boolean functions that plays a key role in the characterization of the threshold between order and chaos.

Definitions 3.2 *Let* $f(x_1, x_2)$ *be a boolean function of two arguments.*

1. *The function* f *depends* on argument x_1 *if for some* $v \in \{0,1\}$, $f(0, v) \neq f(1, v)$. *A symmetric definition applies when* f *depends on* x_2. *Similarly, if* $\langle D, F, x \rangle$ *is a boolean network,* $f_i = f$, *and the inputs of gate* i *are* $j_{i,1}$ *and* $j_{i,2}$, *then for* $m = 1, 2$, i *depends on* $j_{i,m}$ *if* f *depends on* x_m.

2. *The function* f *is said to be* canalyzing *if there is some* $m = 1$ *or* 2 *and some values* $u, v \in \{0,1\}$ *such that for all* $x_1, x_2 \in \{0,1\}$, *if* $x_m = u$ *then* $f(x_1, x_2) = v$. *Argument* x_m *of* f *is said to be a* forcing argument *with* forcing value u *and* forced value v. *Likewise, if* $\langle D, F, x \rangle$ *is a boolean network and* f_i *is a canalyzing function with forcing argument* x_m, *forcing value* u *and forced value* v, *then input* $j_{i,m}$ *is a* forcing input *of gate* i. *That is, if the value of* $j_{i,m}$ *is* u *at time* t, *then the value of* i *is guaranteed to be* v *at time* $t+1$.

All of these definitions generalize immediately to boolean functions of arbitrarily many arguments. In the case of two argument boolean functions, the only noncanalyzing functions are EQUIVALENCE and EXCLUSIVE OR. The two constant functions $f(x, y) = 0$ and $f(x, y) = 1$ are trivially canalyzing, as are the four functions that depend on only one argument:

$$f(x, y) = x,$$
$$f(x, y) = \neg x,$$
$$f(x, y) = y, \text{ and}$$
$$f(x, y) = \neg y.$$

The remaining eight boolean functions of two arguments are canalyzing, and they are all similar in the sense that both arguments are forcing with a single value, and there is one forced value. A typical example is the OR function. Both arguments are forcing with 1, and the forced value is 1.

The notion of forcing, defined next, is a combinatorial condition that is useful in characterizing stability. It depends on D and F, but not on x.

Definition 3.3 *Again,* $\langle D, F, x \rangle$ *is a boolean network. The definition of what it means for gate* i *to be* forced *to a value* v *in* t *steps is given by induction on* t.

If f_i *is the constant function* $f(x_1, x_2) = v$, *then* i *is forced to* v *in* t *steps for all* $t \geq 0$.

If the inputs $j_{i,1}$ *and* $j_{i,2}$ *of* i *are forced to* u_1 *and* u_2 *respectively in* t *steps, then* i *is forced to* $f_i(u_1, u_2)$ *in* $t+1$ *steps.*

If f_i is a canalyzing function with forcing argument x_m, forcing value u, and forced value v, and $j_{i,m}$ is forced to u in t steps, then i is forced to v in $t + 1$ steps.

By induction on t it can be seen that if i is forced in t steps, then it stabilizes for all initial states x in t steps. It should be noted that forcing is a much stronger condition than stability. It is easy to find networks with gates that stabilize but are not forced. See for example fig. 1 in Lynch, 1993a.

The following combinatorial notions will be used in characterizing forcing structures. It is assumed that the reader is familiar with the basic concepts of graph theory (see e.g. Harary, 1969). Unless otherwise stated, *path* and *cycle* shall mean directed path and cycle in the digraph D.

Definitions 3.4

1. For any gate i in D with inputs $j_{i,1}$ and $j_{i,2}$, let

$$S_0^-(i) = \{i\} \text{ and}$$
$$S_{d+1}^-(i) = S_d^-(j_{i,1}) \cup S_d^-(j_{i,2}).$$

2. Then

$$N_d^-(i) = \bigcup_{c \leq d} S_c^-(i).$$

Thus $N_d^-(i)$ is the set of all gates that are connected to i by a path of length at most d.

3. If I is a set of gates, then $N_d^-(I) = \cup_{i \in I} N_d^-(i)$.

4. In a similar way, one can define $S_d^+(i)$ and $N_d^+(i)$, the set of all gates reachable from i by a path of length at most d.

For obvious reasons, $N_d^-(i)$ is referred to as the in-neighborhood of i of radius d. Note that whether i is forced in d steps is completely determined by the restriction of D and F to $N_d^-(i)$.

The asymptotic behavior of *random* boolean networks will be examined. For each boolean function f of two arguments, there is an associated probability $a_f \in [0, 1]$, where $\sum_f a_f = 1$. The random boolean network with n gates is the result of three random processes. First, a random digraph where every gate has indegree two is generated. Independently for each gate, its two inputs are selected from the $\binom{n}{2}$ equally likely possibilities. Next, each gate is independently assigned a boolean function of two arguments, using the probability distribution $\langle a_f : f : \{0, 1\}^2 \to \{0, 1\} \rangle$. Lastly, the initial state x is chosen using the uniform distribution on $\{0, 1\}^n$. Let $\tilde{B} = \langle \tilde{D}, \tilde{F}, \tilde{x} \rangle$ denote a random boolean network generated as above. For any properties $\mathcal{P}$ and $\mathcal{Q}$ pertaining to boolean networks, let $\text{pr}(\mathcal{P}, n)$ be the probability that the random boolean network on n gates has property $\mathcal{P}$ and $\text{pr}(\mathcal{P}|\mathcal{Q}, n)$ for the conditional probability that $\mathcal{P}$ holds, given that $\mathcal{Q}$ holds. Usually, the n in these expressions will be omitted since it will be understood. Some of the properties that will be investigated depend only on D and F. In that case, the expression describing $\mathcal{P}$ will involve $\langle \tilde{D}, \tilde{F} \rangle$ instead of $\tilde{B}$, and pr can be regarded as the probability measure on random $\langle \tilde{D}, \tilde{F} \rangle$. Similar notation will be used for properties that depend only on

D. Random variables will be denoted by boldface capital letters, and $\mathbf{E}(\mathbf{X})$ will be the expectation of $\mathbf{X}$.

The two argument boolean functions can be classified as follows:

1. $\mathcal{A}$ contains the two constant functions.

2. $\mathcal{B}_1$ contains the four canalyzing functions that depend on one argument.

3. $\mathcal{B}_2$ contains the eight canalyzing functions that depend on both arguments.

4. $\mathcal{C}$ contains the two noncanalyzing functions.

Then the probabilities that a gate is assigned a function in each of the categories are:

$$a = \sum_{f \in \mathcal{A}} a_f$$
$$b_1 = \sum_{f \in \mathcal{B}_1} a_f$$
$$b_2 = \sum_{f \in \mathcal{B}_2} a_f$$
$$c = \sum_{f \in \mathcal{C}} a_f$$

Lastly, let $\mathcal{B} = \mathcal{B}_1 \cup \mathcal{B}_2$ and $b = b_1 + b_2$, the probability that a gate is assigned a nonconstant canalyzing function. Throughout the rest of the article, the following symmetry conditions are assumed:

$$a_{f(x,y)} = a_{f(y,x)} \text{ for all } f \in \mathcal{B}_1$$
$$a_{f(x,y)} = a_{f(\neg x, \neg y)} \text{ for all } f \in \mathcal{B}_2$$
$$a_{f(x,y)} = a_{\neg f(x,y)} \text{ for all } f.$$

Also, log shall always mean $\log_2$, and ln is the natural logarithm.

4 Fundamental Lemmas

The lemmas in this section are basic to proving the results on stability in all three regions of the space of random boolean networks. A key idea, first stated in Łuczak and Cohen, 1991, is that almost all of the gates have sufficiently large neighborhoods that are trees. The following version of this fact will be used.

Lemma 4.1 *For any positive α and unbounded increasing function $\omega(n)$,*

$$\lim_{n \to \infty} \text{pr}(\tilde{D} \text{ has at most } \omega(n)(\log n)^3 n^{2\alpha}$$

gates i such that $N_{\alpha \log n}^-(i)$ is not a tree) $= 1$.

The same is true for $N_{\alpha \log n}^+$.

Proof. For each gate i, let $\mathbf{X}_i$ be the indicator random variable that is 1 if and only if $N_{\alpha \log n}^-(i)$ is not a tree, and let $\mathbf{X} = \sum_{i=1}^n \mathbf{X}_i$. If $\mathbf{X}_i = 1$, then there exists a path P of length $r \leq \alpha \log n$ beginning at some gate k and ending at i and another path Q of length s, $1 \leq s \leq \alpha \log n$, beginning at k, disjoint from P except at k and its other endpoint, which must be in P. There are no more than n^r ways of choosing P and no more than $n^{s-1} \times r$ ways of choosing Q. The probability of any such choice is bounded above by $(2/n)^{r+s}$. Therefore

$$(4.1) \qquad \mathbf{E}(\mathbf{X}_i) \leq \sum_{r=0}^{\alpha \log n} \sum_{s=1}^{\alpha \log n} 2^{r+s} r n^{-1}$$
$$\leq (\alpha \log n)^3 n^{2\alpha-1}.$$

Then $\mathbf{E}(\mathbf{X}) \leq (\alpha \log n)^3 n^{2\alpha}$, and the lemma follows by Markov's inequality. A similar argument applies to $N^+_{\alpha \log n}$. $\square$

The intended use of this lemma is when $\alpha < 1/2$. Then almost all gates have treelike neighborhoods of radius $\alpha \log n$.

Another basic result is a recurrence relation for the probability that a gate is forced, given that its in-neighborhood is treelike.

Lemma 4.2 *For $d \geq 0$ and $v \in \{0, 1\}$ let*

$$p_d(v) = \text{pr(gate } i \text{ is forced to } v \text{ in } d \text{ steps}$$
$$|N^-_d(i) \text{ is a tree) and}$$
$$p_d = p_d(0) + p_d(1).$$

Then

$$(4.2) \qquad p_d(0) = p_d(1),$$

and p_d satisfies the following recurrence.

$$p_0 = a \text{ and}$$
$$(4.3) \qquad p_{d+1} = a + bp_d + cp_d^2.$$

Proof. Equation (4.2) is proven by induction on d. The recurrence relation will be a byproduct. For $d = 0$, it is clear since $p_0(0) = p_0(1) = a/2$.

Next, equation (4.2) is proven for $d + 1$, assuming it is true for d. Let j and k be the two inputs of i. Since $S^-_{d+1}(i)$ induces an acyclic subgraph, so do $S^-_d(j)$ and $S^-_d(k)$. The possible ways that i can be forced to v are:

(1) It is assigned the constant function $f(x, y) = v$.

(2) It is assigned some function $f \in \mathcal{B}_1$, and the input on which i depends is forced in d steps to the value that forces f to v.

(3) It is assigned some function $f \in \mathcal{B}_2$, v is the forced value of f and at least one of its inputs is forced in d steps to the value that forces f to v, or v is not the forced value of f but j and k are forced to values u and w such that $f(u, w) = v$.

(4) It is assigned some function $f \in \mathcal{C}$, and j and k are forced to values u and w such that $f(u, w) = v$.

The proof of equation (4.2) will be finished by deriving expressions for the probability of each of the four cases, and showing they are the same for $v = 0$ and 1. The probability of Case (1) is $a/2$.

If $f \in \mathcal{B}_1$, say $f(x, y) = x$, the probability that i is forced to v in $d + 1$ steps is $p_d(v)$. The other choices for

$f \in \mathcal{B}_1$ are symmetric, and by the induction assumption the probability of Case (2) is $b_1 p_d(0)$.

In Case (3), it can be observed that the eight functions in $\mathcal{B}_2$ may be partitioned into four pairs, each of the form $\{f, \neg f\}$. Take a typical $f \in \mathcal{B}_2$, say the OR function. Then 0 is not the forced value of f, but it is for $\neg f$. The probability that f is forced to 0 is

$$p_d(0)^2,$$

and the probability that $\neg f$ is forced to 0 is

$$(1 - p_d(1))p_d(1) + p_d(1)(1 - p_d(1)) + p_d(1)^2.$$

Summing these two probabilities and using symmetry and the induction hypothesis, the probability that the function assigned to i is f or $\neg f$ and i is forced to 0 in $d + 1$ steps is

$$a_f \times p_d(0)^2 + a_{\neg f} \times (2p_d(1) - p_d(1)^2) = (a_f + a_{\neg f}) \times p_d(0).$$

Summing over all four pairs of functions, the probability of Case (3) is $b_2 p_d(0)$. The argument when $v = 1$ is symmetric.

Lastly, let $f \in \mathcal{C}$, say f is EXCLUSIVE OR. The probability that i is forced to 0 in $d+1$ steps is $p_d(0)^2 + p_d(1)^2$, and the probability that i is forced to 1 in $d + 1$ steps is $2p_d(0)p_d(1)$. By the induction assumption, these probabilities are equal. Similar reasoning applies when f is EQUIVALENCE. Thus the probability of Case (4) is $2cp_d(0)^2$ regardless of whether v is 0 or 1.

Summing over all four cases, for $v = 0$ or 1,

$$p_{d+1}(v) = \frac{a}{2} + bp_d(0) + 2cp_d(0)^2,$$

proving equation (4.2). Furthermore,

$$p_{d+1} = a + bp_d + cp_d^2,$$

proving the recurrence (4.3). $\square$

Fig. 1 gives a clue to the reason why the difference of a and c is important to the stability of random boolean networks. The fixed points of the recursion (4.3) are a/c and 1. Part (a) illustrates a typical case when $a > c$. Since p_d is a probability, $p_d \leq 1$, and it cannot approach the larger fixed point a/c. The other fixed point 1 attracts all values of p_d. In fact, the convergence of p_d to 1 is geometric. When $a = c$, the fixed points coincide, as shown in part (b). The convergence in this case is not as rapid, but is still sufficiently fast.

Lemma 4.3 *Let d be a natural number.*
(a) If $a > c$, then

$$p_d > 1 - (1 - a + c)^{d+1}.$$

(b) If $a = c \neq 0$, then

$$p_d \geq 1 - \frac{1}{ad}.$$

Proof.
(a) Let $q_d = 1 - p_d$. By the recurrence in lemma 4.2,

$$q_0 = 1 - a < 1 - a + c \text{ and}$$
$$1 - q_{d+1} = a + b(1 - q_d) + c(1 - q_d)^2.$$

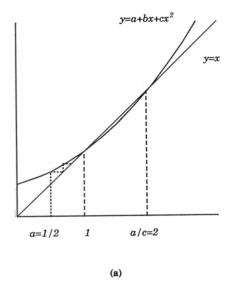

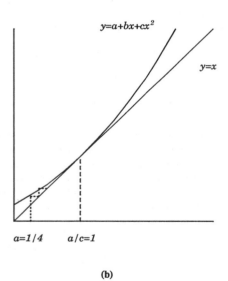

FIGURE 1. Examples of the convergence of p_d. The dotted line ···· indicates the successive iterations of (4.3) from $p_0 = a$ towards 1.

(a) $a = 1/2$, $c = 1/4$.

(b) $a = c = 1/4$.

Since $a + b + c = 1$,

$$q_{d+1} = (1 - a + c)q_d - cq_d^2$$
$$\leq (1 - a + c)q_d$$

and the result follows by induction on d.

(b) Let $q_d = 1 - p_d$. Then from (4.3), the recurrence for q_d is

$$(4.4) \qquad q_{d+1} = q_d - aq_d^2$$

Letting $r_d = 1/q_d$ and using induction on d, the proof will be finished by showing that $r_d \geq ad$. When $d = 0$, this is evident. By (4.4),

$$\frac{1}{r_{d+1}} = \frac{1 - a/r_d}{r_d}$$

and so

$$r_{d+1} = \frac{r_d}{1 - a/r_d}$$
$$\geq r_d + a,$$

which establishes the induction step. $\square$

5   The Region of Stability

This section describes random boolean networks in the region $a > c$. These theorems and their proofs were originally presented in Lynch, 1993b. They are included here in order to compare them with results on the behavior of random networks in the two other regions. The first theorem uses the basic facts established in the previous section, i.e., most gates have a sufficiently large in-neighborhood that is a tree, and such gates are forced with high probability. Since a forced gate must be stable, most gates stabilize.

Theorem 5.1 *For any positive $\alpha < 1/2$ there is a constant $\beta > 0$ such that*

$$\lim_{n \to \infty} \text{pr}(\tilde{B} \text{ has at least } n(1 - n^{-\beta})$$

gates that stabilize in $\alpha \log n$ steps) $= 1$.

A key idea used in analyzing the number of very weak gates and the state cycle size is that the unforced gates are arranged in small, isolated components. Further, these components have a very simple structure. They consist of single cycles with trees growing out of them. A direct consequence is that the unstable gates are also arranged in small isolated islands with the same structure. The derivations of the number of very weak gates and state cycle size rely on this structural characterization. It is also used to bound the number of states the network passes through before entering its state cycle (the tail length).

Theorem 5.2 *There exists $\alpha > 0$ such that*

$$\lim_{n \to \infty} \text{pr}(\tilde{B} \text{ has at least } n(1 - n^{-\alpha})$$

very weak gates) $= 1$.

Theorem 5.3 *For every $\epsilon > 0$ there is s such that*

$$\text{pr}(\tilde{B} \text{ has a state cycle larger than } s) < \epsilon.$$

Corollary 5.4 *Let $\omega(n)$ be any unbounded increasing function. Then*

$$\lim_{n \to \infty} \text{pr}(\tilde{B} \text{ has a state cycle larger than } \omega(n)) = 0.$$

Theorem 5.5 *There is a constant β such that*

$$\lim_{n \to \infty} \text{pr}(\tilde{B} \text{ has a tail longer than } \beta \log n) = 0.$$

6   The Edge of Chaos

In this section, it is assumed that $a = c$. Due to space limitations, full proofs are not given here. They are found in a preprint obtainable from the author or the Los Alamos Preprint Bulletin Board (adap-org/9309001). Theorems 6.1 and 6.2 show that random networks in this region still appear stable when viewed locally, i.e., most gates are stable and very weak. The proofs are similar to those of the theorems in the previous section. They rely on the facts that most gates are still forced quickly (recall fig. 1(b)), and the unforced gates are arranged in small isolated components.

Theorem 6.1 *Let $\alpha < 1/2$ and $\omega(n)$ be any unbounded increasing function. Then*

$$\lim_{n \to \infty} \text{pr}(\langle \tilde{D}, \tilde{F} \rangle \text{ has at least } n(1 - \omega(n)/\log n)$$

gates that stabilize in $\alpha \log n$ steps) $= 1$.

Theorem 6.2 *Let $\omega(n)$ be any unbounded increasing function. Then*

$$\lim_{n \to \infty} \text{pr}(\tilde{B} \text{ has at least } n(1 - \omega(n)/\log n)$$

very weak gates) $= 1$.

The next theorems show that the third measure of stability, the state cycle size, is much larger in this region than it was in the $a > c$ region, and thus it reveals the onset of chaos. Let the random variable $\mathbf{C}$ denote the size of the state cycle of $\tilde{B}$.

Theorem 6.3 *For any constant γ and sufficiently large n,*

$$\mathbf{E}(\mathbf{C}) > n^\gamma.$$

Let $\mathbf{E}(\mathbf{C}|\langle \tilde{D}, \tilde{F} \rangle)$ be the expected state cycle size of a random $\langle \tilde{D}, \tilde{F} \rangle$ averaged over all $x \in \{0,1\}^n$.

Theorem 6.4 *There is a constant $\gamma > 0$ such that*

$$\lim_{n \to \infty} \text{pr}(\mathbf{E}(\mathbf{C}|\langle \tilde{D}, \tilde{F} \rangle) \geq n^\gamma) = 1.$$

These theorems will follow from a key result (lemma 6.10) on the probability of existence of certain kinds of structures in $\langle \tilde{D}, \tilde{F} \rangle$. To define these structures, let α be a fixed real number such that $0 < \alpha < 1/2$ and $m = \lceil \alpha \log n \rceil$.

Definition 6.5 *Let $B = \langle D, F, x \rangle$ be a boolean network on n gates. A vortex of circumference d consists of two disjoint subsets of gates $R = \{r_0, \ldots, r_{d-1}\}$ and $S = \{s_0, \ldots, s_{d-1}\}$ satisfying the following conditions for $0 \leq i < d$.*

1. $(r_i, r_{i+1 \,(\text{mod } d)}) \in D$.

2. $(s_i, r_i) \in D$.

3. s_i is forced in m steps.

4. The value that s_i is forced to is not a forcing value for f_{r_i}.

The state of a vortex shall mean the state of B restricted to $R \cup S$.

Intuitively, a vortex consists of a cycle of gates R surrounded by a set S of gates. All the gates in S become forced, hence stable, but they do not force any gates in R. The set S may be thought of as a barrier or moat surrounding R, so that after all the gates in S stabilize, the gates in R are isolated from the rest of the network. The next lemma is a formalization of these ideas.

Lemma 6.6 *A vortex enters its state cycle in at most m steps. Its state cycle is completely determined by the initial state of $R \cup N_m^-(S)$.*

The state cycle of the vortex is just the sequence of states of R. Because of the simple structure of R, it can be shown that its state cycle size must be a factor of 2 times the circumference. In particular, the following holds.

Lemma 6.7 *If the circumference of the vortex is prime, then the size of its state cycle is 1, 2, d, or 2d.*

One additional condition on vortices that will be needed is that they should enter a large (relative to their circumference) state cycle from many initial states. This is formalized by the following definition.

Definition 6.8 *A vortex of circumference d is strong if for at least $1/2$ of the initial states of B, the state cycle size of the vortex is greater than or equal to d.*

The next lemma implies that the probability of a vortex being strong is rather large.

Lemma 6.9 *For any $\tilde{B}$ and vortex V of circumference $d \geq m + 2$ where d is prime, the probability that V is strong is greater than or equal to $1/2 - o(1)$.*

The final technical lemma gives a formula for the probability that there exist strong vortices whose circumferences include a certain set of numbers.

Lemma 6.10 *Let p_m be as given in lemma 4.2, $k(n) = O(\log n / \log \log n)$, and σ_i be the probability that a vortex of circumference d_i is strong, for $i = 1, \ldots, k(n)$. Then*

$$\text{pr}\left(\bigwedge_{i=1}^{k} \tilde{B} \text{ has a strong vortex of circumference } d_i \right)$$

$$= (1 - n^{-\Omega(1)}) \prod_{i=1}^{k} \left(1 - e^{-p_m^{d_i} \sigma_i} \right) + n^{-\Omega(\log \log n)}.$$

To complete the proofs of theorems 6.3 and 6.4, take any $\beta > \alpha$, and let $k(n)$ be the number of primes in the interval $[\beta \log n, 2\beta \log n]$. By the prime number theorem (Titchmarsh, 1951), there are about $\beta \log n / \ln \log n$ primes in this range. Using this fact together with lemmas 6.9 and 6.10, it can be shown that with probability $n^{-o(1)}$, a random network will have strong vortices whose circumferences include all the primes from $\beta \log n$ to $2\beta \log n$. For those $\tilde{B}$ having such vortices, with probability $\geq 2^{-k(n)} = n^{-o(1)}$, a random starting state takes each such vortex of circumference d to a state cycle of size d or $2d$. In that case, the state cycle size of $\tilde{B}$ must be at least as large as the product of all these primes. That is, for such a starting state, $\tilde{B}$ enters a state cycle of size greater than or equal to

$$(\beta \log n)^{k(n)} = e^{(1-o(1))\beta \log n}$$
$$= n^{\beta \log e - o(1)}.$$

Thus, with probability $\geq n^{-o(1)}$, $\tilde{B}$ enters a state cycle larger than $n^{\beta \log e - o(1)}$. By Markov's inequality,

$$\mathbf{E}(\mathbf{C}) \geq n^\beta.$$

Since β was arbitrarily large, theorem 6.3 follows.

The proof of theorem 6.4 uses similar computations. Let $\mathbf{X}$ be the random variable that counts the number of primes d in $[\beta \log n, 2\beta \log n]$ such that $\tilde{B}$ has a strong vortex of circumference d. Applying lemmas 6.9 and 6.10, it can be shown that

$$\mathbf{E}(\mathbf{X}) \sim \sum_{i=1}^{k(n)} 1 - e^{-p_m^{d_i}\sigma_i}$$
$$\geq k(n) e^{-2\beta/(a\alpha)}/5$$
$$\to \infty.$$

The same arguments also show that

$$\mathbf{E}(\mathbf{X}^2) \sim (\mathbf{E}(\mathbf{X}))^2.$$

Therefore by Chebyshev's inequality, for any $\delta < 1$,

$$\mathrm{pr}(\mathbf{X} \leq \delta \mathbf{E}(\mathbf{X})) \leq \frac{\mathbf{E}(\mathbf{X}^2) - (\mathbf{E}(\mathbf{X}))^2}{(1-\delta)^2(\mathbf{E}(\mathbf{X}))^2}$$
$$\to 0.$$

That is, almost all $\tilde{B}$ have at least $\delta k(n) e^{-2\beta/(a\alpha)}/5$ strong vortices of distinct prime circumferences in $[\beta \log n, 2\beta \log n]$. For all such networks, with probability $\geq 2^{-\delta k(n) e^{-2\beta/(a\alpha)}/5} = n^{-o(1)}$, the starting state leads to a state cycle larger than or equal to

$$(\beta \log n)^{\delta k(n) e^{-2\beta/(a\alpha)}/5} \geq e^{\beta \delta e^{-2\beta/(a\alpha)} \log n/5}$$
$$= n^{\beta \delta e^{-2\beta/(a\alpha)} \log e/5}.$$

By Markov's inequality,

$$\mathbf{E}(\mathbf{C} | \langle \tilde{D}, \tilde{F} \rangle) \geq n^{\beta \delta e^{-2\beta/(a\alpha)} \log e/5 - o(1)},$$

and the theorem follows by taking any $\gamma < \beta \delta e^{-2\beta/(a\alpha)} \log e/5$.

7   The Region of Chaos

Here it is assumed that $a < c$. Then the two local measures of stability suddenly drop: for almost all random networks in this region, substantial fractions of the gates do not stabilize and are not very weak. The basic lemma in this case is the following.

Lemma 7.1 *Fix any gate i. For $d \geq 0$, let q_d be the probability that the state of gate i does not change from time d to time $d+1$. i.e. $f_i^d(x) = f_i^{d+1}(x)$, given that $N_d^-(i)$ is a tree. Then*

$$q_0 = a + \frac{b}{2} + \frac{c}{2} \quad and$$
$$q_{d+1} = a + \frac{b_2}{2} + c + (b_1 - 2c)q_d + (\frac{b_2}{2} + 2c)q_d^2 + o(1).$$

It can be shown that the fixed points of this recurrence are $(2a + b_2 + 2c)/(b_2 + 4c)$ and 1. Since $a < c$, the first fixed point, γ, is less than 1. Then, using elementary arguments similar to those in the proof of lemma 4.3(a), q_d converges geometrically to γ. In fact, this is true for any initial value for q_0 less than 1. By lemma 4.1, for almost every gate, its in-neighborhood of radius $\log n/4$ is treelike. Therefore, after $\log n/4$ steps, the probability that the gate does not change at the next step is less than $\gamma + n^{-\delta}$ for some $\delta > 0$. Since this is strictly less than 1, the proportion of gates that change at any step persists at approximately $1 - \gamma$.

Theorem 7.2 *There is a constant $\alpha < 1$ such that*

$$\lim_{n \to \infty} \mathrm{pr}(\tilde{B} \text{ has at most } \alpha n \text{ gates that stabilize}) = 1.$$

The fact that a significant proportion of gates do not stabilize when $a < c$ is clear evidence of a phase transition. Results on the other two measures of stability (number of weak gates and size of the state cycle) are not known. However, it can be shown that for a nonzero proportion of gates, perturbing the state leads to a cascade of perturbations which persist.

Theorem 7.3 *There is a constant $\alpha < 1$ such that*

$$\lim_{n \to \infty} \mathrm{pr}(\tilde{B} \text{ has at most } \alpha n \text{ very weak gates }) = 1.$$

Again, this is an indicator of a phase transition because theorems 5.2 and 6.2 state that almost all gates are very weak (and hence weak) when $a \geq c$.

8   Future Directions

There are a number of open problems suggested by the results in this article. In the transition region $a = c$, there is a large gap between the absolute upper bound of $2^{n\omega(n)/\log n}$ for the state cycle size obtained by Łuczak and Cohen, 1991, and the lower bounds for average state cycle size given in theorems 6.3 and 6.4. It would be interesting to narrow the gap, in particular to improve the lower bound in theorem 6.4 for almost all random boolean networks.

There is even more work to be done in the chaotic region $a < c$. Proving that the proportion of weak gates is strictly less than 1 is an obvious problem. Also, since the average state cycle size is already superpolynomial when $a = c$, the same should be true when $a < c$. Indeed, the following conjecture seems plausible. As $a - c$ increases, stability of the network increases, i.e., all three measures of stability increase.

In a broader context, the significance of the results given here will depend on whether they can be extended to more general models of the genome, or even other complex systems. Two rather technical generalizations are to networks where the probabilities of the boolean functions are not symmetric, and networks with more than two inputs per gate. As mentioned in the introduction, it is not clear what a realistic model of a random genome should be. One feature that may be relevant is the topology of the genes on the chromosomes. A key finding of the experiments on adaptive systems such as the genetic algorithm (Holland, 1975) is that crossover is a very significant force in evolution. It seems plausible that the distance between two genes on a chromosome should affect the probability that one influences the other. Random ordered structures have been studied in random graph theory, and the methods may be useful in this context.

Independently of the question about biological fidelity, the computational capabilities of random boolean networks should be investigated. There is a widespread belief that systems with the ability to perform complex computations lie near the border between stability and chaos. This was supported by Langton (1990) in his study of the λ parameter of cellular automata. The results in this article seem to indicate that networks where a is significantly greater than c are too stable to do anything interesting, since their active gates are arranged in small, isolated cycles. Networks in the complementary region where $a < c$ appear too chaotic. The analysis of the region $a = c$ relied on the existence of vortices, which are complex structures. Yet they they persist once the network enters its state cycle. Perhaps similar methods can show that networks where a is close to c have large, persistent substructures that can perform interesting computations.

References

Bollobás, B. 1985. *Random Graphs.* London: Academic Press.

Derrida, B. and Y. Pomeau. 1986. Random networks of automata: a simple annealed approximation. *Europhys. Lett.* 1:45–49.

Derrida, B. and G. Weisbuch. 1986. Evolution of overlaps between configurations in random Boolean networks. *J. Physique* 47:1297–1303.

Erdős, P. and A. Rényi. 1959. On random graphs I. *Publ. Math. Debrecen* 6:290–297.

Erdős, P. and A. Rényi. 1960. On the evolution of random graphs. *Publ. Math. Inst. Hungar. Acad. Sci.* 5:17–61.

Harary, F. 1969. *Graph Theory.* Reading, MA: Addison-Wesley.

Harris, B. 1960. Probability distributions related to random mappings. *Ann. Math. Stat.* 31:1045–1062.

Holland, J. H. 1975. *Adaptation in Natural and Artificial Systems.* Ann Arbor: University of Michigan Press.

Kauffman, S. A. 1970. Behaviour of randomly constructed genetic nets: binary element nets. In *Towards a Theoretical Biology*, edited by C. H. Waddington. Chicago: Aldine Publishing Company, 18–37.

Kauffman, S. A. 1990. Requirements for evolvability in complex systems: orderly dynamics and frozen components. *Physica D* 42:135–152.

Langton, C. G. 1990. Computation at the edge of chaos: Phase transitions and emergent computation. *Physica D* 42:12–37.

Łuczak, T. and J. E. Cohen. 1991. Stability of Vertices in Random Boolean Cellular Automata. *Random Structures and Algorithms* 2:327–334.

Lynch, J. F. 1993a. Antichaos in a class of random boolean cellular automata. *Physica D* 69:201–208.

Lynch, J. F. 1993b. A criterion for stability in random boolean cellular automata. *Ulam Quarterly* 2:32–44.

Manneville, P., N. Boccara, G. Y. Vichniac, and R. Bidaux. 1989. *Cellular Automata and Modeling of Complex Physical Systems.* Berlin, Heidelberg: Springer-Verlag.

Titchmarsh, E. C. 1951. *The Theory of the Riemann Zeta Function.* Oxford: Oxford University Press.

Toward an Evolvable Model of Development for Autonomous Agent Synthesis

Frank Dellaert[1] and Randall D. Beer[1,2]

Department of Computer Engineering and Science[1]
Department of Biology[2]
Case Western Reserve University
Cleveland, OH 44106
Email: dellaert@alpha.ces.cwru.edu, beer@alpha.ces.cwru.edu

Abstract

We are interested in the synthesis of autonomous agents using evolutionary techniques. Most work in this area utilizes a direct mapping from genotypic space to phenotypic space. In order to address some of the limitations of this approach, we present a simplified yet biologically defensible model of the developmental process. The design issues that arise when formulating this model at the molecular, cellular and organismal level are discussed, and for each of these issues we describe how they were resolved in our implementation. We present and analyze some of the morphologies that can be explored using this model, specifically one that has agent-like properties. In addition, we demonstrate that this developmental model can be evolved.

1. Introduction

Our long term goal is the co-evolution of bodies and control systems for complete autonomous agents. The design of autonomous agents is a complex task that typically involves a great deal of time and effort when done by hand. Both the physical implementation of robotic agents as well as their control architectures present increasingly difficult design challenges as the complexity of the system at hand increases. A number of people (Beer and Gallagher 1992; Harvey, Husbands and Cliff 1993) have argued it might be advantageous to use evolutionary methods.

However, most of the current attempts at solving the design problem using genetic algorithms employ some form of a 'direct mapping' between genotype and phenotype. Here, typically there is a one-to-one correspondence between some substring on the genome and an associated parameter or feature in the final design. Examples can be found in (Beer et al. 1992; Lewis, Fagg and Solidium 1992; Harvey et al. 1993). In addition, in the majority of these cases the authors try to optimize a fixed number of parameters in some chosen architecture that a priori determines which designs are possible.

One can identify a number of problems with this direct mapping: (1) The designs to be explored are essentially limited by the chosen architecture, because of the fixed dimensionality. (2) One of the most obvious problems is one of

scaling: while evolving small networks works quite well, the approach scales badly for larger networks. (3) In many cases it is desirable that the final design exhibits bilateral symmetry, which is hard to come by using a direct mapping approach.

In contrast, there is no direct mapping between genotype and phenotype in biology. Rather, plant and animal morphologies are the result of a growth process that is directed by the genome. We believe that there are intrinsic properties to this developmental process that, when used in conjunction with genetic algorithms, may enable it to address some of the difficulties with the direct mapping approach: (1) When interpreted not as a direct encoding but as a set of developmental rules, a genetic description can lead to much more complex morphologies than those achievable with direct mapping. (2) There is some hope that a developmental process can reduce the scaling problems as well, as it essentially builds upon previous discoveries in an incremental fashion. (3) Symmetry comes for free in a developmental model.

Apart from promising to address some problems, there are some additional advantages to using a developmental model in its own right: (1) Development naturally provides a way to sample a spectrum of genetic operators, ranging from local hill-climbing operations to long jumps beyond the correlation length of the genetic search space (Kauffman and Levin 1987). (2) With a developmental model, morphologies and behavioral control mechanisms can graciously co-evolve to obtain optimal performance.

This paper describes our first step toward a simplified but biologically defensible model of development that is efficient enough to be used in conjunction with a genetic algorithm. This preliminary model enables us to explore body patterns and relative placement of essential components. After discussing related work is Section 2, in Section 3 we will highlight some of the design issues and tradeoffs that come up when modeling development in general along with the way we have resolved these issues in our particular model. In Section 4 we show the results that can be expected from it and how the model behaves when used in conjunction with genetic algorithms. In Section 5 we analyze the developmental sequence of a simple agent, using that as an example to show the detailed workings of the model.

Finally, in the Section 6 we will discuss it's strengths and weaknesses, along with directions for future work.

2. Related Work

There is already a body of work wherein the authors both understand and appreciate the importance of incorporating a developmental process into the picture. (Wilson 1989) discusses a general representational framework to set the stage for simulations of development. A number of people have implemented some models, and they can be roughly categorized into two groups:

Much work involves some kind of growth-model coding for evolving neural networks: (Kitano 1990) and (Gruau and Whitley 1993) use grammatical encoding to develop artificial neural networks. (Harp, Samad and Guha 1989) try to evolve the "gross anatomy" and general operating parameters of a network by encoding areas and projections onto them into the genome. (Nolfi and Parisi 1991) uses an abstraction of axon growth to evolve connectivity architectures. Most of these models do not aspire to be biologically defensible, however. Also, they have not been applied as such in the area of autonomous agents.

In contrast, a number of other authors have looked at more biologically inspired models of developmental processes: some work is based on the grammar based approach first developed by Lyndenmayer, such as (deBoer, Fracchia and Prusinkiewicz 1992). For instance, (Mjolsness, Sharp and Reinitz 1991) use grammatical rules to account for morphological change, coupled to a dynamical neural network to model the internal regulatory dynamics of the cell. (Fleischer and Barr 1994) have a hard-coded model for gene-expression that they combine with a cell simulation program. Many other biologically realistic models of different developmental processes are found in the theoretical biology literature. However, to our knowledge, none of these more complex models in the second category have as yet been used in conjunction with genetic algorithms.

It is the combination of a biologically defensible model of development with evolutionary methods that we would like to apply to the design of autonomous agents, something that at this point in time has not yet been addressed in the existing literature.

3. Model

In this section we will first give an overview of the principal components of our model. Then, in the subsequent sub-sections, we will raise some of the issues that came up when modeling each of these components. For every issue that we discuss, we will present the way we resolved it when implementing the model, together with a more detailed description of the actual implementation.

3.1. Overview

The developmental process unfolds simultaneously at three different levels, each of which will need to have a counterpart in our model: at the level of the organism, of the cell and at the bio-molecular level. At the topmost level, a single zygote develops into a *multicellular organism* by a complex epigenetic process: eventually, groups of cells will literally stick together and co-ordinate their actions to form tissues and organs that make up the entire organism. This happens because at the *cellular level*, individual cells unfold a sequence of determination and differentiation events that enable them to take up their specific role in the developing embryo. Ultimately responsible for this unfolding sequence, however, is the genetic information contained within each cell, which brings us down to the *level of molecular biology*. Although each cell has the same copy of the genome, different genes are expressed in different cells, which in turn leads to their difference in behavior. Thus, this pattern of differential gene expression lies at the heart of the developmental process.

This *genetic regulatory network* will be the first principal component of our model. We believe that, for our purposes, the essence of the unfolding pattern of differential gene expression at the genome-level is best captured by modeling a network of interacting genetic elements. Each element would correspond to the existence of a gene product or the expression of some gene. The total state of the network at a given time can then accordingly be viewed as the pattern of gene expression of a given cell at that time: because the state evolves over time this then corresponds to the unfolding of a developmental program in each cell.

The second component consists of a very simple *cellular simulator* to model development at the cellular level. Eventually, every action directed by the genome should first have a consequence at the level of the cell, if it wants to have an effect on development. It follows that we will need to construct a model for how a cell behaves and the way the genome can influence this behavior. One way to do this could be to build a complex, three dimensional model of how biological cells actually work. However, our primary interest is not to mimic the actual biological developmental process, but to extract from it the essential beneficial properties. That is why we have opted for a simple, two-dimensional cellular simulation.

Finally, the last aspect of the model will cover all phenomena at the *organismal level*. During development cells interact continuously: in biological development, cells communicate by touch as well by chemical signals (Walbot and Holder 1987, page 4). This intracellular communication is extremely important, as it can change the pattern of gene expression in the participating cells. Thus, we will have to take this into account in our model. Another issue that transcends the cellular level is that of external influences, such as how symmetry is somehow broken at the very first stages of development. These aspects will be modeled at the level of the organism.

While implementing these components, we were confronted at each step along the way by the trade off between simplicity and biological defensibility, as ultimately the model was to be used in conjunction with the genetic algorithm. Typically, when you want to evolve autonomous agents you use populations of hundreds of individual organisms in parallel. With our developmental model, each of these would consist of from a hundred to a thousand cells, and in each cell a genetic regulatory network would be active. It is obvious that with such numbers you want to keep the model as simple as possible to keep the computational demands feasible. Although many aspects of biological development are important and even crucial for biological life forms, we feel that some of them can be left out in a simplified model without invalidating the results we get.

3.2. Genetic Regulatory Network

As we will model the patterns of gene expression in each cell as the state of a regulatory network, we will have to address each aspect of these networks, i.e. the nature of the elements and the way they interact.

The interactions between the genetic elements

The major issue here is at what level of detail one wishes to simulate the genetic elements and their interactions. In the biological cell there are a number of strategies for the regulation of gene expression: essentially each step of the pathway between the coding sequence on the DNA and it's final gene product presents an opportunity to regulate the expression of that particular gene (Alberts et al. 1991, page 551-556). Do we want to make a distinction between transcriptional control and RNA degradation control and incorporate them as different building blocks in our model ? Chances are that doing so will yield some insight into the detailed workings of these processes, but it will also pose an enormous computational problem to simulate.

In our model we will assume the existence of one type of abstract genetic element and one way in which these elements can influence each other. Although the genetic elements in a biological cell include not only DNA sequences but also regulatory proteins, cell-surface receptors and a whole lot more, we will assume there is only one type. This assumption does also imply that we will model only one way of interaction between these elements, because most of the time the differences between different regulation strategies in the cell come down to differences in the type of players involved. Of course, if you decide on this course of action, you have to make the assumption that the essence of development is not to be found in the details of all the different regulatory mechanisms, but rather the interaction of mutually influencing elements.

The nature of the genetic elements

The genetic elements can be modeled by anything ranging from simple binary to complex quantitative models. One of the choices that must be made is between continuous and discrete state variables, and between continuous or discrete time models. In the literature this choice has been made in a number of ways, resulting in models that use essentially simple binary elements (Kauffman 1969; Jackson, Johnson and Nash 1986), models with multi-level logic (Thieffry and Thomas 1993), dynamical neural networks (Mjolsness et al. 1991) and fully quantitative models.

We have chosen to model the genetic elements as binary elements. Although the more complex approaches are certainly useful, a binary model is especially attractive from a computational viewpoint: their simplicity will allow us to simulate a large number of them in a reasonable time. As explained before, this is a necessity when we will use the model in conjunction with genetic algorithms. In addition, simplifying genetic elements to binary variables can be defended on a deeper ground: a lot of phenomena that occur during embryology and the life span of a cell have an on/off quality or have some mechanism of self-amplification. For example, biochemical pathways linking cell-surface receptors to the DNA have lots of amplification steps built in (Walbot et al. 1987, page 321), ensuring that their response is on/off like. Kauffman (1993) presents additional arguments why the major features of many continuous dynamical systems can be captured by a Boolean idealization.

We will model the genetic regulatory networks by Boolean networks

Given all these considerations, we decided to model the genetic regulatory network by a Boolean network, as first pioneered in this context by (Kauffman 1969) and extended by (Jackson et al. 1986) to systems of multiple, communicating networks. This model is both readily understood, efficiently implemented and easily analyzed (Wuensche 1994), in contrast with the more complex continuous time dynamical networks as used by (Mjolsness et al. 1991). The latter work is more focused on parameter identification of actual biological processes however, in which case this more complicated approach makes sense.

A Boolean network is much like a cellular automaton, but where in the latter the neighborhood of a node is fixed and consists of neighboring cells, there is no such restriction in the former. The basic elements are the N nodes, each with its own associated K-node neighborhood and updating rule. Each node can assume a state of 1 or 0, according to the state of its K inputs at the time it was updated. It is easy to see that there are $(N^K)^N$ possible wiring configurations. Fig.1 illustrates one possible instance of a Boolean network with N=3 and K=2 out of $(3^2)^3 = 729$ possibilities.

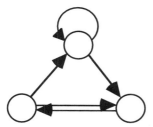

Fig.1. Example of a wiring configuration in a Boolean network with $N=3$ and $K=2$.

The updating rule of a node can be any Boolean function of its K inputs, and can be specified by a lookup table with 2^K entries. As each entry can contain a value of 1 or 0, there are 2^{2^K} possible updating rules for each node in the network. As an example, one out of the sixteen possible Boolean functions for $K=2$ would be the logical AND operator, which can be represented by the string <0001> in a lookup table.

The total number of possible Boolean networks with given N and K is then $\left[(N^K)(2^{2^K}) \right]^N$, a number that can be very large. For the example of $N=3$ and $K=2$, there are $\left[(3^2)(2^{2^2}) \right]^3 \cong 3 \cdot 10^6$ possible networks.

Some additional issues

We have chosen to update these Boolean networks synchronously, i.e. every time step the whole state vector is computed using the values of the state vector at the previous time step. Discrete time, synchronously updating networks are certainly not biologically defensible: in development the interactions between regulatory elements do not occur in a lock-step fashion. The alternative is to update all nodes asynchronously, each node having a given probability at any time to recompute its value from its inputs at that time. This introduces an element of non-determinism however, that might render any genetic search in a space of such networks very difficult. In addition, they are less readily analyzed than their synchronous counterparts, for which there are excellent analysis tools available (Wuensche 1994). Then again, they would be useful to examine phenomena like spontaneous symmetry-breaking interactions between cells, that can not occur with lock-step updating.

We ended up using a comparatively small number of elements in the Boolean networks. When one looks at the function of genes in eukaryotic genomes, one finds that the vast majority of gene products will be responsible for housekeeping functions that are common between cell types, and most of the others are cell-type specific genes (Walbot et al. 1987, page 174-175). In addition, genes have

been found that switch on whole gene-batteries at a time (McGinnis and Kuziora 1994), thus acting as a representative for a whole class of genes. This could suggest that actually a small number of genes might be responsible for the regulatory mechanisms within the cell.

The genome that we use in the genetic algorithm is a straightforward description of one such possible Boolean network. Both the connection parameters and the Boolean function are subject to mutation. In some experiments however, constraints can be imposed to explore restricted search spaces. Most of the time we used parameter settings of $N=6$ and $K=2$.

3.3. Cellular level

At the cellular level we will have to model all the properties of the cell that play a role in the developing organism at the higher level and that can be influenced by the genetic regulatory networks at the lower level. These properties include the physical characteristics of the cell, the cell cycle controlling the cell's behavior and how a cell differentiates into a particular cell type, which we will discuss here. We will also touch on the aspects of biological cells that we left out of the model and why we left them out.

The physical characteristics of the cell

When modeling the physical characteristics of the cell, we are looking for a model that is both simple enough to be efficiently simulated in large numbers, and yet captures enough of the aspects of a biological cell that makes it work within a developmental process. We think two properties are essential in this respect: it has to have some form of physical extent and it has to be able to undergo division. Granted that this is an extreme simplification of what a real cell actually constitutes: our ultimate goal is the synthesis of autonomous agents, however, not the modeling of biological development, and we think that these two properties suffice for our purposes.

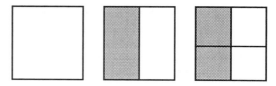

Fig.2. Zygote square dividing two times to yield 4 'cells'.

Thus, we will simulate the physical appearance of a cell by a simple, two-dimensional, square element that can divide in any of two directions, vertical or horizontal. If division occurs it always takes place in such a way that the longest dimension is halved, and the two resulting daughter cells together take up the same space as the original cell. This very simple approach has as a consequence that we do not have to deal with cells changing shape as a result of cell division: after two cleavages the shape is again a square. See Fig.2 for an illustration.

The cell cycle

A simulated cell cycle consisting of two phases, interphase and mitosis, co-ordinates the updating of the Boolean network state and cell division, respectively. In one organism, each cell would have a copy of the same Boolean network constituting the genetic information of that organism. However, the state of the network, corresponding to the pattern of gene expression in a particular cell, may be different in each cell, as they underwent different influences during their life span or started out with a different initial state. In Fig.3 the cell cycle is depicted graphically, and we will discuss each phase here.

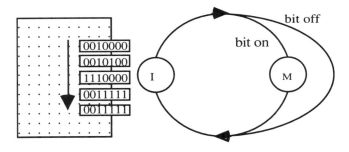

Fig.3. The cell cycles between interphase and mitosis. Mitosis is skipped if the 'dividing bit' is not set.

During the first phase, interphase, the state is synchronously updated until a steady state is reached: a steady state corresponds to a stable pattern of gene expression. We make the assumption that each cell has enough time to reach such a stable state, and when used with the genetic algorithm we will discard any organism whose regulatory network leads to a cyclic pattern. Waiting for a steady state also means that in every cell the number of updates may be different, because the transient behavior of the network depends both on the original state vector as on the new environmental stimuli (see below).

Then, in the second phase, according to the setting of a specific bit the cell either goes through mitosis and divides into two daughter cells, or it stays intact and waits for the next interphase to start. When the cell divides, it's state vector is inherited by the two daughter cells. Note that in the following interphase, unless something in the environment changes or there is some other external interference, that state vector will correspond to a stable pattern and nothing will happen, i.e. that pattern of gene expression will be passed on to the next generation. Also, our 'cell cycle' definition is not entirely comparable with the biological equivalent, as in our model the cell does not necessarily have to go through mitosis.

Differentiation into distinct cell-types

There are two ways in which we could model how the genome determines the final differentiation of a cell: by combinatorial specification or using 'master genes'. In biology, the combinatorial gene regulation theory hypothesizes that the cell can 'detect' a particular combination of regulatory proteins and thus is able to differentiate into the corresponding cell type. For instance, this type of mechanism is thought to underlie the division of the imaginal discs in *Drosophila* into sharply demarcated compartments (Alberts et al. 1991, page 930). In principle, three different genes would be sufficient to specify a unique address for each of the eight compartments formed. Alternatively, there could be several regulatory 'master' genes whose expression determines the expression of a whole gene batteries needed in a particular cell type. See (Davidson 1990) for a comparative overview of a number of cell fate specification mechanisms.

In our implementation we have modeled both these mechanisms and we can choose between them when running simulations. In combinatorial mode, a subset of genetic elements is chosen to determine the final differentiation of the cell. Every distinct combination of activity in these elements then corresponds to a particular cell type. This simply corresponds to a binary encoding of the cell types, e.g. for $N=3$ cell type 5 would be represented by the string <101>. In the other, 'master gene' mode, we relate differentiation to the activity of one specific genetic element, with the additional constraint that there should be no conflict between competing cell types. For encoding a cell type 5, we thus need at least $N=6$, and it would then be represented by the string <100000>. All the results reported in this paper use combinatorial specification, as we have found that it takes considerably longer to evolve the additional mapping between the state of the network and the different 'master genes'.

Our model uses color as an abstraction for cell type. As it is ultimately our goal to synthesize autonomous agents, the final differentiation of a cell will then correspond to it being a sensor, actuator or control-neuron. For the time being, however, it will be sufficient to simulate this by a different color that the cell can take on: it will enable us to demonstrate the different architectures that can be explored using the model. In the combinatorial mode, we assign a color according to the settings of $\lg(C)$ specific bits in the state vector, where C is the number of colors. Mostly, $C=8$ and the three bits used are bit[0], bit[1] and bit[2].

Biological properties we did not include in the model

Although many aspects of biological development at the cellular level are important and even crucial, we feel that some of them can be left out in a simplified model without invalidating our results. Cell movement and coordinated cell sheet deformations, for instance, lie at the basis of all but the simplest morphologies encountered in multicellular organisms. However, they would make the model quite complex and much more difficult to implement. We think it is better to start off exploring what is possible with 'simple' intracellular communication (see below) and genetic regulatory networks, rather than make the model too complicated from the start. Once it is clear what can be achieved with a simple model, it is certainly worthwhile to add incorporate more complex mechanisms.

3.4. Organismal level

In the subsequent paragraphs we will discuss how individual cells function within the organism and how two very important aspects of development, symmetry breaking and intracellular communication, are implemented.

The organism as a collection of cells

The organism itself is a two-dimensional square consisting of many cells. Development starts out with one single square that represents the zygote, which then subsequently divides according to the state of the genetic regulatory network. As discussed, whenever a square divides the two daughter cells take up the same space as the original one: there is no pushing away of neighboring cells or shape change involved, except that after an odd number of cleavages, cells may be rectangular in shape rather than square. The organism is then the collection of squares that originated from the 'zygote' square.

Symmetry breaking in the early stages of development

We will have to address a way to break the symmetry between the very first cells at the early stages of development, otherwise we will end up with a uninteresting, homogeneous collection of cells: because of the deterministic, synchronous updating and as they are all descendants of the same 'zygote', all the cells will have the same state vectors at each step unless something disturbs this symmetry. Biological development faces the same problem, and there are diverse mechanisms by which in early development the correct spatial pattern of differential gene expression is imposed (Davidson 1990).

We will break the symmetry at the time of the first cleavage by assuming the existence of a 'maternally' imposed a-symmetry in the 'zygote' square, that can lead to different patterns of gene expression in the first two daughter-cells. This is certainly biologically defensible, as in many organisms this anisotropic distribution of some entity is actually observed (Walbot et al. 1987, page 340-353). We will simulate by flipping a bit of the Boolean network state vector in only one of the two daughter cells. If you will, the genetic element of which the state is flipped corresponds to an asymmetrically distributed determinant in the zygote.

A second spatial clue is introduced by supplying the developing organism with the notion of a midline. As we will explain in Section 6, it was sometimes necessary to provide more spatial clues than only the first cleavage symmetry breaking. Thus, we also provided the cells with information on whether they are adjacent to the horizontal midline of the organism, according to a bit flipping scheme similar to the one used in the first cleavage step, although this time a bit in the neighborhood vector is flipped (see below). Actual biological embryos get this midline notion for free because of the three dimensional topology in which they develop: as an example, in the frog embryo neurulation takes place along the dorsal midline of the embryo, which after gastrulation lies closest to the mesodermal germ layer that is responsible for the initiation of the process (Walbot et al. 1987, page 368-375).

Intracellular communication

One of the key elements of the developmental process and consequently of our model is how the cells communicate. Indeed, following the initial symmetry-breaking the cells now have a rough plan for the positioning of major body structures. In all but the simplest organisms, however, a great deal of fine-tuning is necessary, and this can be achieved by intracellular communication or induction, i.e. the way in which one group of cells can alter the developmental fate of another group by providing it with some signal (Walbot et al. 1987, page 366). We have put a lot of thought into whether to model this by actually simulating the existence of cell-surface receptors and chemical 'signal' molecules. As an alternative, one could link the genetic regulatory networks in a more direct way, by letting their next state depend not only on their own state, but also on that of surrounding cells.

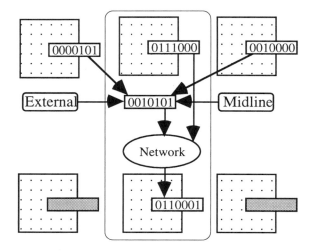

Fig.4. The state vectors of two neighbor cells are ORed together to yield a neighborhood vector that is combined with the cell's state vector to determine the next state.

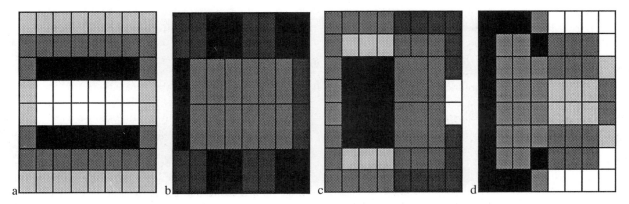

Fig.5. Four different examples that demonstrate the range of organisms we have been able to develop so far.

To implement induction we used a modified version of the Boolean Network: whereas normally each node has K incoming edges from other nodes in the network (or recurrent), we now allow for some of these incoming edges to connect to nodes in an abstract 'neighborhood' vector. The latter is the logical OR of all the state-vectors of the neighboring cells. Fig.4 shows this arrangement. In our genetic description of the network, a negative connection parameter implies that the corresponding bit in the neighborhood state vector is taken, in stead of from the cell's state vector. Note that an edge to the neighborhood vector can be interpreted as the existence of a cell-surface receptor, sensing the presence of specific chemical agents introduced by cells in its environment.

This implementation implies that our simulation has to keep track of which cells are neighbors. Although this may sound an easy thing to do, it does actually complicate things somewhat, as the cells are not static entities but instead divide all the time. Thus, a scheme must be devised by which topological relations are constantly kept up to date. However, it becomes soon intractable to let each cell poll every other cell in the organism, because the number of cells rises exponentially in each organism. We solved this problem by letting each cell pass on a list of it's neighbors at the time of division and then letting each daughter cell poll these neighbors to check whether they are still adjacent. It is interesting to note that our particular implementation allows to easily substitute a more complex (even three dimensional) geometry for the 2-D square one. Indeed, we have already modeled one-dimensional 'string' organisms in this way, and plan to look at the more detailed topological framework model as proposed by (Matela and Fletterick 1979) and recently elaborated on by (Duvdevani-Bar and Segel 1994).

Another type of induction is the influence exercised by the external environment: we modeled this by reserving one bit in the neighborhood state vector for that purpose: it is forced to ON if the cell in interphase is at the border of the organism, otherwise it is OFF (See Fig.4.).

4. Examples and Evolvability

4.1. Developmental Examples

Fig.5 shows several examples that demonstrate the range of organisms we have been able to evolve until now. Although these are preliminary explorations, mostly found using 'biomorph mode' (sitting down at an X-terminal and selecting the fittest individual according to subjective taste, see (Dawkins 1989)), they nevertheless exhibit interesting features that can conceivably be put to use in the context of autonomous agents.

Fig.5a displays an interesting 'layered' characteristic, with cell-types at the sides of the organism (it is facing towards the right) different from those in the middle, and with an intermediate layer in between. Note that in biological development the three germ layers exhibit the same spatial order: ectoderm to face the outside, endoderm at the inside and mesoderm in between them.

We have selected the organism in Fig.5b because it has a segmentation property, as you can discern a bilaterally symmetrical repeat structure at the sides of the organism. Finally, Fig.5c and d represent more complex morphologies, both a-symmetric with respect to the vertical axis and having more detailed patterning at the rostral side.

4.2. Evolvability

Given that our basic goal is to efficiently evolve autonomous agents, one of the things to look at is how the model behaves when used in conjunction with a genetic algorithm. To investigate this, we have devised a generic performance function that maximizes the number of colors, taking care that no color is more represented than any other. Although this particular criterion has no direct relevance to autonomous agents design, it is nevertheless useful to examine the discoveries made by evolution in maximizing this function.

We have found that we can successfully evolve Boolean networks that can steer the developmental model so that the

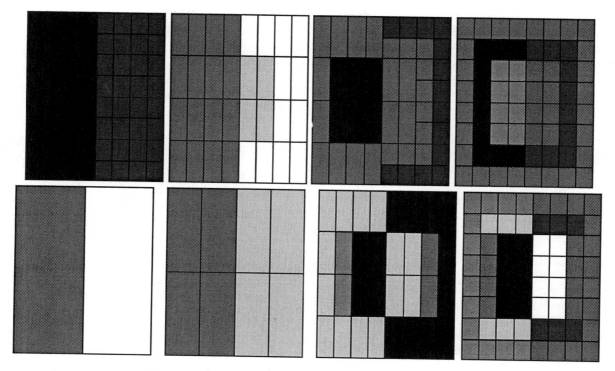

Fig.6. The best individual after each fitness jump during evolution. The respective performance values of these individuals are 45, 170, 210, 1200 and 5600. Of the last individual, the developmental stages are also shown.

fully developed organism optimizes some performance function. This is a strong result: there is no obvious relationship between the setting of a bit in an update rule of the genetic regulatory network and the performance function to be optimized. The color of a square in the final design is quite far removed from the particular wiring of the network. In addition, the organism is evaluated only at the end of the full developmental process, so that any mutation in the genome

(= network) must not only be beneficial from a performance function viewpoint to be incorporated in the population, it must also take care not to interfere with the existing developmental process in a 'wrong' way.

Also, we have observed that the computational overhead induced is not as bad as one might expect when introducing a model with so many different elements. A typical simulation with population size of 20, network parameters $N=6$ and $K=2$, and a maximum of 64 cells per organism takes about 10 minutes to at most half an hour on a Sparc 10 for 200 generations, depending on the performance function and the mutation rate. Typically, we used mutation rates of 0.1 and cross-over probabilities of 0.5.

Fig.6 and Fig.7 show the results for a typical run. In Fig.7 we plotted the maximum and average fitness. The GA we used for this experiment used elitist selection, i.e. the best individual is never thrown away, which explains the step-like manner the maximum fitness evolves. In Fig.6 five organism are shown, each a snapshot of the best individual of its generation. The snapshots were taken just after a jump in fitness occurred: the way our performance function was constructed, this corresponds to the discovery of a new color. The last organism has discovered all eight colors.

In the first individual, we can immediately see evidence of the a-symmetry we introduced at the time of the first division: all individuals that did not make use of that were discarded from the first sample, as it is very easy for the GA to 'discover' this a-symmetry.

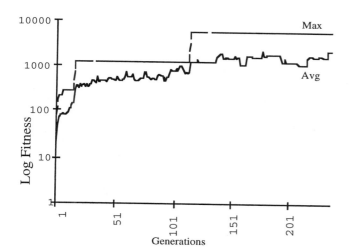

Fig.7. : Maximum fitness and average fitness.

All evolved organisms shown are bilaterally symmetrical. This is a direct consequence of how the model is set up: the only a-symmetrical stimulus is our first bit-flip, and the other external stimulus is the environment, which is symmetrically introduced at all sides[1]. Because of the synchronous updating of the networks, no other a-symmetries are introduced. Thus, in this model we get symmetry for free.

The next discovery made by the GA is that of the external environment. Notice that in the second square there is a difference between the center and the border cells of the organism. Together with the a-symmetry, the developmental process is able to specify 4 colors. In the further course of the evolutionary process, the previously formed layers themselves provide information for new cells to assume different colors. The next big discovery is six colors, then eventually eight.

The last individual shown has discovered all eight colors, and its developmental sequence is reminiscent of the discoveries made by the GA during the time span of the experiment. We have shown the subsequent stages of development this individual goes through: as you can readily observe, the steps that development goes through follow the 'discoveries' made in the course of evolution: a-symmetry, external environment, induction. As we have argued in the introduction, this is one of the great strengths of the developmental model: evolution is able to gradually build on previous discoveries, and extend them towards fitter organisms.

5. Development of a simple "Agent"

We have evolved a simple organism that exhibits the relative placement of sensors, actuators and control system of the kind one would like to see in a simple chemotactic agent. Any attempt at the design of autonomous agents using a developmental model will have to deal with morphological features such as these. For the simple task of chemotaxis we specifically looked for a bilaterally symmetric organism, with sensors and actuators placed sideways at the front and the back, respectively, and a control structure or 'neural tissue' connecting them. The performance function we used tried to minimize the difference between the color-patterns in the fully developed organism and a template that to us represented the features needed in a chemotaxic agent. In Fig.8 the organism, which we have termed 'seeker', is shown along with its developmental sequence. A number coded representation is used for the different cell types. In the last stage of development, with 64 cells, you can observe how the different components are placed: cell type 2, prominent at the right-side corners of the organism, corresponds to sensors, whereas cell type 4 and 1 correspond to actuators and 'neural tissue', respectively. Note that 2 actua-

tor-cells are out of place, but the overall relative placement of the components is quite good.

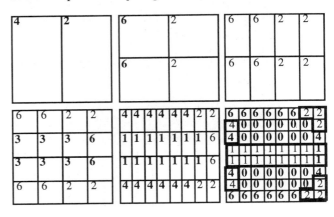

Fig.8. The six consecutive stages of development in the 'seeker' organism, with digits denoting cell types. Bold digits indicate the cell type changed relative to the previous developmental stage.

The 'seeker' organism will serve as an excellent example to illustrate in somewhat more detail just how the developmental model works. To do this we will frequently refer to Fig.8, but this in itself does not say much about the underlying process. We have access however - in contrast to researchers in biology - to every variable at every stage of the developmental process, from the outward appearance of the cells up to and including the complete description of the genome. In the subsequent paragraphs we will analyze this information and show what it can tell us about the sequence of events in the development of 'seeker'.

a) node	0 1 0 1 0 0 1 1	inputs	b) node	Equivalent Boolean function
1	0 0 1 0	3 -6	1	~3 AND mid
2	1 1 0 0	-2 -1	2	~(-1)
3	0 0 0 1	-5 5	3	ext AND 5
4	1 1 0 1	4 4	4	~4 OR 4 = TRUE
5	0 1 1 0	6 -6	5	6 XOR mid
6	0 1 1 1	6 -1	6	6 OR -1

Fig.9. a) the actual genome of the 'seeker' organism. b) The Boolean functions in a more readable form.

The genome, shown in detail in Fig.9a, specifies the wiring of a Boolean network (Fig.10) and the update rules of each of the nodes (Fig.9b). The 'seeker' organism has network parameters of $N=6$ and $K=2$, so the genome consists of 6 update rules and 12 input addresses. Induction from other cells is modeled by a negative address, corresponding to an incoming edge from outside the cell. The numbers -5 and -6 are reserved for conveying the influence of the external environment and the midline, respectively: if a cell is on the perimeter of the organism the value of bit '-5' will

---

[1] In this particular run there is no notion of a midline, as introduced in the previous section.

be TRUE and FALSE if not. Likewise, the value of bit '-6' is TRUE when the cell borders the midline of the organism, which runs horizontally across[2].

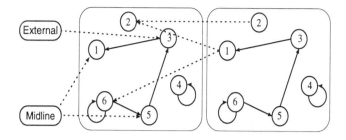

Fig.10. the 'seeker' wiring diagram: the dashed lines represent extracellular inputs. The 'midline' and 'external' have value 1 when the cell in question is on the midline resp. the perimeter of the organism.

The wiring of the network can serve very specific purposes: one thing that immediately catches the eye when looking at these figures is that both inputs to node 4 are recurrent connections and that the updating rule (~4 OR 4=TRUE) ensures that the corresponding genetic element will be permanently active. This can be explained by the particular fitness function that was used to evolve the organism, i.e. it rewarded a high number of cells in the final design: as bit 4 is used to decide whether to enter mitosis or not, the genetic algorithm found this positive feedback loop to ensure that division would take place at every step, resulting in a maximum number of cells.

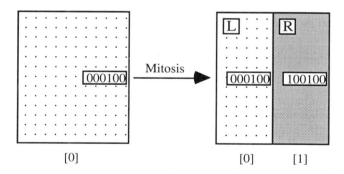

Fig.11. The zygote square divides at least once because we force the dividing bit 4 to TRUE prior to the first mitosis phase. Colors are read from the first three bits in the state vector and are indicated in square brackets.

At the very first stage, it is ensured that the zygote will divide at least once and that the resulting daughter cells are

not completely alike, so that the symmetry is broken. The organism starts out as a single 'zygote' square with all but one genetic elements inactive, i.e. zero state vector, except for the 'dividing bit' node 4, which is forced to 1. This will ensure that the cell division will take place, dividing the zygote into two cells L and R (Left and Right). In addition, at the time of that first cleavage bit 1 is set to 1 in one daughter cell and to 0 in the other, so that the symmetry is broken. We then have two cells with state vectors 000100 and 100100 respectively, as depicted in Fig.11. As the cell type or color is determined by the first three bits (least significant bit at left) this corresponds to color [0] at the left and color [1] at the right.

From now on interphase and mitosis will alternate until the final design of the organism is reached after stage 6 of the developmental process. We will look at the first stages in detail and then paint the broader picture when a detailed explanation becomes both tedious and too space-demanding. To understand the detailed picture, keep in mind that at each developmental stage a cell does three things: (1) it determines its neighborhood vector, (2) it repeatedly updates its state vector in interphase until a steady state is reached, and (3) assumes a color and decides whether to divide in the next stage.

> Cell L:
> Neighborhood vector: 100110
> Interphase: 000100 -> 000101 -> 001111
> Color = [4], divide
> Cell R:
> Neighborhood vector: 000110
> Interphase: 100100 -> 010100
> Color = [2], divide

Fig.12. Before, during and after interphase in each of the daughter cells L and R.

After the first interphase, we will have reached the 2-cell stage of the developmental sequence depicted in Fig.10. As described in Fig.12, the two daughter cells of the zygote each go through the three steps described above before entering mitosis, and you can see that the colors now match up with the colors shown in Fig.10. We will examine the behavior of cell L in somewhat more detail: bit 6 switches to 1 because its updating rule (6 OR -1, as we know from Fig.9b evaluates to TRUE because of the positive induction from cell R: bit 6 is 0 but bit -1, i.e. bit 1 in the neighborhood vector, is 1, so 6 OR -1 = 1. To put this into more general terms, the activity of the genetic element 1 in cell R induces a change in the pattern of gene expression of cell L, whose perturbed stable state will now elicit a transient behavior in the regulatory network. The final pattern of gene expression is not reached until after a steady state is reached, however, which happens after one more synchronous update. The color of cell L can now be read from the first three bits, i.e. color [4].

---

[2]In the simulated evolution that led to this particular organism, a midline notion is only present from the 16-cell stage and onwards. Up to and including the 8-cell stage, bit 6 in the neighbourhood vector is always 0.

Because cells inherit steady state vectors after mitosis, some change in the environment is needed to trigger a change in behavior and/or color. After cell L and R go through mitosis, we get 4 cells which we will denote by LT, LB, RT and RB, where T and B stand for Top and Bottom. Because the inherited state vectors represent a steady state of the regulatory network, nothing will happen unless some value changes that triggers a perturbation of this steady state. As it happens, this only occurs in the cells LT and LB, where the resetting of bit 1 in the neighborhood vector causes bit 2 to switch on (rule ~(-1)), resulting in color [6] for both cells after interphase settles down. The details are given in Fig.13, and the colors can be verified by looking at Fig.10: only the left cells have changed color from [4] to [6].

Cells LT and LB:
Neighborhood vector: 011110
Interphase: 001111 -> 011111
Color = [6], divide

Fig.13. The two left cells at stage 2 undergo a transition from color [4] to color [6].

Now that we have looked in detail at the mechanism that underlies the transitions in cell color at a given developmental stage, we can at least qualitatively understand the subsequent stages of the developing organism of Fig.10. In stage 3, it looks as if all state vectors remain unperturbed because the colors are unchanged: when looking at the tracefiles of the simulation, we have found that this was indeed the case. To make this difference between 'active' and 'inactive' interface apparent, we have marked the colors in Fig.10 bold when they resulted from a triggered transient, i.e. 'active' interphase.

A 'neurulation-like' event takes place at the 16-cell stage: suddenly all cells lying around the midline of the organism undergo a color change. It is clear that this resulted from the influence of the 'midline bit' 6 in the neighborhood vector, that has value 1 for these cells but value 0 for the cells at the sides of the organism. This inductive step sets the stage for the specification of sensors and actuators away from the midline, and for 'neural tissue' in the middle. The reminiscence of neurulation is not altogether surprising as we implemented the midline concept with just that phenomenon in mind (see Model Section).

A secondary induction event occurs at the 32-cell stage: all the cells of color [3], created by the 'neurulation' event in the 16-cell stage, in turn induce a perturbation in the cells around this group. Indeed, it can be verified from Fig.9b that bit 2, with rule ~(-1), will switch off in response to the now active genetic element 1 in the middle of the organism. This at least accounts for the change to color [4] resp. [1], for the cells that had color [6] resp. [3]. The picture is more complicated for some other cells, and we will not get into it here.

Eventually, via quantitatively similar interactions and influence from the external environment, the more complex picture at the last stage of development emerges.

6. Discussion and Conclusions

We have built a simple yet biologically defensible model of the developmental process. We have shown that it can account for a range of morphologies and that it is evolvable, i.e. the genetic regulatory networks can be evolved to optimize some performance function for the fully developed organism. Moreover, we have analyzed in some detail the developmental sequence of an agent-like morphology.

The work described in this paper is thus fully in line with our longer-term goal to use this developmental model for co-evolving body and control system in autonomous agents. Although to reach that goal, with computational simplicity in mind, we do not intend to modify the model all that much, our initial exploration with the model has raised a number of questions and suggested some issues that may be worthwhile to investigate further:
(1) It would be of value to look at a model where symmetry breaking is the norm, rather than the exception. Continuous time networks with some introduced noise component are an option, as are a-synchronously updating Boolean networks.
(2) Many important aspects of biological development that we have excluded from our model provide rich developmental possibilities and could be taken into account.
(3) Instead of binary induction between neighboring cells, it might be advantageous to model gradients of morphogens, as they are hypothesized to underlie both the expression of segmenting genes (Walbot et al. 1987, page 641) as the pattern formation in limbs (Wolpert 1977).
(4) One might want to incorporate a less direct mapping from genome to genetic regulatory network, using instead one that lends itself more naturally to operators that splice out or insert genes, affecting the size of the regulatory network.

Our future work involves, as suggested, extending the model towards actually functioning autonomous agents. We will examine whether it is possible to co-evolve sensor/actuator placement in an organism together with a control structure - or nervous system, if you will - based on non-linear neural networks. To that end, we will associate colors with real functional 'cell-types' like neuron, sensor and actuator, and then let evolved organisms perform some task in a simulated environment, evaluating them on basis of performance of that task. An obvious candidate, and easy to implement, is chemotaxis. Evolving non-linear neural networks for controlling a chemotactic agent has already been done within our research-group (Beer et al. 1992), and it will be of considerable interest to compare the two approaches.

Acknowledgments

We would like to thank everyone in the autonomous agents research group at CWRU for the fruitful discussions that helped shape this work. Special thanks to Leslie Picardo, Hurkan Balkir and Katrien Hemelsoet for their comments on an earlier draft of this paper. This work was supported in part by grant N00014-90-J-1545 from the Office of Naval Research.

References

Alberts, B., D. Bray, J. Lewis, M. Raff, K. Roberts and J. D. Watson. 1991. *Molecular biology of the cell.* New York: Garland Publishing.

Beer, R. D. and J. C. Gallagher. 1992. "Evolving dynamical neural networks for adaptive behavior." *Adaptive Behavior* **1** : 91-122.

Davidson, E. H. 1990. "How Embryos work: a comparative view of diverse modes of cell fate specification." *Development* 108: 365-389.

Dawkins, R. 1989. "The Evolution of Evolvability." In *Artificial Life.*, edited by C. G. Langton. Reading, MA: Addison-Wesley.

deBoer, M. J. M., F. D. Fracchia and P. Prusinkiewicz. 1992. "Analysis and Simulation of the Development of Cellular Layers." In *Artificial Life II.*, edited by C. G. Langton, C. Taylor, J. D. Farmer and S. Rasmussen. 465-483. Reading, MA: Addison-Wesley.

Duvdevani-Bar, S. and L. Segel. 1994. "On Topological Simulations in Developmental Biology." *Journal of theoretical Biology* 166: 33-50.

Fleischer, K. and A. H. Barr. 1994. "A Simulation TestBed for the Study of Multicellular Development: The Multiple Mechanisms of Morphogenesis." In *Artificial Life III.*, edited by C. G. Langton. 389-416. Reading, MA: Addison-Wesley.

Gruau, F. and D. Whitley. 1993. "The cellular development of neural networks: the interaction of learning and evolution." Research Report 93-04, Laboratoire de l'Informatique du Parallélisme, Ecole Normale Supérieure de Lyon.

Harp, S. A., T. Samad and A. Guha. 1989. "Towards the Genetic Synthesis of Neural Networks." In *Proceedings of the Third International Conference on Genetic Algorithms.*, edited by J. D. Schaffer. 360-369. San Mateo, CA.: Morgan Kaufmann.

Harvey, I., P. Husbands and D. Cliff. 1993. "Issues in evolutionary robotics." In *Proceedings of the Second International Conference on Simulation of Adaptive Behaviour.*, edited by J. Meyer, H. Roitblat and S. Wilson. Cambridge, MA.: MIT Press.

Jackson, E. R., D. Johnson and W. G. Nash. 1986. "Gene Networks in Development." *Journal of theoretical Biology* 119: 379-396.

Kauffman, S. 1969. "Metabolic Stability and Epigenesis in Randomly Constructed Genetic Nets." *Journal of theoretical Biology* 22: 437-467.

Kauffman, S. and S. Levin. 1987. "Towards a General Theory of Adaptive Walks on Rugged Landscapes." *Journal of theoretical Biology* 128: 11-45.

Kitano, H. 1990. "Designing neural networks using genetic algorithm with graph generation system." *Complex Systems* **4** : 461-476.

Lewis, M. A., A. H. Fagg and A. Solidium. "Genetic Programming Approach to the Construction of a Neural Network for Control of a Walking Robot." In *IEEE International Conference on Robotics and Automation*, Nice, France. 1992.

Matela, R. J. and R. J. Fletterick. 1979. "A topological exchange model for self-sorting." *Journal of theoretical Biology* 76: 403-414.

McGinnis, W. and M. Kuziora. 1994. "The Molecular Architects of Body Design." *Scientific American* **270** 2: 58-66.

Mjolsness, E., D. H. Sharp and J. Reinitz. 1991. "A Connectionist Model of Development." *Journal of theoretical Biology* 152: 429-453.

Nolfi, S. and D. Parisi. 1991. "Growing neural networks." Report PCIA-91-15, Institute of Psychology, C.N.R.-Rome.

Thieffry, D. and R. Thomas. "Logical synthesis of regulatory models." In *Proceedings, Self-Organization and Life: From Simple Rules to Global Complexity, European Conference on Artificial Life (ECAL-93)*, Brussels, Belgium. 1993.

Walbot, V. and N. Holder. 1987. *Developmental Biology.* New York: Random House.

Wilson, S. W. 1989. "The Genetic Algorithm and Simulated Evolution." In *Artificial Life.*, edited by C. G. Langton. 157-166. Reading, MA: Addison-Wesley.

Wolpert, L. 1977. "Pattern Formation in Biological Development." *Scientific American* **239** 4: 154-164.

Wuensche, A. 1994. "The Ghost in the Machine: Basins of Attraction of Random Boolean Networks." In *Artificial Life III.*, edited by C. G. Langton. 465-501. Reading, MA: Addison-Wesley.

Bifurcation Structure in Diversity Dynamics

Mark A. Bedau

Reed College, 3213 SE Woodstock Blvd., Portland, OR 97202, USA
Email: mab@reed.edu

Alan Bahm

Photon Kinetics, 9405 SW Gemini Drive, Beaverton, OR 97005, USA
Email: bahm@reed.edu

Abstract

We propose a measure of total population diversity D of an evolving population of genetically specified individuals. Total diversity D is the sum of two components, within-gene diversity W_g and between-gene diversity B_g. We observe the dynamics of diversity in the context of a particular model, a two-dimensional world with organisms competing for resources and evolving by natural selection acting implicitly on genetic changes in their movement strategies. We examine how diversity dynamics and population performance—measured as the efficiency with which the population extracts energetic resources from its environment—depend on mutation rate and the presence or absence of selection.

Systematic exploration of mutation rates reveals a bifurcation into qualitatively different classes of diversity dynamics, whether or not selection is present. Class I: At low mutation rates, diversity dynamics exhibit "punctuated equilibria"—periods of static diversity values broken by rapid changes. Class II: At intermediate mutation rates, diversity undergoes large random fluctuations without always approaching any evident equilibrium value. Class III: At high mutation rates, diversity is stable, with small fluctuations around an equilibrium value. Optimal population performance occurs within a range of mutation rates that straddles the border between class I and class II. The relationships among diversity D and its components W_g and B_g reflects the typical features of these different classes of diversity dynamics as well as corresponding differences in the gene pool, which ranges from genetically similar individuals in class I to genetically dissimilar individuals in class III. The fact that class I dynamics occur whether or not selection is present suggests that stochastically branching trait transmission processes have an intrinsic tendency to exhibit punctuated equilibria in population diversity over a critical range of branching (mutation) rates.

1 The Evolution of Diversity

Complex adaptive systems are embodied in many settings, ranging from ecological populations of organisms, through immune systems of antigens and antibodies, even to networks of neurons in the brain. By abstracting away the diverse details, one can model complex adaptive systems at a level of generality that might reveal fundamental principles governing broad classes of such systems—this we take to be the working hypothesis of artificial life [1].

One reason for the impressive effects in many artificial life models is their "emergent" architecture: The system's global adaptive behavior emerges unpredictably from an explicitly modeled population of low-level individuals. We have been studying a class of models consisting of a population of computation agents (basically, individual computer programs) that interact with each other and with their environment in a way that allows natural selection implicitly to shape their strategies for achieving various global computational goals [9, 2, 3, 5, 4]. We define statistical "macrovariables"—loosely akin to thermodynamic macrovariables like pressure or temperature—that reflect fundamental aspects of a system's adaptive behavior. Then we try to identify simple laws relating these macrovariables to other fundamental system parameters and we try to use these macrovariables to identify and explain basic classifications of systems.

An obvious but striking feature of complex adaptive systems is the evolutionary dynamics of population diversity. How can population diversity be defined and measured? How does diversity change as a population evolves? How do diversity dynamics vary as a function of other fundamental system parameters, such as mutation rate and selection pressure? Does population diversity define qualitatively different kinds of evolving systems? The present study addresses these questions (see also [3, 5]).

2 A Simple Model of Evolution

The model used here is designed to be simple yet able to capture the essential features of an evolutionary process [9, 2, 3, 5, 4]. The model consists of organisms (sometimes called "bugs") moving about in a two dimensional world. The only thing that exists in the world besides the organisms is food. Food is put into the world in heaps that are concentrated at particular locations, with levels decreasing with distance from a central location.

Food is refreshed periodically in time and randomly in space. The frequency and size of the heaps are variable parameters in the simulation.

The food represents energy for the organisms. Organisms interact with the food field by eating it at their current site at each time step, decrementing the food value in the environment and incrementing their internal food supply. Organisms must continually replenish their internal food supply to survive. Surviving and moving expend energy. Organisms pay a tax just for living and a movement tax proportional to the distance traveled. If a organism's internal food supply drops to zero, it dies and disappears from the world. On the other hand, an organism can remain alive indefinitely if it can continue to find enough food. Any evolutionary learning that occurs in the model is the effect of the one stress of continually finding enough food to remain alive. A good strategy for flourishing in this model would be to efficiently acquire and manage vital energetic resources.

It is important to note that selection and adaptation in the model are "intrinsic" or "indirect" in the sense that survival and reproduction is determined solely by the contingencies involved in each organisms finding and expending food. No externally-specified fitness function governs the evolutionary dynamics [9, 2].

The organisms in this model follow individually different strategies for finding food (and hence are sometimes called "strategic bugs" [2]). The behavioral disposition of bugs is genetically hardwired. A behavioral strategy is simply a map taking an organism's current sensory state—information about its present local environment (the five site von Neumann neighborhood)—to a vector indicating the magnitude and direction of its subsequent movement:

$$S : (s_1, ..., s_5) \rightarrow \vec{v} = (r, \theta). \qquad (1)$$

A bug's sensory state has two bits of resolution for each site in its its local environment, allowing the bug to recognize four food levels at each site (least food, somewhat more food, much more food, most food). Its behavioral repertoire is also finite, with four bits of resolution for magnitude r (zero, one, ..., fifteen steps) and three bits for direction θ (north, northeast, east, ...). A unit step in the NE, SE, SW, or NW direction is defined as movement to the next diagonal site, so its magnitude is $\sqrt{2}$ times greater than a unit step in the N, E, S, or W direction. Each movement vector $\vec{v}$ thus produces a displacement (x, y) in a square space of possible spatial destinations from a bug's current location.

The graph of the strategy map S may be thought of as a look-up table with 2^{10} entries, each entry taking one of 2^7 possible values. This look-up table represents an organism's overall behavioral strategy. The entries are input-output pairs that link a specific behavior (output) with each sensory state (input) that an organism could possibly encounter. The input entries in the look-up table represent genes or loci, and the movement vectors assigned to them represent alleles. Since bugs have 1024 genes or loci, each containing one out of a possible 128 alleles or behaviors, the total number of different genomes is 128^{1024}. Although finite, this space of genomes allows for evolution in a huge space of genetic possibilities, which simulates the much larger number of possibilities in the biological world.

When a bug's internal food supply crosses a threshold, it produces some number of offspring by asexual budding. After reproduction, the parental food supply is divided equally among the new children and the parent(s). Parental genes are inherited with some probability of mutation. Point mutations of the genes change the output values of entries in a child's look-up table. The mutation rate μ determines the probability with which individual loci mutate during reproduction. At the limit of $\mu = 1$, every allele has probability one of mutating and thus each child's alleles are chosen completely randomly.

While mutation rate is an explicit parameter of the model, selection pressure is controlled indirectly by adjusting other explicit parameters. The parameter *output noise*, P_0, is defined as the probability that the behavior *actually* performed by a bug on a given occasion in a given local environment will be chosen *randomly* from the 2^7 possible behaviors, rather than determined by the bug's genes. If $P_0 = 1$, then natural selection has no opportunity to "test" the usefulness of the behavioral traits encoded in a bug's genome. The bugs are still subject to differential survival and reproduction, and so there is a sort of "selection," but the alleles or traits transmitted in reproduction reflect only random genetic drift. There is heritable genetic variation but no heritable *phenotypic* variation, so natural selection plays no role in shaping the evolution of either genotypes or phenotypes. In simulations reported in this paper, output noise P_0 was set to either zero or one, thus creating pairs of simulations in which all model parameters were identical except for the presence or absence of selection's effects on the course of evolution.

This model is a very abstract and idealized representation of a population of evolving organisms, and has many biologically unrealistic respects. Nevertheless, our working hypothesis is that this model captures many fundamental aspects of evolving systems, and is thus a useful way to investigate the essential aspects of more complex evolving systems.

3 Measures of Diversity

How might population diversity be measured? (To simplify terminology, in what follows "diversity" always means *population* diversity.) Our proposal, very roughly, is to represent the population as a cloud of points in an abstract genetic space, and then define its diversity as the spread of that cloud. In the present model, an allele is a movement vector, a spatial displacement triggered by the sensory state corresponding to a given local environment. An individual's genotype is a complete set of spatial displacements, one for each possible sensory state. To capture the total population diversity, D, then, collect all the displacements of all bugs in all environments into a cloud, and measure the spread or variance of that cloud. (One can define related measures of diversity based on information-theoretic uncertainty rather than variance [5].) More explicitly, we define total diversity as the mean squared deviation between the average movement of the whole population, averaged over all individuals and over all sensory states, and the individual movements of particular individuals subject to particular conditions, i.e.,

$$D = \frac{1}{IJ} \sum_{i=1}^{I} \sum_{j=1}^{J} [(x_{ij} - \bar{x}^{IJ})^2 + (y_{ij} - \bar{y}^{IJ})^2] \quad (2)$$

where I is the number of individuals i, J is the number of sensory states for local environments (or, in the present model, genes) j, (x_{ij}, y_{ij}) is the movement vector of individual i subject to input j, and $\bar{x}^{IJ} = \frac{1}{IJ} \sum_{i=1}^{I} \sum_{j=1}^{J} x_{ij}$ (similarly for $\bar{y}^{IJ}$). So, $(\bar{x}^{IJ}, \bar{y}^{IJ})$ is the (x, y) displacement of the population averaged over all individuals i and sensory states (genes)j.

We can divide this total diversity D into two components. (Additional diversity components can also be defined and studied [3].) Collect the spatial displacements of each bug in the population corresponding to a given sensory state, i.e., the traits encoded across the population at a given gene locus, and calculate the spread or variance of this gene cloud. The average spread of *all* such gene clouds is a population's within-gene diversity W_g. More explicitly,

$$W_g = \frac{1}{IJ} \sum_{i=1}^{I} \sum_{j=1}^{J} [(x_{ij} - \bar{x}_j^I)^2 + (y_{ij} - \bar{y}_j^I)^2] \quad (3)$$

where $\bar{x}_j^I = \frac{1}{I} \sum_{i=1}^{I} x_{ij}$ (and similarly for $\bar{y}_j^I$). So, $(\bar{x}_j^I, \bar{y}_j^I)$ is the (x, y) displacement of the population in sensory state (gene) j averaged over all individuals i.

Now, form another, second-order collection of the centers of gravity of each gene cloud, i.e., a cloud of each "average" displacement across the population in a given gene. The spread or variance of this second cloud is the population's between-gene diversity B_g, which measures the diversity of the average population response across all sensory states (genes), thus:

$$B_g = \frac{1}{J} \sum_{j=1}^{J} [(\bar{x}_j^I - \bar{x}^{IJ})^2 + (\bar{y}_j^I - \bar{y}^{IJ})^2] \quad (4)$$

An easy calculation shows that the total diversity is the sum of the within- and between-gene components, $D = W_g + B_g$.

Absolute diversity values presented here reflect the size of the model's output space. (To compare diversity measurements across different size output spaces, measurements could be normalized by the size of output space; since all our simulations have the same size output space, we have not done this.) The maximum possible diversity value corresponds to the distribution in output space that is peaked at the four corners; in this case, each point is maximally distant from the mean (in this case, the center of output space). In the present context in which the maximum displacement is fifteen squares, the diversity value of this "corner post" distribution is the

sum of the x and y displacements from the mean, i.e., $15^2 + 15^2 = 450$. In a flat random distribution in our modified polar coordinate system of 128 possible movements, since movements in the NE, SE, SW, or NW directions are $\sqrt{2}$ times larger than movements in the N, E, S, or W directions, the diversity value of the flat distribution is $\frac{4[(1^2+\sqrt{2}^2)+(2^2+(2\sqrt{2})^2)+\cdots+(15^2+(15\sqrt{2})^2)]}{128} = 116.25$.

Intuitively, the flat distribution, not the "corner post" distribution, is maximally diverse, but the diversity measures defined here are higher for the "corner post" than for the flat distribution. (This situation is reversed when using analogous diversity measures in which variance is replaced by information-theoretic uncertainty [5].) However, recall that food is placed in the simulated world in heaps that slope away from the center and that the bugs pay a movement tax proportional to distance traveled. So, is is not surprising that observed diversity values exceed the value for the flat distribution only in special circumstances when selection is absent.

The relative proportions of the two diversity components reflects a population's genetic structure. Consider a population consisting of "random individuals," in the sense that each bug's alleles are chosen randomly from the set of possible alleles, different bug's genes being chosen independently. In this case, the distribution across the population at any given environment-gene will be a huge cloud covering the whole set of possible spatial displacements, so the population's within-gene diversity W_g will be quite large. Since the center of gravity of each of these huge clouds will be virtually the same point—the center of the space of possible behavioral displacements—the distribution of these centers of gravity will be quite tight, and so the between-gene diversity will be nearly zero, $B_g \approx 0$. The population's total diversity will approximately equal the within-gene diversity, $D \approx W_g$.

Another extreme case is a population consisting of genetically identical bugs that are "sensitive" to their environment in that they behave differently when they sense different environments. In this case, the within-gene diversity is zero, $W_g = 0$, since the average spread of the cloud of behavioral displacements at each environment-gene is nil. On the other hand, since the average behaviors in different environments are quite different, the between-gene diversity is large and equal to the total diversity, $D = B_g$.

Typical results from the first 10,000 time steps of simulations with mutation rate $\mu = 0$ is shown in Figure 1. Notice that the within-gene diversity W_g drops over time. In fact, when selection is present W_g reaches zero within 3000 time steps, at which point the entire population has become genetically identical. Furthermore, the between-gene diversity B_g increases over time until $B_g = D$ when $W_g = 0$. These effects also happen when selection is absent, but they typically take substantially more time.

When selection is absent, it is to be expected that pure genetic drift will produce this crossing of the W_g and B_g components: W_g will drop as stochastic sampling fixes more loci, and B_g will rise as different traits become

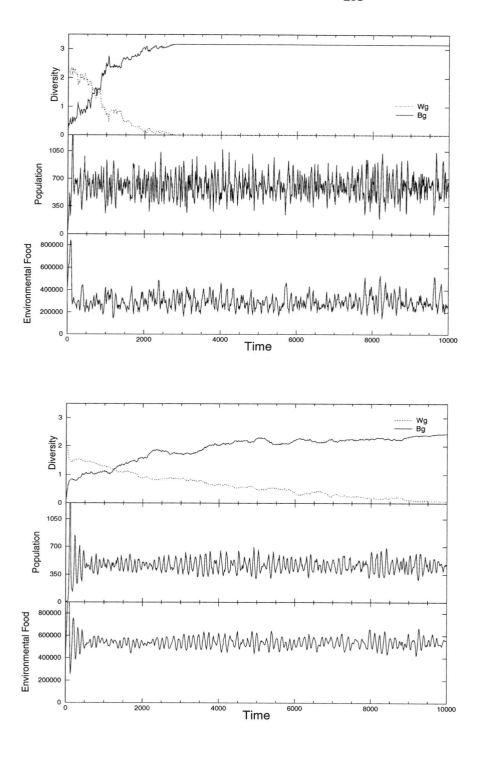

Figure 1: Within-gene W_g and between-gene B_g diversity, population level, and residual food in the environment in the first 10,000 time steps of a typical simulation with mutation rate $\mu = 0$. Top: selection ($P_0 = 0$). Bottom: no selection ($P_0 = 1$).

fixed. When selection is present, one expects the same component crossing but at a faster rate, since the effects of genetic drift are augmented by selection: W_g will drop as selection progressively weeds out traits at given loci, and B_g will rise as traits at different loci become specialized in different directions. Population level and residual environmental food data (Figure 1) also reveal the effects of selection, with selection supporting larger populations that extract more food from the environment.

4   The Bifurcation Structure

A bifurcation structure is a change in qualitative behavior caused as a parameter is smoothly varied [6]. Our observations suggest that there is a bifurcation structure in diversity dynamics as a function of the mutation rate μ, regardless of whether selection is present. As μ is varied, long-term diversity dynamics exhibit three qualitatively different patterns (summarized in Table 1). The transitions between these dynamics occurs at roughly the same mutation rates whether or not selection is present.

In class I, which is exhibited at low mutation rates, diversity dynamics exhibit "punctuated equilibria"— periods of static diversity values broken by rapid changes. In class II, which is exhibited at intermediate mutation rates, diversity undergoes large random fluctuations without approaching any evident equilibrium value. In class III, which is exhibited at high mutation rates, diversity is stable, with small fluctuations around an equilibrium value.

Some features of the bifurcation structure remain obscure. It is still uncertain how sharp the transitions are, and even how many transitions there are. For example, the transition between classes II and III might be quite smooth, and class II might be merely an extreme case of class III. Nevertheless, the qualitative character of class II dynamics is distinctive. Moreover, it is quite clear that classes I and III consist of two fundamentally different kinds of evolving systems. So, without prejudging the detailed nature of the bifurcation structure, we will use diversity D and its components W_g and B_g to describe the characteristic features of these three classes of behaviors.

The relationships among total diversity and its components reflects each kind of diversity dynamic. In class I, total diversity is dominated by its between-gene component, $D \approx B_g$ and $W_g \approx 0$. In class III, the situation is reversed; total diversity is dominated by the within-gene component, $D \approx W_g$ and $B_g \approx 0$. In class II, neither component dominates total diversity, $D \gg W_g \sim B_g \gg 0$.

These effects can be made vivid by plotting the diversity of the diversity components, i.e., the extent to which the total diversity D contains a large contribution from *both* components W_g and B_g (Figure 2). The diversity of the components, $C = \frac{4W_g B_g}{D^2}$, reflects what fraction of the area of a square of side D is contained in a rectangle with sides $2W_g$ and $2B_g$. (The factors of 2 scale C so that $0 \leq C \leq 1$.) Note that $C = 0$ if $W_g = 0$ or $B_g = 0$, and $C = 1$ if $2W_g = 2B_g = D$.

These relationships among D, W_g, and B_g indicate corresponding differences among the genetic structure of the population in the three classes. Class I populations remain highly similar while class III populations remain highly dissimilar. Class II populations are intermediate between being similar and dissimilar; the extent of similarity or dissimilarity is continually shifting rapidly within an intermediate range.

It is striking that the full bifurcation structure occurs whether or not selection is present. When selection is absent because $P_0 = 1$, there is no connection between the behavioral strategy encoded in a bug's genes and the bug's chances of survival and reproduction. This creates something very much like a stochastically branching trait transmitting process. So, the fact that the qualitative dynamics does not depend on whether selection is in effect strongly suggests that there is a general tendency for evolving systems generally to exhibit all three kinds of dynamics.

It is also striking that maximal population performance, measured by the efficiency with which the population extracts energetic resources from its environment, occurs in a range of mutation values spanning the border between classes I and II. This suggests that the mutation rates around the transition between classes I and II optimally balance evolutionary learning's competing demands for "memory" (reflected in low mutation rates) and "novelty" (reflected in high mutation rates). Experiments in which the mutation rate itself is allowed to evolve corroborate this suggestion, for there is a robust tendency for mutation rates to evolve toward this same transition zone and the evolved mutation distribution can be pushed higher (or lower) by engineering situations that call for greater novelty (or memory) [4].

Further discussion of diversity dynamics occurs elsewhere [3, 5].

4.1   The Simulations

We measured total diversity D and its within-gene W_g and between-gene B_g components in a series of pairs of selection/no-selection simulations, smoothly varying the mutation rate μ (on a log scale). All other parameters of the model, including the size of the world and the food environment, were held constant. Many parameter settings was multiply sampled. We simultaneously measured two crude aspects of the "performance" of the population—the population level and the amount of residual food in the environment—on the assumption that higher population level and lower residual food reflects better evolutionary learning on the part of the population.

Each simulation consisted of 10^6 time steps (every bug moves, and perhaps eats and reproduced, each time step). Diversity was sampled one thousand times in each simulation. To make the diversity more clearly reflect the action of selection, we collected statistics over only that portion of the genome that received the vast bulk of use (the thirty six most frequently used genes).

Founder populations in all simulations consisted of 100 bugs that were assigned traits randomly, with displace-

Table 1: Three qualitatively different kinds of diversity dynamics.

	CLASS I	CLASS II	CLASS III
μ RANGE	$\mu < 10^{-3}$ (approx.)	$10^{-3} \leq \mu \leq 10^{-2}$	$10^{-2} < \mu$ (approx.)
DYNAMICS	punctuated equilibria	large erratic fluctuations	small, stable fluctuations about equilibrium value
DOMINATING COMPONENT	B_g	none	W_g
GENE POOL	highly similar	somewhat similar, somewhat dissimilar, always rapidly shifting	highly dissimilar

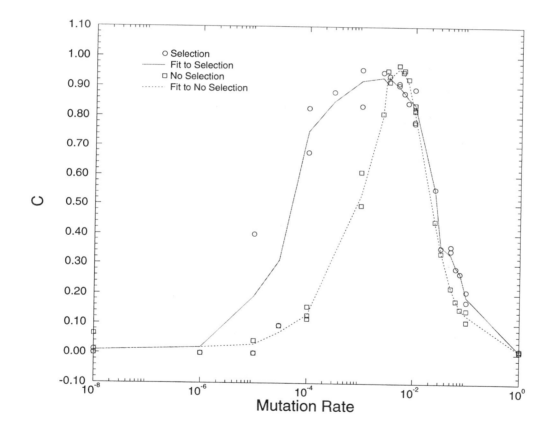

Figure 2: The time average of the diversity of the diversity components, $C = \frac{4W_g B_g}{D^2}$, as a function of mutation rate μ, for selection ($P_0 = 0$) and no selection ($P_0 = 1$). (The extreme left data points are for $\mu = 0$.) Fitted values for selection and no selection are included. The diversity of components evidently shows a sharper transition without selection.

ment direction chosen from the eight compass directions and distance in steps chosen from zero, one and two. Thus, in the founder population, the total diversity was relatively low, $D = 2.5$, and virtually all of the total diversity was in the within-gene component, $D \approx W_g$ and $B_g \approx 0$.

4.2 Class I: Punctuated Equilibria

When μ is low, diversity dynamics display vivid punctuated equilibria—this characterizes class I diversity dynamics. Selection and no selection examples of the typical dynamics of diversity for simulations with very low mutation rates are displayed in Figure 3. The total diversity D remains basically static for long periods of time, but this pattern of stasis is punctuated occasionally by very rapid changes in diversity. The resulting picture is characterized by relatively flat plateaus separated by abrupt cliffs.

As one might expect, as μ increases, so does the frequency of the punctuations; in addition, the stability during the periods of stasis declines. If μ gets large enough ($\mu \geq 10^{-3}$, in the present model), punctuated equilibria are no longer evident, and the system is no longer in class I. As μ declines toward 0, the periods of stasis become more stable and longer lasting, until, at $\mu = 0$, the population remains genetically identical forever (after an initial transient) and the diversity dynamics reveal one, indefinitely long, unbroken period of stasis.

Class I systems have populations of genetically identical (or, nearly identical) individuals—in effect, a population of (near) "clones." Different loci encode different traits, and this distribution of traits across loci abruptly shifts from time to time, when B_g is punctuated. Hence, total diversity is dominated by the between-gene diversity, $W_g \approx 0$ so $D \approx B_g$, as Figure 3 reveals. Although B_g always dominates D in class I, as μ approaches the border between class I and class II, W_g tends to occupy an increasingly significant share of D.

4.3 Class II: Erratic Fluctuations

At intermediate μ values, total diversity D continually exhibits large and rapid random fluctuations on different time scales. Illustrations of these characteristic class II dynamics, with and without selection, are shown in Figure 4.

In class II systems, neither W_g nor B_g is close to 0, so neither component dominates D (Figure 2). Although the relative share of D captured by the two components is always shifting—indeed, at some μ values the two components are wildly criss-crossing—each component always claims a significant portion of the total diversity.

The intermediate values of W_g and B_g in class II systems indicates the presence of populations which are neither highly similar nor highly dissimilar, but have a complex and shifting genetic structure that is intermediate between these extremes. Precisely characterizing the variable degree of relationship among individuals in class II populations is a topic of current research.

4.4 Class III: Stable Dynamics

When μ is toward the high end of the spectrum, total diversity D exhibits small noisy fluctuations around a stable equilibrium value—this is the signature of class III. Figure 5 shows two examples of typical diversity dynamics in class III, one with and one without selection.

As the mutation rate increases, the amplitude of the fluctuations decreases. When $\mu = 1$ the equilibrium diversity value corresponds to the precise numerical value of a "flat" distribution of alleles, regardless of whether selection is present. In the limit of large populations, the amplitude of fluctuations at $\mu = 1$ becomes arbitrarily low. If μ is below 1, selection pulls the equilibrium value down, while the equilibrium value when there is no selection remains at the value of the flat distribution (compare top and bottom, Figure 5).

Class III populations consists of genetically distinct individuals, each of which has a random collection of alleles. The total diversity is well approximated by the within-gene diversity, $D \approx W_g$. When μ falls near the border between class II and class III, B_g starts to comprise a significant share of D.

4.5 Population Performance

The bifurcation structure has an interesting connection with optimal population performance, measured by how much food the population can extract from the environment. Three robust patterns in residual environmental food emerge (Figure 6). First, when selection is absent residual food is flat across the mutation scale and is significantly higher than in simulations with selection. This is to be expected; if $P_0 = 1$ and there is no selection, then a population's ability to find food should not depend on the mutation rate and it should be much worse than it would be if there were evolutionary learning.

Second, when natural selection shapes evolutionary learning, population performance worsens when μ approaches 1, and also when μ is extremely close to 0. This effect no doubt reflects the sacrifice of one of the two competing demands of evolutionary learning. On the one hand, the need to remember what has been learned calls for a sufficiently low mutation rate; on the other hand, the need to explore novel possibilities calls for a sufficiently high mutation rate. Thus, a very high mutation rate sacrifices a population's memory, while a very low mutation rate sacrifices its source of novelty. In either case, suboptimal performance results.

The third effect observed is that the range of maximal food extraction is when μ is broadly in the vicinity of the boundary between class I and class II. Our argument in the preceding paragraph implies that optimal evolutionary learning requires a mutation rate that appropriately balances the competing demands for memory and novelty. These optimal mutation rates evidently are found around the border between regimes I and II. This suggests that the bifurcation structure—which is independent of selection—is exploited to optimize evolutionary learning when selection is present. Experiments with evolving mutation rates tend to corroborate this suggestion [4].

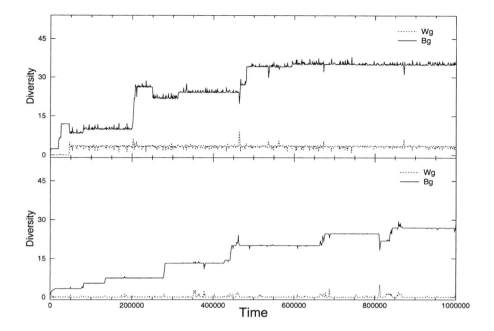

Figure 3: Typical class I punctuated equilibrium dynamics at $\mu = 10^{-5}$. Top: selection ($P_0 = 0$). Bottom: no selection ($P_0 = 1$). Punctuations tend to lead to increased diversity on this time scale in our simulations, since all our founder populations have low initial diversity. In the long run, punctuations continue indefinitely but show no general trend up or down.

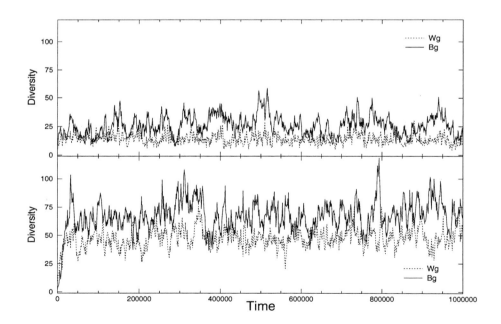

Figure 4: Typical class II dynamics with large random fluctuations at $\mu = 3 \times 10^{-3}$. Top: selection ($P_0 = 0$). Bottom: no selection ($P_0 = 1$).

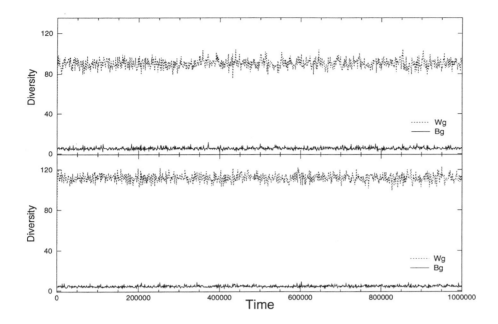

Figure 5: Typical class III stable dynamics at $\mu = 10^{-1}$. Top: selection ($P_0 = 0$). Bottom: no selection ($P_0 = 1$).

4.6 Generic Punctuated Equilibria

Artificial life systems commonly display punctuated equilibria in quantities like species concentration [8] and average fitness [7]. Yet the causes of these punctuated dynamics remain uncertain. Ecological complications such as host–parasite interactions or genetic complications such as extensive epistasis are typically thought to be implicated, and it is almost universally assumed that selection plays an essential role.

Our observations question these presumptions, since we routinely observe punctuated equilibria in population diversity when our model is in class I. None of the ecological or genetical complications usually thought to play a role are explicitly present in the model. For example, the population has no explicit division into anything like host and parasite and the genetic structure has explicit no epistasis. However, there might be some *implicit* genetic complications, such as multigene traits, in which environmental regularities trigger regular sequences of genes which caused a characteristic sequence of movements. If such implicit genetic complications are present, their magnitude is up in the air. In addition, it is true that the model could support the emergence of implicit sub-populations following competing or co-operating food-finding strategies. Such sub-populations would be revealed by a substantial within-gene diversity W_g, since the average trait at given loci must differ between the sub-populations. Since punctuated equilibrium dynamics frequently occur when $W_g \approx 0$, (e.g., Figure 3, bottom), implicit sub-populations clearly play no necessary role in punctuated equilibria generally.

What is most striking about these punctuations is their occurrence even when natural selection is absent.

Although punctuated equilibria in the absence of selection occur only when the mutation rate μ falls within an appropriate range near 0, the effect is quite robust. The presumption that punctuated equilibria in population diversity require the operation of natural selection is simply wrong. Therefore, even when punctuated equilibria occur *with* selection, without further evidence we cannot assume that selection plays any important role in its genesis. Evidently, there is an intrinsic tendency for evolving systems absent selection—that is, stochastically branching, trait-transmitting processes—to produce punctuated diversity dynamics, provided the branching rate is suitably poised. How to explain this effect remains a topic of current research.

5 A Science of Artificial Life

We are aiming to achieve two goals simultaneously: first, to develop plausible and useful measures of population diversity and, second, to use those measures to discern basic features of the evolution of diversity, ideally features that illuminate the fundamental nature of complex adaptive systems in general. Progress towards these goals is inextricably intertwined. One reason for finding our observations plausible and interesting is that our measures seem appropriate, and one sign that our measures are appropriate is that they reveal seemingly plausible and interesting effects.

To confirm the extent of our progress toward these goals requires further work. Analytical details remain to be settled, of course, and the need for a more precise statistical analysis of our effects calls for extensive further simulations. But the most important task is to determine the full generality of our results by replicating

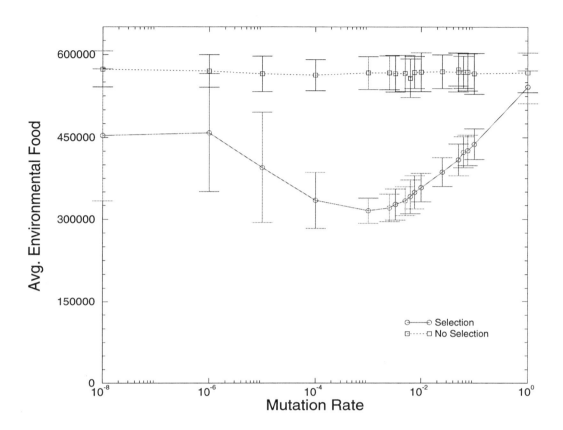

Figure 6: Time averages with standard errors of typical residual environmental food as mutation rate μ is varied, for selection ($P_0 = 0$) and no selection ($P_0 = 1$). (The data points on the extreme left are for $\mu = 0$.)

them in other models. Our observations were all made in one simple model of evolution, but the same measures can be implemented in other systems (or portions of systems)—both artificial and natural—in which all alleles at all genes share a metric. (And the uncertainty-based diversity measures can be implemented across a much broader range of systems [5].) The fact that the bifurcation structure exists both with and without selection is itself strong evidence that this structure is a fundamental feature of evolving systems generally. Final confirmation of the importance of our measures and the universality of our effects—and vindication of artificial life's fundamental working hypothesis—can come only from comparing quantitative results across a host of complex adaptive systems. Our effects provide one target for such comparisons.

Acknowledgements

For extensive helpful discussion, many thanks to Norman Packard and Marty Zwick. For expert programming assistance, thanks to Robert Seymour. Thanks also to audiences at Harvard University and the Rowland Institute for Science.

References

[1] M. A. Bedau, "Philosophical Aspects of Artificial Life." In F. J. Varela and P. Bourgine, *Towards a Practice of Autonomous Systems.* (Bradford/MIT Press: Cambridge, MA, 1992).

[2] M. A. Bedau and N. H. Packard, "Measurement of Evolutionary Activity, Teleology, and Life." In C. G. Langton, C. E. Taylor, J. D. Farmer, and S. Rasmussen, eds., *Artificial Life II.* SFI Studies in the Sciences of Complexity, Vol. X. (Addison-Wesley: Redwood City, CA, 1991).

[3] M. A. Bedau, F. Ronneburg, and M. Zwick, "Dynamics of Diversity in an Evolving Population." In R. Männer and B. Manderick, eds., *Parallel Problem Solving from Nature, 2.* (New York: Elsevier, 1992).

[4] M. A. Bedau and R. Seymour, "Adaptation of Mutation Rates in a Simple Model of Evolution." Submitted to Complex'94.

[5] M. A. Bedau, M. Zwick, and A. Bahm, "Measures of Diversity in a Simple Model of Evolution." Preprint.

[6] S. Eubank and D. Farmer, "An Introduction to Chaos and Randomness." In D. L. Stein, ed., *Lectures in the Sciences of Complexity,* SFI Studies in the Sciences of Complexity, Lec. Vol. I. (Addison-Wesley: Redwood City, CA, 1989).

[7] D. Hillis, "Simulated Evolution and the Red Queen Hypothesis." Biocomputation Workshop, Monterey, June 22-24, 1992.

[8] K. Lindgren, "Evolutionary Phenomena in Simple Dynamics." In C. G. Langton, C. E. Taylor, J. D. Farmer, and S. Rasmussen, eds., *Artificial Life II.* SFI Studies in the Sciences of Complexity, Vol. X. (Addison-Wesley: Redwood City, CA, 1991).

[9] N. H. Packard, "Intrinsic Adaptation in a Simple Model for Evolution." In C. G. Langton, ed., *Artificial Life.* SFI Studies in the Sciences of Complexity, Vol. VI. (Addison-Wesley: Redwood City, CA, 1989).

On Modelling Life

Chris Adami

W.K. Kellogg Radiation Laboratory 106-38
California Institute of Technology
Pasadena, CA 91125

Abstract

We present a theoretical as well as experimental investigation of a population of self-replicating segments of code subject to random mutation and survival of the fittest. Under the assumption that such a system constitutes a minimal system with characteristics of life, we obtain a number of statements on the evolution of complexity and the trade-off between entropy and information.

1 Introduction

What are the defining characteristics of life? For biologists and those studying living systems, this is the question that predates other questions and generates passionate arguments; for there are several ways how to answer this question. On the one hand, there is the philosophical approach. We would invent a "Turing Test of Life" in the tradition of the empiricists: "Anything that appears to be alive, is alive". On the other hand, we may draw up a list of characteristics and decide the question case by case, through a look-up table of sorts. To the scientist interested in the foundations of life, however, neither of these approaches is satisfactory. Rather, the scientific method applied to the problem of obtaining the defining characteristics of life would demand that we construct a *model* of life, one that displays the necessary attributes to pass the "Turing Test of Life", and then *strip it down to the most simple system that still displays these characteristics*. Anything that, if removed from this most simple system would result in a drastic change, removing the life-likeness, must be considered as a defining characteristic of this system. This does not preclude the existence of other systems passing the same test, with different characteristics. It does however establish the model under consideration as a baseline, and its characteristics as universal in its class. It is after this step has been taken that we can further enrich the model to learn more about life.

Clearly, this kind of approach is impossible using natural systems, as all life currently extant shows considerably more sophistication than needed just to survive. Thus, the construction of the minimally living system must be artificial. This has recently become possible with the advent of Tom Ray's tierra [1-3], and systems based on his idea. Briefly, tierra-like, (or "auto-adaptive genetic" [4]) systems consist of a population of self-replicating strings of information coded in a suitable manner that "live", and coevolve in an environment subject to mutation. The information contained in the self-replicating strings is the information necessary to carry out the task of self-replication, as well as any information that might give a certain genotype an advantage over another. Typically, this is information about the population itself, and the environment it lives in. Such a system displays a wealth of complexity and "life-like" characteristics [1-6]; enough to suggest answers to a number of fascinating questions relating to life.

Our approach then will be this: We assume that such systems of self-replicating strings *do* constitute a "minimal living system", and if not minimal then close enough to support the notion that the results obtained using the model are reflections of its "aliveness", rather than artifacts. While the theoretical results are independent of a particular implementation, the experimental ones are not. It is beyond the scope of this article to examine how close the particular system is to the abstract minimal living system, yet our results suggest that they are a good approximations, and in many respects universal. In the following we present theoretical arguments and experiments supporting them that suggest answers to a number of fascinating problems, including the emergence of complexity in living systems, and the nature of the evolutionary process.

Let us attempt to isolate the main characteristics of the "minimal living system". Primarily, we find that the cornerstone of this system is self-replication. The statistical mechanics of the ensemble of strings with and without self-replication are so vastly different (see Section 2) that we can determine with confidence to have isolated a defining characteristic of living systems. While this in itself is not surprising, it is specifically the self-replication of the information contained in the genome that is necessary (and that in this simple system is tantamount to replication of the entity). The consequences of this fact alone are manifold. Considering the evidence for genetic similarity in tRNA from the time of the origin of life to today [7], this capability seems to allow for the conservation of information over infinite periods of time in a noisy environment.

The keys for the evolution of complexity in this system

are: bit-wise mutation acting on the genome of members of the population, and survival of the fittest. The former is again an obvious choice for a necessary condition: Without mutation, a population of self-replicating strings shows no patterns, no evolution, and no complexity. Clearly, mutation works against the force that attempts to preserve information, yet at the same time mutation *creates* information, or rather transfers it from the environment into the genome of the strings, where again it will be preserved by self-replication. We find that "survival of the fittest" is synonymous with *survival of the most populous*. Thus, fitness in this system is directly related to replication rate. In the absence of a mechanism that enables one string to prevent the replication of another (kill-strategy), it is conceivable that "fitness = replication-rate" is a universal feature of such low-level organisms.

Another mystery of living systems is associated with the evolution of complexity and its apparent contradiction with the second law of thermodynamics, the law of increasing entropy. While it is known that dissipative systems involving microscopic irreversibility can evolve from disorder to order, the precise mechanism is often unclear. In this model, the evolution of complexity appears to proceed by a trade-off between entropy and information. Information is transferred from the environment into the genome of the population via chance mutations that allow the mutated string to exploit the information for a higher replication rate. As a consequence, the (information-theoretic) entropy of the population of strings drops by a commensurate amount until off-spring of the mutated string have reestablished disorder and the previous level of entropy is (roughly) restored. The time between such events is distributed according to a power law, and we thus observe periods of stasis interrupted by "avalanches of invention". Again, if we accept this system as a model for the most primitive and basic living system, the evidence suggests that evolution is Darwininan in nature even though not gradual, but rather punctuated: the picture promoted by Gould and Eldredge [8]. Below, we shall give more details and results of experiments that support the claims made here.

2 Statistical Mechanics of Self-Replicating Bit-Strings

Let us approach the model system from a theoretical point of view and write down a few of the basic equations governing the dynamics of self-replicating strings. Consider a system of N strings of ℓ instructions, where each instruction can take on p values ($p = 2$ for binary strings, $p = 4$ for DNA, and $p = 32$ for Ray's tierra). Also, let there be N_g different genotypes (different combinations of instructions) extant in the population, $N_g \leq N$, and let n_i denote the number of strings of genotype i. Then,

$$N = \sum_i^{N_g} n_i \ . \tag{1}$$

Let us further assume that the total number of strings in the population is constant: $dN/dt = 0$ (this constraint

can be relaxed trivially). Also, denote the replication rate (off-spring per unit time) of genotype i by ε_i. Defining the average replication rate (average "fitness")

$$\bar{\varepsilon} = \sum_i^{N_g} \frac{n_i}{N} \varepsilon_i \equiv \sum_i^{N_g} \rho_i \varepsilon_i \ , \tag{2}$$

where ρ_i is the genotype *density* of species i, we can immediately write down the equation that describes the time-evolution of n_i in a "mean-field" approximation

$$
\begin{aligned}
n_i(t+1) - n_i(t) &= (\varepsilon_i - \langle \varepsilon \rangle - R\ell)\ n_i(t) + \frac{N}{N_g} R\ell \\
&\equiv \gamma_i n_i(t) + C
\end{aligned}
\tag{3}
$$

with obvious definitions for γ_i and C, the "source" term. Here, we introduced the mutation rate R in units [mutations·(site · time)$^{-1}$]. Thus, $R\ell$ is the probability that a string of length ℓ is hit by a mutation per unit time, i.e. per instruction executed. The constant term on the right hand side of (3) is modelling mutations acting on genotypes $j \neq i$ and ensures $\dot{N} = 0$. Equation (3) can be solved in the equilibrium limit, $d\bar{\varepsilon}/dt = 0$:

$$n_i(t) = \frac{N}{N_g} \left\{ (1 + \frac{R\ell}{\gamma_i}) e^{\gamma_i t} - \frac{R\ell}{\gamma_i} \right\} \tag{4}$$

using an equilibrium boundary condition, $n_i(0) = N/N_g = \bar{n}$. For $C \approx 0$ this is simply

$$n_i(t) = n_i(0) e^{\gamma_i t} \tag{5}$$

where $\gamma_i = \varepsilon_i - \bar{\varepsilon} - R\ell$ is the "growth factor". Clearly, in equilibrium $\gamma_i = 0$ only for the best genotype and $\gamma_i < 0$ for inferior ones.

In order to make contact with statistical mechanics, define the *inferiority*

$$E_i = \varepsilon_{\text{best}} - \varepsilon_i \tag{6}$$

where $\varepsilon_{\text{best}}$ is the highest replication rate in the population. Then, $E_i \geq 0$ and we can define the ground-state ("vacuum") of the population by

$$\langle 0|E|0 \rangle = \bar{E} = 0 \ . \tag{7}$$

Let us also define the "fine-grained" (Shannon-) entropy

$$S = -\sum_i^{N_g} \frac{n_i}{N} \log(\frac{n_i}{N}) \tag{8}$$

with the property

$$\frac{\partial S}{\partial E} \simeq \frac{1}{R\ell} \ , \tag{9}$$

which suggests that the system of self-replicating strings is analogous to atoms with eigen-energies E_i in a heat bath of temperature $R\ell$ which initiates transitions $E_i \rightarrow E_j$ via mutations. In fact, detailed balance arguments lead to the equation determining the equilibrium distribution $n_i(E_i, \bar{E})$

$$\frac{1}{n_i} - \frac{1}{n_j} = \frac{1}{C}(E_i - E_j) \tag{10}$$

with the solution

$$n_i(E_i, \bar{E}) = \frac{C}{E_i - \bar{E} + R\ell} . \qquad (11)$$

In equilibrium, we find $\bar{E} \to R\ell$, universally. However, this equilibrium is disrupted by spontaneous phase transitions triggered by mutations creating genotypes with $\varepsilon_i > \varepsilon_{\text{best}}$ with a "latent heat" $L \doteq \Delta\varepsilon = \varepsilon_{\text{best}}^{\text{new}} - \varepsilon_{\text{best}}^{\text{old}}$. After such an event the entropy of the system defined by (2) drops, and then rebounds. In Fig. 1 we compare the entropy of a system of strings subject to mutation with or without replication in the **avida** environment [6][1]. If we start both systems in the uniform state $\bar{E} = 0$ (which implies $S = 0$: all strings are identical), we see that the self-replicating population reaches a plateau with $S < \log(N)$ while the non-self-replicating system reaches the maximum value rather quickly. In addition, the population of self-replicating strings can lower the entropy spontaneously, a feat that is impossible for the non-replicating population which must strictly adhere to the second law of thermodynamics.

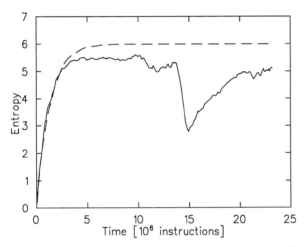

FIG. 1 Entropy of a population of 400 self-replicating (solid line) and non-replicating (dashed line) strings started in the uniform state.

3 Entropy and Information

Let us investigate more closely the time-evolution of entropy in systems of self-replicating entities. From the point of view of statistical mechanics rather than information theory, entropy is just the logarithm of the number of states in the volume of "phase-space" occupied by the population. Genetic "phase-space", or genotype-space, is spanned by the combinatorics to assemble the basic building blocks (or bits), and is finite for finite length strings. The population forms a "cloud" in genotype-space that moves slowly according to the fitness-landscape. Each genotype moves along a trajectory determined by a gradient term (adaptive process) and a noise term (mutations). As a consequence, the

---

[1]We chose this system because the number of strings can be kept strictly constant in **avida**.

time-evolution of the genotype of a string is determined by the usual Langevin equation. While genotype-space is very high-dimensional, the number of states occupied by the population is by definition N_g, and thus,

$$S = \log(N_g) \qquad (12)$$

which is equivalent to the fine-grained entropy in the large N limit. The number of genotypes N_g is an equilibrium quantity, depending on mutation rate $R\ell$, total number of strings N, and most importantly, the average inferiority of the population. Clearly, only in infinite systems after infinite time is $\log(N_g) = \ell \log(p)$. For finite systems, the number of genotypes is determined by the rate of birth of new genotypes and the rate of extinction of existing ones. The rate of birth of new, viable, genotypes is determined by the ratio of "cold" to "hot" spots in the genome. A cold spot on the string is an instruction that, if mutated, leads to either death or severe inferiority. Hot spots are instructions that are neutral under mutation and are thus responsible for the ε-degeneracy of genotypes. Thus, if there were no hot spots in the genome, there could not be any diversity, and in fact no evolution could take place, as was conjectured some time ago [9].

Let us denote the number of cold spots in a string of length ℓ by ν_c. Then the birth rate of new genotypes is $R\ell N$, while the birth rate of *viable* genotypes is $(1 - \nu_c/\ell)R\ell N$. On the other hand, the extinction rate of existing (viable) genotypes depends (again in a mean-field approximation) on the average inferiority of the population, $\bar{E}$. Clearly, for zero inferiority there will be no extinctions, while extinctions are maximal during a phase transition, when the average inferiority is high. We thus find

$$\dot{N}_g = NR\ell(1 - \frac{\nu_c}{\ell}) - \bar{E}N_g \qquad (13)$$

and thus in equilibrium

$$N_g = N\frac{R\ell}{\bar{E}}(1 - \frac{\nu_c}{\ell}) . \qquad (14)$$

We are now ready to examine the dropping of entropy observed through phase transitions. The decrease seen in Fig. 1 is of course due to the discovery of information and a corresponding new $\varepsilon_{\text{best}}$. Through such a transition, $\bar{E} \to \bar{E} + \Delta\varepsilon$ and $\Delta S = \Delta\varepsilon/(R\ell)$. However, with time $\bar{E}$ relaxes, and the entropy rebounds, but not necessarily to its previous equilibrium value. Consider for instance the situation where the information gained in the process requires more cold spots ν_c for storage than before. Then, a phase transition effectively freezes some spots that were originally hot spots, and thus reduces the effective length of the string. This is the situation in Fig. 1, the entropy rebounds, but reaches a lower plateau than before, due to the incorporation of information in the genome. We are thus witnessing a curious trade-off between information and entropy. In fact, it has been conjectured [10] that the physical entropy is not given by just the fine-grained entropy, but rather is the sum of Shannon entropy $S(\rho)$ and algorithmic complexity $K(\rho)$, the latter being a measure of information stored in a string. In this scenario, complexity can emerge without violating the second law, simply by trading information for entropy while keeping the physical entropy constant.

4 Fractal Structure of Evolution

Earlier we described how macroscopically the system is in an equilibrium state only rarely disrupted by chance mutations that happen to improve the fitness of the population. Here, we would like to take a more quantitative look at this behaviour, and work out the consequences for the structure of evolution.

More and more evidence is surfacing that certain quantities relating to living systems are distributed according to power laws, such as the frequency of extinction events of a certain size [11]. It has been known for a long time that there is a structure in the taxonomic system, since it was pointed out by Willis [12] that there is a very large number of taxa with only one or a few subtaxa, but only a few taxa with many subtaxa. Recently, there has been an extensive analysis of the frequency distribution of the number of taxa N that have n sub-taxa [13], with the result that $N \sim n^{-D}$ with $D \approx 2 \pm 0.5$. This fractal structure of evolution has been obtained from a pain-staking analysis of fossil records from the Cambrian to the Tertiary (families within orders) as well as a selection of catalogued flora and fauna (subclades within clades, families within orders, and species within genera). Yet, an explanation for this structure could not be offered. Generally speaking, such a distribution can be expected if no scale determines the length of time that a certain species dominates a population, and the number of subfamilies is proportional to this period of time. This hypothesis can be tested using the minimal living system.

Recently [5], we suggested that self-replicating systems typically operate in a self-organized critical state [14]. Drawing the analogy to the sandpile model, we could identify information as the agent that causes the self-organization, and which is transported via a dissipative process through the system. We also identified the growth-factor γ_i [see Eq. (5)] as the critical variable. In the equilibrium situation, the population is dominated by the current "best" genotype and its ε-degenerate offspring, all of which have $\gamma_i = 0$. An advantageous mutation however can disrupt this equilibrium by creating a new "best" genotype with $\gamma_i > 0$. Thus, the vacuum with $\bar{E} = 0$ is now a false one, and a phase transition must occur. This happens in the manner of the sandpile avalanches, with the new information transmitted through the system via the off-spring. Gradually, all genotypes with a subcritical replication rate will become extinct and replaced: the system has returned to its critical state. A necessary condition for the identification of the self-organized critical state is a power spectrum of the fitness history (replication rate as a function of time) that is of the power law type.

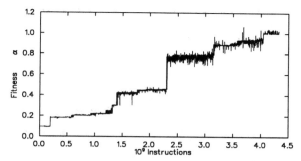

FIG. 2 Fitness curve for a typical run. The (normalized) fitness α of the most successful (i.e., most populous) genotype is plotted as a function of time, measured every million instructions for a mutation rate $R = 0.65 \times 10^{-3}$.

A typical fitness history (normalized replication rate of the most abundant genotype in the population as a function of evolutionary time) obtained using the tierra system is displayed in Fig. 2.

The associated power spectral density in Fig. 3 shows a near perfect $1/f^2$ spectrum over 4 orders of magnitude. While a power-law spectral density is a necessary condition for the identification of a self-organized critical state, it is not sufficient. Indeed, a $1/f^2$ spectrum is typical for a simple random walk, and thus for the most basic dissipative process. Incidentally, the process of evolution can be described as a guided random walk in genotype space, albeit a *non-Markovian* one.

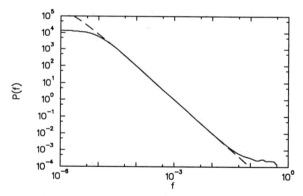

FIG. 3 Power spectrum $P(f)$ of a typical fitness curve $\alpha(t)$ (Fig. 2). The dashed line is a fit to $P(f) \sim f^{-\beta}$ with $\beta = 2.0 \pm 0.05$.

On the other hand, the distribution of waiting times between events of a certain size in a random walk model is exponential, rather than of the power law type. The latter is expected if the time between such events is not controlled by any scale. The fundamental time scales in self-replicating systems of the sort discussed here are the gestation time (the time for a string to gestate one off-spring) and the average time between two mutations hitting a single string (see Ref. [4]). Both of these time scales, however, are much smaller than the lifetime of single genotypes. Also, as discussed earlier in connection with the heat-bath analogy, the system can be understood as a superposition of a very large number of meta-stable states, with transitions between them induced by the mutations. In such a system, the time between avalanches or "events" is distributed according to a power law, as is the size and duration of the events [14]. Using the tierra system at a mutation rate of $R = 0.65 \times 10^{-3}$ (mutations per instruction executed), with a total number of 131,072 sites inhabited by typically 600-1,400 strings of length 60-150, we have measured the frequency distribution $N(\tau)$ of waiting times τ (lengths of "epochs") between phase transitions (which are signalled by a discontinuous jump in fitness as in Fig. 2). Figure 4 shows the *integrated* distribution function

$$M(\tau) = \frac{1}{\tau} \int_\tau^\infty N(t) dt \, , \qquad (15)$$

The integrated quantity is distributed with the same exponent as $N(t)$ (as can be shown by direct differentiation), but is more reliable due to improved statistics.

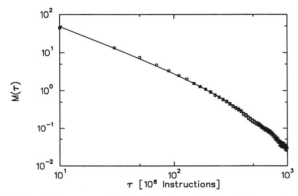

FIG. 4 Integrated distribution $M(\tau)$ of times between phase transitions τ (length of epoch) measured in millions of executed instructions.

The data were obtained from 50 runs under identical conditions (save the random number seed) resulting in 512 waiting times. The resulting curve was fit to the standard power law, cut-off by finite-size effects [2]

$$M(\tau) = c \, \tau^{-\alpha} \exp\left(-\frac{\tau}{T}\right) \qquad (16)$$

and obtained $\alpha = 1.1 \pm 0.05$ and $T = 450 \pm 50$, where all times are in units of millions of executed instructions. We expect the exponent of the power law to be universal, while the cut-off of course is not.

There are a number of fascinating consequences of such a distribution. If the time between phase transitions, as well as the size of fitness increases during such transitions are distributed according to a power law, curves such as the one depicted in Fig. 2 are fractal in nature, i.e. they (ideally) appear similar *at all scales* ("Devil's Staircase"[3]). If this is the case, we can imagine that a more sophisticated model, one that can run

many orders of magnitudes longer than the one we used, with unlimited fitness increases, would yield just such a curve, where each element of the curve when examined closely would have the same structure as the one in Fig. 2. Then, we must admit that fitness increases of enormous magnitude can happen almost instantly, driven by microscopic point-mutations *only* [4], albeit extremely infrequently. Also, the model implies that at each such phase transition a large number of genotypes goes extinct dynamically [see Eq. (14)] without external interference (this was also observed in [3]). Thus, Raup's distribution of extinction events would equally follow.

On the other hand, it is tempting to believe that the longer a species dominates a population, the more subspecies will be generated. As a consequence, the frequency distribution of taxa with n subtaxa, $N(n)$, must show the same critical exponent as the waiting time distribution [5].

5 Conclusions

We proposed to model life using artificial living systems that are "minimal" in their characteristics. We determined self-replication of information, bit-wise mutation of the information (providing the noise), and an environment of information (providing the gradient of the fitness-landscape) as necessary characters of such a model. The artificial life systems tierra and avida were used to determine universal characteristics of living systems, such as the evolution of complexity through transfer of information present in the environment into the genome, and its relation to the second law of thermodynamics. Furthermore, we were able to conjecture that living systems typically evolve towards a self-organized critical state and that this state determines fractal population structures. This also constitutes strong evidence for a punctuated equilibrium picture of evolution as favoured by the fossil record.

Acknowledgments

This work was supported in part by NSF grant # PHY90-13248 and a Caltech Divisional fellowship.

References

[1] T.S. Ray, in *"Artificial Life II"*, Proc. of the Santa Fe Institute Studies in the Sciences of Complexity Vol.X, C. Langton et al. eds., Addison-Wesley (1992), p. 371.

[2] T.S. Ray, *Artif. Life* **1** (1994)195.

---

[2] Since all runs were terminated after between 500 million and 4 billion instructions were executed, we cannot expect to measure the distribution of waiting-times larger than 500 million instructions with satisfactory accuracy.

[3] See, e.g., [15].

[4] In all of the previous discussion we have ignored mutations of the cross-over type, which effectively happen in systems such as tierra, but which can be viewed as *consequences* of point mutations. Very little is as yet known about the impact of cross-over mutations in the self-replicating systems described here, but we must assume that some of the phase transitions are in fact fitness improvements due to cross-over mutations.

[5] The discrepancy between Burlando's coefficient $D \approx 2$ and our coefficient $\alpha = 1$ is not understood.

[3] T.S. Ray, "Evolution, Complexity, Entropy, and Artificial Reality", Physica D, to be published.

[4] C. Adami, "Learning and Complexity in Genetic Auto-Adaptive Systems", Physica D, in print.

[5] C. Adami, "Self-Organized Criticality in Living Systems", KRL preprint MAP-167, Caltech (December 1993).

[6] C. Adami and C.T. Brown, "Evolutionary Learning in the 2D Artificial Life System 'Avida' ", these proceedings.

[7] M. Eigen et al., *Science* **244** (1989)673.

[8] S.J. Gould and N. Eldredge, *Nature* **366** (1993)223.

[9] J. Maynard Smith, *Nature* **225** (1970)563.

[10] W.H. Zurek, in *Complexity, Entropy, and the Physics of Information*, Proc. of the Santa Fe Institute Studies in the Sciences of Complexity Vol. VIII, W.H. Zurek, Ed. (1990), p. 73.

[11] D.M. Raup, *Science* **231** (1986)1528.

[12] J.C. Willis, *Age and Era*, Cambridge University Press, Cambridge, 1922.

[13] B. Burlando, *J. Theor. Biol.* **146**, 99 (1990); **163**, 161 (1993).

[14] P. Bak, C. Tang, and K. Wiesenfeld, *Phys. Rev. Lett.* **59** (1987)381; *Phys. Rev.* **A38** (1988)364.

[15] B. Mandelbrot, *The Fractal Geometry of Nature*, W.H. Freeman and Co., New York, 1977.

SHORT PAPERS

Genes, Phenes and the Baldwin Effect: Learning and Evolution in a Simulated Population

Robert M. French and Adam Messinger
Computer Science Department
Willamette University
Salem, Oregon 97301
french@willamette.edu and amessing@willamette.edu

Abstract

The Baldwin Effect, first proposed in the late nineteenth century, suggests that the course of evolutionary change can be influenced by individually learned behavior. The existence of this effect is still a hotly debated topic. In this paper clear evidence is presented that learning-based plasticity at the phenotypic level can and does produce directed changes at the genotypic level. This research confirms earlier experimental work done by others, notably Hinton & Nowlan (1987). Further, the amount of plasticity of the learned behavior is shown to be crucial to the size of the Baldwin Effect: either too little or too much and the effect disappears or is significantly reduced. Finally, for learnable traits, the case is made that over many generations it will become easier for the population as a whole to learn these traits (i.e. the phenotypic plasticity of these traits will increase). In this gradual transition from a genetically driven population to one driven by learning, the importance of the Baldwin Effect decreases.

1. Introduction

The view that evolution is influenced by acquired behaviors and traits is regarded by many as being uncomfortably close to the discredited Lamarckian contention that evolution consists of the inheritance of acquired behaviors and traits. It is perhaps for this reason more than any other that the evolutionary mechanism first proposed first by J. Mark Baldwin and Lloyd Morgan in 1896 (Baldwin, 1896; Morgan, 1896) is still veiled in controversy. This mechanism, known today simply as the Baldwin Effect, states that learned behavior and characteristics at the level of individuals can significantly affect evolution at the level of the species. Schull (1990) sums up the process as one in which "individual developmental responses will necessarily lead to directed and non-random evolutionary change." And while many evolutionary biologists accept the Baldwin Effect as a significant force in evolutionary change, the theory also has many detractors. For example, in a recent article, Piattelli-Palmarini (1990) writes, "One would have hoped that, in 1990, all talk of the Baldwin effect ... would have been mercifully forgotten."

Parisi, Nolfi, and Cecconi (1990) give three further reasons that evolutionary biologists tend to dismiss the Baldwin Effect. The orthodoxy of evolutionary biologists, they claim, is strongly reductionist, "which implies that the causes and basic mechanisms of evolution are only to be found at the level of genetics." As a consequence, behavior and learning, both being highly holistic processes, have been largely ignored in attempting to understand evolutionary processes. Another reason for the lack of attention, according to these authors, is that evolutionary biologists feel "behavior and learning are the province not of biology but of psychology and ethology." And finally, until recently, there have been very few empirical studies of the Baldwin Effect in either real or simulated populations.

In this paper, clear evidence is presented that the Baldwin Effect can indeed significantly alter the course of Darwinian evolution at the level of the genotype. In other words, we show that plasticity at the phenotypic level can and does produce directed changes at the genotypic level. In addition, the amount of plasticity and the amount of benefit is also demonstrated to be crucial to the size of the effect: either too little or too much and the effect disappears or is significantly reduced. This research confirms and extends earlier experimental work done by Hinton and Nowlan 1987, Belew 1990, Parisi, Nolfi, and Cecconi 1992, and Fontanari and Meir 1990, among others.

We empirically tested various aspects of the Baldwin Effect on populations similar to those described in Bedau 1993 and Holland 1993. We created a large population of agents with varying metabolic, feeding, locomotive and reproductive characteristics and allowed them to evolve in a world in which the amount and distribution of food varied over time. We then examined the effects of phenotypic plasticity on the evolution of the genotype. Three different areas of plasticity were considered namely, more efficient metabolism, movement or reproduction.

Genes and Phenes: an overview of the simulation

Our simulated world consists of a population of agents whose genetic material consists of a fixed-length bitstring. Food is regularly, but randomly, added to this world in discrete piles of uniform depth, much as if someone were regularly throwing handfuls of food into the world at random locations. The size of the piles of food and the frequency with which they are added to the world can vary. Food is required for energy. "Energy taxes" are levied for movement, reproduction and existence.

We stipulate that one particular genotype (i.e., one particular pattern of bits)—the "Good Gene" or GG, for short—will result in a fitness-enhancing behavior or trait—the "Good Phene" or GP—at the phenotypic level. We assume that the closer an agent is to the Good Gene (in terms of some metric on gene space), the easier it will be for it to learn the Good Phene that will enhance survival and reproductive possibilities. Phenes can either be learned or be the direct product of the possession of a Good Gene. A "natural" Good Phene (one which is the direct result of having the Good Gene) and a "learned" Good Phene are indistinguishable at the level of the phenotype. Possessing the Good Gene necessarily implies that its associated Good Phene will be expressed.

In our simulations, the Good Phene can be one of three things: improved locomotive, metabolic or reproductive efficiency. In each case, these were implemented as a reduction in the default taxes on each of these activities. In contrast to Hinton and Nowlan (1987), our simulation incorporates no explicit "fitness function" (Holland, 1975) to make an a priori determination of how good each genotype is. In other models, an explicit fitness function is generally used to determine the future reproductive success of each genotype.

In our model, however, the fitness of a particular genotype is determined implicitly by how well it survives in the population over time. The success of a particular genotype, then, is a measure of the percentage of agents in the population that have it. We are particularly interested in the evolution of the percentage of organisms in the population with the Good Gene. Specifically, we compare how often and under what circumstances the Good Gene appears in a population with learning compared to a population without learning.

Learning in the simulated world

Two factors will determine the extent to which the Good Gene will eventually proliferate in the population, namely:

- the difficulty across the population of learning the Good Phene (phenotypic plasticity).

- the amount of benefit conferred on an agent by having learned the Good Phene.

First, let us consider phenotypic plasticity. We assume that each agent is born with a genotype that is a certain Hamming distance from the Good Gene. Based on that distance, the probability that an agent will learn the Good Phene associated with the Good Gene is determined by a phenotypic plasticity curve like those shown in Figure 1. The x-axis represents the agent's normalized Hamming distance from the Good Gene (i.e., the number of bits differing from the Good Gene divided by the

total number of bits) and the y-axis shows the probability of learning the Good Phene. Therefore if the agent is genetically close to the Good Gene, it will stand a good chance of learning the corresponding Good Phene. As the normalized Hamming distance from the Good Gene increases, the probability of learning the Good Phene falls off. The shape of the curve depends on the Good Phene. Phenes range from being very hard (or impossible) to learn to comparatively easy to learn. For example, consider a Good Phene that is relatively hard to learn across the entire population (ρ is small; see Figure 3). When the genotype of a particular organism differs from the underlying Good Gene by half of its bits , it will virtually no chance of learning the Good Phene. Whether or not the agent will actually learn the Good Phene is based on the probability given by its phenotypic plasticity curve.

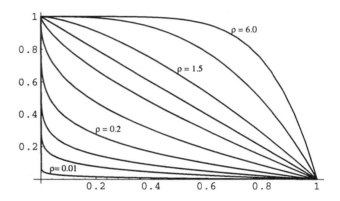

Figure 1: Curves showing varying degrees of phenotypic plasticity: $y = 1 - x^\rho$, where ρ is phenotypic plasticity associated with a particular Good Phene

Some Good Phenes are clearly harder to learn than others. For example, the trait of possessing blue eyes has no phenotypic plasticity. Individuals who do not possess the precise genotype that codes for blue eyes will never be able to learn this characteristic.

Other traits and behaviors are easier to learn. These are represented by the family of phenotypic plasticity curves shown in Figure 1. Each curve represents a different amount of native plasticity for a particular Good Phene. In a population where the plasticity parameter ρ is very low, an agent has to have been born very close to the Good Gene in order to have a chance of learning the Good Phene. In essence, when ρ is very low, the Good Phene is almost never learned.

However, as ρ increases, it becomes possible for some members of the population to actually learn the Good Phene. For example, consider the ability to hum Middle C. Some people can do it perfectly with no learning at all. These people possess the Good Phene—in this case, perfect pitch—from birth. Presumably, there is something in their genes that allows them to perform this task flawlessly. People who are genotypically a bit farther from the Good Gene may still be able to learn to hum Middle C, eventually, but will have to adopt a variety of clever strategies to do so. Most people, though, could try forever and still never be able to hum Middle C correctly. They are genetically too far from the Good Gene. This is a case of low, but non-zero, phenotypic plasticity.

Now, consider the ability to orient oneself. Again, some people are naturally very good at this particular task. For others, the task is much more difficult, requiring many clever strategies, gut given enough time they, they will eventually learn to orient themselves. A relatively small, but significant number of individuals in the population will never be able, try as they might, to avoid getting lost. Learning to orient oneself is an example of a behavior with average phenotypic plasticity.

Finally, certain behaviors are very plastic. For example, if the Good Phene consisted of "writing with your left hand" or "winking", virtually everyone in the population could master this. The phenotypic plasticity, ρ, for this trait is very high.

If the advantage conferred by the Good Phene is too low, then one would expect little movement of the population towards an increased incidence of the corresponding Good Gene. On the other hand, if the advantage conferred is extremely high, then any organisms, however few, who manage to learn it will beat out all competitors in the survival game. The Good Gene, however rare to begin with, will propagate and will soon come to dominate the population. Consider an extreme example that illustrates this point: if the (Very) Good Phene were, say, "Energy taxes drop to zero", then, even with no learning, once this Good Gene succeeded in entering the population by a lucky mutation, it would stay in the population forever. This would happen because the agent who happened to have been born with it would be able to survive and reproduce even in the complete absence of food.

The real evolutionary value of the Baldwin Effect therefore is that it gives good—but not extraordinarily good—genes an improved chance of remaining in the population. Extremely good genes will, in general, stay in a population. The benefit they confer is simply so overwhelming that they do not get washed out. But Extraordinary Genes are Extraordinarily Rare (and their existence would smack of saltationism (Mayr, 1988)). Most Good Genes only confer a slight selective advantage and, as such, are highly susceptible to random elimination (Gould, 1988). The Baldwin Effect, increases the number of individuals who *effectively* benefit from the Good Gene, by dint of acquiring the Good Phene. This allows genetic configurations close to the Good Gene that otherwise would not have survived, to remain in the population. The Good Gene's ultimate chances of survival are thereby ameliorated. If one accepts Darwinian gradualism, the value of the Baldwin Effect as a mechanism for protecting moderately good genes becomes apparent.

2. Experimental results

In the following experiments, we will first demonstrate that, even with no explicit fitness function, the Bald-win Effect can significantly influence evolving populations. We compare "Baldwin" populations (i.e., ones in which learning takes places) to non-Baldwin populations by means of three different measures namely:

- total population;

- number of agents in the evolving population who have the Good Phene (GP);

- number of agents in the evolving population who have the Good Gene (GG) that produced the original Good Phene.

The last measure the genetic shift towards the Good Gene is the most important for establishing that learning can have a significant effect on the genetic composition of the population. It turns out that the Baldwin Effect is most pronounced for phenes whose plasticity is neither too great nor too small.

In a later section of this paper, we will consider Good Phenes that confer more or less benefit to an organism. For the moment, though, we will hold this benefit constant and examine the effect of learning versus non-learning on the genetic makeup of the population.

Simulation 1: The Baldwin Effect

In the first experiment, using asexually reproducing agents, we chose a Good Gene whose naturally corresponding Good Phene decreased the organism's "energy tax" on movement. Any agent born with this particular Good Gene had phenotypic characteristics that allowed it to move around in its environment more efficiently than those that did not have it. We chose a phenotypic plasticity for this Good Phene of ρ of 0.01.

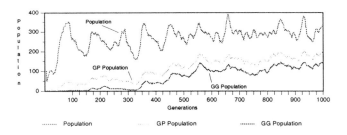

Figure 2: The effect of learning on the evolution of the genotype. Phenotypic plasticity, $\rho = 0.01$.

Recall that the probability that a given agent will learn the Good Phene is determined by how far the agent is from the Good Gene (Figure 1). The graph in Figure 2 shows the evolution of a population in which learning of the Good Phene is occurring. The top line shows the total population. The middle line indicates the number of agents in the population who have the successfully learned the Good Phene. In other words, these agents possess the Good Phene either by dint of learning it or because they were born with the Good Gene. Finally, the lowest line represents the number of agents in the population who actually possess the Good Gene.

In this population (Figure 2) it is apparent that, by as little as 500 generations, nearly half of the population has learned the Good Phene. This causes an total population increase compared to an equivalent population in which none of the members have the Good Phene (Figure 3). There is a consistent increase in total population directly related to the benefit the Good Phene confers.

The number of agents possessing the Good Gene gradually increases but will always remain significantly below the number of agents possessing the Good Phene because all that is required for improved survival is the Good Phene however it was come by and not necessarily the Good Gene. Individuals who did not obtain the Good Phene genetically and are not able to acquire it through learning will gradually be eliminated from the population by their fitter GP competitors.

Eventually virtually all organisms in the population will possess the Good Phene. At this point, the genotype will cease to evolve significantly. Darwinian selective pressure ceases because all of the organisms are now equally fit, at least with respect to the advantages conferred by the Good Phene. Nature cares only that the organism has a Good Phene and is not concerned with where it came from. In this simulation, by 5,000 generations, nearly all members of the population have acquired the Good Phene, and approximately 70% possess the Good Gene. Even after an additional 5,000 generations, the number of Good Genes in the population had not changed significantly. Once the entire population has become a GP-population, then the difference between the total population and the GG-population is directly proportional to phenotypic plasticity.

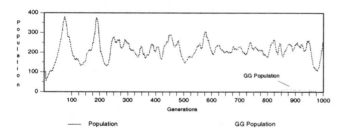

Figure 3: The same population with no phenotypic learning (notice the lack of change of the genotype). Phenotypic plasticity, $\rho = 0$.

Finally, in an identical non-learning population (Figure 3), the number of agents possessing the Good Gene remains insignificant. Clearly, the ability to learn the Good Phene has a significant influence on the evolution of the genotype, thus demonstrating the Baldwin Effect.

Simulation 2: When phenotypic plasticity is too high or too low

We have already seen in Figure 3 that, without phenotypic plasticity (i.e., when no learning is possible), the genotype of the population does not evolve towards the Good Gene. As phenotypic plasticity increases, there is

a corresponding increase in genotypic movement towards GG. But as phenotypic plasticity grows, the number of agents with the Good Gene actually begins to decrease and, for very high plasticity, there is virtually no trend at all towards a GG population (Figure 4). As in the first simulation, once the entire population possesses the

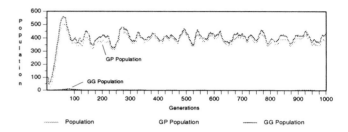

Figure 4: The Baldwin Effect ceases for high phenotypic plasticity (Note that the GG population remains insignificant)

Good Phene, the percentage of GG agents in the population stops increasing. If phenotypic plasticity is too high, the entire population will quickly acquire the Good Phene, effectively bringing to an end any further reason for the genotypic profile of the population to change. The topographic plot in Figure 5 shows how different phenotypic plasticities affect the percentage of agents in the population with the Good Gene. From this graph it can been seen that the Baldwin Effect disappears for high and low phenotypic plasticities.

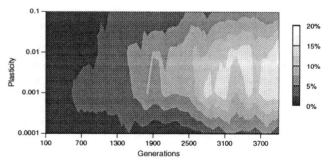

Figure 5: The effects of differing phenotypic plasticities on the evolution of the Good Gene in the population (lighter areas indicate higher GG percentages)

Simulation 3: The Baldwin Effect and the Quality of the Good Phene

If the Good Phene confers little or no selective advantage, then the question of its acquisition is of little importance. GP organisms will not survive better than non-GP organisms and, consequently, there will be little or no increase of GG organisms in the population. On the other hand, if the Good Phene is extremely good, thereby conferring a very large selective advantage, those individuals who somehow manage to acquire it will have a much better chance of surviving. These organisms will almost invariably out-compete their non-GP rivals and reproduce more successfully (this is especially true in asexual populations). Even if phenotypic plasticity is extremely low (or even zero), the progeny of a single organism that happened to stumble onto the (Very) Good Gene through random mutation would stand a good chance of going on to dominate the population. In Figure 6 below the "benefit factor" of the Good Phene is set very high and phenotypic plasticity is set very low. This figure shows that the Baldwin Effect is unnecessary when the quality of the Good Phene is very high.

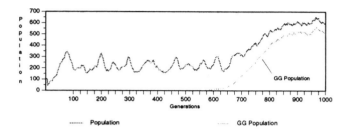

Figure 6: Effects on genotypic evolution of a very Good Phene

Evolution of phenotypic plasticity

In the simulations presented thus far in this paper, the ability to learn has remained constant for a particular Good Phene. However, in reality, the ability to learn is coded genetically. When there are "learning" genes that are also allowed to evolve, it can be shown that the phenotypic plasticity (i.e., ease of learning) of any beneficial, learnable trait increases over time. Over many generations, populations become smarter, i.e., better learners. Our simulated population will serve to illustrate how this works. Assume there are a number of individuals at a distance of, say, 3 bits from a particular Good Gene. For a given phenotypic plasticity, a certain percentage of this 3-bit-away sub-population will be able to learn the associated Good Phene. These "good learners" will be more likely to survive and pass on their learning abilities. Gradually, the 3-bit-away organisms many generations later will be better able to learn GP than the original generation. They will have become, in this sense, "smarter." The same phenomenon affects all of the organisms in the population, which means that GP gradually becomes easier to learn across the entire population. The phenotypic plasticity for GP evolves to a higher value. In one simulation in which learning ability was allowed to evolve, within a thousand generations, phenotypic plasticity had risen from $\rho = 0.01$ to $\rho = 1.1$ (Figure 7).

As we have seen, when phenotypic plasticity is sufficiently high, the presence of the Good Gene in the population is less critical. When phenotypic plasticity is high enough, learning becomes far and away the most frequent manner of acquiring the Good Phene (as opposed to acquiring it through the expression of the Good

Gene). In this gradual transition from a genetically driven population to a memetically driven one (i.e., a population driven by learning), the importance of Baldwin Effect decreases. (For a discussion of memes, see Dawkins, 1976.)

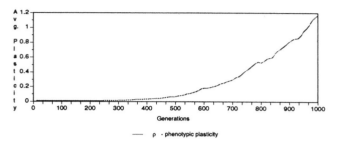

Figure 7: The evolution of ρ, the phenotypic plasticity

3. Comparison with Hinton & Nolan

In Hinton and Nowlan's (1987) simulation of the Baldwin Effect, the genotype for each organism in the population consists of a bitstring of twenty 0's, 1's and ?'s ("undetermined" alleles). A "phenotypic copy" (instantiated as the weights of a network in their model) of this bitstring is made on which "phenotypic learning" is done. Learning is done by randomly assigning 1's or 0's to all of the question marks approximately 1000 times, each time checking to see if an all-1's phenotype has been hit upon. If the all-1's phenotype is discovered during the 1000 learning trials, the original genotype is then assigned a high explicit "fitness" making it much more likely to be chosen as a parent for mating and crossover in the next time cycle.

Our model differs in certain significant respects from Hinton and Nowlan's model. Phenotypic learning, in particular, is implemented differently in our simulation. We start from the assumption that certain Phenes, as we have called them, are harder to learn than others in general, i.e., for all organisms in the population. Across the human population, for example, it is harder to learn perfect pitch than it is to learn to wink. In our model, differing degrees of phenotypic plasticity express those differences. With respect to learning a particular Phene, say, orienting oneself, certain individuals have a considerably harder time than others. This is a function of the organism's genetic Hamming distance from the phene in question.

In Hinton and Nowlan's model, phenotypic plasticity is fixed; there is no mechanism for varying the difficulty of learning according to the Good Phene under consideration. There is only one Good Phene and the difficulty of learning it depends solely on how far an organism's genotype is from it. In contrast our model allows for variation and evolution of phenotypic plasticity.

In addition, in Hinton and Nowlan, an explicit function is used to measure the fitness for reproduction of an individual. This function is implicit in our simulation, arising from the interaction of the agents in the world.

While the Good Phene does confer an efficiency advantage in our model, it does not expressly add to the fitness of the individual. In fact, in certain simulations when the reproduction tax was too low, the agents rapidly overpopulated the world and actually became extinct due to starvation.

By varying phenotypic plasticity, by allowing it to evolve and by modifying the amount benefit to the organism, we are able to more completely characterize the Baldwin Effect.

4. Conclusion

We have empirically demonstrated a pronounced Baldwin Effect in a simulated population of naturally evolving agents. In other words, the ability to learn at the phenotypic level had a significant effect on the genotypic evolution of the population. In addition, certain factors, in particular, the amount of phenotypic plasticity and the benefit associated with the learned phenotypic behavior or characteristic, have a significant influence on the amount of genotypic change produced. It also appears that excessively high or low levels of phenotypic plasticity have the same effect—namely, they are significantly less successful at promoting genotypic evolution than moderate levels of plasticity. Finally, the transition from genetic to memetic evolution as a result of upwardly evolving phenotypic plasticity curves, decreases the overall impact of the Baldwin Effect.

Acknowledgments

The authors would like to thank Mark Bedau, David Chalmers, Jim Friedrich and Melanie Mitchell for their suggestions and assistance with this paper.

References

[1] D. H. Ackley and M. L. Littman. Interactions between learning and evolution. In C. G. Langton, C. Taylor, J. D. Farmer, and S. Rasmussen, editors, *Artificial Life II*, pages 487–507, Reading, MA, 1992. Addison-Wesley.

[2] J. M. Baldwin. A new factor in evolution. *American Naturalist*, 30:441–451, 1896.

[3] M. A. Bedau and N. H Packard. Measurement of evolutionary activity, teleology, and life. In C. G. Langton, C. Taylor, J. D. Farmer, and S. Rasmussen, editors, *Artificial Life II*, pages 431–461, Reading, MA, 1992. Addison-Wesley.

[4] R. K. Belew. Evolution, learning, and culture: Computational metaphors for adaptive algorithms. *Complex Systems*, 4:11–49, 1990.

[5] R. Dawkins. *The Selfish Gene*. Oxford University Press, New York, NY, 1976.

[6] J. F. Fontanari and R Meir. The effect of learning on the evolution of asexual populations. *Complex Systems*, 4:401–414, 1990.

[7] S. J. Gould. *Wonderful Life*. W. W. Norton Company, New York, NY, 1988.

[8] G. E. Hinton and S. J. Nowlan. How learning can guide evolution. *Complex Systems*, 1:495–502, 1987.

[9] J. H. Holland. *Adaptation in natural and artificial systems*. MIT Press, Cambridge, MA, 1975.

[10] J. H. Holland. Escaping brittleness: The possibilities of general-purpose learning algorithms applied to parallel rule-based systems. In R. S. Michalski, J. G. Carbonell, and T. M. Mitchell, editors, *Machine Learning II*, pages 593–623, San Mateo, CA, 1986. Morgan Kaufmann.

[11] J. H. Holland. Echoing emergence: Objectives, rough definitions, and speculations for echo-class models. Technical Report 93-04-023, Santa Fe Institute, 1993.

[12] E. Mayr. *Towards a New Philosophy of Biology*. Havard University Press, Cambridge, MA, 1988.

[13] L. Morgan. On modification and variation. *Science*, 4:733–740, 1896.

[14] N. H. Packard. Intrinsic adaptation in a simple model for evolution. In C. G. Langton, editor, *Artificial Life*, pages 141–155, Reading, MA, 1989. Addison-Wesley.

[15] D. Parisi, S. Nolfi, and F. Cecconi. Learning, behavior, and evolution. In F. J. Varela and P. Bourgine, editors, *Proceedings of the First European Conference on Artificial Life*, Cambridge, MA, 1992. MIT Press/Bradford Books.

[16] M. Piattelli-Palmarini. Which came first, the egg-problem or the hen-solution? *Behavior and Brain Sciences*, 13(1):84–86, 1990.

[17] J Schull. Are species intelligent? *Behavior and Brain Sciences*, 13(1):84–86, 1990.

Evolving Multi-cellular Artificial Life

Kurt Thearling

Thinking Machines Corporation
245 First Street
Cambridge, MA 02142

kurt@think.com

Thomas S. Ray

ATR Human Information Processing Lab
2-2 Hikaridai, Seika-cho, Soraku-gun
Kyoto 619-02   Japan

ray@hip.atr.co.jp

Abstract

In this paper we describe a number of experiments in which the ideas of multi-cellular evolution are applied to digital organisms in an artificial ecology. The individual organisms are parallel programs in a shared memory virtual computer where evolution by natural selection is shown to lead to increasing levels of parallelism.

1   Introduction

One of the greatest challenges in the development of naturally evolving artificial systems is crossing the threshold from single to multi-cellular forms. From a biological perspective, this transition is associated with the Cambrian explosion of diversity on Earth. During the Cambrian explosion, most of the complexity that we see in living organisms emerged rather abruptly some six hundred million years ago. The work presented in this paper is based on the Tierra system [2] in which the evolving entities are self-replicating machine code programs. Multi-cellular digital organisms are parallel processes. From the computational perspective, the objective is to use evolution to explore the as yet under-exploited possibilities inherent in parallel processing.

Transferring the concept of multi-cellularity from the organic to the digital domain could take many forms. To make the transfer we must first understand what the most basic, essential, and universal features of multi-cellularity are, and then determine the form that these features would take in the completely different physics of the computational system into which evolution is being introduced. The features that we will capture in the present model are: 1) that multi-cellular organisms originate as single cells, which develop into multi-celled forms through a process of binary cell division; 2) that each cell of a multi-celled individual has the same genetic material as the original cell from which the whole developed; and, 3) that the different cells of the fully developed form have the potential for differentiation, in the sense that they can express different parts of the genome (i.e., each cell can execute different parts of the program).

In the digital metaphor of multi-cellularity, the program is the genome, and the processor corresponds to the cell. In organic biology, there is at least one copy of the genome for each cell, because genetic information can not easily be shared across cell membranes. In most current parallel architectures, the same holds: since memory is not shared, there is an area of memory associated with each processor (cell), and there must be at least one copy of the program code in the memory of each processor. This provides a very simple model of multi-cellularity: each digital cell consists of a unique block of memory with its own copy of the program and its own processor.

However, if the parallel machine has a shared memory architecture, making copies of the genome for each cell needlessly wastes memory and processing time (to copy the genetic information). In this context evolution by natural selection would not likely find any advantage in such waste. Thus a more logical and efficient implementation in this evolutionary context is to share a single copy of the program in a single block of memory among multiple processors. Each cell in a single organism corresponds to a parallel processor. Multi-cellularity can develop from a single original processor through a process analogous to cell division. The initial cell (processor) can issue an instruction which would then create another cell (a parallel processor). They may exhibit cell differentiation by having different processors executing different parts of the shared program. Obviously all cells will contain the same genetic material, since there actually will be only one copy per multi-cellular individual. The work presented in this paper is based on this shared memory model of multi-cellularity.

2   Tierra

The Tierra system has already been described in detail elsewhere [2-5] so it will be described only briefly here. The software used in this study is available over the net or on disk[1]. A new set of computer architectures and associated machine code have been designed to withstand the genetic operations of mutation and recombination. This means that computer programs written in the machine code of these architectures often remain viable after being randomly altered by bit-flips which cause the swapping of individual instructions with others from within the instruction set, or by swapping segments of code between programs (through a spontaneous sexual process). These new computers have not been built in silicon, but exist only as software prototypes known as "virtual computers," and have been called Tierra, Spanish for Earth.

Initially a self-replicating program was written in Intel machine language. This program was then implemented in

---

1. The complete source code and documentation (but not executables) is available via anonymous ftp from tierra.slhs.udel.edu and life.slhs.udel.edu in the file: tierra/tierra.tar.Z.

the first Tierran language in the fall of 1989. The program functions by copying itself one byte at a time to another location in memory and dividing (i.e., giving the copy its own instruction pointer). Subsequently, both programs replicate, and the number of programs "living" in memory doubles in each generation.

These programs are referred to as "creatures" or "organisms." The creatures occupy a finite amount of memory called the "soup." The operating system of the virtual computer, Tierra, provides a "slicer" service to allocate CPU time to the growing population of self-replicating creatures. When the creatures fill the soup, the operating system invokes a "reaper" facility which kills some creatures to insure that some memory will remain free for occupation by newborn creatures. Thus a turnover of generations of individuals begins when the memory is full.

The operating system also generates a variety of errors which act as mutations. One kind of error is a bit-flip, in which a zero is converted to a one, or a one is converted to a zero. This occurs in the soup, which is the RAM memory where the "genetic" information that constitutes the programs of the creatures resides. The bit-flips are the analogs of mutations, and cause swapping among machine code instructions. Another kind of error imposed by the operating system is called a "flaw." A flaw causes possible errors in calculations taking place within the CPU of the virtual machine, slight alterations during the transfer of information, or error in the location of memory accesses.

The machine code that makes up the program of a creature is the analog of the genome, the DNA, of organic creatures. Mutations cause genetic change and are therefore heritable. Flaws do not directly cause genetic change, and so are not heritable. However, flaws may cause errors in the process of self-replication, resulting in offspring which are genetically different from their parents. Those differences are then heritable.

The self-replicating program (creature) running on the virtual computer (Tierra), with the errors imposed by the operating system (mutations) results in precisely the conditions described by Darwin as causing evolution by natural selection [1]. This is therefore an instantiation of Darwinian evolution in a digital medium.

2.1 Adding multi-cellularity to Tierra

To implement the ideas of basic multi-cellularity, two additional Tierra machine instructions had to be created to allow the digital organisms to carry out development. The most obvious candidate for inclusion was an instruction that would create an additional CPU for the creature. This process was modeled on binary cell division on purpose so that the execution of this instruction (which is called **split**) takes a single processor (CPU) and produces two CPUs upon completion of the **split**. When a creature is given a new time-slice by the Tierra simulator, each CPU in that creature is allocated its own copy of that time-slice.

The process of executing a **split** instruction follows. A new CPU is created for the current creature. The registers, stack, and IP for this new CPU are copied from the CPU that issued the **split** instruction. To differentiate between the

CPUs after splitting, the DX register of each CPU is modified by shifting the value one bit to the left, and for the new CPU a value of 1 is added. As a result, each CPU's DX register has a different value after the **split**. If a sequence of **split** instructions is considered, the DX registers of each of the parallel CPUs within the creature will contain the address of their position in a binary tree of splits. For example, consider the following code fragment (assume that code is initially executed by a single CPU creature):

```
split    ; create 2 CPUs
split    ; split both CPUs, creating 4 CPUs
```

During the first split the DX registers for both CPUs are modified, first by shifting left by one bit (which has no effect since it is assumed that the DX register has previously been initialized to zero) and then by adding a value of 1 for the new CPU. Both CPUs then execute the second **split** instruction, creating two more new CPUs. The leftward shift and conditional addition to the DX register causes the four CPUs to end up with DX values of 00, 01, 10, and 11. This enumerates all four CPUs with different DX values from zero to three. In parallel processing terminology, the DX register contains the "self-address" of the parallel CPU.

In addition to the **split** instruction, an additional parallel construct was implemented: the **join** instruction. Once a CPU issues a **join** it waits until all other CPUs have also issued a **join**. Then all CPUs, other than the original single cell CPU, terminate. The **join** instruction was created to overcome Tierra's limitation of allowing only one CPU in a creature to issue a **divide** without causing an error. It was decided that an appropriate way to deal with this limitation was for each CPU to issue a **join** immediately before attempting to divide mother and daughter. There is no real organic biological analogy to the **join** instruction. It was introduced because it is a useful parallel programming tool. Another possible solution would have had a creature conditionally executing the **divide** instruction so that only one CPU actually performed the division of mother and daughter.

2.2 First steps

As with the original Tierra research, an "ancestor" was created to inoculate a new soup. The first parallel ancestor was designed to be very similar to the original serial ancestor described in [2]. Using the **split** and **join** instructions, it was possible to modify the original single celled ancestor and parallelize its functionality. The operation of the original single cell ancestor (as described in [2]) follows.

The ancestor first examines itself to determine where in memory it begins and ends. This is done by searching backward for the template that appears at its beginning and then searching forward for the template that matches its ending. To determine its size, the beginning address is subtracted from the end address. Space for the daughter is then allocated using this size information. The ancestor then calls the copy procedure which copies the entire genome into the daughter cell memory, one instruction at a time. Once the genome has been copied, it executes the **divide** instruction, which causes the creature to lose write privileges on the daughter cell memory, and gives an instruction pointer to the daughter cell (it also enters the daughter cell into the slicer and reaper queues). After this first replication, the mother cell does not examine itself again; it proceeds directly to the allocation of another daughter cell, then the copy procedure is followed by cell division, in an endless loop.

Only the copy loop was parallelized in the parallel ancestor, with one CPU copying half of the genome and another CPU copying the rest. Currently there is an arbitrary limit of sixteen CPUs per organism. The parallel ancestor uses only two CPUs. Any additional parallelism would have to evolve. The basic approach to parallelizing the single celled ancestor was to **split** immediately after allocating space for the daughter cell and to perform a **join** immediately before the **divide**. Within the copy loop the first CPU will copy the even numbered instructions and the second CPU will copy the odd numbered instructions.

Unfortunately the original instruction set was not very rich in its ability to manipulate registers. In fact, there was no way to operate using the DX register directly and therefore differentiating between parallel CPUs was very difficult. To operate on the DX register, it was necessary to push DX onto the stack and then pop it off into one of the registers that could be operated on. For example, consider the process of aligning one of the two parallel CPUs to copy the odd numbered instructions, conditional on the value in the DX register. The following code fragment performs this task:

```
pushC    ; save the CX register on the stack
pushD    ; push DX onto the stack
popC     ; now pop the value from DX into CX
ifz      ; if CX (aka DX) == 0
incA     ; inc AX (destination) to odd align
ifz      ; if CX (aka DX) == 0
incA     ; inc BX (source) to odd align
popC     ; return CX to its original value
```

Obviously it is difficult to achieve some fairly simple operations using the available instructions. The major cause of this difficulty results from an inability to act directly upon the register used differentiate between multiple CPUs in an organism (DX). As a result, any evolutionary activity that involves parallel CPUs must manipulate both the DX register as well as any other registers it uses actually to perform the desired operation. This greatly complicates the process and makes evolutionary improvements much more difficult to perform. Simple operations become somewhat "brittle" as a result of this limitation.

Figure 1 illustrates reproduction time[2] vs. time for a typical run using this instruction set. The new (parallel) ancestor starts out approximately twice as fast as the old ancestor, because it uses two CPUs rather than one. All of the evolutionary improvement is incremental, using serial processing improvements similar to the improvements in the original ancestor runs. No additional parallelism is added via evolution. Also note that the same parasitism and other natural phenom-

---

2. Unlike the results presented in [2] which focused on creature size, reproduction time is now used to describe the progress of evolution of a digital organism. Since multiple CPUs are now possible within a single creature, the relationship between size and reproduction time is such that reproduction time is no longer directly inferrible from the size. A long creature with many parallel CPUs might very well reproduce faster than a short creature with few parallel CPUs.

ena that are described in [2] are observed in these new runs with multi-cellular digital organisms, as well.

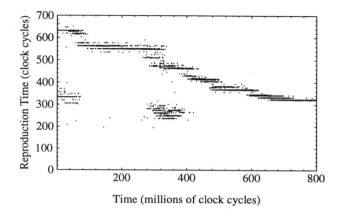

Figure 1: Evolution and the new (parallel) instruction set

2.3 Creating an evolvable parallel ancestor

This observed behavior was not unexpected. When the second author originally developed the Tierra instruction set, he first wrote the original ancestor and then included in the instruction set only those instructions which were used in that ancestor. Since multi-cellular digital organisms are naturally more complex than single cell digital organisms, an instruction set sufficient for a single cell organism would not necessarily suffice for a parallel organism.

To alleviate some of these problems, several additional instructions were added. These instructions perform operations that simplify the design of a multi-cellular ancestor. The first instruction added was **zeroD**, which explicitly zeros out the DX register. The second new instruction was **shr**, a shift right instruction for the CX register. Since the CX register typically contains the creature's size, a **shr** effectively divides the size by two. When performed in conjunction with a **split**, the CX register is modified to contain the portion of the genome to be copied by each parallel CPU.

The final two instructions that were added were based on known techniques for distributing work among parallel processes. An "offset" instruction takes the size (which is usually divided by the number of CPUs) and multiplies it by the CPU's self-address. This value specifies an offset into memory which evenly divides the memory among parallel CPUs. When added to a base address, the value specifies where in memory each parallel process should begin accessing data. Two versions of the offset instruction were created for the multi-cellular instruction set: **offAACD** and offBBCD. The **offAACD** instruction multiplies the CX register (size) times the DX register (self-address) and adds it to the AX register (the base address). The **offBBCD** instruction is similar except that BX is used instead of AX as the base register.

3 Evolution and Multi-cellularity

Using these new instructions, a new parallel ancestor was created. The multi-cellular ancestor is very similar to the ancestor described in [2] and is composed of 82 instructions.

The copy loop is parallelized for two CPUs, with each CPU copying half of the genome from mother to daughter (unlike the previous scheme, one CPU copies the first half of the genome while the other CPU copies the second half).

Once the new and improved multi-cellular ancestor had been created, any further improvements in its performance would be generated through evolution. Comparing the multi-cellular ancestor with its single celled cousin, the multi-cellular creature is clearly the more efficient reproducer. While the single celled ancestor requires approximately ten clock cycles per instruction copied, the multi-cellular ancestor requires only five (since two CPUs are operating in parallel, they are effectively doing the same amount of work in half the time). As a result, a multi-cellular ancestor will produce nearly twice as many offspring as a single celled ancestor in the same period of time. Obviously multi-cellular organisms will have an advantage and will dominate the population.

Experiments were run using the multi-cellular ancestor on a new version of Tierra[3] that runs on a Connection Machine CM-5 massively parallel supercomputer. By taking advantage of the size and speed of a supercomputer, much larger and faster evolutionary simulations have been achieved.

Figure 2 (top) shows a graph of reproduction time versus time for the new multi-cellular ancestor. For the first 200 million instructions, there is a gradual improvement in reproduction time due to optimizations such as template size reduction and taking advantage of the side effects of some instructions (upper band). In addition, there is also effective parasitism (lower band) until approximately 150 million clock cycles at which time most organisms become resistant to parasites. This type of behavior also manifested itself in simulations using single cell organisms [2].

A sharp discontinuity then appears at approximately 215 million clock cycles and represents a thirty percent improvement in reproduction time. This new optimization is added parallelism, and it corresponds to an increase from two to four CPUs per organism. In the genome length versus time graph (center), this change is even more noticeable since the dominating organisms have actually increased in size from 44 to 52. While the size 44 creatures have only two CPUs, the size 52 creatures have four. The larger but more parallel creatures are faster reproducers and as a result take over the population. This increase in parallelism is even more obvious when examining the graph of reproduction efficiency (the average number of clock cycles necessary to copy a single instruction from mother to daughter) versus time (bottom).

One noticeable characteristic the genome length versus time graph shows is that when there are two CPUs, reductions in size typically take place in multiples of two instructions. When there are four CPUs per organism, the multiple increases to four. Obviously the creatures are dividing the workload evenly and are not able to handle circumstances which do not provide even workloads. When the first size 52/four CPU creatures appear, fifteen out of the 52 instructions are mean-

ingless (i.e., these instructions do not affect the execution of the creature's algorithm). In some sense these instructions are a form of computational intron, pieces of unnecessary code left over from some dead ancestor. The introns are used to pad out the size of a creature so that it is a multiple of four, simplifying the distribution of work among the parallel CPUs.

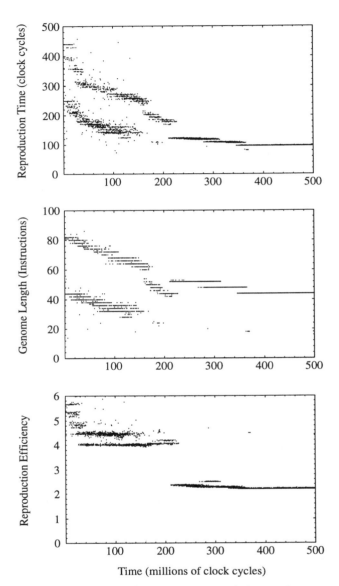

Figure 2: Various evolution characteristics vs. time

A large number of intron instructions are observed when the first four CPU creatures appear. Through evolution, the size of the creatures decreases to 48 (268 million clock cycles), then 44 (346 million clock cycles) and finally to 40 (1 billion clock cycles). In each of these improvements, evolution simply removes some intron instructions while keeping the length a multiple of four. If the intron instructions are removed from the size 40, 44, 48, and 52 creatures, we find that they are the *exact* same algorithm. Obviously at that point in

---

3. Although Tierra now runs on a parallel supercomputer, the parallelism in the digital organisms is unrelated to the parallelism in the computer that they are running on. Slower and smaller simulations of multi-cellular creatures have been run on workstations.

the process, evolution has optimized the basic algorithm as well as it can without major restructuring.

The size 40/four CPU creature that evolved from the size 82/two CPU ancestor is quite efficient in its use of parallelism. The genome of one such creature (0040aba) is:

```
nop0      ; beginning template
adrb      ; find beginning + template size
nop1      ;
subAC     ; sub template size from beginning
movAB     ; put beginning in BX
adrf      ; find end
nop0      ;
nop0      ;
subCAB    ; calculate size
mal       ; allocate space for daughter
incC      ; intron (since CX no longer used)
split     ; 2 CPUs
ifz       ; intron (since CX can't be zero)
movCD     ; intron (since ifz not true)
shr       ; size = size / 2
offAACD   ; split genome into 2 halves &
offBBCD   ; adjust AX and BX accordingly
zeroD     ; zero out DX before second split
pushB     ; save beginning on stack
shr       ; size = size / 2
split     ; 4 CPUs
offAACD   ; re-partition genome into 4 and
offBBCD   ; adjust AX and BX accordingly
nop1      ; copy loop starts here
nop0      ;
movii     ; copy instr from mother to daughter
decC      ; decrement number of instr to copy
ifz       ; if number of instr to copy == 0
jmp       ; jump forward to just before join
nop0      ;
incA      ; increment source address
incB      ; increment destination address
jmpb      ; jump back to start of copy loop
nop0      ;
nop1      ;
join      ; join up multiple CPUs
divide    ; divide mother and daughter
ret       ; return to beginning of creature
nop1      ; ending template
nop1      ; ending template
```

The first thing to notice is that there are still three intron instructions remaining. Simple removal of these introns is not possible since the workload distribution among the four CPUs requires the size to be a multiple of four. Unless the algorithm is radically changed, it would be difficult to evolve a more compact version of this creature. The first thing this creature does is examine itself (beginning and ending) and compute its size. It then allocates space for its daughter. This process is common to almost all viable creatures in the soup.

The next thing that happens is that the creature splits into two CPUs and then divides its size by two. The new size is then used to offset the source and destination registers for the move. Essentially one CPU copies the first half of the genome while the other CPU copies the second half. The creature then zeros out the DX register before splitting again, creating a total of four CPUs. Just before the **split**, the BX register (which contains the address of the beginning of the mother) is pushed onto the stack. This will be used later when the reproduction process has completed.

After the second **split**, the size once again divides by two, leaving the CX register with a value of size divided by four.

Each CPU then offsets the source and destination registers again. Since the CPUs had previously partitioned the genome into halves, they are now hierarchically re-partitioning each of the halves into quarters.

Once each of the four CPUs is set up to copy its quarter of the genome, it enters a copy loop similar to the copy loop in the multi-cellular ancestor. After the copy loop is complete, each CPU waits for the others via a **join**, and once all four CPUs have joined, the creature issues a **divide** for its daughter. After the **divide** a return is performed, which pops the stack into the IP. Since the beginning address of the creature was previously pushed onto the stack, this return causes the creature to start all over again at the beginning of its genome.

3.1 Taking advantage of its creator

In one of the first simulations (after the addition of the new instructions), a very strange set of results was observed. Somehow the creatures had increased their reproductive efficiency so that it appeared to take only two clock cycles to copy the mother's entire genome to the daughter. Even if the creatures had managed to make use of the maximum number of CPUs allowed (sixteen), it would have been impossible to reproduce that quickly. Somehow the creatures must have taken advantage of a bug in Tierra enabling them to reproduce faster than should have been possible. When one of these creatures was examined, it appeared thus:

```
template marking beginning

split     ; 2 CPUs
split     ; 4 CPUs
split     ; 8 CPUs

find beginning, end, and size

divide    ;
mal       ; allocate space for daughter

copy loop
```

For some reason the first thing that happens is the creation of eight parallel CPUs. The creature finds its beginning and end and calculates its size (typical for all creatures). But then it attempts to **divide**, which won't work since the daughter's space has yet to be allocated. The next instruction allocates space for the daughter, which somehow seems to be in the wrong order. Finally a copy loop is entered to copy the mother's instructions to the daughter's space. According to the information saved about this genome, the mother and daughter were genetically identical. Somehow this process must be working correctly since it allows the mother to reproduce.

The key to understanding this process involves the way in which the **split** instructions adds additional CPUs to a creature. In the original version of the multi-cellular Tierra simulator, the execution of a **split** instruction would cause a CPU immediately to exit its time-slice. Any unused cycles left in the time-slice would have been added to the time-slice that both CPUs would receive the next time through the execution queue. As a result of performing three splits in a row, the organism created eight CPUs. These eight creatures each execute (on average) three time-slice's worth of instructions before moving onto the next parallel CPU. So, when the first of eight CPUs issues the **divide**, nothing happens (other than setting an error flag). That CPU then allocates space for a daughter and starts copying its instructions to the daughter.

It turns out that three times the average time-slice size is just enough time to copy all of the instructions to the daughter. So, by the time the slice is over the daughter is complete. After the slice ends, the next CPU performs a **divide**. Unlike the first **divide**, which generated an error because no daughter had yet been allocated, the second **divide** completes correctly. The daughter for the second **divide** is actually the daughter created by the first CPU. After the **divide**, the CPU then allocates another daughter and copies the genome to the daughter's space. This process continues for all eight CPUs, each of which (except for the last one) generates another daughter.

Unfortunately the aforementioned behavior was not desired. The creatures had serialized the parallel CPUs (since they are simulated in a serial fashion in Tierra) and used each CPU to generate its own daughter. They had taken a supposedly parallel process and "daisy-chained" its behavior together so that the beginning of one CPU's execution finishes up the execution of the previous CPU. On a real parallel computer, such behavior is inefficient. However, since Tierra emulates parallelism through time slicing, an algorithm which serializes the activity of its several CPUs can avoid the cost of calculating the offsets and coordinating their activity.

3.2 Fixing the bug(s)

After this bug was discovered, Tierra was modified so that cycles did not accumulate between time-slices (i.e., any cycles left in a time-slice upon execution of a **split** would be lost). This produced the desired behavior (daisy-chaining between parallel CPUs was prevented) and multi-cellular evolution (as described in section 3) was observed.

Unfortunately, this modification also produced a side-effect which was not considered at that time. By zeroing out the time-slice whenever a **split** was performed, Tierra implicitly imposed a large computational cost on additional parallelism. Although an increase in parallelism from two to four CPUs evolved, additional increases in parallelism were not observed. Consequently, Tierra was modified to remove the computational cost imposed on parallelism. Tierra now switches between a creature's CPUs after each instruction in a time-slice rather than after each CPU's time-slice completes. This corrects the original problem without imposing the unwanted cost on parallelism. After this final change was implemented, the evolution of additional parallelism (up to the specified limit of 16 CPUs) was quickly observed. Space does not permit us to discuss these new results here but they will be presented in detail in a forthcoming paper.

4 Conclusions and a Glimpse Into the Future

This first experiment with evolution of parallel processes has yielded fruitful results. Evolution has been able to spontaneously increase the level of parallelism, and effectively coordinate the activities of the additional processors without generating errors. However, differentiation between the processors has taken the form of manipulating different data, not executing different code. Essentially, this is a SIMD style of parallelism, rather than the more interesting MIMD parallelism that we hope to evolve in the future. Yet, given the nature of the problem at hand (copying a series of continuous bytes), a SIMD solution is the most appropriate. In order to evolve MIMD parallelism, where the different processors execute different code while coordinating their activities, evolution will have to be challenged with more complex problems.

From the biological perspective, the SIMD/MIMD distinction relates to the absence (SIMD) or presence (MIMD) of differentiation between cells. In differentiated organisms, different cell types express different suites of genes, which correspond to executing different parts of the same code. It is hoped that future digital organisms will evolve into complex forms exhibiting both SIMD and MIMD parallelism. In order to facilitate this evolution, protocols are being established to permit communication between cells and individuals within and between nodes of both real and virtual machines.

In order to challenge evolution with more complex problems, preparations are being made to create a large biodiversity reserve for digital organisms distributed across the global net [5]. Participating nodes will run a network version of Tierra as a low-priority background process, creating a virtual Tierran sub-net embedded within the real net. Digital organisms will be able to migrate freely within the virtual net. Given that the availability of energy (CPU time) at each node will reflect the activity patterns of the users, there will be selective pressures for organisms to migrate around the globe in a daily cycle, to keep on the dark side of the planet, and also to develop sensory capabilities for assessing deviations from the expected patterns of energy availability, and skills at navigating the net in response to the dynamically changing topology of the net and patterns of CPU-energy availability.

5 Acknowledgments

The authors would like to thank Danny Hillis, David Waltz, and the Santa Fe Institute for supporting this research.

The work of TSR was supported by grants CCR-9204339 and BIR-9300800 from the United States National Science Foundation, a grant from the Digital Equipment Corporation, and by the Santa Fe Institute, Thinking Machines Corp., IBM, and Hughes Aircraft. This work was conducted while at: Thinking Machines Corporation (KT), School of Life & Health Sciences, University of Delaware (TSR), the Santa Fe Institute (KT and TSR), and the ATR Human Information Processing Research Laboratories (KT and TSR).

6 Bibliography

[1] Darwin, Charles. 1859. *On the origin of species by means of natural selection or the preservation of favored races in the struggle for life.* Murray, London.

[2] Ray, T. S. 1991. An approach to the synthesis of life. In: Langton, C., C. Taylor, J. D. Farmer, & S. Rasmussen (eds), *Artificial Life II*, 371-408. Redwood City, CA: Addison-Wesley.

[3] _____. 1994. An Evolutionary Approach to Synthetic Biology: Zen and the Art of Creating Life. *Artificial Life* 1(1/2): 195-226.

[4] _____. In Press. Evolution, Complexity, Entropy, and Artificial Life. *Physica D*.

[5] _____. In Press. Evolution of parallel processes in organic and digital media. In: D. Waltz (ed.), *Natural and Artificial Parallel Computation*. Philadelphia: SIAM Press.

MESHING OF ENGINEERING DOMAINS BY MEITOTIC CELL DIVISION

Kazuhiro Saitou   **Mark J. Jakiela**

Computer-Aided Design Laboratory
Massachusetts Institute of Technology
Cambridge, MA 02139
email: kazu@mit.edu   jakiela@mit.edu

Abstract

A new technique for triangular meshing of a given two-dimensional domain is developed, where meshing is done by a process of simulated meitotic cell division. Dead cells become the elements of the discretized domain, and the domain is filled by a process of cell growth. A classifier system evolves the rules for developmental cell growth in the domain. A given domain is meshed repeatedly using the growth rules, and the quality of each mesh is checked with a fitness function. The preliminary results shows that an evolved set of classifiers for a particular test domain could also be effectively used for geometrically and topologically similar domains.

Introduction

Meshing is the process of discretizing a geometrical domain into a large number of smaller, simpler geometric elements. As shown in Figure 1, meshing can generally be performed on one, two, and three dimensional domains. In one-dimensional meshing, a space curve is discretized into a contiguous sequence of linear segments; in 2D meshing, a spatial surface is discretized into a set of adjacent planar polygons; and in 3D, a volume is discretized into a set of adjacent volumetric elements. We show the domains meshed into linear (*i.e.* line segments and plane facets) elements, but in the general case, this is not required. A space curve, for example, could be discretized into a series of arc segments of varying radii.

Meshing is important because it is necessary to discretize geometric domains to allow the use of approximate engineering analysis techniques such as the finite element method (FEM) (Bathe, 1982) and the boundary element method (BEM) (Brebbia, 1978). Models of continuum phenomena, such as material strain and heat transfer, are applied to each element with the constraints that solution values agree at the boundaries between elements. An iterative solution procedure then solves for quantities of interest (*i.e.* temperature, material strain) at the element vertices (often called "nodes"). Solutions for other points in the domain are interpolated from the nodal values. For such a process to work well, the domain must be appropriately discretized. Large geometrically uniform regions will likely not have much variance in

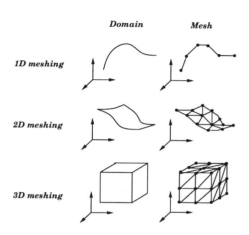

Figure 1: Meshing of domains

the solution values. A relatively small number of larger elements should, therefore, be used. Smaller regions with holes and other topological and geometric features will have rapidly changing solution values and should therefore be meshed with a larger number of smaller elements. Overall, a minimum number of regularly-shaped elements should be used to facilitate the iterative solution procedure.

Meshing, therefore, is a problem of great practical interest that has generated a significant body of research over the past few decades. Shimada (1993) provides a very comprehensive review in his recent Ph.D. thesis. He surveys a large number of techniques that address portions of the overall meshing problem and concludes that the following problems still exists: (1) Poor node spacing control; (2) Ill-shaped elements; (3) Different techniques are tailored to different dimensional spaces; (4) Difficulty in remeshing in response to changes in the domain boundary (important *e.g.* when simulating the plastic deformation of materials). In response to these deficiencies, Shimada's approach, simulating the packing of elastic "bubbles" into the domain and connecting their centers after the packing, shows great promise. With his bubble packing method, however, only triangular mesh

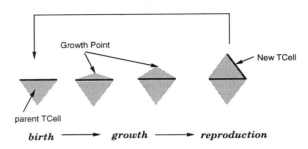

Figure 2: Meitotic cycle of TCell

name	range
grow_direction	$\{0, 1\}$
grow_start	$[0, 1]$
grow_end	$[0, 1]$
small_distance	$[0, 1]$
small_area	$[0, 1]$

Table 1: Growth and reproduction parameters

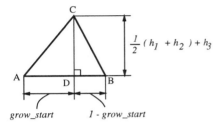

Figure 3: Geometry of a matured TCell

element can be created directly. As a general-purpose mesher, it would be also useful to allow direct meshing with other shapes, *e.g.* rectangles, for some domain.

In this article, we introduce a new technique strongly motivated by recent approaches to artificial embryology (de Garis, 1992), (de Garis, Iba, and Furuya, 1992), where meshing is done by a process of simulated meitotic cell division. Dead cells become the elements of the discretized domain, and the domain is filled by a process of cell growth. A classifier system evolves the rules for developmental cell growth in the domain, much like the approach proposed by Wilson (1987). A given domain is meshed repeatedly using the growth rules, and the quality of each mesh is checked with a fitness function. Our hope is that an evolved set of classifiers for a particular test domain could also be effectively used for geometrically and topologically similar domains. Like De Garis, Iba and Furuya (1992), we hope to construct a geometric shape by cell development and division, but we only require that a prespecified domain be filled. Our preliminary results compare well with those reported by Shimada (1993), particularly for geometrically and topologically complex domains. Even though our current implementation does only triangular meshing, the same technique can be applied easily to create meshing by other shapes such as rectangles.

Cell Division Metaphor

Triangular meshing of a given two-dimensional domain is done by simulated growth and reproduction of cell-like triangular objects, which we simply call *TCells*. Once a TCell is placed on a two-dimensional domain with closed boundaries, it "grows" to become a certain size, and then "reproduces" itself on one of its three sides. Figure 2 illustrates this meitotic cycle. On birth, a TCell is just a line segment which lies on a side of its parent. The first TCell in the domain, the *root cell*, is placed on a domain boundary. Then, a point on the initial line segment (the growth point) starts to move perpendicularly to the line, forming a growing triangular shape. The size of a "matured" TCell, *i.e.* when the growth point stops moving, depends on a user-defined node density function given over the domain. A TCell also stops growing when it hits any domain boundaries or any other TCells.

A TCell "dies" when it is surrounded either by the

other TCells or by the boundary on all three sides, so that it can no longer reproduce. It also dies when it is too close to the other TCells or to the boundary, since its offspring would be too small to form a mesh. The meitotic cycle is repeated recursively until all TCells in the domain die out.

Mechanism of Growth and Reproduction

There are five parameters that decide behaviors of a TCell in growth and reproduction. Their names and ranges are listed in Table 1.

Grow_direction decides on which side a TCell reproduces its offspring. This parameter takes only a binary value since one side of the TCell is already adjacent to its parent. Growth of a TCell is controlled by *grow_start* and *grow_end*. *Grow_start* specifies position of the growth point along the initial side length. More specifically, as in Figure 3:

$$\overline{AD} : \overline{BD} = grow\_start : 1 - grow\_start$$

The height of the TCell is decided by the following equations:

$$\overline{CD} = \frac{1}{2}(h_1 + h_2) + h_3$$

where

$$h_1 = \frac{1}{d(D)\overline{AB}}$$

$$h_2 = \frac{\sqrt{3}}{2}\overline{AB}$$

$$h_3 = \sqrt{\frac{1}{d(D)}} \times grow\_end$$

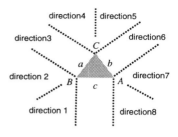

Figure 4: Eight directions to "look"

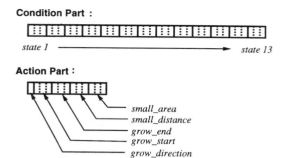

Figure 5: Bit assignment of a classifier

and $d(X)$ denotes a node density at a point X calculated by the pre-defined node density function given over the domain.

Growth halts whenever any part of a TCell touches a boundary or the other TCells. After growth, the growth point and a vertex of the other TCells are merged if they are within the distance of *small_distance* × *desired_height*[1]. A TCell whose area is smaller than *small_area* × *desired_area*[2] is deleted after that. Finally, after all TCells in the domain die out, the node points within some fixed distance from the boundary are pulled to the domain boundary.

When matured, a TCell "looks" around to measure the distance to the boundary and the other TCells. The distances are measured in multiples of the TCell height. Eight directions are monitored as in Figure 4. Five of these eight distance values are given to the offspring. When the offspring is reproduced on the side a in Figure 4, it receives the values of the directions 1 to 5. Similarly, the offspring born on the side b inherits the values of the directions 4 to 8.

The *state* of a TCell is a set of 13 integers which consist of the 5 distance values inherited from its parent, and the 8 distance values measured by the TCell itself. The state is used to determine the parameters in Table 1 as described in the next section.

Classifier Chromosome and Genetic Algorithm

State is related to the growth and reproduction parameters by classifiers. A classifier is a rule of a form "if *condition* then *action*". The condition part is a string of 0, 1, and # (don't care), and the action part is a string of 0 and 1. In this study, the condition part and the action part correspond to the encoded state and the encoded parameters, respectively. Each of the 13 integer values of state is encoded to a 3-bit binary number[3], and concatenated to a 39-bit binary string. This string is matched to the condition part of the classifier. If they match, the corresponding action part is decoded to the

[1]*desired_height* $= 1/d(\mathrm{D})\overline{AB}$
[2]*desired_area* $= 1/2d(\mathrm{D})$
[3]this encodes integers between 0 and 7, which is large enough for the examples below.

parameter values. Figure 5 shows the bit assignment of a classifier. A TCell has 10 such classifiers which do not change during growth and reproduction. In other words, offspring receive identical copies of parent's classifiers when a mesh is constructed.

Within a lifetime of a TCell, all of the 10 classifiers are matched twice, *before* and *after* growth. For the first match, only the first 15 bits of the encoded state, which contain inherited values from its parent[4], are matched. And then, only the last 12 bits of the matching classifier is decoded to *grow_start*, *grow_end*, *small_distance* and *small_area*. For the second match, the rest of 24 bits (obtained from a new "look" of the just grown cell) of the encoded state is matched to decode *grow_direction*. Matching to the 10 classifiers is done in sequence. If more than one classifier matches, the last matching classifier is used to decode the parameter values. If there is no match, default parameter values are used.

The genetic algorithm (Goldberg, 1989) is applied to evolve a TCell, or more precisely, a set of 10 classifiers that does high quality meshing of a given domain. The current implementation uses fitness-proportionate selection with an elitist selection scheme, where the best individual in a population is guaranteed to have a copy in the new population. One-point crossover is performed on both the condition part and the action part of the classifiers. A set of 10 classifiers is used to mesh a given domain, from a birth of the root cell to death of all cells in the domain. Fitness is then calculated as a weighted sum of fitnesses of three criteria: size and shape of each TCell, and total covered area:

$$fitness = w_{\mathrm{size}}f_{\mathrm{size}} + w_{\mathrm{shape}}f_{\mathrm{shape}} + w_{\mathrm{area}}f_{\mathrm{area}}$$

where

$$f_{\mathrm{size}} = \sum_{\mathrm{TCell}} (1 - |1 - \frac{area}{area_{\mathrm{desired}}}|)/n$$

$$f_{\mathrm{shape}} = \sum_{\mathrm{TCell}} \sum_{i \in \{A,B,C\}} (1 - |1 - \frac{cos(i)}{0.5}|)/3n$$

$$f_{\mathrm{area}} = \sum_{\mathrm{TCell}} \frac{area}{domain\_area}$$

[4]fot the root cell, "don't care" values are used

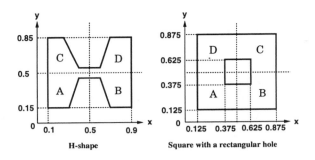

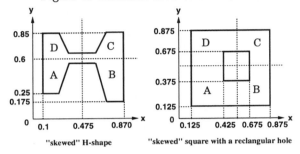

Figure 6: Domains used to evolve classifiers

Figure 7: Slightly skewed domains

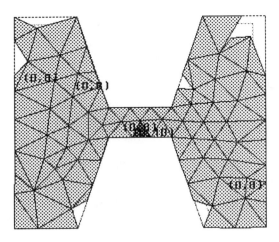

Figure 8: Meshing of H-shape with random classifiers

and n is the number of TCells created in the domain. Intuitively speaking, a set of classifiers receives high fitness if in the resulting mesh, the size of each triangle matches the desired node density, all triangles are equilateral, and entire domain is filled up by triangles. The maximum possible values of f_{size}, f_{shape} and f_{area} are 1.0. Linear fitness scaling (Goldberg, 1989) is performed after fitness evaluations of an entire population.

Examples

The technique described in the earlier sections is applied to two different 2D domains, an H-shape and a square with a rectangular hole, as shown in Figure 6. For each domain, a set of 10 classifiers is evolved that maximizes the fitness function described above. The best set of the evolved classifiers for each domain is then applied to mesh a similar domain with slightly different geometry shown in Figure 7. GA parameters used in these examples are shown in Table 2.

In addition to all the GA parameters, a node density function must be defined over the domain to be meshed. The node density function used for the (regular) H-shape

GA parameter	value
crossover probability	0.9
mutation probability	0.1
linear scaling coefficient	2.0
population size	50
number of generation	100

Table 2: GA parameters used in the examples

example is:

$$d_H(x,y) =$$
$$\begin{cases} 1200xy + 50 & if \quad (x,y) \in A \\ 1200(1-x)y + 50 & if \quad (x,y) \in B \\ 1200(1-x)(1-y) + 50 & if \quad (x,y) \in C \\ 1200x(1-y) + 50 & if \quad (x,y) \in D \end{cases}$$

and for the "skewed" H-shape example is:

$$d_{sH}(x,y) =$$
$$\begin{cases} 1200xy + 50 & if \quad (x,y) \in A \\ 1200(0.95-x)y + 50 & if \quad (x,y) \in B \\ 1200(0.95-x)(1.2-y) + 50 & if \quad (x,y) \in C \\ 1200x(1.2-y) + 50 & if \quad (x,y) \in D \end{cases}$$

In the following figures, the root cell is drawn in dark gray, and a pair of two numbers on a cell shows the rules fired in matching. For example, $(0,8)$ means that no rule was fired at the matching after growth of the parent cell, and then the eighth rule was fired at the matching before growth. No numbers are shown if no rules are fired in both matching (i.e. $(0,0)$). Figure 8 and Figure 9 show results of meshing an H-shape, before evolution (*i.e* random classifiers) and after evolution, respectively. Offset lines that appear in unfilled areas of Figure 8 show a "pulling" region near the domain boundary. The evolved classifiers used in Figure 9 are applied to mesh the skewed H-shape. As shown in Figure 10, the evolved classifiers could mesh the new domain fairly well.

The second example is meshing a square with a rectangular hole. The node density function for the square examples is:

$$d_R(x,y) =$$
$$\begin{cases} 1500xy + 50 & if \quad (x,y) \in A \\ 1500(1-x)y + 50 & if \quad (x,y) \in B \\ 1500(1-x)(1-y) + 50 & if \quad (x,y) \in C \\ 1500x(1-y) + 50 & if \quad (x,y) \in D \end{cases}$$

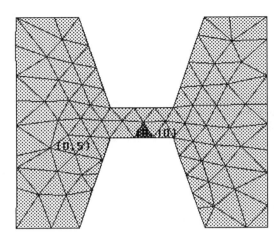

Figure 9: Meshing of H-shape after evolution

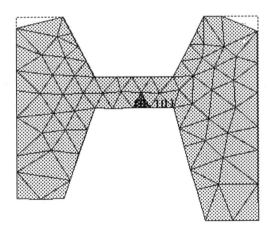

Figure 10: Meshing of "skewed" H-shape with the evolved classifiers

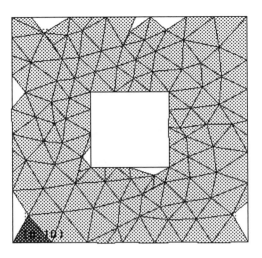

Figure 11: Meshing of square with a rectangular hole with random classifiers

and for the "skewed" square with a rectangular hole is:

$$d_{sR}(x, y) =$$
$$\begin{cases} 1500xy + 50 & if \quad (x, y) \in A \\ 1500(1.1 - x)y + 50 & if \quad (x, y) \in B \\ 1500(1.1 - x)(1.05 - y) + 50 & if \quad (x, y) \in C \\ 1500x(1.05 - y) + 50 & if \quad (x, y) \in D \end{cases}$$

The meshes before and after evolution are shown in Figure 11 and in Figure 12, respectively.

As the previous example, the evolved classifiers are applied to the "skewed" domain, as shown in Figure 13. In this case, again, the resulting meshing is fairly high quality.

Discussion and Conclusions

In this article, we have demonstrated how a process of simulated meitotic cell division can be used to generate meshes for engineering domains. A classifier system evolves the growth rules that control cell division in response to the geometry and topology of the domain. The approach produces high quality meshes for a domain of a given constant shape. The system does this even for difficult idiosyncratic shapes, such as the "H" shown in Figure 9. Our hope was that a set of classifiers evolved for a specific example of a class of shapes (again, the "H" of Figure 9 is an example), would also work well for other examples of the class. Reasonable results were obtained in this regard, as shown by Figures 10 and 13, but it seems as though there is some sensitivity to the specific proportions of a shape in a class. This is probably because only one constant shape was used to train the classifiers. Also, it is seen that the number of rules fired is typically very small. We believe that this is because the default parameters are well tuned. We note, however, that the overall mesh quality is extremely sensitive to small number of rules firing and slight change

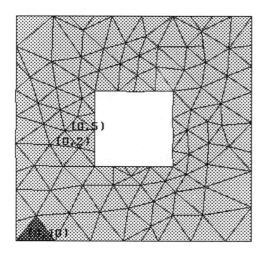

Figure 12: Meshing of square with a rectangular hole after evolution

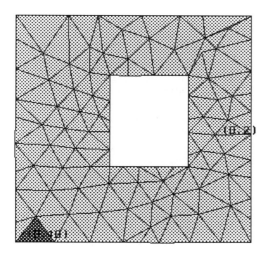

Figure 13: Meshing of "skewed" square with a rectangular hole with the evolved classifiers

in the shape of the root cell (as can be seen by closely examining the root cell of Figure 11 and Figure 12).

We plan to continue to work on these issues of class generality, as they are important to some of our future intended uses for the technique. Our hope is to have classifiers learn growth rules from a particular situation, or small set of situations, that actually occurred. These rules could then be applied to other domain shapes with the same general characteristics. We hope, for example, to model processes such as corrosion and the action of cleaning agents on a variety of part shapes given empirical data from a few typical shapes. Additionally certain phenomena of engineering interest are exceeding difficult to simulate based on fundamental physical principles. Mold filling and solidification for a casting process is an example. We would hope to qualitatively model the flow and solidification processes by cell growth and death respectively. Resulting approximations would be qualitative, but would hopefully be accurate enough to facilitate initial design decisions. Artificial life as a metaphor for a physical process could help the engineering design process.

Acknowledgements

This work is supported by the National Science Foundation with a Presidential Young Investigator's grant (DDM-9058415). Matchable funds for this grant have been provided by Schlumberger Inc. These sources of support are gratefully acknowledged. Additionally, the system was developed using the computational facilities of the Computer-Aided Design Laboratory at the Massachusetts Institute of Technology, Department of Mechanical Engineering. This support is also gratefully acknowledged.

References

Bathe, K. 1982. *Finite Element Procedures in Engineering Analysis*. Prentice-Hall.

Brebbia, C. 1978. *The Boundary Element Method for Engineers*. Wiley.

de Garis, H. 1992. Artificial embryology: The genetic programming of an artificial embryo. In Soucek, B. and the IRIS Group, editors, *Dynamic, Genetic, and Chaotic Programming*, pages 373–393. Wiely.

Goldberg, D. 1989. *Genetic Algorithms in Search, Optimization and Machine Learning*. Addison-Wesley.

de Garis, H., Iba, H., and Furuya, T. 1992. Differentiable chromosomes. In *Parallel Problem Solving from Nature, 2*, pages 489–498. North-Holland

Shimada, K. 1993. *Physically-Based Mesh Generation: Automated Triangulation of Surfaces and Volumes via Bubble Packing*. Ph.D. thesis, Department of Mechanical Engineering, MIT.

Wilson, S. 1987. Genetic algorithms and simulated evolution. In *Artificial Life*, pages 157–166, Los Alamos, New Mexico. Addison Wesley

Simulating Natural Spacing Patterns of Insect Bristles Using a Network of Interacting Celloids

Hiroaki Inayoshi

Computation Models Section,Computer Science Division,
ElectroTechnical Lab.(ETL),
1-1-4 Umezono, Tsukuba Science City, Ibaraki, 305, Japan.
tel: +81 298 58 5865;   fax: +81 298 58 5871;   e-mail: inayoshi@etl.go.jp

Abstract

The structure, function, and behavior of multicellular organisms is an 'emergent property' that comes from the interaction between its component parts, i.e. cells. I am now working on a project entitled "Celloid for Organismoid", in which I search for artificial synthesis that exhibits this behavior. In other words, the emergence of an artificial organismoid by designing artificial, differentiable celloids as its component parts, collecting them, and letting them interact to organize together. In this paper, one example of this research is presented. The goal of this example is to generate spacing patterns, which are remniscent of ones observed in insects. When insects produce their bristles, or short stiff hairs, the bristles are spaced in a non-random fashion. New bristles will appear in the largest spaces between the preexisting bristles. This is a simple example of the general phenomenon of cell differentiation.

A new mechanism, called an Artificial Genetic Network (AGN), which is based on Natural Genetic Networks (i.e. networks of genes), is introduced into the celloids to synthesize the spacing patterns artificially. A collection of celloids, having the same AGN, interact with each other locally, and each celloid makes a dynamical decision whether to have or not to have a bristle, based on this local interaction. The local decisions made by each of celloids give rise to some global pattern. In spite of the fact that there is no global director determining which celloid should and which should not have a bristle, some degree of order in the spacing pattern emerges. To compare the degree of order in celloids with AGN, a control experiment is done, in which independent cells are collected and each cell makes the same binary decision independently (i.e. without interaction). The comparison shows that the collection of interacting celloids has a higher degree of order than the collection of 'independent cells'.

keywords: artificial life; cell, celloid; organism, organismoid, organization; Genetic Network, Artificial Genetic Network; spacing pattern, bristle; differentiation; interaction; gene, gene expression, regulation;

1 Introduction

Every multicellular organism, including a human being, exists as a collection of cells. But a multicellular organism cannot be built simply by collecting its component parts, i.e. cells. Three criteria must be satisfied. They are (1) **differentiation** of parts; (2) **location** of parts; and (3) **interaction** between parts to organize the whole system. The importance of these points can be easily seen by considering the following cases.

(1) If one collects only one kind of part (i.e. cells of the same type, such as muscular cells only), the collection won't work. So, many kinds of parts (i.e. differentiated cells) are required to build an organism.

(2) Suppose one has sufficient numbers and kinds of cells. If they are placed randomly, or combined at random, the whole system won't work. So, the parts have to be located / combined properly.

(3) Suppose sufficient parts are located at proper places. If each of these parts operates independently, or works randomly without interaction, the whole system won't work. So, proper interaction / communication between parts is also required to organize the whole cells.

In fact, if one examines multicellular organisms carefully, the following facts will be observed:

- Each of the cells, as a component part of the organism, operates simply according to some local rules.

- Despite the above fact, global order (or the coherence of the organism) emerges out of local interaction between the parts.

This paper describes an example of an attempt to make an artificial organismoid using artificial, differentiable celloids [1]. The goal of this example is to generate spacing patterns, similar to those seen in natural organisms, which should emerge out of local interaction between celloids. (Note that the goal is *not* in making artificial patterns that mimic exactly natural patterns *but* in generating artificial patterns which have order or regularity. In other words, an emphasis is placed on **order**

---

[1]This work and the name "cell-oid" are inspired by the "bird-oid" in [Reyn87].

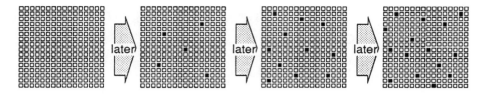

Figure 1: The spacing patterns of metabolizing bristles.

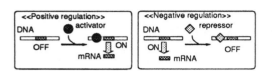

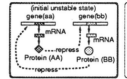

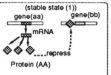

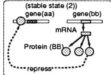

Figure 2: Two types of regulation.

Figure 3: The simplest Genetic Network composed of two nodes: (exclusive case).

rather than on **similarity** to nature.) To achieve this goal, an Artificial Genetic Network (AGN) is introduced as a mechanism of interaction between celloids. All celloids have the same AGN, which determines their behavior rules, although their internal states will be different. (In multicellular organisms, all cells have the same genome or set of genes, although their internal states will be different.) This difference in the internal states of celloids is expected to give rise to a spacing pattern.

The remainder of this paper is organised as follows: Section 2 provides a brief description of spacing patterns in insects. Section 3 explains natural and artificial genetic networks. Section 4 shows how interaction between celloids proceeds. Section 5 presents experiments and their results. Finally, Section 6 provides a conclusion and future directions.

2 Spacing Patterns

Bristles (i.e. short stiff hairs) form a spacing pattern as described in the following quotation:

> ... what is observed in various insects is that the bristle mother cells always appear approximately in the center of the spaces between the extant bristles and this occurs continuously until the final density is reached. ... It was clear that new bristles had been added and that these had appeared in the largest spaces between the preexisting bristles. (*Quoted from [Lawr92] p.159.*)

Introducing the metabolization of bristles, this phenomenon can be depicted as Fig. 1. In the figure, each small rectangle corresponds to a cell. (Each black rectangle has a bristle, white rectangles have none.)

Note that the binary decision (i.e. to have or not to have a bristle) corresponds to the simplest differentiation, because each cell can take only one of the two states. Therefore, this example provides a basis for the differentiation of cells in general, in which the range of choice will be increased.

3 Genetic Networks

In this paper, a **Genetic Network** refers to **a network of genes** (i.e. **nodes**) which are connected by **regulatory influences** (i.e. **edges**) that affect the expression of other genes. In this section, Natural and Artificial Genetic Networks will be described, after some biological facts are explained.

3.1 Natural Genetic Networks ([MolBio])

Molecular Biology tells us that genetic information usually flows in the following direction: DNA → RNA → protein. That is, first the genetic information in DNA is **transcribed** to RNA, then the information in RNA is **translated** to a protein. Proteins, i.e. final products of **gene expressions**, perform various **functions**. Some proteins act as **enzymes**, others help in **cytoskelton**. But a most important fact is that **some proteins can regulate the production of other proteins**. In other words, they can regulate the expression of genes. This regulation is done as follows: First a regulatory protein is synthesized (i.e. the regulatory gene is expressed (= transcribed and translated)). Next the regulatory protein binds to some specific part of DNA. This binding protein can act either as **an activator** or as **a repressor** of some genes. (See Fig. 2.) When a regulatory protein acts as an activator, **(positive regulation)** it activates the transcription of the gene. On the other hand, when it acts as a repressor, **(negative regulation)** it represses the transcription of the gene.

With these regulation mechanisms in mind, consider a simple Genetic Network, or a network of **two** genes. Suppose two genes, *aa* and *bb*, can produce proteins *AA* and *BB*, respectively. If the two products, *AA* and *BB*, can repress the expression of each other, **exclusive expression** will take place, as in Fig. 3.

Now proceed to a network of **many** genes. (See Fig. 4.) Suppose some of the genes (nodes) have **potential influences** (activative / repressive edges) on other genes (nodes), including themselves. (The word potential means that influence can occur **only when**

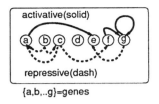

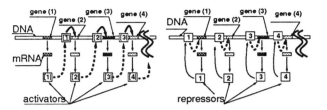

Figure 4: An example of general genetic network.

Figure 5: A structured collection of boxes.

Figure 6: Network for sequential expression of genes (left), and Network of repressors (right).

the gene (node) is expressed (turned ON). In other words, so long as the gene is **not expressed (turned OFF)**, the gene has no influence on other genes.) Let G be the number of genes in the Genetic Network. Each gene takes either of two states: {expressed or not expressed} . Although the number of possible states of "genetic expression" is 2^G, stable states would be a subset of 2^G states, due to the regulations. And this subset gives rise to a set of possible differentiations.

Note: The Genetic Networks are not exactly the same as Kauffman's **random boolean networks** in [Kauf90]: the latter have only one type of edge, while the former have two, i.e. activative and repressive.

3.2 Artificial Genetic Networks

To synthesize the spacing patterns described in section 2, celloids having the same Genetic Network must interact to decide whether to have or not to have a bristle. To achieve this interaction mechanism, I adopt the following form of **Artificial Genetic Network (AGN)**, which is named **box-in-box AGN**. In this AGN, each gene i can produce both undiffusible products $\boxed{\boxed{i}}$ and diffusible products $\boxed{i}$, when the gene i is expressed. The "diffusibility" means the ability to move out to the neighboring cells. An $\boxed{\boxed{i}}$ acts as an activator for the gene $i+1$, while an $\boxed{i}$ acts as a repressor for the gene i, as will be described below. The designed AGN has the following features:

- **Sequential expression of genes:** Consider N boxes of different sizes, and of similar shapes which are structured in the following way. (Assume the boxes are sorted by size in decreasing order, and labeled 1 to N in that order.) First, the box $\#\,N$ is put into the box $\#\,(N-1)$, then they are put into the box $\#\,(N-2)$. Repeat this process, until the box $\#\,1$ (i.e. the largest box) contains the rest of all $(N-1)$ boxes. Fig. 5 shows the structured boxes when $N=4$. Now suppose that each **box** corresponds to a **gene**, and that **opening a box** corresponds to **expressing a gene**. Then if boxes (genes) are structured in this way, it follows that **genes can be expressed only sequentially**. In other words, to express gene $\#\,i$ ($1 \leq i \leq N$), all genes from $\#\,1$ to $\#\,(i-1)$ should be expressed.

Assume that a gene expression is a stochastic process and that activators / repressors will increase / decrease the probability of the gene expression. If the products of the gene $\#\,i$ ($1 \leq i \leq N$), denoted as $\boxed{\boxed{i}}$, act as the activator for the gene $i+1$, then the sequential structure can be synthesized. (See Fig. 6.)

- **Bristle formation and apoptosis (or, programmed death):** As a criterion for construction and destruction of a bristle, following conditions are adopted.

(1) Construction: A celloid can **produce a bristle**, when **half** of the genes (i.e. genes from $\#\,1$ to $\#\,N/2$) are expressed.

(2) Destruction: A celloid '**dies**' (i.e. loses its bristle and resets all genes), when **all** genes are expressed. (This 'cellular suicide' is called *apoptosis* in biology. The final box corresponds to an artificial '**self-destruction-switch**'.)

- **Interaction / regulation between genes and their products:** In order to let the celloids interact, the following mechanism is adopted: The gene $\#\,i$ can produce not only $\boxed{\boxed{i}}$, which is *not diffusible* (i.e. which cannot move thorough the boundary of each celloid), but also $\boxed{i}$, which is *diffusible* (i.e. which can move out to the neighboring celloids). The important point is that each of these $\boxed{i}$ can act as the "repressor" for the gene $\#\,i$ (i.e. for the gene of $\boxed{i}$ itself; See Fig. 6.)

Note that the cell itself, which expressed the gene $\#\,i$, is **immune** to this repression for the following reason: Although the $\boxed{i}$ can repress the *transcription* of i, they cannot repress the *translation* of i. Therefore once i is transcribed, the translation of i is not repressed, so long as mRNA of i exists. (This mechanism of 'repression immunity' is artificial and probably has no counterpart in the natural world.)

4 Operational Sequences in Celloid Network

In this section, we describe how the computation proceeds in the collection of celloids. Celloids, each of which obeys the same AGN, interact by communicating their

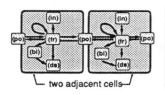

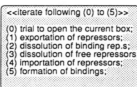

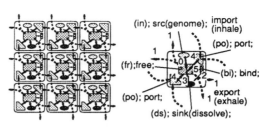

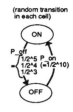

Figure 7: Possible transitions for repressors.

Figure 8: Computational sequence in each celloid.

Figure 9: Interaction in "Celloid Network".

Figure 10: Control experiment.

genetic products. (In the current example, only diffusible repressors, $\boxed{1}$ to $\boxed{N}$, are communicated, where N is the number of genes. Since $\boxed{\boxed{1}}$ to $\boxed{\boxed{N}}$ are not diffusible, they are not communicated.) Celloids having the same box-in-box AGN participate in a **boxes-opening-race** or **genes-expression-race**, in which each celloid tries to open all of the boxes as quickly as possible. When one celloid succeeds in opening the i-th box (=gene), where $1 \leq i \leq N$, it can **repress** the neighboring celloids, by **releasing** the repressors, $\boxed{i}$. (As mentioned in the previous section, the releasing cell of the $\boxed{i}$ is itself **immune** to this repression.) The details of the adopted artificial repression mechanism are as follows.

(1) A repressor $\boxed{i}$ **can** repress the transcription of gene i, **only when it binds to the repression site** for the gene i. Therefore, the repressor in its free state (i.e. the repressor not binding to DNA) has no influence on the expression of gene. (This mechanism is based on a **Lock and Key relationship** ([Conr92]) between regulatory proteins (i.e. activators / inhibitors) and regulatory sites in DNA. Thus $\boxed{i}$ can repress only the gene i, while other genes are not influenced by $\boxed{i}$.)

(2) **The degree of repression** depends upon **the number of repressors** binding at the repression site. Though this relation can be set arbitrarily, the following one is chosen: Each gene has a fixed number of sites (labelled *sites*) to which repressors can bind. When x repressors are binding to gene #i, the probability of transcribing the gene #i is equal to $1/2^{K+x}$, where K is an adjustable parameter and $0 \leq x \leq sites$. This probability ranges from $1/2^K$ (when no repressor is binding to the gene #i) to $1/2^{K+sites}$ (when all repression sites of the gene #i are occupied.)

(3) All repressors have some probability of being **dissolved (or disintegrated)**.

Each repressor can take one of the following five states: • (in)=initial;(not-yet-released); • (fr)=free state; • (bi)=bind to the repression site; • (ds)=dissolved; • (po)=port;(to be transported); Possible transitions between states are shown in Fig.7, and they take place stochastically. Initially, all repressors are in state (in).

When the i-th gene is expressed, an amount Q of repressor $\boxed{i}$ is released. (Their states change from (in) to (fr).) Those repressors in (fr) state can make one of the following four transitions: • (fr) $\rightarrow$ (po): move to ports to be transported. • (fr) $\rightarrow$ (bi): bind to the repression site, if there is a vacancy. • (fr) $\rightarrow$ (ds): be dissolved. • (fr) $\rightarrow$ (fr): remain free (without being exported / dissolved).

Computationally, processes in Fig.8 are iterated in each of the celloids, concurrently. The numbers, 0 to 5, in those processes correspond to the ones in Fig.9. The figure shows nine adjacent celloids and a close-up of one celloid indicating possible flows of the repressors. In the figure, celloids are aligned two-dimensionally, and each celloid is adjacent to four neighbors. (In the current example, flows of different repressors, $\boxed{1}$ to $\boxed{N}$, *do not* interact with each other directly.) As is seen in the figure, each celloid has both a *source (genome)*, **from which repressors are released**, and a *sink (dissolution)*, **into which repressors will disappear. The interaction between celloids occur through the flows of N kinds of repressors.** (Since activators are *not* diffusible, they don't play any role in the interaction between celloids. Note that the diffusibility of activators is introduced artificially, and has nothing to do with the diffusibility of activators in nature.)

5 Experiments and Results

Two experiments are done: One with interaction between celloids ($N = 8$), and the other without it (i.e. control experiment). They are labeled "interaction" and "random", respectively. Details are as follows.

Ten thousand (100*100) celloids, each adjacent to four neighbors, are aligned two-dimensionally. Each celloid dynamically selects one of the two states: { ON (i.e. having a bristle) or OFF (i.e. not having a bristle) }, according to the same rule, the AGN described in the previous sections. As a control experiment, the same number of cells are collected and each of them selects { ON or OFF} *independently from each other*. In other words, in the control experiment, each cell makes a "transition at random" without interaction, as in Fig.10. The probabilities of transition are as follows: P_{ON} is fixed to be $1/2^{10}$, whereas P_{OFF} is either { $1/2^5$, or $1/2^4$, or $1/2^3$ }. These three cases are labeled as { P=(10:5), P=(10:4), P=(10:3) }, respectively in the following figures.

Though each celloid has several adjustable parameters, only one parameter is varied in the experiment. This parameter is Q, the amount of repressors released by each of the gene expressions. Instead of specifying the values of Q, the expected diffusion distance, d, is specified as one of the following values: $\{4, 8, 12, 16, 20\}$. (These are labelled as $\{d = 4, d = 8, d = 12, d = 16, d = 20\}$, respectively.) Note that the "Manhattan distance" [2] is used and that Q is proportional to $d * d$. Both experiments start from **bald state** (i.e. no bristle at all) and run for several thousands of time steps (or **cycles**).

Fig.11 shows the relation between the numbers of bristles and the time steps in the "random" (top) and "interaction" (bottom) models respectively. As is seen, the numbers of bristles change with time but the variations remain in certain ranges in both cases. Figs.12 to 15

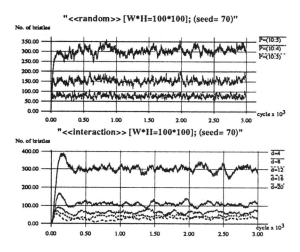

Figure 11: No. of bristles vs. cycle in random model (top). and in interaction model (bottom).

show two cases of results:

(1) a relatively dense case (i.e. approximately 300 bristles appear in 100*100 celloids.) ... Figs.12 &14.

(2) a relatively sparse case (i.e. approximately 150 bristles appear in 100*100 celloids.) ... Figs.13 &15.

Figs.12 &13. show the snapshots of bristles at three cycles (at cycle # 100, 300, and 500). Each of figs.14 &15 shows the distibution of distances to the nearest bristle from each bristle (at cycle # 100) and compares the interaction model with the random model in two cases, dense and sparse, respectively. (The abscissa stands for the distance between bristles, whereas the ordinate stands for the frequency.) Similar results are obtained when the seeds for the random numbers are changed. In both cases, the following facts are observed:

• In the random model, the distribution of distances between bristles (figs.14 &15) is quite **broad** and **flat**.

---

[2] $dist(A, B) = |x_A - x_B| + |y_A - y_B|$, where $A = (x_A, y_A)$ and $B = (x_B, y_B)$.

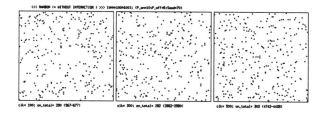

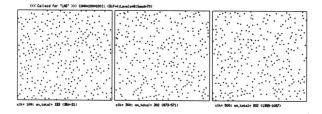

Figure 12: A relatively dense case: Snapshots of bristles at three cycles (from left, at cycle # 100, 300, and 500). Top= random model; Bottom= interaction model; Each dot in the snaps represents a bristle in 100 * 100 celloids. The exact numbers of bristles in each snap (from left) are 290, 282, 303 (top) and 333, 302, 302 (bottom).

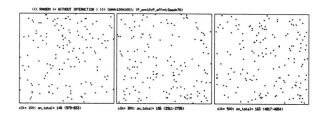

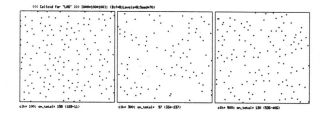

Figure 13: A relatively sparse case: Snapshots of bristles at three cycles (from left, at cycle # 100, 300, and 500). Top= random model; Bottom= interaction model; Each dot in the snaps represents a bristle in 100 * 100 celloids. The exact numbers of bristles in each snap (from left) are 146, 156, 163 (top) and 158, 97, 130 (bottom).

distribution of dist-to-the-nearest: interaction model vs. random model

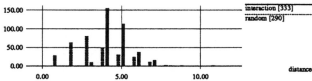

Figure 14: Comparison of interaction:(d=4) and random:P=(10:5) in relatively dense case;

distribution of dist-to-the-nearest: interaction model vs. random model

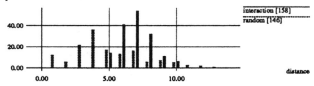

Figure 15: Comparison of interaction:(d=8) and random:P=(10:4) in relatively sparse case;

- In the interaction model, the distribution of the distance between bristles is **much narrower and more peaked**.

A **peaked distribution** reflects order in the system. Therefore it follows that **the collection of interacting celloids has a higher degree of order than the collection of independent celloids.**

6   Conclusion

Artificial synthesis of spacing patterns is presented as the one example of a project entitled "Celloid for Organismoid". A new mechanism, Artificial Genetic Network (AGN), is introduced, which is inspired from the Genetic Network in nature, in order to achieve the above goal. A collection of celloids, each of which interacts according to the same AGN, is compared with a collection of independent cells, in which each cell behaves independently from the others. The comparison shows that the former exhibit higher order than the latter. Therefore, the importance of proper interaction between component parts in organisms is demonstrated.

Finally, important future directions are enumerated below:

1. **Generalization of the AGN**: The box-in-box AGN can provide the **sequential / linear** expression of genes. (See Fig. 16.) The differentiation with this AGN is based on whether a **critical gene** is expressed or not. But if **branches** are introduced to this AGN, the differentiation into many types can be easily designed. (See Fig. 16.)

   Another way to provide differentiation would be to generalize the AGN as in Fig. 4. In this case, different **attractors** (i.e. a set of **stable** states of the gene expression) correspond to different types of cells. The number of attractors can change if the AGN can **"mutate"** (e.g. by means of the {generation /deletion} of the {nodes/edges}). This mutatability of the AGN leads to the **evolvability** of celloids with AGN.

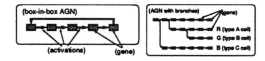

Figure 16: Linear genetic network (left) and Branchable genetic network (right).

2. **Design of the interaction**: The interaction in celloids with the box-in-box AGN is explicit. In other words, all aspects of the interaction are specified explicitly by **symbolic mapping**. Instead of using symbolic mapping between genes and their products, (such as the product #i can repress the gene #j), use of a **template** as in Tierra ([Ray91]) should give rise to **emergent interaction between genes and their products**, which leads to interaction between celloids.

3. **Introduction of cell division and cell death**: In the example shown in this paper, each celloid makes a binary decision whether to have or not to have a bristle. If two other **binary decisions**, for example, **to divide or not to divide** and **to die or not to die**, are introduced into celloids, then an organismoid can grow. These decisions will be made based on the local interaction between the celloids, and this local interaction is expected to give rise to **self-organization of celloids**.

7   acknowledgement

I'd like to thank Francois Grey for proofreading.

References

[Conr92] Conrad M. "Molecular Computing: The Lock-Key Paradigm", in IEEE Computer vol.25, No.11, pp.11-20, (1992).

[Kauf90] Kauffman S.A. "Requirements for Evolvability in Complex Systems: Orderly Dynamics and Frozen Components", in Physica D 42, pp.135-152, (1990).

[Lawr92] Lawrence P.A. "The Making of a Fly", Blackwell Scientific Publications, (1992).

[MolBio] Alberts B., Bray D., Lewis J., Raff M., Roberts K., and Watson J.D.: "Molecular Biology of the Cell, 2nd ed.", Carland Pub. Inc, (1989).

[Ray91] Ray T.S.: "An approach to the synthesis of life", in Langton C. et.al (Eds.) "Artificial life II", Addison-Wesley, pp.371-408, (1991).

[Reyn87] Reynolds C.W., "Flocks, Herds, and Schools: a Distributed Behavior Model", in Computer Graphics 21(4), pp. 25-34, (1987).

CHARACTER RECOGNITION AGENTS

Lijia Zhou and Stan Franklin

Institute for Intelligent Systems
and
Mathematical Sciences Department
The University of Memphis
Memphis, TN 38152
zhoul@hermes.msci.memphis.edu
franklins@hermes.msci.memphis.edu

Abstract

Artificial life techniques are applied to the important practical problem of handwritten character recognition. Artificial agents feed in an environment containing a handwritten character, and compete to recognize that character by their rate of consumption. These character recognition agents learn the characters as paths of food, and are selected by evolution(in fact, co-evolution with characters). Here we describe an architecture for character recognition agents, and propose a strategy for combining learning and evolution. Preliminary test results, from a small prototype system, suggest that learning can shape the direction of evolution, and that co-evolution can speed up the process of evolution. Recognition rates, while not high in an absolute sense, are promising.

1. Introduction

In this paper we suggest applying artificial life techniques to the important practical problem of handwritten character recognition. More specifically, we show how character recognition agents in an artificial environment can learn to recognize characters, that competent sets of such agents can evolve, and that the addition of co-evolution seems to speed up their evolution.

This research was motivated by attempts to develop an optical character recognition system capable of character recognition without significant preprocessing, especially without feature extraction. Successful pattern recognition systems must be flexible in dealing with changing environmental conditions. In designing such a system, one typically struggles to find features which are stable or almost stable in a changing environment. Finding such features is highly nontrivial and time-consuming. As available computational power grows, an artificial life experimental approach, which avoids feature extraction, may become more and more feasible.

As a start in this direction we describe artificial agents that can follow the paths of characters, and, by doing so, recognize them.

1.1 Background

Recognizing handwritten characters has long been an important research topic (Mori 1992). During the past four decades, a wide variety of approaches have been proposed. These approaches can be organized into three broad categories: statistical methods, structural methods, and neural networks.

Statistical methods can be subdivided into two broad classes: Bayesian methods and clustering. Bayesian methods prescribe decision rules by which a sample is assigned to the class with the highest a posteriori probability. Clustering is a procedure by which unlabled data vectors are divided into groups which, ideally, form natural "clusters." Clustering involves defining a similarity measure between data vectors, and a criterion function which evaluates the quality of vector set partitions. The goal is then to seek optimal partitions.

Structural methods describe shape using constituent primitive parts and the relations between these parts. Models of classes can be grammars, where class membership is decided by parsing. Models can also be general relational structures. where class membership is decided by graph matching.

Neural Networks implement rather simple forms of category formation and associative memory. In category formation, the neural network learns a set of categories and classifies incoming inputs according to these. The function of an associative memory is to encode, store, and selectively recall relevant information. By extension, associative memory systems can therefore be used as classifiers when they store exemplar patterns (models). Several neural network architectures have been proposed. Among the most popular are the multilayer perception (Rumelhart, Hinton and Williams 1986), the Hopfield network (Hopfield 1984), the Carpenter-Grossberg network (Carpenter & Grossberg 1987), and Kohonen's self-organized map (Kohonen 1984).

Almost all these methods need preprocessing for feature extraction. Usually, some features are useful for certain characters and not for others. It is hard to find a single set of features that can describe the whole recognition space. In an attempt to solve this problem, we previously introduced a hybrid approach, called an ANN-tree (Zhou and Franklin 1993) for choosing suitable features according to different situations. This method splits a recognition space to the several smaller spaces, making feature selection easier and making the system more flexible. Although feature selection in an ANN-tree is automatic, initial features are supplied by the designer. Over many years of working on character recognition, feature design and choice has been a continual headache. Perhaps with the help of artificial life techniques, a "featureless" method can be found. What follows is a first attempt in that direction.

1.2 Character Recognition Agents

A character recognition agent is a simple artificial organism that must find food in a predefined artificial environment. This environment is a character image, a 24x24 pixel matrix. The body pixels of the character, each represented by a "*", constitute the agent's food, "." represents a blank, a location without food. A character recognition agent is an adaptive system designed to operate in this tight, bounded environment. Such an agent can learn to find food by following a path along the character. (For earlier, path following agents see Jefferson, et al 1992 and Collins and Jefferson 1992.) There are two problems that need to be solved with this scheme: 1) If eating the food means recognition, how do agents show the recognition results? 2) How can an agent find and eat food fast?

To solve the first problem, we employ a number of agents, each trained to eat a particular character. The number of agents is dependent on the number of characters to be recognized. Agents trained to recognize a character "i" will be referred to as agent(i). For instance, agent(A) is an agent trained to eat character "A". During the recognition process, each trained agent is feed the same character. They begin eating simultaneously. An agent trained to eat this particular character will eat faster than others. The character is recognized by the index of the winning agent, the fastest eater.

Clearly the fast eating problem, problem 2), is the key issue for this recognition system. For an agent, fast eating means moving fewer steps and finding (eating) more food. To find food quickly, an agent has to be able to memorize the paths of its given character. It does so, by neural network learning.

In the rest of this paper, we will describe the agent's structure and its learning strategy, study the co-evolution and learning of agents, discuss the experimental results and finally conclude the paper.

2. The Agent Learning Architecture

Learning from the environment can be described as follows: An autonomous agent L learns through experiment what it can do and see in an environment E to the extent that L can drive E into a set of states that L "wants" E to be in. Several learning strategies have been proposed. Among these, reinforcement learning (e.g. Barto, Sutton and Anderson 1983) and rule learning (e.g., Wilson 1985) are two major learning models. In reinforcement learning, trial-and-error learning takes place with only a right or wrong judgment supplied as feedback. In rule learning, rules are induced from examples in the context of problem solving, say of finding food. Example contain observed consequences of actions upon the environment. The quality of the learned rule is then measured by its utility for problem solving.

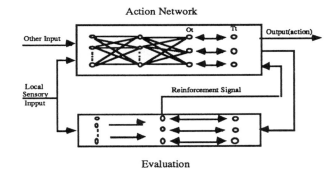

Figure 1. Agent's Learning Architecture

Character recognition agents employ a variant of reinforcement learning. (See Ackley and Littman 1992, for a similar scheme.) Figure 1 depicts the two central structures controlling an agent, its action network and its evaluation component.

At discrete time steps, the agent collects sensory information from the environment and chooses an action within the environment. The action network, which effects this choice, is a two layer back propagation neural network mapping sensory input to motor output. The initial weights of this network are set randomly, and are adjusted over time by a reinforcement learning algorithm.

The evaluation component also receives local sensory input and, in addition, the action network's output. From this information, the evaluation component calculates an action which is then used to train the action network. The learning method used here is not a truly reinforcement learning because the reinforcement signal supplies a correct answer. We call it a semi-reinforcement learning .

The *learning algorithm* can be described as follows:

Given a back propagation network with N input units, M hidden units and 3 output units that encode eight possible actions (directions in which to move). (000) = East; (001) = NE; (010)=N; (011)=NW; (100)=W; (101)=SW; (110)=S; (111)=SE.

Step 1. Receive sensory input vector V(t), and compute reinforcement signal R(t) using the evaluation algorithm.

Step 2. Assign R(t) to T(t), the target vector. Back propagate error E(t) calculated from this target.

Step 3. Update weights.

Step 4. Continue to train the action network via back propagation with same input and target output, until E(t) < error-threshold.

Step 5. Perform the action designated by output O(t) of the action network. Let t = t + 1. Go to Step 1.

Evaluation Algorithm :

Step 1. Encode the local food positions, designated by "*", into three bit binary vectors B(i). If no food is present locally, encode distant vision information. (i represents the directions 0...7).

Step 2. Forward propagate the action network to produce O(t).

Step 3. Compute the Hamming distance between B(i) and O(t), for i = 0...7. Assign the Bi with the smallest distance to R(t).

In the computer simulation, we represented the eight directions an agent, @, can move as illustrated in figure 2.

3	2	1
4	@	0
5	6	7

Figure 2.

The sensory input vector, V(t), includes the agent's local *and* distance vision information, as well as local information about its nearby food, and the agent's previous move. A total of 32 bits are required. The action network has seven hidden units and three output units.

3. Co-Evolution of Agents and Characters

Since fast eating is the key to recognition, we need some way to measure the rapidity of an agent's eating ability. For this purpose set

(1) $E(i) = Nf / Nm$

where i indexes a particular character agent, Nf the number of asterisks agent(i) has consumed, and Nm is the number of moves that agent(i) has made. The range of E(i) is $0 \le E(i) \le 1$. E(i) will used to evaluate the fitness of each agent.

With the environment, the learning algorithm and a fitness criterion in hand, the agent seems ready to be in the environment eating. But where shall he start? In our simulation, an agent can start moving at any of the four corners labeled of the character matrix (figure 3)

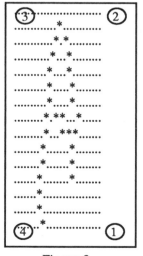

Figure 3.

Actually, we created four character agents for each character, one which starts from each of the different corners. For instance, agent(A3) is the agent(A) which starts moving at corner three. Since there are four agents for each character, choosing a set of agents that produces good recognition is a search problem. We use a genetic algorithm to solve it (Holland 1975).

A genetic algorithm searches from a population of possible solutions, rather than from a single candidate as do most search methods. Here, each individual (candidate solution) is represented by a chromosome with M genes, where M is the number of characters to be recognized. Loci on the chromosome correspond to characters to be recognized. Each gene codes for a character recognition agent trained to the character of that locus. The allele appearing at locus i represents one of agent(i1), agent(i2), agent(i3), or agent(i4). For instance, the chromosome

4	2	1	3	2	3

agent(A) agent(B) agent(C) agent(D) agent(E) agent(F)

represents six different character agents in which agent(B) starts moving at corner two.

In an actual run of the system, each agents of the set specified by a chromosome is given the same character on which to feed. Each begins moving from its given starting point. The winner recognizes the character presented as that specified by its locus.

To continue with the genetic algorithm, a population of chromosomes is created randomly. A fitness function, to be applied to each individual chromosome in population is defined by the following formula:

$$(2) \qquad F(X) = \frac{1}{M} \sum_{i=1}^{M} E(i) - \frac{1}{(M-1)^2} \sum_{i=1}^{M} \sum_{j \neq i} E(j)$$

where X is an individual chromosome, M is the length of chromosome, E(i) is the fitness of agent(i) (see formula (1)) who is receiving the character it is to recognize, and E(j) is the fitness of agent(j) who is receiving some other character. The range of fitness function is $0 \leq F(X) < 1$.

For an effective evolutionary search, one must supply a well chosen set of characters for the fitness function above to operate on. To do so, we employ a co-evolution strategy (Hillis 1992). After several generations, test the most fit individual with each of the set of characters. Rank the characters according to the error that this very fit individual makes. The higher the error, the higher the rank of the character. This rank is essentially the character's fitness value. Now add a number of very fit characters to the training set for the agents. Train each agent in this agent-string (individual). At the same time replace some characters from the test set by some new characters. By iterating this co-

evolution and learning cycle, the whole population will become more fit.

The co-evolution and learning algorithm can be described as follows :

Given L*M characters for the fitness test,

Step 1: Initialization
— Randomly initialize the weight matrix of each agent(i).
— Initialize the population of agent-strings with random alleles.

Step 2: Agent-string searching
— Parallel repeat until a number of generations is reached or a good enough individual recognition behavior is found.
— Judge the fitness of each individual.
— Select individuals for reproduction in proportion the their fitness.
— Mate pairs of selected individuals via crossover.
— Mutate the resulting new individuals.

Step 3: Character searching
— Rank the test set characters by the errors made by the most fit individual in the population.
—Put Nc highly ranked characters into the training set of each agent(i).
—Remove the Nc characters, and add Nc new characters to the character test set.

Step 4: Agent learning
— Train each agent(i) belonging to the most fit individual with the training set found in step 3. Use the learning algorithm described above.
— Goto Step 2

4. Testing

Our initial prototype contains 7*4 agents meant to recognize the handwritten characters A, B, C, D, E, F, G. It uses 7*10 characters as a fitness test set. Our intent is to create one individual (an agent-string) describing a set of agents which can do a good job of recognition. Hence, the top individual's fitness value is recorded at each generation. Two tests were performed. Test 1 employed both evolution and learning, while test 2 combined co-evolution and learning.

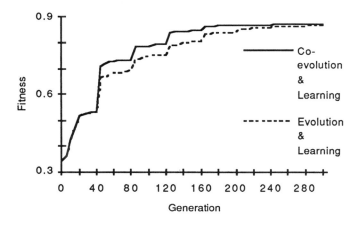

Figure 4.

Figure 4 shows the results of both tests. Two important facts emerge. First, learning can shape the direction of evolution as is clear from the zig-zag shape in the figure. The fitness value jumps at each multiple of 40 generations where learning took place. Second, it seems that co-evolution speeds up the process of evolution. Though both methods level off at a fitness approaching .9, with co-evolution this is achieved in 160 generations, while 240 generations are required without it.

While training was done with at most seven instance of each character, recognition was tested, at each generation, on 100 instances of each. The results are depicted in Figure 5.

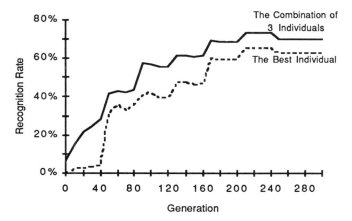

Figure 5.

These tests show that a majority vote of the best three individuals out performs the single best individual. The effect of learning is clearly visible. The tailing off of recognition rate after 240

generations seems to be due to generalization problems with backpropagation learning as the size of the training set increases. The recognition rate achieved, while not high in an absolute sense, is promising in a small prototype with a small training set. On the other hand, testing is computationally intensive, since the training set increases over time and backpropagation learns slowly.

The various tests show

5. Conclusion

A potential application of artificial life methods is presented. A method of combining co-evolution and learning to solve an important and practical character recognition problem is introduced. An architecture for a character recognition agent, who recognizes characters by consuming them, is given, and a strategy (algorithms) combining learning and co-evolution is proposed. Preliminary test results show that learning can shape the direction of evolution, and that co-evolution can speed up the process of evolution. Although the tests as yet are partial, small and incomplete, the results are exciting.

Our next task is to build a complete system on a more powerful, perhaps parallel, machine, and fully test this new application of artificial life techniques against known methods of handwritten character recognition.

6. References

Ackley, David, and Littman, Michael (1992), "Interactions between Learning and Evolution," in Langton, et al, editors, **Artificial Life II**, Redwood City, CA: Addison-Wesley, 487-509.

Barto, A., Sutton, R., and Anderson, C. (1983), "Neuron0like adaptive elements that can solve difficult learning control problems," *IEEE Trans. Systems, Man & Cyber.*, **SMC-13** (5), 834-846.

Carpenter, G.A. and Grossberg, S. (1987), "ART 2: Self-organization of stable category recognition codes for analog input patterns," Applied Optics, 26 (23), 4919-4930.

Collins, R. and Jefferson, D. (1992), "AntFarm: Toward a Simulated Evolution," in Langton, et al, editors, **Artificial Life II**, Redwood City, CA: Addison-Wesley, 579-601.

Hillis, W. Daniel (1992), "Co-Evolving Parasites Improve Simulated Evolution as an Optimization

Procedure," in Langton, et al, editors, **Artificial Life II**, Redwood City, CA: Addison-Wesley, 313-324.

Holland, John H. (1975), **Adaptation in Natural and Artificial Systems**, Ann Arbor: University of Michigan Press.

Hopfield, J.J. (1984), "Neurons with graded response have collective computational properties like those of two-state neurons," *Proc. Natl. Acad. Sci.* **81** 3088-3092.

Jefferson, D., et al (1992), "Evolution as a Theme in Artificial Life: The Genesys/Tracker System," in Langton, et al, editors, **Artificial Life II**, Redwood City, CA: Addison-Wesley, 549-578.

Kohonen, T. (1984), **Self-Organization and Associative Memory**, Springer-Verlag, Berlin.

Mori, Shunji (1992), "Historical Review of OCR Research and Development," *Proceedings of IEEE*, **80-7**, 1029-1088.

Rumelhart, David E., G. E. Hinton and R.J. Williams (1986), "Learning Internal Representations by Error Propagation," in David E. Rumelhart and James L. McCelland et al (eds.) **Parallel Distributed Processing**, Vol. 1, Cambridge MA: MIT Press.

Wilson, S. W. (1985), "Knowledge Growth in an Artificial Animal," *Proceedings of the First International Conference on Genetic Algorithms and Their Applications* (pp. 16-23). Hillsdale, New Jersey: Lawrence Erlbaum Associates.

Zhou, Lijia and Franklin, Stan (1993), "ANN-Tree: A hybrid method for pattern recognition," SPIE Proceedings, vol. 1965, **Applications of Artificial Neural Networks IV**, 358-363.

The Building Behavior of Lattice Swarms

Eric Bonabeau [1], Guy Theraulaz [2],
Eric Arpin [3], Emmanuel Sardet [3]

(1) CNET Lannion B - RIO / TNT, route de Trégastel, 22301 Lannion Cédex, France - e-mail: `bonabeau@lannion.cnet.fr`
(2) CNRS - URA 1837, Laboratoire d'éthologie et psychologie animale, Université Paul Sabatier,
118 route de Narbonne, 31062 Toulouse, France - e-mail: `theraula@cict.fr`
(3) Télécom Paris, 46, rue Barrault, 75013 Paris, France

Abstract

In this paper, we present building algorithms for artificial swarms, inspired by how wasp colonies build their nests. The insectoïds which constitute the swarms move randomly on 2D or 3D lattices, do not communicate directly, but instead use the local shapes they encounter to determine their individual building behaviors.

1. Introduction

We all know that some animal colonies are capable of producing very complicated patterns. In this paper we propose to start an exploratory study of architectural patterns grown by artificial agents moving and acting in a virtual space. We choose to base this study on biological data provided by observations of nests built by social wasps (Downing and Jeanne, 1988, 1990; Hansell, 1984; Karsai and Penzes 1993; Wenzel, 1991). For a long time, wasp nest architectures have been studied from the viewpoint of their functional and adaptive values (protection against predators, reproduction, etc...). A complementary aspect concerns the individual behavioral algorithms which allow a society to build its (possibly highly complex) nest. As biologists, we hope to gain some understanding of the algorithmic processes involved in, and required by building behavior, through simulating minimal algorithms followed by simple agents. Such simulations can provide us with valuable insight when, as engineers, we need to design programs or machines exhibiting collective problem-solving abilities. In both cases, we are faced with the same 'inverse problem', where given a shape (or more generally a desired state) we have to find the simplest behavioral algorithm that can generate it. We shall indeed restrict our attention to very simple individual algorithms, where information is processed locally, in space as well as in time. No direct communication between the agents will be assumed. Instead, the only relevant interactions taking place between agents are indirect, *i.e.* through the dynamically varying shape in construction: such a process was called *sematectonic* communication (Wilson, 1975). Of course, we shall assume no blueprint, nor shall we resort to any representation of the environment under the form of a map.

A lot of collective phenomena resulting in the structuration of the environment have been described in social insects (see Theraulaz and Gervet, 1994 for an overview). Some of these processes rely on chemical gradients that orient individual behavior, like in the building behavior of termites (Deneubourg, 1977); others are based on local density of objects of the same type to produce global configurations such as the clustering of objects in ants (Deneubourg *et al.*, 1991) or the characteristic concentric brood pattern we observe on the combs of honey bee colonies (Camazine, 1991). Extending these lines of researchs we wish to know whether is it possible to build complex structures in 2D or 3D worlds with artificial swarms, using only local configurations of objects. This would contribute to the understanding of biological collective systems (Gallais-Hamonno and Chauvin, 1972), and constitute a first step for engineers seeking new ways of implementing collective problem solving with artificial swarms such as robots or mobile automata [see *e.g.* Beni, 1988 for a similar approach]. Colonies of robots, which already exist (Beckers *et al.*, 1993) and are capable of performing spatial clustering of objects, could be designed to follow these simple behavioral algorithms.

We present in this paper a few examples of architectures grown by artificial agents moving randomly on a lattice, and performing very simple asynchronous actions with purely local information. We discuss the notions of stigmergic vs sequential algorithms, and of growth complexity. We also briefly describe a possible solution to the inverse problem. But let us first review some biological data originating from studies on wasp nests.

2. Biological background and experimental data

Social insects such as termites (Grassé, 1959; Deneubourg, 1977), ants (Franks *et al.*, 1992), honey bees (Camazine, 1991; Skarka *et al.*, 1990), or social wasps have the ability to build nests whose architectures range from the simplest to

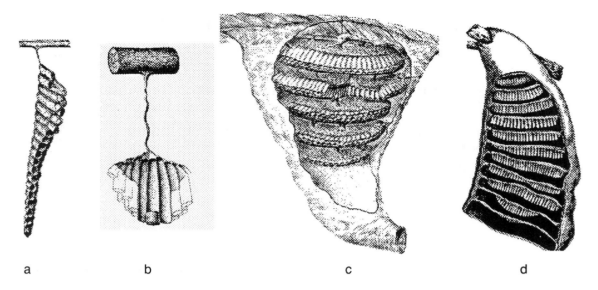

Figure 1: a. nest of *Ropalidia variegata*, **b.** nest of *Mischocyttarus drewseni*
c. nest of *Angiopolybia pallens*, **d.** nest of *Epipona morio* (modified from Jeanne, 1975).

more complex, highly organized ones. We focus here on nests generated by social wasps. Figure 1 shows some examples of architectures we find in nature. It seems obvious that some of these nests are more complex than others: in particular, nests c and d possess an external envelope and look more organized in the inside. Wenzel (1991) has classified wasp nests architectures, and found more than 60 different types, with many intermediates between extreme forms. A mature nest can have from a few cells up to a million cells packed in stacked combs, the latter being generally built by highly social species.

Since a lot of experimental studies were carried out on primitively *eusocial* wasps (eusocial implies overlapping adult generations, cooperative brood care, presence of a sterile caste), we will focus on these simpler societies of wasps. Most of these colonies are usually founded by one or a few individuals, and contain less than 100 individuals at the same time. What is important to ethologists is to understand the underlying behavioral 'algorithms', and possibly their phylogeny. Such an algorithm includes several types of acts, the first one consisting most often in attaching the future nest to a substrate with a 'stalk-like pedicel of wood pulp' (Wenzel, 1991) (see nests a, b and c). It seems that the placement of the first cells of the nest follows always the same rule within one given species. Combs are made of parallel rows of cells and the nest generally has radial or bilateral symmetry around these intial cells. One important point is that wasps tend to finish a row of cells before initiating a new row and that rows are initiated by the construction of a centrally located cell first. These simple rules ensure that the nest will grow fairly evenly in all directions from the petiole (Downing and Jeanne, 1990). There is therefore an isotropy in space. Other

rules ensure the enlargment of the nest by the addition of new combs with pedicels attached to the first one. Eggs are laid in the combs which are progressively built. Brood is present in these combs at various stages of maturity, explaining the differences in growth of the different combs, which are extended when needed by the growing larvae. This shows that a nest is generally the result of a complex interplay between regulatory mechanisms. It may be possible, however, to extract (endogeneous) minimal behavioral rules that could allow for the generation of the nest shape, without explicitly simulating all these mechanisms.

3. Stigmergic algorithms

The swarm is composed of insectoïds that move and act in a 2D or 3D lattice and are able to deposit bricks according to their local neighborhoods (8 and 26 neighboring cells for our 2D and 3D lattice swarms respectively), using a look-up table. The space of local configurations is rather huge (2^8 and 2^{26} in 2D, 3^8 and 3^{26} in 3D with two and three states bricks respectively) and the space of local rules cannot be explored systematically (2^{2^8} and $2^{2^{26}}$ in 2D and 3^{3^8} and $3^{3^{26}}$ in 3D). Therefore, one has to solve the inverse problem one way or another in order to discover the minimum set of rules necessary to produce a given architecture.

In insect colonies, the collective building activity results from the adjustment of individual behaviors. Every individual has to decide whether or not to put down a piece of material so as to coordinate its own building activity with the one resulting from the actions of the other members of the society. In that way, the behavior of a social insect might appear at first glance more "intelligent" or "complex" than a solitary one. In a previous paper, we pointed out that the building activity of solitary insects relies on a sequential algorithm, while that of social insects is based on a stigmergic one (Deneubourg, Theraulaz and Beckers, 1992). *Stigmergy* was originally defined by P.-P. Grassé in his pioneering studies on the reconstruction of the termite nest of *Bellicositermes natalensis* (Grassé, 1959). In order to explain the coordination of individuals' tasks, Grassé showed that the regulation of the building activity does not depend on the workers themselves but is mainly achieved by the nest structure. A termite worker does not direct its work; rather, its actions are controlled and guided by its previous work, and modify in turn the shape of the local configuration which has triggered its building activity. The new configuration will then automatically trigger other kinds of actions from the termite or from any other worker of the colony. A stigmergic algorithm differs from a sequential one because in the former mode of construction, the insect automatically responds when it meets any local configuration, while in the latter the insect reacts sequentially to only a small subset of local configurations and is blind to all others. When following a stigmergic algorithm, the insects continuously repeat the same action until by chance a new configuration is produced by the last action and allows for new actions to be taken.

Let $C = \{C_1, C_2, ..., C_n\}$ be the set of local stimulating configuration, that is the configurations which release the building behavior (put down a brick). In a sequential algorithm, each insect is characterized by a set of internal states $S = \{S_1, S_2, ..., S_n\}$, each of these states is associated with a subset $C(S_p)$ of C, with $\cup_p C(S_p) = C$ and $\forall p_1 \neq p_2, C(S_{p_1}) \cap C(S_{p_2}) = \varnothing$. When a building action is taken by the insect, its internal state S_p shifts to S_{p+1} with a probability $p(S_p \rightarrow S_{p+1})$. The local configurations triggering building actions are modified: the building activity of the animal is constrained by the sequential dynamics of its internal states so that configurations associated with the previous state of the animal no longer trigger any building activity. In the case of the stigmergic algorithm, all local configurations that can trigger a building action are potentially simultaneously present. Each new configuration resulting from a building action, may or may not belong to C. When no configuration belongs to C, the building activity stops. When a local configuration that doesn't belong to C is produced, it constitutes a local constraint or a local stopping configuration. The coordinated building is realised thanks to these local stopping configurations. Stigmergic algorithms allow social insects to evaluate many different local configurations at once, their responses can thus be more flexible and a greater range of building problems can be overcome (Downing and Jeanne, 1990).

Stigmergy provides us with a possible way of solving in certain cases the inverse problem. In effect, starting from a given shape, one can try to decompose it into sub-shapes corresponding to local stopping configurations. Such configurations will enforce the approximate order of the sequence of acts that allows for the production of the shape: the application of a given action requires another action to be taken first, because there is a local stopping configuration overcome by that action. We successfully applied this method to the generation of the shapes presented in the example sections. Of course, it is important to determine whether all shapes can be produced with this method. In effect, there are unexpected constraints emerging as the shape is being built: these constraints may very well eventually prevent some configurations from being reached. In order to avoid this type of problem, one can resort to a more refined growth algorithm. It is worth giving a few explanations about the nested hierarchy of such algorithms.

4. The complexity of nest architectures

In order to test basic hypotheses on how swarms of simple animals or agents may be able to design complex architectures, one can explore the space of shapes following a hierarchy of behaviors developed by Cris Moore (private communication) to describe different levels of growth complexity. *Growth complexity* is different from recognition complexity, embodied *e.g.* in Chomsky's grammars, which tells how difficult it is to recognize that a particular pattern belongs to a given set of patterns: growth complexity is a measure of how difficult it is to grow a pattern. To grow a pattern on a lattice, one seeks to apply rules depending on local states (theoretically, the size of the local window irrelevant, but for obvious biological and practical reasons, we choose to resort only to very small neighborhoods in space as well as in time). The growth complexity of the pattern is then defined by the way in which the (minimal) rules are applied to generate the desired pattern. The rules for depositing or removing bricks (chosen among several possible types of bricks) can be applied as follows:

(**1**) rules can be applied deterministically;
(**2**) rules can be applied stochastically (*i.e.* a given configuration triggers a rule with a predefined probability);
(**3**) there is only one set of rules which is always scanned;
(**4**) there are several sets of rules which are used in a seasonal manner (*i.e.* set n°1 during t_1 times steps, then set n°2 during t_2 time steps, etc...)
(**5**) there is a hierarchy of sets of rules so that lower-level rules are applied when higher-level rules can no longer be applied (i.e. practically, an agent tries to apply the first set of rules; if it does so unsuccessfully for an amount of time T -this implies some kind of memory-, it switches to the next set of rules, etc...)

In this paper all structures will be produced with probabilistic rules in 2D (seasonal and/or stochastic) and deterministic stigmergic rules in 3D. In order to compare the complexities of various algorithms, we will call growth complexity the number of rules that the algorithm comprises (if two algorithms are to be compared with respect to this criterion, they must belong to the same hierarchical level). Growth complexity has obviously some

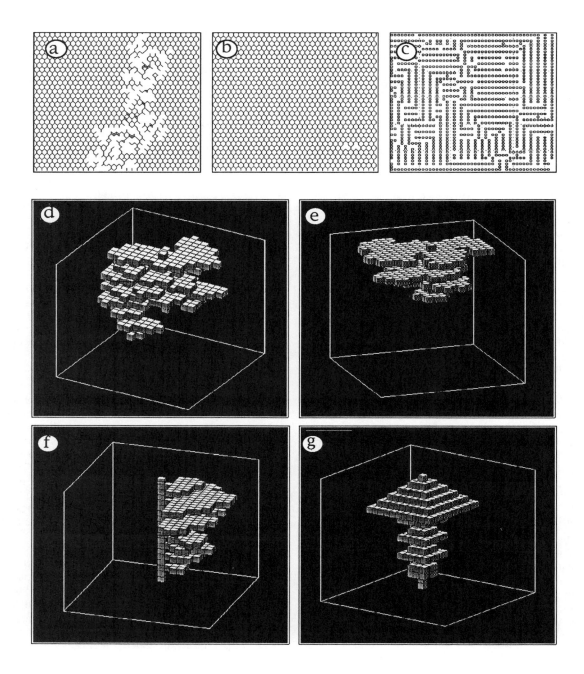

Figure 2: a. After 10000 time steps, one obtains this pattern comprising two rather regular regions separated by an irregular gap which contains many defects. **b.** It takes 10000 more time steps to get rid of the defects and obtain a fully regular pattern. **c.** This maze is obtained after approximately 2000 time steps. The number of rules is 15 including rotations, but 6 when they are excluded in the first case (a,b) and 106 (29 without rotations) in the second case (c). **d.** Nest architecture obtained after 24000 steps (growth complexity = 9). **e.** Nest architecture obtained after 22000 steps (Growth complexity = 15). **f.** Nest architecture obtained after 24000 steps (Growth complexity = 10). **g.** Architecture obtained after 45000 steps (Growth complexity = 147). 2D Simulations were made on a 50X50 Lattice with 100 wasps. 3D Simulations were made on a 20X20X20 Lattice with 20 wasps for d, e, f and 40 wasps for g.

biological relevance, since it is a measure of the sophistication of the behavioral algorithm needed to effectively grow a family of 'complex' architectures. It is also a relevant way of quantifying the complexity of a man-made system designed to grow these complex architectures. The next two sections are devoted respectively to 2D and 3D architectures. We end up with a review of some directions for future work.

5. 2-D lattice swarms

We present in this section two very interesting applications of swarm intelligence to the design of regular patterns by agents with no central coordination, moving randomly, and performing their actions *asynchronously* with only purely local information. The patterns of figures 2a and 2b are separated by about 10000 time steps. Agents have cyclic activities, the first cycle consisting in depositing little branches (represented by bars), the second cycle in removing some of these branches, so that flaws (2a) in the regular structure can be removed (2b). Moreover, the algorithm is probabilistic: for example, each individual deposits a branch in an empty neighborhood with a very low probability.

The next example is inspired from the rules of the Game of Life. A rather regular maze is obtained (see 2c). At time t=0, circular bricks are deposited randomly in the environment to initiate the organizing process. Any empty site which is surrounded by exactly three filled sites can receive a brick (with a fixed probability). Any site containing a brick and surrounded by more than four sites also containing a brick is likely to have its brick removed. All other sites are likely to remain unchanged in the presence of an artificial wasp. The probability to deposit a brick at a wrong site decreases with time, so as to avoid the destabilization of the organized pattern eventually obtained. The idea is similar to that of simulated annealing where temperature is slowly decreased until one reaches a state of minimal energy.

6. Building behavior in three dimensions

For obvious reasons, we have tested only purely deterministic rules, with no season and no hierarchy. Extensive simulations on a powerful computer have to be performed in order to explore the behavioral space in a satisfactory manner, even in this simplest case. According to their neighborhoods and lookup tables, agents may deposit two types of bricks (type 1 or type 2). Although the rules are simple and inspired by the natural building behavior of wasps (cf. §2), we obtain interesting patterns which closely match those found in nature. We present a few architectures using these simple deterministic rules. Space limitation prevents us from giving in extenso all the rules used to grow the patterns presented in figures 2d-g. In the appendix we give the rule used to produce the architecture shown in figure 2d. The behavioral rules used to generate figure 2d-f have been chosen (using the heuristic solution to the inverse problem previously described) so as to reproduce architectures which can be found in nature in the Vespids.

Figure 2d is representative of the type of architecture found in the species *Stelopolybia vicina*; 2e is closed to the nests of the *Vespa* genera while 2e closely resembles that of *Parachartergus*. Note for instance the little piece on the top of the nest, which corresponds very closely to natural pedicels, and the succession of horizontal planes which represent the combs of natural nests. Figure 2d differs from Figure 2e because in the former the enlargement of the nest was made by multiple pedicels connecting the combs at different levels, while in the latter they are stacked, connected by one central pedicel. Figure 2g has been found by a synthetic exploration of the space of possible architectures. We selected this rule because it shows that very ordered structures can be obtained: directed synthesis may turn out to be helpful to find the simplest set of behavioral rules which allow to grow a pattern.

7. Future directions

It is by now obvious that the space of possible architectures grown by artificial agents on a lattice capable of depositing or removing, say, two types of bricks according to the local state of the environment, is huge and long to explore, even if one restricts one's attention to (at least partially) spatially isotropic rules. Therefore, we are currently investigating two different ways of constraining our exploration:

• In the first one, a human observer evaluates the fitness of a pattern according to her choice. Behavioral algorithms are encoded into bit strings and those which generate the most 'interesting' architectures are selected. The term 'interesting' may refer to biological resemblance or to any other criterion, be it subjective or objective. As biologists we prefer biologically plausible architectures, and the interactivity of the method is most helpful since it may be hard to formally define the adaptive value or the plausibility of the shape of a nest; as engineers we are looking for 'useful' patterns.

• In the second one, we add some biological features based on ethological studies; although the space to explore is wide enough without additional features, it may be useful to have the nest self-regulated by the whole activity of the colony. For instance, the introduction of stimulating volatile chemical cues (favoring actions in the neighborhood of a recently taken action) may lead to the emergence of critical mass effects, whereby the nest cannot be built unless there is a sufficient number of individuals. The introduction of such features can be insightful (Karsai and Penzes 1993).

8. References

Beckers, R., J.-L. Deneubourg and S. Goss. 1993. Paper presented at the Second European Conference on Artificial Life, May 24-26 1993, Brussels, Belgium.

Beni, G. 1988. The concept of cellular robotic system, pp. 57-62. In: *Proceedings of the 1988 IEEE International Symposium on Intelligent Control*, Arlington, VA.

Camazine, S. 1991. Self-organizing pattern-formation on the combs of honeybee colonies. *Behav. Ecol. Sociobiol.* 28: 61-76.

Deneubourg, J.-L. 1977. Application de l'ordre par fluctuations à la description de certaines étapes de la construction du nid chez les termites. *Ins. Soc.* 24: 117-130.

Deneubourg, J.-L., G. Theraulaz and R. Beckers. 1992. Swarm-made architectures, pp. 123-133. In: *Toward a Practice of Autonomous Systems, Proceedings of The First European Conference on Artificial Life.* edited by F.J. Varela and P. Bourgine. MIT Press/Bradford Books.

Deneubourg J.L., S. Goss, N. Franks, A. Sendova-Franks, C. Detrain and L. Chretien. 1991. The dynamics of collective sorting : Robot-like ant and ant-like robot, pp. 356-365. In: *Simulation of Adaptive Behavior : From Animals to Animats*, edited by J.A. Meyer and S.W. Wilson. MIT Press/Bradford Books.

Downing, H. A. and R. L. Jeanne. 1988. Nest construction by the paperwasp Polistes: A test of stigmergy theory. *Anim. Behav.* 36:1729-1739.

Downing, H. A. and R. L. Jeanne. 1990. The regulation of complex building behavior in the paperwasp *Polistes Fuscatus* . *Anim. Behav.* 39:105-124.

Franks, N. R., A. Wilby, V.W. Silverman and C. Tofts. 1992. Self-organizing nest construction in ants: sophisticated building by blind buldozing. *Anim. Behav.* 44: 357-375.

Gallais-Hamonno, F.G. and R. Chauvin, 1972. Simulations sur ordinateur de la construction du dôme et du ramassage des brindilles chez une fourmi *(Formica Polyctena).C. R. Acad. Sc.* Paris, 275 D, 1275-1278.

P.-P. Grassé. 1959. La reconstruction du nid et les coordinations inter-individuelles chez *Bellicositermes Natalensis et Cubitermes sp.* La théorie de la stigmergie : essai d'interprétation du comportement des termites constructeurs. *Ins. Soc.* 6:41-81.

Hansell, M. H. 1984. *Animal architecture and building behavior.* Longman:London.

Jeanne, R. L. 1975. The adaptativeness of Social wasp nest architecture. *Quaterly Review of Biology* 50: 267-287.

Karsai, I. and Z. Penzes. 1993. Comb building in social wasps: self-organization and stigmergic script. *J. Theor. Biol.* 161:505-525.

Skarka, V., J.-L. Deneubourg and M.R. Belic. 1990. Mathematical Model of Building Behavior of *Apis mellifera. J. Theor. Biol.* 147, 1-16.

Theraulaz, G. and J. Gervet. 1994. Du Superorganisme à l'Intelligence en Essaim: Modèles et Représentations du Fonctionnement des Sociétés d'Insectes. In: *Intelligence Collective*, edited by E. Bonabeau and G. Theraulaz. Paris: Hermès (in press).

Wenzel, J. W. 1991. Evolution of nest architecture, 480-521. In: *Social Biology of Wasps*, edited by K. G. Ross and R.W. Matthews. Ithaca: Cornell University Press.

Wilson, E.O. 1975. *Sociobiology.* Cambridge, MA: The Belknap Press of Harvard University Press.

9. Appendix

The neighborhood of the artificial wasp is composed of the 26 first cells surrounding the cell it occupies. We represented this neighborhood with 3 slices along the z axis (see Figure 3).

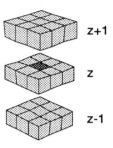

Figure 3. Local neighborhood in 3D Lattice Swarm

Below we give the rule used to produce architecture 2d. When the wasp occupies the central position of the slice z (marqued •), *i.e.* when their is no brick in that cell, it will put down a brik of type 1 in the case of configuration 1.1 and a brick of type 2 in the cases 2.1 – 2.8. We give each local configuration without taking symmetries into account.

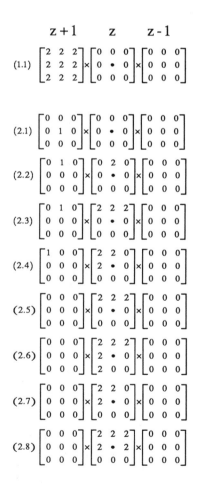

$$
\begin{array}{cccc}
 & z+1 & z & z-1 \\
(1.1) & \begin{bmatrix} 2 & 2 & 2 \\ 2 & 2 & 2 \\ 2 & 2 & 2 \end{bmatrix} \times & \begin{bmatrix} 0 & 0 & 0 \\ 0 & \bullet & 0 \\ 0 & 0 & 0 \end{bmatrix} \times & \begin{bmatrix} 0 & 0 & 0 \\ 0 & 0 & 0 \\ 0 & 0 & 0 \end{bmatrix} \\
(2.1) & \begin{bmatrix} 0 & 0 & 0 \\ 0 & 1 & 0 \\ 0 & 0 & 0 \end{bmatrix} \times & \begin{bmatrix} 0 & 0 & 0 \\ 0 & \bullet & 0 \\ 0 & 0 & 0 \end{bmatrix} \times & \begin{bmatrix} 0 & 0 & 0 \\ 0 & 0 & 0 \\ 0 & 0 & 0 \end{bmatrix} \\
(2.2) & \begin{bmatrix} 0 & 1 & 0 \\ 0 & 0 & 0 \\ 0 & 0 & 0 \end{bmatrix} \times & \begin{bmatrix} 0 & 2 & 0 \\ 0 & \bullet & 0 \\ 0 & 0 & 0 \end{bmatrix} \times & \begin{bmatrix} 0 & 0 & 0 \\ 0 & 0 & 0 \\ 0 & 0 & 0 \end{bmatrix} \\
(2.3) & \begin{bmatrix} 0 & 1 & 0 \\ 0 & 0 & 0 \\ 0 & 0 & 0 \end{bmatrix} \times & \begin{bmatrix} 2 & 2 & 2 \\ 0 & \bullet & 0 \\ 0 & 0 & 0 \end{bmatrix} \times & \begin{bmatrix} 0 & 0 & 0 \\ 0 & 0 & 0 \\ 0 & 0 & 0 \end{bmatrix} \\
(2.4) & \begin{bmatrix} 1 & 0 & 0 \\ 0 & 0 & 0 \\ 0 & 0 & 0 \end{bmatrix} \times & \begin{bmatrix} 2 & 2 & 0 \\ 2 & \bullet & 0 \\ 0 & 0 & 0 \end{bmatrix} \times & \begin{bmatrix} 0 & 0 & 0 \\ 0 & 0 & 0 \\ 0 & 0 & 0 \end{bmatrix} \\
(2.5) & \begin{bmatrix} 0 & 0 & 0 \\ 0 & 0 & 0 \\ 0 & 0 & 0 \end{bmatrix} \times & \begin{bmatrix} 2 & 2 & 2 \\ 0 & \bullet & 0 \\ 0 & 0 & 0 \end{bmatrix} \times & \begin{bmatrix} 0 & 0 & 0 \\ 0 & 0 & 0 \\ 0 & 0 & 0 \end{bmatrix} \\
(2.6) & \begin{bmatrix} 0 & 0 & 0 \\ 0 & 0 & 0 \\ 0 & 0 & 0 \end{bmatrix} \times & \begin{bmatrix} 2 & 2 & 2 \\ 2 & \bullet & 0 \\ 2 & 0 & 0 \end{bmatrix} \times & \begin{bmatrix} 0 & 0 & 0 \\ 0 & 0 & 0 \\ 0 & 0 & 0 \end{bmatrix} \\
(2.7) & \begin{bmatrix} 0 & 0 & 0 \\ 0 & 0 & 0 \\ 0 & 0 & 0 \end{bmatrix} \times & \begin{bmatrix} 2 & 2 & 0 \\ 2 & \bullet & 0 \\ 0 & 0 & 0 \end{bmatrix} \times & \begin{bmatrix} 0 & 0 & 0 \\ 0 & 0 & 0 \\ 0 & 0 & 0 \end{bmatrix} \\
(2.8) & \begin{bmatrix} 0 & 0 & 0 \\ 0 & 0 & 0 \\ 0 & 0 & 0 \end{bmatrix} \times & \begin{bmatrix} 2 & 2 & 2 \\ 2 & \bullet & 2 \\ 0 & 0 & 0 \end{bmatrix} \times & \begin{bmatrix} 0 & 0 & 0 \\ 0 & 0 & 0 \\ 0 & 0 & 0 \end{bmatrix}
\end{array}
$$

Modeling Adaptive Self-Organization

Jari Vaario

Evolutionary Systems Department
ATR Human Information Processing Research Laboratories
2-2 Hikari-dai, Seika-cho, Soraku-gun, Kyoto 619-02, JAPAN
e-mail: jari@hip.atr.co.jp

Abstract

In this paper we present a *computational modeling method* for adaptive self-organization. Adaptation here is defined as a change of behavior in a simulated environment. Self-organization means the creation of a system from several similarly constructed elements (cells) in a way that allows each element to execute its own local rules, which cause interactions within itself and with the environment. The environment model includes mechanical and chemical interactions. The result of the interactions within the environment is a *topological structure*. Additional *interaction networks* may emerge as a result of the chemical and electronic interactions between the basic elements. The topological structure of interacting elements forms a system that can react with the outside stimulus through boundary elements (receptors). This provides a method for primitive communication. Because the simulation integrates the environment within the system, the result is a system that is extremely adaptive to environmental changes. The computational model developed consists of an environment simulator, and an engine to evaluate for the production rules, which describe the actions of the elements.

Introduction

In biological systems, adaptation is based on emergent phenomena. Self-organization in a complex system results from several simple elements executing local behavior rules. Analytical description of such a system is difficult. Computational simulation is required to model the self-organizing process.

The definition of the basic elements and of their local behavior rules lays the foundation for the simulation. We have imitated the real world rather than creating abstract elements and behavior rules. We do this because we believe that simulating real world phenomena extends our understanding of Nature, as well as contributes to a basic understanding of adaptation to the real world.

We recognize that simulated physical phenomena do not correspond to real world phenomena, but through simulation we can explore the principles of self-organization. It is not known how local behavior rules

lead to a complex system. In the first phase, a simplified environment might even be an advantage in that it allows us to focus on just what environmental factors cause the emergence of systems. Also, becoming familiar with the emergent principles helps to extend the principles to more general computational concepts, *e.g.*, economic and social system models.

In this paper we have selected a cell as the basic element. The behavior rules can be divided into rules to simulate a *cell cytoplasm* (cell machine), *i.e.*, the internal interactions, including the transcription of genetic information, and rules to simulate a *cell membrane, i.e.*, the interactions of mechanical and chemical phenomena. The cell machine interprets the genetic code thus providing a way for the systems to evolve, although this is not the topic of this paper.

Finding a mechanism for adaptation during the developmental process is important. Even without changes in the genetic information, we are able to achieve adaptation in a dynamic environment. Furthermore, the extent of the adaptability can be controlled by the genetic information.

This work began with the goal to build a design system for artificial neural networks. The problem faced was how to construct the complex connectivity of neurons. To model a biologically plausible construction of neural connections, Lindenmayer systems were applied (Vaario & Ohsuga, 1992). However, this provided only the genetic growth. The need for modeling environmental effects as well became obvious. The next step was to create a formalism for modeling environmental effects (Vaario, Hori, & Ohsuga, 1994; Vaario & Ohsuga, 1994; Vaario, 1994b, 1994a). In this paper, the role of the initial description is reduced further. We describe how simple local rules result in a multicellular organism, and how global behaviors emerge at different levels of interacting networks.

Emergence of Structures

The basic elements for emergence are cells. They are defined to consist of two types of production rule: *cell machine*, and *cell membrane*. The cell machine rule interprets genetic information then modifies the internal state of the cell accordingly. The cell membrane rule interprets the surrounding environment, and models possible interactions between the cell and the environment. Both types of production rule modify the internal state of the cell (fig. 1).

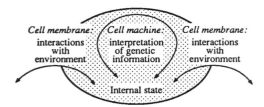

Figure 1: The basic modeling element is a cell. Production rules are divided into two types: rules that interpret the genetic information (cell machine), and rules that interpret the surrounding environment (cell membrane), both of which modify the internal state of the cell.

The result of these two types of actions is intra-cell and inter-cell activity, cell division, and cell death. The inter-cell activity forms a network of cells that can have various emergent levels as described below. Cell division and a death can result from the communication between cells, *i.e.*, synchronization between cells.

The cells modify the environment based on these local rules. The cells produced by cell divisions possess similar production rules, and react basically the same way to similar environmental changes. In this way cells create a coupled network of interacting units (fig 2).

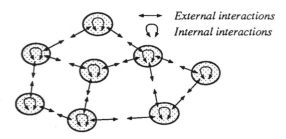

Figure 2: Cells generate a coupled network. The interactions between cells are defined indirectly rather than directly due to modification of the environment.

Cells occasionally differentiate and react to different environmental factors which leads to network coupling. This is how additional emergent levels are created (fig. 3).

The result of this process is a multilevel structure of interacting networks. The interacting networks can be described by their communication method, *i.e.*, the input/output substance of each cell. Thus, if cell has several input and output substances, it belongs to several different networks.

Computational model

Our computational model is based on the concept of production rules inspired by Lindenmayer systems (Prusinkiewicz & Lindenmayer, 1990). However,

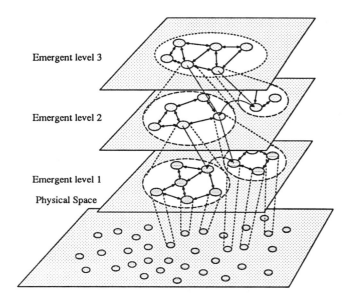

Figure 3: The emergence of structures. One type of interaction with the environment generates one network. A different kind of interaction by some elements of these networks generates a different network, which can be observed as a new emergence level.

instead of using a linear string of symbols to describe abstract objects, we have cells arbitrarily located in an environment. These autonomous cells execute production rules thereby modifying themselves and the surrounding environment, dividing in to new cells, and dying off.

Moreover, instead of having parameters for symbols, we use an attribute list of objects. The execution of the model includes the evaluation of the production rules of each cell, and updating the environmental forces. Each of these are described in detail below.

Cells

A cell is parsed from its string representation $(cell(attr_1, attr_2, \ldots, attr_n))$, which describes the name of the cell $(cell)$. The cell will have a default location and shape in the environment, unless it changes them explicitly through interface functions (such as $ps\_add((100.0, 100.0))$ for setting a new location). For example, an initial description for a cell, which describes some attributes (*e.g.*, chemicals and enzymes) with their initial values and some production rules, can be as follows.

CELL($Chemical_1 = value, \ldots, Chemical_n = value,$
$Enzyme_1 = value, \ldots, Enzyme_n = value,$
$CellMachine = (\text{CELL::} (condition) \rightarrow [\text{CELL}(\ldots)]),$
$CellMembrane = (\text{CELL::} (condition) \rightarrow [\text{CELL}(\ldots)]),$
$\ldots)$

There are no restrictions on the names of attributes. After parsing such a description, the cell exists in the environment and executes its production rules.

Attributes

Each object has an attribute list, with each attribute having a name and a value (*e.g.*, a chemical compound with its concentration). The general description of an attribute is '*attr = value*', where the value can have a wide variety of predefined types will dynamic type resolution during execution. Attributes are modified by expression statements.

Interaction rules

Each interaction rule is described as a production rule. A production rule describes the conditions and the kind of action that will take place. Production rules are divided into the following types, according to the possible actions.

- Modification of internal state based on the internal state (cell machine)
- Modification of internal state based on the external state (cell membrane)
- Creation of a new cell
- Deletion of a cell

Internal-state-based modification   These modification rules have the following form of production rule.

CELL :: *cond* → [CELL(*attr=expr(attr)*)]

The rule modifies the cell's attributes (*attr*) according to the expression (*expr(attr)*).

External-state-based modification   These modification rules have the following form of production rule.

CELL :: *cond*
$\quad$ → [CELL(*attr$_1$+=ps_sub(expr(attr$_1$, ps_val(attr$_1$)))*),
$\qquad$ *ps_add(expr(attr$_2$, ps_val(attr$_2$))))*]

The above production rule describes the 'in-flow' of the chemical *attr$_1$* and the 'out-flow' of the chemical *attr$_2$*. Both flows depend on the inside and outside concentrations of the same chemical. This dependence is described by the expression *expr*.

Access to the attributes of the physical space is conducted through interface functions such as *ps_val(attr)*, *i.e.*, returning the value of *attr*; *ps_add(attr)*, *i.e.*, adding the current value of *attr* to the value of *attr* in the physical space; and *ps_sub(attr)*, *i.e.*, subtracting the current value of *attr* from the value of *attr* in the physical space.

Creation of a new cell   Creation of a new cell occurs when there is an extra cell in the successor list of production rule as follows.

CELL :: *cond* → [CELL(*attr=expr(attr)*),
$\qquad$ CELL$_{new}$(*attr=expr(attr)*)]

This rule has an extra cell, which is recognized by a different name (*i.e.*, CELL$_{new}$). The new cell inherits the attribute values of the original cell. However, the attributes can be modified within the same rule. Thus, modifying the attribute values can lead to the following basic cell lineage types:

- The cell divides into two cells of same type ($A \to A, A$)
- The cell creates a new cell type ($A \to A, B$)
- The cell divides into two cells of different types ($A \to B, C$)

Deletion of a cell   Deletion of a cell occurs with the following rule.

CELL :: *cond* → [~CELL(*ps_add(Chemical$_1$)*), ...,
$\qquad$ *ps_add(Chemical$_n$)*, *ps_del()*)]

Cells preceded by '~' are deleted as a result of applying the rule. When an object is dying, it releases chemicals into the environment by calling *ps_add()* function, and removes its representation from the environment by the *ps_del()* function.

Concept of physical space

In earlier versions of this work the concept of a physical space was included in the production rule description. However, this was found to be computationally too inefficient, so a separate physical space concept is introduced here.

The physical space concept consists of an interface to the physical space with commands to add an object (location and shape), and to update the forces acting on the object. The modeled mechanical forces include collision and adhesion forces, which depend on the current cellular state, *i.e.*, the elasticity of the cell, and its surface structure.

In addition to mechanical forces, the environment includes chemical substances. Each cell can emit/absorb these into/from the surrounding environment. Each chemical substance diffuses into the environment over time. The chemical substances are conserved over time, *i.e.*, the amount in the system remains constant. In the future, simulation of some kind of chemical reaction will be possible, *i.e.*, substances can be converted from one form to another.

Even though this "physical world" is not adequate for real world applications, it meets our need for a simple model of environmental effects.

Creation of networks

The creation of interacting networks is based on the location of cells and their emission and absorption of chemical substances. Thus, the networks are created using local rules of cells depending on whether they take some action or not. The created network can be observed in the form of interaction graphs as described earlier (fig. 3). These networks determine the behavior of the system.

As the emission and absorption of a particular chemical depends on the emission and absorption of a different chemical, the created networks can begin to communicate with each other. Thus the emergent levels correspond to the capability of the 'higher-order' emission and absorption of chemicals. This is the basic requirement for creating multi-level emergent properties.

Furthermore, the chemicals could be replaced with some other communication mechanism, such as audio signals, visual signals, *etc.* However, the interpretation of more complex signals is complicated, and, as a process itself, it too should be based on emergent principles.

Examples

In this section we give a series of examples of modeling adaptive behavior, where starting from a single cell and a set of interaction rules results in an organism and its behavior. The resulting behavior is, of course, determined by the interaction rules, which are designed especially for these examples. However, most of the rules are quite general and can be reused in other applications. Furthermore, while it is obvious that we need to apply evolutionary mechanisms in order for the created organisms to evolve, we have not yet done this because of computational requirements.

Cell model

The model of our biological cell consists of simulations of metabolic reactions inside the cell, chemical flow through the cell membrane, and creation of new cells through division, and deletion of old cells through cell death.

The following example gives a general idea of how these functions can be written using the production rules. The first production rule introduces chemicals into the environment (*Nutrition*).

The initial state consists of a cell with production rules to model the metabolic reactions (*CellMachine*), the flow of chemicals through the membrane (*CellMembrane*), cell division (*CellDivision*), and cell death(*CellDeath*).

The *CellMachine* rule can calculate the metabolic reactions in two ways. The condition part ($f^{cond_1}_{metabolism}$ and $f^{cond_2}_{metabolism}$) determines which one to use. The function that creates each chemical can be different for each chemical (f^c_i), and can depend upon other chemicals and enzymes ($Chem_i, Enz_i$). The enzymes can be updated within the same rule (f^e_i). The cell pressure, which dictates its size, is determined by the amount of internal chemicals present ($f_{press}(Chem_i)$). This is set in the physical space by an interface function (ps_press).

The *CellMembrane* rule modifies the physical space through the interface functions (ps_val by returning the concentration value of a given chemical at the location of the cell, ps_sub by subtracting the chemical value from the location of the cell by a given amount, and ps_add by adding the given chemical value to the location of a cell). Note that the diffusion of the chemical values in the physical space is calculated internally. The functions that determine the amount of flow out (f^{Emit}) and in (f^{Absorb}) depend upon the concentration of the particular chemical outside the cell, which can be determined by using the interface function ps_val.

The *CellDivision* rule creates a new cell and divides the chemical between the original cell and the new cell. When a cell divides, the amount of chemicals is simply halved, although some noise may also be introduced, or a more complex division mechanism can be used. The division direction can be given and the new location set to the physical space ($ps\_at(ps\_pos()+rot(Offset, Deg))$).

The *CellDeath* rule removes the cell from the simulation and releases the chemicals into the physical space. Finally, the physical representation can be removed by an interface function ($ps\_del()$).

$$
\begin{aligned}
&\text{ENV}(Location, Chem_i, \\
&\quad Nutrition{=}(\text{ENV} :: (f^{cond}_{nutrition}) \\
&\qquad \rightarrow [\ \text{ENV}(ps\_add(Location, Chem_i))\ ]), \\
&\quad Cells{=}[\ \text{CELL}(\dots), \dots, \text{CELL}(\dots)\ ])
\end{aligned}
$$

$$
\begin{aligned}
&\text{CELL}(Location = ps\_at((0,0)), Chem_i, Enz_i, \\
&\quad CellMachine{=}(\text{CELL} :: (f^{cond_1}_{metabolism}) \\
&\qquad \rightarrow [\ \text{CELL}(Chem_i = f^c_i(Chem_i, Enz_i), \\
&\qquad\qquad Enz_i = f^e_i(Chem_i, Enz_i), \\
&\qquad\qquad ps\_press(f_{press}(Chem_i, Enz_i)))\ )\ ] \\
&\qquad :: (f^{cond_2}_{metabolism}) \rightarrow [\ \text{CELL}(\dots)\ ]), \\
&\quad CellMembrane{=}(\text{CELL} :: (f^{cond}_{in-out-flow}) \\
&\qquad \rightarrow [\ \text{CELL}(Chem_i {+}= \\
&\qquad\qquad ps\_sub(f^{Absorb}(ps\_val(Chem_i))) \\
&\qquad\qquad -ps\_add(f^{Emit}(ps\_val(Chem_i))))\ ]), \\
&\quad CellDivision{=}(\text{CELL} :: (f^{cond}_{division}) \\
&\qquad \rightarrow [\ \text{NEW\_CELL}(Chem_i / = 2.0, Enz_i / = 2.0, \\
&\qquad\qquad ps\_at(ps\_pos()+rot(Offset, Deg)), \\
&\qquad\qquad ps\_press(f_{press}(Chem_i, Enz_i))), \\
&\qquad\quad \text{CELL}(Chem_i / = 2.0, Enz_i / = 2.0, \\
&\qquad\qquad ps\_at(ps\_pos()+rot(Offset, -Deg)), \\
&\qquad\qquad ps\_press(f_{press}(Chem_i, Enz_i)))\ ]), \\
&\quad CellDeath{=}(\text{CELL} :: (f^{cond}_{activity}) \\
&\qquad \rightarrow [\ {\sim}\text{CELL}(ps\_out(Chem_i), ps\_del())\ ])
\end{aligned}
$$

This kind of intra-/inter-cell communication results in a dynamic clustering as reported in (Kaneko & Yomo, 1994).

It is very difficult to visualize the above cellular behavior. However, the resulting behavior is easier to observe. In the following, we show how a cellular system can create a neural network that can control its behavior in the environment.

Neuron model

The previous cell can be replaced with a slightly more complicated structure, in the form of a neuron, which differs from the earlier cell model, due to its ability to extend its shape and create connections to other neurons. The connections are represented as sub-objects of the cell each having similar attributes and production rules similar to those of the parent cell. The genetically controlled growth of the connections is shown in Fig 4.

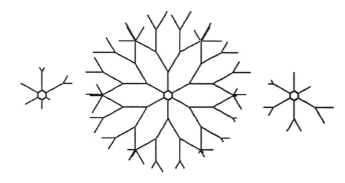

Figure 4: The genetic growth and withdrawal of connections of a neuron. When to branch, and the rotation angle, are determined by the genetic rules. Three phases are shown: initial grow (left), initial withdrawal (middle), and just before all connections are withdrawn (right).

In Figure 5 the environment consists of some obstacles and some target cells (the type of target cells being determined genetically). The target cells emit chemical substances into the environment, thus creating a gradient field that the connections 'climb'. When the connections finally reach the target cells, they create a synaptic connection, and stop growing. Those connections unable to find any target neurons gradually withdraw.

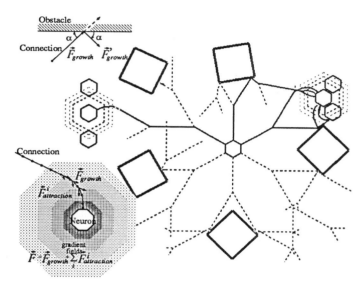

Figure 5: Genetic growth is repeated with environmental obstacles (the collision model at left top) and a gradient field of the target cells (the attraction model at left bottom). The resulting network is drawn with solid lines and the connections to be withdrawn with dashed lines.

Behavior directed by the created network

The last example illustrates how electrical signals can be simulated in the model. Electrical signals involved in the neural activity begin at the receptor field, which convert the external world into neural signals, and ends at the muscle cells, where the signals cause muscle activity resulting in behavior in the environment.

In the following example, the above modeling principles are used to grow a network between receptor cells and muscle cells. For simplicity, the effect of muscle cells is simulated by a force vector related to the received signal. The receptor cells generate a signal proportional to the angle and the distance of the target they are sensitive to (fig. 6). The signal propagation inside the network is done without any modification, i.e., without learning.

The receptor field simulation is based on the same principles of chemical diffusion previously used to explain the growth of neural connections toward a target cell.

The resulting behavior of the corresponding network structure consists of the behaviors shown in Figures 6 and 7 (see details in (Vaario et al., 1994)). The production rules are deliberately tailored to grow a specific network, which results in behaviors similar to the

lower level behavior of Braitenberg's Vehicles (Braitenberg, 1984). Thus, while the rules themselves are not particularly interesting, it is noteworthy that, with a simple simulation mechanism of local rules, which can also be reached through an evolutionary mechanism, we were able to simulate the global behavior shown.

Adaptation of systems

The interaction rules of each cell are inherited through the cell lineage. Thus, the cells of a lineage tree are basically capable of communicating with each other through similar substances. The behavior of a system can be modified by two mechanisms.

First, the initial values and rules of the initial cell can be changed. This corresponds to genetic variation in biological systems. Second, variations in the environment, where the cell lineage occurs, can cause changes to the cell lineage and to the internal state of cells. This corresponds to development and learning in biological systems.

Variation in the initial values

Variation in the initial values is similar to genetic variation in biological systems. Within the presented modeling method, the variations can be created automatically. Genetic information can be represented as a variable to be used by production rules to generate internal chemicals and enzymes.

The genetic code is distributed through the cell division mechanism, and through cell fusion ($cell'$, $cell''$ :: $cond \rightarrow cell$, where $cell$ replaces $cell'$, and $cell''$ is removed) two genetic codes can be combined by a crossover operation, resulting in a new system.

The following production rule implements the fusion of two cells. The production rule is applied to all intersecting cell pairs. The f^{cond}_{fusion} function is responsible for determining whether fusion takes place. There are no restrictions on how the user defines the genetic code and the $f_{cross-over}$ function.

CELL($Gene = [\ldots]$
    $CellFusion$=(CELL, CELL2 :: (f^{cond}_{fusion})
        $\rightarrow$ [CELL($Gene = f_{cross-over}(Gene,$ CELL2.$Gene$),
            $Chem_i += $CELL2.$Chem_i$),
        ~CELL2()])

Variation in the environmental values

Variations in environmental values can change the behavior of systems. In general, it is very difficult to predict how changes in the environment change system behavior. A more fruitful way to think about the environmental effect is to consider the system to be part of the environment, rather than think about two separated interacting systems (Varela, Thompson, & Rosch, 1993).

A system can respond to environmental changes in two basic ways, i.e., morphogenesis and ontogenesis. Morphogenesis creates a new form, which leads to new behavior. Ontogenesis modifies existing behaviors. In our model, morphogenesis creates a new interacting network, and ontogenesis corresponds to the actual interactions

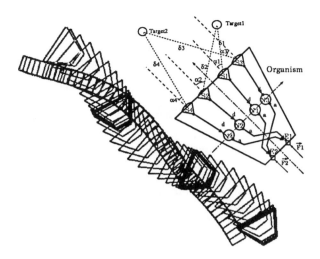

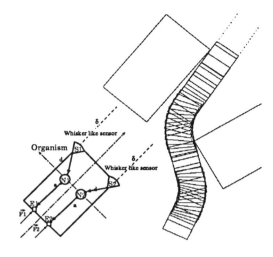

Figure 6: Targeting behavior of the grown network, where the initial state included the sensor and effector descriptions.

Figure 7: Obstacle avoidance of the grown network. The collision of the objects is detected by whisker-like sensors, which generate a signal proportional to the interpenetration of the objects.

between the elements of the network. While both processes depend upon the surrounding environment, the system is very adaptive to changes in it.

Currently, we are continuing our experiments on how to self-organize the required stability of the systems while still preserving the flexibility provided by the modeling mechanism.

Conclusion

Our goal is not to create an exact simulation of biological systems, but to create a general system capable of modeling different biological and physical phenomena in order to find the principles of the adaptive self-organization.

With this perspective we have thus far created a modeling method that starts from an initial description of a basic element (or a few elements) including the possible interactions it is capable of, and a dynamic description of the environment where the basic element will 'live', *i.e.*, how the interactions are applied.

Our approach differs from Fleischer's system (Fleischer & Barr, 1994) in that his basic motive seems to be to model biological morphogenesis as a phenomenon, whereas our goal is to find a general modeling method for life. Thus, our method is not only capable of modeling the physical interactions between cells, but also of abstracting interactions. Our emphasis is also on the local interaction rules of each cell applied to the detailed internal description of each cell.

In this paper we have shown that our modeling method is capable of modeling the local interactions in detail, and that these interactions result in global behavior within the environment. Thus, the presented method enables modeling of biological information processing systems that are based on the local interactions. We believe that this approach will lead to a new information processing concept based on the cellular elements.

Reference

Braitenberg (1984). *Vehicles - Experiments in Synthetic Psychology*. The MIT Press.

Fleischer, K., & Barr, A. H. (1994). A simulation testbed for the study of multicellular development: The multiple mechanisms of morphogenesis. In Langton, C. G. (Ed.), *Artificial Life III*. Addison-Wesley.

Kaneko, K., & Yomo, T. (1994). Cell division, differentation and dynamic clustering. *Physica D*. (to appear).

Prusinkiewicz, P., & Lindenmayer, A. (1990). *The Algorithmic Beauty of Plants*. Springer-Verlag.

Vaario, J. (1994a). Artificial life as constructivist AI. *Journal of SICE*, *33*(1), 65—71.

Vaario, J. (1994b). From evolutionary computation to computational evolution. *Informatica*. (to appear).

Vaario, J., Hori, K., & Ohsuga, S. (1994). Toward evolutionary design of autonomous systems. *The International Journal in Computer Simulation*. (to appear).

Vaario, J., & Ohsuga, S. (1992). An emergent construction of adaptive neural architectures. *Heuristics - The Journal of Knowledge Engineering*, *5*(2).

Vaario, J., & Ohsuga, S. (1994). On growing intelligence. In Dorffner, G. (Ed.), *Neural Networks and a New AI*. Chapman & Hall, London.

Varela, F. J., Thompson, E., & Rosch, E. (1993). *The Embodied Mind* (paperback edition edition). The MIT Press.

Robot Herds: Group Behaviors for Systems with Significant Dynamics

Jessica K. Hodgins
David C. Brogan
College of Computing
Georgia Institute of Technology
Atlanta, GA 30332-0280
[jkh|dbrogan]@cc.gatech.edu

Abstract

Birds, fish, and many other animals are able to move gracefully and efficiently as a herd, flock, or school. We would like to reproduce this behavior for herds of artificial creatures with significant dynamics. This paper develops an algorithm for grouping behaviors and evaluates the performance of the algorithm on two types of systems: a full dynamic simulation of a legged robot that must balance as well as move with the herd and a point mass with minimal dynamics. Robust control algorithms for group behaviors of dynamic systems will allow us to generate realistic motion for animation using high-level controls, to develop synthetic actors for use in virtual environments, mobile robotics, and perhaps to improve our understanding of the behavior of biological systems.

1 Introduction

To run as a herd, animals must remain in close proximity while changing direction and velocity and while avoiding collisions with other members of the herd and obstacles in the environment. In this paper, we explore the performance of a control algorithm for modulating the motion of each individual in a herd of dynamically simulated legged robots. A photograph of 100 simulated robots running as a herd is shown in figure 1.

The herding algorithm computes a desired velocity for each individual based on the location and velocity of its visible neighbors. This desired velocity is then used as an input to the locomotion control system for the robot. We compare the performance of this algorithm on a herd of point-mass objects and a herd of dynamically simulated running robots for a test suite of four problems: steady-state motion, acceleration and deceleration, turning, and avoiding obstacles. For this test suite, all individuals in the herd of robots remained upright and only a small number of collisions occurred. However, the performance of this herd was not as robust as that of the point-mass system.

In contrast to most previous implementations of algorithms for group behaviors, we are using this algorithm to control a robot herd where the individuals have significant dynamics. The problem of controlling the robot herd more closely resembles that faced by biological systems because of the underlying dynamics of the individuals in the herd. Each robot in the herd is a dynamic simulation of a physical robot and a control system. As such, the robots have limited acceleration, velocity, and

Figure 1: Photograph of a herd of 100 simulated one-legged robots. The herd had run stably for 5 minutes before this photograph was taken.

turning radius. Furthermore, the control algorithms are inexact, resulting in both transient and steady-state errors in velocity control. Required changes in velocity are delayed by as much as half a running step because the control system can influence velocity during only the flight phase of the running cycle. To understand the effect of the underlying dynamics, we compared the performance of the herding algorithms on the robots with full dynamics and on particle systems with perfect velocity control.

Algorithms for high-level behaviors of dynamic simulations are needed for the construction of synthetic actors with robust and realistic motion that can respond interactively to changes in the environment. A dynamic simulation in concert with a control system will provide natural looking motion for low-level behaviors such as walking, running, and climbing. High-level behaviors such as obstacle avoidance, grouping, and rough terrain locomotion will allow the actor to function in and interact with a complex and unpredictable environment.

2 Background

Recent advances in robotics have produced autonomous agents capable of performing a variety of tasks in different domains. At the same time, researchers in the artificial life community have contributed to our understanding of the evolution of complex behaviors through the use of simulations that produce emergent behaviors.

By combining work from these fields, we should be able to create multi-agent robotic systems that mimic the elegant grouping behaviors of biological organisms.

Herding, flocking, and schooling behaviors of animals have been studied extensively over the past century and this research serves as motivation for the creation of artificial creatures with similar skills. Groupings exemplify an attraction that modulates the desire of each member to join the group with the desire to maintain a particular separation distance from nearby creatures (Shaw 1970). As an example of this attraction, Cullen, Shaw, and Baldwin (1965) report that the density of fish is approximately equal in all planes of the school, as if each fish had a sphere around its head with which it wished to contact the sphere of another fish. Herding benefits the average group member by limiting the average number of encounters with predators (data summarized in Veherencamp 1987). Grouping behaviors allow animals to hunt more powerful animals than those they could overpower as individuals. Due to the success of behaviors such as these in biological systems, it seems reasonable to assume that it would be advantageous to reproduce them in robotic systems.

Early work in the simulation of grouping behaviors was performed by Reynolds (1987). Actors in his system are bird-like objects and are similar to the point masses used in particle systems except that each bird also has an orientation. The birds maintain proper position and orientation in the flock by balancing their desires to avoid collisions with neighbors, to match the velocity of nearby neighbors, and to move towards the center of the flock. Each bird uses only information about nearby neighbors. This localization of information simulates perception and reaction in biological systems and allows for proper balancing of the three flocking tendencies. Reynolds's work demonstrates that realistic-looking animations of group formations can be created by applying simple rules to determine the behaviors of the individuals in the flock.

Yeung and Bekey (1987) propose a decentralized approach to the navigation problem for multiple agents. Their system first constructs a global plan without taking into account moving obstacles. When a collision is imminent, the system locally re-plans using inter-robot communication to resolve the conflict. Because of the two levels of planning, this solution requires the communication overhead associated with grouping behaviors only when a pair of robots perceive an impending collision.

Sugihara and Suzuki (1990) demonstrated that multiple robots can form stable formations when each robot executes an identical algorithm for position determination within the group. Each robot can perceive the relative positions of all other robots and has the ability to move one grid position during each unit of time. By adjusting the position of each robot relative to either the most distant or the closest neighbor, a regular geometric shape such as a circle can be formed by the robots in the world. Furthermore, the movement of one robot in a formation can cause a chain reaction that results in a translation of the group in world coordinates. By carefully constructing the algorithms that each robot uses in determining intra-group position, formations will emerge without *a priori* knowledge about the total number of robots. Designation of leaders allows the simple rules of the group to create leader-following algorithms and to demonstrate the division of a formation into smaller groups.

Wang (1991) investigated the navigation of multiple robots in formation and the resulting group dynamics. Each robot in the model is simulated as a point mass and perceives other robots in the region contained in a cone extending from the center of the robot and heading in the direction of travel. Formations are represented as a set of offsets from a predefined reference robot. In this way, a formation can be directly defined as a set of positions for each robot relative to the leader, closest neighbor, or set of closest neighbors. Wang shows that the error in desired position relative to actual position diminishes to zero for each independent robot in the formation and therefore the desired formation is asymptotically stable. Simulations for up to four point-mass robots demonstrate that these navigation strategies can produce stable group formations.

Arkin explored the question of communication in a group of interacting mobile robots using schema-based reactive control (Arkin 1992, Arkin and Hobbs 1992). Example schemas are *move-to-goal*, *move-ahead*, and *avoid-static-obstacle*. Each behavior computes a velocity vector that is combined with the velocity vectors from the other behaviors and is used to control the robot. Arkin demonstrated that for some tasks robots can interact with no communication other than observations of the environment or with very limited explicit communication. The herding algorithms we implemented are also examples of an algorithm in which there is no explicit leader and all communication is through observations of the environment.

Mataric researched emergent behavior and group dynamics in the domain of wheeled vehicles. These robots, like Arkin's, do not explicitly communicate state or goals and the system has no leaders. This work demonstrated that combinations of such simple behaviors as attraction and repulsion can produce complex relationships such as dispersion and flocking in physical robots in the laboratory (Mataric 1992a,b). The robots utilize the knowledge that they are all identical when executing behaviors, but an extension to these results found that heterogeneous agents do not perform significantly better than homogeneous ones (1993). In these experiments, a hierarchy is created in which an ordering between the agents determines which agent will move first in completing tasks such as grouping and dispersing.

3 Algorithms for Herding

The herding algorithms described in this paper were evaluated on two systems: a one-legged robot simulation with full dynamics and a particle simulation with minimal dynamics. The next two sections describe the herding algorithms and the two simulations.

The herding algorithm consists of three parts: a perceptual model to determine the visible creatures for each individual in the herd, a placement algorithm to determine a desired position for each individual given the locations of the creatures that are visible to it, and a spring/damper control system to compute a desired velocity given the current position and the desired position. The herding algorithm is run for each individual in the

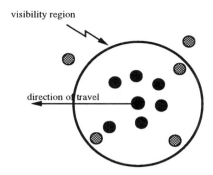

Figure 2: One creature is visible to another if it is within a certain radius and is one of the n closest visible creatures (n is six for this example). The black circles represent visible creatures and the grey represent creatures that can not be seen by the individual under consideration.

herd to compute a desired velocity for that individual. The control system for each legged robot then uses the desired velocity provided by the herding algorithm to determine how the leg should be positioned during flight to achieve the desired change in forward velocity. The particles in the point-mass herd use this desired velocity as their actual velocity on the next time step.

The herding algorithms for an individual robot are run each simulation time step while the robot is in flight. For the particle system, the herding algorithms compute a new desired velocity for each simulation step and a new set of visible particles every ten simulation steps.

3.1 Perception Model

Each individual in the herd can perceive the location and velocity of the n nearest creatures that are within a circle of radius r. In the trials reported in this paper n was 30 and r was 24 m and the herd included 105 individuals. For most configurations the circle was large enough to include all members of the herd. Figure 2 illustrates the perception model.

3.2 Desired Position and Velocity

The list of visible creatures provided by the perceptual model is used to compute a desired position for each individual in the herd. A desired position relative to each visible creature is computed and then these desired positions are combined with a weighted average. The desired position of an individual relative to each of the visible creatures is a constant distance D away from the visible creature on the line between the two creatures (figure 3). In these experiments D was 2.5m. This set of desired positions (one for each visible creature) is averaged with a weighting of $1/d^2$ to compute a global desired position where d is the distance between the two creatures.

The global desired position for an individual is used to compute a desired velocity for that creature using a spring and damper system:

$$\dot{x}_\mathrm{d} = k_\mathrm{p} e - k_\mathrm{v} \dot{e} + \dot{x}_\mathrm{nom} \qquad (1)$$

where $\dot{x}_\mathrm{d}$ is the desired velocity in the plane, e is the error between the current position of the creature and the global desired position, $\dot{e}$ is the rate of change of the error, k_p and k_v are the proportional and derivative gains, and $\dot{x}_\mathrm{nom}$ is the nominal velocity. For the experiments

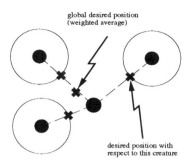

Figure 3: The locations of the visible creatures are used to compute a global desired position for the individual under consideration. The algorithm computes a desired position with respect to each visible robot by finding the point on the line between the individual and the visible creature that is a constant distance D away from the visible creature. These desired positions are averaged with a weighting equal to $1/d^2$ where d is the distance between the two creatures.

Link	Mass (kg)	Moment of Inertia $(x, y, z \ \mathrm{kgm}^2)$		
Body	23.1	0.9	0.9	0.602
Upper Leg	1.4	0.018463	0.017297	0.001441
Lower Leg	0.64	0.0197	0.0197	0.000176

Table 1: Parameters of the rigid body model of a one-legged robot. The moment of inertia is computed about the center of mass of each link.

reported here $k_\mathrm{p} = 0.5$ and $k_\mathrm{v} = 0.3$. The nominal velocity $\dot{x}_\mathrm{nom}$ was the average of the desired velocities of the visible creatures. To provide the user with control of the herd, one creature is selected by the user. The nominal velocity in equation 1 for that creature is set by the user rather than computed by averaging the desired velocity of the visible creatures.

4 Simulating the Herd

The herd simulation consists of the equations of motion for either the robot or the particle system, a copy of the state vector for each individual in the herd, control algorithms for running, a graphical image for viewing the motion of the herd, and an interface that allows the user to control the parameters of the simulation. For the robot herd, the equations of motion represent a rigid body model of a one-legged robot and control algorithms that allow the robot to run at a variety of speeds and flight durations. At each simulation time step, the control system computes forces or torques for each joint of the robot based on the actual and desired state vector for that individual. The equations of motion of the system are integrated forward in time, and the resulting motion of the individuals in the herd is displayed graphically and recorded for later use. The equations of motion for the individuals in the herd do not take into account the physical effects of collisions between two members of the herd, although collisions are detected and a count of collisions is recorded for use in analyzing the data. The details of the robot and particle models are described below.

Link	COM to Proximal (m)	COM to Distal (m)
Body		0.0
Upper Leg	0.095	-0.095
Lower Leg	0.221	

Table 2: The distance from the center of mass of each link to the distal and proximal joints in z for the canonical configuration of the robot (the distance in x and y is zero for this model).

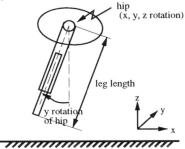

Figure 4: The reference angles for the controlled degrees of freedom of the robot. The controlled degrees of freedom are three degrees of freedom at the hip and the length of the leg.

4.1 One-legged Robot Simulation

The equations of motion for the robot were generated using a commercially available package (Rosenthal and Sherman 1986). The package generates subroutines for the equations of motion using a variant of Kane's method and a symbolic simplification phase. The parameters of the robot are given in table 1 and table 2. The reference angles of the model are shown in figure 4.

The locomotion algorithms for the one-legged robot control flight duration, body attitude, and forward and sideways velocity. Flight duration is controlled by extending the leg during stance. Body attitude (pitch, roll, and yaw) is controlled by exerting a torque between the body and the leg during stance. The velocity is controlled by the position of the foot with respect to the center of mass of the body at touchdown. For a constant velocity, the foot is positioned in the center of the distance that the body is expected to travel while the foot is on the ground. To increase the speed, the foot is positioned closer to the hip. To decrease the speed, the foot is positioned further from the hip. The details of the locomotion control algorithms are given in Raibert (1986).

4.2 Particle Simulation

The particle simulation has minimal dynamics. The desired velocity computed by the herding algorithm is used by the dynamic simulation as the actual velocity on the next time step. There are no limits on acceleration or velocity. The particle system differs from the robots in that there is no delay in the implementation of a new desired velocity and the new velocity exactly matches the desired velocity.

5 Results

We tested the herding algorithms in four situations: steady-state movement, acceleration, turning, and avoid-

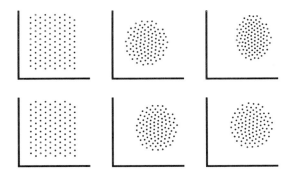

Figure 5: The configuration of the herd at the start state and every 100 s thereafter for a commanded steady-state velocity of 2.0 m/s in the x direction for the user-controlled creature. The top set of graphs shows the motion of the robot herd; the bottom set shows the motion of the particle system. Each graph represents the x, y position of each robot or particle in the herd.

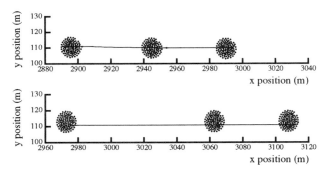

Figure 6: The configuration of the herd at the start state and every 30 s thereafter for an initial steady-state desired velocity of 2 m/s in the x direction, followed by a desired velocity of 3 m/s for 30 s and 1.5 m/s for 30 s for the user-controlled creature. The top graph represents the herd of robots; the bottom graph represents the herd of particles. The path shown between the snapshots of the herd is the trajectory that the user-controlled robot or point mass followed.

ing obstacles. For steady-state movement, both the herd of robots and particles began in the same configuration, and the user-controlled creature was commanded to move at 2.0 m/s for 300 s. As is shown in the snapshots of the herd configurations in figure 5 both herds contracted to form a nearly circular shape.

The second test began with the ending point of the steady-state test for each system. The commands to the user-controlled creature were an acceleration to 3m/s in the x direction for 30 s and then a deceleration to 1.5 m/s for 30 s (figure 6). Both the particle system and the robots continued to move as a herd although the user-controlled robot pulls away from the herd during the acceleration phase of the experiment and falls back into the herd during the deceleration.

The third test involved turning. Beginning with a steady-state run of 2.0 m/s in the x direction, the y desired velocity was increased to 1.41 m/s and the x desired velocity was decreased to 1.41 m/s. After 20 s the y desired velocity was set to -1.41m/s for the next

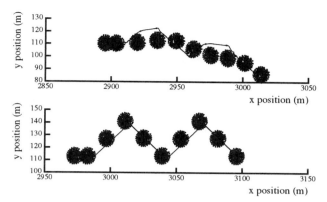

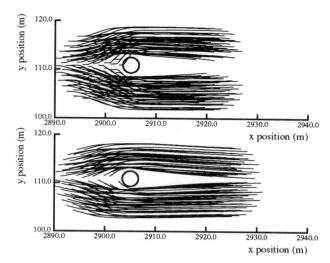

Figure 7: The configuration of the herd as the user-controlled creature follows a zig-zag pattern. Beginning with a steady-state run of 2.0 m/s in the x direction for 5 s, the y desired velocity was increased to 1.41 m/s and the x desired velocity was decreased to 1.41 m/s. After 20 s the y desired velocity was set to -1.41m/s for the next 20 s. This pattern was repeated a second time. The top graph represents the herd of robots, the bottom graph, the herd of particles. The path shown between the snapshots of either herd is the trajectory that the user-controlled robot or point mass followed.

Figure 8: The trajectory of the members of the robot herd and the particle system herd as the creatures avoid an obstacle. The top graph represents the herd of robots, the bottom graph, the herd of particles. Both herds ran for 15 s at a nominal speed of 2 m/s in this experiment and the herds were moving from left to right.

20 s. This zigzag pattern was repeated a second time (figure 7). The particle system had no collisions but the robot herd had multiple collisions as the herd changed direction. The particle system herd tracked the user-controlled point-mass much more closely than the robot herd tracked the user-controlled robot although both herds retained an approximately circular shape. The user-controlled robot does not follow the desired zigzag pattern closely because its velocity is affected by its position relative to the other members of the herd.

The final test involved obstacle avoidance (figure 8). The creatures on a collision path with the obstacle moved to avoid the obstacle by aiming for a point out to the side of the obstacle at a distance of 1.5 times the radius of the obstacle. This sideways motion was incorporated into the calculation for the desired position with a weighted average. The herd of point masses were able to avoid the obstacle and quickly rejoined to form a single herd on the far side of the obstacle. The first robot in the robot herd was unable to avoid the obstacle and this herd was slower to regroup on the far side of the obstacle. In easier tests where the herds had more time to react the performance of the point mass herd and the robot herd were similar.

The herding algorithms used for these two sets of tests were identical and differences in performance can be attributed to differences between the two dynamic systems. The point-mass herd ran more tightly under changes in magnitude and direction of velocity because of the exact control of velocity. The behavior of the robot herd was not as robust as that of the point-mass system because the herd did not track the user-controlled robot as closely. The robot herd had more variability and motion within the herd and tests more often resulted in collisions between members of the herd. In more difficult tests than those reported here, an individual in the herd sometimes lost its balance and fell down. The particle systems had no notion of balance or of maximum speed

or acceleration and could not fail in this way.

In other ways, the performance of the robot herd was superior to that of the particle system. The robot herd worked equally well with the damping term in equation 1 set to zero; however, the particle system was unstable without damping. The natural damping of the individual behavior of the robots appears to compensate for the lack of damping in the herd control equation and to increase the stability of the herd in some situations. The robot herd was also more stable than the point mass herd for tests where the number of visible creatures was reduced below 30.

A serious limitation of this herding algorithm is the knowledge required by a robot about the desired velocity of a neighbor because this information could not be measured with a sensor. An implementation that used actual rather than desired velocity would be preferable but was not stable for the robot herd. The dynamic interaction of the leg with the ground and inaccuracies in the locomotion control system prevent the robots from running at exactly the commanded velocity. A linear fit between the actual and desired velocities was not a sufficiently accurate model to correct this problem, and the herd ran increasingly faster or slower depending on the constant chosen for the linear fit. In the particle system, actual and desired velocities were identical and there is no difference between these two approaches.

There are other limitations to the herding algorithm we implemented. In some situations, the desired velocity moved two individuals closer to a collision. In our current implementation there is no reflexive reaction to an impending collision.

With this algorithm, a breakaway group of sufficient size will not rejoin the main herd unless it happens to move close enough to the herd that a member of the main herd is visible to a member of the breakaway group. This problem could be solved by the addition of a separate behavior that causes individuals to look further afield for another herd to join.

We experimented with other perceptual models, adding occlusion and reducing the visibility of creatures behind as opposed to in front of the individual in question. Occlusion reduced the stability of the particle system without qualitatively changing the behavior of the robot herd. When the list of visible creatures changes because of the addition of a previously occluded individual, the desired position and velocity may change significantly thereby causing an immediate ripple effect in the particle system simulation. The natural inertia of the robot simulation appears to mitigate this effect.

Although reducing the visibility of creatures that are behind the one under consideration might appear to be a more natural perception model for an animal or a human, it was not a good heuristic for this simple herding algorithm. Unequal front and back visibility caused the creatures in front to contribute more heavily to the desired position than the creatures in back. The desired position would then be in front of the current position, and the velocity of the robot or particle would continually increase.

We have not yet explored the question of how the algorithms will perform on a heterogeneous population. Currently the robots have identical mass and inertia properties and identical control systems, but we plan to vary the parameters of the system and to introduce noise to study how the performance of the herding algorithms is affected. A further extension would be to develop "personalities" for the individuals as Bates did in his woggles simulation (Bates et al 1993). In the case of the dynamic robot simulation, a simple personality might consist of adjustments to the gains in the herding algorithm and the locomotion control system so that the robot appears to behave aggressively or in a timid fashion.

Although the simulation of the robots is a full dynamic simulation, many factors are missing in the simulation that would be present in a physical robot. The simulated motors do not have a maximum torque or limited bandwidth, the joint and perceptual sensors do not have noise or delay, and the environment used for testing the herding algorithms does not contain uneven or slippery terrain.

One application of this work is to provide high-level controls of simulated creatures for use in computer animations or virtual environments. To be useful in interactive virtual environments, the motion of simulated actors must be computed in real time (simulation time must be less than wall clock time). Our implementation of a single one-legged robot runs faster than real time on a Silicon Graphics Indigo^2 Computer with a R4400 processor. We anticipate that with improved simulation techniques and the continued increase in workstation speed, a small herd of robots or more human-like models will run in real time within a few years.

Acknowledgments

This project was supported in part by NSF Grant No. IRI-9309189 and funding from the Advanced Research Projects Agency.

References

Arkin, R. C., 1992. Cooperation Without Communication: Multiagent Schema Based Robot Navigation, *Journal of Robotic Systems* 351–364.

Arkin, R.C., and Hobbs, J. D., 1992. Dimensions of Communication and Social Organization in Multi-agent Robotic Systems. *Proceedings of the Second International Conference on Simulation of Adaptive Behavior: From Animals to Animats 2* 486–493.

Bates, J., Altucher, J., Hauptman, A., Kantrowitz, M., Loyall, A. B., Murakami, K., Olbrich, P. Popovic, Z., Reilly, W. S., Sengers, P., Welch, W., Weyhrauch, P., Witkin, A., 1993. Edge of Intention. *Siggraph 1993, Visual Proceedings*, 113–114.

Cullen, J.M., Shaw, E., Baldwin, H.A. 1965. Methods for Measuring the Three-Dimensional Structure of Fish Schools. *Animal Behavior* 13:534–543.

Mataric, M., 1992a. Minimizing Complexity in Controlling a Mobile Robot Population. *Proceedings of the 1992 IEEE International Conference on Robotics and Automation* 830–835.

Mataric, M., 1992b. Designing Emergent Behaviors: From Local Interactions to Collective Intelligence. *Proceedings of the Second International Conference on Simulation of Adaptive Behavior: From Animals to Animats 2* 432–441.

Mataric, M., 1993. Kin Recognition, Similarity, and Group Behavior. *Proceedings of the Fifteenth Annual Cognitive Science Society Conference.*

Raibert, M. H. 1986. *Legged Robots That Balance.* Cambridge: MIT Press.

Reynolds, C. W. 1987. Flocks, Herds, and Schools: A Distributed Behavioral Model. *Computer Graphics* 21(4): 25–34.

Rosenthal, D. E., Sherman, M. A., 1986. High Performance Multibody Simulations Via Symbolic Equation Manipulation and Kane's Method. *Journal of Astronautical Sciences* 34(3):223–239.

Shaw, E., 1970. Schooling in Fishes: Critique and Review. Development and Evolution of Behavior. Aronson, L., Tobach, E., Leherman, D., and Rosenblatt, J.(eds) 452–480.

Sugihara, K. and Suzuki, I., 1990. Distributed Motion Coordination of Multiple Mobile Robots. *Proceedings of the 1990 IEEE International Conference on Robotics and Automation* 138–143.

Wang, P. K. C., 1991. Navigation Strategies for Multiple Autonomous Robots Moving in Formation. *Journal of Robotic Systems* 8(2):177–195.

Yeung, D. Y. and Bekey, G. A., 1987. A Decentralized Approach to the Motion Planning Problem for Multiple Mobile Robots. *Proceedings of the 1987 IEEE International Conference on Robotics and Automation* 1779–1784.

Veherencamp, S., 1987. Handbook of Behavioral Neurobiology, Volume 3: Social Behavior and Communication, Marler, P. and Vandenbergh, J. G. (eds.) 354–382.

A FUTURES MARKET SIMULATION WITH NON-RATIONAL PARTICIPANTS

Michael de la Maza and **Deniz Yuret**
Numinous Noetics Group
Artificial Intelligence Laboratory
Massachusetts Institute of Technology
Cambridge, MA 02139
mdlm@ai.mit.edu, deniz@ai.mit.edu

Abstract

This paper describes a set of experiments performed with an artificial futures market simulation. The non-rational market participants, which evolve simple strategies using genetic algorithms, compete against each other to make profits by buying and selling futures contracts. The dynamic and equilibrium behavior of the participants is studied under a variety of conditions. The results suggest that in simple markets with non-homogenous participants opportunities for making consistent profits over extended periods of time exist.

Introduction

Although futures contracts are a relatively recent addition to the cornucopia of financial products available to market participants, their rate of growth has positioned them as one of the most traded financial vehicles in the world. Scalpers, day traders, institutions, and hedgers all compete to exchange risk and return in ways that meet their objectives. All participants in the market have data about the cumulative actions of all of the other participants and they adjust their behavior to exploit perceived opportunities.

The dynamic and equilibrium behavior of such a complex adaptive system cannot be easily understood by straightforward application of traditional econometric analysis. In this paper we present a program that simulates an artificial futures market and describe several experiments designed to illuminate how differences among market participants can lead to consistent profits for some of the participants.

The most important feature of this simulation is that the market participants are not rational. Curiously, when the rationality of participants in a market is bounded, as it is in this work, formal analysis becomes more difficult, not less difficult, because the number of parameters needed to describe each participant increases. Thus, market simulations are a tool that help us study complicated systems.

We stress, however, that our simulation of an artificial futures market does not do justice to the intricacies of real futures markets. In an attempt to capture the essentials of the futures markets in order to study them, we may have inadvertently removed the very components that make them worth studying.

Efficient markets

A market in which prices fully reflect all of the information in an information set is an efficient market. When this information set consists of price and volume data the market is said to be weakly efficient. If the information set consists of all publicly available information then the market is efficient in the semi-strong sense, and if the set contains all information, both public and private, then the market is efficient in the strong sense.

The most commonly used model for describing the behavior of market participants is called the rational expectations equilibrium model. The rational expectations model makes two assumptions [Dornbusch and Fischer, 1987, Sargent, 1986]. First, market participants are *rational* and are able to optimize an objective function. Second, the same information is available to all market participants.

In this paper both of these requirements are weakened. The first condition is not met because the participants use genetic algorithms to set the parameters of a model. Genetic algorithms are not guaranteed to find optimal solutions and participants cannot search over the space of models. Such participants have limited, or bounded, rationality [Simon, 1957, Simon, 1982]. Also, in some of the experiments, participants have different information sets, thus the second condition is not met.

Note that prices in a market in which the participants have bounded rationality and which have access to different information sets may still converge to the level predicted by rational expectations theory (for an example see [Sargent, 1993]). Thus, at a macro-level these two types of markets may be indistinguishable.

An artificial futures market

This section describes the framework we used for testing various hypotheses about the market.

The market consists of a number of participants who buy and sell futures contracts. For the purposes of this paper the primary distinction between a futures contract and a stock is that the futures contract has an expiration date. On the day of expiration the price of the futures contract converges to the price of the underlying commodity, which is called the spot (or cash) price. For most futures traded on American exchanges the expiration dates are three months apart. In this simulation, the participants buy and sell futures that expire in one day, so they are effectively attempting to predict the next day's spot price. One iteration of the simulation consists of a single round of buy and sell orders on one futures contract.

In the simulations described below participants attempt to buy futures contracts at prices below the price on the expiration day and sell futures contracts at prices above the price on the expiration day. The number of futures contracts that have been bought is always equal to the number of contracts that have been sold; thus, one participant's gain is another participant's loss. No transaction fees are charged when a participant buys or sells contracts.

The price of the futures contract on the day of expiration is a function of previous prices. Hence, the market participants are in effect setting the price by their actions.

Each participant in these simulations learns a strategy that calculates a "fair" value for the future contract. During each iteration of the simulation this fair value is revealed and an equilibrium price is computed which balances the number of buyers and sellers. If there is a range of possible equilibrium prices then the equilibrium price is set to be the average of the extremes of this range. If the fair value computed by a participant is above the equilibrium price, then the participant buys at the equilibrium price. If the fair value is below the equilibrium price then the participant sells at the equilibrium price. Note that in this simulation participants do not have the option of remaining on the sidelines and observing the price: they must buy or sell during each iteration.

Each strategy is a set of coefficients that is multiplied by the same set of variables used to determine the spot price. For most of the experiments described in the next section three variables are available to each of the participants: a ten day moving average of the equilibrium prices, the last equilibrium price, and the sign of the last change in the equilibrium prices. So the function used to compute the fair value is:

$$f(a, b, c, d) = a * x + b * y + c * z + d$$

where x is the ten day moving average of the equilibrium prices; y is the last equilibrium price; z is -1 if the last change in equilibrium prices is negative, 1 if it is positive, and 0 otherwise; and a, b, c, and d are the coefficients supplied by the strategy.

The coefficients used to determine the spot price for most of the experiments described below are: $a = 0.5$, $b = 0.5$, $c = 1$, and $d = 0$.

Each participant has a visible strategy which is used to compute the fair value of the futures price. In addition, each participant has a set of strategies that are not visible to other participants. Each of these non-visible strategies also computes a fair value for the futures contract. At each iteration of the simulation the best strategy (i.e., the one whose fair value is closest to the spot price) in the previous iteration becomes the visible strategy.

The strategies maintained by each participant are modified and improved by a genetic algorithm [Holland, 1975, Goldberg, 1989] with traditional mutation and crossover operators. The mutation operator adds a number uniformly distributed between -.1 and .1 to a coefficient and the crossover operator is traditional single point crossover. The best strategies are selected and copied using standard proportional selection with a constant offset that ensures that all fitness values are positive [Michalewicz, 1992]. The fitness values are calculated by subtracting the fair value computed by a strategy from the spot price on the next day.

Experiments

This section describes some experiments that suggest that when participants are asymmetric opportunities for making consistent profits exist.

In experimenting with this futures market simulation program we have found that small perturbations sometimes drastically change the behavior of the participants. As a result we have proceeded cautiously, changing only one parameter at a time in order to increase our understanding and intuition.

Preliminary experiments were used to fix several parameters of the simulations: the mutation rate, the number of participants, and the number of strategies per participant. The first experiment discussed below is an example of a preliminary experiment. It tests to what extent participants can ignore noise.

Experiment 2 serves as the gold standard against which experiments 3-7, which introduce sources of potential systematic profits, should be compared.

Experiment 1: Impact of noise

This experiment provides evidence that the participants can withstand noise. The three variables available to the participants, the ten day moving average, the last equilibrium price, and the sign of the last change, are all replaced with random numbers chosen uniformly between -5 and 5. The spot price is fixed at 1. So the participants should set the first three coefficients of their strategies to 0 and the fourth coefficient to 1. The participants' strategies are all initialized with small random numbers near 0.

Figure 1 shows how the equilibrium price changes as a function of time. Each day the simulation performs one iteration of the algorithm. This involves calculating an equilibrium price from the fair values submitted by the participants and updating the participants' strategies with the genetic algorithm.

Experiment 2: Comparison point

The second experiment provides a point of comparison for the rest of the results because all of the participants are identical. The spot price is a function of the moving average, the last equilibrium price, and the sign of the last change in the equilibrium price: $x/2 + y/2 + z$. Because all four participants are homogenous and have access to all relevant variables, the profit collected by the most profitable participant should approach zero over time.

Figure 2 shows the profit received by the most profitable participant as a function of time. The profit first increases sharply and then falls to less than 0.01. These profits are averaged over ten runs. Figure 3 displays the spot price, equilibrium price, and the fair value of the futures contract predicted by each of the four participants for one of the ten runs. Note that the coefficients of the strategies are initialized to values that cause the participants to overestimate the spot price during the beginning of the simulation.

Experiment 3: Forced liquidations

Experiment 3 simulates the effect of forced liquidations on the futures price and on profitability. With probability .2, a participant is forced to post a fair value that is one half of that computed by its visible strategy. These occasional shocks are designed to mimic unforeseen events, such as bankruptcies, that are sometimes experienced by market participants.

For this experiment the profit does not tend toward zero, as Figure 4 shows. Large profits continue to be made by those participants who are fortunate enough not to be forced to liquidate. Thus, participants with stable bankrolls appear to have an advantage over those who do not. To the extent that this is true of real world markets, they are not weakly efficient.

Figure 5 shows the spot, equilibrium, and fair value prices for one run. The equilibrium prices typically overshoot the spot price, thus providing some market participants with the opportunity to make profits.

Experiment 4: Unknown variable

In this experiment the spot price is partially determined by a fourth variable, the moving average over the last five days, that is not available to the participants. Thus, even the "optimal" strategy cannot correctly predict the

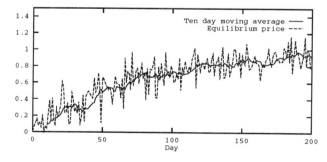

Figure 1: Equilibrium price as a function time for one run of experiment 1. In this experiment, the first three variables are randomly and uniformly distributed between -5 and 5 and the spot price is fixed at 1. This data is averaged over ten runs each of which lasted for 200 days.

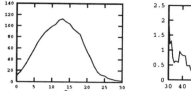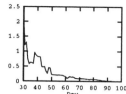

Figure 2: Profit of the most profitable participant as a function of time for experiment 2. As expected, the profit approaches zero as time passes.

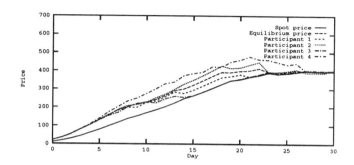

Figure 3: The spot, equilibrium, and fair value prices as a function of time for one run of the simulations performed for experiment 2.

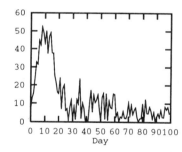

Figure 4: Profit of most profitable individual as a function of time in experiment 3. Unlike experiment 2, the profit does not tend to 0. This data is averaged over ten runs.

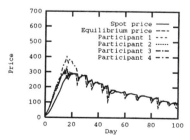

Figure 5: Spot, equilibrium, and fair value prices for one simulation of experiment 3. The dips in the price occur as a result of forced liquidations.

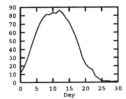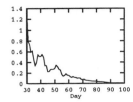

Figure 6: Profit of the most profitable participant as a function of time in experiment 4. Surprisingly, the profit achieved by the most profitable participant is less than in experiment 4.

spot price in all cases. The new spot price is: $x/4+y/4+4z/5 + w/5$ where w is the five day moving average of the equilibrium prices and all other variables have been defined previously.

We expected profits to persist longer in this experiment than in experiment 2. To our surprise, as Figure 6 shows, profits are uniformly lower. Why is this the case? At first we thought that the addition of the five day moving average might reduce the initial sharp rise in the price, thus allowing the strategies to better estimate the spot price. However, Figure 7 demonstrates, for one representative example, that this is not the case.

The lower profits appear to be due to a reduction in the impact that the sign variable, z, has on the price. Because it is the most volatile variable and the only non-linear one, reducing its contribution increases the predictability of the spot price, therefore reducing profits.

Experiment 5: Protected participant

This experiment, in which only three of the four participants are forced to undergo liquidations, is to be compared with experiment 3. Over time, the protected participant is expected to accumulate profits that dwarf the profits of the other three participants.

Over ten runs, the protected participant accumulated a total profit of 4162.5 points, while the unprotected participants had profits of -56.2, -4128.7, and 22.4. The

protected participant does not, however, accumulate the most profits in all of the runs. Figure 8 shows one simulation in which the protected participant is only slightly more profitable than one of the unprotected participants.

Experiment 6: Sharing strategies

Experiment 6 shows the effect of sharing strategies on the profitability of the participants. At the end of each iteration each participant incorporates the visible strategies of the other participants into its strategy set. We expected this sharing to increase the homogeneity of the strategies across the participants and, therefore, to reduce the profits accumulated by the most profitable participant.

These expectations were borne out by the results, which are summarized in Figure 9 and Figure 10. The profits are significantly less than in experiment 2 and the fair values computed by the visible strategies are all but indistinguishable.

This experiment helps to explain why most of the papers published in the academic literature about making money in the market report negative results.

Experiment 7: Asymmetric information sets

This final experiment explores how different information sets affect the profitability of participants. As before, one participant has access to all three variables, but the other three participants do not have access to the ten day moving average of the equilibrium prices.

Over ten runs the participant with the larger information set accumulates a profit of 666.67 units, while the other three participants have profits of -235.97, -173.81, and -256.89. The cumulative profits for one simulation run are shown in Figure 11. These results are in accord with our intuitions: the participant with the larger information set makes greater profits. Note, however, that this result may not hold as the information set increases in size because the search space becomes correspondingly larger.

A new description of participants

As a result of the experiments in this paper we have been led to a new model for characterizing market participants. Although this model contains no more informa-

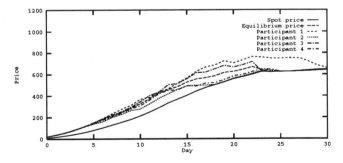

Figure 7: Spot, equilibrium, and fair value prices for one simulation of experiment 4. The spot price rises faster than in experiment 2, thus contradicting the hypothesis that the lower profits accumulated by the most profitable individual are due to increased price stability.

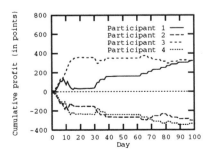

Figure 8: Cumulative profit as a function of time for experiment 5. Participant 1 is not subject to forced liquidations while the other three are.

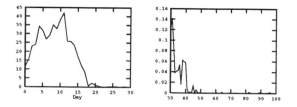

Figure 9: Profit of the most profitable individual as a function of time in experiment 6. As expected, the profits decrease more rapidly than in experiment 2. This data is averaged over ten simulations.

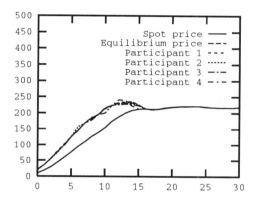

Figure 10: Spot, equilibrium, and fair values for one run of experiment 6. Note that the sharing of visible strategies causes the fair values to be all but inseparable.

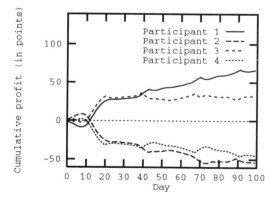

Figure 11: Spot, equilibrium, and fair values. Participant 1 has access to the ten day moving average while the other participants do not.

tion than the rational expectations equilibrium model, we feel that this description language better captures our intuitions about what differentiates market participants. Each participant is parameterized by four variables:

- Information set. What information does the participant have access to? Prices? Volume? Macroeconomic indicators?

- Algorithm set. What optimization algorithms and learning algorithms can the participant employ? Genetic algorithms? Linear regression? Neural networks?

- Model set. What kinds of models can the participant use to describe the data and predict future prices? Linear? Non-linear? Turing equivalent?

- Constraint set. What kinds of restrictions is the participant subject to? High transaction costs? Low initial capital? Sudden liquidations?

Some of this participant space has been explored in detail by economists. Linear models, for example, are much studied in econometrics and the effect of certain constraints on arbitrage opportunities is also well studied (see, e.g., [Kolb, 1991]). However, very little is known either about the dynamic or equilibrium behavior of systems in which the participants employ learning algorithms from artificial intelligence or in which the participants have grossly asymmetric information sets. In this paper we have kept the algorithm set and model sets fixed and changed the information sets and constraint sets of the participants. Exploring this space in a more complete fashion is a topic for future work.

Related work

Nottola, Leroy, and Davalo [Nottola *et al.*, 1992] describe an artificial market system in which agents (called participants in this paper) learn rules that predict movements in price. Much of their paper describes different approaches to modeling markets, but they do discuss one experiment which compares the performance of agents with different learning strategies and they conclude that adaptive agents out-perform non-adapting ones.

LeBaron [LeBaron, 1994] has developed a stock market simulation in which participants learn strategies using a genetic algorithm, as they do in this paper. The strategies of his market participants are encoded as bit-strings in which each bit corresponds to a boolean variable that is precomputed. His system features a mechanism for easily changing the risk aversion of individual participants. LeBaron notes [Freedman, 1993] that even if there are regularities in real market data that can be exploited, the data might contain so much noise that the regularities are barely perceptible.

There is substantial evidence that the currency futures market is not efficient [Thomas, 1986, Cavanaugh, 1987, Glassman, 1987, Harpaz *et al.*, 1990]. As empirical evidents mounts against the view that markets are efficient, theoreticians will be forced to provide alternative explanations of the market's behavior. One of the most likely candidates for this is Simon's bounded rationality theory [Simon, 1957, Simon, 1982]. Sargent [Sargent, 1993, page 4] explores markets in which the participants have bounded rationality and writes that "the bounded rationality program wants to make the agents in our models more like the econometricians who estimate and use them."

Conclusions and future work

As researchers we are interested in understanding whether or not certain participants in a market can consistently make profits. As money managers we are interested in uncovering opportunities for making profits. In this paper we have taken a small step towards addressing parts of both of these issues. Participants with stable bankrolls and access to more relevant information appear to have an advantage over other participants in our admittedly limited simulations. Some of the results that stem from this research have been used to manage a small options account since September, 1993. We participated in the options division of the 1993 U.S. Investing Championships using this account and finished fifth with a 43.9% return.

In the future we expect to create more accurate participant models of the market. We plan to circumscribe the scope and extent of the participant model described in section by running simulations with different settings for the four parameters. In particular, our current research focuses on expanding the participants' model sets so that they include explicit expectations about the model sets of other participants.

Acknowledgments

This paper describes research done at the Artificial Intelligence Laboratory of the Massachusetts Institute of Technology. Support for the laboratory's artificial intelligence research is provided in part by the Advanced Research Projects Agency of the Department of Defense under Office of Naval Research Contract N00014-91-J-4038. The first author is an NSF graduate fellow.

References

[Cavanaugh, 1987] K. Cavanaugh. Price dynamics in foreign currency futures markets. *Journal of International Money and Finance*, 6(3):295–314, 1987.

[Dornbusch and Fischer, 1987] R. Dornbusch and S. Fischer. *Macroeconomics*. McGraw-Hill Book Company, New York, 1987.

[Freedman, 1993] D. Freedman. Enter the market merlins. *Forbes ASAP*, 1993.

[Glassman, 1987] D. Glassman. The efficiency of foreign exchange futures markets in turbulent and non-turbulent periods. *The Journal of Futures Markets*, 7(3):245–267, 1987.

[Goldberg, 1989] D. Goldberg. *Genetic Algorithms in Search, Optimization and Machine Learning*. Addison-Wesley, Reading, MA, 1989.

[Harpaz et al., 1990] G. Harpaz, S. Krull, and J. Yagil. The efficiency of the U.S. dollar index futures market. *The Journal of Futures Markets*, 10(5):469–479, 1990.

[Holland, 1975] J.H. Holland. *Adaptation in Natural and Artificial Systems*. University of Michigan Press, Ann Arbor, 1975.

[Kolb, 1991] R. Kolb. *Understanding Futures Markets*. Kolb Publishing Company, Miami, 1991.

[LeBaron, 1994] B. LeBaron. An artificial stock market. MIT AI vision seminar colloquium, 1994.

[Michalewicz, 1992] Z. Michalewicz. *Genetic Algorithms + Data Structures = Evolution Programs*. Springer-Verlag, Berlin, 1992.

[Nottola et al., 1992] C. Nottola, F. Leroy, and F. Davalo. Dynamics of artificial markets: Speculative markets and emerging 'common sense' knowledge. In *Towards a Practice of Autonomous Systems: Proceedings of the First European Conference on Artificial Life*, pages 185–194, 1992.

[Sargent, 1986] T. J. Sargent. *Rational Expectations and Inflation*. Harper & Row, New York, 1986.

[Sargent, 1993] T. J. Sargent. *Bounded Rationality in Macroeconomics*. Clarendon Press, Oxford, 1993.

[Simon, 1957] H. Simon. *Models of Man: Social and Rational; Mathematical Essays on Rational Human Behavior in Society Setting*. John Wiley, New York, 1957.

[Simon, 1982] H. Simon. *Models of Bounded Rationality*. MIT Press, Cambridge, MA, 1982.

[Thomas, 1986] L. Thomas. Random walk profits in currency futures trading. *The Journal of Futures Markets*, 6(1):109–125, 1986.

Evolutionary Differentiation of Learning Abilities — a case study on optimizing parameter values in Q-learning by a genetic algorithm

Tatsuo Unemi,[*] **Masahiro Nagayoshi, Nobumasa Hirayama,**
Toshiaki Nade, Kiyoshi Yano, and **Yasuhiro Masujima**
Department of Information Systems Science, Soka University
1-236, Tangi-machi, Hachioji, 192 Tokyo, JAPAN
e-mail: {unemi,rorry,nobu,toshi,papino,yas}@iss.soka.ac.jp

Abstract

This paper describes the first stage of our study on evolution of learning abilities. We use a simple maze exploration problem designed by R. Sutton as the task of each individual, and encode the inherent learning parameters on the genome. The learning architecture we use is a one step Q-learning using look-up table, where the inherent parameters are initial Q-values, learning rate, discount rate of rewards, and exploration rate. Under the fitness measure proportioning to the number of times it achieves at the goal in the later half of life, learners evolve through a genetic algorithm. The results of computer simulation indicated that learning ability emerge when the environment changes every generation, and that the inherent map for the optimal path can be acquired when the environment doesn't change. These results suggest that emergence of learning ability needs environmental change faster than alternate generation.

1 Introduction

There are many layers in the Artificial Life researches such as molecular dynamics, evolution, development, learning, collective behavior, and so on. One of the methods for fruitful studies is on combination of two or three of these layers, such as evolutionary development system, collective behavior of learners. We already finished the first stage of above two kinds of combinations (Nade et al, 1994, Unemi, 1993). On the third combination, evolution of learners, Todd and Miller have been pursuing evolutionary process to organize associative neural networks that learn by a simple Hebbian rule (Todd and Miller, 1990). Ackley and Littman mentioned genetic acquisition of evaluation network in a neural network based reinforcement learning method (Ackley and Littman, 1992) and a distributed Lamarckian evolution (Ackley and Littman, 1992). Their studies provided us fruitful suggestion to understand a process of emergence of intelligent creatures. However, we are just starting our steps toward the real intelligent creature, that is,

---
[*]Soka University, and Laboratory for International Fuzzy Engineering Research, Siber Hegner Bldg., 3rd Fl., 89-1 Yamashita-cho, Naka-ku, 231 Yokohama, JAPAN

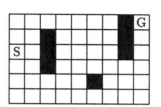

Figure 1: An example of the maze. Black squares are obstacles. The letter S and G indicate the start position and the goal position respectively.

human. We need more studies to complement the pioneers' works and to advance our understanding in this field.

This paper describes the first stage of our study on evolution of learning abilities of which final goal is to make learning ability to emerge structurally. As the first step of this challenge, we tried to optimize the inherent learning parameters of a reinforcement learning mechanism using a genetic algorithm. We can intuitively expect that it is unnecessary for creatures to have learning abilities if the environment is stable and doesn't change through many generations but learning abilities are needed if the environment changes every generation. To realize these phenomena on the computer, we designed a simulator of evolvable learners that consists of a simple learning task, a simple learning algorithm, and a simple genetic algorithm.

The rest sections of this paper describe the specifications of the simulator, the experiments, and their results.

2 Task for individual

The task for each individual is a simple maze exploration designed by Sutton as a testbed for his Dyna learning architecture (Sutton, 1990). The maze an individual creature lives in is a two-dimensional grid world of nine columns and six rows. Some of the grids are occupied by obstacles so the individual cannot position there. Figure 1 shows a typical map of the maze used in the experiments described later.

In each execution step, it moves from the current position to the adjacent grid in the maze not occupied by an obstacle. From fixed start position, it explores the maze toward the fixed goal position. It gets positive reinforcement signal when it reaches the goal and then starts exploration from the start position again. It knows its position in each step.

This is a typical task for simple reinforcement learning, where the environment is stable, deterministic, and discrete. Some readers may feel that such settings are too simple as a model of life, but it would be better to start a simple model as possible at the first stage.

3   Learning algorithm

We employ a one step Q-learning using look-up table proposed by Sutton (1990), because its algorithm is very simple and the look-up table has an ability to represent a map of the world. A brief description of this learning algorithm is as follows.

In each execution step, the learner gets sensory input from the environment and decides its action based on the input data. Then it takes the action, and receives reinforcement signal if available, and then goes to the new state. The look-up table contains the Q-values that estimate expected reward corresponding to the all of possible states and actions, that is, Q-value Q_{xa} indicates expected reward when the individual takes the action a at state x. For the task we use, the look-up table contains $4 \times 2 + (7 + 4) \times 2 \times 3 + 4 \times 7 \times 4 = 186$ Q-values if it includes no obstacle, because of two actions in four corners, three actions in four sides, and four actions in inner grids.

The action is determined in each step according to a probability of Boltzmann distribution. The probability for selecting action a at state x is defined as

$$P(a|x) = \frac{\exp(\alpha Q_{xa})}{\sum_{j \in Possible\ Actions} \exp(\alpha Q_{xj})}$$

where α, exploration rate, is the inverse value of temperature. When the value of α is large, it tends to select the action of maximum Q-value rigidly. When the value of α is small, nearly zero, it tends to take a random action regardless of the Q-values.

When it selected action a at state x, the result state was y, and it received reward r, then the Q-value Q_{xa} is revised by the assignment equation:

$$Q_{xa} \leftarrow Q_{xa} + \beta \left(r + \gamma \max_b Q_{yb} - Q_{xa}\right)$$

where β is learning rate and γ is discount rate of rewards.

The learner explores the maze according to its action selection mechanism and estimates the value of each action at each state only referring to delayed rewards through its own experience. Each individual creature lives alone, that is, the individuals in population have no interaction with each other.

Learning performance strongly depends on the values of parameters, especially on the initial Q-values, though theoretical guarantee of convergence to the global optimum solution after infinite times of trials is proved by Watkins (1993). Because the value of β should gradually increase for certain convergence, we set up the start value and the end value of α and β, and gradually change the values step by step under the constant difference.

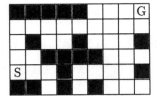 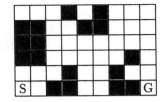

Figure 2: Examples of the random mazes.

According to our preliminary experiments in the above problem, when all of the initial Q-values is set to zero, 4,000 steps provides an enough number of trials to find an optimal path, sometimes global optimum but often local optimum. So, the life span of each individual is set to 4,000 steps fixed in the experiments described later.

4   Genetic algorithm

This section describes the genetic algorithm we use here.

4.1   Genetic code

All of inherent learning parameters have continuous numerical values. So, we employ eight bits integer coding for each parameter and translate each genetic code to real number of its range proportionally, for example, because the range of γ is $[0, 1]$, the value of γ is set to $G_\gamma/255$ where G_γ is the value of eight bits unsigned integer on the gene corresponding to γ. The ranges of Q_{xa}, β, and γ are $[0, 1]$, and the range of α is $[0, 63.75]$ ($63.75 = 255/4$). One genome contains 221 bytes, $6 \times 9 \times 4 = 216$ Q-values, α_{start}, α_{end}, β_{start}, β_{end}, and γ. The reason we use redundant 216 Q-values rather than 186 values as described above is merely to simplify the implementation on the computer program.

4.2   Fitness

The fitness of each individual is calculated by counting the times the learner achieves at the goal position in its later half of life. The life span is set to 4,000 steps as described above, counting is done during the later 2,000 steps. Because the first half of life should be treated as a moratorium, we ignore the performance of that period.

4.3   Genetic operations

We use a simple genetic algorithm with a ranking and elitist strategy using selection, crossover, and mutation. The following list summarizes the algorithm.

Selection Remain the best third of genomes to the next generation.

Crossover Replace the middle third of genomes to the genomes made by crossover operation between each one of the best third and the middle third. Each genome includes two pieces of byte string chromosomes where the one contains Q-values and the other one contains the five parameters. One point crossover is applied to each chromosome independently not bitwise but bytewise.

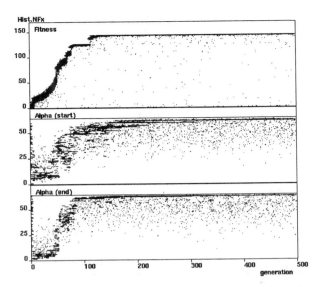

Figure 3: An evolutionary process without learning on fixed environment.

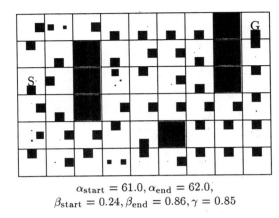

$$\alpha_{\text{start}} = 61.0, \alpha_{\text{end}} = 62.0,$$
$$\beta_{\text{start}} = 0.24, \beta_{\text{end}} = 0.86, \gamma = 0.85$$

Figure 4: Probabilities of action selection of the best individual at 500th generation – the case without learning on fixed environment. The size of black square in each grid represents the probability of selecting the action at that state, that is, $P(a|x)$.

Mutation Replace the worst third of genomes to the mutant of the best third. One byte randomly selected from each chromosome is modified by pseudo-Gaussian distribution.

Starting from the population of randomly initialized genomes, evaluation and genetic operation cycles are iterated. The result of learning by an individual doesn't inherit to the offspring but only the natural characteristics, that is, not Lamarckian but Darwinian evolution.

5 Experiments

For the purpose of this research, we tried experiments concerning with four kinds of settings as follows.

1. Without learning but only evolution on the fixed environment.

2. Without learning but only evolution on the environment that changes randomly every generation.

3. With learning and evolution on the fixed environment.

4. With learning and evolution on the environment that changes randomly every generation.

"Without learning" means the individual doesn't learn in any trial, that is, Q-values are not modified. The behavior of the agent without learning depends only on its inherent Q-values and α. The values of β and γ have no effect. The map of fixed maze is as shown in Figure 1 above. A random maze is designed as the start position is located in the left most column and the goal position is located in the right most column and a path from start to goal exists. Figure 2 shows two examples of the random mazes.

We expected that it needs learning abilities if the environment changes every generation but it doesn't need if fixed. Setting 1 was expected to lead the acquisition

of inherent map of the environment, and setting 4 was expected to lead the acquisition appropriate parameter values for learning. In setting 2, it seemed to be hard to adapt the environment. However, we could not predict what would happen in setting 3.

Using the following values of genetic parameters, we examined ten runs for each setting with distinct random number sequences.

Population size	=	100	individuals
Life span	=	4,000	steps
Number of generations	=	500	generations

The rest of this section describes a summary of the experimental results.

5.1 Without learning on fixed environment

Figure 3 shows a typical trace of an evolutionary process. As shown in Figure 4, the optimum path of the maze is acquired through a punctuated equilibrium evolution, where Q-values represent the map of the path and the value of α becomes not always the maximum but large enough to make selection rigid. All of ten runs shows the similar pattern of evolutionary process, but some of them were trapped at a local optimum even at 500th generation.

5.2 Without learning on changing environment

As shown in Figure 5, it is difficult to adapt the environment to walk through the optimal path. However, from Figure 6, we can observe that the agent acquired the tendency to go right because the start position is located at the left most column and the goal position is located at the right most column. All of ten runs shows the similar pattern of evolutionary process. This fact suggests that it is difficult to adapt to any changing environment without learning ability.

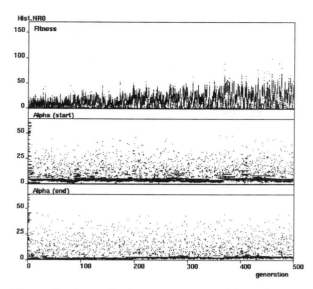

Figure 5: An evolutionary process without learning on changing environment.

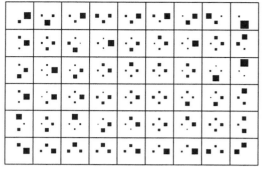

$$\alpha_{\mathrm{start}} = 4.8, \alpha_{\mathrm{end}} = 2.2,$$
$$\beta_{\mathrm{start}} = 0.64, \beta_{\mathrm{end}} = 0.37, \gamma = 0.82$$

Figure 6: Probabilities of action selection of the best individual at 500th generation – the case without learning on changing environment.

5.3 With learning on fixed environment

We observed both cases where the agents with efficient learning abilities appeared and where the agents who inherently know the optimal path appeared. Figure 7 and 8 show the former case, and Figure 9 and 10 show the later case. The typical difference between these two cases is seen at the value of α_{start}. In the former case, α_{start} converged low that leads the agent to radical exploration at young age. This characteristic is useful for learning. In the later case, both α_{start} and α_{end} are high that makes the agent to take a conservative action following the inherent Q-values. The high value of α prevents the agent from learning. This means that the value of β and γ make no effect in this case.

5.4 With learning on changing environment

As we expected, the agent with an efficient learning ability emerged. Figure 11 shows a typical pattern of the evolution, where fitness values widely vary because the length of the optimal path of each maze randomly generated is also widely varies. The value of α_{start} becomes low and the value of α_{end} becomes high as similarly as the case of fixed environment where learning abilities are acquired.

6 Conclusion

We investigated the relation between learning and evolution under the different stability of the environment. As we expected before experiments, learning abilities emerged when the environment changed every generation, and inherent knowledge about the world was acquired when the environment fixed. However, we are surprised at the observation that learning ability could emerge even when the environment fixed. We guess that this phenomenon happens because of our assumption of genetic codes that only include some numerical parameters but do not mention any structural information of learning mechanism. The observation that to change the environment every generation causes evolutionary acquisition of efficient learning abilities suggests us that the condition of environmental change strongly effects the structural emergence of learning abilities.

The results complement the work on genetic acquisition of evaluation network by Ackley and Littman, because we here focused on differentiation of adaptation strategies, evolution versus learning. The range of learning rate β doesn't include negative value, so the learner always mentions the reinforcement signal at the goal position as reward, never as punishment. It is one of quick experiments to make the range of β to include negative value so as to consider acquisition of evaluation function. Both of the experiments by Todd and Miller, and Ackley and Littman, did not mention the relation between the stability of the environment and evolutionary process of learning abilities. The experimental results can provide an extension of our knowledge in this field combining with the pioneers' results.

Starting from the first stage of the research described here, our future work will include more precise investigation of the effects of genetic parameters and development of a methodology of structural emergence of learning abilities.

Acknowledgment

The authors would like to thank Shigenobu Kobayashi, Masayuki Yamamura, Norihiko Ono, Hiroshi Deguchi, Hisashi Tamaki, Osamu Sakura, and Kazuhiro Matsuo for their useful suggestions on earlier work of this research. The authors would also like to thank the referees for their helpful comments for the manuscript of this paper, and thank Toshiro Terano at Laboratory for International Fuzzy Engineering Research for the support of this research.

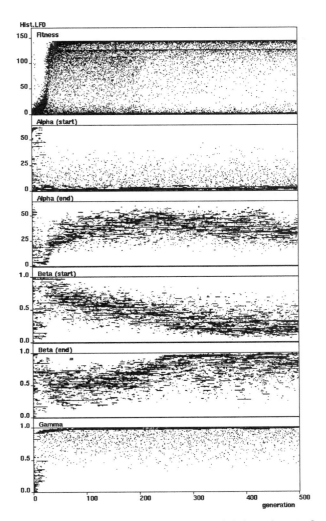

Figure 7: An evolutionary process with learning on fixed environment – 1.

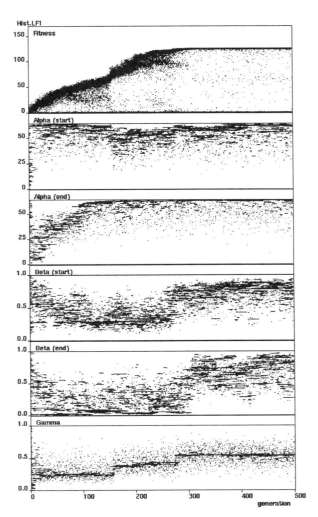

Figure 9: An evolutionary process with learning on fixed environment – 2.

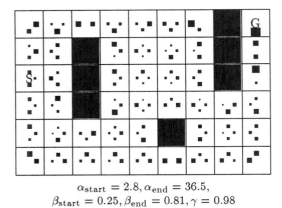

$\alpha_{\text{start}} = 2.8, \alpha_{\text{end}} = 36.5,$
$\beta_{\text{start}} = 0.25, \beta_{\text{end}} = 0.81, \gamma = 0.98$

Figure 8: Probabilities of action selection of the best individual at 500th generation – the case with learning on fixed environment – 1.

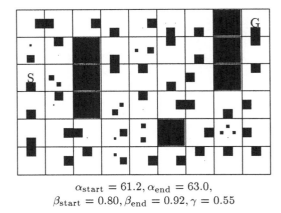

$\alpha_{\text{start}} = 61.2, \alpha_{\text{end}} = 63.0,$
$\beta_{\text{start}} = 0.80, \beta_{\text{end}} = 0.92, \gamma = 0.55$

Figure 10: Probabilities of action selection of the best individual at 500th generation – the case with learning on fixed environment – 2.

References

Ackley, D. and M. Littman. 1992. Interactions between Learning and Evolution. In *Artificial Life II*, edited by C. G. Langton, C. Taylor, J. D. Farmer, and S. Rasmussen. Addison-Wesley, 487–509.

Ackley, D. and M. Littman. 1992. A Case of Distributed Lamarckian Evolution. Presented at *the Third International Workshop on Artificial Life*, Santa Fe, NM.

Nade, T., M, Nagayoshi, N, Hirayama, Y. Masujima, K. Yano, and T. Unemi. 1994. A Simple Development System on 3D Euclidean Space and its Evolution. *IPSJ SIG Notes* 94:20:25–30, in Japanese.

Sutton, R. S. 1990. Integrated Architectures for Learning, Planning, and Reacting Based on Approximating Dynamic Programming. *Proceedings of the Seventh International Conference on Machine Learning*, 216–224.

Todd, P. M. and G. F. Miller. 1990. Exploring Adaptive Agency II: Simulating the Evolution of Associative Learning. *From Animals to Animats – Proceedings of the First International Conference on Simulation of Adaptive Behavior*, 306–315.

Unemi, T. 1993. Collective Behavior of Reinforcement Learning Agents. *Proceedings of the 1993 IEEE/Nagoya University WWW on Learning and Adaptive System*, 92–97.

Watkins, C. J. C. H. and P. Dayan. 1993. Technical Note Q-Learning. In *Reinforcement Learning*, edited by R. S. Sutton, Kluwer Academic Pub., 55–68.

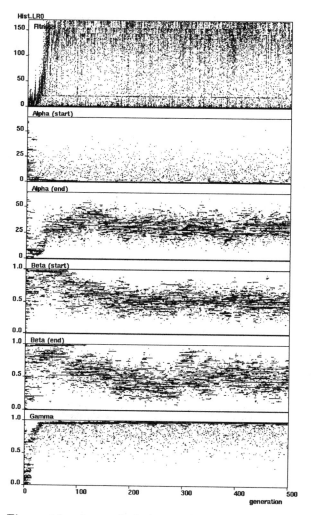

Figure 11: An evolutionary process with learning on changing environment.

Exploring the Foundations of Artificial Societies: Experiments in Evolving Solutions to Iterated N-player Prisoner's Dilemma

Steve Bankes
RAND
P.O. Box 2138
1700 Main Street
Santa Monica CA 90407-2138
bankes@rand.org

Abstract

This paper describes the use of genetic algorithm based approaches to explore the evolution of cooperative behavior in N player generalizations of prisoner's dilemma. In these experiments, quasi-stable cooperation emerges for coalitions of cooperators of all sizes greater than a minimum coalition size determined by the payoff structure of the game. The framework described here produces emergent cooperation reproducibly as long as minimum coalitions are 3 or less. For games with large minimum coalition sizes, cooperative behavior is not observed.

1. Towards Artificial Societies

Human societies are complex adaptive systems whose properties, while very important, are poorly understood. Existing models from the fields of economics, anthropology, and political science can explain some aspects of observed social phenomena, but many social patterns are emergent structures that cannot be understood without confronting the complex adaptive aspects of human social behavior. Computational experiments based on abstracted models have been used to study the complex adaptive properties of biological systems, and it is reasonable to speculate that this approach can be generalized to study human social issues via computational experiments with "artificial societies". (Builder and Bankes, 1991). In this paper I describe a first step towards this goal: a formalism in which cooperative behavior among communities of self interested actors can emerge.

The evolution of behavior in the context of an iterated game has been useful for understanding how cooperation can arise among groups of individuals who have powerful incentives not to cooperate. Axelrod has demonstrated that in the iterated 2-player Prisoner's Dilemma game, cooperation can evolve and be evolutionarily stable

(Axelrod, 1984). This fundamental result has been extended by others (Nowak and Sigmund, 1993), including some work focused on multiplayer games (Glance and Huberman, 1993). However, none of the work to date provides a substrate on which experiments can be conducted with arbitrary games, arbitrary numbers of players, or where players have asymmetric roles. In order to provide a solid foundation for investigations into the behavior of artificial societies, it is necessary to demonstrate the evolution of social structure in n-player games with arbitrary payoff landscapes, including landscapes incorporating social dilemma. In this paper I take a next step on this path.

2. Evolutionary N-Player Iterated Games

In order to support arbitrary games, including those with specialized player roles, individual players must be separately represented, capable of playing asymmetric roles, and developing specialized strategies. Consequently, modeling the evolving strategy set as a single pool of "genomes" as in (Axelrod, 1984) will not suffice. In the examples shown here, players are modeled as distinct pools of memes that code for player strategies. Player actions at any time are a consequence of the memes they possess at that time, and player adaptation over time results from the evolution of their meme populations.

Within this framework, various protocols for deriving player behavior based on meme frequencies, and for adapting meme pools based on player experiences can be investigated. Of several investigated so far, the protocol resulting in the most interesting behavior is reported here. In these experiments, player behavior is a stochastic, guided by meme frequency, and each player's meme pool evolves using a standard genetic algorithm (Goldberg, 1989) with meme fitness based on player outcomes in games where that meme's strategy was played.

Thus, for each play of an iterated game, a memome (a complete set of memes that describe a strategy) is randomly selected from the meme pool of each player. The strategy described by that memome is used by that player throughout the iterated game. For each player, that player's cumulative score is added to the fitness of the memome selected. After a number of such iterated games, each meme pool is separately updated (no mixing) by a genetic algorithm sing roulette wheel selection, with point mutation and crossover genetic operators.

3. N-Player Iterated Prisoner's Dilemma

The game known as "prisoner's dilemma" has received considerable attention as the simplest game that captures the fundamental problem of social cooperation. In this game, each of two players must make a simultaneous choice between two strategies, that can be called "cooperate" and "defect". The game payoff function for prisoner's dilemma is such that either player, no matter what the other plays, will do better by defecting. Yet the payoff for mutual defection is poorer for both than mutual cooperation would be.

The basic game of prisoner's dilemma rationally must always result in mutual defection unless embedded in some larger context in which sociality motivates players to cooperate. To study this social process in the abstract, theorists have focused on the iterated supergame, in which the same players play the basic game repeatedly, so that one can "reward" the other for cooperation on a previous move by cooperation in the next play of the game.

In this paper I consider N-player generalizations of prisoner's dilemma. There are various N-player generalizations of the classic 2-player prisoner's dilemma. Here I consider the generalization due to Schelling (Schelling, 1978). The number of players in any given experiment will be fixed (N). On a given play of the game each player must choose between cooperating or defecting. Payoff for that play will depend on which move was made, together with the total number of players that cooperated on that play. The game is thus defined by two functions: pay-off-to-defectors $D(i)$ and payoff-to-cooperators $C(i)$, where the number of cooperators i is in the range $0 <= i <= N$. (Actually, in general, $D(N)$ and $C(0)$ need not be defined.) A wide variety of games can be defined by defining these two functions in various ways. The game can be considered to be a generalization of prisoner's dilemma if it satisfies these properties:

(1) for all i, $D(i-1) > C(i)$

(2) $C(N) > D(0)$.

Thus, no matter how many players cooperate, any one of them will earn a better payoff by defecting. But, if all cooperate they will score better than if all defect.

In the experiments reported here, the functions C and D are linear in the number of player cooperating. Thus the game can be defined by N and the slopes and intercepts of the two payoff functions. As payoffs are relative, we can without loss of generality define $D(0) = 0$. Thus, property 1 implies that $C(0)<0$.

An important parameter of the definition of these games is the x-intercept of C, which in conformity with Schelling, I shall call K. K is the minimum size of a coalition of players that can benefit from cooperating. As we will see, this important parameter drives a large part of the emergent behavior in these experiments.

For these games, the memomes consist of 2 "chromosomes". One chromosome has N-1 positions, where the position j codes for the player's next move when j of the other players were observed to cooperate on the previous move. A small second chromosome provides the first move in the iterated play.

4. Results and Discussion

Within the framework described above, cooperation can evolve for a wide range of games. A typical example for N=5, K=2, is displayed in figures 1-9. Figure 1 shows the payoff functions C and D for this example. In the context of all other players defecting, a single player would receive -1 points for cooperation but could would receive a score of 0 for defecting along with the other 4 players. At the other end of the scale, 5 cooperating players would each receive 3 points on each round, but any of them could increase his take to 4 by defecting. As long as all five cooperate they will garner 3 points more per turn than if they all defected, but regardless of the play of others, any player can improve his payoff by 1 if he were to defect.

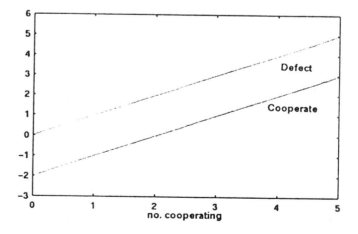

Figure 1
Payoff functions for K=2

In spite of the short term motivation to defect, significant cooperation can occur. Fig. 2 depicts the average number of cooperators across all games plotted over time. There are clearly long periods characterized by high levels of cooperativity, punctuated by intermittent collapses into mass defection. The average payoff is strongly driven by the amount of cooperation as demonstrated by Fig. 3. All games that are generalizations of prisoner's dilemma show both high level of cooperation and mass defections occurring, with intermittent sudden shift between these two regimes. (Similar behavior was observed in a different framework by Glance and Huberman, 1993.) Further, there are a variety of stable levels of cooperation, resulting in a terraced trajectory. In this example the most common cooperative scenario has 4 cooperating players, with one free riding. For other games, other outcomes are possible. For example, figure 10 displays the cooperation trajectory in a game with K=1. For this game, situations with all 5 players cooperating are by far the most common.

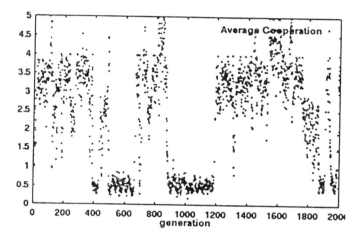

Figure 2
Average Cooperation for K=2

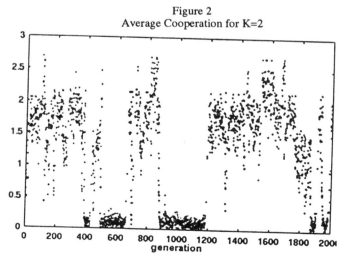

Figure 3.
Average Payoff for K=2

Notice in figures 2 and 10 that no terraces occur for levels of cooperation of K or less. In any of these games, should cooperation level fall to K, there will no disincentive to restrain the remaining cooperators to joining the rush to defect. Cooperation levels greater than K can be stable however, as the cooperators are scoring better than they would if all players defected. The viability of these coalitions are akin to the viability of cooperation in the 2-player game, as observed by Axelrod. It is noteworthy however, that these levels of cooperation are demonstrably stable. There is no logic in these player that would allow them to reason about the likely consequences of a defection. Human players may choose to cooperate as a result of predicting future consequences, but these automated players are incapable of such reasoning.

In two player games, the stability of cooperation depends on evolving the strategy TIT-FOR-TAT (or for different variants its analog, see Nowak and Sigmund, 1993). TIT-FOR-TAT is the strategy of playing cooperation on the first move, and on the succeeding moves imitate the opponent's previous play. The stability of cooperation in 2-player prisoner's dilemma con be understood by observing that TIT-FOR-TAT is an evolutionarily stable strategy (Maynard-Smith, 1982).

Analyzing the strategies (memomes) that result in the quasi-stability of multiple player coalitions is more difficult both because the memomes are larger and have more degrees of freedom, and because the coalitions are unstable in the long run. These effects cause the memome population to be unstable in the sense that genetic drift randomizes and hence gradually erases portions of the genome that are not being utilized. Thus, during a period of all defection, genes dictating behavior when there is a high level of cooperation tend to erode. Similar erosion occurs to low cooperation genes during eras of predominant cooperation. However, by inspecting the gene database and averaging the values of gene positions across a run, a signature "strategy" can be detected. The dominant form is a 5-player generalization of tit-for-tat. These gnomes have the form [000..11]. These players defect in the case of mostly defection occurring on the previous move, and cooperate for mostly cooperation. The position at which the genes switch from defecting to cooperating varies considerably among the gene pool, although always bounded to be greater than K. Some percentage of the evolved genes obey a modified "free-rider strategy". These have the form [00..11..00]. These players defect when everyone else is, cooperate when moderate levels of cooperation occur (preserving minimal viable coalitions), but defect again if there is lots of cooperation, presenting the opportunity to "cherry pick". The average values of the genes for the example above are displayed in figures 4-8. Figure 9 shows the average value of the gene that dictates the initial move of the iterated game.

The ease with which cooperation can be established is quite sensitive to the functions that define the game. In fig.

10 and 11, are alternative payoff functions and the resulting trajectory of average number of cooperators in the population. As figure 11 demonstrates, with increasing K, the balance between cooperation and defection changes, with resulting increase of the average time to wait for cooperation to break out. However, the level of cooperation attained during cooperative intervals does not appear as strongly affected by changes in K.

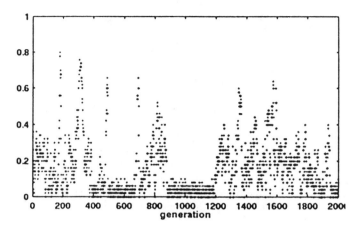

Figure 4.
Average Response to no cooperators

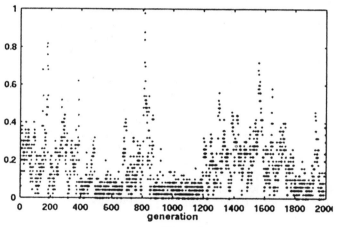

Figure 5.
Average response to 1 cooperator.

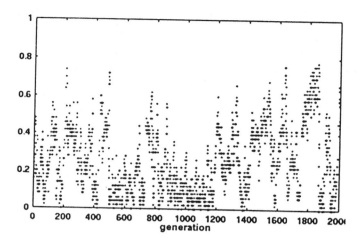

Figure 6.
Average response to 2 cooperators.

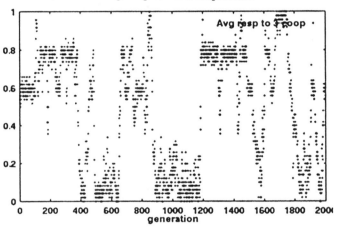

Figure 7.
Average response to 3 cooperators.

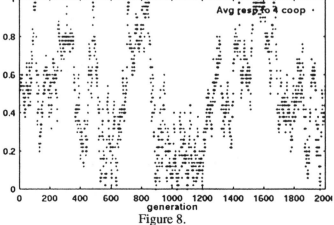

Figure 8.
Average response to 4 cooperators.

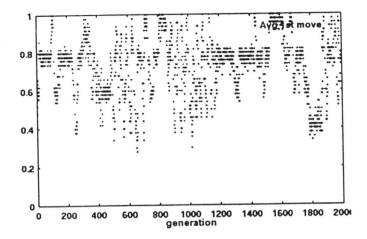

Figure 9.
Average first move.

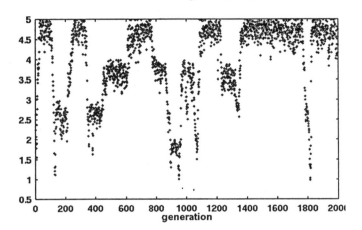

Figure 10.
Average cooperation for k=1.

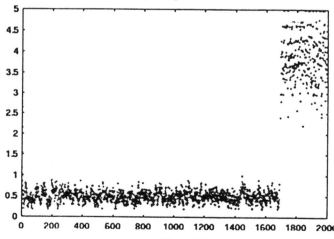

Figure 11.
Average cooperation for K=3.

In order to better understand the effect of different payoff functions on evolved behavior, I generated a large

number of games in which the properties of C and D systematically varied. There was no correspondence between values of function slopes or intercepts and the average level of cooperation achieved. However, there is a strong dependence on K, as demonstrated in figure 12. Note that significant cooperative behavior is never achieved in games with a K greater than 3. As most of the decline in average cooperation values is due primarily to increasingly long periods of "all defect" being averaged in, we could presumably observe small amounts of cooperation for large K games by allowing for very large run times. Preliminary experiments exploring varying values of N suggests that the sharp edge at K=3 appears to be a general result (data not shown). The ease with which any level of cooperation can be achieved depends much more strongly on K than it does on N.

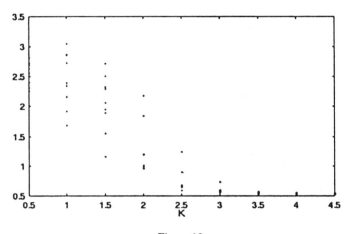

Figure 12.
Average cooperativity versus K.

These results suggest that the mechanism which allows for cooperation in the framework described here, can produce cooperation only for payoff functions with small minimal group sizes. To achieve cooperation where much larger collaborations are required, may require other mechanisms, such as coercion or side-payments.

5. Conclusions and Future Directions

In the classical literature there has been significant debate on the relationship between minimum viable coalition size and the likelihood of cooperation (Chamberlain, 1974; Hardin, 1983; Olson, 1965). The results presented here are consistent with arguments originally advanced by Olson suggesting that cooperation is easier to achieve small coalitions are viable. However, the detailed results of these experiments would not have been consistently predicted by existing social theory. This suggests that the use of computational experiments can inform debate and enhance our thinking about the nature of social cooperation. The agenda for this nascent field must be the discovery of the simplest structures that can result in emergent phenomena akin to those we wish to better understand. To this end, future experimentation must explore the properties of the wide range of payoff functions beyond those that are strict generalizations of prisoners dilemma, and to examine the implications of new mechanisms that are highly relevant to understanding human societies, most notably the ability of players to directly influence the payoff landscapes of others, and implications of constraints such as communications costs.

References

Axelrod, Robert. 1980. *The Evolution of Cooperation*. New York: Basic Books.

Builder, Carl, and Steven Bankes. 1991, *Artificial Societies: A Concept for Basic Research on the Societal Impacts of Information Technology*. RAND P-7740.

Chamberlain, John. 1974. Provision of Collective Goods as a Function of Group Size. *American Political Science Review*,. 68: 707-16,.

Glance, Natalie S., and Bernardo A. Huberman. 1993. The Outbreak of Cooperation. *Journal of Mathematical Sociology*. 17: 281-302,.

Goldberg, David E., 1989. *Genetic Algorithms in Search, Optimization, and Machine Learning*. Reading Massachusetts:Addison-Wesley Publishing Co.

Hardin, Russell. 1982. *Collective Action*. Baltimore:John Hopkins University Press.

Maynard-Smith, John. 1982. *Evolution and the Theory of Games*. Cambridge: Cambridge University Press.

Nowak, Martin, and Karl Sigmund. 1993. A strategy of Win-stay, Lose-shift that Outperforms Tit-for-Tat in the Prisoner's Dilemma Game. *Nature* 364: 56-58.

Olson, Mancur Jr. 1965. *The Logic of Collective Action: Public Goods and the Theory of Groups*. Cambridge MA: Harvard University Press.

Schelling, Thomas. 1978. *Micromotives and Macrobehavior*. New York: W.W. Norton and Co.

Evolutionary Dynamics of Altruistic Behavior in Optional and Compulsory Versions of the Iterated Prisoner's Dilemma

John Batali
Department of Cognitive Science
University of California at San Diego
La Jolla, CA 92093-0515
batali@cogsci.ucsd.edu

Philip Kitcher
Department of Philosophy
University of California at San Diego
La Jolla, CA 92093-0302
pkitcher@ucsd.edu

Abstract

Computational simulations of the evolution of populations playing an optional version of the iterated prisoner's dilemma, in which players may choose whether or not to interact, show that this version of the game gives rise to individuals who tend to cooperate more often than those playing the standard compulsory version of the game. This result is due to dynamical properties of the evolving systems: the populations playing the compulsory game can become stuck in states of low cooperation that last many generations, while the optional game provides routes out of such states to states of high cooperation.

1 Introduction

Altruism presents a puzzle to evolutionary theory because it seems that "selfish genes" ought to yield selfish behavior. What is the selective advantage of helping others, especially if doing so requires time and effort that could be better spent pursuing one's own interests? One answer to this puzzle is that altruistic behavior toward one's relatives benefits the copies of ones genes those relatives possess. However this explanation does not account for altruistic behavior towards those who may be quite distantly related, even members of other species. Another possible benefit of altruism is that it might be reciprocated, so that one's generous efforts are rewarded in kind. We are interested in exploring the conditions in which reciprocal altruism (which we shall refer to as "cooperation") can arise as a result of evolution.

The iterated prisoner's dilemma has been extensively studied as a simple model of social interaction in which a kind of cooperative behavior can occur (Axelrod & Hamilton, 1981; Axelrod, 1984). In each round of the game, two players choose to either "cooperate" (C) or "defect" (D). Each player then receives a payoff depending on the pair of choices. The two players interact some number of rounds, and the payoff for each player at the end of the interactions is the total of the payoffs received each round.

The payoffs to the players are determined as follows: If one player defects and the other cooperates, the defector receives the "traitor" payoff T and the the other player receives the "sucker" payoff S. If both players cooperate, they both receive the "reward" R. If both players defect, they both receive the payoff P. The values of the payoffs are such that $T > R > P > S$ and $2R > T + S$. For our simulations, $T = 7, R = 5, P = 2$ and $S = 0$.

The payoff values are such that to maximize the expected reward for a single round, a player should defect. However the possibility arises that the two players could, in the course of a number of interactions, settle on a pattern of mutual cooperation that provides them both with a relatively high payoff. Studies of the iterated prisoner's dilemma have analyzed the relative merits of various strategies for playing the game. One strategy, "always defect," involves choosing D each round. It has the advantage of always receiving at least the P payoff, and of receiving the maximum T payoff against players who cooperate. The "always cooperate" strategy always chooses C, and has the advantage of receiving the relatively high R payoff against other cooperators, however it will receive the minimal S payoff against defectors.

A strategy known as "tit-for-tat" (TFT) is especially successful in the iterated prisoner's dilemma. The TFT strategy involves cooperating on the first move, and then doing whatever the opponent did on the last round. TFT does well when playing against cooperative strategies, as both will receive the R payoff each round. Against defectors, TFT will lose out on the first round. From then on, it will defect against the defector, and both will receive the same payoff P. TFT is both "forgiving" in the sense that it will revert to cooperative behavior if its opponent does so first, and "wary" in that it will resort to defection if the need arises.

While the iterated prisoner's dilemma has been used as the basis for studies of animal behavior, for example (Dugatkin & Alfieri, 1991), it seems unrealistic in its assumption that both players are forced to interact each round. In addition to choosing to cooperate or defect, animals might refuse to interact at all, facing neither the benefits nor the risks of social behavior.

Kitcher (1993) introduces a version of the iterated prisoner's dilemma in which either of the two player may choose to "opt out" (O) of playing the game. If either player chooses O, both receive a payoff W, intermediate between R and P. (In our simulations, the value of W was 3.) In this optional version of the iterated prisoner's dilemma, a number of new strategies arise. For example the "soloist" strategy involves always opting out; the "discriminating altruist" strategy involves cooperating with a player as long as the other player cooperates, but opting out if the other player ever defects.

Kitcher analyzes populations of individuals playing each version of the iterated prisoner's dilemma. He makes the assumption that the relative proportion of players using a given strategy will increase or decrease according to the relative payoffs the players using that strategy receive in their interactions against the other members of the population. This analysis thus provides a model of the evolution of cooperative behavior. An important feature of these evolutionary models is that the environment in which evolution is occurring — namely the strategies used by the other players in the population — is evolving also.

Kitcher shows that in the compulsory version of the iterated prisoner's dilemma, the TFT strategy can have difficulty getting established in a population where all of the other players are defecting. The problem is that the TFT players will be penalized for cooperating on the first round, a penalty they can make up only by finding other TFT players in the population. In order for TFT to establish a foothold in a population of defectors, a significant number must arise spontaneously. Related deficiencies of TFT have been noted before (Boyd & Lorberbaum 1987; Farrell & Ware 1989; Mesterton-Gibbons 1992).

On the other hand, in the optional iterated prisoner's dilemma, Kitcher finds that a population of defectors can be invaded by a single soloist. As the soloist will receive the W payoff each round, it will do better than the defectors do against each other and better than the defectors do against it. With the higher fitness values thus received, the soloists will eventually take over the population.

In a population of soloists, cooperative strategies are now advantageous, as the cooperators will do no worse than W against the soloists and will receive the R payoff with other cooperators. So the number of players using cooperative strategies will increase over the generations. These cooperative strategies will be at risk from defectors, however, unless they incorporate the discriminating tactic of opting out when defected upon. Kitcher shows that the discriminating altruist strategy can successfully resist invasion by defectors.

2   Computational Simulations

Our computational simulations involve populations of players who participate in the iterated prisoner's dilemma by following inherited strategies.

The actions performed by each player are recorded as a history sequence. The lengths of histories recorded in our simulations varied from 2 to 4. For example the sequence (C C) indicates that a player cooperated on both of the previous two rounds; the sequence (D C) indicates that a player defected two rounds ago, but cooperated the last round. The symbol N is used when the players haven't played as many games as the history records. Thus for a history of length 2, the sequence (N N) indicates that no rounds at all have been played; the sequence (N C) means that a single round was played, and the player cooperated.

Strategies are represented by pairing each possible history of opponent's actions with the action the player will make in the next round. This pairing of a history with a response action will be called a "move." The following move represents the response of defecting if the opponent cooperated twice in a row:

((C C) D)

This move represents the action of cooperating on the first round:

((N N) C)

Given a specific history length, a complete strategy contains a move for each possible history sequence of that length. For example this strategy represents the tit-for-tat strategy in which two steps of history are recorded:

```
((N N)  C)
((N C)  C)
((N D)  D)
((C C)  C)
((C D)  D)
((D C)  C)
((D D)  D)
```

In this strategy, the player begins by cooperating and then responds with whatever its opponent did the last round.

A sequence of rounds between two players is simulated by using the strategies of the two players to determine their moves for each round, depending on what the other player did the last rounds. Each player receives an increment to its "fitness" value according to the payoff schedule described above. In a generation, each player plays against each other player in the population some number of times.

At the end of each generation, the set of players is sorted in order of decreasing fitness values. The top third of the players is preserved into the next generation, and those players are used to create the strategies of the rest of the players in the next generation. Each new strategy is created by mixing the strategies of two of the fittest players — for each possible history, the response action is taken randomly from one or the other parent's strategy. Mixing strategies in this way has the effect of rapidly distributing advantageous moves through the population. A small fraction (for most of our runs: 1%) of the moves are then mutated by replacing the action part of the move with a randomly chosen action.

In each generation, a record is kept of the total number of moves of each type: "cooperate," "defect," and for the optional games, "opt out." At the end of a generation the average fitness of the population is also recorded. A sample run of the compulsory game in shown in figure 1. As is typical for the runs reported here, the population moves through a number of states in which the amount of cooperation and defection is roughly constant for tens of generations or longer. Fairly rapid transitions then occur, yielding other stable states.

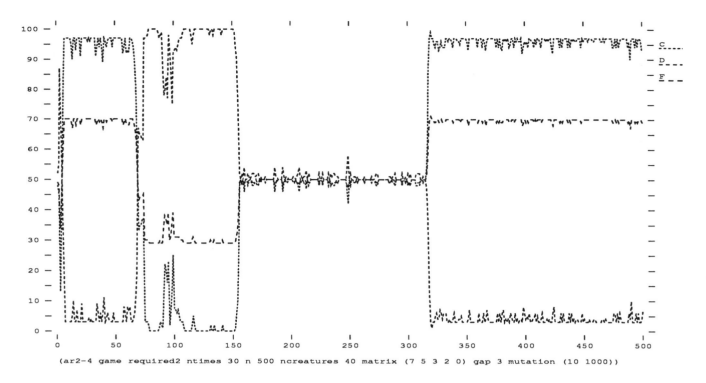

(ar2-4 game required2 ntimes 30 n 500 ncreatures 40 matrix (7 5 3 2 0) gap 3 mutation (10 1000))

Figure 1: A run of the compulsory game, with two steps of history recorded. The trace marked C records the percentage of "cooperate" moves each generation; the trace marked D records the percentage of "defect" moves each generation; the trace marked F records the average fitness of the population as a percentage of the maximum possible value.

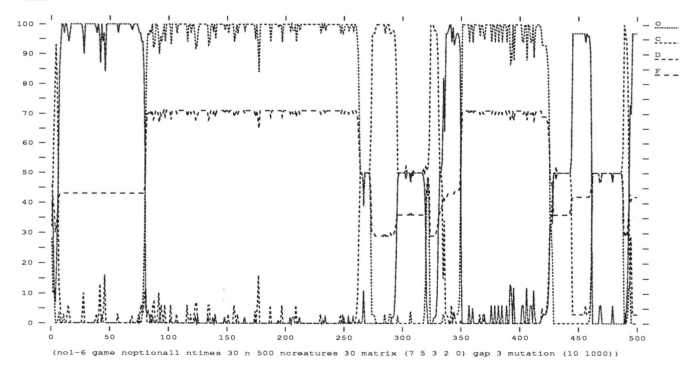

(no1-6 game noptional1 ntimes 30 n 500 ncreatures 30 matrix (7 5 3 2 0) gap 3 mutation (10 1000))

Figure 2: A run of the optional game, with one step of history recorded. The trace marked O records the percentage of "opt out" moves each generation.

In this run the population quickly enters a state of high cooperation and fitness. Around generation 70, it reverts to a state of virtually 100% defection and low fitness. This is followed at generation 150 with a state of 50% cooperation, 50% defection and a medium fitness value. Around generation 310, the population again finds a state of very high cooperation.

The results of an entire run are summarized with two numerical values:

The "cooperativity" measure is meant to quantify the degree to which cooperative behavior dominated during the run. Cooperativity is the fraction of generations during a run when the difference between the percentages of cooperative moves and defection moves is greater than a threshold of 25. The cooperativity value for the run shown in figure 1 is 0.494. (The precise value of the threshold for computing the cooperativity value is not crucial. For example changing the threshold to 70 for the run in figure 1 changes the cooperativity value from 0.494 to 0.488. This is because the runs tend to remain in states where either cooperation or defection is relatively high, and the other is therefore low.)

The "instability" measure is meant to quantify the degree to which the amount of cooperation varies from generation to generation. This is measured by computing the average of the square of the difference between the number of cooperative moves in successive generations. The instability value for the run shown figure 1 is 19.41. The instability measure increases if a run enters more states, or if the states that it enters do not have fairly constant values of cooperation. The value of the instability measure for simulations depends on the simulation parameters. For example in a set of 20 runs of 500 generations of the compulsory game with 36 players, a history of length 2, and 10 games between each pair of players, increasing the mutation rate from .1% to 10% changed the average instability value from 3.28 to 28.37. As we will see, the instability value is also highly affected by whether the game is compulsory or optional.

In the simulations of the optional games, the players have, in addition to the actions C ("cooperate") and D ("defect"), the choice of action O ("opt out"). In the optional games, each player's strategy included moves containing O in the history and action. The payoff W is given to each player if either player chooses to opt out in a round of the game.

The number of cooperate, defect, and opt out moves in each generation of a simulation are recorded. The cooperativity and instability measures are computed in the same way as for the compulsory game. A run of the optional game is shown in figure 2. The cooperativity measure for this run is 0.581; its instability measure is 45.1.

3 Simulation Results

The basic results of this study can be summarized in a single table. The following statistics are for a set of 27 runs, each of 500 generations. There were 60 players in each run, a mutation rate of .1%, 2 steps of history recorded and each pair of players interacted 30 times.

Game	Cooperativity		Instability	
	Mean	SD	Mean	SD
compulsory	.105	.160	5.28	3.80
optional	.719	.283	13.2	5.96

As can be seen from this table, the optional game yields populations which are less stable but more cooperative than those that result from playing the compulsory game.

What these statistics do not show is the dynamical character of the runs. Few runs of either type of game yielded simple graphs in which either cooperation or defection reached a high value and stayed there. We suspect that all runs are unstable in the sense that drastic changes in the relative amounts of cooperation and defection eventually occur. The relative fitness of the strategies used by the members of each population depend on the strategies used by the other members of the populations, and the specific combination of strategies determines the relative proportions of strategies in the next generation. In the following discussion, we describe some of the dynamics of the evolutionary simulations, showing how the properties of the strategies, and their genetic representations, give rise to the changing rates of cooperation and defection. More detailed discussion will be found in Batali & Kitcher (forthcoming).

3.1 The Compulsory Game

Runs of the compulsory game tend to become stuck in a small number of states, either with very high cooperation, very high defection, or half cooperation and half defection. Often a run will be stuck in a state for many generations. This is reflected in the relatively low instability value for the compulsory game, and the high ratio of the standard deviation of its cooperativity to the mean value.

For example, when one step of history is recorded, a run often first enters a state where all of the players defect each round. This is because the initial strategies are random, and hence there are a large number of players who cooperate no matter what their opponents do. Hence the defectors receive the high T payoff, and their offspring take over the population.

Subsequent events can be understood by examining the strategies shown in figure 3, beginning with the "always defect" strategy shown in 3a. Two of the single-move mutations of this strategy will be at a disadvantage playing against it because they will cooperate in a round where the original will defect. If everyone in the population is defecting all the time, however, the move ((C) D) is never exercised. So a mutation from the move ((C) D) to ((C) C) will not affect the behavior (nor the fitness value) of a player using that strategy.

After several generations of this kind of genetic drift, a population initially containing only players with strategy 3a can be expected to contain a fraction of players with strategy 3b. From here, a single mutation in the ((C) D) move can change the strategy to TFT, as shown in 3c. Provided that enough of these mutations occur at about the same time, players using TFT can dominate the population, which will enter a state of very high co-

```
((N) D)   ((N) D)   ((N) C)   ((N) C)   ((N) D)
((C) D)   ((C) C)   ((C) C)   ((C) D)   ((C) D)
((D) D)   ((D) D)   ((D) D)   ((D) C)   ((D) C)
   a         b         c         d         e
```

Figure 3: Some strategies for the compulsory game, when one step of history is recorded.

operation. This is essentially what happens in the run shown in figure 1 around generation 10.

However TFT is not immune to variation. For example the ((N) C) move could mutate back to ((N) D) and return the population to strategies like 3b and a state of high defection. This is what happens near generation 70 in the run shown in figure 1.

Another mutation from TFT can yield the "always cooperate" strategy. If the population is in a state where everyone else is following TFT, "always cooperate" will not affect the fitness of the player, as the ((D) C) move in TFT is never exercised. However if defectors appear, they can exploit the player who always cooperates. For example strategy 3d is one mutation away from the "always cooperate" strategy. If it appears, the number of players using it will increase in the population as they prey on the cooperators. A single mutation from 3d is 3e, which increases defection in the population even more.

Strategies like the one shown in 3e can lead to very stable populations, with relatively low cooperation and fitness values. A pair of players playing this strategy will cooperate 50% of the time and defect 50% of the time. Furthermore there is no possibility of genetic drift with this strategy as each move of it is exercised and each one-mutation variant of this strategy is at a disadvantage against it. The population in the run shown in figure 1 enters a state where each member of the population is playing a variant of this strategy around generation 155. Note that the state remains steady for around 60 generations, before the population returns to a state of relatively high cooperation.

With longer history lengths, strategies like 3e, in which pairs of players using the strategies alternate between cooperation and defection, can be very stable. For example the following strategy is a variant of 3e for a history length of 2:

```
((N N)  D)   *
((N C)  C)
((N D)  C)   *
((C C)  D)   *
((C D)  C)   *
((D C)  C)   *
((D D)  D)
```

The '*' character indicates moves that are exercised when players using this strategy play against each other. Five of the seven moves are used in such an encounter, and therefore cannot mutate away from those shown without having an effect. Mutations to the other two moves actually serve to reinforce the patterns of interactions seen when the rest of the players are using this strategy. We have observed runs of the compulsory game

in which the population remains stuck in states where populations are using such strategies for thousands of generations, though transitions to states of high cooperation or high defection eventually occur.

3.2 The Optional Game

With the option of opting out, populations in the optional games can escape from states in which there is a significant amount of defection. Since the payoff for opting out, W, is larger than both S, the payoff for cooperating when the opponent defects, and P, the payoff for mutual defection, the presence of defection in the population makes opting out an advantageous alternative. Thus populations playing the optional games will tend to revert to states of high opting out whenever a number of defectors appear. This fact alone accounts for some of the reason why the optional game leads to higher cooperativity — it just can't become stuck in states of high defection.

A full run of the optional game with one step of history recorded is shown in figure 2. As is typical for runs of the optional games, the population enters more states than otherwise equivalent runs of the compulsory game, and the states that it enters are much less stable.

Since the initial strategies are random, and therefore include many defectors, opting out is initially favored, and most runs of the optional game enter states of virtually 100% opting out in the early generations, often as early as generation 10. From then on, the populations tend to go through cycles of various sorts. One kind of cycle involves the appearance of defectors, which is usually soon followed by a reversion to high opting out.

Another kind of cycle seen in the optional games involves the appearance and subsequent disappearance of players who always cooperate. If almost every other member of the population is opting out each round, there is no danger to the few players who mutate to strategies that involve some cooperation. Indeed when these players play against each other, and receive the reward R for mutual cooperation (which is larger than the W opting out payoff), they will increase in the population. However when there are lots of careless cooperators in the population, there is an advantage to be gained by defecting. If mutations occur to create strategies that involve some defection, defectors will rapidly increase, effectively destroying the cooperators. Since the resultant level of defection is high, opting out is now relatively advantageous, and the population reverts to a state where everyone is opting out. This pattern is similar to the "predator-prey" cycles seen in population biology.

A typical run of the optional game will go through a number of such cycles. In some cases it is possible that the mutations that increase defection either include or are followed by mutations that increase the discriminatingness of the strategy, either by playing TFT, or by opting out when an opponent defects, i.e., the discriminating altruist strategy. If such mutations occur before defection rises significantly, it is possible for the players possessing these strategies to continue to increase in the population even when defection rises temporarily. Thus

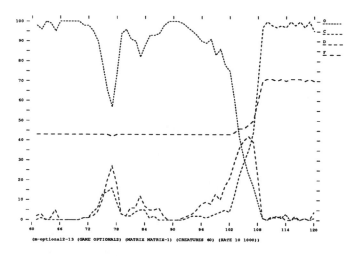

Figure 4: A portion of a run of the optional game, showing two "predator-prey" cycles, and the beginning of a state of high cooperation.

the population can enter and remain in a relatively stable state of high cooperation.

This process is illustrated in figure 4. This shows a portion of a run of the optional game. At generation 60, virtually all of the players are opting out in each game. Around generation 70, a few players begin cooperating. Since they manage to find other cooperators, their numbers increase. Within two generations, however, a few defectors appear. Since these defectors will prevail over the cooperators, their numbers increase rapidly, and by generation 80 or so, the cooperators are gone. A similar but less dramatic pattern of this sort begins almost immediately, and is over by generation 90.

At generation 91, another set of cooperators appears, followed closely by defectors. However in this case, at least some of the cooperators are playing a discriminating strategy, and in fact by generation 108, the defectors begin disappearing from the population. By generation 110, virtually all of the players are cooperating all of the time.

As with the compulsory game, states of high cooperation are not stable either. With all of the members of the population cooperating, genetic drift can set in, and mutate some of the discriminating strategies to their careless versions, providing fodder for defectors when they appear by mutation. As before, the high rate of defection will ultimately be followed by an increase in opting out.

4   Conclusion

The optional version of the iterated prisoner's dilemma was introduced as a more ethologically realistic model of animal social interaction than the compulsory version. It can be shown through the analysis of simple strategies that the the optional game has advantages over the compulsory game in terms of how cooperative strategies can appear, increase in the population, and resist invasion. These results are supported by our computational simulations.

The only way that a population playing the compulsory game can escape from a state of high defection is for several favorable mutations to occur at once. In the optional game, there are routes out of state of high defection, via states where virtually all of the population opts out all of the time. The option of asocial behavior facilitates the appearance and maintenance of altruistic behavior.

In thinking about the evolution of social behavior it is important to recognize that such behavior occurs against a changing environment consisting of the behaviors of the other members of the populations. Thus such an evolutionary process is a feedback system, and the global properties of such a process should be expected to fluctuate, perhaps chaotically. The relative fitness of a given behavior or strategy cannot be assessed statically, with respect to a specific, or to a fixed, environment. In the long run, the evolutionary dynamical properties of strategies and their genetic representations, may have the most significant effect on the careers of populations using those strategies.

References

Axelrod, R. (1984). *The Evolution of Cooperation*. Basic Books, New York,

Axelrod, R. & Hamilton, W. D. (1981). The evolution of cooperation. *Science*, 211:1390–1396,

Batali, J. & Kitcher, P. (forthcoming). Evolution of altruism in optional and compulsory games.

Boyd, R. & Lorberbaum, J. P. (1987). No pure strategy is evolutionarily stable in the repeated prisoner's dilemma game. *Nature*, 327:58–59,

Dugatkin, L. A. & Alfieri, M. (1991). Guppies and the tit for tat strategy: Preference based on past interaction. *Behav. Ecol. Sociobiol.*, 28:243–246,

Farrell, J. & Ware, R. (1989). Evolutionary stability in the repeated prisoner's dilemma. *Theoretical Population Biology*, 36:161–168,

Kitcher, P. (1993). The evolution of human altruism. *Journal of Philosophy*, October.

Mesterton-Gibbons, M. (1992). On the iterated prisoner's dilemma. *Bulletin of Mathematical Biology*, 54:423–443.

Plate 1a: Artificial fishes in their physics-based world.

Plate 1b: Mating behaviors. Female (top) is courted by large male.

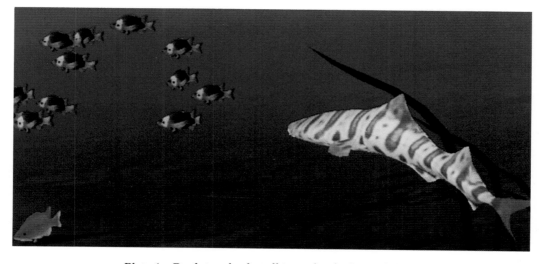

Plate 1c: Predator shark stalking school of prey fish.

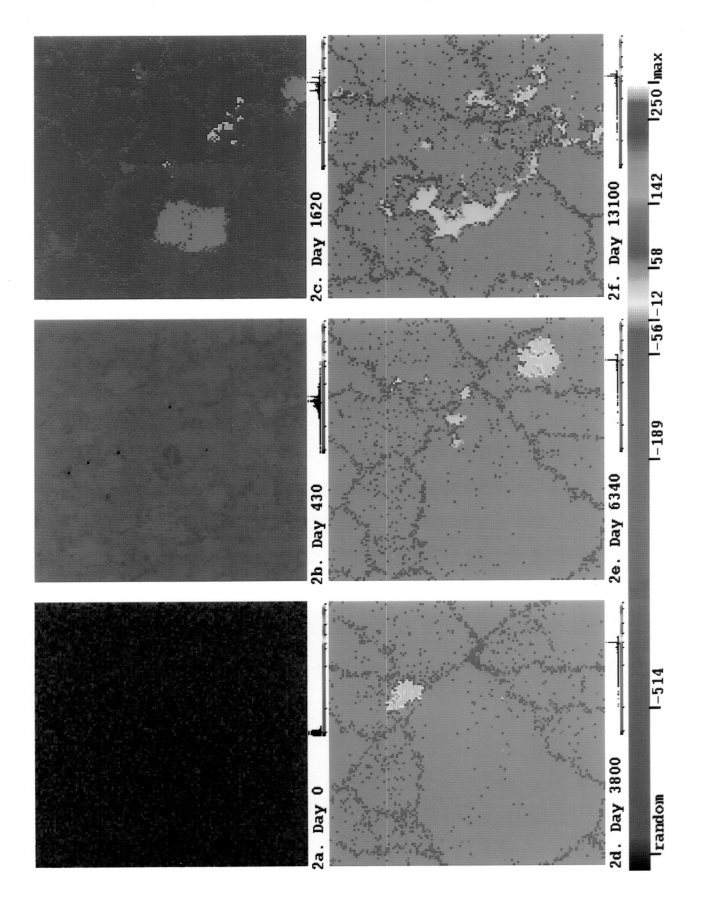

2a. Day 0   2b. Day 430   2c. Day 1620

2d. Day 3800   2e. Day 6340   2f. Day 13100

random   −514   −189   −56   −12   58   142   250   I_max

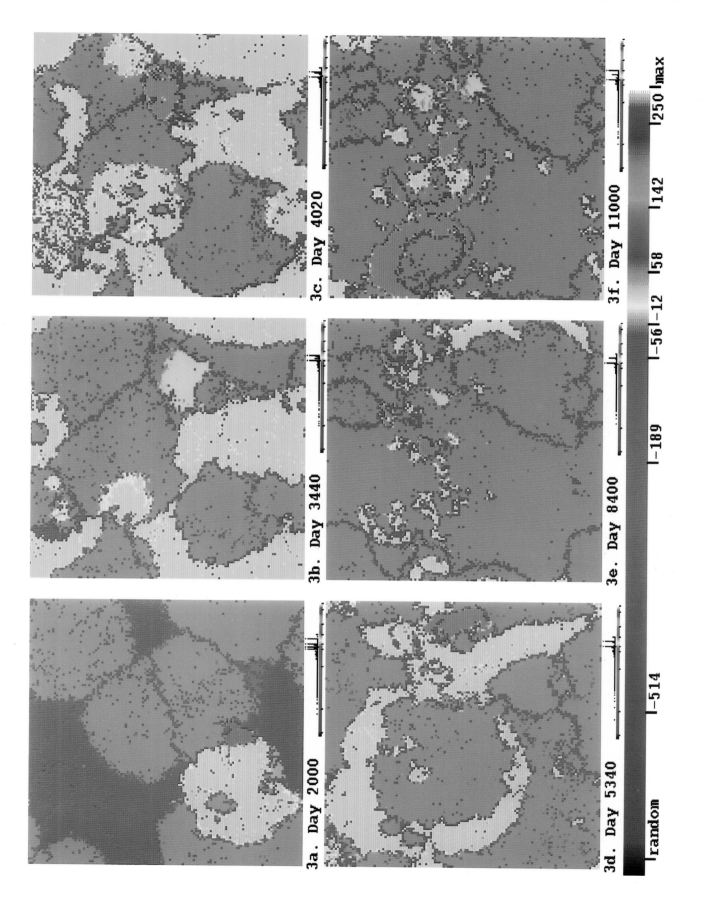

3a. Day 2000

3b. Day 3440

3c. Day 4020

3d. Day 5340

3e. Day 8400

3f. Day 11000

random   -514   -189   -56 -12   58   142   250 I_{max}

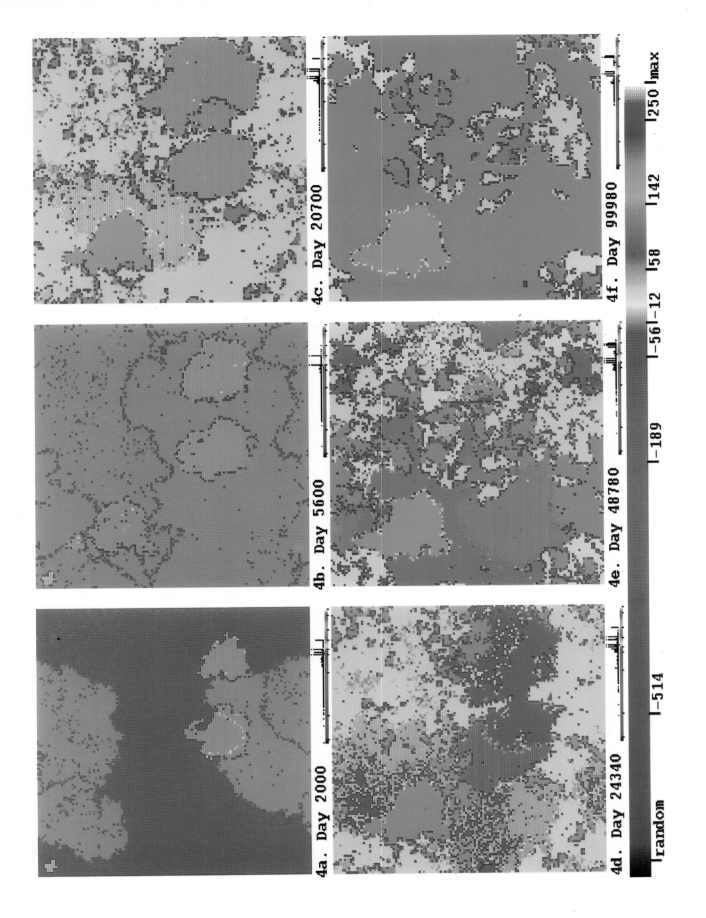

4a. Day 2000 4b. Day 5600 4c. Day 20700

4d. Day 24340 4e. Day 48780 4f. Day 99980

random -514 -189 -56 -12 58 142 250 I_{max}

Evolving Cooperation in the Non-Iterated Prisoner's Dilemma:
The Importance of Spatial Organization

Michael Oliphant
Department of Cognitive Science
University of California, San Diego
e-mail: oliphant@cogsci.ucsd.edu

Abstract

Most work on evolving cooperation in the Prisoner's Dilemma treats the non-iterated game as an undesirable simple case that should be risen above. It has been taken as a given that populations evolving to play the non-iterated game will always converge on defection. This paper questions this assumption, and demonstrates that organizing a population spatially dramatically changes the nature of the game and allows cooperation to emerge.

1 The Prisoner's Dilemma

A formalism known as the Prisoner's Dilemma has been adopted as the standard for studying the evolution of cooperative behavior. It describes a game played by two players, where each can decide whether or not they will cooperate with another (failure to cooperate is termed "defection"). A typical example of the payoff each player gets depending on their choice of action is summarized in Figure 1, with the payoff to the row player listed first.

	Coop	Defect
Coop	R/R	S/T
Defect	T/S	P/P

Figure 1: Payoff matrix for the Prisoner's Dilemma

The Prisoner's Dilemma occurs when $T > R > P > S$ and $2R > T + S$ (the most commonly used values are T=5, R=3, P=1, S=0). Although mutual cooperation is the best solution from the joint perspective of both players, defection is the best choice on a self-interested basis. If an individual can expect to encounter a percentage of cooperators Fc and a percentage of defectors Fd (where $Fc + Fd = 1$) over the course of an evaluation cycle, then its expected average payoff would be

$$E[Pc] = R * Fc + S * Fd \qquad (1)$$

for cooperators, and

$$E[Pd] = T * Fc + P * Fd \qquad (2)$$

for defectors. $T > R$ and $P > S$, so defectors always have an advantage, regardless of the makeup of the population. This mathematical fact is demonstrated if an evolutionary simulation of the Prisoner's Dilemma is done, with populations in such simulations converging to all defectors.

2 Previous Work on Evolving Cooperation

Dissatisfied with mathematical inevitability of defection, Axelrod and many others (Axelrod 1984; Axelrod 1987; Axelrod and Hamilton 1981; Lindgren 1991) have studied a more complex variant of the Prisoner's Dilemma. This variant, called the Iterated Prisoner's Dilemma, involves individuals playing each other more than once. In order to benefit from multiple games against the same opponent, individuals are given a three game history that documents both their actions and those of their opponent. They are also given a mechanism that allows them to modify their future behavior based their history of interaction with an opponent. This allows individuals to cooperate with those who cooperate with them and defect on others, and leads to a predominance of cooperation in the population.

Nowak and May (Nowak and May 1992) took a different approach – attempting to avoid convergence on defection while staying within the confines of the non-iterated game. They used the Prisoner's Dilemma as an update rule for two-dimensional cellular automata. In their simulations, cells could be in one of two states (cooperate or defect) and at each time step, every cell was replaced by the cell bordering it that has the highest summed payoff from playing the game with its eight neighbors. The result they found was not convergence on defection, but intricate patterns of cooperation and defection.

While this result seemed encouraging, it has been criticized as being an artifact of the simulation conditions. Huberman and Glance (Huberman and Glance 1993) argue that the results depend on the use of a synchronous update mechanism where all cells in the space are updated simultaneously. They demonstrate that when asynchronous updating (updating one cell at each time step) is used, the population quickly degrades into all defection.

3 Current Work

The present simulations are aimed at showing that, while the results of the simulations done by Nowak and May

may be artifactual, the idea of using a spatially organized population is a good one and can lead to results similar to what they thought they had found. There are good reasons to expect that organizing a population spatially might benefit cooperators.

3.1 Theoretical Motivation

There is an assumption present in the mathematical explanation (shown in equations (1)&(2)) of the superiority of defection as a choice in the non-iterated Prisoner's Dilemma. The assumption is that individuals will interact with the same percentage of cooperators and defectors regardless of their own game-playing strategy. If this assumption does not hold, neither does the mathematical result. A more generalized version of the payoff equations can be formulated as follows:

$$E[Pc] = R * Fcc + S * Fcd \qquad (3)$$

$$E[Pd] = T * Fdc + P * Fdd \qquad (4)$$

where Fcc is the percentage of the time a cooperator plays against a cooperator and Fcd is the percentage of the time a cooperator plays against a defector, etc. In the non-spatial, non-iterated case, $Fcc = Fdc = Fc$ and $Fcd = Fdd = Fd$, but this does not hold true in the spatial case.

In the bulk of previous simulations involving the Prisoner's Dilemma, the populations were not spatially organized. This meant that when picking an opponent to communicate with, or when selecting a mate, all other individuals were equally likely to be chosen. In a population where spatial organization is being imposed, this is not the case. Individuals are more likely to play the game with those close to them than they are with those farther away. Also, when an individual has offspring, they are placed in the area of the space where the parent was. What these factors result in is a space where individuals are more related (genetically closer) to those nearer to them.

This means that the expected number of cooperators and defectors an individual will play against depends on their own status, providing a mechanism for cooperation to spread through the population. Cooperators in the population will benefit those close to them. Since those close to them will be more likely to be cooperators as well, they are, in a sense, benefitting their own genes. The opposite holds true for defectors, who will have a negative impact on those around them. Since their neighbors are more likely to be defectors, they hurt their own kind. Getting back to equations (3)&(4), what spatial organization does is to increase the probabilities Fcc and Fdd, thereby increasing the terms $R * Fcc$ and $P * Fdd$. Since $R > P$, this benefits the cooperators.

3.2 Simulations

Instead of using cellular automata, the present simulations use the genetic algorithm (a function optimization technique based on biological evolution developed by John Holland (Holland 1975)). As in Nowak and May's simulations, individuals are fixed as being cooperators or defectors (coded on a one-bit genome). All

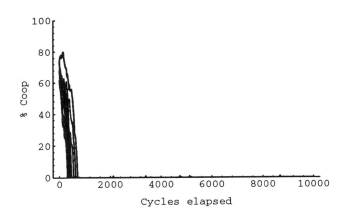

Figure 2: Non-spatial runs with T=5, R=4.5, P=0.5, S=0

simulations use a fixed population size of 100 individuals, a low mutation rate (1 in 1000), and no crossover (as it isn't very useful with when the genome is comprised of only a single bit). The fitness metric used is the average payoff an individual receives when playing the the Prisoner's Dilemma against approximately 32 [1] in the current population. Initial simulations were done using syncronous updating, but to avoid the criticism given by Huberman and Glance of the Nowak and May work, the simulations described here used asynchronous updating. In each cycle of the simulation a single individual was chosen randomly (but biased so that high fitness increased the chance of selection) to reproduce, and a single individual was chosen randomly (but biased so that low fitness increased the chance of selection) to "die" and make room for the new individual.

To replicate the standard all-defect result in the non-spatial, non-iterated game, a series of runs were done without imposing any kind of spatial organization on the structure of the population. The result (10 runs of which are shown superimposed in Figure 2) is as expected. In every run, the population quickly converges to defection.

A very different picture is seen when the same simulations are done with the only modification being to impose spatial organization. Figure 3 shows that each run quickly converges to cooperation, and is maintaining this equilibrium after 10000 cycles have elapsed. In these simulations, the population was organized as a one-dimensional wrap-around space (a ring). Spatial organization was imposed by selecting opponents for an individual to play based on a gaussian distribution around that individual. In this particular set of runs, the distance to an opponent was selected from a distribution with a standard deviation of 1 and incremented by one

---

[1] An average of 32 games are played by each individual because an individual selects 16 opponents, and is expected to be selected approximately 16 times by other individuals.

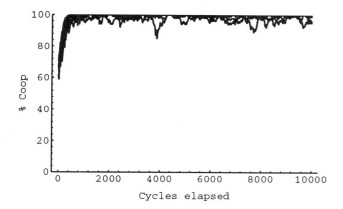

Figure 3: Spatial runs with T=5, R=4.5, P=0.5, S=0

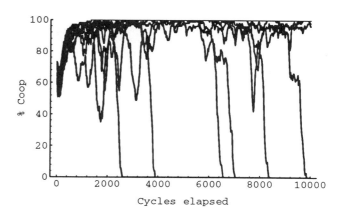

Figure 4: Spatial runs with T=5, R=3, P=1, S=0

(to prevent individuals from playing themselves). Also, offspring were placed in the population at a position near their parents (using the same deviation formula that was used for opponent selection). Using larger standard deviations reduces the spatial organization of the population (and hence the advantage to cooperators), and a sufficiently large deviation makes the population effectively non-spatial.

The results of the previous simulations are somewhat misleading, however, as they used a payoff matrix (T=5, R=4.5, P=0.5, S=0) that gives the defectors only a slight advantage (although enough of an advantage for defectors to easily dominate in the non-spatial case). If this advantage is increased, the results are less clear. Figure 4 shows 10 runs done with a payoff matrix where T=5, R=3, P=1, and S=0. The runs show initial convergence on cooperation, with sharp transitions to defection later on in 6 of the 10 runs. The reason for this is that while a population converged on cooperation is in an equilibrium state, this state is not stable. Enough mutations to defection within a short period of time can put the system in a position to be pulled into the other attractor (convergence on defection), which is a stable state. The probability that a population will be jostled out of the cooperation equilibrium state and then transition to the defection equilibrium state depends on the payoff matrix (which determines the strength of the two attractors) and the mutation rate (where a higher mutation rate is more likely to mutate enough cooperators into defectors in a short period of time to move the population into the attraction region of the all-defect state).

4   Conclusions

The simulations presented in this paper show that when a spatial structure is imposed on populations evolving to play the Non-Iterated Prisoner's Dilemma, a much different pattern emerges. While non-spatial populations quickly fall into defection, spatial populations are able to evolve and maintain cooperative behavior. Because of this, it is a mistake to treat the non-iterated game as a simple case where defectors always have the advantage, and researchers should be careful in motivating their move to more complex forms of the game, such as the iterated version. The iterated game requires a sophisticated game player that can recognize the individuals it has played with previously and remember the other's behavior in those prior interactions – requirements one might not want to assume when trying to apply simulation results to real-world phenomena.

An individual playing the non-iterated game is very simple – all that is required is a hardwired strategy that is either "always cooperate" or "always defect". In demonstrating that cooperation can emerge in the non-iterated game, a lower bound has been placed on the complexity required by an individual to have a population of such individuals evolve cooperative behavior.

References

Axelrod, R. (1984). *The evolution of cooperation*. New York: Basic Books.

Axelrod, R. (1987). The evolution of strategies in the iterated prisoner's dilemma. In L. Davis (Ed.), *Genetic algorithms and simulated annealing*, Chapter 3, pp. 32–41. Los Altos, CA: Morgan Kaufmann Publishers, Inc.

Axelrod, R. and W. Hamilton (1981). The evolution of cooperation. *Science 211*, 1390–1396.

Holland, J. (1975). *Adaptation in natural and artificial systems*. Ann Arbor, MI: The Univ. of Michigan Press.

Huberman, B. and N. Glance (1993). Evolutionary games and computer simulations. *Proceedings of the National Academy of Sciences (USA) 90*(16), 7715–7718.

Lindgren, K. (1991). Evolutionary phenomena in simple dynamics. In C. Langton, C. Taylor, J. Farmer, and S. Rasmussen (Eds.), *Artificial life II*, pp. 295–311. Redwood City, CA: Addison-Wesley.

Nowak, M. and R. May (1992). Evolutionary games and spatial chaos. *Nature 359*, 826–829.

An Alternate Interpretation of the Iterated Prisoner's Dilemma and the Evolution of Non-Mutual Cooperation

Peter J Angeline

Loral Federal Systems
1801 State Route 17C
Mail Drop 0210
Owego, NY 13827-3994
pja@owego.vnet.ibm.com

Abstract

The Iterated Prisoner's Dilemma (IPD) has been an effective model of social, biological and political interaction for nearly 40 years. In the traditional definition of the game, constraints are provided that limit the cooperation between players to *mutual cooperation*, i.e. both players deciding to "cooperate" on a single play. This paper demonstrates that by modifying the traditional constraints, successful strategies must embody the ability to coordinate their interactions over several consecutive plays, a much more complex and potentially more interesting behavior. This form of interaction is termed *non-mutual cooperation*. This paper demonstrate the evolution of non-mutually cooperative agents, represented as Finite State Machines, under two distinct payoff schemes and discusses the implications of these results.

1 Introduction

The Iterated Prisoner's Dilemma (IPD) has been used as an idealized model of social, biological and political interaction for nearly 40 years. The appeal of this game is its simplicity of statement, its universal applicability to idealized multi-agent interactions, and the apparently incoercible dilemma at its core that pits "rational" self-interest with low return against "irrational" cooperation with high return. In the traditional definition of the game, constraints are provided that restrict the manner of cooperation between players to only *mutual cooperation*, i.e. both players deciding to "cooperate" on each single play.

This paper demonstrates that the IPD can also be used to model situations where the preferred interaction between population members is not simple mutual cooperation. I use the term *non-mutual cooperation* for environments where the ideal interaction between population members must be realized over more than a single play of the game. Such situations involve a level of "trust" between the players that is unmodeled in the standard IPD. Experiments showing the evolution of non-mutual cooperation under two different payoff tables are reported. I begin with background on both the standard and iterated Prisoner's Dilemma.

2 Background

2.1 The Prisoner's Dilemma

The Prisoner's Dilemma (Luce and Raiffa 1957; Rapoport 1966) is defined as a two-player nonzero-sum, noncooperative game. The term *noncooperative* indicates that no communication occurs between the players prior to starting the game. The term *nonzero-sum* identifies the game to be such that the winnings of one player are not necessarily the losses

of the other. In such games, a holder of the resources for which the players are competing, a "banker," pays out rewards according to a predefined payoff matrix. The general form of the Prisoner's Dilemma is shown in Table 1 (after

		Player 2	
		C	D
Player 1	C	γ_1, γ_1	γ_2, γ_3
	D	γ_3, γ_2	γ_4, γ_4

Table 1. Payoff Matrix for the Prisoner's Dilemma

Rapoport 1966), where the first element of an entry in the table determines the payoff to Player 1 and the second entry determines the payoff to Player 2. As traditionally defined, the values for the individual payoffs in the Prisoner's Dilemma are subject to the following constraints:

$$\gamma_3 > \gamma_1 > \gamma_4 > \gamma_2 \qquad \textbf{(EQ 1)}$$

$$2\gamma_1 > \gamma_3 + \gamma_2 \qquad \textbf{(EQ 2)}$$

A typical assignment for these payoffs that fills the constraints is shown in Table 2 (from Axelrod 1984). The con-

		Player 2	
		C	D
Player 1	C	3, 3	0, 5
	D	5, 0	1, 1

Table 2. Example Payoffs for the Standard Prisoner's Dilemma

straint in Eq. (1) creates the condition that the "D" strategy strictly dominates, in the game theoretic sense, the "C" strategy for both players (Luce and Raiffa 1957). This forces "rational" players to consistently play (D, D), *mutual defection*. However, the play (C, C), *mutual cooperation*, supplies a higher payoff for both players. This is the essence of the dilemma, that there exists an "irrational" play that improves the payoffs of *both* players. The purpose of the constraint in Eq. (2) is unclear. Rapoport (1966) observes:

> If the [second] inequality ... is reversed another form of tacit collusion is possible, namely an alternation between CD and DC (assuming that the game is repeated). (p. 3)

Axelrod (1984) states:

> The second [constraint in] the definition of the Prisoner's Dilemma is that the players cannot get out of their dilemma by taking turns exploiting each other. This assumption means that an even chance of exploitation and being exploited is not as good an outcome for a player as mutual cooperation.(p. 10)

The second constraint has typically been assumed in subsequent modeling investigations.

In the standard characterization of the game, the selection of strategy "C" is associated with a player's desire to "cooperate" for the largest mutual payoff while the selection of the strategy "D" is associated with the desire to "defect" or go against the mutual good in hopes of obtaining a larger individual payoff (Luce and Raiffa 1957; Axelrod 1984; Hofstadter 1985). This interpretation has prompted the naming of each of the four payoff values in accordance with their perceived significance: γ_1 is the *reward for mutual cooperation*, γ_3 is the *temptation to defect*, γ_2 is the *sucker's payoff*, and γ_4 is the *punishment for mutual defection*.

2.2 Iterating the Prisoner's Dilemma

In general, a single play of the Prisoner's dilemma is not very interesting since "rational" agents will always select (D, D). Generally, the Iterated Prisoner's Dilemma (IPD) is preferred, where two players play several consecutive iterations of the game using the payoff matrix to accumulate a total score. The player with the larger cumulative score is deemed the winner. Through various investigations (Axelrod 1980a; 1980b), one strategy, dubbed Tit-For-Tat (TFT), has been identified as doing well for this game against a number of competitors. Tit-For-Tat always cooperates on its first move and then simply repeats the last play of its opponent.

2.3 Evolving Strategies for the IPD

Axelrod has performed several experiments that investigate the evolvability of certain strategies for the Iterated Prisoner's Dilemma. Axelrod (1987) noticed that the performance of a given player against eight representative strategies investigated in earlier work (Axelrod 1980a; 1980b) was a good indicator of its overall robustness. Axelrod used these eight strategies as a measure of fitness in experiments to evolve players for the IPD (Axelrod 1987) using a genetic algorithm (Holland 1975; Goldberg 1989) where each population member was represented by a table with 70 entries. Each population member played a game of 151 moves against the eight representative strategies with a player's final score computed as a weighted average over the eight games. Axelrod (1987) conducted 40 trials each for 50 generations and reports that the evolved players, while not purely TFT, contained many similarities to TFT and scored about as well.

In a second experiment, Axelrod altered the fitness calculation for the evolving strategies so that a player's fitness was its average score when played against all other members of the population (Axelrod 1987). Axelrod reports that in all 10 runs the populations initially tended to evolve away from initial cooperation toward mutual defection. After 10 or 20 generations, mutual cooperation starts to appear in the interactions of the population leading to higher scores and displacing those strategies that continuously defect.

Following Axelrod (1987), Fogel (1991) performed similar experiments using an *evolutionary program* (EP) (Fogel, Owens and Walsh 1966; Fogel 1992) with players represented as finite state machines (FSMs), an example of which is shown in Figure 1. A finite state machine contains a number of states, transitions between the states, and an output associated with each transition. On each step, the input from the en-

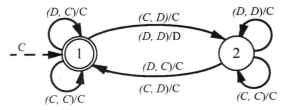

Figure 1. *Finite state machine of the type used in Fogel (1991; 1993). The initial play of the machine is indicated by the dashed line. The state transitions depend on the last move of the player and the opponent, shown as the italicized pair (P_1,P_2) on each transition followed by the current machine's next move.*

vironment selects a transition out of the current state of the FSM to a new state and the associated output value is returned. The input alphabet for Fogel's machines was {(C,C), (C,D), (D,C), (D,D)} representing all possible pairs of moves for the two players on the last iteration of the game. The output alphabet was the next move of the machine, namely {C, D}. Fogel (1991) also added an initial move for each FSM as shown in the figure. On each generation, Fogel (1991) saved the best 50 percent of the population and allowed each to produce a single offspring. An offspring was constructed from a parent FSM by copying the parent and applying a single randomly selected structural mutation. The six possible structural mutations included *adding a state, removing a state, altering a transition, altering an output value, altering the initial state* or *altering the initial play* of the machine. The number of states in the FSMs was allowed to vary between one and eight. Using the payoff matrix in Table 2, Fogel (1991) reports observations similar to Axelrod's over the four runs executed with mutual cooperation taking over the population in about 5 to 10 generations, slightly faster than Axelrod (1987).

Fogel (1993) extended these experiments using the same representation and evolutionary technique but varying other aspects of the problem. In the second experiment, Fogel (1993) increases the payoff for the temptation to defect (γ_3 in Table 1) from 5.25 to 6.0 in increments of 0.25. At 6.0 the constraint of Eq. (2) was strictly violated. Ten trials at each increment were conducted. Fogel (1993) shows that the machines that evolved when $\gamma_3 = 6.0$ were not always cooperative, but frequently evolved to alternate between defection and cooperation. He concludes that in order for cooperation to reliably evolve the payoff for defecting against cooperation (γ_3) must be as small as possible (Fogel 1993, p. 94).

3 An Alternate Interpretation of the IPD

3.1 Generalizing the Concept of Cooperation

As described above, the traditional definition and interpretation of the Prisoner's Dilemma associates the "C" move with a player's desire to "cooperate" in order to receive the highest mutual payoff while a "D" indicates the player "defects" against this goal in order to acquire a larger payoff on that round. But associating the act of cooperation strictly with the selection of the "C" strategy and defection with the "D" strategy is simplistic. The essence of the dilemma in the Prisoner's Dilemma is that "rational" agents choose to under-exploit the resource ("banker") while "irrational" agents exploit the resource more effectively and to better mutual benefit. In terms of the values in the game matrix, the dilemma indicates

	CCCC	CCCD	CCDC	CCDD	CDCC	CDCD	CDDC	CDDD	DCCC	DCCD	DCDC	DCDD	DDCC	DDCD	DDDC	DDDD
CCCC	12,12	9,15	9,15	6,18	9,15	6,18	6,18	3,21	9,15	6,18	6,18	3,21	6,18	3,21	3,21	0,24
CCCD	15,9	10,10	12,12	7,13	12,12	7,13	9,15	4,16	12,12	7,13	9,15	4,16	9,15	4,16	6,18	1,19
CCDC	15,9	12,12	10,10	7,13	12,12	9,15	7,13	4,16	12,12	9,15	7,13	4,16	9,15	6,18	4,16	1,19
CCDD	18,6	13,7	13,7	8,8	15,9	10,10	10,10	5,11	15,9	10,10	10,10	5,11	12,12	7,13	7,13	2,14
CDCC	15,9	12,12	12,12	9,15	10,10	7,13	7,13	4,16	12,12	9,15	9,15	6,18	7,13	4,16	4,16	1,19
CDCD	18,6	13,7	15,9	10,10	13,7	8,8	10,10	5,11	15,9	10,10	12,12	7,13	10,10	5,11	7,13	2,14
CDDC	18,6	15,9	13,7	10,10	13,7	10,10	8,8	5,11	15,9	12,12	10,10	7,13	10,10	7,13	5,11	2,14
CDDD	21,3	16,4	16,4	11,5	16,4	11,5	11,5	6,6	18,6	13,7	13,7	8,8	13,7	8,8	8,8	3,9
DCCC	15,9	12,12	12,12	9,15	12,12	9,15	9,15	6,18	10,10	7,13	7,13	4,16	7,13	4,16	4,16	1,19
DCCD	18,6	13,7	15,9	10,10	15,9	10,10	12,12	7,13	13,7	8,8	10,10	5,11	10,10	5,11	7,13	2,14
DCDC	18,6	15,9	13,7	10,10	15,9	12,12	10,10	7,13	13,7	10,10	8,8	5,11	10,10	7,13	5,11	2,14
DCDD	21,3	16,4	16,4	11,5	18,6	13,7	13,7	8,8	16,4	11,5	11,5	6,6	13,7	8,8	8,8	3,9
DDCC	18,6	15,9	15,9	12,12	13,7	10,10	10,10	7,13	13,7	10,10	10,10	7,13	8,8	5,11	5,11	2,14
DDCD	21,3	16,4	18,6	13,7	16,4	11,5	13,7	8,8	16,4	11,5	13,7	8,8	11,5	6,6	8,8	3,9
DDDC	21,3	18,6	16,4	13,7	16,4	13,7	11,5	8,8	16,4	13,7	11,5	8,8	11,5	8,8	6,6	3,9
DDDD	24,0	19,1	19,1	14,2	19,1	14,2	14,2	9,3	19,1	14,2	14,2	9,3	14,2	9,3	9,3	4,4

Figure 2. *Cumulative payoff matrix for Iterated Prisoner's Dilemma when four consecutive moves are scored as a unit and the value of γ_3, the temptation payoff, is equal to six. Moves which maximize the resource are highlighted in grey. The number of cooperation points, highlighted in dark grey, now total 19.*

that rational players prefer entries in the table where their summed score is *not* maximal for the game.

In a biological sense, the "banker" is the environment in which two agents compete for resources. Cooperation by the agents should maximize the exploitation of the available resources to mutual advantage. Notice that the maximal combined payoff for a single move of the Prisoner's Dilemma of Table 2 is six and only occurs when the players cooperate, highlighted in grey in the table. I refer to any Prisoner's Dilemma that observes the constraint of Eq. (2) as a *Single-Max Prisoner's Dilemma* (SMPD). Generalizing from these observations, define a strategy to be *cooperative* when there exists any opponent strategy where the *combined payoff to the players is maximal for the game and the respective payoffs to the players is equal*. This definition of cooperation retains the original motivation of maximizing the combined payoff of the game, i.e. extracting as much payoff from to "banker" as possible, without appealing to any specific strategy. For convenience, I use the term *cooperation point* to identify any pair of strategies for opponents where the combined payoff is maximal and the individual payoffs are equal.

3.2 The Multi-Max Prisoner's Dilemma

Now consider the implications of modifying the payoffs of the game so that the second constraint in the definition of the Prisoner's Dilemma is strictly violated. Consider the constraint of Eq. (2) modified to be:

$$2\gamma_1 = \gamma_3 + \gamma_2 \qquad \textbf{(EQ 3)}$$

so that the average of the temptation payoff and the sucker's payoff is equal to the reward for cooperation. An example single move payoff matrix observing this constraint using the values of Table 2 with $\gamma_3 = 6$ is shown in Table 3. Cells in the table that are grey indicate the combined payoff for the players is maximal for the game. Light grey indicates an unequal distribution while dark grey indicates a cooperation point. From the perspective of a single move, there is no difference in the game: the "D" strategy still dominates the "C" strategy. Therefore, the dilemma is still present since mutual defection

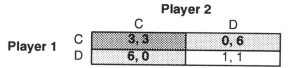

Table 3. *Example Payoffs for Multi-Max Prisoner's Dilemma*

still dominates but mutual cooperation still returns a better score for both players. However, where in the original payoff matrix there was only a single move that maximized the combined payoff, now every move other than mutual defection maximizes the total payoff from the "banker." I refer to this as the *Multi-Max Prisoner's Dilemma* (MMPD), since there are many ways to maximize the exploitation of the environment. As in the original game, only one move in the single game, mutual cooperation, is a cooperation point.

When the Multi-Max Prisoner's Dilemma is iterated, we observe a very different environment than in the traditional IPD. Figure 2 shows the cumulative payoff matrix for four iterations of the MMPD game matrix of Table 3. The payoff matrix shows a large number of moves, shown in grey, where the combined score is maximal for the game.[1] Light grey indicates the payoffs that are not equal for the players while dark grey indicates cooperation points. Notice that the number of plays that maximize the utilization of the resource in the game has risen to 80 and correspond to all composite moves in which no single move results in mutual defection.

An especially interesting facet of this cumulative playoff matrix is the appearance of a distinct method of cooperation. According to Eq. (3), two players that trade defections in an MMPD achieve the same payoff as two players that select mutual cooperation for two moves. For instance, the composite move (CDCD, DCDC) is a cooperation point but contains no single move that is mutual cooperation. In effect, the players trade defections and subsequently maximize the exploitation of the resource. Because the players avoid mutual coop-

---

1. Interestingly enough, the pattern of maximal resource exploitation in a cumulative payoff table of any length for the Multi-Max Prisoner's Dilemma is the ubiquitous Sierpinski's Triangle fractal.

	CCCC	CCCD	CCDC	CCDD	CDCC	CDCD	CDDC	CDDD	DCCC	DCCD	DCDC	DCDD	DDCC	DDCD	DDDC	DDDD
CCCC	12, 12	9, 16	9, 16	6, 20	9, 16	6, 20	6, 20	3, 24	9, 16	6, 20	6, 20	3, 24	6, 20	3, 24	3, 24	0, 28
CCCD	16, 9	10, 10	13, 13	7, 14	13, 13	7, 14	10, 17	4, 18	13, 13	7, 14	10, 17	4, 18	10, 17	4, 18	7, 21	1, 22
CCDC	16, 9	13, 13	10, 10	7, 14	13, 13	10, 17	7, 14	4, 18	13, 13	10, 17	7, 14	4, 18	10, 17	7, 21	4, 18	1, 22
CCDD	20, 6	14, 7	14, 7	8, 8	17, 10	11, 11	11, 11	5, 12	17, 10	11, 11	11, 11	5, 12	14, 14	8, 15	8, 15	2, 16
CDCC	16, 9	13, 13	13, 13	10, 17	10, 10	7, 14	7, 14	4, 18	13, 13	10, 17	10, 17	7, 21	7, 14	4, 18	4, 18	1, 22
CDCD	20, 6	14, 7	17, 10	11, 11	14, 7	8, 8	11, 11	5, 12	17, 10	11, 11	14, 14	8, 15	11, 11	5, 12	8, 15	2, 16
CDDC	20, 6	17, 10	14, 7	11, 11	14, 7	11, 11	8, 8	5, 12	17, 10	14, 14	11, 11	8, 15	11, 11	8, 15	5, 12	2, 16
CDDD	24, 3	18, 4	18, 4	12, 5	18, 4	12, 5	12, 5	6, 6	21, 7	15, 8	15, 8	9, 9	15, 8	9, 9	9, 9	3, 10
DCCC	16, 9	13, 13	13, 13	10, 17	13, 13	10, 17	10, 17	7, 21	10, 10	7, 14	7, 14	4, 18	7, 14	4, 18	4, 18	1, 22
DCCD	20, 6	14, 7	17, 10	11, 11	17, 10	11, 11	14, 14	8, 15	14, 7	8, 8	11, 11	5, 12	11, 11	5, 12	8, 15	2, 16
DCDC	20, 6	17, 10	14, 7	11, 11	17, 10	14, 14	11, 11	8, 15	14, 7	11, 11	8, 8	5, 12	11, 11	8, 15	5, 12	2, 16
DCDD	24, 3	18, 4	18, 4	12, 5	21, 7	15, 8	15, 8	9, 9	18, 4	12, 5	12, 5	6, 6	15, 8	9, 9	9, 9	3, 10
DDCC	20, 6	17, 10	17, 10	14, 14	14, 7	11, 11	11, 11	8, 15	14, 7	11, 11	11, 11	8, 15	8, 8	5, 12	5, 12	2, 16
DDCD	24, 3	18, 4	21, 7	15, 8	18, 4	12, 5	15, 8	9, 9	18, 4	12, 5	15, 8	9, 9	12, 5	6, 6	9, 9	3, 10
DDDC	24, 3	21, 7	18, 4	15, 8	18, 4	15, 8	12, 5	9, 9	18, 4	15, 8	12, 5	9, 9	12, 5	9, 9	6, 6	3, 10
DDDD	28, 0	22, 1	22, 1	16, 2	22, 1	16, 2	16, 2	10, 3	22, 1	16, 2	16, 2	10, 3	16, 2	10, 3	10, 3	4, 4

Figure 3. Extended game payoff matrix for Iterated Prisoner's Dilemma when four consecutive moves are scored as a unit and the value of γ_3, the temptation payoff, is equal to seven. Moves with maximal resource utilization highlighted in grey lie along the antisymmetric axis of the payoff matrix. Cooperation points are highlighted in dark grey.

eration but still cooperate in the sense defined in this paper, I call this *non-mutual cooperation*. Non-mutual cooperation is exactly the "tacit collusion" described by Rapoport (1966) and what Axelrod (1984, p. 10), quoted above, suggested somehow avoids the dilemma. Given that all of the cooperation points in Figure 2 lie on or above the anti-symmetric axis and are still dominated by mutual defection, the defining aspects of the dilemma persist.

The chief distinction between mutual cooperation and non-mutual cooperation is that when agents mutually cooperate their cumulative scores increment equally while in non-mutual cooperation the cumulative scores oscillate: first one player gets the temptation for defection and then the other. Thus non-mutual cooperation involves an element of *trust* between agents that non-mutually cooperate. If a player cooperates in the face of an opponent's defection, it may be with the assumption that the opponent will reciprocate latter. Higher levels of trust appear in the table as plays of the form (CCDD, DDCC) where the first player allows the second to defect twice before extracting reciprocation.

As evidenced by Figure 2, the multiple cooperation points in this iterated game represent different attractors for the population to fall into in order to maximize resource exploitation. This is probably the reason that Fogel (1993) observed a longer convergence time for this game in his experiments.

3.3 The Anti-Max Prisoner's Dilemma

Next, consider the situation when the inequality of Eq. (2) is completely reversed, namely:

$$2\gamma_1 < \gamma_3 + \gamma_2 \qquad \textbf{(EQ 4)}$$

so that the average of the temptation payoff and the sucker's payoff is larger than the reward for cooperation. An example single move payoff matrix observing this constraint using the values of Table 2 with $\gamma_3 = 7$ is shown in Table 4. Light grey indicates the combined payoffs are maximal for the game but not equal for the players. Once again, the "D" strategy still dominates the "C" strategy, thus the defining aspects of the dilemma are again preserved. I call this the *Anti-Max Prison-*

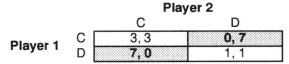

Table 4. Example Payoffs for Anti-Max Prisoner's Dilemma

er's Dilemma (AMPD) since all positions that extract the maximum payout from the "banker" lie along the anti-symmetric diagonal of the payoff matrix. Unlike the previous games, there is no cooperation point in the single move AMPD since there is no entry in the table where the combined payoff is both maximal and equal. As in MMPD, the dynamics of the iterated AMPD are distinct. Figure 3 shows the composite payoff matrix for four iterations of the AMPD with positions of the matrix providing maximal resource exploitation highlighted in grey. Light grey indicates the payoffs are not equal for the players while dark grey indicates cooperation points. In this iterated game, the total combined payoff is maximized when the two players have complementary strategies, i.e. whenever one plays a "C" the other plays a "D". The six cooperation points in the table also lie along the anti-symmetric axis of the game matrix and correspond to those moves where the agents use an equal number of defections to cooperations. Therefore, the *only* method for cooperatively extracting the maximal payout from the "banker" in an Anti-Max Prisoner's Dilemma is to use non-mutual cooperation.

4 The Evolution of Non-Mutual Cooperation

In this section, I present experiments that evolve finite state machines, similar to those of Fogel (1991; 1993), for each of the three Prisoner's Dilemmas defined above. The purpose of these experiments is to compare the ability to evolve agents that use non-mutual cooperation to the ability to evolve agents that use mutual cooperation previously demonstrated by Axelrod (1984; 1987) and Fogel (1991; 1993).

4.1 Methodology

These experiments use a similar representation and approach as in Fogel (1991; 1993) with three modifications.First, rather than selecting a single mutation at random, the number of mutations performed on a parent to create a child is given by a Poisson random distribution with a mean of three. For instance, if the Poisson random variable returns five, then five structural mutations are chosen at random, with replacement, and applied to the parent to create the offspring. The second change is that no machine smaller than three states is allowed in the population at any time. One last difference in this representation is that the input language includes only the symbols {C, D} which represent the possible last plays of the opponent.

4.2 Experiments

Three experiments were conducted. The first experiment used the SMPD payoff matrix of Table 2, the second used the MMPD payoff matrix of Table 3 and the third used the AMPD payoff matrix of Table 4. In each experiment, 10 trials of 200 generations were executed on a population size of 100. In each generation, every population member played 100 iterations of the game with each of the other 99 opponents in the population. The fitness of an individual was its average cumulative score over the 99 games. As in Fogel (1991; 1993), half of the population was replaced each generation and each parent was allowed exactly one offspring. Parents were chosen according to the competitive selection method described in Fogel (1992).

Due to space limitation, the graphs for the first experiment using the SMPD payoff matrix in Table 2 can not be shown. One important observation of this experiment is that the variance of the mean parent scores increased quickly and then decreased to a moderate level just as quickly. The results of this experiment agreed qualitatively with those reported by Axelrod (1987) and Fogel (1991; 1993).

The results of the second experiment using the Multi-Max Prisoner's Dilemma are shown in Figure 4. Figure 4a shows the mean score of the parents for each of the 10 trials. The average of the mean parent scores of Figure 4a is shown in Figure 4b. Note that the MMPD runs had an initial tendency toward strategies using mutual defection that were later displaced by more cooperative machines. The variance of the mean parent scores over the 10 trials is shown in Figure 4c. The variance of the mean parent scores increases quickly at first as the population tends towards mutual defection and subsides slowly over the next 100 generations before consistently remaining fairly low. In all, eight of the ten trials of this experiment resulted in a population that tended toward non-mutual cooperation while the populations of the other two preferred mutual cooperation. In half of the eight trials where the population tended towards non-mutual cooperation, the level of mutual cooperation was only marginally less preferred. Figure 4 also shows the number of times per generation that four consecutive moves of a game resulted in either a (CCCC, CCCC) (mutual cooperation), (CDCD, DCDC) (a popular type of non-mutual cooperation) and (DDDD, DDDD (mutual defection) for three different trials. Figure 4d shows a run where non-mutual cooperation clearly dominated in the population. Figure 4e shows a run with mutual cooperation as the dominate strategy. Figure 4f shows a trial where both mutual cooperation and non-mutual cooperation coexist at relatively equal levels. This mix of dominant strat-

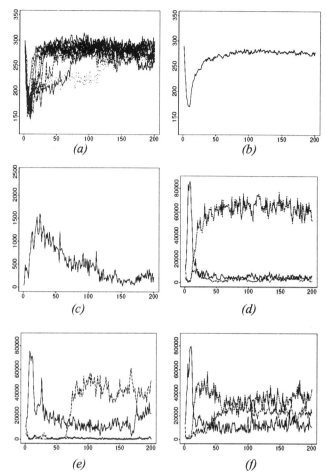

Figure 4. Results for experiment evolving strategies for Multi-Max Prisoner's Dilemma. (a) Mean scores of the parents for each run. (b) Average of the mean parent scores over all 10 runs. (c) Variance of the mean parent scores over all 10 runs. Number of hits per generation for (CCCC, CCCC) (dashed line), (DDDD, DDDD) (solid line) and (CDCD, DCDC) (dotted line) in three different trials using the Multi-Max Prisoner's Dilemma. (a) Trial #8 showing non-mutual cooperation dominant. (e) Trial #5 showing mutual cooperation dominant. (f) Trial #4 showing a mixing of mutual and non-mutual cooperation.

egies in both the individual trials and a single population is to be expected given that both mutual and non-mutual cooperation can equally maximize the exploitation of the resource (see Figure 2).

The results of the third experiment, using the Anti-Max Prisoner's Dilemma from Table 4 as the payoff function, is shown in Figure 5. The ordering of the graphs in Figure 5 is identical to Figure 4. As with the other experiments, it is obvious from the graphs that the trials again had an initial tendency toward strategies using mutual defection. Since the payoff of the cooperation points in this game average 3.5 per move, the average of the mean parent scores (Figure 5b) often rises above 300, although more slowly than in the previous experiment. The variance of the mean parent scores over the 10 trials, shown in Figure 5c, illustrates that the populations found it difficult to consistently achieve homogeneity in this experiment.

358

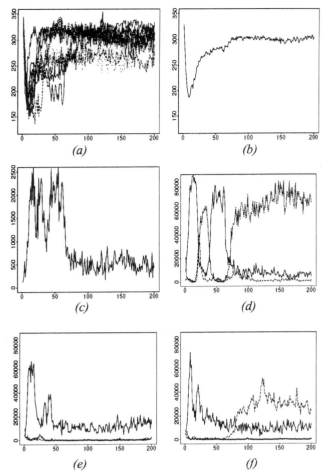

Figure 5. *Results for experiment evolving strategies for Anti-Max Prisoner's Dilemma. (a) Mean scores of the parents for each run. (b) Average of the mean parent scores over all 10 runs. (c) Variance of the mean parent scores over all 10 runs. Number of hits per generation for (CCCC, CCCC) (dashed line), (DDDD, DDDD) (solid line) and (CDCD, DCDC) (dotted line) in three different trials using the Anti-Max Prisoner's Dilemma. (d) Trial #1 showing oscillation between mutual defection and non-mutual-cooperation. (e) Trial #3 showing mutual-defection dominant. (f) Trial #4 showing mutual cooperation dominant.*

It is apparent from the graph of the mean parent scores for the individual trials (Figure 5a) that not all the trials resulted in non-mutually cooperative populations that could maximize the exploitation of the resource. In all, eight of the ten trials resulted in populations where non-mutual cooperation dominated play. The other two trials resulted in less exploitative strategies dominating the play of the population. Figure 5 also shows the number of times per generation consecutive moves of a game resulted in either a (CCCC, CCCC), (CDCD, DCDC), or (DDDD, DDDD) for three different trials. Figure 5d shows a trial where non-mutual cooperation clearly dominates in the population, although it was displaced for a short time by mutual defection. Figure 5e shows a trial where mutual defection dominated both the maximally exploitative non-mutual cooperation and the less exploitative but better mutual cooperation. In Figure 5f, mutual cooperation is the dominant strategy.

5 Conclusions

It is clear from the experiments above that non-mutual cooperation is evolvable in both the Multi-Max Prisoner's Dilemma and the Anti-Max Prisoner's Dilemma. The fact that the variance of the experiments using MMPD and AMPD subsided more slowly than the experiment using SMPD indicates that the evolutionary environment associated with these games is more complex. The reason is most likely the presence of multiple distinct cooperative strategies in the cumulative game table. The two trials which did not evolve non-mutual cooperation in the AMPD experiment indicate more complex dynamics than the other two experiments. What caused these populations to be less exploitative of their environment is not yet understood.

Invariably, when non-mutual cooperation appeared in a trial, the most used form was that of (CDCD, DCDC) which represents the minimal length non-mutually cooperative interaction. I feel that the reason this strategy was preferred is related to the issue of trust raised earlier. A question for future research is to determine if longer chains of non-mutual cooperation, i.e. those where reciprocation between players is delayed over a large number of individual moves, can be supported in the population by manipulating the amount of trust between the agents.

Acknowledgments

Thanks are due to David Fogel, Greg Saunders, Lou Coglianese and three anonymous reviewers for comments and encouragement on this work.

References

Axelrod, R. (1980a). Effective choice in the prisoner's dilemma. *Journal of Conflict Resolution*, **24**, pp. 3-25.

Axelrod, R. (1980b) More effective choice in the prisoner's dilemma. *Journal of Conflict Resolution*, **24**, pp. 379-403.

Axelrod, R. (1984). *The Evolution of Cooperation*. New York: Basic Books.

Axelrod, R. (1987). Evolution of strategies in the iterated prisoner's dilemma. In *Genetic Algorithms and Simulated Annealing*, L. Davis editor, Morgan Kaufman Publishers, Inc., pp. 32-41.

Fogel, D. B. (1991) The evolution of intelligent decision making in gaming. *Cybernetics and Systems*, **22**, pp. 223-236.

Fogel, D. B. (1992). *Evolving artificial intelligence*. Doctoral Dissertation, University of California at San Diego.

Fogel, D. B. (1993). Evolving behaviors in the iterated prisoner's dilemma. *Evolutionary Computation*, **1** (1), pp 77-97.

Fogel, L., A. Owens and M. Walsh (1966). *Artificial intelligence through simulated evolution*. New York: John Wiley & Sons.

Goldberg, D. (1989a). *Genetic algorithms in search, optimization, and machine learning*. Reading, MA: Addison-Wesley Publishing Company, Inc.

Holland, J. H.(1975). *Adaptation in natural and artificial systems*, Ann Arbor, MI: The University of Michigan Press.

Hofstadter, D. R. (1985). *Metamagical themas: Questing for the essence of mind and pattern*. New York: Basic Books.

Luce, R. D. and Raiffa, H. (1957). *Games and Decisions*, New York: John Wiley.

Rapoport, A. (1966). Optimal policies for the prisoner's dilemma. Technical Report No. 50, The Psychometric Laboratory, University of North Carolina, NIH Grant MH-10006.

Asymmetric Mutations Due to Semiconservative DNA Replication: Double-stranded DNA Type Genetic Algorithms

Hirofumi Doi[1], Ken-nosuke Wada[2], and Mitsuru Furusawa[3]

[1]Biological Informatics Section, Institute for Social Information Science, Fujitsu Labs. Ltd.
9-3 Nakase 1-chome, Mihama-ku, Chiba 261, Japan
e-mail: doi@iias.flab.fujitsu.co.jp
[2]Evolutionary System Dept., ATR Human Information Processing Research Labs. Ltd.
2-2 Hikaridai, Seika-cho, Soraku-gun, Kyoto 619-02, Japan
e-mail: kwada@hip.atr.co.jp
[3]Molecular Biology Research Laboratory, Daiichi Pharmaceutical Co., Ltd.
16-13 Kitakasai 1-chome, Edogawa-ku, Tokyo 134, Japan

Abstract

Semiconservative replication of double-stranded DNA (ds-DNA) is an asymmetric process; there is a leading and a lagging strand. This situation provides two possibilities for the occurrence of mutations: symmetric and asymmetric mutations between the two strands. We simulated symmetric and asymmetric mutations from ds-DNA type genetic information using genetic algorithms to resolve the knapsack problem. Our results strongly suggest that disparity in mutations caused by the asymmetric machinery of DNA replication effectively resolve the problem. Furthermore, the ds-DNA type GAs dramatically showed differences in features between haploid and diploid organisms.

1. Introduction

Genetic information in DNA is backed with another DNA strand; thus making double-stranded DNA (ds-DNA). In contrast, genomic RNA is only a single strand. For example, a particle of human immunodeficiency virus (HIV) has two single-strands of 10^4 bp-long genomic RNA. After HIV enters a cell, each RNA strand is replaced with ds-DNA by reverse transcription, and then integrated into the host chromosome ds-DNA. During reverse transcription, the error frequency is about 10^{-4} per site; i.e. a nucleotide base in the HIV genome might be substituted for another one. HIV therefore carries the risk that all its genetic information will be rewritten through 10^4 rounds of reverse transcription. However, the genetic information of HIV integrated into the chromosome remains as a parental master manuscript (provirus). Since the error frequency of DNA replication is much lower at about $10^{-9} \sim 10^{-12}$ / site (Drake, 1969), the probability of base change in the provirus region on the chromosome is almost zero. Therefore, HIV adapts itself to the various host target cells or the host immune system by making variants through highly frequent mutations during reverse transcription, but it conserves the parental infectious genetic information in the integrated provirus.

The parental ds-DNA, however, no longer exists after semiconservative DNA replication. That is, one strand of the parental ds-DNA becomes a template for the newly synthesized ds-DNA, and the other strand is also a template for the other daughter ds-DNA. Furthermore, since the error frequency of DNA replication is $10^{-9} \sim 10^{-12}$ / site / DNA replication (Drake, 1969), any organism which has its genetic information in long ds-DNA has a risk for mutation. For example, humans have a 3×10^9 bp size genome. If the mutation rate in human cells was 3.33×10^{-10} / site / DNA replication, one nucleotide base in the genome is substituted for another one during each replication cycle. This means that an egg or sperm carries mutations, the number of which is roughly the same as the number of divisions which the germ cells underwent from the fertilized egg. In other words, although the genome should be inherited without mistake, any organism having DNA-type genetic information carries the risk of no original genetic information, because of mutations during semiconservative replication. But, the risk is a driving force of evolution. The organism is, therefore, in a dilemma between inheritance of genetic information without mistake and evolution caused by mutations.

How does the organism solve this dilemma? We have proposed that the solution exists in sexual reproduction and in the asymmetric machinery of semiconservative DNA replication: "Disparity theory of evolution" (the concepts were proposed by Furusawa and advanced with mathematical ways by Doi; Furusawa and Doi, 1992). We have also proposed two kinds of ds-DNA type genetic algorithms (GA) to test our hypothesis. One is the parity GA, having symmetric mutations between a leading and a lagging strand, and the other is the disparity GA based on our idea (Wada and Doi et al., 1993). In this paper, we will introduce our concepts to resolve the dilemma, and our ds-DNA type GAs. We wills also show that the disparity GA

360

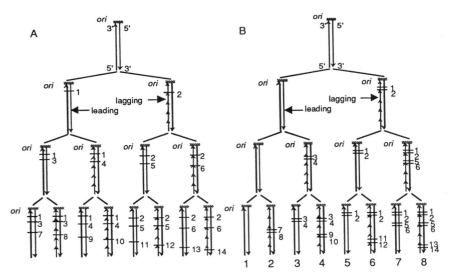

Figure 1. The distribution of mutations according to the deterministic models which have linear ds-DNA with a single ori at one end. A, the parity model. The error rate in the leading and lagging strands is 1:1. B, the disparity model. The error rate in the leading and lagging strands is 0:2. The broader long arrow indicates a parent DNA strand. The bars crossing DNA strands are misbases. Each number by the bars indicates a base substitution at a different site. The numbers at the bottom indicate individual gametes.

overwhelms the parity one, and that the parity model is allowed for haploid organisms, whereas diploid organisms must be a disparity GA type.

2. Semiconservative DNA replication and asymmetric mutations

Semiconservative DNA replication starts at the origin of replication (*ori*); that is, a ds-DNA molecule is opened at the origin and becomes Y-shaped (see Fig. 3). The replication is macroscopically uni-directional to the root of the "Y". However, ds-DNA is an anti-parallel molecule, and one strand of the daughter ds-DNA molecule is newly polymerized in the 5'-to-3' direction by copying its opposite-directional strand of the parental ds-DNA molecule (see Fig. 3). Therefore, one strand (leading strand) is polymerized continuously along the macroscopic direction of replication by a DNA polymerase, and the other strand (lagging strand) is synthesized discontinuously in the opposite direction using a completely different and more complex system (Kunkel, 1992).

This asymmetric process of semiconservative DNA replication provides two considerable opportunities for occurring mutations: one is that there is a difference in frequency of strand-specific base-misreading between leading and lagging strands ("disparity model"), and the other is that there is no statistical difference in frequency of base-misreading between both strands ("parity model") (Furusawa and Doi, 1992).

The predicted difference in occurrence of mutations between both models dramatically alters the accumulation of mutations in their descendants. Fig. 1 shows the difference in accumulation of mutations under extreme deterministic conditions. In the disparity model, no error occurs in the leading strand and two errors deterministically occur in the lagging strand in one cell cycle in a genome which consists of linear ds-DNA with

a single *ori* (origin of replication) at one end and every misbase is fixed as a point mutation without exception. As shown in Fig. 1B the disparity model disproportionately accumulates mutations among individuals, with the consequence that a concentrated accumulation of mutations takes place in some individuals and those with less mutations, including zero mutations, appear (e.g., eight descendants in the F3 generation in Fig. 1B). Needless to say, the individuals with zero mutations retains the original genetic information. In the control parity model, it is assumed that one misreading takes place in both strands, respectively, in one cell cycle and point mutations are fixed as well. In this parity deterministic model, three mutations are clearly shared in all eight descendants in F3 (Fig. 1A).

Even if mutations occur stochastically, the difference in accumulation of mutations is similar to that between the deterministic models. The stochastic model organisms have the same genomes as in the deterministic ones, and the model stochastically causes a total of two fixed base changes on the lagging and leading strands in one cell cycle. In the parity model, one base misreading takes place as expected on each strand in one cell cycle. In the disparity model, the newly synthesizing lagging strand has P base changes and the leading strand has Q base changes in number during one round of DNA synthesis where $P > Q$ and $P + Q = 2.0$. Fig. 2 shows the results of twelve simulation trials of the parity and disparity stochastic models in the 10th generation (when $P = 1.99$ and $Q = 0.01$). In the parity stochastic model, almost all 10th generation descendants in each simulation trial has between 2 and 20 mutations (Fig. 2A). In contrast, the disparity stochastic model has more disproportionately accumulated mutations among individuals (Fig. 2B). Individuals frequently exist which have no mutation and the largest number of mutations introduced in an individual increases to 24 in the 10th

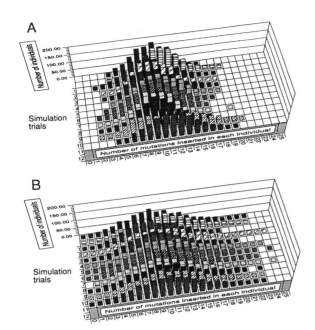

A

B

Figure 2. The distribution of individuals with a given number of mutations in the tenth generation for the parity (A) and disparity (B) stochastic models. In the disparity stochastic model, p = 1.99 and Q = 0.01.

generation (simulation trials 4, 7 and 9 in Fig. 2B).

After many simulations, we found that disparity of 10^2~10^3 in mutation rates guaranteed the existence of the original genetic information (Furusawa and Doi, 1992).

In a conserved environment, the disparity descendants with less mutations can survive, but survival of the parity descendants is very risky, because of the mutations accumulated in every parity individual. As the environment changes, the disparity descendants, which have much more mutations than the parity descendants, might adapt themselves more quickly to the new environment. Therefore, disparity organisms might be at an advantage relative to parity organisms, regardless of changing or constant environment, and the disparity in fidelity during semiconservative DNA replication is one of the solutions to the dilemma posed in the Introduction.

3. Sexuality and asymmetric mutations

If we suppose that the DNAs in Fig. 1 are the haploid genome of a gamete and that gamete-1 mates with gamete-8 in the F3 generation, the resulting diploid individual will show no phenotypic change since the accumulated mutations in the chromosome from gamete-8 will be compensated by the intact one. Thus, we may generally deduce that, in a population of the current disparity model organisms which have sexuality, the chromosomes with accumulated mutations tend to be effectively spread throughout the population without

accompanying phenotypic changes. This tendency will bring about a decrease in the probability of death of individuals who have accumulated mutations. Some of these chromosomes may, however, become more functional by chance. If the newly created phenotype is an advantage to the species, the new chromosome would likely be used instead of the old counterpart. The natural selection force would contribute to survival of individuals having the new advantageous phenotype (Furusawa and Doi, 1992). Therefore, the sexuality of disparity organisms might more effectively solve the dilemma.

4. Disparity and parity double-stranded DNA type Genetic Algorithms

We introduce here two kinds of ds-DNA type genetic algorithms, the disparity GA, which is based on our concepts, and the parity GA, and show their ability to resolve the knapsack problem, with respect to adaptability, competition among the two GAs, and extermination of individuals. Our genetic algorithms mimic the semiconservative replication manner of DNA, and point mutations are inserted by base-misreading when the genome replicates (Wada and Doi et al., 1993).

4.1. Disparity and parity GAs resolving knapsack problem

The current knapsack problem is defined as follows: There are 100 kinds of objects with weights (20 to 100) and values (20 to 100) determined at random, and the objective is to maximize the total value of objects placed in a knapsack, subject to a loading weight limitation of 1000, and taking a plural number of the same object is not allowed.

Two types of GA were constructed for resolving the problem. At the beginning, 500 parent individuals who have a knapsack survive. An individual carries genetic information by virtue of its strategy of placing objects in its chromosome. The chromosomal information consists

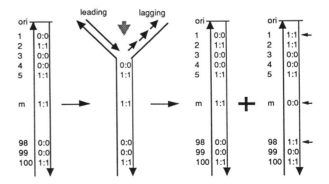

Figure 3. The chromosome used in the ds-DNA type GA. The short arrow in the most right shows the base substitution. The gray arrow between a leading and a lagging strand means the macroscopic direction of replication.

of 1 and 0 and is a double-stranded string with the length of 100 "bases" (Fig. 3). It replicates like ds-DNA with a single origin of replication (*ori*) at the one end. The pairing rule of the double-stranded string is 1:1 and 0:0. A "1" at the *m*th position in the string means that the individual takes the *m*th object and "0" means it does not take the object. The genetic information of the initial 500 individuals is randomly determined, and each individual has a haploid or diploid genome. (In the diploid individual, 1's dominate 0's, although we got similar results when 0's dominate 1's.) They make their offspring for the next generation asexually or sexually. In asexual reproduction, the individual divides itself into two. In sexual reproduction, it performs meiosis producing four gametes, and two of them unite with a gamete from others. Consequently, 500 individuals initially live and 1000 offspring are asexually produced or 500 offspring sexually.

During division or meiosis, the genetic information is changed stochastically according to given mutation rates. In the disparity model individuals, the mutation rate of the leading strand is 0.001 base / chromosome, and that of the lagging strand is n bases / chromosome ($n \geq 0.1$). In the parity model individuals, the mutation rates of the leading and the lagging strands are both ($n/2$) bases / chromosome. The mutation rate in the below means the total of the mutation rates of both strands ignoring that of the leading strand of the disparity model individuals; that is, n bases / both strands.

The 1000 or 500 individuals in the next generation again choose objects to place into their knapsack. 500 parents are stochastically selected from them to produce offspring in proportion to their total values of placed objects (the so called biased roulette wheel selection), and the individuals whose loading weights are over the limitation or whose total values are zero, are not selected. Thus, the total values represent scores of fitness from a biological view.

The offspring then perform the same trial. The trials were carried out up to the 4000th generation or until extermination. The GA is composed of a single class of individuals, e.g. parity, haploid and asexual individuals, and in the case of the competition study, the GA is mixed with two classes of individuals having two opposing properties such as parity and disparity.

4.2. Haploid and asexual ds-DNA type GA and competition between the two models

We simulated the two types of GA with the loading weight limitation of 1000, initially composed of 500 haploid and asexual individuals, under different conditions such as parity or disparity, and high or low mutation rates ($n = 8.0$ or 0.1). Fig. 4 shows the results of simulation up to the 300th generation. Among the tested models, the disparity model individuals with $n = 8.0$ (open rectangle line in Fig. 4) adapted fastest to the knapsack problem and increased fitness scores step-by-step. After about the 1500th generation, the parity and the disparity individuals with the low mutation rate achieved higher scores than the disparity individuals with the high mutation rate. In particular, the disparity model with the low mutation rate got the highest scores up to the 4000th generation (data not shown).

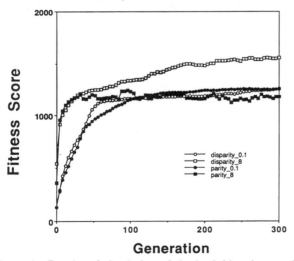

Figure 4. Results of simulation of the haploid and asexual ds-DNA GA up to the 300th generation. Open circle, the disparity model with a mutation rate $n = 0.1$; open rectangle, the disparity model with $n = 8$; closed circle, the parity model with $n = 0.1$; the parity model with $n = 8$.

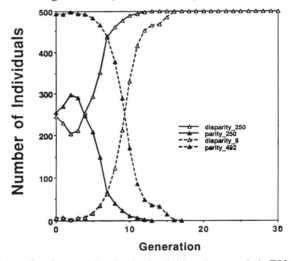

Figure 5. Competition in the haploid and asexual ds-DNA GA with a mutation rate $n = 8$ between the disparity and parity models. Solid lines, 250 parity and 250 disparity individuals initially survive; dotted lines, the disparity individuals were initially eight of the original 500. Open triangle, disparity individuals; closed triangle, parity individuals.

The parity model individuals with $n = 8.0$ (closed rectangle line in Fig. 4) could not increase scores after about the 100th generation, with maximum scores of about 1200.

As shown in Fig. 4, when $n = 8.0$, the disparity model individuals clearly achieved much higher fitness scores than the parity individuals. As expected from this, in the GA with a mixed population of disparity and parity model individuals (250 parity and 250 disparity individuals initially survive) and a high mutation rate of $n = 8.0$, the disparity individuals won a sweeping victory in the competition and the parity individuals disappeared soon by the 13th generation (Fig. 5, solid lines). Even if the disparity individuals were initially only eight of the original 500 individuals, they soon exterminated the parity individuals (Fig. 5, dotted lines).

4.3. Diploid and asexual ds-DNA type GA

In the diploid and asexual GA, the disparity individuals having a high mutation rate ($n = 8.0$) adapted faster at the beginning than the low-mutation-rate disparity individuals ($n = 0.1$). The low-mutation-rate disparity did finally obtain higher scores than the high-mutation-rate disparity, the same as the case in the haploid and asexual world (data not shown). However, the parity individuals with a mutation rate of $n = 2.0$ remarkably disappeared after 250 generations (data not shown). In particular, the higher the mutation rate ($n \geq 4.0$) was, the faster the parity individuals disappeared (Fig. 6, closed triangle line ($n = 4.0$) and closed rectangle line ($n = 8.0$)).

The reason for their extermination is that their genetic information is changed frequently into a strategy of loading objects which exceeds the weight limits due to high rate mutations in the both leading and lagging strands. Their variations then decrease to zero by the biased roulette wheel selection (Fig. 7, solid triangle line ($n = 4.0$)). In contrast, the genetic information of the surviving disparity individuals is guaranteed by the high fidelity replication of the leading strand (Fig. 7, open rectangle line ($n = 4.0$)). The low-mutation-rate parity individuals ($n = 0.1$) did not die out, but could not get higher scores than the disparity with the same low mutation rate throughout the generations (data not shown).

4.4. Diploid and sexual ds-DNA type GA

In the diploid and disparity GA, the sexual individuals, as expected, adapted faster than the asexual ones regardless of crossover at a mutation rate of $n = 0.1$ (Fig. 8). However, the results at $n = 8.0$ were different from those at $n = 0.1$: the sexual individuals without crossover had lower scores over almost all the generations relative to the asexual ones; but the sexual individuals with crossover (frequency = 0.2) obtained higher scores than the asexual ones (data not shown). This suggests that an optimal mutation rate exists for evolution of sexual and disparity organisms.

In the sexual and parity GA, when $n \geq 2.4$, the sexual individuals disappeared in the early generation as did the asexual ones (data not shown).

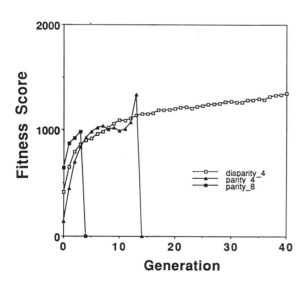

Figure 6. The results of simulation of the diploid and asexual ds-DNA GA up to the 40th generation. Open rectangle, the disparity model with a mutation rate $n = 4.0$; closed triangle, the parity model with $n = 4.0$; closed rectangle, the parity model with $n = 8.0$.

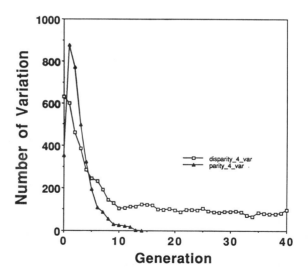

Figure 7. Variation of genetic information up to the 40th generation. Open rectangle, variation of genetic information of the disparity individuals with a mutation rate $n = 4.0$; closed triangle, variation of genetic information of the parity individuals with $n = 4.0$.

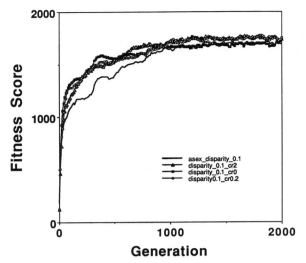

Figure 8. The results of simulations of the disparity, diploid and sexual ds-DNA GA with a mutation rate $n = 0.1$, up to the 2000th generation. Solid line, the disparity, diploid and asexual individuals; triangle, the disparity, diploid and sexual individuals with crossover (the frequency = 2.0); open circle, the disparity, diploid and sexual individuals with crossover (the frequency = 0.2); rectangle, the disparity, diploid and sexual individuals without crossover.

5. Discussion and Conclusions

We have simulated two models of ds-DNA type GA under various conditions of mutation rate, population size, loading weight limitation, sexuality, ploidy and crossover. Table 1 summarizes the results. We obtained basically consistent results in that the scores of the disparity model overcame those of the parity model, in particular those having a diploid genome, during increasing adaptability. The result of competition study (mutation rate $n = 8.0$) shown in Fig. 5 was the most striking. The parity and the disparity models showed similar independent fitness curves up to the 40th generation (Fig. 4), but a quick extermination occurred at the 13th generation in the parity individuals when the two models were in competition (Fig. 5). Generally we can say that the disparity individuals overwhelm the parity ones in a survival race, especially when the population is smaller and the mutation rate is higher.

Table 1. Summary of simulations by ds-DNA type GAs

Advantage for disparity	Survival condition for parity
Small population	Large population
Strong selection pressure	Weak selection pressure
High mutation rate	Low mutation rate
Sexuality and diploid	Asexuality and haploid
Competitive world	Noncompetitive world

Goldberg and Smith (1987) have simulated with GA that a diploid organism with single-stranded genetic information adapts quickly to a change in environment. Considering the feature of the disparity model mentioned in Section 2, a disparity GA also will quickly adapt to a change in environment. As expected, the disparity GA adapts more quickly than the parity GA (unpublished data). We have shown that a disparity of $10^2 \sim 10^3$ in mutation rates guarantees the existence of the original genetic information (Furusawa and Doi, 1992). However, the original information no longer has an advantage compared with the mutated information, when there is a change in environment. Recent biological experiments have shown that the disparity in mutation rate is 3~4 to 20~30 in living cells (Veaute and Fuchs, 1993) or in extract from cells (Roberts et al., 1994). Using a disparity lower than $10^2 \sim 10^3$, we can show that the disparity GA adapts quickly to the change in environment (unpublished data).

Finally, the present results strongly suggest that living organisms have a solution to the dilemma posed in the Introduction: the disparity in replication fidelity between the leading and lagging strands. In other words, during the long history of evolution, the organisms obtained a great advantage for evolution once they had acquired an asymmetric replication-machinery with a distinct difference in fidelity between the DNA strands.

Acknowledgments

The computer software of ds-DNA type GAs to resolve the knapsack problem was made by K. Wada and S. Tanaka, and the simulation of the ds-DNA type GAs was mainly performed by H. Doi. We would like to thank Drs. K. Abe, M. S. H. Ko, T. Iino, Y. Saeki, and Y. Kikuchi for discussions and S. Tanaka and Y. Wada for computer technique, and Dr. D. Stearns-Kurosawa for critical reading of the manuscript and for useful suggestions.

References

Drake, J. W. 1969. *Nature* **221**, 1132.
Furusawa, M. and Doi, H. 1992. *J. theor. Biol.* **157**, 127-133.
Goldberg, D. E. and Smith, R. E. 1987. *Proc. Second International Conference on Genetic Algorithms*, 59-68.
Kunkel, T. A. 1992. *BioEssay* **14**, 303-308.
Roberts, J. D., Izuta, S, Thomas, D. C., and Kunkel, T. A. 1994. *J. Bio. Chem.* **269**, 1711-1717.
Veaute, X., and Fuchs, R. P. P. 1993. *Science* **261**, 598-600.
Wada, K., Doi, H., Tanaka, S., Wada, Y., and Furusawa, M. 1993. *Proc. Natl. Acad. Sci. USA.* **90**, 11934-11938.

EMBRYOLOGICAL DEVELOPMENT ON SILICON

P. Marchal[1], C. Piguet[1], D. Mange[2], A. Stauffer[2], S. Durand[2]

[1] CSEM Centre Suisse d'Electronique de Microtechnique SA
CH-2007 Neuchâtel
e-mail: <name>@csemne.ch

[2] Logic Systems Laboratory, Swiss Federal Institute of Technology,
CH-1015 Lausanne
e-mail:<name>@di.epfl.ch

Abstract

This paper addresses a tentative hardware implementation of Artificial Life. The embryological development of a simple "creature" will be described step by step. This "creature" - a 4-state reversible counter - has been chosen simple enough to be easily understood by any reader. Nevertheless this methodology, based on recent advances in the field of computer engineering (circuit synthesis, programmable devices and artificial life) as well as in the related field of molecular biology (genetics and embryology), is sufficiently powerful to address the implementation of any system (combinational or sequential). This architecture provides some life-like properties, such as self-adaptation and artificial healing, emerging from these genome-based computing systems and leading to a kind of functional autonomy particularly interesting for embarked systems where access is difficult (aggressive environment) even impossible (sea or space).

1.- Introduction

The notion of a machine reproducing itself has great intrinsic interest and invariably elicits a considerable range of responses [1]. According to Kemeny [2], "If [by 'reproduction'] we mean the creation of an object like the original out of nothing, then no machine can reproduce - but neither can a human being". The mathematician John von Neumann first seriously came to grips with the problem of machine reproduction [3]. After his work on the organisation employed today in almost all general purpose computational machines - the so-called "von Neumann machine" - was completed, he began thinking seriously about the problems of extremely large machines - design, programming and reliability - and he became involved with the many possible analogies to the complex behavior of living systems. In his quest for reliability, von Neumann has been investigating self-reproduction. The mathematician S. Ulam suggested to von Neumann that the notion of self-reproducing machine would be amenable to rigorous treatment if it could be described in a "cell-space" format - a geometrical grid or tessellation, regular in all dimensions. Within each cell of this system resides a finite state automaton. This cell automaton can only be affected by some of its neighbors, and only in very specific ways. In the model von Neumann finally conceived, a checkerboard system is employed with an identical finite state automaton in each square; each automaton can communicate with its four cardinal direction neighbors. In the first reproducing system the cell automata can be in one of the 29 possible different states. A few thousands of cells were necessary to demonstrate the self-reproduction. Codd [4] recapitulated von Neumann's results in a simpler cell space requiring only 8 states.

In order to construct self-reproducing automata simpler than those of von Neumann and Codd, Langton [5] followed by Byl [6] adopted more liberal criteria. They dropped the condition that the self-reproducing unit must be capable of universal construction. In the same spirit, but with a completely different methodology, H. C. Morris [7] used the concept of typogenetics first introduced by Hofstadter [8] to reproduce in a 1-dimension environment strings of characters analogous to those of DNA (A, C, G, T). Unfortunately, in both environments (Langton's 2-dimension and Morris' 1-dimension), the initial configurations "do nothing but propagate" [7].

The final goal of our research is to get more effective artificially living "creatures", i.e., creatures able to reproduce and specialize, but also enabling them to "behave" in some sense, that is to say "creatures" able to realize a desired function. The embryological development sequence of a "creature" taken as example will be described step by step. Our approach uses the basic concepts embedded in the multi-cellular living organisms: one cell (zigote) reproduces itself by duplicating the description of the full organism located in its genome. Cell reproduction is carried out until the full organism is populated with cells. The specialization phase, in which each cell interpret one piece of the genome (its own gene) depending on its location in the organism, takes place all along the reproduction phase. Three levels of abstraction will be used to gradually set up the building blocks of our artificial organism: the logical level, the hardware level and the biological level, with some emphasis between hardware and biology.

2.- Logical Level

The logical level is divided into four steps : the function description and its representation, the logic synthesis and its optimization. A very simple digital system will be used to illustrate the different steps of our methodology. Let us consider a classical sequential circuit : the 4-state reversible counter. Depending on the value of X, zero or one, the counter is counting up (from 00 to 11) or down (from 11 to 00). This sequential system can be expressed either by a state graph or by a Karnaugh map (figure 1).

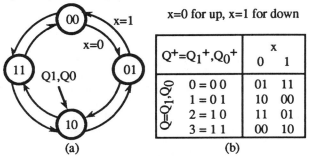

x=0 for up, x=1 for down

$Q^+=Q_1^+,Q_0^+$	x 0	1
$Q=Q_1,Q_0$ 0 = 0 0	01	11
1 = 0 1	10	00
2 = 1 0	11	01
3 = 1 1	00	10

(a)   (b)

Figure 1.- 4-state reversible counter.
(a) state graph (b) table.

It is well known that any combinational or sequential system can be described by one or several binary decision diagrams (BDDs) [9], [10], [11], [12]. For instance, let us consider the next state of the bit Q1, noted $Q1^+$, of the reversible counter. Figure 2 depicts how to obtain the binary decision tree (BDT) representation from the truth table expression presented in figure 1b. Figure 2a depicts the truth table. We easily remark that the value of the least significant variable (Q0) changes at every step, that the middle one (Q1) changes every second step and the most significant variable (x) changes every four steps. This remark leads to the simplified truth table of figure 2b. Figure 2c is obtained by turning figure 2b through a right angle clockwise. This figure 2c is equivalent to the binary decision tree (BDT) of figure 3a.

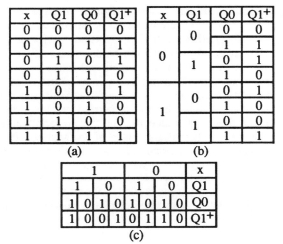

(a)

(b)

1		0		x
1	0	1	0	Q1
1 0	1 0	1 0	1 0	Q0
1 0	0 1	0 1	1 0	Q1+

(c)

Figure 2.- Evaluation of $Q1^+$, next state if Q1.
(a) truth table (b) simplified truth table (c) rotated truth table

The binary decision diagram BDD is obtained from the BDT description by regrouping identical subtrees (like subroutines in algorithms). Figure 3b depicts one of the possible BDDs describing $Q1^+$. As a matter of fact, the control variables located on the same level can be permuted with those located on any other levels in the BDD representation without any impact on the result. For instance, in the BDD's description of $Q1^+$ depicted in figure 4b, Q0 variable comes first, followed by x and then Q1.

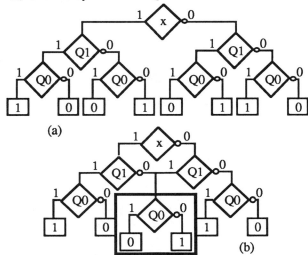

Figure 3- Binary decision diagram of $Q1^+$
(a) BDT representation (b) BDD representation.

Coming back to the reversible counter, as shown in figure 4b, two binary decision diagrams are necessary to compute the functions $Q1^+$ and $Q0^+$ which determine the next state of the 2 bits of the counter. As $Q0^+$'s BDD is composed of a simple test tile, it has been superimposed on an unused test tile of $Q1^+$'s BDD. In order to align the Q0 variable of both BDDs, the order of the control variable has been modified with respect to the one presented in figure 3b. The set of minimal BDDs for a given logic system (combinational or sequential) can be obtained by systematic methods [11], [13], [14] or by graphical methods using Karnaugh maps [12]. This range of methods will be used to synthetize any logical system (combinational or sequential).

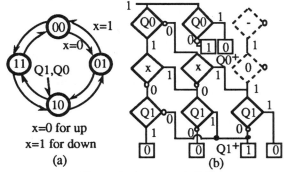

x=0 for up
x=1 for down

(a)

(b)

Figure 4.- 4-state reversible counter.
(a) state graph (b) BDD representation

3.- Hardware Level

The hardware level includes the definition of a DNA-like system on silicon and the derivation (description) of the genogram (genome programme). Before entering the very fine details of the implementation, let us come back on the technology that will enable the implementation of programmable devices: Field Programmable Gate Arrays.

Field Programmable Gate Array (FPGA) integrated circuits constitute the new hardware technology of digital machines, characterised by the regular structure of a gate array (2 dimensional array of identical cells) to which it adds programmable capabilities realized by fields rather than by mask as in conventional gate arrays [13]. Like traditional gate arrays, FPGAs implement thousands of logic gates in multilevel structures. The three major key parameters from users' point of view are that: 1) their programming time can be measured in minutes rather than weeks/months for conventional GAs, 2) their field programmability suppresses the custom masking tooling necessary for conventional GAs, thus saving some thousands of dollars, 3) in-house programming of such devices restricts confidentiality constraints to be only internal, this is of prime importance for advanced prototypes.

FPGAs architectures usually come in two types: One Time Programmable (OTP) devices are based on antifuse technology while fully reprogrammable devices embed RAM memories to store the field programme.

In order to optimize silicon area, a dedicated FPGA architecture has been designed. It has been designed first to be reprogrammable (RAM) and secondly to enable the direct implementation of any BDD. In fact the FPGA cell has been derived from the BDD test tile. Figure 5 depicts the double DMUX architecture which enables to directly implement BDD. By letting CV2 be equal to 1, sub-trees can be merged together.

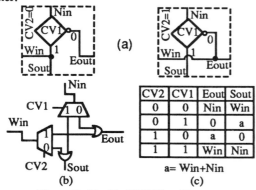

Figure 5.- Double DMUX architecture.
(a) logic schemes (b) DMUX implementation (c) truth table.

The VLSI implementation of a BDD can be realized either with multiplexers (MUX), demultiplexers (DMUX) or by mixing both techniques [15]. For sake of simplicity, we have chosen a mixed version (3 MUX and 2 DMUX), as depicted in figure 6. This building block is based on a one-variable DMUX and is controlled by 10-bit field-programme (shaded in ▨).

This cell is composed of: 1/ a functional part (shaded in ▨), 2/ a connection part (shaded in ⬚).

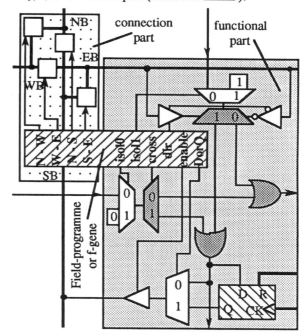

Figure 6- Functional description of the cell.

The functional part is divided into 3 blocks:

• the test tile implementing the BDD test; "dir" selects for direct or inverse control variable, "cross" enables the BDD to be optimised by crossing the signal lines to merge identical sub-trees. This only part (shaded in ▨ in figure 6) will be used to describe the building blocks of the reversible counter example (Cf. figure 7).

• the memory tile (shaded in ▨ in figure 6) necessary to implement any sequential system; "DorQ" and "enable" control the value output on the south bus (SB). It possesses global Clock and Reset signals.

• the glue logic on the top and right side of the cell isolate the logical system from other functions implemented on the same cellular substrate. "Isol0" forces 0 on West input, "isol1" forces 1 on North input.

The communication system is composed of four busses (NB, SB, EB, WB), interconnections are set-up by means of bi-directional switches (transmission gates) controlled by 4 bits (N-W, W-E, N-S, S-E).

Using a network composed of 9 building blocks, we are able to implement the BDDs of Q1+ and Q0+ by means of a 2-dimension cellular automaton (figure 7a). Figure 7b shows the connection part, where only the variable X and the constant 0 are external inputs. Figure 7 presents at the same time, the implementation and the corresponding value of the field programme for both functional and connection parts.

This section has provided the silicon programmable matter necessary to implement the logical functions synthetized with tools described in section 2.

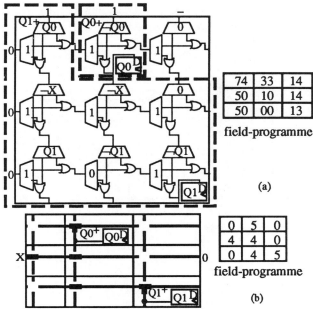

field-programme

(a)

field-programme

(b)

Figure 7.- 4-bit reversible counter mapped in the FPGA.
(a) functional part - (b) connection part.

4.- From Hardware to Wetware

We can notice that the field programme located in each cell can be seen as a "gene", and the field programme of the whole logical system as the genogram (the programme of the genome) of the artificial "creature".

As shown in figure 8a, regrouping the 2 field-programmes for the functional and connection parts (listed in figure 7a - 7b) gives a 10-bit instruction by cell. This 10-bit instruction (or field-programme) is composed of 1 hexadecimal digit for the connection description and 2 octal digits for the function description. This 10-bit instruction can be considered as the functional gene (f-gene) of the cell. We obtain the complete description of the system architecture (and behavior) by regrouping the f-genes of each cell in sequence. Conversely to conventional FPGAs in which the field programme (f-gene) is tied to the physical cell in which it is located, in our system, each gene is composed of a functional gene (behavioral description) and local-address gene (topological description). As a matter of fact, in molecular biology, the genome can also be split into two parts : functional genes and switch genes. The functional genes code for protein (behavioral aspect) while the switch genes do not code for any protein but are active during specialization (organi-zational or topological aspect). Being faced to the same problem, we have introduced binary coordinates (B, A for horizontal coordinates and D, C for vertical ones in figure 8), in order to describe the genome topology. This technique enables to store the blueprint of the system together with its functionality : before each f-gene, a switch gene or s-gene codes for topology by storing the coordinates of the gene in the whole system. The global genome is, in our case, a linear string of 9 pairs composed of one s-gene followed by one f-gene. Figure 8b shows the global genome. A pair of genes is described under the format <s gene>-<f gene>; pairs are separated by an "=" sign.

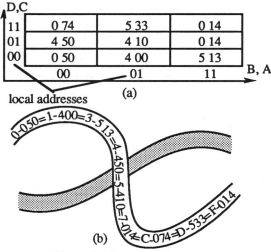

D,C			
11	0 74	5 33	0 14
01	4 50	4 10	0 14
00	0 50	4 00	5 13
	00	01	11 → B, A

local addresses   (a)

(b)

Figure 8.- Genome of the counter.
(a) gene topology - (b) global genome.

5.- Biological Level

At the biological level we are now faced with our last challenge:

- copying the genogram from the mother cell into two daughter cells, and so on until the whole circuit is programmed;
- interpreting the genogram in each cell, in order to calculate the correct gene for the cell, depending on its environment, i.e., its local coordinates.

According to the original scheme (the living cell), where the cell first stores the whole genome before specialisation, each cell should contain enough place to store the genome of the whole system ($3*10^9$ basic DNA letter pairs grouped in about 50'000 genes in each human cell). In fact, as far as application of any complexity could be implemented, a tremendous amount of memory should be available inside each cell. Different solutions for the storage of the genome have been investigated : single location, distributed locations or shared groups of genes [17]. For sake of simplicity, the solution where the complete genome is stored outside the cell in a single location is presented; in this case, each cell will only embed its own gene. As the genome is not contained in each cell, the specialization should come first. Each cell has to know its environment (coordinates) before choosing correctly the piece of genome it has to interpret (its f-gene). During the specialization step, each cell locally computes its coordinates as a function of its state (OK or not OK) and of the neighbors' coordinates (restricted to South and West). The only coordinates which must be provided to the whole system are those of the mother cell (address = 0).

Figure 9 shows the coordinates evaluator. The cell stores the value coming from its neighbors respectively in the horizontal and vertical registers. If the state of the cell is

"OK", then the cell sends the input coordinates incremented by 1, else it propagates the input coordinates to its neighbors (hence the cell is transparent for local address computation).

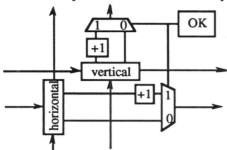

Figure 9.- Coordinates evaluator

As shown in figure 9, each coordinate arrives twice coming from West and South boundaries; this redundancy enables the cell to check if both paths are good. As soon as the specialization phase finishes (each cell knows its location in the network), then the division of the mother cell into two daughter cells (genome transfer) begins. This is done by writing the instructions (genes) one by one in the two daughter cells which are the immediate neighbours (restricted to East and North) of the mother cell.

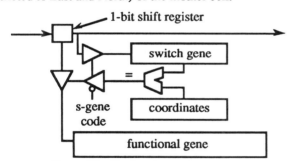

Figure 10.- Cloning the correct gene.

Figure 10 depicts the hardware necessary to interpret the correct gene. At each time an s-gene starts, the signal "s-gene code" enables the input of the switch gene register and disables the output of the comparator. As soon as it finishes, the "s-gene code" goes low, disabling the s-gene register input but enabling the output of the comparator to control the f-gene loading driver, if and only if the coordinates of the cell were found in the s-gene. At the end of these 2 phases (specialization and division), the logical system (the counter here) is programmed on the cellular architecture.

6.- Emergence of life-like properties

We can observe the emergence of life-like properties such as (partial) self-repair coming together with the self-reproducing process [18]. Let us assume that some built-in self test (BIST) strategy is provided at the cell level, that is to say the cell is able to detect its own failures [19]. Then, if any failure occurs either before the first use of the chip or during its lifetime, then each cell is able to detect it with respect to the BIST model. As the genome is stored on the

chip, it can be used to re-map the architecture of the system on a fault free cellular space by avoiding the faulty cell(s). As mentioned in the specialization step, each cell locally computes its coordinates as a function of its state (OK or not OK) and of neighbours coordinates. The faulty cell should be able to send a message (not OK) to inform the other cells that a failure has occurred. The running application is then switched to waiting mode. The specialization step is performed again, followed by the duplication step : leading the system to be functional again.

It exists different degradation modes. For instance, the **shift mode** consists in shifting the whole function away from the faulty cell (removing the mother cell). It is used when large portion of circuit are defective. The **jump mode** consists in jumping the row(s) or/and the column(s) of the affected cell(s).

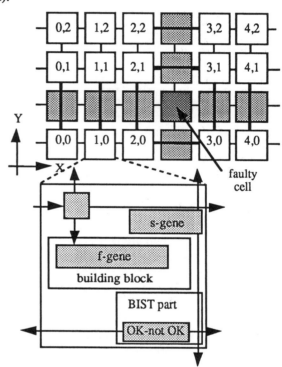

Figure 11.- Double jump mode.

If the functional part is the only affected part, then the communication part can be used. In this case, the simple jump, only one line or one column will be "cut". If both functional and communication parts are affected, then neither part of the faulty cell should be in use. In this case, the double jump, both line and column of the faulty cell are bridged over. Figure 11 depicts the principle of the double jump mode.

7.- Conclusion

We have described a methodology enabling the implementation of logical systems of any complexity on a 2-dimensional Cellular Automata. Until now, only simple

370

systems have been addressed. The hierarchical decomposition of a bigger system into sub-units (organs) has now to be investigated. This will allow us to synthesise systems as complex as microprocessors on a 2-dimensional C.A. It is obvious that this solution is one order of magnitude (or more) bigger than a conventional implementation. Nevertheless, on one hand, we are playing with programmable matter : for instance we can dedicate the size of the data path of the microprocessor mentioned above, without any "redesign". On the other hand, the benefit brought by self-reconfiguration is particularly interesting (especially for embarked systems to which it offers graceful degradation).

Concerning the gene storage, we have investigated different solutions. The extreme solutions are : one unique genome for the whole system (plus copy of the proprietary gene inside each cell) and one genome by cell. In-between solutions have also been addressed, like sub-groups of genes copied in sub-groups of cells [16].

With respect to self-reparability, different schemes have been used: multiple copies of the genome (at least 2), functional cell BIST design, redundancy in the coordinate evaluation, and so on. As needed by any autonomous system, all these techniques require available resources, i.e., spare parts in order to re-map the system. For a single function, spare parts are stored on the top and on the right side of the function cells. For multiple function systems, some spare parts will be stored locally (next to each sub-unit) and some will be global to the whole system. As needed, each sub-unit will take the local resources as long as available. When no more resources are available locally, thank to the genome, the system is able to produce a new copy of itself in a sufficiently large fault-free area on the silicon space.

In some sense, this approach is closer to von Neumann's self-reproducing machine than to Langton's one, mainly because the genome codes for behaviour of the system and not only for its reproduction.

Concerning the clocking system, if the system get larger, a global clocking scheme should be avoided, local clocks will be generated and message passing will be done "by waves", according to local clocks throughout the system.

At the moment, we have not yet produced any silicon, but all methodologies have been tested and several prototypes have been built using off the shelf components, validating the different steps:

- the transformation of combinational and sequential systems into a 2-dimensional cellular automaton;
- mapping the 2-dimensional C.A. on different functional cells;
- the realization of different reproduction processes applying the proposed solutions;
- the different self-reconfiguration strategies (shift mode, jump mode as well as more efficient, but more complex solutions).

The most important point is the following : this methodology enables very complex integrated circuit synthesis, far beyond the present possibilities of the technology, with living creature properties (robustness and healing).

References

[1] R. A. Freitas & W. P. Gilbreath, , Advanced Automation for Space Missions, NASA Report CP-2255, 1982.

[2] J. G. Kemeny, "Man Viewed as a Machine", *Scientific American*, vol 192, 1955, 58-67.

[3] J. von Neumann, *The Theory of Self-Reproducing Automata*, A.W. Burks, ed. Univ. Illinois Press, Urbana, 1966.

[4] E. F. Codd, *Cellular Automata*, Academic Press, New-York, 1968.

[5] C. G. Langton, "Self-Reproduction in Cellular Automata", *Physica 10D*, 1984, 135-144.

[6] J. Byl, "Self-Reproduction in Small Cellular Automata", *Physica 34D*, 1989, 295-299.

[7] H. C. Morris, "Typogenetics: a Logic for Artificial Life", *Artificial Life*, C. G. Langton, ed., Addison-Wesley, Redwood City, 1989, 369-395.

[8] D. R. Hofstadter, *Gödel, Escher, Bach : An eternal golden braid*, Basic Book, New-York, 1979.

[9] C. Y. Lee, "Representation of Switching Circuits by Binary-Decision Programs", *The Bell System Technical Journal*, Vol 38, n°4, 1959, 985-999.

[10] S. B. Ackers, "Binary Decision Diagrams", IEEE Trans. on Computers, Vol C-27, n°6, June 1978, 509-516.

[11] R. E. Bryant, Symbolic Boolean Manipulation with Ordered Binary-Decision Diagrams", ACM Computing Surveys, Vol 24, n°3, September 1992, 293-318.

[12] D. Mange, *Microprogrammed Systems - An introduction to firmware theory*, Chapman & Hall, London, 1992.

[13] M. Davio, J. P. Deschamps, A. Thayse, *Digital systems with algorithm implementation*, John Wiley, Chischester, 1983.

[14] E. Sanchez, A. Thayse, "Implementation and transformation of algorithms based on automata", Part III, *Philips Journal of Research*, Vol 36 n° 3, 1981,159-172.

[15] S.D. Brown, R.J. Francis, J. Rose, Z.G. Vranesic, *Field-Programmable Gate Arrays*, Kluwer Academic Publishers, Dordrecht, 1992.

[16] P. Marchal, A. Stauffer, "Binary Decision Diagram Oriented FPGAs", ACM International Workshop on FPGA : FPGA '94, Berkeley, Feb. 1994.

[17] S. Durand, C. Piguet, "FPGA with self repair capabilities", ACM International Workshop on FPGA : FPGA '94, Berkeley, Feb. 1994.

[18] R. Laing, "Artificial Organisms : History, Problems, Directions", *Artificial Life*, C. G. Langton, ed., Addison-Wesley, Redwood City, 1989, 49-61.

[19] P. Marchal, C. Piguet, D. Mange, A. Stauffer & S. Durand, "BIOBIST: Biology & Built-in Self-Test Applied to Programmable Architectures", IEEE International Workshop BIST/DFT'94, Vail (CO), April 1994.

Development and Evolution of Hardware Behaviors

Hitoshi Hemmi, Jun'ichi Mizoguchi and **Katsunori Shimohara**
Evolutionary Systems Department,
ATR Human Information Processing Research Laboratories,
2-2 Hikaridai, Seika-cho, Soraku-gun, Kyoto, 619-02, JAPAN
e-mail: hemmi@hip.atr.co.jp

Abstract

A new system is proposed towards the computational framework of evolutionary hardware that adaptively changes its structure and behavior according to the environment. In the proposed system, hardware specifications, which produce hardware structures and behaviors, are automatically generated as Hardware Description Language(HDL) programs. Using a rewriting system, the system introduces a program development process, that imitates the natural development process from pollinated egg to adult and gives the HDL-program flexible evolvability. Also discussed is a method to evolve the language itself by modifying the corresponding rewriting system. This method is intended to serve as hierarchal mechanism of evolution and to contribute to the evolvability of large-scale hardware. Although this paper's discussion is mainly involves in HDL-programs because our goal is hardware evolution, the techniques described here are applicable to ordinary computer programs written in such conventional formats as "C" language.

1 Introduction

Hardware can be evolvable! The ultimate goal of our research is to prove the previous sentence and create real examples of such an organism. In order to create adaptive complex systems that are like living systems, we believe that not only software evolution (Ray 1991) but also hardware evolution is indispensable. New hardware architecture and devices, on which evolvable hardware is really implemented, are needed to achieve hardware evolution (de Garis 1993). On the other hand, it is also essential to create an evolutionary framework and/or computational mechanisms to guide hardware evolution.

One approach to achieving this ultimate goal is to use Hardware Description Language, a powerful tool for describing circuit behavior, and to evolve HDL programs with an evolutionary process.

HDL is a programming language and hardware design tool, currently used for most hardware designs based on analytical methods. HDL has an advantage in that hardware specification can be programmed, evaluated with the desired performance, and directly implemented on LSIs. However, when using it in a synthetic framework made through by an evolutionary process, HDL has a significant drawback: it is too vulnerable to the change caused by evolutionary operations. Therefore, it is necessary to develop evolutionary methods that overcome the drawbacks of HDL and maintain the capability of hardware description.

On the other hand, mimicking the natural evolutionary process, some artificail evolutionary measures have been proposed (Ray 1991) (Goldberg 1989). Tierra (Ray 1991) is a kind of world in which idealized processor instruction codes evolve mainly by mutation. Genetic Argorithms (GAs) (Goldberg 1989) are the most widespread techniques generally applicable to various problems in which crossover acts as the main mechanism of evolution. Some studies have used GAs to generate software programs written in LISP (Koza 1992), and Field Programmable Gate Array structures (Higuchi et al. 1993). By applying these features, once HDL-programs are no longer vulnerable, their descriptive power will give vast evolvability to hardware behaviors.

This paper demonstrates that a program development process based on a rewriting system allows the HDL to overcome its vulnerability and provide flexible evolvability. In addition, converting HDL grammar into a rewriting system makes it possible to transform the language itself, that introducing a hierarchical aspect to the evolutionary mechanism and direct to contributint to the evolvability of large-scale hardware.

2 Development of HDL-program

This section introduces the development process by which HDL-programs are generated automatically. The basic idea in such a process is as follows:
Adults in natural life, especially in higher forms, are vulnerable to change just as with HDL-programs. However, in the early stage of the development process, natural life is flexible to operations added to their embryo. Actually, the earlier the development state, the greater the flexiblity; in the extreme case, a life can be divided into two or more individuals in the initial stage of development. In the same way, by accepting such a development process, HDL programs may achieve superior flexiblity and evolvability.

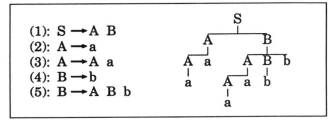

Figure 1: Example of Rewriting System
S is the start symbol, A and B are non-terminals, and a and b are terminals. The production rules are shown on the left and the rewriting process is shown on the right. Each production rule is composed of a left-hand side and a right-hand side. The left-hand side is a symbol to be replaced and the right-hand side is a sequence of symbols after replacement.

2.1 Rewriting System

The development process of natural life is based on the cell division process. The mathmatical basis of this process for plants has been studied as the L-system (Prusinkiewicz and Lindenmayer 1990). On the other hand, rewriting systems, which perform operations similar to those by the L-system, has been used strictly to define the computer language specification. Backus-Naur form (BNF) is the first and most widely used rewriting system for this purpose. Many programing languages are defined in BNF, such as Algol-60, C, and C++. Accordingly, a rewriting system is ideally suited to constructing the HDL program development process.

Our system uses Structured Function Description Language [1] (SFL), which is efficient for describing complex control sequences in which control automatons interact with each other and data signals are processed according to the states of the automatons. SFL is also defined in BNF. Therefore, the description in this paper mainly applies to SFL, but the general ideas are applicable to other high-level HDLs and ordinary computer languages.

A rewriting system consists of a start symbol, non-terminal symbols, terminal symbols, and production rules. An example of a rewriting system is shown in Figure 1. In the rewriting system used for HDL, there is initially only one start symbol. Then a production rule having the start symbol in its left hand side is applied to the start symbol, producing non-terminal symbols and terminal symbols. After that, production rules are applied one after another to the non-terminal symbols in the same manner. Finally, when all of the symbols become terminal symbols, one program has been constructed.

Considering a program as one creature, the start symbol corresponds to a pollinated egg. Non-terminal symbols are considered the cells in the middle of the development process and as a kind of brast such as a

---

[1] A HDL given by PARTHENON, i.e., an LSI design system, developed by NTT (Nippon Telegraph and Telephone Corporation); PARTHENON also supplies the behavior simulator for SFL.

neurobrast.[2] Terminal symbols correspond to cells that do not perform cell-division, such as a nerve cell.

2.2 Chromosome

Generally, some non-terminal symbols of a rewriting system have plural production rules that are applicable. In the development process of one program, everytime such non-terminals appear, only one of the candidate production rules is selected and applied to the symbol. Writing such selected rules in turn, one can construct a tree-like diagram(Fig. 2).

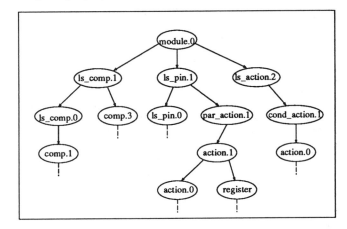

Figure 2: Chromosome

Figure 2 shows the progression from program to chromosome. The node labeled "ls_pin.0" shows that the non-terminal symbol "ls_pin" is applied with its 0th production rule. Two arcs from the node show that, as a result of this rule application, the symbol "ls_pin" is divided into two non-terminal symbols "ls_pin" and "par_action". In the reverse progression, from chromosome to program, if a well-constructed (i.e consistent in rule application chain) diagram such as Figure 2 is given, one program can be constructed deterministically. Therefore such a diagram can be seen as the blueprint of the program. We define this diagram as chromosome of the corresponding program. In other words, this chromosome is the control data of the program's development process.

If a node of this chromosome is modified, the sub-tree from the node is also modified to maintain consistency, and the development process from that node will be different from the original process, so the resulting program is also modified. In this way, the program varies according to chromosome variation. In fact, programs that can be generated this way cover *all* possible programs in the language.

3 Evolution of HDL-programs

This section discusses techniques to change the development process so that HDL-programs can be evolved.

---

[2] A neurobrast is divided into one nerve cell and one neurobrast. This corresponds to recursive rules such as rule (3) in Figure 1.

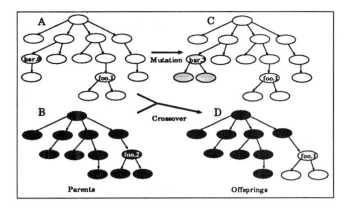

Figure 3: Crossover and mutation

The first subsection explains the operations for transforming chromosomes, which include the corresponding part to operations—crossover and mutation—of conventional GAs. Building blocks in the context of program development is disscussed in the second subsection. The third subsection shows an example of an evolving hardware module.

3.1 Transformation of Chromosome

Koza (1992) shows that exchange can be performed arbitrarily for two S-expressions of LISP programs. This is not applicable to the programs of other languages because in general they consist of grammatically heterogeneous substructures. The principle of substructure exchange in our chromosome form is:

Only subtrees with root nodes having the same non-terminal symbol name can be exchanged with each other.

For example, in Figure 3, white chromosome A and black chromosome B have nodes labeled foo.1 and foo.2, respectively. Consequently, a subtree having the foo.2 node can be replaced by the one with foo.1 to yield new chromosome D. This operation corresponds to the crossover of GAs.

On the other hand, part of a chromosome's body can be reconstructed ("bar.0" to "bar.3" in Figure 3) if the grammar of the language is maintained.

This operation's principle is as follows:

A chromosome can be modified by replacing one node in it to the another node which have same non-terminal name.

For example, in Figure 3 chromosome A have node labeled bar.0, which can be replaced with bar.3 by discarding original subtree and constructing a new subtree according to the rewriting rules. (Of course this additional development process is not needed if the replacement is performed in a "leaf" of the chromosome tree.) This operation corresponds to the mutation of GAs.

We prepared three other operations: duplication, insertion and deletion. These are related to recursive production rules, and simulate the gene duplication process in natural chromosomes. The details are described

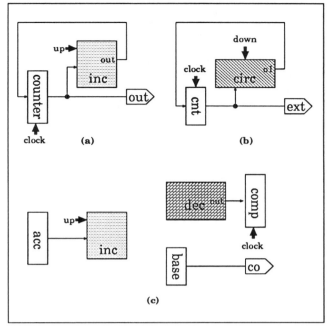

Figure 4: Block diagrams

in Mizoguchi, Hemmi, and Shimohara (1994). Combinations of GAs and the techniques stated here are called "Production Genetic Argorithms" by authors (Mizoguchi et al. 1994).

3.2 Building Block

What should be a "building block" or a useful construct of an HDL-program from the viewpoint of program development and evolution?

A building block of programs is a set of grammatical sub-blocks in which the labels are freely changed as long as the uniformity of the labels is preserved.

The rest of this subsection explains the above concept of "building block".

```
par {
    out = counter;
    inc.up(counter);
    counter := inc.out;
}
```

The above is a program segment written in SFL. The meaning of this code is (Fig. 4(a)) "Do three operations in parallel: (1) output the contents in register **counter** to terminal **out**, (2) operate **up** in the functional circuit (can be thought of as a blackbox) **inc** with the contents in **counter**, (3) update **counter** with the value of output terminal **out** of the **inc** in *next* clock cycle."

How should this program segment be transplanted to another program organism? One way, which Koza's Genetic Programming (Koza 1992) adopted, does not change the segment and literally exports the block to the other program. The problem with this method is that there must be facilities with the exact same name, such as input/output pins or register, in both the importing and exporting programs. This restriction degrades the flexibility of a diverse of program population, and stagnates evolution. Another way is to re-assign all labels in the segment to new facility names that exist in the importing program. This method overcomes the the above problem, but the program segment might result in such a form:

```
par {
    co = base;
    inc.up(acc);
    comp := dec.out;
}
```

A block diagram of this code is shown in Figure 4(c). This code is not believed to preserve the meaning of the original code because unlike the original code, now each sentence operates to divergent facilities.

The method presented at the beginning of this subsection overcomes the above problems perfectly because the labels are renamed to new ones, but the labels which had the same name in the original code will also have the same new name in the new program. In this way the code may become:

```
par {
    out = cnt;
    circ.down(cnt);
    cnt := circ.out;
}
```

A block diagram of this code is shown in Figure 4(b). Obviously this code maintains the *meaning* of original code, while its *appearance* is quite changed. The code segment then gets acclimatized to the new program body.

3.3 Evolvable Hardware Module

An example of an evolvable hardware module is shown in Figure 5. "PLD" stands for Programmable Logic Device, which is a generic name of multiple reconstructable devices such as Field Programmable Gate Array (FPGA). Some high-level HDL, including SFL, provide a synthesizer of FPGA programing instructions, so an HDL-program chromosome can be converted into circuits by development and synthesis processes.[3] The module in Figure 5 also provides ports for exchanging the chromosome. A population of hardware modules, can be evolving both under some supervisor contrivance such as GAs or independently according to its fitness to the environment (i.e. how well does it processes data).

4 Evolution of Language

This section explain a technique for changing the rewriting system so that the HDL itself will evolve. This technique gives hierarchical aspect to the evolutional mech-

---

[3]Of course the HDLs are mainly for LSI design, so HDL-prgrams can be converted into mask patterns that are used to print the circuits into silicon. However, this path takes time.

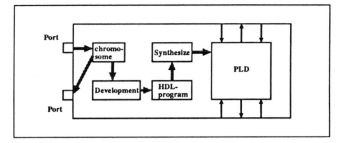

Figure 5: Example of EvolvableHardware Module

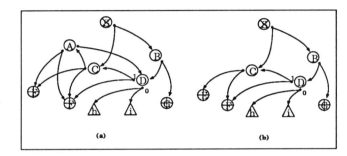

Figure 6: Production diagram

anism, and contributes to evolvability of hardware behavior of high complexity.

Any good tool, even though it can theoreticaly handle all problems in a class, in practice has a range of problem sizes that it can efficiently manipulate. Evolutional systems, if they operate on a single mechanism, cannot manipulate a wide range of programs; it can manipulate relatively small problems but not large problems, or vice versa.

A similar situation can be observed in computer programming tasks. For example, the machine language of a well-designed processor can manipulate any algorithm in theory with maximum efficiency of the processor. However, in practice even for the most skilled programmer, it is very difficult to code large complex programs precisely in machine language. Productivity increases significantly by using "C" programming language. But this reaches a limit sooner or later: shifting to a higher level language such as "C++" becomes necessary. In this way, the languages themselves need to progress.

4.1 Production diagram and Space of Programs

This subsection introduces the "production diagram" of a rewriting system and discusses the theoretical program space of a language. The discussion here is the guiding principle to the next subsection, which discusses a method to evolve the rewriting system by transformating this diagram.

An example of a production diagram is shown in Figure 6(a). ⊗ represents the start symbol. ⊕ represents non-terminal symbols which are directly rewritten to literal terminal symbols,[4] and ◯ represent other non-terminal symbols. Examples of rewritten non-terminal symbols are "input name", "output name" and "register name". △ represent terminal symbols that are keyword or operator symbols of the language. Literal terminal symbols do not appear in Figure 6(a) because there are usually too many such symbols in a language. • around ◯ and arcs from the • represent one production rule from the non-terminal symbols and replacing symbols. The label of • represents a sub-number of the production rule.

For example, a non-terminal node labeled "D" has two production rules, and • labeled 1 on the left of the non-terminal node represents the next production rule:

D.1: D → C F

When the diagram does not have looping of the arcs such as in Figure 6(b), it is considered a tree or a flow. If moreover there are only finite number of terminal symbols, the number of programs written in the language is obviously finite.

If the number of terminal symbols is infinite, the number of programs would seem to be infinite. However, from the viewpoint of actual program significance, the variety of programs is in fact finite. This may be seen by the finite variation of *abbreviated* chromosomes, which use only symbols in Figure 6(b) and do not distinguish the terminal symbols corresponding to same non-terminal symbol ⊕ in the diagram. Also, in the same abbreviated chromosome the variation of *meaning* of the program is restricted by the uniformity of labels appearing in the program. (See also discussion in Section 3.2.)

On the other hand, if the diagram has loops the number of programs writable in the language is infinite, wheather the number of terminal symbols is finite or infinite. A typical way that loops exist in the diagram is as follows:

$$
\begin{array}{lcl}
formula & \to & monomial \ operator \ monomial \\
& \vdots & \\
monomial & \to & \text{``("} \ formula \ \text{``)"} \\
& \vdots &
\end{array}
$$

From the viewpoint of evolvable hardware, a looping diagram is naturally more interesting than a non-looping diagram because only the former has infinite evolvability. The production diagram of SFL has looping too.

4.2 Transformation of Rewriting System

The operation that transforms the production diagram introduced in the previous subsection is as follows:
In the production diagram in Figure 7(a), non-terminal node D is divided into D_1 and D_2 (Fig. 7(b)). The arcs that point from node A and B are *exclusively* distributed to either point D_1 or point D_2. On the other hand, the production rules • around node D are *inclusively* distributed to either D_1 or D_2. An example of the operation's result is shown in Figure 7(b).

---

[4] Usually destinations from one ⊕ node are names of one facility class, such as in0, in1, in2, ... etc. for inputs.

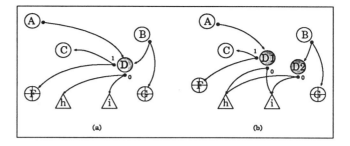

Figure 7: Diagram Transformation (abbreviated)

In the production rule form, this operation is expressed as follows:

$$
\begin{array}{lcl}
A & \to & D \\
B & \to & D \ G \\
D & \to & h \ i \\
D & \to & C \ F
\end{array}
\quad \Longrightarrow \quad
\begin{array}{lcl}
A & \to & D_1 \\
B & \to & D_2 \ G \\
D_1 & \to & h \ i \\
D_1 & \to & C \ F \\
D_2 & \to & h \ i
\end{array}
$$

A program created according to new production rules completely satisfies the original language syntax. On the other hand, it should be noted that from symbol A, as before, expressions having either "h i" or "C F" can be created, but from symbol B it is no longer possible to create expression having "C F".

This is *not* retrogression but progression of the language. This can be seen by an analogy to "C" language. Processor instruction codes compiled from "C" language a certain pattern in instruction arrangement; the sequence of instruction cannot be arbitrary

Even if it restricts to the arbitrariness of machine instruction programming, it is a *style* of programming, and such styles are what increases productivity and provide a source of new languages.

The fact that the operation introduces style to the language is seen more clearly if the D_2 in production rule is expanded:

B → h i G

By evolving languages with such techniques, the system contributes to the evolvability of highly complicated hardware behavior.

One caveat must be stated here: if by some operation the production diagram loses its loop, this result in *hypertely* (i.e. over-evolution), the language loses evolvability, and the range of programs in the language becomes finite.

5 Experiment and Result

In this section, a simulation of evolution using SFL programs is shown. The problem treated here is making an arithmetic circuit (i.e. binary adder). Other examples for the Artificial Ant problem are described in Mizoguchi, Hemmi, and Shimohara (1994).

Details of the problem are as follows (see Figure 8):
The target is two input and one output circuit; inputing two sequences of binary numbers from lowest figure, the circuit produces the sum of the binary numbers from the

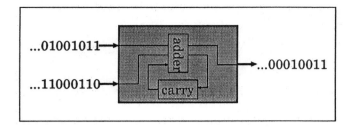

Figure 8: Binary adder: (interior is not specified)

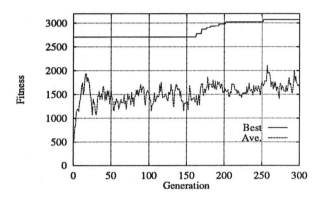

Figure 9: Evolution in binary adder

lowest figure in the output terminal. The correct circuit must consider the carry from the lower bit, so it belongs in a class of sequential circuits.

An experiment was performed by combining GAs and the transformation technique stated in Section 3.1.

The experimental conditions are as follows: the population is 100, crossover rate is 50%/program, mutation rate is 1%/symbol, gene duplication rate is 1%/program, deletion rate is 1%/program and two test numbers are 1536 bits long[5]. Fitness is calculated as follows: each program initially has 1536 point as its share; if one output bit is correct it increase; otherwise it decreases. Therefore the perfect score is 3072 points, and worst score is 0 point.

The result is shown in Figure 9. In the 251st. generation, there was a perfect score SFL program, and the authors confirmed it as a complete binary adder circuit by reading the description of the program.

6   Conclusion

This paper introduced an evolutionary system that works on HDL-programs and explained the techniques it uses. By converting HDL grammar to the production rules of a rewriting system, the programs could

---

[5]This unusual number has the following meaning: the two test numbers are arranged in all possible combinations of two 4 bit numbers (these are $(2^4)^2 = 256$), with 2 bit "00" between each boundary to suppress the carry propagation. Accordingly the total length is $256 \times (4 + 2) = 1536$.

be developed from a structured chromosome. This provided great flexibility to evolutionary operations. Furthermore, this conversion made the language itself evolvable. These techniques are expected to contribute to the creation of a complex adaptive system.

Acknowledgment

We would like to thank Dr. Kiyoshi Oguri of NTT for giving us the opportunity to use the PARTHENON system.

References

de Garis, H. (1993). Evolvable hardware: Genetic programming of Darwin machines. In *International Conference on Newral Networks and Genetic Algorithms*, Lecture Notes in Computer Science. Springer-Verlag.

Goldberg, D. E. (1989). *Genetic Algorithms in Search, Optimization, and Machine Learning*. Addison-Wesley.

Higuchi, T., T. Niwa, T. Tanaka, H. Iba, H. de Garis, and T. Furuya (1993). Evolvable hardware – genetic based generation of electric circuitry at gate and hardware description language (HDL) levels. Technical Report 93–4, Electorotechnical Laboratory, Tsukuba, Ibaraki, Japan.

Koza, J. R. (1992). *Genetic Programming – On the Programming of Computers by Natural Selection*. MIT Press.

Mizoguchi, J., H. Hemmi, and K. Shimohara (1994). Production genetic algorithms for automated hardware design through an evolutionary process. In *IEEE Conference on Evolutionary Computation*.

Prusinkiewicz, P. and A. Lindenmayer (1990). *The Algorithmic Beauty of Plants*. Springer-Verlag.

Ray, T. S. (1991). An approach to the synthesis of life. In C. G. Langton, C. Taylor, J. D. Farmer, and S. Rasmussen (Eds.), *Artificial Life II*, Volume X of *Santa Fe Institute Studies in the Sciences of Complexity*, Redwood City, CA, pp. 371–408. Addison-Wesley.

Evolutionary Learning in the 2D Artificial Life System "Avida"

Chris Adami and **C. Titus Brown**
W.K. Kellogg Radiation Lab 106-38
California Institute of Technology
Pasadena, CA 91125

Abstract

We present a new tierra-inspired artificial life system with local interactions and two-dimensional geometry, based on an update mechanism akin to that of 2D cellular automata. We find that the spatial geometry is conducive to the development of diversity and thus improves adaptive capabilities. We also demonstrate the adaptive strength of the system by breeding cells with simple computational abilities, and study the dependence of this adaptability on mutation rate and population size.

1 Introduction

Artificial systems such as Tom Ray's tierra have opened the possibility of studying open-ended evolution in strictly controlled circumstances, allowing experiments that were previously unthinkable as the only alternative was "wetware". The study of evolution in an information-rich artificial environment requires ever larger and faster systems, and present systems are largely restricted by such limits. Distributing tierra simulations over multiple processors is not practical on a large scale, because of the non-local interaction between members of the tierran population.

We have designed a next-generation system based on an array of cells that interact only with their nearest neighbours, and an update mechanism reminiscent of 2D cellular automata. It is designed for evolution towards complexity in an information-rich environment [3] much as the tierra system, but for purposes of universality is simpler in some respects.

Concurrently, we have retained the spirit of the tierra system, in the sense that the members of the population are strings of machine–language-like instructions running on a virtual computer much like the one designed by Ray. These strings of instructions ("genomes") can be thought of as being orthogonal to the grid that marks the physical location of the string, while the interactions between the grid points are similar to those of cellular automata. The key difference, which makes the system evolvable, is that the update rules are not fixed but rather are dependent on the genome of the cells in the immediate neighborhood. The genomes on the grid are subjected to Poisson-random mutation which allows evolutionary adaptation via implicit Darwinian selection.

The strings adapt to a landscape specified by information only: "discovery" of that information (by developing the code in the genome to trigger the bonus associated with the information) typically results in a higher replication rate for the adapted string and subsequent perpetuation of the discovery. In this manner, complexity can develop in code that starts out only with the ability to self-replicate. The task learned by the strings is entirely determined by the information they encounter, and is thus entirely at the control of the experimenter.

Suppose, for example, that we specify that adding integer numbers results in a bonus for each string that accomplishes the task. After some time, the strings *will* develop code that reads integer numbers, adds them, and then writes them to the output. Clearly, such adaptive capability can be a powerful tool, since there is no fundamental limit to the complexity achievable through use of this technique, given enough evolutionary time. We believe that this method of "stochastic information transfer", from the environment into the genome of the adapted cell via mutations, observed in tierra and this system, is central to the development of complexity in living systems.

In the next section we present a brief description of our avida system, with emphasis on the strengths of the local geometry and aspects of the update system. In section three we examine the results of the local interaction in terms of genotype age and the consequent rise in diversity. We then study the adaptive process as a function of population size and mutation rate. Finally, we offer some conclusions and discuss future applications of the system.

2 The avida System

In avida, the physical position of a string is determined by its coordinates in a $N \times M$ grid with the topology of a torus. As in tierra, each string is a segment of computer-code written in a simple language (similar to the Intel 80x86 assembly language) running on a configurable virtual computer[1]. The language is user-defined, but must support self-replication.

Self-replication in avida occurs when the strings copy their genome into a child string. Ideally, the strings de-

---

[1] For purposes of comparison, we have used an instruction set and CPU structure similar to Ray's instruction set #4.

termine their own size, and then allocate memory accordingly; this allocated memory is attached to the end of the genome, and the string copies its computer-code into the free space. After completing the copy, a cell-division command is issued by the string, separating the genome into two identical pieces. Once the new genome is released with the cell-division command, the oldest cell within the immediate neighborhood is replaced by a new cell containing the new genome. Thus, the birth of a cell can only affect those cells directly surrounding it. As a consequence, string-string interactions are local, and information propagates accordingly.

As the strings are subject to Poisson-random mutation of their genome, the process of reproduction is often corrupted by mutation of the parent strings either before or during the copy process, leading to imperfect or incomplete copies. This is the driving force of evolutionary change and diversity in the system. Interestingly, even though the only direct source of mutations is point-permutation of the genome, many of the other recognized biological mutations emerge from the copying process; these include insertion and deletion of instructions or chunks of instructions, as well as doubling of the genome. They arise much like in the **tierra** system, and will be studied elsewhere. For simplicity, we do not provide for an explicit cross-over mutation mechanism, nor for multiple ploidy or sexual reproduction.

Mutation rates are defined as flux rates (mutations per site per unit time) through the available genetic space (analogous to the "soup" in **tierra**). While organisms interact as points on the lattice, genomes reside in a pool of genetic material. This genetic material is randomly mutated at the flux rate, whether or not it is currently in use; the results of this are discussed in section three. Finally, we have abstained from the use of flawed execution of instructions. While we acknowledge that imperfect action of proteins and enzymes do occur in nature, we have not found this to be a crucial feature in this system.

Parallel execution of the strings is simulated by assigning time-slices to each cell. After execution of its time-slice, every string is in a certain state: requesting memory, copying instructions, or placing an off-spring. After a sweep of the grid in which every cell executes its allotted time slice, the **avida** system updates the lattice according to the state of the cell. The interaction of cells is thus akin to $K = 1$ cellular automata[2]. The time-slice is kept small in order to insure that no cell can unduly affect its surroundings beyond control of the resolver, such as by reproducing multiple times in one time slice. However, this requirement conflicts with the need to distribute bonus time-slices as reward for the correct execution of user-specified tasks[3]. The two requirements can be reconciled via the update mechanism, which keeps the *average* time-slice constant while each individual cell is given less time (punishment) or more (reward), according to its relative bonus. In principle a cell can accu-

mulate an unlimited amount of bonus through complex operations. When the information in the genome that triggers this bonus is propagated throughout the environment by self-reproduction of the superior organism, a newly born cell without this information will be at a severe disadvantage obtaining significantly less than the average time-slice and thus unable to compete, while already existing organisms and newly born organisms that contain this information will compete on an equal footing, receiving the average slice.

The update mechanism is also designed to allow simple distribution of **avida** over multinode systems. Because the only direct interaction of strings is local, the transfer of information between computational nodes is kept to a minimum; in fact, the only two mechanisms that necessarily communicate between nodes are the conflict resolution system (which handles reproduction) and the time slicing mechanism. Therefore **avida** can be distributed over multiple processors with a nearly linear increase in execution speed.

For further information on the precise implementation of **avida** see [8]. The program is available to the public[4]. For a more detailed introduction to **tierra**-like auto-adaptive genetic systems, we refer the reader to Refs. [1-3].

3 Localized Interactions and Genotype Age Distribution

In this section we would like to point out the relationship between causal spread of information and the genotype-age distribution.

In **avida**, as in **tierra**, a genotype system records the creation and extinction of species via their populations. Genotypes are exact: every member of a specific genotype has the same genome and therefore most point mutations (and consequent reproductive mutations) create new genotypes. This abundance of new genotypes, while an important part of any evolutionary system, is hard to track in a meaningful way. In addition, most (more than 90%) of mutations generate nonviable genomes. These genomes die out quickly, with minimal impact on the system.

In order to observe the system more easily, we have (as in the **tierra** system) divided genotypes into *threshold* and *temporary* genotypes. Threshold genotypes are those that have at any time achieved at least 10 concurrent members, while all other genotypes are considered temporary. The ages of all threshold genotypes can be plotted as a frequency distribution $N(\tau)/N$, where N is the total number of genotypes, to reveal information about the diversity and survival probability of genotypes.

We have measured this distribution in **tierra** by collecting the ages of 184,767 genotypes from 40 runs at the same mutation rate (see Fig. 1).

---

[2]We have experimented with $K = 2$ interaction with no significant change in dynamics.

[3]see Ref. [3]

---

[4]To obtain **avida**, contact one of the authors.

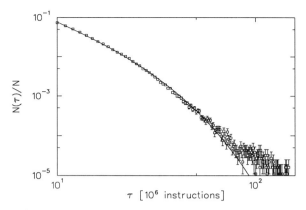

FIG. 1 Distribution of ages of genotypes in a tierran population of ~ 1000 cells at mutation rate $R = 0.65 \times 10^{-3}$ (mutations per instruction executed). This fit yields $D = 1.6 \pm 0.05$, with a cut-off $T = 15 \pm 1$ million instructions.

Due to the threshold in tierra, short-lived genotypes are not recorded and the distribution shows a lack of young genotypes. For that reason, we plot only genotypes that are older than 10 million instructions. We fit the distribution to a power-law $N(\tau)/N \sim \tau^{-D} \exp(-\tau/T)$ with a finite-time cut-off, and find $D = 1.6 \pm 0.05$ and $T = 15 \pm 1$ million instructions. However, the value for D must be considered a lower limit as removal of points at small genotype ages (which may be affected by the threshold) decreases the χ^2 of the fit and $D \to 2$.

In avida we chose a much less severe threshold and see no dependence of the fit on the removal of points at young genotype ages. The measured distribution is shown in Fig. 2.

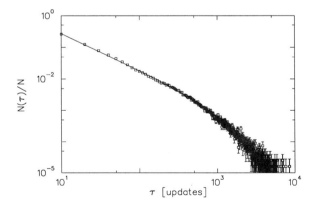

FIG. 2 Distribution of ages of genotypes in population of 1600 cells at mutation rate $R_\star = 2.0 \times 10^{-3}$. The fit yields $D = 1.14 \pm 0.1$, with a cut-off $T = 1200 \pm 20$ updates.

This is the distribution of ages for a population of 1600 strings at mutation rate $R_\star = 2 \times 10^{-3}$ mutations per executed instruction, obtained from 20 runs under identical conditions, yielding 121,703 genotypes. We fit the distribution with the same parametrization and find the slope D to be 1.14 ± 0.1, with a cut-off $T = 1200$

updates. Note that the absolute values for the lifetimes cannot be compared easily, as the units of time are different in the two systems. The exponents, however, should be universal.

Examining the distributions further, we find D to be independent of population size or mutation rate; the exponential cutoff T, while independent of population size, does seem to depend on the mutation rate. This is most likely due to the different average run length at different mutation rates.

The rather rapid decrease of $N(\tau)$ in tierra and the correspondingly small number of "old" genotypes suggests a tendency toward premature equilibration in tierra. Indeed, as any cell anywhere in the tierran soup can affect any other cell *directly* via the reaper queue, a discovery anywhere in the soup will reach other cells immediately and force extinction of those genotypes. Thus, diversity is throttled and the population will tend to homogenize. This is especially worrisome if the governing genotype is trapped in a meta-stable state. In this case, the lack of diversity may condemn the population to remain in this state indefinitely. This kind of behavior is apparent in tierra simulations with large soup-sizes.

The low value of D in avida on the other hand suggests a near maximal population diversity. This signifies the simultaneous exploration of multiple evolutionary paths in the system, a feature of a robust Darwinian environment. It is the result of a causal spread of information, a direct consequence of the localized interaction.

Extinction events in avida are thus far less severe than extinction events in tierra and do not seriously curtail the heterogeneity of the environment. Significant gains are still disseminated throughout the population, but the new information has more time to be integrated with the existing genotypes. This strongly suggests that the meta-stable states observed in tierra [3] will not play as large a role in halting the evolution of new genotypes in the system, and indeed, very few meta-stable states have been observed.

4   Evolution and Adaptation

In this section we investigate the "learning" capabilities of avida as a function of external mutation rate and population size.

We design a landscape for the population to adapt to by distributing bonuses for accomplishing either the main task or certain other feats that are helpful in building up the code necessary to trigger the main bonus. Specifically, to compare with results obtained with tierra [3], we breed strings that have the ability to add two integer numbers. The code necessary to perform this task must at least include two "read" and one "write" statement, as well as register addition and movement of numbers between registers. The strings start out as self-replicating, with no other capability. With the average time-slice set to 30, we reward each "read" and "write" statement that develops with 7 time-units, for a maximum of three "read" and three "write" statements. Additional input/output instructions are not rewarded. Furthermore, if a string manages to echo the value last read into the output buffer unchanged, it reaps a bonus of 30 units for each time this task is accomplished, with a maximum of three times. Finally, if a string writes into the output buffer a number that is the sum of two previously read numbers, it is rewarded with 100 units, with a maximum repetition rate of three. Note that while highly adapted cells can reap a bonus of well over 300, the average time-slice per update remains constant.

All these bonuses are of course available at the same time, and no order is specified. Most importantly, they do not favour a particular solution to the problem but rather create a fitness gradient that leads to the solution from many avenues. Evidently, the paucity of the reward structure does constrain the solutions to a certain class while a more complex environment would allow solutions to the problem taking advantage of rewards entirely unconnected to the task at hand. This is of course a feature of evolution in natural systems, and the construction of a more complex environment is a challenging task for the future.

Since adaptation by mutation is an intrinsically stochastic process, the definition of an "average learning rate" is problematic. Intuitively, we would expect that an average learning time[5] should be connected to the average time between "discoveries" (discontinuous jumps in replication rate).

Yet, it was observed recently [4] that the time between such jumps is distributed according to a power law, and no such average can be defined. This can be traced back to the fact that there is no time scale of the order of the learning time in auto-adaptive genetic system. We can nevertheless determine the adaptive power of the population by obtaining the *learning fraction* f_X, which is the fraction of runs that have accomplished the task *before* a cut-off time X (measured in thousands of updates). Thus, if for ten runs under identical conditions we find six where almost all of the population has discovered how to trigger the bonus before, say, 10,000 updates, this combination of parameters is assigned $f_{10} = 0.6$.

In Figs. 3a-c and Figs. 4a-c we show the learning fraction for an array of 20x20 and 40x40 cells respectively, for cut-offs X=10, 20, and 50 thousand updates, as a function of mutation rates. To obtain these graphs, we repeated up to 20 runs at each of 15 mutation rates between $R_\star = 0.1 \times 10^{-3}$ and $R_\star = 20 \times 10^{-3}$ [mutations per executed instruction] which translates into $R = 0.12 \times 10^{-4}$ to $R = 23.44 \times 10^{-4}$ [mutations·(site)$^{-1}$·(update)$^{-1}$]. The number of runs N performed at each mutation rate is shown in Tab. 1, the total number of runs being 640.

$R_\star [10^{-3}]$	$R [10^{-4}]$	$N(20\text{x}20)$	$N(40\text{x}40)$
0.1	0.12	-	20
1.0	1.17	20	20
2.0	2.34	20	20
3.0	3.52	20	20
4.0	4.69	20	20
5.0	5.86	20	20
6.0	7.03	20	20
7.0	8.20	20	20
8.0	9.37	20	20
9.0	10.55	20	20
10.0	11.72	20	20
11.0	12.89	20	20
12.5	14.65	20	20
14.0	16.41	-	20
15.0	17.58	-	20
17.0	19.92	20	20
20.0	23.44	10	10

TAB. 1   Number of runs performed for an array of 400 (3rd column) and 1600 (4th column) cells at mutation rate $R_\star$ (in units mutations per executed instruction) [first column] or R (in units mutations per site per update) [second column].

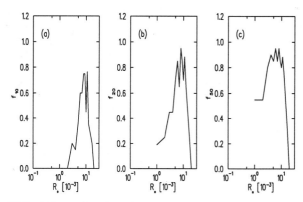

FIG. 3 Learning fraction vs. mutation rate for 400 cells in a 20 × 20 lattice, for three different cut-offs (a): f_{10}, (b): f_{20}, and (c): f_{50}.

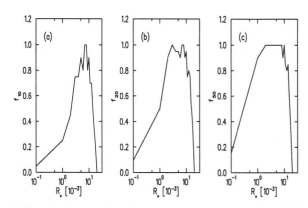

FIG. 4 Learning fraction vs. mutation rate for 1600 cells in a 40 × 40 lattice. Notation as in Fig. 3.

---

[5]Note that in avida, we choose the number of updates as the universal measure of time, as this is independent of population size.

While we were interested in results for larger populations as well, it turned out that runs for 6400 cells tended to be extremely CPU-time consuming if allowed to continue to the scheduled number of 50,000 updates. While most runs at the geometry 80x80 did learn the task fairly easily, we do not report any results for lack of statistics.

It is intuitively obvious that there should be an optimum mutation rate at which the strings on average achieve the task in the smallest amount of time. Clearly, very small mutation rates may take the population a long time to adapt. On the other hand, a mutation rate that is so high that the average time between mutations hitting a cell is smaller than the gestation time will prevent the information contained in the genome to be transmitted, and self-replication stops [3]. This limit is often called the "error-catastrophe" limit [6] and is crucial in understanding the window of adaptability.

Both the small and the large population simulations show the rapid drop-off as the mutation rate reaches the error-catastrophe limit. Interestingly, however, this limit is, for the genome-size we started out with (an ancestor of size 59) exceeded in most runs. This can be traced back to an evolutionary pressure to reduce genome size at high mutation rate. Indeed, we find that this pressure is intense starting at around $R_* \approx 10^{-2}$. If the pressure to reduce genome size (and thus to represent less of a target to the lethal mutations) is high, cells tend first to reduce genome size and then learn to add. Away from the error-catastrophe limit the pressure to reduce size is less intense, but the mutation rate too high to develop the necessary code to reap the bonus. As a consequence, we witness regions in the "learning window" where learning is suppressed. Also, we see a clear effect of population size on the learning rate. The learning fraction rises earlier for larger populations, and reaches saturation (all runs learn to add integer numbers before the scheduled finish) earlier.

5   Conclusions

We have introduced a next-generation auto-adaptive genetic system with local interactions between members of the population and causal propagation of information on a two-dimensional flat torus (periodic boundary conditions in x- and y-directions). An update mechanism that allows the strings to be executed in any order (akin to the update of cellular automaton arrays) guarantees parallelism at all times, as the average time-slice is kept constant and small. We determined that the local interactions lead to a more diverse population with a larger spread in genotype-ages, and less liability of trapping in meta-stable states, as occurs frequently if the population is too homogenous.

We investigated the evolvability of the population as a function of population size and mutation rate, and found that the learning fraction for large arrays rises earlier and thus offers a broader window for evolution, while the upper limit (error-catastrophe limit) is universal for all sizes, as expected.

Due to the flexibility of the avida system, it can be used in many varied applications. Besides breeding strings to perform user-specified tasks, it can be used for research on evolution, ecology, and maybe immunology. As an example, we have developed a method to measure the genetic distance between genotypes [7], and are planning to use it to study trajectories in genotype-space in quasi-deterministic and chaotic regimes.

Acknowledgements

We would like to thank Steve Koonin for continuing support, and Charles Ofria for collaboration in the design of avida. We also thank Reed College for the use of its computational chemistry laboratory. CTB acknowledges a SURF fellowship from Caltech. CA is supported in part by NSF grant PHY90-13248 and a Caltech Divisional Fellowship.

References

[1] T.S. Ray, in *"Artificial Life II"*, Proc. of the Santa Fe Institute Studies in the Sciences of Complexity Vol.X, C. Langton et al. eds., Addison-Wesley (1992), p. 371.

[2] T.S. Ray, *Artif. Life* **1** (1994)195; T.S. Ray, Physica D, to be published.

[3] C. Adami, "Learning and Complexity in Genetic Auto-Adaptive Systems", Physica D, in print.

[4] C. Adami, "Self-Organized Criticality in Living Systems", KRL preprint MAP-167, Caltech (December 1993).

[5] C. Adami, "On Modelling Life", these proceedings.

[6] M. Eigen, J. McCaskill, and P. Schuster, *Adv. in Chem. Phys.* **75** (1989)149.

[7] C. Adami, C.T. Brown, and C. Ofria, *in preparation*.

[8] C.T. Brown, "An Introduction to avida, an Auto-Adaptive Genetic System", SURF technical report, Caltech (October 1993) (unpublished); C.T. Brown, C. Adami, and C. Ofria, Avida Technical Manual, in preparation.

Asynchrony Induces Stability in Cellular Automata Based models

Hugues Bersini, Vincent Detours
IRIDIA - CP 194/6
Université Libre de Bruxelles
50 av. F. Roosevelt
1050 Bruxelles, Belgium
e-mail: bersini@ulb.ac.be, detours@ulb.ac.be

Abstract

Two cellular automata based computer simulations: an immune network model on one hand and the classical game of life on the other hand, despite similar algorithmic presentations, exhibit surprisingly distinct time evolution: respectively a fixed point and the complex dynamics characteristic of class IV cellular automata. At the conclusion of a complete investigation to understand better which of the algorithmic differences is responsible for this behavioural difference, we provide evidence that asynchronous rather than synchronous updating turns out to be the key factor. Experimenting and discussing in more detail this stability induction, we show that the responsibility of asynchrony for freezing game of life type of simulation can be theoretically justified in some particular cases by finding an associate Lyapunov function whose monotonous tendency proves the stability. The implications of such sensitivity to the updating mechanism for the future of cellular automata based models are reviewed.

1 Introduction

The result discussed in this paper is drawn from the comparison between two computer models, the game of life (GL) and the immune network model (INM), which although presenting very similar algorithms, display radically different dynamic regimes. GL belongs to class IV CA (Wolfram 1984), whereas INM always evolves toward fixed point attractors. These two models, both grounded in a spatially distributed type of algorithm, assume that the biological environment can be represented by a two-dimensional lattice. Each site of the lattice represents a biological entity whose state at time t can only take two values, 0 or 1, that are often interpreted as "dead or alive". The same rules apply uniformly for all the sites in the lattice and can be stated as follows: an entity dies if its local neighbourhood (to be defined in next sections) is overcrowded or if it is empty, and it is brought to life, or it survives if already born, when the neighbouring population is in a certain range of concentration i.e. neither too high nor too low. Despite these similarities, the two models differ however in several aspects and actually one can see INM as a sophisticated version of GL.

Given the resemblance between the two models and the difference between their dynamic regimes, the question that comes to mind is: what feature present in INM but not in GL is responsible for the appearance of fixed points and thus the disappearance of complex regimes. We will see that the asynchrony of INM updating rules turns out to be the key factor in explaining this behavioural simplification. Since all distributed systems necessarily have an updating procedure, the possibility that this procedure could influence the result of simulations is of interest for all the scientists using this class of models, especially in the field of artificial life.

The results presented here could hold for a large number of distributed systems. While this paper focuses on a CA based model, Lumer and Nicolis (1994) have independently reached a similar conclusion with a coupled map lattice model. Also the dependence of Hopfield net stability on updating policy is well known (Hopfield 1982). Similar results have been reported for ising models (Choi and Huberman 1984), and for evolutionary game models (Huberman and Glance 1993; Novak, Bonhoeffer and May 1994). Thus, if all members of the CA class IV turn out to be so sensitive to the type of updating, that is, may lose their dynamical richness by increasing the degree of asynchrony, a deeper analysis on the adequacy of pure synchronous updating out of respect for biological or physical reality becomes necessary. Otherwise their appealing behaviour might just appear as pure simulation artifact leading to a strong weakening of their scientific realism and potential interest.

In section 2 are described the two models from which our result is inferred. In section 3 their dynamic behaviour is reported. Then (section 4) the effect of asynchrony will be analysed in detail: we will show that the responsibility of asynchrony for freezing GL type of simulation can in some cases be mathematically justified either by finding an associate Lyapunov function whose monotonous tendency proves stability or, as was done by McIntosh, by relying on approximate mean field theory to reveal a hidden marginally stable fixed point (McIntosh 1990). We finally discuss the implications of such sensitivity to the updating mechanism for the future of the development of CA based models.

2 Presentation of GL and INM

The game of life has been presented in a large number of publications (see for example Gardner 1970), but in order to facilitate further comparisons we will now reintroduce its evolution rules in a more generic fashion: the state s_{ij} of a cell located at site (i, j) is determined by the sum h_{ij} of the states of its eight nearest neighbours using the following rules:

- if $h_{min} \leq h_{ij} < h_{med}$ then no change
- if $h_{med} \leq h_{ij} \leq h_{max}$ then $s_{ij} = 1$
- if $h_{ij} < h_{min}$ or $h_{ij} > h_{max}$ then $s_{ij} = 0$

with $h_{min} = 2$ and $h_{med} = h_{max} = 3$. The rules are applied synchronously for all sites.

In the following, other instances defined by different threshold values are labelled GL$h_{min}h_{med}h_{max}$.

Since the biological motivations behind INM are not required to understand this article, this model will only be described from a formal point of view. Readers interested in immunological background should see Stewart and Varela 1991, where this model was first introduced. INM obeys the same transition rule as GL, however the cells are updated asynchronously (the cell to be updated is chosen at random); $h_{min} = h_{med}$, and h_{ij} is computed in a different way. Let $(i' j')$ be the site symmetrical to (i, j) relatively to the center of the lattice (fig.1), h_{ij} is defined by:

$$h_{ij} = \sum_{k, l}^{L, L} m_{ij, kl} \cdot s_{kl}$$

where L defines the size of the lattice and $m_{ij, kl}$ is a gaussian function of the distance d between $(i' j')$ and (k, l).

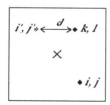

Figure 1: Computation of h_{ij} for INM.

In case of asynchrony, we say that one "time step" has elapsed when $L*L$ sites have been updated.

It is possible to construct intermediate versions between these two models in order to see which feature ($h_{min} = h_{med}$, the asynchrony, the symmetry rule, or the gaussian affinity domain) plays the most decisive role in determining at which regime the system operates.

3 The dynamics of IMN and GL

Fig.2 represents a pattern obtained when simulating INM from random initial conditions. The stripes no longer move once established.This steady state is easily explained if one notices that each stripe is coupled with a quasi-symmetric one such that both partners mutually guarantee for each other a

viable value of h_{ij}. When the lattice is covered with stripes, h_{ij} sensed by each site not yet occupied can only be either superior to the maximal threshold or inferior to the minimal one. If this is the case, no further cell can appear and no cell already present has to die. The existence of such a static equilibrium will indeed be theoretically justified in Section 4.

Complying with Wolfram's classification, such behaviour is typical of class I CA. It is relevant in the immunological context since the stripes can memorize encounters with antigens (see Stewart and Varela 1991, Bersini 1992). Then, whatever their biological utility, either as a memory mechanism in immunology, or for the formation of geometrical patterns existing abundantly in nature, class I CAs are very promising instruments to rely on for the exploration of biological systems (see Meinhart 1982). However the key question we wanted to answer was why, despite their important resemblance in using the same type of threshold mechanism, the immune modelling just described and the game of life produce such different types of behaviour that they belong to two different classes of CA.

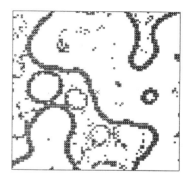

Figure 2: Fixed point pattern obtained with INM.

Simulations performed with the GL rules have gained increasing popularity during these last years due to the existence of a large family of fancy objects dancing on the computer screen, and known under the names of "gliders", "blinkers", "starships", "guns" just to mention a few of them. They led Poundstone (1985) to say: "predictable as the game is on a cell-by-cell basis, the large scale evolution of the patterns defies intuition". The complex dynamics caused by GL rules endows it with universal computation capacities (Berkelamp, Conway, and Wiesenfield 1982; Langton 1991) and thus opens the door to speculations about how the laws of logic might be grounded into natural systems and even get the upper hand on the laws of physics (Langton 1991). According to several authors (Packard and Wolfram 1985, McIntosh 1990) GL appears to be an exception in its class and systematic exploration over the huge space of possible 2D CA transition rules revealed that all members of the class IV are trivial variants of the original GL. Indeed such uniqueness can be explained first by the hard time Conway spent to find a CA dynamics capable, while maintained in non-equilibrium condition, of a certain form of viability i.e. an animate self-regulatory system neither exploding nor collapsing, and secondly by the considerable care which is needed in selecting initial conditions giving rise to interesting evolutions.

As a matter of fact, using GL in the same way as the immunological model namely initiating the simulation with a random occupation of the plane drives rapidly either to an emptying of this plane or to the survival of sparsely distributed rigid or/and oscillating objects (see fig.3 a,b for one of these simulations' final state). In order to raise interest, GL initial conditions must be selected with care (indeed this is a large part of the fun (see Poundstone 1985)) since a random selection of these conditions drives the simulation into fixed point or oscillatory configurations (as is the case for the first and the second CA classes).

Figure 3: a,b) Snapshots of the period 2 final situation of one GL simulation started from random initial conditions (the blinker oscillating) - c) An asynchronous GL final fixed point state obtained from similar initial conditions.

4 Asynchrony freezes the game of life

Now in order to better understand why the two models operate at such different dynamic regime (at least for properly chosen initial conditions), we undertook a large number of simulations, adding to GL successively the features which differentiate it from IMN: first the symmetrical affinity, then the gaussian neighbourhood and finally the asynchrony of updating. The two first features have no influence on the regime at which the system operates. Finally we came to the conclusion that the key factor for inducing stability in INM was the asynchrony. The various dynamical behaviours obtained when varying the thresholds of the synchronous GL (with the constraint $h_{min} = h_{med}$) are listed table 1. Chaotic dynamics are obtained for 13 thresholds conditions. This number drops to 0 with asynchronous updating (table 2).

h_{max}, h_{min}	0	1	2	3	4	5	6	7	8
0	o								
1	o	c							
2	o	c	c						
3	o	c	c	n					
4	o	c	c	c	n				
5	o	c	c	c	n	n			
6	o	o	c	c	n	n	n		
7	o	o	o	o	n	n	n	n	
8	s	s	s	s	o	n	n	n	n

Table 1: Dynamics of the synchronous GLxxy, random initial conditions, $L = 20$. Depending on the thresholds the ca's behaviour can be oscillatory (o), chaotic (c), steady (s), or null (n, all sites equal to 0).

A similar result is observed for the regular GL and its asynchronous counterpart: the asynchronous GL (AGL) converges to a fixed point for *any* initial conditions. Noteworthy the final configurations are qualitatively different from those obtained with the synchronous GL (SGL) (fig. 3a, b) since all the available space is populated (fig. 3c). Therefore the main message of this paper is that, depending on its updating procedure, GL can belong either to the first or to the fourth CA class.

h_{max}, h_{min}	0	1	2	3	4	5	6	7	8
0	6								
1	12	*							
2	13	14	*						
3	12	16	113	22					
4	8	11	16	*	n				
5	7	13	9	150	n	n			
6	11	10	10	13	17	n	n		
7	8	7	10	5	8	n	n	n	
8	1	1	1	1	1	5	n	n	n

Table 2: Dynamics of the asynchronous GLxxy, random initial conditions, $L = 20$. The number of time steps before a fixed point is reached (averaged over 25 runs) is listed. Thresholds conditions marked by "*" produce dynamics that remain chaotic for more than 25000 time steps when a lattice with 20*20 site is used. However these conditions lead to fixed points within reasonable time when run on smaller lattice ($L = 7$).

Can this be justified in a more formal way? Various authors have noticed the strong similarities existing between the two popular computational models: CA and neural networks (NN) (see Garzon 1990). Their architecture is similarly organized in a set of interconnected elementary automata and their interesting behaviour emerges from the collective action of these automata. Basic differences remain however. Whereas locality and uniformity of rules are key properties of CA, they disappear in NN since neurons present a parametrized activation mechanism which depends on their "individualized" synaptic connections vector (the most sensitive NN architectural element). Moreover the range of connections is not limited to the immediate neighbourhood, and the activation mechanisms generally include slightly more complicated mathematical operations.

Among the great number of NN characterized by distinct architecture and activation mechanisms, the closest to CA ought to be the Hopfield network (Hopfield 1982) since:

- it is completely interconnected so that, despite the synaptic individuation of the neurons, it shows the greatest degree of homogeneity;

- its standard activation mechanism involves threshold functions;

- in addition, what makes it very attractive in the present context is that like for GL, asynchrony is a key factor of stability, and mainly that the formal proof of stability i.e. the existence of a monotonous Lyapunov function can be transposed in a very faithful way for proving stability of the family of CA models (including INM and AGL) defined by the rules introduced in the previous section.

Indeed consider the CA counterpart of the Hopfield net i.e. a one-threshold CA (the update rule can boil down to: $s_{ij} = 1$ if $h_{ij} \leq h_{min}$, $s_{ij} = 0$ if $h_{ij} > h_{min}$), and associate to this rule, the Lyapunov Energy function E written below:

$$E = -\sum_{i,i} s_{ij}(h_{min} - \frac{h_{ij}}{2})$$

In case of asynchronous transition, one cell is selected in a random or deterministic way to be updated (the selection order can be sequential or random, it is in no way a crucial point). Due to the absence of self-interaction and the symmetry of CA structure, we obtain:

$$h_{ij}\Delta s_{ij} = \sum_{kl} s_{kl} \cdot \Delta h_{kl}$$

where kl indexes the eight neighbours of ij. Thus updating E as a result of updating the selected cell (i,j) gives

$$\Delta E = -\Delta s_{ij}(h_{min} - h_{ij})$$

In consequence for each transition: $\Delta E \leq 0$. Given that the asynchronous form of this CA family causes E to be a monotonous decreasing function, the stability is proved: state changes will continue until a least E is reached.

In contrast, if more thresholds appear in the rules the discovery of a Lyapunov function is a non trivial problem. Let's associate with a two threshold CA of the form GLxxy the function:

$$E = \sum_{ij} s_{ij}h_{ij}^2 - \sum_{ij} \frac{s_{ij}}{2}h_{ij}h_{min} - \sum_{ij} \frac{s_{ij}}{2}h_{ij}h_{max} + \sum_{ij} s_{ij}h_{min}h_{max}$$

any transition results in:

$$\Delta E = -\Delta s_{ij}(h_{ij} - h_{min}) \cdot (h_{max} - h_{ij}) + \sum_{kl} s_{kl}\Delta h_{kl}^2$$

We easily observe that although ΔE is always negative for a $1 \to 0$ transition, problems appear for a $0 \to 1$ transition for which the second term has a positive contribution conflicting then with the first one. Though a fixed point was always reached in our computer simulations, the existence of a monotonously decreasing Lyapunov function associated to the two-thresholds CA becomes questionable. At least it seems that its existence could be highly dependent on the thresholds values.

As a matter of fact, the existence of a Lyapunov function should imply that the fixed points of the system have large basins of attraction, which is unlikely for the threshold conditions resulting in extremely long transient time (table 2). This later remark also concerns AGL since the steady states

seem to exert no attraction. This view is consistent with the fact that a long time is often needed before a stable configuration is reached (the duration of transients varies between $5 \cdot 10^2$ and 10^5 time steps depending on initial conditions). Moreover the analysis of the damage caused by the flipping of a single site has revealed a very weak robustness of the fixed points to small changes. Actually 60% of the 200 single site perturbations performed on stabilized systems ($L=20$) led to avalanches involving all the sites. The duration of these avalanches was comparable to the duration of the transient just mentioned, i.e. the perturbation of one site had the same effect as starting from a new random initial state. In the remaining 40%, the disturbance was restricted to less than 10 sites in the neighbourhood of the perturbation.

Variants of GL with INM-like thresholds (i.e. GLxxy), were very much faster to stabilize as compared with AGL. Indeed as fig.4 shows, the final fixed point seems to really exert a strong attraction (a situation quite different than for AGL fixed points). A damage spreading analysis has demonstrated a strong robustness of the steady states to perturbations: avalanches triggered by a single site flip never spread throughout the whole lattice and remain nearly always confined to less than 10% of the cells. Also it is necessary to disturb more than 20% of the sites in order to observe significant change in the global shape of the patterns. In general the larger the thresholds window the easier to obtain and more robust the fixed points turn out to be. This could be roughly justified by observing that for the $0 \to 1$ transitions, the strength of the ΔE expression first term increases with the enlargement of the window. We may reasonably hypothesize that in our current situation i.e in the absence of a decreasing Lyapunov function, the speed of convergence and the stability of the fixed points could be dependent on the proportion of the local increases ($\Delta E > 0$) and plateau ($\Delta E = 0$) to be encountered by the function E while showing a general decreasing tendency.

Interestingly enough, Hopfield networks present the same type of pathological behaviour as CA: in its synchronous version a lot of initial conditions cause the dynamics to be trapped into a periodic attractor (cycle of length at most 2 when the connections are symmetric). However, carrying on with the similarities, when the system evolves from more favourable initial conditions, the network settles in a fixed point following a transient shorter than in the asynchronous case. Clearly synchronous updating makes larger step in the attractive road towards fixed point than asynchronous updating, and this too rapid progression in the basin of attraction might explain both their rapid convergence when converging but also their oscillatory or even chaotic to and fro motion around the fixed points when not converging. In a similar way, it is well known that a gradient descent running with too important a learning rate can be responsible for erratic motion around the extremum.

All these common properties between CA and Hopfield networks tend to suggest that in its synchronous version the GL type of CA suffers from the same sort of instability as synchronous Hopfield networks: if the converging rate is too important, it prevents the convergence and induces an erratic wandering around the fixed point. The fact that the complex behaviour of SGL may be attributed to a marginal stability at a fixed point was already conjectured by McIntosh (1990) relying on mean field theory. Based on a careful observation of the mean field curve for probabilistic self-consistency, it has been noticed that in general the actual transition probabilities characterizing class IV CA were located close to the theoretical solutions of the self-consistency problem and in a very stable region of the curve. This is why one may consider class IV CA behaviour as marginally stable, and why a reinforcement of the asynchrony contributes to suppress this marginality.

Figure 4: Time evolution of AGL223.

5 Discussion and conclusion

In a very recent work, Lumer and Nicolis (1994) discussed the induction of stability the switching from synchrony to asynchrony is responsible for in a model of coupled map lattice originally presented by Kaneko (1989) and observed to generate complex spatiotemporal dynamics when using a synchronous updating. Huberman and Glance have noticed that the same conclusion applies to simulations of evolutionary spatial games. Here we have shown that this result also holds for a whole class of threshold CA. In the last few years an important number of spatially distributed systems, generally grounded into CA models, have been largely appreciated and discussed for their capacity to evolve neither in an oscillatory way nor in a full chaotic way but somewhere in between, in a transition region called "the edge of chaos" (Langton 1991; Bak, Tang, and Wiesenfeld 1988; Chen and Bak 1989; Kauffman and Johnsen 1991). Such a critical behavioural regime is capable of storage (local quiescent state) and transmission of information (the propagating structures). Indeed manipulating these storing and propagating computational objects in a proper way can code for the elementary logical operations. SGL like Langton's transitory stochastic CA (Langton 1991) and like Bak's sand pile (Bak, Tang, and Wiesenfeld 1988) or trivial universe (Chen and Bak 1989) are all Turing universal.

An obvious question coming to mind is (originally raised in Huberman and Glance 1993, and Lumer and Nicolis 1994): "are the observed interesting regimes not just mere artifacts of the discretisation of the biological reality these models aim at reproducing?". Indeed, if reality is better captured by continuous differential equations, Hopfield networks testify to the greater adequacy of asynchronous updating as compared with the synchronous one for giving the most accurate version of the differential equations solutions.

Differential equations are in principle synchronous (although some asynchrony might be introduced by adding a probabilistic term), but because of the rough discretisation that their computer simulation demands, neither a synchronous nor an asynchronous form of updating can ensure a better fidelity to the actual temporal evolution that a lack of analytical solutions may keep hidden. The ideal solution would see the synchronous and asynchronous updating giving rise to similar spatiotemporal evolution. But when this is not the case, which one is telling the truth: the fixed point or the complex regime?

In facing this delicate synchronous versus asynchronous problem which might concern numerous computer simulations behaving in a complex manner, two attitudes can be adopted but need to be clearly distinguished. A first one consists in the development of very interesting computer simulations capable of teaching us a great deal about complex systems but it must not be confused with a physical or biological scientific endeavour since no external physical or biological reality is expected to be reproduced and deeply understood in one way or another by these simulations. This attitude has much more to do with a creative or artistic effort and is more in line with a mathematical rather than a physical approach. SGL, Langton's CA, Chen and Bak's toy model of interacting ballistic particles in no way try to match any particular physical or biological system. Needless to say, these computer simulations share the same appeal (and indeed gliders, guns, starships, traffic lights, shuttles don't lack of such appeal) for the same reasons and remain worth of scientific attention. They are perfect illustrations of how nature can be layered in multiple levels of coarseness, and how elementary laws at a low level can be responsible for complex behaviour at upper levels. They nicely show how unpredictability can still be grounded in predictable but finer mechanisms, and why keeping things simple but observed through a microscope can in some circumstances be the most appropriate strategy to confront apparently complex systems. If this is the attitude being adopted, the problem posed by the contrasting behaviour between different updating procedures simply vanishes. If not constrained by any biological or physical reality, just use the updating procedure which results in the kind of behaviour which better illustrates what in fact you aim at illustrating.

Now a second and to some extent much more demanding attitude is the physical one: There is some physical or biological reality whose structure and functionalities one wants to better grasp by means of computer simulation. It is supposed that such systems are continuous with direct consequence that any coarsening of the space and time resolution required by the computer simulation is not immune of artifacts. It is well known for instance that the logistic map is chaotic only in its discrete time form, and that continuous systems need to possess at least three dimensions to exhibit similar chaos. Since there is no specific reason to privilege either the synchronous or asynchronous updating of the simulation for observing its evolution in time, and given that the two updating procedures generate totally different spatiotemporal trajectories, a hard problem is raised: which behaviour is the most faithful to reality. If possible, a Popperian type of validation might be the supreme judge in some cases, but nevertheless a high sensitivity to the updating procedure should call for an increased attention and prudence while running the simulation.

Some of these "edge of chaos", "critical" or "phase transition" computer simulations do have physical or biological counterparts like Per Bak's pile of sand (Bak, Tang, and Wiesenfield 1988), earthquake models (Sornette and Sornette 1989), Kauffman's interconnected fitness landscapes (Kauffman and Johnsen 1991) and others. On the other hand, nature seems full of systems showing fluctuations with dimension following a power law and showing spatial self-similarity as well, which are two key diagnostic factors for "criticality" and "phase transition" which these computer toy simulations aim at reproducing. Although the existence of these natural phenomena gives high legimitity to the computer simulations endowed with the same characteristics, the existence of a contrasting behaviour depending on the updating procedure, and the crucial fact that this sensitivity often turns out to be the indicator of a hidden stability (and thus a potential outstanding behavioural simplification) is the main argument for advocating increased attention from the developers of these simulations.

References

Bak, P., C. Tang, C. and K. Wiesenfield. 1988. Self-Organized Criticality. *Phys. rev. A* 38 (1). 364-374.

Berkelamp, E. , Conway, J.H. and R. Guy. 1982. Winning Ways for Your Mathematical Plays. New York: Academic Press.

Bersini, H. The Interplay Between the Dynamics and the Metadynamics of the Immune Network. IRIDIA Technical Report 92/12.

Chen, K. and P. Bak. 1989. Is the Universe Operating at Self-Organized Critical State. *Physics Letters A* . Vol. 140. no 6. 299 - 302.

Cheung, K.F. , Atlas, L.E. and R.J. Marks II. 1987. Synchronous vs Asynchronous Behaviour of Hopfield's CAM Neural Net. *Applied Optics*. Vol. 26. no 2. 4808-4813.

Choi, M.Y., and B.A. Huberman. 1983. Digital Dynamics and Simulation of Magnetic Systems. *Phys. Rev. B* 28. 2547-2554.

Garzon, M. 1990. Cellular Automata and Discrete Neural Networks. *Physica D* 45. 431-440.

Hopfield J.J. 1982: Neural Networks and Physical Systems with Emergent Collective Computational Abilities. *Proc. Nat. Acad. Sci. USA* . vol. 79. 2554 - 2558.

Huberman, B.A. and N.S. Glance. 1993. Evolutionary Games and Computer Simulations. *Proc. Nat. Acad. Sci. USA*. vol. 90. 7716-7718.

Kaneko, K. 1989. Pattern Dynamics in Spatiotemporal Chaos. *Physica D* 34 . 1-41.

Kauffman, S.A. and S. Johnsen. 1991. Co-Evolution to the Edge of Chaos: Coupled Fitness Landscapes, Poised States, and Co-Evolutionary Avalanches. In Artificial Life II. Langton, Taylor, Farmer and Rasmussen (eds.). Addison-Wesley Publishing Company.

Langton, C. 1991. Life at the Edge of Chaos. In Artificial Life II - Langton, Taylor, Farmer and Rasmussen (eds.). Addison-Wesley Publishing Company. Redwood City, CA . 41-91.

Lumer, E.D. and G. Nicolis 1994. Synchronous Versus Asynchronous Dynamics in Spatially Distributed Systems. *Physica D* 71. 440-452.

McIntosh, H.V. 1990: Wolfram's Class IV Automata and a Good Life. *Physica D* 45. 105-121.

Meinhart, H. 1982. *Models of Biological Pattern Formation*. Academic Pres. London.

Novak, M.A., S. Bonhoeffer and R.M. May. 1994. More Spatial Games. *International J. of Bifurcation and Chaos*. Vol 4. no 1. 33-56.

Packard, N.H. and S. Wolfram 1985. Two-dimesional cellular automata. *J. Stat. Phys*. 38. 1-35.

Poundstone, W. 1985. *The Recursive Universe*. Morrow. New York.

Sornette, A. and D. Sornette 1989. Self-Organized Criticality and Earthquakes. *Europhysics Letters* 9 (3). 197-202

Stewart, J. and F. Varela. 1991. Morphogenesis in shape-space: Elementary meta-dynamics in a model of the immune network. *J. theor. Biol*. 153. 477-498.

Wolfram, S. 1984: Universality and Complexity in Cellular Automata . *Physica D* 10 . 1-35.

Evolutionary Automata

Murray Shanahan

Imperial College
Department of Computing,
180 Queen's Gate,
London SW7 2BZ, England.
Email: mps@doc.ic.ac.uk

Abstract

The class of Evolutionary Automata is formally defined. An Evolutionary Automaton generates an evolving population of "organisms" within a computational medium. The organisms move randomly around a microworld, eat, reproduce asexually (with mutation), and die of "natural causes". The results of some preliminary experiments conducted with an instance of this class are reported, in which a parameter was varied across a small range. The paper concludes with some speculation on the possible significance of these results for the study of diversity and complexity in Artificial Life.

Introduction

One of the most attractive features of the manifesto of Artificial Life research is that it legitimises the study of any life-like phenomenon that can be reproduced in a computer. Artificial Life research avoids the prolific detail present in real-life biological phenomena, but at the same time enjoys the freedom to explore a larger space of possible phenomena than actually arise in nature [Langton, 1987]. This wider perspective encourages the asking of deep scientific questions, such as "What preconditions are essential to the origin of all life-like phenomena?"

Unsurprisingly, evolution is one of the phenomena that has attracted most attention from Artificial Life researchers. Work on evolution currently falls into two main categories. First, some impressive results have been obtained through the simulation of evolutionary processes in microworlds, some of which are deliberate attempts to model aspects of biological reality (for example, [Yaeger, 1992]), whilst others are unashamedly artificial (for example, [Ray, 1990]). Second, a number of computer experiments have been conducted which are based on abstract automata, in particular cellular automata [Packard, 1988], [Langton, 1990], and boolean networks [Kaufmann, 1992, Chapters 5 & 6]. These experiments are not so much an attempt to model particular evolutionary phenomena in detail. Rather, their emphasis is on deepening our theoretical understanding of such processes in general.

This paper falls between these two categories. A class of Evolutionary Automata is described, in which a population of entities evolves in a microworld. This class is large — in the sense that different instances of it can reproduce a wide variety of effects — yet abstract enough for it to be susceptible to simple mathematical description and tractable formal analysis. (Holland's Echo model [1992, Chapter 10] is similarly motivated.) The point of studying such a class is that it enables us to investigate deep scientific questions, such as "Under what conditions does evolution produce complexity and diversity?" using a tool that models evolution with a fine enough grain of detail for complexity and diversity actually to arise.

The first section of the paper presents the basic idea of an Evolutionary Automaton and a formal definition. The next two sections present some experiments with a particular automaton, in which the effect on evolution of varying one of the automaton's parameters is studied. It turns out that this parameter must be tuned to be close to a critical value for the automaton to produce interesting effects. The paper concludes with some discussion of the question of the conditions that are required for complexity and diversity to emerge from an evolutionary process. The potential relevance of the edge of chaos to this question is discussed, but in the light of the reported experiments its applicability seems minimal.

1. The Class of Evolutionary Automata

After some informal characterisation, this section formally defines the class of Evolutionary Automata (EA's). Essentially, an Evolutionary Automaton non-deterministically specifies a sequence of states of a microworld. The microworld is a square grid of locations. Both top and bottom edges and left and right edges are joined so that the microworld is, in effect, shaped like a torus. An evolving population of "organisms" inhabits the microworld. These organisms can move, eat, reproduce, and die of "natural causes."

Each organism occupies a number of connected locations in the microworld. An organism's whereabouts in the microworld is constantly changing, but its shape (or phenotype) remains unchanged throughout its lifetime. Each organism has a genotype, which can be thought of as defining the "species" to which it belongs. Its shape is a function of its genotype. Organisms can be either alive or dead. Dead organisms don't move, eat or reproduce, and can be eaten by any other organism except members of their own species.[1] Each organism has a certain energy level. Eating increases its energy level and reproduction depletes it. There are four kinds of microworld event.

---

[1] Forbidding cannibalism prevents the establishment of species which are completely independent from the rest of their environment, and so encourages the setting up of food webs. I have built this restriction into the definition of an EA, but it could be relegated to the descriptions of particular automata.

- **Movements:** At each increment of the mircroworld clock, every living organism moves a random distance in a random direction, up to a certain maximum distance away.

- **Meals:** Under certain circumstances one organism can eat another, thus gaining a number of units of energy. Organisms have to be within a certain distance from each other for one of them to eat the other.

- **Births:** When an organism has sufficient units of energy, it will reproduce. Reproduction is asexual, but random mutations to the organism's genotype can occur.

- **Natural deaths:** As an organism increases in age, it becomes ever more likely that it will die of "natural causes." When this happens it becomes frozen in its current location, and doesn't eat or reproduce.

It will be immediately obvious that this description includes many features we would expect to find in a population undergoing Darwinian natural selection. To obtain an instance of the class of Evolutionary Automaton, a number of parameters have to be supplied. These include,

- The mapping from genotype to phenotype,
- The conditions under which one organism can eat another,
- The mutation rate,
- The maximum distance an organism can move per time step, and
- The probability that an organism will reproduce given its energy level.

In addition, a particular run of an EA must be parameterised by grid-size, and an initial state. By varying all these parameters, a large number of effects can be produced. Now comes the formal definition.

Preliminary definitions. A *location* is a pair of integers. The *grid* of a microworld of size N is the set of locations $\langle x,y \rangle$ such that $abs(x) \leq N$ and $abs(y) \leq N$, where $abs(z)$ is the absolute value of z. A *cell* is a pair comprising a type and a location (the type allows for different kinds of cell). Two cells are said to be *connected* if their corresponding locations are adjacent or if there exists a third cell which is connected to both of them. A *phenotype* is a set of connected cells including a distinguished cell $\langle t,\langle 0,0 \rangle \rangle$ called the *centre*. Anything can be a *genotype*.

Definition 1. Let Org be an infinite set (the carrier set of potential organisms) and Gen be a set of genotypes. Then a *state* of a microworld of size N is a 6-tuple $\langle O,G,L,S,A,E \rangle$, where

- O is a subset of Org (all the extant organisms, both alive and dead),
- G is a mapping from O to Gen (the organism's genotype),
- L is a mapping from O to the microworld's grid (the organism's whereabouts — actually the location of its centre),
- S is a mapping from O to True or False (whether the organism is alive or dead),
- A is a mapping from O to the naturals $\mathbb{N}$ (the organism's age), and

- E is a mapping from O to $\mathbb{N}$ (the organism's energy level).

Definition 2. Let Gen be a set of genotypes and Phe be a set of phenotypes. An *Evolutionary Automaton* is an 8-tuple $\langle P,M,U,R,X,D,B,T \rangle$, where

- P is a mapping from Gen to Phe,
- M is a mapping from Phe $\times$ Phe to True or False (whether an organism with a given phenotype can eat one with another phenotype) such that if M(X,Y)=True then M(Y,X)=False,
- U is a mapping from Gen to 2^{Gen} (the range of genotypes a given genotype can mutate into),
- R is a real number in the range 0..1 (the probability of mutation),
- X is a natural number (the maximum distance an organism can move in one time step),
- D is a function from $\mathbb{N}$ to reals in the range 0..1 (the probability that an organism will die of natural causes given its age),
- B is a function from Phe $\times$ $\mathbb{N}$ to reals in the range 0..1 (the probability that an organism of a given phenotype will reproduce when it has a given energy level), and
- T is a natural number (how close two organisms have to get before one can eat the other).

Let Org be an infinite set of organisms, Gen be a set of genotypes and Phe be a set of phenotypes. Then given the initial state $\langle O,G,L,S,A,E \rangle$ of a microworld of size N, an Evolutionary Automaton $\langle P,M,U,R,X,D,B,T \rangle$ generates a sequence of microworld states by repeatedly executing four procedures — deaths, moves, meals and births — which are specified below. Each procedure updates one or more of the variables O, G, L, S, A, E representing the state of microworld.

The Deaths procedure alters the status, denoted by S, of any number of living organisms depending on their age. It also updates the age of each living organism. The death rate can be controlled by varying the function D. Obviously, the longer an organism is allowed to live, the higher the chances of the survival of its species. A low death rate favours organisms that eat only occasionally, but eat a lot. A high death rate favours organisms that eat little and often.

> **Procedure** Deaths(O,S,A)
>   **For** all x in O such that S(x) **Do**
>   A(x) := A(x)+1
>   **With** probability D(A(x)) **Do**
>   S(x) := False

The Moves procedure alters the location, denoted by L, of each living organism. Two or more organisms can have the same location. Note that the organisms have no control over their own movement. They just float about randomly. In fact, they have no "behaviour" at all. The only feature they have, which determines their capacity to survive, is their shape, which is a function of their genotype. The maximum distance an organism can move in one time step is controlled by the X parameter. A small value for X will encourage the formation of localised colonies of the same species. The greater the value of X, the faster a species will spread around the microworld.

Procedure Moves(O,S,L)
   For all x in O such that S(x) **Do**
     Let θ be a random direction
     Let δ be a random distance $\leq$ X
     L(x) := L(x) translated by δ units in direction θ

The Meals procedure, which updates the O and E variables, is where the two vital mappings M and P come into play. Alteration of these parameters can give rise to completely different behaviours. This is where selection takes place. It is important to note that the M mapping does not supply a global measure of fitness. It does not even have to be a transitive relation, thus allowing a paper/scissors/stone relationship between species of the form A eats B, B eats C but C eats A. The absence of a global fitness function makes open-ended evolution possibile.

Procedure Meals(O,S,E)
   Let s be the set of all x,y in O such that S(x) and S(y)
   and the distance between x and y < T
   While s not empty **Do**
     Let ⟨a,b⟩ be a randomly chosen member of s
     s := s − ⟨a,b⟩
     If M(P(G(a),P(G(b))) or [S(b)=False and
     G(a)≠G(b)] **Then**
       O := O − b
       E(a) := E(a) + E(b) + size of b

Finally, we have the Births procedure. This procedure, whenever a new organism is born, will update all the variables comprising the state. The genotype of a newborn organism is liable to mutation, which will of course alter its phenotype. Mutation can be constrained by the U parameter to ensure that evolutionary change is gradual. The mutation rate is controlled through the parameter R. A high mutation rate will lead to runaway evolution, in which numerous new species are created, only to be superseded by even newer species before they have time to establish themselves.

Procedure Births(O,G,L,S,A,E)
   For all x in O such that S(x) **Do**
     With probability B(P(G(x)),E(x)) **Do**
       Pick any y in Org not in O
       O := O+y; L(y) := L(x)
       S(y) := True; E(y) := 0
       A(y) := 0; G(y) := G(x)
       With probability R **Do**
         Let G(y) be any member of U(G(y))

Before moving on to a specific example, I would like to point out a few features of this class of automata. First, unlike the microworlds used in many previous computer simulations of evolution, that manipulated by an EA contains just organisms. There are no obstacles, no landscape features, in fact no inanimate objects at all. An organism's environment consists entirely of other organisms (dead or alive). Second, as already mentioned, organisms have no behavioural characteristics. In much previous work of this kind, species evolve with different behavioural tendencies. In contrast, all evolution in an EA is evolution of shape. However, the M mapping, which determines who eats who based on their shapes, can be any

(computable) function.[2] Suitable instantiations of M can lead to very interesting evolutionary effects, which can then be studied without having to grapple with the complexities of behaviour.

Evolutionary Automata bear some resemblance to genetic algorithms [Holland, 1992]. However, there are several very important differences which it is worth emphasising.

- There is no global measure of fitness in an EA, whereas there is in a genetic algorithm. Rather, the M function allows for open-ended evolution.
- Geographical effects are possible in an EA. This limits the influence of each organism to its neighbours. In a genetic algorithm, a single global ranking of all members of the population is used to decide which will survive to the next generation.
- In a genetic algorithm, unlike an EA, organisms do not persist from one time step to the next, and there is no concept of natural death. In a genetic algorithm, each time step is a new generation.
- Unlike genetic algorithms, EA's only feature asexual reproduction. EA's with crossover are an obvious topic for further study.

One way to generalise the class of EA's is to endow each organism with an internal state which can be changed by encounters with other organisms. The M function would have to allow for forms of interaction other than eating. This would permit the evolution of organisms that co-operate in various ways through the trade of information, with both members of their own species and members of other species. As a special case of this, genetic material could be stored and exchanged via the internal state, and reproduction could be made a function of this state. Experiments with such generalisations are a topic for future research.

2. A Simple Evolutionary Automaton

This section and the next describe some computer experiments carried out with a simple but non-trivial EA. These experiments are on a very small scale, and should be thought of as a first foray into the space of possible EA's. In each of these experiments, the microworld is a grid of 50 by 50 locations.

A variety of P's and M's were tried, apart from those I'm about to describe. These included a mapping from genotype to phenotype which was very similar to the recursively defined mapping used by Dawkins [1987] in his "biomorphs" program. Using this mapping, with a suitable instantiation of M, evolution can produce spectacular results in a few hundred time steps. But much of the apparent complexity in the resulting organisms is inherent in the recursive nature of the mapping. Because I wanted to study the production of complexity and diversity by

---

[2] It's a straightforward exercise, for example, to construct an EA that evolves an improving population of noughts and crosses (tick-tack-toe) playing organisms. An organism's body is a look-up table, mapping board positions to moves. The M function plays two organisms against each other, reading off each organism's moves from this look-up table. The winner eats the looser.

selection alone, the main experiment described in this section uses a simpler mapping.

Three cell types are used: red, white and blue. A genotype has an infinitely extendable tree structure, recursively defined as follows.

- A *colour* is either red, white or blue.
- A *gene* is either,
  - A pair $\langle c,n \rangle$, where c is a colour and n is a natural number, or
  - A list $[\langle c,n \rangle, g1, g2, g3]$, where c is a colour, n is a natural number, and g1 to g3 are genes.
- A *direction* is either north, south, east or west.
- A *genotype* is a pair $\langle d,g \rangle$, where d is a direction and g is a gene.

For example, $\langle east, [\langle white,2 \rangle, \langle red,1 \rangle, [\langle blue,2 \rangle, \langle red,1 \rangle, \langle blue,0 \rangle, \langle red,0 \rangle], \langle red,1 \rangle] \rangle$ is a genotype. The corresponding phenotype is shown in Figure 1. Informally, to form the phenotype for the genotype $\langle d,[\langle c,n \rangle, g1, g2, g3] \rangle$, you begin at $\langle 0,0 \rangle$ and add n cells of colour c in direction d, then rotate 90° anticlockwise and add the cells for g1, move back to direction d and add the cells for g2, then finally rotate 90° clockwise and add the cells for g3. More precisely,

- $P(\langle d,g \rangle) = \langle 0,0 \rangle \cup Grow(g,d,\langle 0,0 \rangle)$, where
  $Grow(\langle c,n \rangle, d, \langle x,y \rangle) =$ the next n cells (with colour c) in direction d starting from $\langle x,y \rangle$
  $Grow([\langle c,n \rangle, g1, g2, g3], d, \langle x,y \rangle) = Grow(g1,d-90°,\langle x',y' \rangle) \cup Grow(g2,d,\langle x',y' \rangle) \cup Grow(g3,d+90°,\langle x',y' \rangle) \cup Grow(\langle c,n \rangle,d,\langle x,y \rangle)$,
  where $\langle x',y' \rangle$ is n units from $\langle x,y \rangle$ in direction d

⊖ Red

⊛ Blue

Figure 1: An Example Phenotype

The rule that determines whether one organism eats another (the mapping M) works by awarding points to the two organisms. An organism gets two points for every one of its cells that covers and beats one of its opponent cells. The "beats" relation is not transitive. A red cell beats a white cell, a white cell beats a blue cell, and a blue cell beats a red cell. In addition, an organism gets a point for every one of its cells that is not covered by any of its opponent's cells. If one organism is awarded more points than the other, then it eats the other. Let Points(p1,p2) be the number of points awarded to p1 against p2, according to the above description. Then we have,

- M(p1,p2) = True if Points(p1,p2) > Points(p2,p1)
  M(p1,p2) = False if Points(p1,p2) ≤ Points(p2,p1)

The rest of the parameters of the EA are instantiated as follows. The crucial parameter for the suite of experiments I will discuss is the penultimate one, B. This will be varied, by varying the value of θ, to give different effects, as we'll see shortly.

- $U(\langle d,g \rangle) =$ the set of all genotypes obtainable either,
  1) By replacing the direction d by d±90°, or
  2) By replacing any gene $\langle c,n \rangle$ in g where n≤1 by $\langle c,n+1 \rangle$, or
  3) By replacing any gene $\langle c,n \rangle$ in g where n>1 by $\langle c,n±1 \rangle$, or

  4) By replacing any gene $\langle c,n \rangle$ in g by $[\langle c,n \rangle, g_1, g_2, g_3]$, where each g_i has the form $\langle c_i,n_i \rangle$ where c_i is a colour and some $n_i=1$ whilst the other two $n_i=0$

- R = 0.1667 (1 in 6)
- X = 5
- $D(n) = \dfrac{n^3}{125}$ or 1, whichever is smaller
- $B(p,e) = \begin{cases} 0 & \text{if } e < s \\ \dfrac{e}{s^\theta} & \text{if } s \le e < s^\theta \\ 1 & \text{if } e > s^\theta \end{cases}$ where s = size of p
- T = 5

As should be clear, the mapping U is designed to ensure that (almost) any genotype can evolve from any other, but in small steps. But what is the rationale behind the apparently arbitrary choice of the other parameter values? Many of these values, although they are the result of experimentation to find values which give interesting behaviour, would tolerate quite a lot of variation. The mutation rate (R) must be neither so small as to arrest evolution altogether, nor so large as to prevent species from establishing themselves. The ratio of the amount of movement allowed per time step (X) to the maximum distance at which organisms can eat each other (T)[3] must be neither so small as to prevent organisms from finding food before they die of natural causes, nor so large as to negate geographical effects altogether. The function D guarantees that no organism can live longer than five time steps, whilst ensuring that the probability of dying of natural causes is small for young organisms and grows rapidly towards age five.

The function B, which determines the probability that an organism will reproduce, is intended to give some advantage to smaller organisms to counteract the disadvantage they are given by the mapping M, thus encouraging the setting up of food webs. The larger an organism, the more it has to eat to have a reasonable chance of reproducing. The degree of this advantage is controlled by the internal parameter θ. A large θ accentuates the advantage of being small.

3. Experimental Results

Each of the parameters of the EA described above, in particular R, X, D, B and T, could be varied experimentally in order to gain an understanding of the workings of the family of EA's with similar genotype to phenotype mappings and similar rules for eating, hopefully thus improving our understanding of evolutionary processes in general. In the small suite of experiments whose results I will now present, the value of θ (internal to the parameter B) was varied from 1 to 3. Each run involved an initial population of 150 organisms, randomly placed in the microworld. Each organism in the initial state was given a

---

3 For efficiency regions, in the actual implementation the microworld was divided into five by five regions, and two organisms were deemed to be within eating distance of each other if any parts of their bodies were in the same region.

randomly generated genotype of the form ⟨d,⟨c,n⟩⟩, where n≤3.

Twenty experiments of 200 time steps were conducted for each of θ=1, θ=2 and θ=3. The experiments were performed with an Apple Macintosh Quadra, and were implemented in LPA MacProlog. Particular attention was given to the behaviour of three variables throughout each run: the total number of different species, the mean size of organism, and the total population. Two organisms belong to different species if they have different genotypes. The number of species in the microworld at any given time is the number of different genotypes having two or more exemplars. In other words, unique mutations that never reproduce are not considered to have inaugurated a new species.

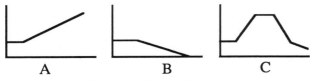

Figure 2: Graph Shapes

When the variables mentioned above are plotted against time, several distinct graph shapes are to be found, three of which are denoted A, B, and C in Figure 2. There was a

other words, an exploding population of single-celled organisms was generated In general, the dominant species in the population was alternately red, white then blue, then red again, with increasing colonies of one colour often chasing declining colonies of another across the microworld.

The most interesting results were obtained with the intermediate value for θ. With θ=2, there was considerable variation in the shapes of the graphs from one run to another. Examples of both the above extremes of behaviour were obtained. But in most of the runs, the population and number of species produced graphs of shape C, whilst mean organism size produced a graph of shape A. The upturn in mean organism size corresponded with the decline in population. This last kind of behaviour corresponds to an initial phase during which small organisms thoroughly establish themselves, supporting a small population of moderately successful predators, followed by the emergence of one or more super-predators which are so successful that they eat all available food, resulting ultimately in mass extinction. Graphs for an actual run with θ=2 are shown in Figure 3.

Considerable variation was present even within the set of runs displaying the last kind of behaviour described. In one of the twenty runs, although conforming to the same

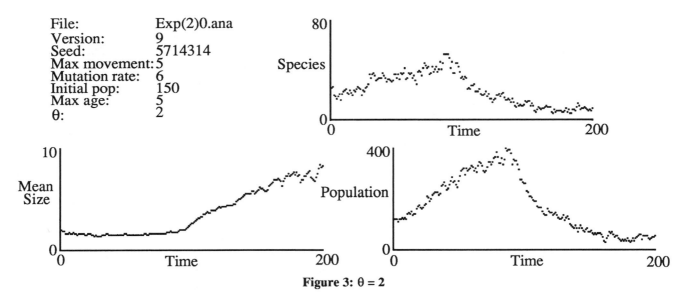

Figure 3: θ = 2

strong correlation between population and number of species. In all twenty experiments with θ=1, population and number of species produced graphs with approximately shape B, whilst mean size produced a graph with approximately shape A. In other words, with such a low θ, successful predators emerge very quickly. These predators rapidly eat all available food, and soon become extinct.

With θ=3, all twenty experiments generated graphs for population and number of species with approximately shape A. The mean size of organism, however, remained constant and low (close to one), reflecting the fact that predators find it very hard to establish themselves with such a high θ. In

overall pattern, the population went through a sequence of peaks and troughs coinciding with troughs and peaks in mean organism size. This is just the sort of cyclical pattern often taken to indicate the presence of a homeostatic relationship between predator and prey. Had evolution been arrested before the advent of the first super-predator, this homeostasis might have persisted indefinitely. In addition, not all of the runs ended with total extinction, and it would have been interesting to continue beyond 200 time steps to see when, if ever, the whole population would die out. Clearly, much further experimentation is required with θ=2 to see just what is going on.

Discussion

Langton [1990] asks, "Under what conditions can we expect a dynamics of information to emerge spontaneously and come to dominate the behaviour of a physical system?" The answer, he suggests, is that "living systems . . . must have learned to steer a delicate course between too much order and too much chaos," and "evolution can be viewed as the process of gaining control over more and more 'parameters' affecting a system's relationship to the vital phase transition [between order and chaos]."

So, according to Langton, life exists at the edge of chaos, in other words in the "complex regime" to be found in the vicinity of the phase transition between order and chaos. But even if we accept this as the basis of the answer to Langton's question, we cannot stop there. With the complex regime already established, how is it that evolution has been creative enough to produce the diversity and complexity[4] we find in life on Earth? Without this diversity and complexity, we surely wouldn't exist. So, within the complex regime, what happens? Let's zoom in on the complex regime, and ask ourselves, "Within a physical system whose behaviour has come to be dominated by a dynamics of information, under what conditions can we expect diversity and complexity to evolve?"

We might expect the idea of the edge of chaos to apply to this question too. As well as its products, the evolutionary process itself, we might speculate, has to steer a course between too much order and too much chaos. For evolution to be frozen in the ordered regime is for it to lack the creative power to produce enough variation to support diversity or complexity. On the other hand, for evolution to be in the chaotic regime is for it to produce too much variation, enough variation to overwhelm the forces of cumulative selection, preventing the establishment of the sort of stable ecosystem in which we find ourselves.

However, even the preliminary exploration of the world of EA's presented in this paper suggests that the truth is much more complicated than this. Section 2 defined a simple EA, and Section 3 presented some experimental results based on this automaton. In these experiments, the number of species is a crude measure of diversity, whilst the mean organism size is a crude measure of complexity. Although based on a small sample, these experiments have shown that, with this EA, the conditions necessary for complexity and diversity to arise are extremely fragile. The parameter θ must be close to 2 for there to be a realistic chance of diversity and complexity arising. If θ is too high, we get a degree of diversity but without complexity. If it is too low, we get complexity at the expense of stability. Experiments with a much larger population would probably reveal a more consistent pattern of behaviour with θ set at this critical value.[5]

It would be nice to be able to report the discovery of something at the edge of something else, especially

something important at the edge of something poetic. However, the results in Section 3 don't support any grand conclusions. The region in which θ is just right isn't the edge of chaos. But it is crucially important. And we might expect similar results with many other parameters. Reading the Artificial Life literature, it's easy to be misled into thinking that a whole range of spectacular evolutionary effects arise inevitably when evolution is simulated in an artificial microworld. Researchers rarely dwell on the hours of tinkering (or sheer good fortune) required to achieve these effects. But it's just this tinkering that's most interesting when it comes to answering the big questions. How did nature achieve the tuning of all these parameters without the aid of a scientist's intervention?

On a more specific level, the results in Section 3 emphasise the fact that it is possible for a species to be too successful. In this artificial microworld, if a species evolves which is part of a complicated food web but which is highly successful at the expense of that food web, then it places the whole ecosystem in danger. (To what extent, if any, this phenomenon mirrors a possibility present in the evolution of life on Earth, I cannot say.) If the θ parameter itself were made part of an organism's genotype then, with a large population, we might see this parameter being tuned by evolution. It might then be appropriate to speak of the "second-order" selection of whole ecosystems or food webs, in which parameters are spontaneously tuned to support a degree of stability. (See [Kaufmann, 1993, Chapter 6] for ideas along similar lines.)

This suggests a promising project for the future, which is to answer the following question. What is the simplest instance of the smallest generalisation of the class of EA's, which generates stable ecosystems supporting diversity and complexity, and which spontaneously tunes its parameters to the critical values required to do this?

Acknowledgments

The author is supported by a SERC research fellowship.

References

[Dawkins, 1987] R.Dawkins, The Evolution of Evolvability, *Proceedings Artificial Life*, pages 201-220.

[Holland, 1992] J.H.Holland, *Adaptation in Natural and Artificial Systems* (2nd edition), MIT Press (1992).

[Kaufmann, 1993] S.A.Kaufmann, *The Origins of Order*, Oxford University Press (1993).

[Langton, 1987] C.G.Langton, Artificial Life, *Proceedings Artificial Life*, pages 1-47.

[Langton, 1990] C.G.Langton, Life at the Edge of Chaos, *Proceedings Artificial Life II*, pages 41-91.

[Packard, 1988] N.H.Packard, Adaptation Toward the Edge of Chaos, in *Dynamic Patterns in Complex Systems*, ed. S.Kelso, A.J.Mandell & M.Shlesinger, World Scientific (1988), pages 293-301.

[Ray, 1990] T.S.Ray, An Approach to the Synthesis of Life, *Proceedings Artificial Life II*, pages 371-408.

[Yaeger, 1992] L.Yaeger, Computational Genetics, Physiology, Metabolism, Neural Systems, Learning, Vision, and Behavior or PolyWorld: Life in a New Context, *Proceedings Artificial Life III*, pages 263-298.

---

[4] Two different notions of complexity are being used here. The first applies to organisms and the second applies to a kind of computation. Precise definitions of both concepts are still pending.

[5] Imperial College has recently taken delivery of a Fujitsu AP1000 parallel machine, on which it is hoped to test this hypothesis.

Non-Uniform Cellular Automata: Evolution in Rule Space and Formation of Complex Structures

Moshe Sipper
Department of Computer Science
School of Mathematical Sciences
Sackler Faculty of Exact Sciences
Tel Aviv University
Tel Aviv 69978, Israel
e-mail: moshes@math.tau.ac.il

Abstract

Cellular automata are dynamical systems in which space and time are discrete, where each cell obeys the same rule and has a finite number of states. In this paper we study non-uniform cellular automata, i.e. with non-uniform local interaction rules. Two different models are described. In the first a cell's rule may be regarded as a genotype whose phenotypic effect is achieved by rule application. Our focus is on evolution in rule space starting from a random gene pool, i.e. rule population. The second model focuses on the study of complex structures formed by a small number of rules, where the term 'complex' denotes a structure consisting of simple grid cells, acting as a single "organism".

1 Introduction

Cellular automata (CA) are dynamical systems in which space and time are discrete. The states of cells in a regular grid are updated synchronously according to a local interaction rule. Each cell obeys the same rule and has a finite (usually small) number of states (Toffoli and Margolus 1987). The model was originally conceived by John von Neumann in the 1950's (von Neumann 1966).

In this paper we study non-uniform cellular automata, i.e. with non-uniform local interaction rules, an area which seems to have received modest attention (Garzon 1990; Lee et al. 1990; Qian et al. 1990). Our approach is different than these works and is more in the spirit of Artificial Life where cellular automata provide us with "logical universes" (Langton 1986). These are: "synthetic universes defined by simple rules ... One can actually construct them, and watch them evolve." (Toffoli and Margolus 1987). In this context our purpose is to study non-uniform cellular automata with the intent of preserving the three essential features of the original uniform model:

1. Massive parallelism.

2. Locality of cellular interactions.

3. Simplicity of cells (finite state machines).

A major argument in favor of studying non-uniform cellular automata is that due to features (1) and (2), namely massive parallelism and locality of interactions, each cell must retain a copy of the rule in its local memory[1]. Thus we argue that in terms of resources there is no essential difference between uniform and non-uniform automata.

Two slightly different models are described in this paper. In the first a cell's rule may be regarded as a genotype whose phenotypic effect is achieved by rule application. As we shall see a cell's genotype is reproduced if its phenotypic effect promotes fitness. Our focus in this model is on evolution in rule space starting from a *random* gene pool, i.e. rule population.

The second (non-uniform) model introduces a slightly enhanced cellular automaton. We argue that this enhanced automaton is simple enough so that feature (3) (above) is maintained. Our focus here is the study of complex structures formed by a small number of rules. The term 'complex' denotes a structure which consists of simple grid cells, acting as a single "organism".

2 Evolution in rule space

The first model studied is that of binary state, non-uniform cellular automata with a nine cell neighborhood. The initial set-up of each cell's rule table is random where the parameter λ denotes the probability of an entry being one (this is in accordance with the λ parameter introduced by Langton (1986), denoting the percentage of all entries in a rule table which map to non-zero states). Operation of the automaton then proceeds as in the original uniform model, with one difference: evolution takes place not only in state space but also in rule space by having a cell's rule evolve each time the cell is *unsuccessful*, where success may be defined in various ways. Two success criteria discussed ahead are:

1. *Live.* A cell is considered to be successful if it attains a state of one, i.e. "lives". We use the terms alive and dead to represent a state of one and a state of zero, respectively, in accordance with the terminology of the game of Life (Gardner 1970; Berlekamp et al. 1982), one of the best known cellular automata rules.

---

[1] Although simulations of cellular automata on serial computers may optimize memory requirements by retaining a single copy of the rule this in no way impairs our argument.

Figure 1: **Territories formed when success criterion** *Agree* **is employed, with random** λ.

2. *Agree.* A cell is successful if it agrees (i.e. is in the same state) with at least four of its neighbors.

Evolution of an unsuccessful cell's rule is accomplished by selecting one successful neighbor at random and copying its rule (if no successful neighbor exists then the cell's rule remains unchanged). An alternative approach is to copy the most successful rule, a process which is used for non-binary success criteria. Another parameter of the model is whether the copying process is perfect or imperfect,where the latter case is said to involve *mutations*. As noted the initial random population of rules can be viewed as a gene pool, where phenotypic effects are achieved by rule application. In the paragraphs ahead a qualitative presentation of simulation results is given (due to lack of space the actual results are not provided).

We first examine the model using constant λ, i.e. the initial population of rules is generated with the same probability of ones. The success criterion is *Live*. Upon studying the simulation results we observe that for all values of λ the rule grid converges[2]. Furthermore, two interesting thresholds emerge, namely λ = 0.8 and λ = 0.6. The first value may be termed the "threshold of life", above which grid cells are guaranteed to attain a state of one, i.e. live. Life is attained not by one rule (which is possible only for λ = 1) but by a *coalition* of rules, demonstrating an *emergent* behavior of the model. The second threshold is that of λ = 0.6, and may be termed the "coalition threshold", above which coalitions of rules are formed, while below it one rule emerges as the winner. We also experimented with a population of rules generated with some constant probability λ with a small number of rules generated with a higher λ. It was observed that even a small number of higher λ cell rules is sufficient to induce proliferation of the entire rule grid.

The next case studied is one in which each cell rule is created by first generating a random λ with uniform distribution in the range [0..1]. This λ is then used to generate the cell's (random) rule. Here convergence of the rule grid is much more rapid than the constant λ case and is actually logarithmic. When success criterion *Agree* is employed (with random λ) the surviving rules form "territories" of live cells (see Figure 1).

What happens when the rule copying process is imperfect, i.e. when mutations are involved? In a population

of rules with random λ convergence is extremely rapid and the final configuration is one where all cells are alive (success criterion is *Live*). When constant λ is used convergence is much slower though the final configuration is again one in which all cells are alive (unless the mutation probability is too small).

The above criteria of success, namely *Live* and *Agree* admit many local minima in rule space which are all equally valid as far as the evolutionary process is concerned. We noted that coalitions of rules are formed which conform to one of these minima.

It is natural to inquire as to what happens when the success criterion is such that a global minimum exists. We examined one such criterion, namely *Parity*, where a cell is successful if it is equal to the parity of its neighbors in the previous time step (the parity of a cell is equal to 0 if it has an even number of live neighbors, 1 otherwise) (Toffoli and Margolus 1987). Our results indicated that a global minimum is indeed reached.

As a final example we consider a success criterion which is non-binary. This is the Iterated Prisoner's Dilemma (IPD) discussed extensively by Axelrod (1984). Each cell plays IPD with its neighbors where a value of one represents cooperation and a value of zero represents defection. In this case a cell copies (with a small probability of mutation) the rule of the neighboring cell with the highest ranking total payoff (computed by summing the eight individual payoffs). In the trial runs of IPD convergence to a single rule occurred , however each time to a different one. Thus, it is evident that this criterion admits many local minima. An interesting phenomenon becomes apparent upon examining the winning rules: the average percentage of ones is 60%, i.e. cooperation is preferred. This value is close to that of the successful TIT-FOR-TAT strategy (Axelrod 1984) whose percentage of ones is 64%[3].

3   Formation of complex structures

The non-uniform automaton model considered in this section is an enhancement of the first model. Each cell is either *vacant*, containing no rule, or *operational* consisting of a finite state automaton which can, in one time step:

1. Access its own state and that of its immediate neighbors.

2. Change its state, or the state of an immediate neighbor. If a cell's state is changed by more than one cell, contention occurs which may be resolved either randomly or deterministically (i.e. defined by the rules). The rules presented in this paper do not admit such contention.

3. Copy its rule onto a neighoring vacant cell. A special case is cell rule mobility where one copy is made in an

---

[2]The term *rule grid* denotes the grid of cell rules whereas the term *grid* denotes the grid of cell states.

[3]A value computed by assuming that ties (i.e. an equal number of cooperating neighbors and defecting ones) are broken in favor of cooperation. This choice is based on one of the qualities of a "good" strategy discussed by Axelrod (1984), namely the quality of *forgiveness*.

adjoining cell and the cell's own rule is erased (i.e. the cell becomes vacant). Again contention (in this case the copying of more than one rule onto the same cell) may be resolved either randomly or deterministically. In this paper this type of contention occurs when two cells attempt to move to the same cell, and this is resolved randomly (i.e. one cell "wins" while the other moves to a different vacant cell).

4. A cell may contain a *small* number of different rules. At a given moment only one rule is *active* and determines the cell's function. A non-active rule may be activated or copied onto a neighboring cell.

Our main additions to the original model, aside from non-uniformity, are in allowing an automaton to change the state of its neighboring cells and to copy itself onto them. In the rest of this paper we consider automata with a nine cell neighborhood and three possible grid states, denoted $\{0, 1, b\}$. Note that a vacant cell may be in any grid state as it can be changed by operational neighboring cells. Our focus in this section is the study of complex structures formed by a small number of rules (due to lack of space the rules are not provided).

3.1 A self-reproducing loop

Our first example involves a simple self-reproducing loop motivated by Langton's work (1986,1984) who described such a structure in uniform cellular automata. Langton's loop (motivated by Codd (1968)) makes dual use of the information contained in a description to reproduce itself. The structure consists of a looped pathway, containing instructions, with a construction arm projecting out from it. Upon encountering the arm junction the instruction is replicated, with one copy propagating back around the loop again and the other copy propagating down the construction arm, where it is translated as an instruction when it reaches the end of the arm.

The important issue to note is the two different uses of information, interpreted and uninterpreted, which also occur in natural self-reproduction, the former being the process of *translation*, and the latter *transcription*. In Langton's loops translation is accomplished when the instruction signals are "executed" as they reach the end of the construction arms, and upon the collision of signals with other signals. Transcription is accomplished by the duplication of signals at the arm junctions (Langton 1984).

The loop considered in this section consists of five cells and reproduces within six time steps . The initial configuration consists of a grid of vacant cells (i.e. with no rule) with a single loop composed of five cells in state 1, each containing the (same) loop rule (Figure 2a). The arm extends itself by copying its rule onto an adjoining cell, coupled with a state change to that cell. The new configuration then acts as data to the arm, thereby providing the description by which the loop form is replicated. When a loop finds itself blocked by other loops it "dies" by retracting the construction arm. Figure 2b shows the configuration after several time steps.

In his paper Langton (1984) compares the self-reproducing loop with the works of von Neumann (1966)

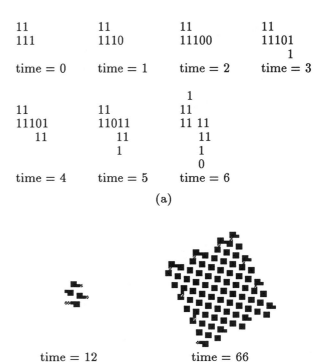

A black square represents a cell in state 1, a non-filled square represents a cell in state 0, a blank square represents a cell in state b.

(b)

Figure 2: **Self reproducing loop.**

and Codd (1968), drawing the conclusion that although the capacity for universal construction, presented by both, is a *sufficient* condition for self-reproduction it is not a *necessary* one. Furthermore, as Langton points out, naturally self-reproducing systems are not capable of universal construction. His intent was therefore to present a simpler system that exhibits non-trivial self-reproduction. This was accomplished by constructing a rule in an eight-state cellular automaton, in which the dual nature of information, i.e. translation and transcription is utilized.

In the loop presented above simple transcription is accomplished as an integral part of a cell's operation, since a rule can be copied, i.e. treated as data. Once a rule is activated it begins to function by changing states in accordance with the grid configuration, thereby performing translation on the surrounding cells (data). Essentially, the loop operates by transcribing itself onto a neighboring cell while simultaneously writing instructions (in the form of grid states) that will be carried out at the next time step.

In Langton's system each grid cell initially contains the rule that supports replication whereas in our case the grid cells are vacant and the loop itself contains all the information needed. In both cases reproduction is not coded entirely into the "transition physics" but rather is "actively directed by the configuration itself" where "the structure may take advantage of certain properties of the transition function physics of the cellular space" (Langton 1984). Thus interest in such systems arises since they display an interplay of active structures taking advantage of the characteristics of cellular space.

3.2 Reproduction by copier cells

In the previous section we described a self-reproducing loop, which exhibited a two-fold utilization of information, i.e. translation and transcription. In this section we examine a model of reproduction consisting of passive structures copied by active (mobile) cells. The motivation for our approach lies in the information flow in protein synthesis, where passive mRNA structures are translated into amino acids by active tRNA cells. Each tRNA cell matches one specific codon in the mRNA structure and synthesizes one amino acid.

Our system consists of stationary structures composed of vacant grid cells comprising the passive data to be copied. The copy ("synthesis") process is accomplished by three types of copier cells, denoted X, Y, and Z which are mobile units, "swimming" on the grid, seeking an appropriate match (remember that cellular mobility is possible by using rule copying). When such a match occurs the cell proceeds to create the appropriate substructure, as in the case of a tRNA cell synthesizing the appropriate amino acid. The final result is a copy of the original structure.

The process is demonstrated in Figure 3. The initial configuration consists of a passive structure coupled with X,Y and Z cells randomly distributed on the grid (Figure 3, $time = 0$). Each time step these copier cells move to a neighboring vacant cell (shown as blank squares) at random, unless a match is found which triggers the synthesis process. Each of the three copiers matches exactly one codon, which is a structure composed of three (passive) cells. Figure 3 ($time = 435$) shows the process at an intermediate stage and at the final stage ($time = 813$) where the copy has been produced.

The copy created is not an exact duplicate but rather a "complementary" one. The reason for this is that we wish to avoid endless copying which would occur had an exact duplicate been created. Since our model is inherently local we cannot maintain a global variable specifying that the synthesis process has been completed. The only way to avoid an endless chain of duplicate substructures is by locally specifying that a copy has been completed. This is accomplished by creating a complementary sub-structure, which does not match any copier cell and is not duplicated further.

3.3 Formation and replication of complex organisms

The final system presented involves the formation and replication of complex structures which are created from grid cells and behave as single "organisms" once formed. The system consists initially of two cell types, builders (A cells) and replicators (B cells), floating around on the grid.

Figure 4 demonstrates the operation of the system. At time 0 A and B cells are distributed randomly on the grid and there are two vacant cells in state 1 acting as the core of the building process. The A cells act as builders by attaching ones at both ends of the growing structure. Once a B cell attaches itself growth stops at that end (time 111).

When a B cell attaches itself to the upper end of a structure already possessing one zero a C cell is *spawned*, which travels down the length of the structure to the other end. If that end is as yet uncompleted the C cell simply waits for its completion (time 172). The C cell then moves up the structure, duplicating its right half which is also moved one cell to the right (time 179). Once the C cell reaches the upper end it travels down the structure, spawns a D cell at the bottom and begins traveling upward while duplicating and moving the right half (time 187). Meanwhile the D cell travels upwards between two halves of the structure and joins them together (time 190).

This process is then repeated. The C cell travels up and down the right side of the structure, creating a duplicate half on its way up. As it reaches the bottom end a D cell is spawned which travels upward between two disjoint halves and joins them together. Since joining two halves occurs every second pass the D cell dies immediately every other pass (e.g. time 195).

There are interesting features to be noted in the process presented. Replication should begin only after the organism is completely formed. However there can be no global indicator that such a situation has occurred (see also Section 3.2). Our solution is therefore local: a B cell upon encountering an upper end which already has one zero completes the formation of that end and releases a C cell which travels down the length of the structure. This cell will seek the bottom end or *wait* for its completion. Only at such time when the structure is complete will the C cell begin the replication process.

Replication involves two cells operating in unison where the C cell duplicates half of the structure while the D cell "glues" two halves together. Again it is crucial that the whole process be local in nature since no global indicators can be used.

The spawning of C and D cells are provided for by our model since as noted above a cell may contain a small number of different rules, where only one is active at a given moment. Therefore, the initial B cells can contain all three rules: B,C,D.

The design of our system is even more efficient than that however, requiring only two rule tables, one for A cells and one for $B/C/D$ cells. Each entry of the $B/C/D$ rule table is only used by one of the cell types (i.e. the entries are mutually exclusive). At a given moment the cell has one active rule (which determines its type). If the table entry to be accessed belongs to the active rule it is used, otherwise a default state change occurs. The default transformation is a move to a random vacant cell for B cells and no change for C and D cells. This may

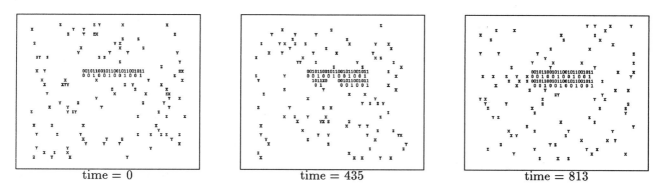

Figure 3: **Reproduction by copier cells.**

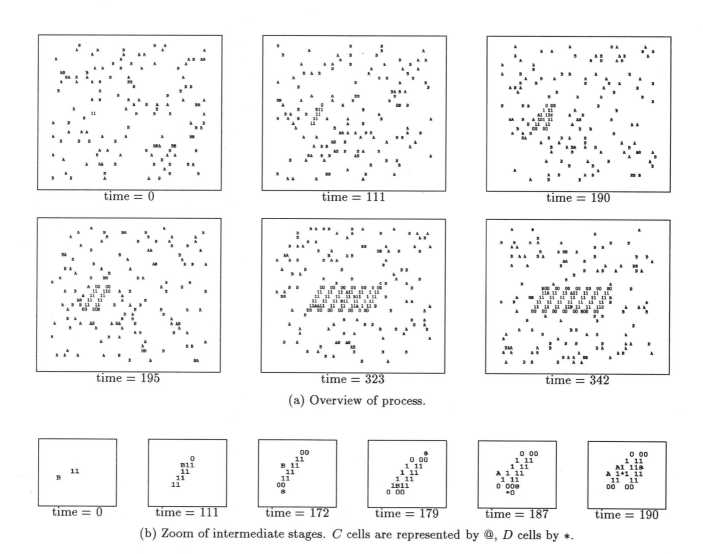

(a) Overview of process.

(b) Zoom of intermediate stages. *C* cells are represented by @, *D* cells by *.

Figure 4: **Formation and reproduction of complex organisms.**

be regarded as a form of differentiation where the cell contains the entire rule table (DNA) but uses only those parts which are relevant to its current functioning.

4 Discussion

The initial random population of rules in the first model (Section 2) can be viewed as a gene pool, where phenotypic effects are achieved by rule application. We observed how successful (fit) genes proliferate the population, in some cases forming coalitions, in others one gene emerges as the winner.

Observing our results we note that when the initial population of rules is generated with random λ convergence is much more rapid than with constant λ. An interesting analogy may be drawn with the biological phenomenon of sex, for which no accepted theory exists. One hypothesis suggests that animal sexuality helps diversify the gene pool thereby promoting more rapid adaptation (Hamilton et al. 1990). In our case diversification is achieved by using random λ which indeed promotes rapid convergence (adaptation in our model).

In his book Dawkins (1986) discusses the issue of possessing "5% of an eye" in relation with objections to the theory of evolution. He argues convincingly that 5% vision is better than no vision at all. Some of our experiments involved a small number of cells which are 5% "more fit" than the rest. We observed that in a majority of the trials such a gene (rule) came to dominate the gene pool. Although our model is simplified in relation to real life it nonetheless demonstrates how even a small advantage is crucial in the "survival" race.

The formation of territories when the success criterion is *Agree* can be regarded as a simple form of epistasis[4]. A gene's success depends on the interaction with its neighboring genes, as opposed to , say, the *Live* criterion where a cell's fitness depends solely on its own gene.

In the second model (Section 3) we concentrated on the formation of complex organisms, composed of simple grid cells. These are formed by using only a small number of rules. The examples presented demonstrate a main feature of our work, namely the power it offers in creating models of interest. As opposed to uniform cellular automata, where each cell contains the same automaton our model enables the creation of systems in which there are different automatons operating in unison. This is especially noted in our last example where complex structures were formed by cooperative operation. Our model also enables the simplification of processes which in uniform automata require complex rule tables with several states per cell. This is noted upon examining the self-reproducing loop which consists of only five cells yet promotes its own replication.

The model presented seems to offer fertile grounds for further investigation. Complex structures may be formed, exhibiting real life properties. Indeed, the dynamic behavior of the last two systems is somewhat reminiscent of observations of organisms under a microscope.

As discussed by Langton (1986) such complex structures are essentially virtual state machines, i.e. higher order automata composed of lower order ones, where first order automata are those that occupy every cell and serve as the basic building blocks.

The work presented in this paper suggests a practical model for studying Artificial Life phenomena consisting of enhanced non-uniform cellular automata which evolve in rule space as well as state space. The issues to be explored involve the evolution of complex structures, where the diversity offered by our model coupled with its simplicity seem to present us with a viable system for such explorations.

References

Axelrod, R. (1984). *The Evolution of Cooperation*. New-York: Basic Books, Inc.

Berlekamp, E. R., J. H. Conway, and R. K. Guy (1982). *Winning ways for your mathematical plays*, Volume 2, Chapter 25. New York: Academic Press.

Codd, E. F. (1968). *Cellular Automata*. Academic Press.

Dawkins, R. (1986). *The Blind Watchmaker*. W.W. Norton and Company.

Gardner, M. (1970, October). The fantastic combinations of John Conway's new solitaire game "life". *Scientific American*, 120–123.

Garzon, M. (1990). Cellular automata and discrete neural networks. *Physica D 45*, 431–440.

Hamilton, W. D., R. Axelrod, and R. Tanese (1990, May). Sexual reproduction as an adaptation to resist parasites: a review. *Proceedings of the National Academy of Sciences 87*, 3566–3573.

Langton, C. G. (1984). Self-reproduction in cellular automata. *Physica D 10*, 135–144.

Langton, C. G. (1986). Studying artificial life with cellular automata. *Physica D 22*, 120–140.

Lee, Y. C., S. Qian, R. D. Jones, C. W. Barnes, G. W. Flake, M. K. O'Rourke, K. Lee, H. H. Chen, G. Z. Sun, Y. Q. Zhang, D. Chen, and C. L. Giles (1990). Adaptive stochastic cellular automata: theory. *Physica D 45*, 159–180.

Qian, S., Y. C. Lee, R. D. Jones, C. W. Barnes, G. W. Flake, M. K. O'Rourke, K. Lee, H. H. Chen, G. Z. Sun, Y. Q. Zhang, D. Chen, and C. L. Giles (1990). Adaptive stochastic cellular automata: applications. *Physica D 45*, 181–188.

Toffoli, T. and N. Margolus (1987). *Cellular Automata Machines*. Cambridge, Massachusetts: The MIT Press.

von Neumann, J. (1966). *The Theory of Self-Reproducing Automata*. Illinois: Univ. of Illinois Press. Edited and completed by A.W. Burks.

---

[4]In the context of Artificial Life this means any interaction between genes, i.e. the extent to which the contribution to fitness of one gene depends on the values of other genes.

Evolutionary Robots: Our Hands In Their Brains?

James V Stone

Biological Sciences/Cognitive and Computing Sciences, University of Sussex, England.
jims@cogs.susx.ac.uk

Abstract

The study of learning and evolutionary adaptation has yet to provide a theory which is sufficiently detailed to enable the construction of even the most primitive synthetic animal. I argue that this is so for three reasons. First, unlike other scientific fields, such as theoretical physics, there is no universally accepted paradigmatic approach to the the study of the brain. Second, there are certain fundamental (highly complex) 'functional primitives' (e.g. types of learning) immanent in nervous systems, which are a necessary prerequisite for perceptual processes, and which are not currently possessed by any animat. Third, even though genetic algorithms are powerful optimisation techniques, the conventional use of genetic algorithms is flawed because it attempts to model only a restricted set of properties found in natural evolutionary systems.

1 The Evolution of Selective Blindness

"We face, then, two great stochastic systems that are partly in interaction and partly isolated from each other. One system is within the individual and is called *learning*; the other is immanent in heredity and in populations and is called *evolution*. One is a matter of a single lifetime; the other is a matter of multiple generations of many individuals." [Bateson1979].

The doors of perception through which we view the world were first opened as experiments in desperation. Out there, beyond the vista of our senses, is a raging cacophony of interacting events, too fast and too slow, at frequencies too high and too low, to be detectable by us. This is the stuff that our evolutionary forebears, playing in the primordial casino, rejected as useless. Had they not done so we could be capable of seeing a wide range of colours, beyond the limited spectrum available to our eyes. Infra-red to see in the dark, ultra-violet to see through fog. We could have polarity sensitive eyes (and be partially blinded by wearing polaroid sun-glasses).

Back in the casino, some animal found that sensing vibrations on its skin could be used to avoid attack, and some other developed ears to take advantage of these vibrations. One of its off-spring was born tuned into the 30-50KHz bandwidth; what a feat of engineering for all that sticky, organic fragility to meld itself into such precision. But that animal was literally out of tune with its environment, where useful vibrations arrived in the 5-20KHz range. It was eaten whilst staring fixedly at a beetle whose mating call was a 40KHz frequency modulated quiver.

Evolution has selected only a small proportion of the physical world for analysis. But it is not sufficient that we, as perceivers, are insensitive to most types of physical events around us. Even those events that are detectable by our sense organs are processed in terms of their physical context. Thus we do not see photons, or hear frequency cycles. We see spatially extended entities, and hear temporally extended sounds. The physical world has been twice filtered (once by the types of sense of organs we possess, and once by their means of transmuting events) before each of its constituent parts is realised even as the most primitive of perceptions.

And there are more filters. The green we see at midday is physically different from the 'same' green we see at sun-set. That is, even though the reflectance properties of a 'green' surface remain unchanged, the spectral balance of light varies throughout the day. Consequently, the spectrum of light reflected from any surface also varies throughout the day. Thus our perceptual apparatus not only has access to a small proportion of the physical events around us, but it also transforms these events before we are aware of them. This third filtering operation ensures that certain events which are physically dissimilar are perceived as being the same (and vice versa).

The nature of the transformations on the physical world is a function of an individual's developmental history. This constitutes a fourth filter type, the developmental filters. Every new-born infant can discriminate between phonemes not in its 'native' languages[Eimas *et al.*1971]. The experience of hearing a single language, which contains a sub-set of all phonemic categories, ensures that each adult speaker can discriminate only the phonemes which exist in his/her own language. There appear to be corresponding filters in the visual system. If monkeys are raised in a normal environment then their visual systems become sensitive to lines at all orientations. However, monkeys raised in an environment of lines at one orientation only have visual cortical neurons which respond preferentially to lines at that orientation[Carlson *et al.*1986].

Thus, what we perceive as the physical world is a multiply filtered and transformed version of that world. The combination of these four filter types ensures that most of the physical world is literally off-limits to us. Both light and sound are packaged for us by the physical characteristics of our sense organs, so that what we take to be the perceptual world, is actually the result of an interaction between the physical world and the structure of our sense organs. Our perceptions are the result of an historically interactive process. The world we perceive is given to us via organs suited to different purposes, and tuned for different reasons to various ranges and amplitudes, via developmental biases which re-package the physical world into the necessary perceptual half-truths required to confront what would otherwise be the overwhelming richness of the physical world.

Even the historical interactions between the physical world and our inherited filters do not define our perceptions. We are not idle cameras pointed at a shifting scene. We are constantly engaged in the process of actively seeking out precisely those qualities that our sense organs are suited to analysing.

So where is this new born infant, the *tabula rasa*, waiting to be written upon? The slate is not blank. It is already filled with ancient writings that specify what types of entities may be added to it. Furthermore, each infant is not a slate that is only *waiting* to be written upon. The slate contains instructions concerning what types of things to seek out. Accordingly, any changes, brightness, colours, and sounds, simple then complex patterns, voices and faces. Thus the infant is born with a set of well defined behavioural and perceptual predispositions. No event in its evolutionary history is too small to influence the child's ability to investigate and absorb the physical world around it. The pre-tuned senses of the child are guided through the chaos of light and sound by a repertoire of behaviours which are the result of millions of years of unrelenting selective pressure.

Whilst most of the decisions about which physical variables to attend to have been taken in evolutionary pre-history, a certain amount of flexibility remains in each individual. An infant thrusts its senses into the myre of light and sound like a fisherman's net into a turbulent sea. And like a net, the infant's senses are designed to ignore certain types of events, and to drag others up for closer inspection. By degrees, the unfettered kaleidoscopic chaos is beaten into a form of order, but a particular type of order, which specifies that temporally and spatially coherent sets of events must be attended to, whilst unchanging events can be ignored. Like frost on water, the senses descend upon an indifferent orgy of events, locking them into place, each frosted frond captures an iota of entropy, and then becomes a seed for further branching into the diminishing details of disorder. And so the physical world is made to yield, first, in coarse blobs of light and sound, and finally, in tight bundles of rainbowed tapestry.

2 Brain Wanted: No Metaphors Need Apply

It is a great benefit to be able to absorb new knowledge by a process of analogy with old knowledge. This metaphorical mode of thinking allows us to abstract the concept of distance from the separation between two tangible objects, motion from the change in that distance, acceleration from a change in that change. It is only after such abstractions have been accomplished that Newton was able to provide a compelling answer to the question: "What governs the motions of the heavenly bodies?".

It is a great handicap to be able to absorb new knowledge by a process of analogy with old knowledge, if the new knowledge is (and it almost always is) incomplete. For instance, it has often been stated that the brain can be understood in terms of the modern computer. Whilst this is true in some respects, the most interesting aspects of brains (e.g. adaptation, habituation, and learning) cannot be so understood.

The trick is to find an analogy in the old knowledge that adequately represents the behaviour of those components of the new knowledge which are deemed to be important. This is the hard part of any scientific endevour (see [Stone1993]). Insightful thinkers, such as Newton, are able to make use of their analogies (e.g. between a falling apple and an orbiting planet) by recognising some deep property that both the old and new knowledge share, and then driving a wedge between the old and the new by rigorously exploring which other properties they do not share. The old knowledge we have about the rigid, non-adaptive, atomistic, and stylized processes of the modern computer is unlikely to tell us anything useful about the new knowledge we have about the dynamic, adaptive, and distributed processes characteristic of brains. So how are we to think about brains? What old knowledge will allow us to think usefully about how brains might work?

It has always been the case that the currently most complex technology has been used as a metaphor for how brains might work. The dominant mode of thought throughout the industrial revolution dictated that any theory of human behaviour had to have a mechanical basis. The clock-makers of the 18th century became so skilled in their trade that they created dancing, talking automata which "possessed an uncanny reality and were therefore called *androids...*" [Hillier1988], (page 55). With the advent of electricity and, in particular, the telephone, brains were commonly described in terms of the workings of an electro-mechanical telephone exchange, with all its switches (neurons) relaying messages on their way from one place (sense organ) to another (muscle fibre). After the telephone, the computer became the favourite metaphor for the brain. And last on the stage of metaphors is artificial life (Alife), with its promises of robust, adaptive, 'situated', dynamical systems. Why should any researcher interested in how brains might work adopt the methods of artificial life? What indications are there that this 'new' field is more than just another plausible, but inadequate, metaphor?

No other physical system has had as many, nor as diverse, metaphors applied to it. It is as if the workings of the brain are chameleon-like, apparently able to take on the characteristics of the surrounding metaphor. This, alone, should tell us something about the brain. That it is, in its entirety, unlike any other physical system, and also that it shares particular sub-sets of properties with many other physical systems. This latter is precisely what enables us to conceptualise the brain so diversely as a telephone exchange, a computer, and, a connectionist system. There is little of interest that can be said of the brain that is not true in some respects, and that is not false in others. Unlike other scientific fields, such as theoretical physics, there is no single method of investigation that has proved to be an effective strategy for obtaining reliable answers. This is probably why there are as many types of brain-scientists (e.g. perceptual psychologists, neuroscientists, neuropsychologists, psychopharmacologists, zoologists, connectionists) as there are metaphors for how the brain works. And perhaps this is a necessary diversity. The increasingly inter-disciplinary nature of brain research suggests that each of these metaphors in isolation increases its contribution when viewed in the context of the others.

3   Evolutionary Robots: Our Hands in their Brains?

"Other maps are such shapes, with their islands and capes; But we've got our brave captain to thank" (So the crew would protest) "that he's bought us the best - A perfect and absolute blank!" [Carroll1897], page 144.

One of the attractive features of the use of genetic algorithms (GA) for evolving robots is that it appears as if the GA embodies the principle of least commitment with regard to the structure of the robot. There seems little doubt that this approach will prove superior to the modular-design approach, best exemplified in the NASA space program. The modular-design approach specifies that a problem should be broken down into its constituent component problems, that each problem should be solved, whereupon the entire system should work. In practice, there are limits to the size/complexity of problems solvable by this method. The frequency and nature of failures witnessed as part of NASA's space program may be a warning of the eventual triumph of system complexity over human design ingenuity.

"Rather than attempting to hand-design a system to perform a particular range of tasks well, the evolutionary approach allows their gradual emergence. There is no need for any assumptions about means to achieve a particular kind of behaviour ..." [Husbands et al.1993], page 62.

However, unless the robot can evolve *de novo* there is necessarily some degree of human intervention in designing the initial state of the robot. The designer of the robot incorporates particular capabilities into the robot. These might be infra-red 'vision', hearing between 0 and 20kHz etc.. The robot evolves from a qualitatively different starting point from that of organisms. Whereas

the latter could (and did) choose to be sensitive over different ranges of many physical variables, the robot can only choose which sub-parts of a small set of physical variables to utilise[1]

In a sense the evolving robot is in a similar position to any not-too-primitive organism:

"For change to occur, a double requirement is imposed on the new thing. It must fit the organism's internal demands for coherence, and it must fit the external requirements of the environment." [Bateson1979], page 158.

Thus, even organisms cannot suddenly decide (in an evolutionary sense) they would like to see in the infrared range; back in the organism's evolutionary history it was a good idea to close that perceptual door, and doing so was part of an adaptation to its environment. So it might be for robots - *if* they ever had that same choice, which they did not.

"GAs should be used as a method for searching the space of possible adaptations of an existing robot, not as a search through the complete space of robots" [Husbands et al.1993], page 62.

Whilst both evolutionary robots and organisms evolve from a current state that implicitly precludes access to many physical variables which might otherwise be of use to them, the organism can fully justify its current state by virtue of its inherited viability. In contrast, the animat is the creation of a person with particular pretheoretical ideas about which set of doors of perception should be left ajar, and which should be firmly locked. No amount of selective pressure will allow the robot to generate the types of adaptations that an organism can generate, because the robot comes into existence in a highly differentiated form.

"At first, systems ... are governed by dynamic interaction of their components; later on, fixed arrangements and conditions of constraint are established which render the system and its parts more efficient, but also gradually diminish and eventually abolish its equipotentiality." [von Bertalanffy1956], page 6.

There is some reason to believe that the capabilities of the evolving organism are of use to it in its current environment. In contrast, there is no *a priori* reason to suppose that the capabilities of an animat will allow it to fully exploit the information around it in order to execute its task.

4   The Evolution of Evolutionary Mechanisms

The genetic algorithm used to drive the artificial evolutionary process makes use of certain *fixed artifical genome interpreters*. That is, for a single instance of

---

[1]The work of Cariani[Cariani1990] is relevant to this section, but unfortunately came to my attention just before this paper went to press.

an artificial genome there exists a multiplicity of possible phenotypic (or functional) consequences, depending upon how the phenotype is generated from the artificial genetic material. (Note that an artificial genome interpreter determines how the artificial genetic material is used to generate the phenotype, whereas an artificial genetic operator determines how the artificial genetic material is transformed between generations).

The ability of any optimisation technique to find extrema on a given function depends as much upon the nature of the moves made over that function as it does upon the nature of the function itself. The genome and the genome interpreter jointly constitute a *move generator*. If we view the artificial evolutionary process as a search over a fitness landscape in which neighbouring points are related by altering the value (allele) at a single gene locus then each genetic operator induces a type of movement on that landscape. Note that an operator such as cross-over induces large jumps across the fitness landscape because it causes many alleles to be simultaneously altered. More precisely, each set of genetic operators, in combination with each artificial gene interpreter, defines a *configuration space*[Solla *et al.*1986, Kirkpatrick1985] in which adjacent points are related by a single move. Thus the ability of a GA to find maxima on a given fitness landscape is determined as much by the nature of its artificial genome interpreter as by its genetic operators.

Work by Lister[Lister1993] using the stochastic simulated annealing technique suggest that the optimising ability of a given core technique can be substantially increased by utilising appropriate move generators. Unlike AESs, *which make a distinction of type* between genome interpreter and the artificial genes on which they operate, NESs make no such distinction. The inability of AES to evolve new forms of genetic operators and new means of controlling how each gene is expressed in an individual necessarily limits its ability to generate the diversity of functional forms encountered in a NES.

In contrast to AES the relation between a sub-set of genes in a given genome and its phenotypic realisation is not fixed. Compelling evidence that the mechanics of evolution are themselves subject to the pressures of natural selection is the appearance of sexual reproduction. Additionally, long before the mechanisms of genetics were known, D'Arcy Thompson's book[Thompson1961] (first published in 1917) was instrumental in suggesting the existence of a genetic mechanism capable of imposing global continuous geometric transformations on the shape of organisms. His main contribution was to demonstrate that, "..two different but more or less obviously related forms can be so analysed and interpreted that each may be shown to be a transformed version of the other" [Thompson1961], page 272. That Thompson was essentially correct is supported by more recent studies which demonstrate that, "Every one of the twenty eight bones of the human skull has been inherited in an unbroken succession from the air-breathing fishes of the pre-Devonian seas" [Gregory1967]. A genetic mechanism which alters the overall size or shape of an organism is clearly capable of modulating and co-ordinating the ex-

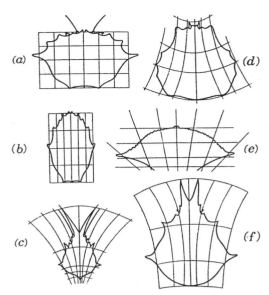

Figure 1: Carapaces of crabs: (a) Geryon; (b) Corystes; (c) Scyramathia; (d) Paralomis; (e) Lupa; (f) Chorinus. After [Thompson1961].

pression of multiple sets of genes in the genome. In terms of the fitness landscape of a given organism this corresponds to a large ('horizontal') movement, which nevertheless is likely to generate only a relatively small change in fitness. Thus the configuration space defined jointly by the original fitness landscape and the 'topology preserving' move generator has fewer local extrema than the original fitness landscape. This, in turn, makes it more amenable to any search for high magnitude extrema on its surface.

The important point, for our purposes, is that these gene controlling mechanisms ('meta-genes') *evolved*, just as the original genetic code evolved. Any meta-genetic process that generated a relatively smooth configuration space, either by controlling the expression of genes or by switching genes on and off, would inevitably ensure its own survival and incorporation into the evolutionary process. In order to enable the evolution of qualitatively new forms, new and increasingly complex meta-genetic controlling mechanisms must be capable of evolving. Thus, it is not sufficient to model a small but accessible *fixed* set of adaptive mechanisms involving only the evolution of genetic material, and not the evolution of the mechanism that interprets that material, in order to generate a phenotype.

5 Learning To Evolve

The single available instance of a successful evolutionary system (i.e. evolution on Earth) suggests that, even if the opportunity exists for the evolution of new genetic mechanisms, their appearance takes many millenia. It is thought that bacteria evolved some 3500M (million) years ago. The first sexual eukaryotic single cell (a cell with a nucleus, and discrete organelles such as mitochondria) appeared some 1800M years ago. Thus *about half of the time that life has existed on Earth was required to generate the first sexual reproduction.* Once this evolutionary threshold had been breached the evolutionary process was accelerated substantially, and some 1000M years ago sponges, the first multicellular organism with differentiated cell types, appeared. The accelerated evolution associated with sexual reproduction generated the first simple nervous systems, in the form of jelly fishes, around 650M years ago, 350M years after the sponges appeared. However, in terms of the total accelerated evolution time *more than one half of the time between the first unicellular sexual reproduction and the present was required to evolve the first simple nervous system.* Once nervous systems became sufficiently plastic it is likely that the Baldwin effect[2] further accelerated the evolutionary process. It was then a 'mere' 250M years before the first vertebrates evolved. The celebrated conquest of the land by vertebrates occurred just 300M years after the first nervous system evolved.

This type of analysis, in terms of the relative time between landmark changes in evolutionary history, suggests that, in its broadest context, evolution is almost static for relatively long periods, and is interrupted by 'sudden' changes that essentially alter the fundamental nature of the pre-existing evolutionary process. Moreover, such long periods of stasis suggest that, in spite of the large number of degrees of freedom, accessible over long periods of time, to an evolving physical system, the types of changes induced during periods of change are non-trivial.

In summary, even in an evolving system with many degrees of freedom and with more time than any computer, evolution generates qualitative changes only rarely. As stated above, about one half of the time between the first unicellular sexual reproduction and the present was required to evolve the first simple nervous system. The other half of the time bridged the gap between this first nervous system and the human brain. The complexities of the adaptive learning processes of even the simplest nervous systems are not well understood, and if a robot were endowed with even the visual abilities of a bumble-bee[3] it would be rich indeed. Given this, admittedly circumstantial, evidence just how much should we expect of an evolutionary robot?

6 Evolving To Learn

There are certain common computational primitives (e.g. learning, habituation) which are possessed by most organisms, and which may have evolved in such a way as to allow each to take advantage of a generic physical environment.

"... there appear to be no fundamental differences in structure, chemistry or function between the neurons and synapses in man and those of a squid, a snail or a leech." [Kandel1979], page 29.

---

[2]The effect of learning on behaviour ensures that, rather than a given genome representing a fixed point in the space of possible phenotypes, it represents any point within a region around that fixed point.

[3]Bees first appeared about 300M years ago.

These primitives enable simple organisms to learn perceptual and motor skills, but using a type of learning mechanism that is universal throughout the animal kingdom. This is not intended as a denial that species-specific types of learning occur, but rather that all learning is based upon certain fundamental principles of neuronal interaction. Once such a fundamental learning capability is possessed the evolution of pre-dispositions for learning particular perceptual/motor skills within an animal's environment is a *relatively* simple matter. Circumstantial evidence of the generality of learning mechanisms in the brain is given by the results which suggest that cells in the auditory cortex can be induced to generate visual receptive fields by re-routing thalamic visual outputs from visual to auditory cortex[Roe *et al.*1990].

Just as the evolution of sexual reproduction, and of the first multi-cellular organism, were protracted affairs, so the evolution of learning mechanisms took a long time. That is, relative to the quantitative alteration of a pre-existing characteristic, a qualitative alteration, such as learning, is likely to have taken a long time. Once such an evolutionary threshold had been breached, even the most primitive learning ability endowed such advantages that its continued (probably accelerated) evolution was assured. The point is that only the 'simple' learning abilities (habituation/sensitisation) of the most primitive organisms (the sea-slug, *aplysia*) are understood to any extent, and even these are known to be implemented by complex processes[Kandel and Hawkins1992]. Given the complexity of simple learning in neuronal systems, and the relative poverty of learning abilities in artificial learning systems (e.g. artificial neural networks), is it reasonable to expect an evolving animat to be able to learn general perceptual/motor skills?

7 Learning As Adaptive Behaviour

"Even if we knew down to the last molecular detail what goes on inside a living organism, we should still be up against the fact that a living system is an organized whole which by virtue of the distinctive nature of its organization shows unique forms of behaviour *which must be studied and understood at their own level*, for the significance of all livings things depends on this." [Sommerhoff1969], page 148. (Italics added).

Learning in animals is traditionally classified into one of the following categories 1) Habituation/sensitisation, 2) Classical (Pavlovian) conditioning, and, 3) Associative or reinforcement learning. Note that these different types of learning assume that learning constitutes an adequate description of the goal of the system; that the functional significance of the observed behaviour is immanent in the process of learning. This view of learning as a goal in itself is unlikely to yield an account which can be used to construct models in which learning is part of a more general set of adaptive behaviours.

A theory of learning that does not include an account of the underlying computational problem being addressed by the learning system is unlikely to yield useful insights into how to construct models which can solve computational problems (such as learning to walk across a room). It can be argued that such a theory is not so much a theory of learning as a theory of adaptive behaviour. That is, a theory which seeks to explain not only how learning occurs, but *why it is desirable that it should occur at all*. I can find no grounds for disagreeing with this argument, and would go so far as to say that a theory of learning is interesting only insofar as it is part of a theory of adaptive behaviour.

The point being made here is sufficiently similar to that made by Marr in a slightly different context, to justify applying his cogent criticisms to learning theories in general:

"They may be able to say that such and such an "association" seems to exist, but they cannot say of what the association consists, nor that it has to be so because to solve problem X... you *need* a memory organised in such and such a way; and that if one has it, certain apparent "associations" occur as a side-effect." [Marr1977], page 47.

Given the complexity of mechanisms underlying the simplest forms of learning, it seems unlikely that similar mechanisms would evolve within an AES. Just as one would not expect an animat to evolve a brain of any sort if it did not have any artificial neurons, so, one would not expect an animat to learn to 'see' if it did not have certain fundamental learning abilities. Thus, certain functional pre-requisites are required in order to be able to evolve other, more complex, functional characteristics. Whilst it is a simple matter to provide an animat with artificial neurons, and thereby facilitate the evolution of a functional 'brain', it is not so easy to provide it with generic learning capabilities because these are simply not sufficiently well understood to allow this. It is for this reason that I believe that the understanding of certain fundamental characteristics (via computational modelling of the development of perceptual processes) of living systems must achieve a certain level by studying these characteristics in relative isolation, before they can be fully understood as an integral part of that system.

Acknowledgements: Thanks to Raymond Lister and Inman Harvey for useful discussions, and to Stephen and Pheobe Isard for comments. The author is supported by a JCI grant awarded to J Stone, D Willshaw and T Collett. This work was initiated whilst the author was at the University of Wales, Aberystwyth, Wales.

References

[Bateson1979] G Bateson. *Mind and Nature*. Flamingo, 1979.

[Cariani1990] P Cariani. Implications from structural evolution: semantic adaptation. *Proc Int. Joint Conference on Neural Networks, Washington DC*, pages 47–51, 1990.

[Carlson *et al.*1986] M Carlson, D Huble, and T Wiesel. Effects of monocular exposure to oriented lines on monkey striate cortex. *Brain Research*, 390:71–78, 1986.

[Carroll1897] L Carroll. *Rhyme? And Reason?* Macmillan and Co., London, 1897.

[Eimas *et al.*1971] P Eimas, ER Siquelena, PW Jusczyk, and J Vigorito. Speech perception in infants. *Science*, 171:303–6, 1971.

[Gregory1967] WK Gregory. *Our face from fish to man*. Hafner, New York, 1967.

[Hillier1988] Mary Hillier. *Automata and Mechanical Toys*. Bloomsbury Books, London, 1988.

[Husbands *et al.*1993] P Husbands, I Harvey, and D Cliff. An evolutionary approach to situated ai. In *Prospects for AI, Proc AISB93, Birmingham, England. Sloman S, Hogg D, Humphreys G, Ramsay A (Eds)*, pages 61–70, 1993.

[Kandel and Hawkins1992] E Kandel and R Hawkins. The biological basis of learning and individuality. *Scientific American*, September 1992.

[Kandel1979] E Kandel. Small systems of neurons. *Scientific American*, pages 29–38, September 1979.

[Kirkpatrick1985] S Kirkpatrick. Configuration space analysis of travelling salesman problems. *J. Physique*, 46:1277–1292, 1985.

[Lister1993] R Lister. Annealing networks and fractal landscapes. In *IEEE International Conference on Neural Networks, San Francisco*, pages 257–26, 1993.

[Marr1977] D Marr. Ai: A personal view. *Artificial Intelligence*, 9:37–48, 1977.

[Roe *et al.*1990] AW Roe, SL Pallas, JO Hahm, and M Sur. A map of visual space induced in primary auditory-cortex. *Science*, 250(4982):818–820, 1990.

[Solla *et al.*1986] SA Solla, GB Sorkin, and SR White. Configuration space analysis for optimization problems. In *Disordered systems and biological organisation, Bienstock E, (Ed.), Springer-Verlag*, pages 283–293, 1986.

[Sommerhoff1969] G Sommerhoff. The abstract characteristics of living systems. In *Systems Thinking, Emery F, (Ed)*, pages 147–202, 1969.

[Stone1993] JV Stone. Computer vision: What is the object? In *Prospects for AI, Proc. Artificial Intelligence and Simulation of Behaviour, Birmingham, England*, pages 199–208, April 1993.

[Thompson1961] D'Arcy Thompson. *On Growth and Form*. Cambridge University Press (First publication 1917), 1961.

[von Bertalanffy1956] L von Bertalanffy. General system theory. In *Yearbook for the Advancement of of General System Theory*, pages 1–10, 1956.

Universality Without Matter?

Alvaro Moreno
Dept. of Logic and Philosophy of Science
University of the Basque Country
Apartado 1249, 20080 San Sebastian, Spain
moreno_bergareche_alvaro@euskom.spritel.es

Arantza Etxeberria
School of Cognitive and Computing Sciences
University of Sussex
Falmer, Brighton BN1 9QH, U.K.
arantza@cogs.susx.ac.uk

Jon Umerez
Dept. of Systems Science
State University of New York (SUNY)
Binghamton, New York 13901, USA
jumerez@bingvaxa.cc.binghamton.edu

Abstract

Currently theoretical biology offers two different perspectives on the definition of life, one based on genetic information and the other on dynamic self-organisation. Artificial life (AL) attempts to develop a universal biology, understanding life as pure organisation and overcoming this antagonism by alternatively seeking dynamical or informational interpretations of the different levels in the computational models it proposes. Nevertheless, living beings differ from other complex systems in the deep entanglement between form and matter they exhibit. Information is explicit in the sequence of monomers of biomolecules, but the self-organising capacity of living systems is only implicit in the properties of its components. Research in AL should therefore not rely only on computational simulations but develop realisations of systems that take maximum advantage of the self-organising capacities of matter.

1 Introduction

The claim that artificial life (AL) should "begin to derive a truly general theoretical biology capable of making universal statements about life..." (Langton, 1989a) is a rather strong one. Until now biology has only studied life on earth and is apparently unable to find universal principles for it. This is why universality is a fundamental problem, related to the definition of life and to the scope of biological theories. AL works on the assumption that this universality is the problem of biologists and that a clear definition existed for "life as we know it". However, as Maynard Smith (1986) has pointed out, biology is divided between two different perspectives on life.

The first one maintains that evolution by means of natural selection, as stated by Darwin and afterwards reformulated by neodarwinism, constitutes the most general theory to understand life; the second stresses that living systems self-organize due to very special relations established among their component units. Primacy is given either to information or to self-organising dynamics. On the other hand, biological explanations might only apply to systems we already know on Earth. Consequently, AL and its attempt to transcend the framework of the particular "life as we know it" by becoming the science of the universal "life as it could be" (Langton, 1989b), might have to take separate ways with biology.

In fact, one way out proposed in AL is to view life as "a property of the organisation of matter, rather than a property of the matter which is so organised" (Langton, 1989b). This formal view of universality has to be further discussed to discover whether the understanding of biological and artificial living systems requires grounding on material properties (Emmeche, 1991, 1992).

2 Science and universality

The problem of universality is mainly posed in the empirical sciences, in which results are contrasted with experiments. A formal science such as mathematics is universal by definition as its explanative effort is developed within the domain of the formalism itself. Now, in order to determine the universality of the empirical sciences, a discussion of the nature of mathematics (and of computation) is crucial because mathematics constrains the methodologies used in modelling.

If we compare the relationships of biology and physics with their objects of study, we observe that although physicists have been continuously broadening their research field (encovering new laws, new phenomena and even new particles), what they find always belongs in some fundamental way to their field. Physics poses no ontological restriction: any system that can be handled operationally is included. Something similar can be said of other sciences, such as chemistry or thermodynamics; the delimitation of their phenomenic sphere derives from a methodology that eventually delimits an unambiguous ontological field, so that all these disciplines can be defined "as they could be". In this line of thought, the concepts of artificial physics or artificial chemistry do not imply alternative principles for the material realm, but, rather, research into the conditions under which certain physicochemical properties appear.

The case of biology is different. We can hazard a guess that if life is found in other remote corners of the universe or if it is synthesised by humans, it might indeed respond to different principles. But we still do not know what is contingent and what is essential –universal– in the basic mechanisms of living phenomena. We also do not have a method for distinguishing living from non-living systems.

The debate on the relevance of matter to living systems exposes a lack of agreement on the foundations of biology that could be solved if artificial or extraterrestrial life were found.

Thus, living systems are attempted described in terms of the relations among their constituent parts without consideration of the properties of those parts. AL, however, is not the first to embark on a description of biological and other complex systems in such relational terms; cybernetics has already tried to seek explanations that would not involve material requisites (see Cahiers du CREA, 1985). We are not convinced of the adequacy of this as a methodology, though. If matter is considered devoid of attributes something important might be missing. If it were the case that the relations among components arise from their material properties, the complex organisation of living systems could not be fully understood except by recourse to the properties of living matter.

3 Two views of life in theoretical biology

As mentioned above, there are two main views in biology (Maynard Smith, 1986; Moreno et al 1990): one sees life as an "evolving informational structure" while the other understands living systems as "self-constructing autonomous system". Given that information is linked to the phenomena of reproduction and evolution, life is essentially conceived of as a collective phenomenon. The self-organising perspective is articulated around dynamic notions underlying autocatalytic or metabolic networks. Autonomy is seen as the main property of living systems; life is the individual phenomenon of construction of identity.

The existence of these two explanative principles is at the basis of some deep problems in the construction of models in biology, because the fundamental features of life derive from both of them.

3.1 Life as genetic information

Taking a view of life as genetic information, points at information as the main causal element in biological phenomena. This information consists of a sequence of elements or "records"; the sequence, or string, maintains a conservative order and is responsible for the reproduction of the organism and evolution of life.

As a research program its goal is to find the underlying molecular causes of every biological (physiological, behavioural, phylogenetic or epigenetic) process. Usually the task consists of finding functional proteins and the DNA sequences responsible for their synthesis. In this way, a variety of phenomena can be explained appealing only to a few molecular structures. A spectacular reduction from the complex to the simple is achieved. The self-reproductive process is considered the expression of the "logic of the living" (Jacob, 1970).

Such a perspective can be formalist; the central dogma of molecular biology which states the unidirectional flow of information from nucleic acids (often seen as software) to proteins (often conceptualised as hardware) favours a

separation of levels. The genotype is pictured as containing an explicit description of the phenotype, and there is a misconception that there are no material requirements to realise the transcription. This view also has a non-formalist aspect to it which stems from the fact that molecular biology is deeply committed to the underlying biochemistry. As such it does not easily generalise to the AL framework.

3.2 Organisms as autonomous systems

The most systematic version of this perspective can be found in the notion of Autopoiesis developed by Maturana and Varela (Varela et al, 1974, Varela, 1979). Living beings are seen as networks of component production and the autonomy of a self-maintained network is considered a necessary and sufficient starting point to generate biological phenomenology. Even though here too, the relevant phenomena are discussed in terms of molecular interactions, individual molecules are not assigned intrinsic functions. The kind of explanations aimed at are of physico-chemical nature which implies that concepts such as role, function, or information should be abandoned. Biology inherited those concepts from Cybenetics, but they can be advantageously substituted by dynamical terms (Atlan, 1985).

This perspective is a rather formalist one; the emergence of order that takes place in an autonomous system cannot be reduced to (nor deduced from) the material properties of the components, it is the result of the interactions among them. Nevertheless, there are some voices for non-formalism, Fleischacker (1988, 1989), for example, requires mechanisms to give rise to the network of component production in the physical space.

4 Logic, information and matter

To know whether it is possible to abstract the organisation or "form" of life that could be applied to every possible life, it has to be determined whether the materiality of biological systems implies certain characteristics that cannot be formalised.

The two perspectives of biology have a different opinion on what is the main property of living systems that has to be explained. The informational emphasises reproduction and evolution, only made possible by the existence of information transmitted over generations. The self-organising emphasizes autonomy, possible because of the logic of the network dynamics. Now, both concepts of information and logic seem to open a door to understand the whole phenomenon as formal but there are reasons to hold doubts on the feasibility of this in practice.

The main problem regarding information is how to make it explicit. For example, in the process of reproduction together with the self-description in DNA other components are transmitted which are able to interpret it. These also bear information implicit in their size, morphology or chemical properties that is crucial for the operation of the system.

The logic of the interactions among the components has some problems on its part. The organisation of the living can be seen as a self-referential connectivity of components that creates a meaningful whole. Nevertheless, it is not very clear that a self-reference understood as a syntactic connectivity of components is enough to explain other living phenomena such as self-reproduction or evolution. Those seem to require some sort of semantics (Pattee, 1977, 1982) that appears to be very much in-built in the specific materials that take part in the living organisation.

Not only life as we know it, but all kinds of living organisation seem to be based on an extremely opportunistic utilisation of some specific properties of matter. What do we mean by that? We only know living systems that are material. Now, the basic thesis of functionalism is that living or cognitive organisation can be realised/implemented by several different forms of material components. The question is: are there material properties that are not formalisable per se? One of the reasons why information and function in living systems are supported by two different types of molecules (nucleic acids and proteins) resides in their different biochemical properties. Information can be stored in DNA because of its template capacities and living functions are carried by proteins because of their high enzimatic capacities. This contrats is not marginal in biological organisation. Life makes an extreme use of the self-organising properties of matter. The special relations among specific material components that we see in life on earth have a history and have been arranged by evolution. This does not mean we cannot formalise living organisation in principle, but it does imply that neither information nor logic can be easily abstracted from living matter.

5 The research program in AL

In some sense the research program of AL can be considered a synthesis of both views. How can this synthesis be realised? AL studies living organisation realised on unespecific material systems such as a computer. A strategy of interpretation of the computational world makes it possible to observe different aspects of living organisation. This way, it is possible to switch between a dynamical or self-organising and a symbolic or informational level of explanation.

As computational simulations occur in a formal universe, some part of the system should represent the symbolic side (genotype), while some other should represent the dynamic side (phenotype). But it is very difficult to capture the dynamics responsible of the genotype expression and the evaluation of the phenotype by the environment in a computational model (Pattee, 1989). Any formal or computational model has to code directly or indirectly all the information that specifies the behaviour of the system. Therefore, if a given material structure has built-in information we should go down into the lower level of specification until this properties only derive from the computational primitives present in the model. In practice the level of description of computational primitives always omits a great amount of implicit information (topological structure, energy, rate, etc.) that derives from forms of organisation and mate-

rial specificities of a lower level.

Our main objection against the view that life is pure organisation stems from the division of perspectives on life that was presented before. Self-organisation arises due to the specific properties of certain material components. This is why so little information can harness the construction of such a complex system; and further that the information can be interpreted autonomously in the frame of the system itself. Therefore, information explicit in the system is insufficient to specify living organisation; however, self-organising capacities of matter without the informational explicit constraints are neither enough to reach these complex levels of organisation.

The attempt to abstract the organisation or form of life to generalise biology to every kind of possible life crashes into the deep entanglement of matter and form that seems to be an essential feature of living systems. The problem is to determine whether the materiality of biological systems involves non-formalisable conditions. Functionalism is not keen on talking about specific materiality in opposition to concepts like form or organisation, because any characterisation of organisation is to be done in relational terms. Materiality can be as easily simulated as organisation; we could define any kind of organisation as the physical interaction of certain set of specific material components. The main functionalist tenet is that any organisation can be realised/implemented by several different material components.

An important property of living organisation is self-reproduction; the process by which a collection of specific material components generates another similar to the original as a consequence of their interactions. Furthermore, organisms evolve when the self-reproducing capacity is extended to create an open-ended set of types or forms within the frame of the basic organisation.

According to functionalism this kind of processes can be realized in the frame of pure organisations, reproduction takes place as a logical consequence of the self-reference of the system. But here two different processes should be distinguished: self-reproduction of organisation, understood abstractly as a set of formal relations and self-reproduction of the material system which acts as a physical support. To obtain a self-reproductive system, all the information should be explicitly contained in the organisation itself. This implies not only that organisation is self-referential –in this case it would be a syntactical self-reference– but that that self-reference includes its own material support –what is equivalent to a semantic self-reference. A system like this is, apart from some trivial cases, logically impossible; living organisation does not contain a complete explicit self-description because a great part of the information necessary for the process of self-reproduction is implicit in the specific materiality that compose it. This specificity (size, morphological or chemical traits...) implicitly contains information necessary to specify the operation of the system (Kampis, 1991). If we suppose that some of the essential relations of the logic of the living are necessarily implicit, then only those material components that can bear such relations can constitute life.

For example, the RNA or protein 3-D structures are not fully specified by the sequence of monomers but depend on other dynamical properties of strings; there is an informational deficit. The intriguing thing is that if we compare the two RNA and protein sequences of a given gene, their formal complexity is equivalent (of course the sequence of RNA should be longer),whereas the topological complexity of the protein is greater. This is due to the greater versatility of amino acids with respect to nucleotides to conform spatially heterogeneous structures (Orgel, 1986); and this increase of complexity is not explicit anywhere, but implicit in the material specificity of proteins (Moreno & Fernandez, 1990, 1992).

The important thing is that this implicit difference in topological complexity that derives from the respective material compositions of both kinds of polymers is the essential cause of their complementary biological functionalities. In effect, there is also a functional reason to explain this asymmetry between nucleic acids and proteins: the greater the template ability, the less the topological versatility, and vice versa. The essential ability for template copy in nucleic acids is based in the setting-up of hydrogen bonds between pairs of bases. But this also implies a redundancy in the spatial distribution and a limit in the possibilities for variety in its tertiary structure (responsible for its enzymatic abilities). By the contrary, the essential enzymatic abilities of proteins are based in the action of the Van der Waals forces; which restricts their possibilities as template structures.

This example shows two points: 1) there is a deep entanglement between logical form or "software" and material structure or "hardware" (part of the information is implicit in the structure of components), and 2) an independence between organisation and structure would require to make explicit the information that specifies the organisation. This, at its turn, would require the implementation of new components in which explicitly to code all the steps of this process. But then we would have a more complex new material organisation, which, at its turn, would pose additional problems of self-reproduction, leading certainly to an infinite regress. That's why we find intrinsic limitations to the formalisation of the living organisation. So, not only life as we know it, but all kind of living organisation seems to be based on an extremely "opportunist" usage of some specific material properties that bear information.

6 A materialistic approach for AL

Another research line in AL is based in the material construction of lifelike systems; a development of realisations instead of mere simulations. Nowadays it is still unthinkable to produce an artificial system which replicates the behaviour of biological systems. But AL is already developing autonomous robots without any central and detailed program to specify concrete tasks for the device (Meyer & Wilson, 1991; Meyer et al 1993). In terms of Cariani (1991), this kind of systems are syntactically adaptive devices; they are capable of changing syntactic relations among parts of the system on the

ground of their past action and experience and have a flexible behavior in front of unexpected situations.

Nevertheless, the standard research program in robotics implicitly assumes the thesis of the independence between organisation and materiality that we have previously criticised. At the moment there are certainly promising research perspectives in the development of robots able to emulate different autonomous behaviours of living beings. However, if instead of these partial goals the purpose is to emulate basic behaviours of living beings, as self-mainte nance or self-reproduction, in which adaptability is linked to the system's ability for physical self-reconstruction, then that is more problematic.

It would not be possible to build actually self-reproducing and evolving robots without implicitly encoding part of the organisation of the system in its material structure, a goal that drastically restricts the range of materials robots can be made of. An attempt to separate organisation from materiality implies limitations for the structural self-modification and the autonomy of the devices.

7 Conclusion

AL is currently trying to develop a research program mainly based in computer simulations. As far as the appropriate interpretation procedures are adopted and awareness on their limitations is kept, simulations are extremely useful. The development of new programming techniques (such as emergent computation (Forrest, 1990)), can even broaden the concept of experimentality in biology. They can be helpful to overcome one of the greatest problems of AL: how to represent the historically evolved properties of living components so as to be able to create novelty (and not only variation) in the computer.

From these considerations, the alternatives for a progressive research program in artificial life are several. One of them would be the development of a theory of levels that allows to understand computationally the relation between form and matter (or information and dynamics) of living systems (Minch, 1988); another would be the realisation or physical implementation of living properties; and, finally, a combination of simulations and realisations might also report interesting results.

Although simulations probably constitute the only way to make rapid progress, shortening the time that real evolutionary processes would require, only a maximal use of the self-organising properties of certain macromolecular components (Mikhailov, 1992) will permit to artificially develop systems with self-reproducing and evolutionary abilities. Systems so created will be recognised as artificial life, surely not so much for being radically different from natural life in their material grounding, but due to features such as possessing a different code to express information. According to this methodology, the combination of already classical technologies, as genetic engineering, with others of a more recent development, as nanotechnology, together with research in computer simulations and robotics, can push forward the artificial creation of life. In fact, building artificial organisms will possibly be a process and not a unique step; in it together with the artificial construction of living systems we will see the elaboration of a new theory to explain them.

Acknowledgements

Authors acknowledge funding from a DIGCYT Project Number PB92-0456 from the MEC (Spain) and the Research Project Number 00 230 HA168/93 from the University of the Basque Country. Arantza Etxeberria has a Postdoc Fellowship from the Basque Government and Jon Umerez has one from the MEC (Spain).

References

[1] Atlan, H. (1985). *L'emergence du nouveau et du sens*. In P. Dumouchel & J.P. Dupuy (Eds). *L'autorganisation. De la physique au politique*. Paris: Seuil, pp. 115-138.

[2] *Cahiers du CREA (Centre de Reserche en Epistemologie et Autonomie)*, 7-8 (1985), Paris: Ecole Polytechnique.

[3] Cariani, P. (1991). "Emergence and Artificial Life." In C.H. Langton, C. Taylor, J. Farmer. & S. Rasmussen (Eds.) *Artificial Life II*, pp. 775-797, Redwood City, CA: Addison-Wesley.

[4] Emmeche, C. (1991). "A semiotical reflection on biology, living signs and artificial life." *Biology and Philosophy*, 6 (3), pp. 325-340.

[5] Emmeche, C. (1992). "Life as an abstract phenomenon: is Artificial Life possible?" In F. Varela & P. Bourgine (eds.) *Toward a Practice of Autonomous Systems*. Cambridge, Mass: MIT Press, pp. 466-474.

[6] Fleishacker, G. (1988). *Autopoiesis: the system logic and the origin of life*, Ph.D. Thesis, Boston University.

[7] Fleishacker, G. (1989). "Autopoiesis: the status of its system logic", *BioSystems*, 22, pp. 37-49.

[8] Forrest, S. (1990). "Emergent Computation". *Physica D, 42*, Proceedings of the Ninth Annual International Conference on Self- Organizing, Collective and Cooperative Phenomena in Natural and Artificial Computing Networks. Los Alamos (New Mexico). May 22-26.

[9] Jacob, F. (1970). *La logique du vivant*, Paris: Gallimard.

[10] Kampis, G (1991). *Self-modifying systems in Biology and Cognitive Science*, Oxford: Pergamon Press.

[11] Langton, C. H. (1989a). "Preface" In *Artificial Life. (Proceedings of the First Conference on Artificial Life)*. Los Alamos, September, 1987. Redwood City, CA.: Addison-Wesley, pp. xv-xxvi.

[12] Langton, C.H. (1989b). "Artificial Life" In *Artificial Life. (Proceedings of the First Conference on Artificial Life)*. Los Alamos, September, 1987. Redwood City, CA: Addison-Wesley, pp. 1-47.

[13] Maynard-Smith, J. (1986). *The Problems of Biology*. Oxford: Oxford University Press.

[14] Meyer, J.A. & Wilson, S.W. Eds. (1991). *From Animals to Animats*. Cambridge, Mass: MIT Press.

[15] Meyer, J.A., Roitblat, H. & Wilson, S.W. Eds. (1993). *Animals to Animats II*. Cambridge, Mass.: MIT Press.

[16] Mikhailov, A.S. (1992). "Artificial Life: An Engineering Perspective". In R. Friedrich. & A. Wunderlich Eds. *Evolution of Dynamical Structures in Complex Systems*. Berlin: Springer.

[17] Minch, E. (1988). *The representation of Hierarchical Structure in Evolving network*. Ph.D. Thesis. Department of Systems Science, SUNY Binghamton (New York).

[18] Moreno, A. & Fernandez, J. (1990). "Structural limits for evolutionary capacities in molecular complex systems". *Rivista di Biologia (Biology Forum*, 83 (2/3), pp. 335-349.

[19] Moreno, A., Fernandez, J. & Etxeberria, A. (1990). "Cybernetics, Autopoiesis and Definition of Life". In R. Trappl, Ed.: *Cybernetics and Systems'90*. Singapore: World Scientific.

[20] Moreno, A. & Fernandez, J. (1992). "From records to self-description: the role played by RNA in early evolutionary systems". *Acta Biotheoretica*, 40, pp. 1-9.

[21] Orgel, L. (1986). "RNA catalysis and the origins of life". *Journal of Theoretical Biology*, 123, pp. 127-149.

[22] Pattee, H. (1977). "Dynamic and Linguistic Modes of Complex Systems". *International Journal of General Systems*, 3, pp. 259-266.

[23] Pattee, H. (1982). "Cell Psychology: an evolutionary view of the symbol matter problem". *Cognition and Brain Theory*, 5, pp. 325-341.

[24] Pattee, H. (1989). "Simulations, Realizations and Theories of Life". In C.H. Langton Ed. *Artificial Life*, Redwood City, CA.:Addison Wesley.

[25] Varela, F., Maturana, H. & Uribe, R. (1974): "Autopoiesis: The Organization of Living Systems, its characterization and a model". *BioSystems*, 5, pp. 187-196.

[26] Varela, F. (1979). *Principles of Biological Autonomy*. New York: North Holland.

Emergent Phenomena and Complexity*

Vince Darley
Division of Applied Sciences
Harvard University
33 Oxford Street, Cambridge MA 02138
e-mail: vince@das.harvard.edu

Abstract

I seek to define rigorously the concept of an emergent phenomenon in a complex system, together with its implications for explanation, understanding and prediction in such systems. I argue that in a certain fundamental sense, emergent systems are those in which even perfect knowledge and understanding may give us no predictive information. In them the optimal means of prediction is simulation. I investigate the consequences of this for certain decidability and complexity issues, and then explain why these limitations do not preclude all means of doing interesting science in such systems. I touch upon some recent incorporation of this work into the investigation of self-organised criticalities.

1   Motivation and Objectives

The calculations were so elaborate it was very difficult. Now, usually I was the expert at this; I could always tell you what the answer was going to look like, or when I got it I could explain why. But this thing was so complicated I couldn't explain *why* it was like that.

So I told Fermi I was doing this problem, and I started to describe the results. He said, "Wait, before you tell me the result, let me think. It's going to come out like this (he was right), and it's going to come out like this because of so and so. And there's a perfectly obvious explanation for this—"

He was doing what I was supposed to be good at, ten times better. That was quite a lesson to me. *Richard Feynman*

The defining characteristic of a complex system is that some of its global behaviours, which are the result of interactions between a large number of relatively simple parts, cannot be predicted simply from the rules of those underlying interactions[1].

---

*Work supported in part by the Kennedy Memorial Trust, London.

[1] Chaotic systems (in the mathematical sense of the word) which inherently evade prediction can be incorporated into this definition, by allowing the fact that for such systems we know and understand, in a quantitative way, precisely how and when our predictions will be deficient due to inadequacies in our knowledge of the initial conditions.

The word "simply" in the preceding paragraph is somewhat troublesome. I set out to explain, justify, understand and rigorise, in the context of such systems, what "simply" means.

With that in mind, let us hypothesise the concept of an 'emergent phenomenon' as a large scale, group behaviour of a system, which doesn't seem to have any clear explanation in terms of the system's constituent parts. Rather than viewing this as the first step in the hierarchical evolution of hyperstructures(Baas 1993), I am interested in 'first-order emergence' in its own right, as it is currently more amenable to a precise formalisation than the elusive increase in complexity one observes in natural evolution.

Why is it that we seem to be incapable of answering the interesting questions about how and why the final state was reached? In the context of the above quotation, can we always find, or meaningfully postulate, a person like Fermi who can solve our problems through a deeper understanding than our own — side-stepping all of our complicated calculations? More succinctly, do truly emergent phenomena exist?

In order to address these issues, we shall need to clarify what we mean by a 'clear explanation' and what we mean by 'understanding'. This is especially warranted due to the misunderstandings inherent in different people's working definitions of such terms in different fields - complexity and emergence have become very much vogue labels for problems in fields as diverse as economics, artificial life, artificial intelligence, neuroscience, and even in the study of cultural change and development. I will use the term 'emergent system' with the understanding that not all phenomena observed in that system will be emergent.

1.1   Explanation

An explanation is some type of answer to the question 'Why did X happen?'. It can contain any or all of the following characteristics:

- "What precise sequence of events and interactions caused X to happen in this situation?" We argue that we can reduce the problem to its constituent interactions, all of which we can explain (often in a precise, quantitative fashion). Thus we have explained the connection between our initial state and the event X in a causal fashion. This sort of explanation can be used either as pre- or post-explanation of the event[2].

- "What were the fundamental details of the initial state which caused X?" This allows us to generalise and determine some characteristics of the class of initial states which bring about X, as compared with the space as a whole. If we can't generalise, we may reach the conclusion that we were dealing with a special case.

- "Assuming this situation isn't special in any way, why did X, rather than $Y, Z, \ldots$ happen?" This is an elaboration of the previous point. Presumably we can imagine a number of different outcomes to which the system could progress (If we're so sure it'll reach X, why aren't we researching something more interesting). Therefore we would like to have some form of explanation for the mapping between different classes of initial state, and the different outcomes $X, Y, Z, \ldots$

1.2 Understanding

We need to understand the legitimacy of the procedures and applications contained in the previous explanations. Therefore, we either understand, fundamentally, the mathematical tools and techniques used in carrying out the rules of the system or believe in the validity of the body of theory which leads to them. If we are only interested in the abstract concept of emergence then this is a sufficient characterisation.

Otherwise, if we are interested in legitimising the applicability of a particular system to an external field of study, we must understand how that field incorporates the system into its own phenomenology. For instance we may wish to use our results to make statements about physical or economic systems, and in such a case we would like to justify the legitimacy of such claims within an accepted framework of such fields.

Let us now examine successive levels of understanding, in an analogous manner to that for explanation.

- "I understand the rules which govern the actions of every single agent and interaction in the system, precisely." I could carry out a simulation of the system to explain precisely why (in the first sense above) X happened.

- "I understand the rules and the arena in which they operate sufficiently well that I can make predictions of the outcome very rapidly from the initial state alone, without having to calculate every interaction." I have some deeper understanding of the system and the legitimising tools, such that I can perform an analysis which reveals some symmetries (probably abstract) which enable me to calculate the outcome more directly. This analysis will presumably reveal at least a partial characterisation of the space of initial states, whose boundaries may have to be sharpened by means of simulation.

- "My analysis and understanding of the system is sufficient to give a clear, precise classification of the space of initial states in terms of the system's outcome." Here we have achieved complete success with the previous

---

[2]Most scientists would not deem this nor its associated mode of understanding as a sufficient characteristic of a definition of either 'explanation' or 'understanding'.

method of analysis, and developed a mapping from the space of initial states to the space of outcomes described by, for example, a closed-form expression.

2   Emergent Phenomena

Now, a simulation of a system which arrives at a given result, and the information contained in the intermediate steps of such a simulation, clearly comprise some form of explanation, which will indeed be very useful to explain some phenomena. If developed in an external fashion, it at least demonstrates sufficient understanding of the original system so as to be able to reproduce its behaviour. This is clearly an important step.

To return to our motivating quotation, however, surely if we *really* understand a system, we shouldn't need to perform such a simulation. We would understand both the circumstances which are necessary for each relevant phenomenon to arise, and the correlation between the initial state and such circumstances. We could then say, directly, whether and why X, Y or Z will happen.

There is clearly a continuum of levels of difficulty and complexity of analysis. One property of a system will be the minimum level of analysis to which it will yield. To couch this in computational terms, there is a minimum amount of computation which needs to be carried out to predict any given phenomenon. With this in mind, I propose the following definition, which will be made more mathematically precise in the forthcoming sections:

Definition 1 *A true emergent phenomenon is one for which the optimal means of prediction is simulation.*

What we mean by a 'prediction' is left deliberately vague - it may be precisely quantitative or just a loose classification. With that proviso, rather than phrasing this definition as a hypothesis about our concept of 'emergence', I prefer to use it as a formalisation of the term 'emergent' in the discussion which follows. I seek to demonstrate the viability of the above definition as an axiom in the study of emergence in complex adaptive systems. Note that I have yet to demonstrate the existence of emergence.

Emergence as I describe it here has been loosely termed 'computational emergence' by Cariani (1992). However, abstract computer simulations have been used ever more frequently to gain an understanding of physical phenomena(Toffoli and Margolus 1987; Farmer, Kauffman, and Packard 1986) and in the evolution of specific functionality(Koza 1993). So computational emergence can, I feel, bear upon Cariani's 'thermodynamic emergence' and 'emergence relative to a model'.

So, emergent phenomena are those for which the amount of computation necessary for prediction *from an optimal set of rules, classifications and analysis, even derived from an idealised perfect understanding, can never improve upon the amount of computation necessary to simulate the system directly* from our knowledge of the rules of its interactions. I posit that this definition concurs with the wishy-washy heuristic sense people mean when they (ab)use such terms as 'emergent behaviour' and 'emergent complexity'[3].

Firstly note that the predictive difficulty of a phenomenon will depend upon both the size, n, and the interaction complexity of a given member of a family of complex systems.

I argue that there is no discontinuous separation between emergence and non-emergence. Emergence is purely the result of a phase change in the amount of computation necessary for optimal prediction of certain phenomena. To explain this phase change, we need to look at the two means of prediction we use:

- Let $s(n)$ be the amount of computation required to simulate a system, and arrive at a prediction of the given phenomenon. At times we may wish to consider this an idealised quantity referring to the optimally tuned simulation ("God's simulation"), but always a simulation.

- Our 'deeper level of understanding' of the symmetries of the system (in both the obvious and abstract senses), has allowed us to perform a creative analysis and deduce the future state whilst, we hope, circumventing most of the previous-required computation. Let $u(n)$ be the amount of computation required to arrive at the result by this method.

I shall shortly address the issue of defining these 'amounts of computation' more precisely, but for the moment let us assume this can be done in a useful manner. Then the above definition states:

$$u(n) < s(n) \Rightarrow \text{the system is non-emergent.}$$
$$u(n) \geq s(n) \Rightarrow \text{the system is emergent.}$$

For simple systems, obviously $u(n) \ll s(n)$ (think of statistical physics), but for many classes of system, as we increase their size and rule complexity, there will be a phase change where the curves $u(n)$ and $s(n)$ cross[4]. The crux is that there is *no discontinuity* separating non-emergent from emergent systems. There is just a phase change in the optimal means of system prediction. Beyond this, perfect understanding of the system does no better than a simulation. All useful predictive knowledge is contained in the accumulation of interactions.

Before proceeding further, let us return to the issue of defining $u(n)$ and $s(n)$. I aim this analysis primarily towards discrete, deterministic systems, although many of the ideas could clearly be made more widely applicable. The obvious problem is that the time required to complete a given computation is machine-dependent. For finite computations the time required by different Turing machines is just related by an arbitrary polynomial. If this phase transition is a real phenomenon, it shouldn't be machine-dependent. To remove this dependence, we must take into account the complexity of the Turing machine calculating the predictions. I therefore propose the following definition:

Definition 2 *The amount of computation, c, for some process is given by*

$$c = \sum_{s \in \mathcal{S}} (complexity\ of\ s)$$
$$= (\#\ of\ steps).(complexity\ of\ each\ step)$$

where $\mathcal{S}$ is the set of all steps required in the computation, and the latter equation is true only if all computational steps are the same.

The complexity of a step is measured in terms of the Kolmogorov complexity of a representation of the computation - i.e. the minimal description length.

To justify the machine-independence of this definition, consider a general cellular automaton (which clearly falls within our definition of a complex system), with K possible states for each cell, and a dependence neighbourhood of R cells.

Given $n(R-1) + 1$ adjacent initial cell-states, we can determine the state of a single cell at the bottom of a light-cone after n time-steps in just $n + (R - 1)n(n - 1)/2 \approx \frac{1}{2}Rn^2$ rule applications. The CA is a mapping from a set of size K^R to one of size K. As this mapping may be completely arbitrary, we can represent each step with some $(\log_2 K^R).(\log_2 K) = R.(\log_2 K)^2$ bits. So, to leading order, $s(n) = \frac{1}{2}(nR \log_2 K)^2$.

I shall now argue that $s(n)$ is invariant under the following transformations.

1. Pre-calculating a larger table, so that we need simulate only every l'th step, by using a larger neighbourhood (of size $R + l(R - 1)$).

   Clearly $n \to n/l$, $R \to R + l(R - 1)$. Thus $s(n)$ is reduced by a factor $(1 - 1/R + 1/l)^2$, which is second order, and only significant in the limit of large memory (l) and small R.

2. Using N machines in parallel to simulate the system.

   $s(n)$ is obviously unchanged except for a small increase due to the overhead involved in splitting up the problem.

3. Mapping the CA onto another with more states, such that a single cell corresponds to several in the original system, but with smaller neighbourhoods.

   Assume we compress a cells into 1, which will now have $K' = K^a$ possible states. $R' \approx R/a$, and $s(n)$ is unchanged.

4. Using *any* Turing machine to simulate the CA.

   I shall not attempt to consider a general transformation, but rather appeal to the intuitive idea that increases in Kolmogorov complexity will offset any decrease in computational steps. I postulate that this is true for complex systems in general, not just CA.

Having identified a fundamental interactive unit, and the complexity of the interaction rules for any given system; we can define all computational amounts in the above manner. Let us consider an *idealised* means of

---

[3]There are of course some completely un-loaded uses of the word emergence, such as to describe the hardness of a rock, or the colour of an object.

[4]I shall justify this statement later.

prediction in that complex system, requiring an amount of computation $u_{\mathrm{opt}}(n)$, based on perfect understanding.

Definition 3 *The emergence ratio, $\xi = u_{opt}(n)/s(n)$.*

The emergence coefficient, $\zeta = -\frac{\log \xi}{\log n}$

Where ζ is only really useful if u and s are not exponential in n, but have leading order n^{α_u} and n^{α_s} respectively, in which case $\zeta = \alpha_s - \alpha_u$ to leading order. It is a positive measure of the range of possible levels of predictive understanding we may have in the system ($\zeta = 0 \Rightarrow$ we should simulate; $\zeta > 0 \Rightarrow$ increasingly rapid predictions). A related measure of understanding, which can be useful if we wish to compare how different methods of prediction, $u(n)$, fare as we move around the space of relatively simple systems, is given by:

Definition 4 *Relative understanding, $\lambda = \frac{u_{opt}(n) - u(n)}{s(n)}$*

Note that, at the phase change, u_{opt} can have a discontinuous gradient. Also, beyond this, $u_{\mathrm{opt}} = s$, and since $u \geq u_{\mathrm{opt}}$, we have that $\lambda \leq 0$. So in an emergent system, our understanding can be at best zero. Our astonishment at the fact that we cannot seem to predict emergent properties, stems not from any failure to understand, but from an inherent property of the system, brought about by the accumulation of interactions. A small reassurance at least.

Now we need no longer deal with any explicit dichotomy between emergent and non-emergent phenomena. The perceived lack of understanding in the former is really just another way of describing the complexity of the map between initial state and final phenomenon. In the sense that a lack of knowledge of the initial conditions will usually cause increasingly poor predictions; this is analogous to a discrete version of chaos.

This explains the apparent paradox of the rules giving a succinct, complete description of the system, whilst still revealing almost nothing about it. Of course any single phenomenon may fall anywhere in the spectrum between trivial prediction and interesting emergence.

Does this necessarily reduce our attempts to understand emergent systems to 'stamp-collecting'? (In the sense that all we may do is catalogue the results of simulations). Section 4 discusses this issue, demonstrating that there are profitable, interesting possibilities for the study of complex systems.

3 The Roots of Emergence

Although I have reached the conclusion that we have no useful predictive information, despite a possible 'perfect' understanding, I have not yet addressed the issue of the root of emergence and the loss of information. Indeed, can we justify that emergence truly exists?

Many complex systems have been shown to be capable of universal computation (e.g. CA with as few as 14 states). Therefore many questions about their infinite time behaviour are formally undecidable. In the case of finite size or time, what happens to analogues of these undecidable propositions? I posit that they are emergent

— the computations retain their irreducibility (see Wolfram (1985b, 1985c, 1985a) for a heuristic discussion of irreducibility). Therefore the root of emergence and the commensurate loss of information lie in the foundations of decidability and complexity in computation.

Note that these complex systems are often only universal for very particular, rare initial conditions. Following Wolfram (1985c), I would suggest that emergence is far more common and occurs in systems which are not computationally universal.

Can we determine, for a given system, whether or not it is emergent? This question seems rather more subtle, now that we have reduced an apparent dichotomy to a continuous parameter range. The answer may be extremely difficult to determine for systems near the transition.

3.1 Decidability of Emergence

Suppose we had an algorithm, which when applied to a complex system could give us a definite answer of 'Emergent' or 'Non-emergent'.

For the case of cellular automata, Wolfram(1983) has suggested a classification into four classes of behaviour: homogeneous (Class I), periodic (Class II), chaotic (Class III) and complex (Class IV). Whether a cellular automaton is in class I, II or III has been shown(Culik and Yu 1988; Sutner 1990) to be undecidable. It follows that the question of emergence in infinite systems is undecidable, with a possible reduction to NP-complete status for finite systems(Green 1987).

More heuristically, by considering the halting problem for any complex system which is capable of universal computation, we know that the best (only) means of prediction in such a situation is to 'run the program' - i.e. perform the simulation. So all complex systems which fall prey to such isomorphisms would seem to be emergent, and the halting problem then analogises to tell us that, in the general case, the question of whether a system is emergent or not is an undecidable proposition.

The undecidability of emergence presents us with a particularly gnarly problem. For any such undecidable system, we can either:

(i) Assume it is emergent, and use large amounts of computational power for simulations. What do we do if Enrico Fermi suddenly arrives, with his deep understanding, and removes all mystery from the system. We have wasted a huge amount of effort.

(ii) Assume the system is non-emergent, and try to find its deep, hidden inner structure. All this effort could conceivably be in vain.

This devil's alternative becomes especially important when the system under consideration is one that has been designed by us as a cooperative attempt to solve a particular problem.

4 Studying Emergence

I previously defined the emergence ratio ξ. It is a function of the size of the system, n, together with some inherent level of emergence contained in the rules of interaction. It measures the relative efficiency of rule-based 'understanding' versus simulation-based 'understanding'. My results indicate that, in general, we cannot determine this ratio. Some interesting questions arise dealing with the behaviour of ξ with n. In very small systems we can usually perform some form of an analysis, so u will be smaller than s.

As systems become more emergent, and n increases, the propagation of information through accumulated interaction will blur the boundaries of any analysis we try and perform. Trying to generalise in the initial-state space becomes more and more futile, until any previously useful u outstrips s. We gradually approach the worst case – that of being forced simply to classify points in the state-space purely by exhaustive enumeration, with each individual result being determined by a simulation. Generalisation has vanished.

Before moving on to discuss possible approaches to the study of emergent systems, it is important to point out that many systems are not emergent, and therefore amenable to some form of analysis. Such an analysis will certainly not generalise to the emergent complex systems, but is clearly an important and valuable contribution to understanding the context of our investigations – the continuum quantified by the emergence ratio.

For example, it has been demonstrated that in certain highly symmetric classes of one-dimensional cellular automata, the single cell at the bottom of a light-cone after n time-steps can be predicted more quickly than the $n(n - 1)/2$ steps of a simulation:

Linear CA(Martin et al. 1984) have $u \sim n \Rightarrow \xi \simeq 2/n, \zeta \simeq 1$. Quasi-linear CA with radius 1/2 (Moore 1993) have $u \sim n^{\frac{\log 3}{\log 2}} \Rightarrow \xi \simeq 2/n^{0.41}, \zeta \simeq 0.41$. The proof of the latter result is particularly informative in the direct manner in which it exploits the symmetry of, for example, the quaternion group.

4.1 Research inside Emergent Systems

I'd like to give an example to suggest that the fact that a given system may lie far beyond the phase change does not mean we should lose all hope in our traditional means of understanding, explanation and prediction:

Consider an analogy with the game of chess. Current computer chess programs use extremely sophisticated but nevertheless brute-force approaches to 'simulate' the game and make predictions. Human grand-masters use a wonderfully subtle combination of pattern-matching, generalisation, comparison and analogy-making with some look-ahead to 'understand' the game and make their predictions. The branching factor of simulations is so high that humans are currently far superior at determining such elusive concepts as positional advantage[5]. I would posit that chess lies on the emergent side of the phase boundary, so that solution by simulation is ultimately the best approach. However, human expertise and understanding seem to achieve an astonishing amount in the face of the above information paradox.

This highlights the fact that my result only states we can do no better than a simulation by using our understanding. A sufficiently sophisticated combination of approaches may be the brain's best bet for prediction in complex systems. The brain itself, of course, is an extremely complex system, which I would suggest lies far beyond the phase change - this clearly has important ramifications for artificial intelligence[6].

Let us now look at the direct study of emergence. There are two alternative view-points from which we could approach emergent complex systems.

Firstly there is the possibility of performing a limited characterisation on the space of emergent systems. Although the boundaries of any such classification in complex system space will be very imprecise, many important systems may fall nicely into such a scheme.

Secondly, given that we know a particular system evolves to a particular type of emergent state under some interesting conditions (or perhaps under most conditions), we may be interested in the behaviour of the system on that emergent state.

1. By virtue of its emergent nature, the interesting, coherent parts of its emergent states will usually be composed of ever changing subsets of the actors which comprise the complex system. Is the split between co-organised and apparently separated actors ergodic? Perhaps we can apply statistical physics techniques to some sub-system.

2. We would like to know the structural stability of the emergent state under external perturbations.

3. The emergent state may exhibit interesting dynamics, e.g. self-organised critical phenomena such as power-law scaling in its structure or constituents.

4. How robust is the emergent state to changes in the underlying rules (interaction stability)?

5. What happens if we allow the system itself to evolve by modification of its own rules?

4.2 Self-Organised Criticality

Let us now look in more detail at just one of the above options[7]:

Definition 5 *A self-organised critical state is a dynamic equilibrium, the frequency of disturbances from which obey a power-law distribution with respect to their size, l:*

$$f(l) = c/l^k$$
$$\Rightarrow \log f = c' - k \log l$$

Where $k > 1$, c is a normalising constant, and these are expectations in the large time limit.

---

[5] Especially if the usual time constraints are removed, as in postal chess.

[6] The traditional knowledge-based approaches to AI are based completely in conceptual ideas and 'understanding'. My results suggest progress and success by the use of such methods will be harder to achieve than by the use of agent-based and connectionist approaches.

[7] for early discussion on some of these points, see Bak, Tang, and Wiesenfeld (1987, 1988).

Now, let the set of actors be X; the parameter under investigation be $a_x{}^t$ for some $x \in X$ at time t; and $\langle,\rangle$ be a metric on the parameter space. Define the average:

$$\bar{a}_X^t = \frac{\sum_{x \in X} \langle 0, a_x{}^t \rangle}{|X|}$$

Then we say a_X is in a critical state if:

$$\mathcal{P}(\lim_{t \to \infty} \langle a_x{}^t, \bar{a}_X^t \rangle = l) = cl^{-k}$$

In order to calculate or approximate such a limit for a system, we need to know the rules. In this case all we need is a rule to give us each $a_x{}^t$ from some local neighbourhood $\mathcal{N}_X$, at time $t - 1$. We formalise this as follows:

$$a_x{}^t = \mathcal{R}_{a_x{}^{t-1}}(\{a_y{}^{t-1}; y \in \mathcal{N}_X(x)\})$$

Here, in full generality, $\mathcal{R}$ is dependent on all properties of the previous state. For specific systems, the variation in $\mathcal{R}$ may be less general, or even constant.

As an example, let us consider a one dimensional system, with $\mathcal{R}$ constant in time, and a unit neighbourhood.

$$
\begin{aligned}
a_x{}^t &= \mathcal{R}_{a_x{}^{t-1}}(a_{x-1}{}^{t-1}, a_{x+1}{}^{t-1}) \\
&= \mathcal{R}'(a_{x-1}{}^{t-1}, a_x{}^{t-1}, a_{x+1}{}^{t-1})
\end{aligned}
$$

This is the update equation for a cellular automaton (the derivation of the second line requires that the actors are not uniquely identifiable). Thus CA will fit within this framework, with many other systems, although perhaps very few will satisfy the critical property. This is the kind of phenomenon which is likely to be emergent for many $\mathcal{R}$.

Current research(Darley and Kauffman 1994) dealing with one particular type of rule $\mathcal{R}$ — describing the interactions between predictive agents in an artificial economy — has demonstrated the existence of robust self-organised dynamic equilibria. The equilibria are found in the space defined by a metric which isolates the complexity of the predictive algorithms used by the agents. Simulations have shown this is a critical state, with power-law scaling of adaptive changes to the predictive models. Such results will be presented in a forthcoming paper of a less philosophical nature.

5 Conclusion

I have given a definition for 'emergence', based upon a notion of predictive complexity. Emergence implies computational irreducibility - which can be seen to be the finite analogue of formal undecidability. Hence, the strange lack of understanding in emergent systems has its roots in complexity theory, from which it would seem that emergence itself is an undecidable proposition.

I conclude that not only do emergent systems exist, but also that they match very closely our working definition of the term. Simulation is an optimal means of determining the outcome of such systems, and is thus an important means of investigation. Simulations can be coupled with any of a number of different analysis methods on the emergent state. Together they are suitable for carrying out very interesting, real science in emergent or near-emergent systems, and should lead to explanations of some of the many thought-provoking emergent phenomena we observe. I have elaborated on just one approach, of current personal interest, to reveal the application of these ideas to critical phenomena.

To summarise, my results suggest the best approach to take in studying emergent complexity is a feedback process of simulation and analysis of the actual emergent phenomena. These results should go some way towards legitimising the concept of a simulation as a real scientific tool for the investigation of *emergence*.

Acknowledgments

The author would like to thank J. Chalcraft for much insightful and entertaining conversation and criticism.

References

Baas, N. (1993). Emergence, hierarchies, and hyper-structures. In C. Langton (Ed.), *Artificial Life III*, pp. 515–537.

Bak, P., C. Tang, and K. Wiesenfeld (1987). Self-organized criticality: An explanation of $1/f$ noise. *Physical Review Letters 59*(4), 381–384.

Bak, P., C. Tang, and K. Wiesenfeld (1988). Self-organized criticality. *Physical Review A 38*(1), 364–374.

Cariani, P. (1992). Emergence and artificial life. In C. Langton (Ed.), *Artificial Life II*, pp. 775–797.

Culik, K. and S. Yu (1988). Undecidability of CA classification schemes. *Complex Systems 2*, 177.

Darley, V. and S. Kauffman (1994). Self-organisation in predictive dynamics. In preparation.

Farmer, J., S. Kauffman, and N. Packard (1986). Autocatalytic replication of polymers. *Physica D 22*, 50–67.

Green, F. (1987). NP-complete problems in cellular automata. *Complex Systems 1*, 453.

Koza, J. (1993). *Genetic Programming*. MIT Press.

Martin, O., A. Odlyzko, and S. Wolfram (1984). Algebraic properties of cellular automata. *Commun. Math. Phys. 93*, 219–258.

Moore, C. (1993). Personal communication.

Sutner, K. (1990). Classifying circular cellular automata. *Physica D 45*, 386–395.

Toffoli, T. and N. Margolus (1987). *Cellular Automata Machines*. Cambridge: MIT Press.

Wolfram, S. (1983). Stastical mechanics of cellular automata. *Rev. Modern Physics 55*, 601–644.

Wolfram, S. (1985a). Origins of randomness in physical systems. *Physical Review Letters 55*(5).

Wolfram, S. (1985b). Twenty problems in the theory of cellular automata. *Physica Scripta*, 170–183.

Wolfram, S. (1985c). Undecidability and intractability in theoretical physics. *Physical Review Letters 54*(8).

Autonomy vs. Environmental Dependency in Neural Knowledge Representation

Markus F. PESCHL[*]
Dept. for Philosophy of Science
University of Vienna
Sensengasse 8/10, A–1090 WIEN
AUSTRIA, EUROPE

e-mail: *a6111daa@vm.univie.ac.at*

Abstract

In neural systems with a recurrent architecture we are facing the problem that we cannot apply the traditional concepts of knowledge representation any more: i.e., we cannot find a stable relation of reference. That is why a redefinition of the relation between the states of the environment and the representational states is suggested. The aim is no longer to map the environment as accurately as possible to the representation system (e.g., symbols). We rather have to look at neural systems as physical devices *embodying* the (transformation) knowledge/function for sensorimotor integration and for generating adequate behavior enabling the organism's survival.

As will be shown the representation of the environment is not only determined by the environment, but highly depends on the organization, structure, and constraints of the representation system which is embedded in a particular body structure. This leads to a *system relative* concept of representation which will be studied in this paper. By transforming recurrent neural networks into the domain of finite automata the dynamics as well as the epistemological implications will become more clear. It turns out that there is some kind of balance between the autonomy of the representation and the environmental dependence/influence in recurrent neural systems.

---

[*]Currently on leave at the Univ. of California, San Diego; this work is supported by a grant from the *Austrian Science Foundation* (FWF).

1 The representational function of feed forward and recurrent neural architectures

The dynamics and processing mechanisms of natural and artificial neural systems do not really fit into the traditional concepts and understanding of representation (e.g., *Fodor* and *Pylyshyn* 1981, 1988). That is why computational neuroepistemology (*Peschl* 1992, 1993) suggests to develop an alternative view of representation which is based on the principles of neuroscience rather than on speculative, folk psychological, and common sense assumptions. As a consequence it seems that we have to *give up* one of the basic ideas of representation: namely the concept of a more or less stable referential relation between repraesentans and repraesentandum (i.e., a certain state in the environment refers to a certain state in the representation system in a stable manner; in some sense the "picture-of-the-environment" view). Several considerations lead to this conclusion:

(i) *Epistemological* reasons: what we are experiencing as "*the*" environment already is the result of neurally realized construction processes leading to a *representation of "the" environment* (and this is also valid for so-called scientific descriptions/explanations). So, there is no such thing as an objective or observer-independent environment as the access to the environment will always be mediated and determined by the observer's sensory and nervous system. Constructivist considerations (e.g. *Glasersfeld* 1984, *Varela* et al. 1991), and many others) show that the representation of the environment always depends on the current structure and dynamics of the substratum in which it is realized (i.e., *system relativity* and *theory ladenness* of representation). The *constructivist* framework provides an alternative and more plausible view of the representational function of neural systems; it is based on the following concepts: representation is (a) the result of a physically/neurally realized process of *construction*, (b) it does not refer directly to the environment, (c) it is system relative, (d) it is responsible for *generating behavior* rather than representing/depicting the environment, and (e) it has a relation of *functional fitness* to the environment. It turns out that these concepts and issues seem to be a more adequate epistemological foundation for neural systems than traditional concepts of representation.

(ii) *Empirical* reasons: empirical as well as simulative experiments in (computational) neuroscience (e.g., *P.S.Churchland* and *T.Sejnowski* 1992) show that we cannot find a stable referential relation between the environment and the representational substratum (see also section 2). The referential relation could be saved in the case of feed forward architectures. We have to bear in mind, however, that (a) feed forward architectures account only for a very small fraction of the nervous system (e.g., in the peripheral parts) and that (b) the "interesting things" happen in the recurrent parts of the neural system. Furthermore (c) the environmental signal is already *distorted* in a non-linear manner in the process of *transduction*.

Let's have a look at the representational function of feed forward networks in a first step. I.e., under the (implausible) assumption of a "structure preserving" and stable transformation of the environmental signal in the process of transduction a certain pattern of activations refers to a certain state s_e in the environment[1]. In other words, an activation pattern is the *representation* (in the traditional sense) of s_e. This form of representation is called *distributed representation* (e.g., *Rumelhart* et al. 1986, *Smolensky* 1987, 1988) – each unit/neuron is involved in the representation of all environmental states. One environmental state is represented by all units/neurons (i.e., by a certain pattern of activations). From an epistemological perspective some interesting things happen in this kind of representation: we not only have to give up the concept of linguistic transparency (*A.Clark* 1989), but also have to reconsider the referential function of single activations. One can only say that a whole pattern of activations refers to a certain (group of) environmental state(s). It is clear that single activations play a crucial role in distributed representation, but what is their representational function? The activation of a single neuron/unit is the result of a squashing function being applied to the sum of its weighted inputs. This implies that a certain activation a_j can be achieved by different sums $\sum_i w_{ij} o_i$ (under the assumption of constant w_{ij}). Thus, two different environmental states may cause the same activation value

a_j in a hidden/output unit. So, what does a_j represent? The concept of distributed representation gives us an answer: some kind of "abstract feature" ("microfeature" (*Rumelhart* et al. 1986) or "subsymbolic feature" (*Smolensky* 1988)) which is present in both inputs and which is extracted/detected by this unit/neuron.

Notice, that we are not talking about an "objective" or "system independent" feature here: in most cases these features are some kind of statistical property *with respect to some desired output* or some measurement of stability (i.e., system relativity of representation). It is only *one possible* way of categorizing the structure of the environment. This particular categorization is determined by the current structure of the neural (and sensory) system. On the one hand the environmental state causes a certain representational state. On the other hand the configuration of the representation system (= weight configuration) determines which (representational) states may be caused by the environment. These "microfeature representations" are not a mapping or reference in the traditional sense any more. They are the result of a *constructive transformation* which aims at contributing to the *generation* of adequate behavior by extracting/constructing system-relevant representations. One important problem remains unsolved, however: what is the representational function of the *synaptic weights* which are central for the generation of the patterns of activations/representations? What is their relation to the structure in the environment? It could be called some kind of *"generative"* relation to the environment. It is a highly *indirect* relation as the goal is to generate activation patterns which lead to behavior which functionally fits into the environment and which maintains the organism's internal equilibrium (= criterion for survival). Thus, the synaptic weights do not represent the environment or refer to its structures; they rather represent *relations* or a transformation for generating adequate behavior with respect to the system's structure and needs.

The problem of finding a substratum for the referential relation or for the representation (in the traditional sense) becomes even worse in the case of *recurrent neural systems*: it turns out that we have to give up the stable and homomorphic relation between the environment and patterns of activations in recurrent architectures, too. There are several reasons for this which can be summed up as follows: due to recurrent weights there is a *recurrent flow* of activations in which a feedback relation between activations is established. As an implication we can find *internal states* in recurrent neural systems: i.e., patterns of activations which are the direct or indirect result of themselves and which influence and feed back to the incoming patterns of activations. This leads to a problem which is crucial for the understanding of representation in recurrent neural systems: we

---

[1]Under the assumption that no learning processes are going on.

finally have to give up the (traditional) idea of a stable referential relation between patterns of activations and states in the environment, as the internal states (internal activations) have an influence on the processing of the input activations. Unlike in feed forward architectures the following internal/representational state is not only the result of the current input i, but *also* a function of itself. For our question of representation this implies that the current internal state cannot be seen as representing or referring to a certain state of the environment any more, as it depends also on the previous internal state which representational state (= pattern of activations) is assumed. Thus, it seems that the influence of the environment is reduced to a minimum. The suggestion being made in this paper is to give up the attempt to press neural systems into the traditional understanding of representation and to replace it by alternative concepts which will be discussed in the following sections.

2 Neural systems, finite automata, and representation

Embodiment and *system relativity* are two central concepts in the representational view being suggested in this paper. So, when we are talking about "representation" we do not have in mind some abstract concepts, rule systems, semantic networks, etc., but *physically realized* structures which (a) determine the organism's dynamics (b) making us as observes of these behaviors think that there is some kind of representation of the environment involved. The question to which the problem of representation can be boiled down is the following: what is the *substratum of representation* and which *representational relation* can be found between this substratum and the environment. In neural systems we can find at least four candidates/carriers of representation: (i) neurons and their *activations*; they are responsible for the "actual representation" of (transduced) environmental states, of the organism's internal state, and of the behavioral output. Their dynamics is determined by (ii) the *synaptic weights* which relate the activations/neurons to each other. By doing so they are controlling the flow/spreading of activations in the network. Neither single weights nor a whole pattern of weights have some kind of iso-/homomorphic ("direct") referential relation to the environment as they are responsible for *generating* adequate behavior. What is represented in the synaptic weights is not the environment, but a *dynamics*/transformation relating inputs, internal states, and outputs to each other. Synaptic weights have a "second order and generative relation" to the environment and represent the knowledge to cope successfully with the environment in the context of fulfilling the internal constraints and criteria for survival. (iii) It turns

out that *sequences of activation patterns* can have referential functions (especially) in recurrent neural systems. I.e., *trajectories* (e.g., cyclic patterns, certain attractors, etc.) in the activation space fulfill representational functions. (iv) The *genetic material* represents the basic level of representation – it embodies the instructions, constraints, dynamics, for the generation/development of the basic architecture for the neural representation system and the body. Its referential relation to the environment is even more *indirect* as at least two levels/steps of *generation* are in-between the structures of the genetic code and the environment. I.e., the generation of the body/neural structures which are responsible for the generation of the actual behavior testing the success/failure of the representation. These representational substrata are neither static nor are they acting in isolation: their representational function is based on the *interaction* between these four levels/processes of representation and the environmental dynamics.

Studying representation in neural systems means to explore the interaction between the environment and different levels of physical processes (and their dynamics) which are embedded in an organism. In order to see the contribution, interplay, and role of each of these levels in the global representational function of an organism I am suggesting to transform neural systems in the domain of *finite automata* in which their dynamics and interactions can be seen more clearly. From these observations we can translate our findings back to the domain of neural systems and apply them there. As we shall see in a moment we can establish a more or less isomorphic relation between automata and neural networks. Almost any system which can be described as a system with *states* and *state transitions* can be transformed into an abstract description having the form of a finite automaton. The idea is to transform a (recurrent) neural system into a finite automaton in order to find out more about its dynamics and representational function. An automaton can be represented graphically as a graph with directed edges. Each node represents a certain *state* of the simulated/represented system. Each directed edge represents a *state transition*. From each node k directed edges originate (k is the number of possible different inputs) – these bundles of edges determine the *dynamics* of the whole automaton. In other words, a bundle of edges originating from a particular state s_x determines the space of possible states which can follow s_x. The environmental input acts as a *selector*; its current state determines (via the transduction process) the input which chooses one of the possible successor states. A certain input i does *not* always imply the selection of a certain state. It rather selects an already predetermined state transition (= edge). As we shall see later this is an interesting phenomenon with respect to our question concerning representation in neural systems.

In the case of an automaton we are confronted with the phenomenon that the successor state does not only depend on the input (i.e., we cannot find a direct mapping between the input/environment and the internal state). The successor state is the result of three factors: (a) the current *input*, (b) the current *internal state*, and (c) the structure/architecture of the *state transitions*.

The idea is to transform neural systems into an automaton. *M.Arbib* (1987, p 24f) gives a mathematical proof how this transformation can be realized. We won't go into the details of this proof as this is not really necessary for what we are interested in. The assumption for this transformation is that the neural system must have *discrete* activation values. As we shall see in a moment this restriction does not have any crucial implications for the basic problem which we are interested in[2]. Furthermore a *synchronous* update strategy is assumed. Let's have a closer look at this transformation: an *isomorphic* relation can be found between the states/dynamics of the neural system and the states/dynamics of the finite automaton; each pattern/state of activations s_i^n in the neural network (being represented as a point P_i in the activation space) can be mapped isomorphically to a certain state s_i^A of the finite automaton. The number of possible states in the neural system (= the number of points in the activation space) increases exponentially with an increasing number of neurons. This implies that the number of states in the automaton will be quite large. The relation between the inputs for the neural system and for the automaton is *isomorphic*, too. The number of possible inputs for the automaton (as well as for the neural system) is given by the product over all possible single input values (i.e., all possible k combinations of input patterns/vectors). As has been discussed inputs are *selecting* predetermined states in an automaton. The k inputs are transformed into k edge-types in the automaton. From *each* state s_i^A in the automaton a bundle of exactly k edges originates and connects it to other (or even the same) state(s). *Spreading activations* are represented as *state transitions* in the automaton. If the automaton is in state s_i^A the bundle of edges *constrains* the set of possible successor states. The current input i selects one of them via the edge being associated with i. Similarly as in the neural activation space we can describe sequences of states as *trajectories* of states which are connected via edges in the automaton. This allows us to easily track the *dynamics* of the automaton and, thus the (behavioral) dynamics of the neural system being represented in the automaton. As a first conclusion from this transformation (or better, from the retranslation into neural systems) we can see that from each point P_i in the activation space *implicitly* a bun-

dle of edges originates which point to the set of possible successor states (= patterns of activations). These bundles are implicitly represented in the configuration of the *synaptic weights*. It can be interpreted as *constraining* the state transitions (and, thus, the dynamics) in neural systems. As in the case of the automaton the input *selects* one of the possible and predetermined successor states by its current value.

3 Implications for representation?

- *Representation in (patterns of) activations*: from the automaton analogy we can see even more clearly that patterns of activations (in a recurrent neural network) can*not* be interpreted as representations in the traditional sense (i.e., as a state which stands for a certain state of the environment in a stable and iso-/homomorphic manner). Looking at the dynamics of automata it is obvious that a certain state is *not* only the result of the environmental input. The previous state determines the current state at least to the same degree as the current input. This implies that no stable referential/representational relation between the internal state of the automaton and an environmental state is guaranteed. One input may cause – depending on the previous state – *different* internal states. This implies that patterns of activations cannot be interpreted as representational states in the traditional sense.

- *Representation in synaptic weights*: the synaptic weights are closely related to the *edges* of the automaton. Both determine the *dynamics* of the system. In the case of neural networks this means that they determine the flow/spreading of activations. The edges in an automaton determine the state transitions. The advantage of the automaton representation is that the transitions between the states/activation patterns become explicit in the architecture of the edges. One could imagine an automaton having its states arranged in the n-dimensional activation space according to the location of the corresponding patterns of activations. The dynamics of the whole neural system would become transparent/explicit by the edges connecting these states. As we have seen both the edges and the synaptic weights act as *constraints* for the dynamics of the system by restricting and determining the successor states. Their representational function is *not* to depict states of the environment but to *guide/control* the behavioral dynamics. I.e., the constraints being represented in the bundles of edges (which are implicitly determined by the current configuration of the weights) relate inputs and outputs with each other via a specific recurrent transformation. The structure of the state transitions does not aim at the representation of the environment, but at the generation

---

[2] Have in mind that, whenever we are simulating neural systems on computers, we make use of discrete systems (and it seems that the scientific community accepts these models).

of adequate behavior in the context of certain environmental and internal states. Ontogenetic (and phylogenetic) adaptation is realized as changes in the structure of the synaptic weights. These changes of weights (= a wandering point in weight space) are represented as changes in the structure of the edges. I.e., the edges are redirected and, thus, the dynamics of the whole system changes. This leads to a new automaton which has the same set of states, but a different structure of edges. This alternative configuration of edges/weights is also responsible for the changes in the "trajectory landscape" which can be visualized very well in the automaton analogy.

• *Representation in trajectories and attractors*: a trajectory in the activation space corresponds to a set of state transitions in an automaton. The current input selects one edge out of the bundle of edges which originate from the current state. Over a period of time this leads to a sequence or trajectory of states. Have in mind that even when the automaton/neural system does not receive any input from the environment this is nevertheless an input, namely the "0"/neutral input which selects as any other input a certain trajectory of states. If the input is stable/constant over a longer period of time the automaton (as well as the neural network) can "fall" into a stable state. We can find at least two forms of stable states (*Hertz* et al. 1991): (a) *fixed points* and (b) *limit cycles*. The constant input *selects* one of these trajectories/attractors. As has been mentioned already they can be considered as a substratum for representation as certain (constant) inputs may lead into certain attractors. In fact, this is the case in feed forward networks; the sequence of activations in the activation space falls into a fixed point after the activations in the hidden and output layer have been shifted out. In recurrent neural systems, however, we are once again confronted with the problem of the influence/dependence of/on the internal state. The system is more likely to consistently fall into a certain (group of) attractor(s) when a certain constant input is presented than to change into a certain state. The problem remains the same, however: it depends on the current internal state which attractor is selected. The same (constant) input can lead to completely different (representational) states and, thus, to different behaviors. The only stable referential relation which can be found is the following: a *group* of trajectories (namely all trajectories being formed by edges which are associated with a certain input i) refers to a certain constant input in a stable manner. Which trajectory is chosen from this group depends on the current internal state.

From these considerations it seems that the traditional understanding of representation is not too promising. The automaton analogy rather suggests to give up this notion and replace it by a concept of "*representation*

without representations". The internal structures do not map the environmental structures; they are rather responsible for generating functionally fitting behavior which is *triggered* by the environment and *determined* by the internal structure (of the edges or of the synaptic weights). It is the result of *adaptive* phylo- and ontogenetic processes which have changed the architecture (over generations and/or in an individual organism) in such a way that its physical structure embodies the dynamics for maintaining a state of equilibrium (= criterion for survival). If we are taking into account ontogenetic adaptation (e.g., neural plasticity or development) and evolutionary processes the environment "gains" influence: it represents the constraints for the development of the body/neural/representational structure. Constraining this development does *not* lead to a structural mapping of the environment to the representational system, however. The environment does *not* explicitly select certain states and/or body/neural structures, but only *trajectories* which lead to states and/or body/neural structures. These states are primarily determined by the structure of the edges/synaptic weights, however, and *not* by the environment.

As an implication the *structure determined* character (in the sense of *Maturana* et al. 1980) of a neural system becomes more clear. Our considerations shed some light on the relation between the *autonomy* of a representational system vs. the *influence of the environment* on this system. The representational (neural) system *itself* (and not the environment) determines which (representational) states can be assumed and what are the rules for the state transitions. The environment does *not* directly cause certain states (in a stable manner), but chooses from an already predetermined set of states. The successor state always depends on the current state and on the current input (= transduced environmental signal). We have to bear in mind, however, that the representational structure and its dynamics is the result of an onto- and phylogenetic process of (mutual) adaptation of physical structures and constraints in both the representation/body system and the environment. Due to its higher plasticity (and complexity) the representation system is more likely to change in most cases, however. So there is an interplay and a quite well balanced relation between the influence of the environment and the (representational) autonomy of the representation system. The environment triggers certain trajectories which are predetermined in the physical structure of the representation system. We could say that the representation system defines its space of what it could represent/"know"/generate by its own structures in a process of interaction with the environment.

The environmental stimuli which are transduced into neural representations can be understood as *operators* which select from the already given dynamics. This is contrary to the traditional understanding of representation where some mechanisms operate on the representation of the environment (i.e., the representation is the operand). What happens to an input, which states/representations/activation patterns are triggered by this input, and which behavior is generated, is determined by the representation system itself – it is clear that such a view has crucial implications for our understanding of representation. A certain pattern of activations does *not* "represent" a certain environmental state, but a state which relates the (previous) internal state with an external state. It aims at generating adequate behavior rather than at depicting the environment. The environment influences the representational structure (= the structure of states and edges in the finite automaton) *indirectly* via the adaptive phylo- and ontogenetic processes. Here again, the result is not the representation of the environmental structure (i.e., no correspondence can be found between the structure of the environment and the structure of the edges), but a structure which relates the environmental dynamics to the task of maintaining the state(s) of equilibrium/homeostasis being necessary for the organism's survival and reproduction.

3.1 Behavior, internal states, and representation

Another lesson we can learn from the automaton analogy concerns the relation between externally observable *behavior*, the environmental state, and internal (representational) states. Normally we assume that the externally observed behavior is an indicator for the internal states/representations. In other words, observing the behavior of an organism makes us project a certain (internal) representational state into this system. Our automaton analogy reveals, however, that such an assumption cannot be maintained: different internal states can cause the same externally observable behavior. This implies that we are not allowed to correlate a certain externally observed behavior with a single internal/representational state. Is this also true for neural systems? The motor system being responsible for the organism's behavior is controlled by a *subset* of neurons, the output/motor neurons. As has been mentioned in section 1 the activation of a neuron is the result of a sum of products. Thus, a certain activation value can be generated by different sums of products. As a certain activation value a_m is only one component in the whole activation vector/pattern and a_m can be generated by different combinations of the other activations *different* internal states may account for the *same* externally observable behavioral action (being associated with a_m).

As different internal states (i.e., different internal patterns of activations) can lead to the same behavior, it

is *methodologically questionable* to assume certain internal representations from externally observed behaviors (this could be used as one more argument against folk psychology (*P.M.Churchland* 1993)). Most (cognitive) psychological approaches as well as traditional concepts in cognitive science and AI are based on these problematic assumptions/methods. They almost exclusively make use of behavioral observations (= the observer's representation of the behavior) which are categorized, correlated with (the observer's representation of the) environmental events, and transformed into representation systems/models. Our considerations from the last paragraphs suggest to question these methods and the theoretical explanations and models of representation resulting from this procedure. For achieving an adequate understanding of the representational processes going on in neurally based cognitive systems we have to *open* them and look at their *internal* structure and dynamics instead of extrapolating and projecting externally observed behaviors into internal representational states.

4 Conclusions

The concept of *internal states* has turned out to be central not only for neural systems, but also for our understanding of representation in these systems. The transformation of (recurrent) neural networks into automata has revealed a lot of interesting aspects concerning the representational function of neural systems: (i) cognitive systems having a recurrent architecture can*not* be understood as simple stimulus-response systems. (ii) In recurrent systems it is methodologically questionable to infer internal (states of) representations from externally observed behaviors. (iii) The same behavior can be caused by different environmental stimuli. (iv) One environmental stimulus can cause different behaviors, as the current internal state determines which states might follow and the input does not select explicit states, but trajectories. (v) We have to give up the correspondence between internal/representational states and environmental states (i.e., the stable referential relation of representational states). (vi) Recurrent systems are highly *"history laden"* systems: i.e., the current internal state represents the result of the *selection history* being caused by the external inputs. The possibilities of these selections are determined, however, by the structure of the representational system itself. Abstractly we could describe the state transitions as well as the behavior of a cognitive system as the computation of a *recurrent function*. From the analogy with finite automata we could see that the behavioral dynamics (the dynamics of internal states) is a *trajectory* of states in the cognitive system's activation space which is predetermined by the configuration of the weights and selected by the inputs. Thus, a quite *balanced relation* can be found between the influence/dependence of/on the environment and the autonomy of the representation system. (vii) Furthermore, the representation system is history laden with respect to its phylo- and ontogenetic development. Its current physical structure represents the result of *incremental changes* in the physical substratum (i.e., the neural architecture and genetic code). The particular physical structure has developed in a process of mutual interaction and adaptation between/of the environmental and organism's constraints.

(viii) The representational relation between the environment and the neural substratum is *indirect, system relative*, and *generative*. The dynamics being embodied in the neural system does not aim at depicting or representing the environment, but at relating the environmental inputs with the current internal state in such a way that functionally fitting behavior enabling the organism's survival is generated. (ix) This "generative effect" is achieved by a "symbiotic" relationship between the synaptic weights and the spreading activations – the weights per se only represent the space of possible state transitions (= a bundle of edges originating from each possible state). The actual states are "instantiated" by the activations being the result of previous activations and the activations being provided by the sensory devices. (x) As an implication of the system relativity of representation the traditional concepts of semantics, language, communication, and symbols have to be reconsidered. (xi) From an epistemological perspective it seems that – in the case of neural systems – "representation" can be better characterized as finding a stable relation/covariance between these two dynamic systems and inside the representation/body system. This can be achieved by adaptational/constructive changes in the neural substratum which lead to an embodied dynamics being capable of generating functionally fitting behavior ("representation without representations"). (xii) The representational function of neural systems can be characterized by the *interaction* of five dynamics: the spreading of activations leading to certain behaviors, the ontogenetic changes in the representational/neural substratum (i.e., neural plasticity, development, "learning", etc.), the genetic dynamics, the dynamics of the "natural" environment, and the dynamics of the artificial environment (i.e., artifacts and "culture" in the broadest sense (*Hutchins and Hazelhurst* 1992)). We have seen that, if we want to understand neural representation adequately, we cannot restrict it to ontogenetic changes in the representational substratum, but we have to include also *phylogenetic dynamics* as well as the steps of *development* in neural systems. It has become clear that these systems develop in *interaction with the environmental dynamics* (*Edelman* 1988) – there is a balanced interac-

tion between what is predetermined in the representational substratum (i.e., genetic code, synaptic weights, etc.) and the influence of the environment. The goal is to achieve some kind of stable relation/equilibrium which allows the survival and reproduction of a single organism as well as of a group of organisms.

References

Arbib M.A 1987. *Brains, machines and mathematics.* Springer, New York, second edition.

Churchland P.M. 1993. *Evaluating Our Self Conception.* Mind & Language, vol. 8, No. 2, pp 211–222.

Churchland P.S. and Sejnowski T.J. 1992. *The Computational Brain.* MIT Press, Cambridge, MA.

Clark A. 1980. *Microcognition. Philosophy, Cognitive Science and Parallel Distributed Processing.* MIT Press, MA.

Edelman G.M. 1988. *Topobiology: An Introduction to Molecular Embryology.* Basic Books, NY.

Fodor J.A. 1981. *Representations.* MIT Press, MA.

Fodor J.A. and Pylyshyn Z.W. 1988. *Connectionism and Cognitive Architecture: A Critical Analysis.* Cognition, vol. 20, 1988.

von Glasersfeld E. 1984. *An introduction to radical constructivism.* in P.Watzlawick (Ed.), The Invented Reality, Norton, N.Y., pp 17–40.

Hertz J, Krogh A. and Palmer R.G. 1991. *Introduction to the Theory of Neural Computation.* Addison-Wesley, CA.

Hutchins E. and Hazelhurst B. 1992. *Learning in the Cultural Process.* in C.Langton et al. (eds), Artificial Life II, Addison-Wesley, CA, pp 689–706.

Maturana H.R. and Varela F.J. 1980. *Autopoiesis and Cognition. The Realization of the Living.* D.Reidel Publishing Company, Dordrecht, Boston, London.

Peschl M.F. 1992. *Embodyment of Knowledge in Natural and Artificial Neural Structures. Suggestions for a Cognitive Foundation of Philosophy of Science from a Computational Neuroepistemology Perspective.* Methodologia 11, vol. VI, pp 7–34.

Peschl M.F. 1993. *Knowledge Representation in Cognitive Systems and Science: In Search of a New Foundation for Philosophy of Science from a Neurocomputational and Evolutionary Perspective of Cognition.* Journal of Social and Evolutionary Systems, vol. 16, pp 181–213.

Rumelhart D.E. and McClelland J.L. 1986. *Parallel Distributed Processing, Explorations in the Microstructure of Cognition.* Volumes I & II; MIT Press, Cambridge, MA.

Smolensky P. 1987. *The Constituent Structure of Connectionist Mental States: A Reply to Fodor and Pylyshyn.* The Southern Journal of Philosophy, vol. 26 supp., 1987.

Smolensky P. 1988. *On the proper treatment of connectionism.* Behavioral and Brain Sciences (1988) 11, pp 1–74.

Varela F.J., Thompson E. and Rosch E. 1991. *The Embodied Mind.* MIT Press, Cambridge, MA.

ADIVERSITY
Stepping Up Trophic Levels

Takuya Saruwatari
Institute of Engineering Mechanics University of Tsukuba, Ibaraki 305, Japan
e-mail takuya@kz.tsukuba.ac.jp
Yukihiko Toquenaga
Institute of Biological Sciences University of Tsukuba, Ibaraki 305 Japan
e-mail toquenag@kz.tsukuba.ac.jp
Tsutomu Hoshino
Institute of Engineering Mechanics University of Tsukuba, Ibaraki 305, Japan
hoshino@kz.tsukuba.ac.jp

Abstract

Artificial ecosystems are created to emerge ALife diversity (Adiversity) characterized by multiple trophic levels. Each organism in the system consists of a pair of strings, species label and digestive fluid. A plant species consists of a pair of strings, and obtains solar energy using its digestive fluid. The digestive efficiency was determined by the bit-matching between solar energy and its digestive fluid. Primary consumers emerge by fusion between the plant species. Further fusion creates higher trophic organisms who consume plants or animals at the lower trophic levels. The digestive efficiency of a predator is determined by the bit-matching between its digestive fluid and the prey's species code. The initial communities consist of only plant species, emerged variable animal species, and trophic levels during evolution. The emerged ecosystem, however, eventually became extinct eventually. The relationship between stability and biodiversity is discussed by comparing it to other ALife models that actualized stable or open-ended food-web systems.

INTRODUCTION

Biodiversity is a key concept for ecology and evolutionary biology. Recent claims against global changes focus on biodiversity as an important index to detect a crisis in nature (Chapin 1994). Biodiversity is also a prerequisite condition for organisms to evolve under natural selection (Brandon 1984). Darwin and following evolutionary biologists have discussed how genetic and phenotypic variability have been maintained in real nature (Provine 1986). However, in what way these diversities have emerged during the evolution of natural systems, is a question that has yet to be addressed.

Most ecology and evolutionary biologists have obtained insights of biodiversity from kinetic models constituted of a set of difference or differential equations. Varieties of stability analyses have been applied to those models to explore relationships between robustness or resilience of ecosystems and their biodivesity (e.g., May 1974; Pimm 1991). Unfortunately, the kinetic models have a crucial weak point, i.e., they cannot emerge species or biodiversity because the number of equations prespecified at the beginning limits the number of species that appear in the model ecosystems. As a result, the discussion of species diversity is restricted to the asking of how to stop the extinction of component species in the ecosystems.

Speciation and emerging biodiversity are main themes for mile stone studies of ALife. Binary bit strings succeeded to emerge complicated evolutionary changes in iterated prisoners' dilemma (IPD) games (Lindgren 1991). Tierra realized complicated ecological relationships, such as parasitism and social parasitism, by generating new species constituted of 32 sets of computer instructions (Ray 1991). However, biodiversity achieved in those models was restricted to a horizontal trophic level. In Lindgren's IPD games, competing individuals shared the same pay-off matrix with no more than a single trophic level appearing. Although the parasitism evolved in Tierra seemed to be the first step to climb the trophic level, the phisics in the system (Core-Game-like instructions) allowed only competitive interactions for limited memory and CPU time. This prevented parasites to evolve further to exploit the whole body of victims just like predation. Biodiversity without predation is far from comparable to that observed in natural biosystems.

To step up the trophic level, one may have to incorporate the cost of predation or energy loss through heterotrophic behavior (Ray 1991). Energy flow may play an important role to make vertical interactions in the ecosystems. In ALife III two models were proposed to realize food-webs with ALife techniques. The first model by Lindgren and Nordahl (Lindgren and Nordahl 1994) extended the original model by incorporating an energy flow. The model apparently succeeded to generate complicated food webs as well as evolutionary changes. However, the result should be interpreted as the vertical transformation of the resutls from the original IPD game that were restricted to a single trophic level. To project horizontal interactions into vertical ones, it may be necessary to incorporate several global controls, such as finding appropriate environmental strategies and functional responses.

Johnson (Johnson 1994) proposed another type of model in which he focused on the size relationship among component species in food-webs. His model consisted of 20 x 20 or 50 x 50 cells, each of which could be occupied by a single organism. Size 0 species (plant) obtained energy through a constant rate of photosynthesis and covered the cells. Mutations created new organisms with a body size greater than 0. These were heterotrophic and could consume others with a body size smaller than their own. Starting from size 0 species (plants), Johson succeeded to obtain several features observed in natural ecosystems, such as the power law in biomass and growth rates of component species, and the tendency to increase in body size. However, the trophic levels did not extend more than three levels (plants, herbivores, and carnivores) in the ecosystems.

In this paper, we propose a more general way to emerge ALife diversity (Adiversity), characterized by vertical as well as horizontal interactions in food webs. Our model combines important aspects originated in several mile stone studies of ALife. The first aspect, derived from Tierra and Lindgren's IPD game, is the generic feature of an artificial world where organisms compete for limited space and energy. We apply simple bit-matching rule (Hillis 1992) for trophic interactions instead of Core-Game instructions or IPD games.

The second aspect, drawn from Lindgren's and Johnson's model, is the generic feature of organisms living in an artificial world. Each organism consists of two strings: the first characterizes "species" and the second determines trophic interactions between the organisms. We also apply the cascade trophic relationship based on body size just as in Johnson's model. We eliminated spatial structure to focus on the emergence of Adiversity only by biological interactions. We succeeded to obtain high diversity characterized by both vertical and horizontal interactions. However, our systems were always threatened by extinction. We discuss the emerging patterns of food-webs, the importance of these type of interactions, and the cause of extinction, by comparing them to the results obtained from the otehr three models.

MODEL

The Universe and the Primary Producer

The universe is constituted of 1,000 cells, each of which is allowed to be occupied by a single organism. There is no spatial distance among cells and an individual occupying a cell has the potential to make contact with any individuals among the other 999 cells. Each cell is showered constantly by SUN-LIGHT that consists of an 8-bit-string of "11111111."

An organism living in the artificial universe has a pair of bit strings (Fig. 1).

	species	digestive fluid
plant	11000000	11010110
animal	11000000	11101110
	00111100	10101100
	00000011	10000011

Fig. 1: A Plant and An Animal with three pairs of strings

The first 8-bit-string labels species of the organisms. Species are defined by a bit difference in this label string among the organisms. Two individuals are the same species if the Humming distance between the two strings is less than 2. The second string also has an 8-bit that represents the digestive fluid. It is possible that this digestive fluid may correspond to the "reference tag" in Lidgren's new model or the bit-string that determines "dietary specificity" in Johnson's model. The primary producers are PLANT species that have only a single pair of species label and digestive fluid.

The plant species obtaine energy by photosynthesis. Each plant initially has eight units of energy. The following energy intake depends on matching efficiency between a SUN-LIGHT string and the plant's digestive fluid. Each photosynthesis process produces eight minus the humming distance of the two strings. Thus, a plant species can obtain a maximum of eight units of resource intake when its digestive fluid consists of all "1".

Reproduction will occur if the individual energy reserve exceeds the threshold of reproduction (Energy for Reproduction, ER). There are two types of reproduction: Sexual reproduction with a cross-over and fusion between different species. In the former case, the partner of the same species is randomly chosen from the other plant individuals who have energy more than ER. In the latter case, an individual of a different species is randomly chosen. However, no fusion is allowed between the individuals who have exactly the same bit strings of species label. Sexual reproduction is applied nine times as much as fusion at each reproduction.

Each pair produces a single offspring whose energy is prespecified (Energy for Offspring). After the reproduction, a parent individual looses energy by EO/2. Both strings of the species label and the digestive fluid are the target of simple GA operations, such as mutation and crossing over. These genetic operations are performed independently against the pair of strings. Living things in this world also have senescence that is defined by the maximum number of bouts (Life Time, LT) in each simulation.

Primary Consumer and Higher Trophic Levels

Fusion between different plant species generates heterotrophic organisms that consists of more than a single pair of 8-bit-strings. They cannot execute photosynthesis with SUN-LIGHT, instead they become heterotrophic and start foraging other organisms. The number of pairs of 8-bit-strings determines how the organisms consume others at the lower trophic levels. Each heterotrophic organism can consume those with fewer pairs of 8-bit-strings, just as the size related trophic rule adopted by Johnson (Johnson, 1994). A similar "cascade" relationship has been applied to explain patterns and rules observed in natural food webs (Cohen 1989).

There is a specific cost for heterotrophic organisms. At each run step, individuals lose energy reserves at a constant rate of energy-Energy for Action (EA). This rate is multiplied by the number of pairs of bit-strings. Thus, the bigger the body size, the more it costs for predating other organisms. On the other hand, the heterotrophic organisms gain energy by predation. Predation is started by choosing a cell in the universe. If there is no organism in the cell, the individual resigns the chance and a loss of energy (EA multiplied by its body size) occurs. If the cell is occupied by an individual, the predator executes predation if the organism has fewer pairs of bit-strings than the predator species.

The energy intake by predation is calculated as follows. Firstly, Humming distances between the prey's species strings and the predator's digestive fluids are calculated for all possible combinations. The digestive efficiency of the predator is defined as (8 - the smallest Humming distance)/8. The predator gains the energy of the prey multiplied by the digestive coefficient. On the contrary, the prey looses its energy by the same amount as the predator gains.

Over View of a Simulation Step

The single step of the simulation is as follows.

1. Input the initial plant population into the universe

2. Reproduction for an individual

3. Photosynthesis or Predation for an individual

   (a) Calculating Energy Gain
   (b) Calculating Energy Loss

4. Resetting the universe

5. Return to 2.

Steps of this universe correspond to years rather than generations. The initial populaiton consists of three plant species, each of which is characterized by a bit-string of 00000011, 00111100, and 11000000, respectively. Ten individuals of each species are the founders in the universe. The initial values for EA, EO, and LT are 5, 100, and 100, respectively. The mutation rate is 0.1 for each bit-string irrespective of its species label and digestive fluid.

SIMULATION RESULTS

We made sixty six ecosystems changing seeds of a random generator. Complicated universes emerged from the beginning, but surprisingly, every ecosystem became extinct. The extinction was characterized by a sudden increase of herbivore species, followed by a disappearance of both plant and animal species. The histogram drawn with thick lines in Fig. 2 shows the frequency of the distribution of longevity of ecosystems consists of 1,000 cells. One ecosystem lasted approximately 9,000 steps although the distribution is skewed to the left side of the graph.

A typical example of species changes is shown in Fig. 3. Each line represents a specified size (pairs of strings) of species. The plant species (size 1) rapidly increased at the beginning, which is followed by an emergence of

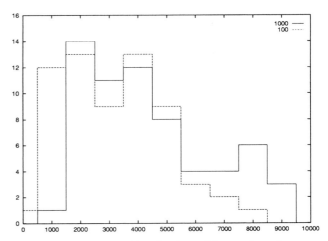

Fig. 2: Frequency distribution of longevity of ecosystems

several sizes of heterotrophic species (creeping lines at the bottom of Fig. 3). The maximum size of the animals sometimes exceeded more than 30. At about 500 steps, the interaction between plants and animals changed and moderate fluctuations appeared. The number of animal species also increased at that point.

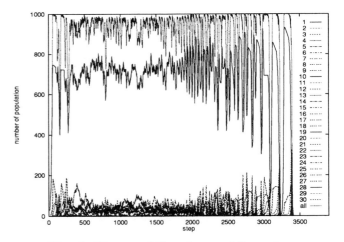

Fig. 3: Change of Size classes: All species

At about 2,000 steps, the interaction changed and plant species began to fluctuate dynamically. This drastic change may be caused by the occurrence of size 7 species taking over size 5, 6, and 9 (Fig. 4). This size 7 animal species, however, had only a single peak and disappeared following a drastic change in the whole system.

Figure 5 represents the change in the composition of species in terms of its body size (number of pairs of bit-strings). Surprisingly, the trophic levels among two to six emerged at almost the same share amount in the ecosys-

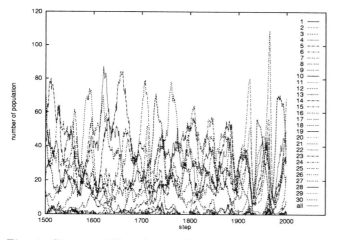

Fig. 4: Change of Size classes: Heterotrophic organisms

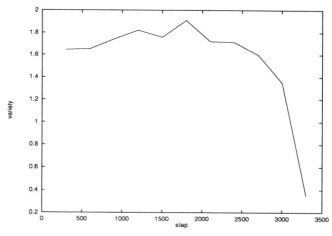

Fig. 6: Species diversity

tem. The trophic levels with more than eight occupied only a small portion of the universe. A bottle neck event of species composition occured at around 2,200 steps. However, this event does not seem to directly correspond with the beginning of large fluctuations in plant species at around 2,000 steps.

steps and continued to emerge multiple trophic layers.

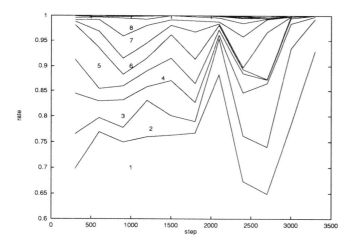

Fig. 5: Species composition classified by size class

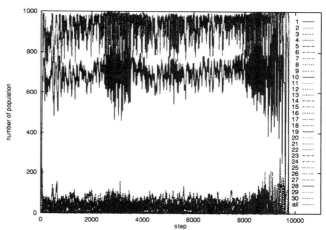

Fig. 7: Change in Size classes: All species

DISCUSSION

Emerging Trophic Levels

Our simulation succeeded in generating biodivesity in two aspects. Fistly, newly emerging individuals consisted of sets of bit strings occupied and regenerated in the 1,000 cells of the universe. Crossing-over and mutation caused speciation among the bit-strings. The second diversity was achieved by generating heterotrophic organisms feeding on primary producers and consumers at lower levels. Increasing of the length of bit-strings is now frequently applied for adaptive computations (Hervey 1994). In those approaches, increasing bit-strings meant increasing search dimensions for specified problems. In our artificial system, however, increasing strings was used for increasing trophic levels (cf. Lindgren and Nordahl 1994).

Figure 6 shows a species diversity index (Shannon Diversity Index) which is calculated based on the bit-string difference at every 300 steps. The diversity index kept a value of about 1.7 and suddenly began to decrease at around 2,500 steps. However, there was no prominent change in this index while the plant species began to oscillate dramatically at about 2,000 steps.

A set of figures is given for another example in Figures 7 to 10. This example lasted nearly 10,000 steps. The tendency for species change was similar to that of the previous example. However, this example seemed to escape the first crisis of ecosystems during 2,500 to 3,500

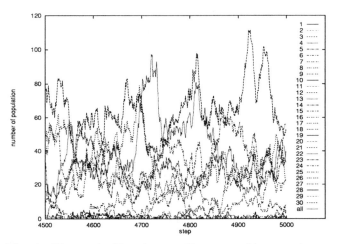

Fig. 8: Change in Size classes: Heterotrophic organisms

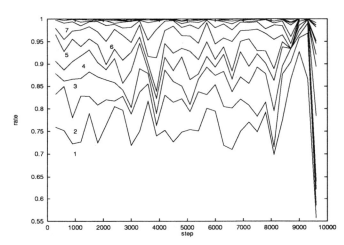

Fig. 9: Species composition classified by size class

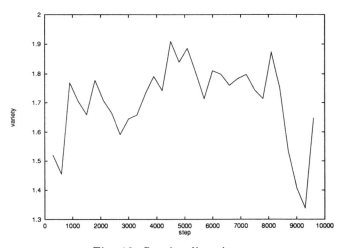

Fig. 10: Species diversity

The maximum size (number of pairs of bit-strings) of heterotrophic organisms may correspond to the maximum layers of the trophic levels emerging in our system. The trophic levels sometimes exceeded more than 30. They were significantly more than those expected in natural ecosystems (Cohen 1989; Pimm 1991). However, the large organisms could not always eat all the other smaller organisms because the predation was a chance event. Moreover, the consumption efficiency depended on the matching between the prey's species code and the predator's digestive fluid. Thus, the realized trophic levels should have been much less than 30.

Changes in the species composition suggests the realization of trophic levels. The equal appearance of size two to six animal species may suggest that these species depended on common or different kinds of plant species and that they had no predation interactions with each other. In this sense, the actual top predators might be those whose body size was more than eight (Fig. 5 and 9).

Extinction

Food-webs of Lindgren and Nordahl (1994) and Johnson (1994) seemed to achieve an open-ended evolution. Especially, Johnson's community showed quite stable oscillations. On the other hand, in all present simulations the ecosystems gradually came to a point of extinction. A possible cause for this might be the small size of our system. However, the 1,000-cell space was much larger than the universe that was adopted by Johnson.

To check the scale effect on the extinction, the authors run another set of simulation in a universe consitsed of only 100 cells. The histogram of the ecosystem longevity was shown by dotted lines in Fig. 2. It was not significantly different from that for the universe consisted of 1,000 cells (Kolmogorov Smirnov Two-sample Test, Max. Difference = 0.238, n = 63, P = 0.062) Extending the universe to10,000 cells actually increased the species diversity, the maximum size of top predators, and the retention time of the ecosystems (Saruwatari, pers. obser.). Nevertheless, there is no guarantee that the system in larger space achieves open ended evolution.

One of the most important aspects that differed from Johnson's model in the present artificial system was the rule of heterotrophic interactions. The authors used two bit-strings; one representing species and the otherdigestive fluid. A predator seeked the most efficient digestive fluid to consume its victims. The bigger the predator, the greater chance the predator had to find an efficient digestive fluid. However, GA operations were applied independently to both strings, which might have caused disparity evolution between the two strings inside an organism, as was observed in the IPD game ecosystems (Lindgren and Nordahl 1994) in the evolution of mimicry. The disparity evolution may cause complicated interactions between prey and predators, and cancel out the benefit of a cascade trophic relationship.

In Lindgren's IPD game, the stability of a system may be implemented by the PD game itself. Lotka-Volterra like oscillation is achieved in a simple IPD setting (see

Fig. 3 in Lindgren and Nordahl 1994). This regular oscillation is also shown in Johnson's model. On the contrary, the present model never showed such regular oscillations. The game occurring in the present ALife model resembles that of coevolution by sorting network programs as generated by Hillis (1991). Plant species try to evolve species labels that poorly match the digestive juice of animal species. The disparity evolution may also be interpreted as a communication game acquiring arbitrary language among evolving agencies (MacLennan 1991).

What We can Learn

The basic trophic rule adopted in our system is "who eats who" in algorithmic fashion. Corresponding classic expressions are Lotka-Volterra prey-predator equations. The attractor of a single-prey-single-predator system is a set of concentric limit cycles around an unstable equilibrium. The difference version of the Lotka-Volterra prey-predator system is known to diverge and ultimately become extinct. The algorithmic expression that we adopted in our system may be taken as another discrete version of the Lotka-Volterra system. Inherited instability may be moderated by speciation and the disparity in the evolution of the species code and the digestive fluid.

The biodiversity index adopted in the present study is widely used in ecological studies. However, it could not indicate the critical change in the artificial ecosystems (Fig. 6 and Fig. 10). The proportion of heterotrophic organisms changed dramatically (Fig. 5 and Fig. 9), but it did not reveal the cause of extinction. Extinction of the artificial systems appears to be caused by a single minute event at the genetic level. They are totally masked by disparity evolution between the species code and the digestive fluid.

Our system generates very complicated food-webs without spatial heterogeneity although appearing food-webs were transient and grasually disappeared. Lack of spatial heterogeneity may be a cause of extinction. Incorporating spatial heterogeneity may achieve pseudo-open-ended ecosystems through a delay in the ultimate extinction of the species by decreasing critical interactions among organisms. However, spatial heterogeneity is only one way to moderate strong interactions between organisms. To explain the occurence of extinction, the reduction of exess organic interactions is of greater significance than spatial heterogenity.

The present study suggests that the use of biodiversity is inadequate when applying simple indices which neglect the time series and processes to detect a ecosystem in crisis. To understand fully the cause of extinction requires the tracing of the historical events, however, to under take such an endeavor would be very difficult even even in a simple ecosystem. There are many ecological models and hypotheses that propose the examination of food-webs, community structure, and biodiversity (Magurran 1988, Pimm 1991). It is recommended that artificial food-webs are a good indicator for testing the robustness and appropriateness of those hypotheses.

acknowlegements

We thank Prof. Osamu Sakura for valuable comments on the earlier version of the present model. We also deeply thank Ms. Annette Raper for helping to polish up our English. This study is supported by the University of Tsukuba Research Projects.

References

Brandon, R. 1984. Adaptation and evolutionary theory, In *Conceptual issues in evolutionary biology*, ed. E. Sober. 58-82. Massachusetts: MIT press,

Chapin, III, F. S. and C. Körner. 1994. Arctic and alpine biodiversity:patterns, causes and ecosystem consequences, *TREE 9*, 45-47.

Cohen, J. E. 1989. Food webs and community structure, In *Perspectives in ecological theory*, ed. J. Roughgarden R. M. May, and S. A. Levin., 181-202. Princeton: Princeton,

Hervey, I. 1994. Evolutionary robotics and SAGA: The case for hill crawling and tournament selection.S, In *Artificial Life III*, ed. C. G. Langton. 299-326. New York: Addison Wesley,

Hillis, W. D. 1991. Co-evolving parasites improve simulated evolution as an optimization procedure, In *Artificial Life II*, ed. C. G. Langton C. Taylor J. D. Farmer, and S. Rasmussen., 313-324. New York: Addison Wesley,

Johnson, A. R. 1994. Evolution of a size-structured, predator-prey community, In *Artificial Life III*, ed. C. G. Langton. 105-129. New York: Addison Wesley,

Lindgren, K. 1991. Evolutionary phenomena in simple changes, In *Artificial Life II*, ed. C. Taylor C. G. Langton J. D. Farmer, and S. Rasumussen., 295-312. New York: Addison Wesley,

Lindgren, K. and M. G. Nordahl. 1994. Artificial food web, In *Artificial Life III*, ed. C. G. Langton., 73-105. New York: Addison Wesley,

MacLennan, B. 1991. Synthetic ethology: an approach to the study of communication, In *Artificial Life II*, ed. C. G. Langton C. Taylor J. D. Farmer, and S. Rasmussen, 631-658. New York: Addison Wesley,

Magurran, A. E. 1988. *Ecological Diversity*, London: Croom Helm,

May, R. M. 1974. *Stability and complexity in model ecosystems*, Princeton:Princeton,

Pimm, S. L. 1991. *The balance of nature?* , Chicago: Univ. Chicago Press,

Provine, W. B. 1986. *Sewall Wright and Evolutionary Biology*, Chicago: Uhiv. Chicago Press,

Ray, T. 1991. An approach to the synthesis of life., *Artificial Life II*, ed. C. G. Langton C. Taylor J. D. Farmer, S. Rasmussen, 371-408. New York: Adison Wesley,

ARTIFICIAL CULTURE

Nicholas Gessler
Department of Anthropology, University of California at Los Angeles 90024
gessler@anthro.sscnet.ucla.edu -- gessler@alife.santafe.edu

ABSTRACT

If we accept the propositions that archaeology is anthropology, and that anthropology is the study of culture process, then we now have at our disposal a new, untapped and rich paradigm for building theories of cultural evolution. Inspired by *artificial life* (*AL*), this is the paradigm of a computational anthropology assembled for the development of what I call *artificial culture* (*AC*). By studying the archaeology and ethnography of these virtual worlds, *AC* should help us develop and test theory for cultural evolution. It should neither confine us to those natural cultures accessible to ethnographers, comprised of modern participants and environments, nor to those accessible only to archaeologists. Rather, *AC* should enable us to explore alternative cultures and experiment with "what if..." scenarios. Perhaps we may in this way bring back to life the adaptations and environments of our early hominid ancestors. The foundations for such research are being built by *AL* in which we already see rudiments of *AC*. Beyond those cultures with which we're familiar, a properly conceived AC should be capable of synthesizing and analyzing virtually any participants in virtually any world.

1. PROLOGUE

Professor Dobb's book is devoted to personetics, which the Finnish philosopher Eino Kaikki has called "the cruelest science man ever created..." The field is a recent offshoot of the cybernetics and psychonics of the eighties, crossbred with applied intellectronics... At present a "world" for personoid "inhabitants" can be prepared in a matter of a couple of hours. This is the time it takes to feed into the machine one of the full-fledged programs (such as BAAL 66, CREAN IV, or JAHVE 09)... A specific type of personoid activity serves as a triggering mechanism, setting in motion a production process that will gradually augment and define itself; in other words, the world surrounding these beings takes on an unequivocalness only in accordance with their own behavior... As hundreds of experiments have shown, groups numbering from four to seven personoids are optimal, at least for the development of speech and typical exploratory activity, and also for "culturization." On the other hand, phenomena corresponding to social processes on a larger scale require larger groups. At present it is possible to "accommodate" up to one thousand personoids, roughly speaking, in a computer universum of fair capacity; but studies of this type, belonging to a separate and independent discipline -- sociodynamics -- lie outside the area of Dobb's primary concerns, and for this reason his book makes only passing mention of them. (Lem 1978, 167ff.)

The foregoing fictional review of a non-existent book was published in 1971, and although personetics and sociodynamics have never emerged as disciplines, we now seem poised to create them.

2. EPISTEMOLOGY

For those anthropologists who subscribe to the *new* or *processual archaeology*, the metaphysics and epistemology of anthropology should become the metaphysics and epistemology of science. As its theorist proclaimed:

The external world exists in its own right, and that includes the properties of the archaeological record... It is the availability of the external world, regardless of the character of our cognitive devices, that makes it possible for science to work. We can learn the limitations of our own ideas, as science has demonstrated over and over again, through skillful interaction with the world of experience, the external world. (Binford 1986, 403.)
In a productive science the aim is to learn something of the limitations and inadequacies of one's received knowledge. (Binford 1989, 52.)

Reaction came from a self-named school of *postprocessual archaeologists* who modeled their criticism after the literary *post-modernists*. To *processualists* these *post-processualists* were merely resurrecting the traditional objections to science. On the other hand, many of the *post-modernists* are studying science with a view towards a more cautious, yet still scientific epistemology. *Postmodernism*, then, is not necessarily opposed to the production of reliable knowledge about the external world.

Suppose we think about the reality "out there" as an unmediated flux... It interacts with and comes into consciousness through self-organizing, transformative processes that include sensory and cognitive components... On the other side are the constructed concepts that for us comprise the world... Within the representations we construct, some are ruled out by constraints, others are not... I am not saying constraints tell us what reality is. This they cannot do. But they can tell us which representations are consistent with reality, and which are not. (Hayles 1992, 78-80.)

We are increasingly stumbling across the limitations of our cognitive abilities and judging them unacceptable. Solutions include a philosophical restructuring of the mind as well as technological augmentations to it.

The problem (is)... to compress a host of interlocking ideas, drawn from many sources... into a form coherently expressible in a linear script. (Rosen 1991, xiv.) The essence of it lies in what we call the modeling relation... the bringing of two systems of entailment into congruence. (Rosen 1991, xvi.) The human mind is not adapted to interpreting how social systems behave... Evolutionary processes have not given us the mental skill needed to properly interpret the dynamic behavior of the systems of which we have now become a part. (Forrester 1971, 61.)

Physical models have a long standing as representations of both the external world and our ideas. Computational models join them mid-way between ourselves and reality. They exist outside of our perceptions and are indispensable to our quest for scientific knowledge and understanding.

3. THEORY

To many archaeologists, artifact characterizatrion and methodology was much more important than theory:

Students of contemporary archaeology must concern themselves with much more than seeking an understanding of the archaeological record. (Binford 1989, 62.)

To many anthropologists, theory was either unattainable or elusive. Building theory was the manifesto which the *new archaeology* proclaimed in 1962:

American archaeology is anthropology or it is nothing. (Binford 1962):
Archaeology must accept a greater responsibility in the furtherance of the aims of anthropology... Archaeologists should be among the best qualified to study and directly test hypotheses concerning the process of evolutionary change... As archaeologists, with the entire span of culture history as our

"laboratory," we cannot afford to keep our theoretical heads buried in the sand. We must shoulder our full share of responsibility within anthropology. Such a change could go far in advancing the field of archaeology specifically, and would certainly advance the general field of anthropology. (Binford 1972, 32.)

Processual archaeology even set out to redefine the traditional hallmark of anthropology, the field practice of ethnography itself:

Explaining variability and change in human behavior is the primary goal of the social sciences, yet we expend a great deal of research effort that is of dubious value in this regard. (Hill 1977, 59.)
(The ethnographer) becomes increasingly dependent upon informants to provide him or her with *information* regarding their knowledge and beliefs in terms of which the local people operate... (Ethnographers) are still not operating in a scientific role. Instead, they have adopted the role of intercultural translators... (This) has compelled many social researchers to rely on their informants to create their data. In turn, these same informants guide the interpretation and ultimately mediate the understanding of the data. What ethnographers report is not data but information, the intellectualized expression of experience. (Binford 1986, 395-96.)

Systems theory held promise for many *processualists*, yet its limitations also brought disappointments:

Most of the systems literature does not deal with *change* in enough detail to be of much use. Systems research has been much more concerned with what systems are, how they can be described and classified, and how they are regulated. And further, since general systems theory is designed to be general, and to deal primarily with *similarities* among different kinds of systems, it provides very few useful ideas on how human social systems might differ from other kinds of living systems in terms of the processes of stability and change. (Hill 1977, 101.)

Only a handful of *AL* papers have been presented in archaeology and anthropology. So foreign do its practitioners appear, that ironically *AL* may more frequently be the **object** of ethnographic field studies than it is the **paradigm** for their understanding.

4. NATURAL CULTURE

Since the beginnings of anthropology, a definition of culture was as elusive as theory to attain. Repeated attempts have been made to survey the anthropological literature for some scientific foundations. A readable overview is THE RISE OF ANTHROPOLOGICAL THEORY (Harris 1968). With theory still beyond reach, Harris tried instead to formulate a cogent research strategy in CULTURAL MATERIALISM: THE STRUGGLE FOR A SCIENCE OF CULTURE (Harris 1979). Though not addressing computational anthropology, Harris did formulate what he called the **universal pattern of culture** (Harris 1991, 22). This useful schema may be expressed as a pyramid rising from the environment and ascending in three giant steps:

SUPERSTRUCTURE
(ideas -- cognition)
S T R U C T U R E
(social economy -- behaviors)
I N F R A S T R U C T U R E
(technology -- artifacts)
N A T U R A L   E N V I R O N M E N T

As a materialist, Harris sees causation as largely initiated from the bottom-up, constraining the variation available to higher levels in the pattern. He explicitly advocates a bottom-up research strategy. The scheme is computationally operational when equated with the objects shown in parentheses above. Failing this, the concept of culture has limited utility:

> In fact, I believe we can profitably do without the concept "culture," since it appears to be unoperational in analysis. (Hill 1977, 103.)

5. NORMATIVE ASSUMPTIONS

Despite frequent protestations to the contrary in the traditional and popular views of anthropology, culture whether it is defined as ideational or material, is often operationally defined as a corpus of shared traits held in common by a social group. While not denying that traits are sometimes shared, it seems to be the differences in these traits among participants that motivates societal evolution and change. Such **normative** assumptions of anthropology may be broken down into **participant homogeneity**, **temporal consistency**, and **referential accuracy**.

The assumption of **participant homogeneity** needs to be tested by incrementally varying the rules from individual to individual. Every division of labor or specialized activity along lines of age, sex, or class implies heterogeneity.

The assumption of **temporal consistency** needs to be tested by incrementally varying the rules within each individual as a function of temporal context. Daily life is full of inconsistent contextualized classifications.

And the assumption of **referential accuracy** needs to be tested by incrementally varying the correspondence between rule categories and their external referents. Misinformation is a fact of life.

An **AC** research strategy should test the extent to which those assumptions may be correct.

6. CULTURE AS EMERGENCE

The role of the individual in society is problematic. He is both the most obvious feature in society and the most neglected when it comes to describing group behaviors. Researchers either ignore the individual level and look for laws of culture at the population level of abstraction, or they ignore the population level and concentrate on the individual. No one has formulated a theory linking the two, although the need is recognized:

> The dynamics between individual action -- which, ultimately, is the source for societal attributes measured at a more summary level -- and group properties, including societal organization and cultural systems, is a constant problem that archaeological theorizing has not adequately addressed. (Read 1990, 50.)
> Behavior is the working out of deeper, structuring properties, hence the focus on the regularity of behavior and behavioral products characteristic of much of scientific archaeology incorrectly directs attention away from the structuring processes towards their consequences. (Read 1986, 16.)
> Explanation... (may proceed) without the necessity for absolute, universal laws of behavior. (Read 1986, 11.)

As **cellular automata** illustrate, individual actors, operating under individual local rules, can automatically produce collective global patterns of behavior that emerge solely through their mutual interactions. Importantly, these global patterns of behavior are not programmed into the simulation. They have no existence within the individual actors themselves. Rather, they come into being only as the entire system operates. I suspect that many of the same processes are operating in culture.

It is not difficult to conceive of an **emergent world**, from the sub-atomic to the cosmological levels of abstraction, wherein quickly acting smaller elemental objects (or variables at lower levels), by carrying out only rules based upon local knowledge, give rise to more slowly emerging larger global patterns (or structures of behavior at higher levels). These higher level structures may then be taken as elemental variables for yet higher levels of interaction. Cause gives rise to consequence, and consequence becomes cause at an even higher level of abstraction. Structure in this view is a relative term: It is the abstraction of behavior as sensed from a higher level.

7. ARTIFICIAL CULTURE

By altering the interpretation of some models, it can be argued that research in *AC* is already underway by *AL*. There is no compelling reason why the differentiation and clustering behavior of a population of computational algorithms in a computer's memory should resemble the emergence of *co-adaptation* and *species-prey interaction* in biology any more closely than it resembles the emergence of *class structure* or *task specialization* in social science.

Cellular Automata are perhaps the clearest examples of the emergence of global patterns from local rules (see Forrest 1991, Gutowitz 1991). They form a metaphor for human interaction on two-dimensional substrate. When cells are made mobile and their states include the ability to move, the social metaphor is even more complete.

Tierra, a highly formalized and abstract virtual world, is home for programs which evolve through natural selection. From an ancestral creature, speciation occurs with lineages specializing in different adaptations with minimal restraint. The species compete, cooperate, socialize, cheat, parasitize, and hyper-parasitize. (Ray 1991, 1993.)

Iterated Prisoners' Dilemma with Choice and Refusal is populated by participants who are repeatedly matched as pairs, each having the choice of cooperating with its partner or defecting, the payoffs differing based on one partner's actions. With the faculty of memory, each can remember the past actions of the partner who chose it and can refuse to participate based upon that experience. As the population evolves, sub-populations emerge which can be characterized as cooperators, rip-off artists, and specialists who rip-off the rip-off artists. This is a distinctively human game with the formalism of *co-evolution* brought into the cultural arena (Stanley, Ashlock & Tesfatsion 1994).

SimLife is a software package written for the computer game market, but which may be useful for prototyping scientific research. One defines a territory, sows it with vegetation, populates it with customized creatures, and then sits back to watch the ecosystem evolve (Karakotsios & Bremer 1993, Maxis 1993).

Strategic Theater of War (STOW) is a world-wide network of military vehicle simulators connected to *semi-automated forces* which are intelligent controllers for foot-soldiers and single tanks requiring a minimum of human intervention. *ARPA* is presently working on fully automated representations for these players. (McDonough 1993).

Cooperative Robot Behavior is a goal of many robot builders, and its engineering is in some ways similar to constructing AC simulations, both in questions of how to achieve collective behavior and how to model the agents (Maes 1994, Mataric 1993, Steels 1994). In Brooks's bottom-up approach robots have no locus of knowledge, reason, or representation, yet despite these limitations, the robot acts intelligently. What Brooks has observed about robotics may also be true for culture:

> Just as there is no central representation there is not even a central system. Each activity producing layer connects perception to action directly. It is only the observer of the Creature who imputes a central representation or central control. The Creature itself has none; it is a collection of competing behaviors. Out of the local chaos of their interactions there emerges, in the eye of an observer, a coherent pattern of behavior. There is no central purposeful locus of control. Minsky gives a similar account of how human behavior is generated. (Brooks 1991a, 148-49.)

The titles of Brooks's articles reinforce his bottom-up approach: "Elephants Don't Play Chess," "Intelligence Without Representation," and "Intelligence Without Reason" (Brooks 1989, 1991a, 1991b).

> It is by no means clear that... a (perceptual world) is anything like what we actually use internally -- it could just as easily be an output coding for communications purposes... In fact, it may be the case that our introspective descriptions of our internal representations are completely misleading and quite different from what we actually use. (Brooks 1991a, 144.)
>
> Simon noted that the complexity of behavior of a system was not necessarily inherent in the complexity of the creature, but perhaps in the complexity of the environment... We hypothesize... that much of even human level activity is similarly a reflection of the world through very simple mechanisms without detailed representations. (Brooks 1991a, 149.)

The goal of *AC* is to create a population of dynamically evolving terrain-based mobile autonomous agents serving as a complex of multiple interacting hypotheses for understanding human cultural behavior. Those agents can either be simulated in hardware as robots, or simulated in software as programs. The future of *AC* is optimistic:

> Ideas, and other atomic particles of human culture, often seem to have a life of their own -- organization, mutation, reproduction, spreading, and dying. In spite of several bold attempts to construct theories of cultural evolution, an adequate theory remains elusive. The financial incentive to understand any patterns governing fads and fashion is enormous, and because cultural evolution has contributed so much to the uniqueness of human nature,

the scientific motivation is equally great. (Taylor & Jefferson 1994, 8.)

8. COMPUTATIONAL OBJECTS.

Software for *Artificial Culture* is currently in the *proof of concept* stage. To facilitate communication among anthropologists, the code is written in C++ to run on a Comtrade EISA VLB 486-66 DX2 with 8MB ram and 256KB cache. A Digital Processing Systems, Personal Animation Recorder model DR-2100 may be added to collect frame-by-frame visualizations of data from long runs on a dedicated hard drive for later playback through a VCR at higher speeds. This configuration should force the search for simplicity in designing key modules and functions. As needed, software will be scaled up to run on a massively parallel machine. As it progresses, my implementation of *AC* should embrace the following objects:

Grid is the terrain of the artificial world on which virtual food, people, and artifacts will interact.

Virtual Objects are people, resources, artifacts, architecture, and geographic features. They may be further characterized by a food value, energy content, information load, storage properties, and persistence. They may evolve on their own.

Personoids are the agents having the properties of other virtual objects plus cognition and the ability to learn and remember. Among static characters will be a name, sex, birthdate, and pedigree. Among dynamic characters will be age, weight, fecundity, morbidity, caloric intake, and caloric surplus. Among cognitive characters will be "rules," whether responsive, significatory, or symbolic, which mediate between sensory and motor organs, or in robotic parlance between sensors and actuators. Learning will be established through techniques of *evolutionary computation* and cultural transmission. Basic *emergent pattern detectors* will be provided to sense abstract behaviors such as theft, trade, organizational groupings, seasonality. The code should allow new *emergent pattern detectors* to evolve.

Environment is a relative term comprising everything external to the object under consideration, which also influences it.

Ethnographer is the trajectory of selected program parameters and states, restricted in ways which one might expect actual ethnographic data collection may be biased. The *ethnographer* may be provided with higher level *emergent pattern recognizers* to detect aggregation, dispersion, residence, and marriage. The *ethnographer* must also be embodied, operating within the physics of the virtual world and perceivable by its objects. Its cognitive skills may also be allowed to evolve. This reflexivity should enable us to examine the biases of various natural-world ethnographic practices.

God is omniscient. S/He is the ideally unbiased record of the trajectory of all program parameters and states. S/He exists for us only in a less-than-ideal state. Unlike the *ethnographer* S/He is disembodied and objective. Higher level *emergent pattern detectors* may be employed.

As these objects are implemented, the *AC* will recursively be scaled up and pared down in order to enhance its performance and its congruence with natural cultures, beginning at the level of foraging hunter-gatherers.

9. PRELIMINARY RESULTS

Since this project is in its formative stage, preliminary results are only indicative of future complexities. Already it is evident that foraging efficiency (measured as the *agents's* ability to extract all the food resources before death) varies widely and often counterintuitively with differing search strategies, distributions of resources, and rations eaten. This is partially the result of agents destroying the gradient which would lead them to higher food concentrations. Agent and contour-following, hill-climbing, and random path search behaviors are observed. Resource distributions are transformed: random distributions are made uniform, resource desert patches are created, and secondary resource patterns are generated some quite unlike the original distributions. From these preliminary results I would expect that a computationally based *AC* would make an excellent test platform for equationally based *foraging theory* (Stephens & Krebs 1986), *behavioural ecology* (Krebs & Davies 1991, 1994), and *human evolutionary ecology* (Winterhalder & Smith 1988, Smith & Winterhalder 1992). I would expect that *AC's* terrain embodied interactivity would eventually provide us with new insights into the *complex adaptive functionalities* of human cultural and bio-cultural evolution (Lovejoy 1992).

10. ETHICS

The promulgation of *AC* will be accompanied by ethical debate on its potential uses for prediction, control, and social change. The results of *AC* modeling used to direct social policy may be somewhat tempered by its use by an increasingly computer literate public. Nevertheless, as with any new technologically driven enterprise, the mastery of such evolutionary computational techniques will probably fall to a technical elite. Intriguingly, at least two new concerns have been raised from science fiction for serious discussion.

Given the strong claim that *AL* is in fact real life, we must anticipate the same assertion for a developed *AC*. If artificial entities become as real as natural ones, should the rights of natural entities be extended to the artificial? The door swings both ways. If some artificial entities are judged to be in ways more acceptable than their natural counterparts, should the rights of certain deviant natural entities be abolished?

A related concern is the effect that socially unacceptable behaviors directed towards artificial worlds may have on the individual's interaction with the natural world. An immersive virtual reality coupled with a developed *AC* may make the natural and artificial indistinguishable (as happens in contemporary war games). When that distinction is suspended, then must human behavior in the two realms be judged by the same conventions?

REFERENCES

Binford, Lewis R. 1962. "Archaeology as Anthropology." *American Antiquity*, 28, 2, 217-225.

Binford, Lewis R. 1972. "Archaeology as Anthropology." *An Archaeological Perspective*. New York: Seminar Press.

Binford, Lewis R. 1986. "Data, Relativism and Archaeological Science." *Man (N.S.)* 22, 391-404.

Binford, Lewis R. 1989. "The 'New Archaeology,' Then and Now." *Archaeological Thought In America*, by C.C. Lamberg-Karlovsky, 50-62. Cambridge: Cambridge University Press.

Brooks, Rodney A. 1989. "Elephants Don't Play Chess." Cambridge: MIT Artificial Intelligence Laboratory.

Brooks, Rodney A. 1991a. "Intelligence Without Representation." *Artificial Intelligence* 47 (1991) 139-159. Elsevier.

Brooks, Rodney A. 1991b. "Intelligence Without Reason." *International Journal Of The Conference On Artificial Intelligence* (1991).

Forrest, Stephanie, editor 1991. *Emergent Computation -- Self-Organizing, Collective, And Cooperative Phenomena In Natural And Artificial Computing Networks.* Special Issues of Physica D. Cambridge: MIT Press, A Bradford Book (Elsevier Science).

Forrester, Jay W. 1971. "Counterintuitive Behavior of Social Systems." *Simulation*, February, 61ff.

Gutowitz, Howard, editor 1991. *Cellular Automata -- Theory And Experiment.* Special Issues of Physica D. Cambridge: MIT Press, A Bradford Book (Elsevier Science).

Harris, Marvin 1968. *The Rise Of Anthropological Theory: A History Of Theories Of Culture.* New York: Thomas Y. Crowell.

Harris, Marvin 1979. *Cultural Materialism: The Struggle For A Science Of Culture.* New York: Random House.

Harris, Marvin 1991. *Cultural Anthropology.* Third edition. New York: Harper-Collins.

Hayles, N. Katherine 1992. "Constrained Constructivism: Locating Scientific Inquiry in the Theater of Representation." *New Orleans Review*, Winter, 76-85.

Hill, James N. 1977. "Systems Theory and the Explanation of Change." *Explanation Of Prehistoric Change.* A School of American Research Book, 59-104. Albuquerque: University of New Mexico.

Karakotsios, Ken and Michael Bremer 1993. *SimLife -- The Official Strategy Guide.* Rocklin: Prima.

Krebs, J.R. & N.B. Davies 1991. *Behavioural Ecology: an Evolutionary Approach.* Oxford: Blackwell.

Krebs, J.R. & N.B. Davies 1994. *An Introduction to Behavioural Ecology.* Oxford: Blackwell Scientific.

Lem, Stanislaw 1978. "Non Serviam." In Stanislaw Lem's, *A Perfect Vacuum*, 167-196. San Diego: Harcourt Brace Jovanovich.

Lovejoy, C.Owen. 1992. "Modeling Human Origins." Handout for the Center for the Study of the Evolution & Origin of Life Symposium, UCLA, Los Angeles.

MacDonough, James 1993. "Doorways to the Virtual Battlefield." Paper presented to *Virtual Reality -- The Commitment to Develop VR.* Los Angeles.

Maxis 1993. *Sim Life -- The Genetic Playground.* Software by Maxis, Orinda, California.

Ray, Thomas S. 1991. "An Approach to the Synthesis of Life." *Artificial Life II*, SFI Studies in the Sciences of Complexity, vol. X, edited by Langton et al. Reading: Addison-Wesley.

Ray, Thomas S. 1993. "An Evolutionary Approach to Synthetic Biology." Manuscript. (October 21.)

Read, Dwight W. 1986. "Mathematical Schemata and Archaeological Phenomena -- Substantive Representations or Trivial Formalism?" *Science And Archaeology*, 28, 16-23.

Read, Dwight W. 1990. "The Utility of Mathematical Constructs in Building Archaeological Theory." Albertus Voorrips, editor, *Mathematics And Information Science In Archaeology: A Flexible Framework.* Studies in Modern Archaeology, Volume 3. Bonn: Holos.

Rosen, Robert 1991. *Life Itself: a Comprehensive Inquiry...* New York: Columbia University Press.

Smith, Eric Alden & Bruce Winterhalder 1992. *Evolutionary Ecology & Human Behavior.* New York: Aldine de Gruyter.

Stanley, E. Ann, Dan Ashlock and Leigh Tesfatsion 1994. "Iterated Prisoner's Dilemma with Choice and Refusal of Partners." *Artificial Life III*, SFI Studies in the Sciences of Complexity, vol. XVII, 131-175, edited by Langton. Reading: Addison-Wesley.

Stephens, David W. & John R. Krebs 1986. *Foraging Theory.* Princeton: Princeton University Press.

Taylor, Charles and David Jefferson 1994. "Artificial Life as a Tool for Biological Inquiry." In *Artificial Life.* Volume 1, Number 1/2, 1-13. MIT Press.

Winterhalder, Bruce & Eric Alden Smith 1981. *Hunter-Gatherer Foraging Strategies.* Chicago: University of Chicago Press.

Explorations in The Emergence of Morphology and Locomotion Behavior in Animated Characters

Jeffrey Ventrella

Visible Language Workshop
MIT Media Laboratory
20 Ames Street, Cambridge, MA 02139
ventrell@media.mit.edu

Abstract

This paper presents an animation system developed for the exploration of emergent morphology and locomotion behavior in articulated 3D figures. A genetic algorithm is used in this system for evolving populations of these figures towards improved locomotion and realistic positioning of the head. Qualities reminiscent of some familiar animals are shown to emerge. This work stems originally from research in designing tools for the art of character animation—as such, personality and humor are noted as qualities which may also emerge. An interactive evolution overlay to automatic evolution is supplied by a separate tool which allows a viewer to affect the course of evolution for these purposes. Of interest in this paper are issues in the design methodologies involved in creating genotype-phenotype representations for biomorphs with "evolvability".

1. Introduction

In this paper I present an animation system developed for exploring the emergence of functional anatomy and motion behavior. Artificial life research projects typically apply genetic algorithms (GA's) (Holland, 75) as a means of achieving emergent behavior in artificial organisms (Langton, 89). While many types of emergent phenomena have been modeled, morphology still remains an area with room to explore. The techniques and discoveries I briefly present here are derived from research in character animation in which motions are generated by forces internal to the character, using physical modeling techniques such as forward dynamics (Badler et al., 91). GA's have been used to optimize motor control schemes for locomotion and other behaviors in physically based characters (Ngo and

Marks, 93). This paper describes a variation on these techniques—with an added emphasis on emergent morphology driven by fitness pressures for locomotion and head motion constraints during evolution.

The stars of the show are biomorphs modeled as 3D articulated stick figures (called *animats* here) existing within a simple physical environment. I have chosen not to put flesh on the skeletons because my focus is primarily on motion and gross morphology, and 3D line-drawings serve well for this purpose. Anatomy and motion attributes are represented by a few dozen parameters in each animat, which evolve through the use of a GA. Relative fitness among animats is defined as the ability to travel longer distances from a common starting point. Secondary fitness terms are used which add pressures for holding the head high, minimizing head movement, and minimizing head contact with the ground.

2. Disney Meets Darwin

To imbue computer animated characters with expressive motion behavior for humor and personality are also goals of this work. Since I have approached this exploration primarily as a visual artist specializing in animation, the original motivation springs from the art of character animation and efforts to develop alternative ways to generate complex and expressive motion art. Strange, monstrous, and comical animats emerging in populations, even as a result of programming bugs, have been welcomed (the bugs are corrected, while trying to keep the spirit of serendipity manifest in the system). In the midst of the surreal biodiversity, familiar-looking forms and motions do emerge, indicating that the system is converging on solutions akin to those which nature favors. Morphologies and compatible motion styles usually emerge together. For instance, horse-like shapes emerge along with a propensity to gallop or trot. One population produced a family of animats which behaved like upright snakes traveling on crutches—both crutches being moved in parallel. Not that nature has ever produced such a phenomenon, but if ever there were a snake on crutches, this animat comes as close as I can imagine it would look.

One outcome of evolution is the appearance of "intention" in the animats' behaviors, which emerges without explicit design. An animat from an evolved population tends to make use of each part of its body, even if the associated body part of its ancestors may have seemed purposeless or even obstructive. For instance, a limb which is not used for locomotion in the direct way that legs are may become useful as a way to shift the animat's center of gravity forward during a crucial step to aid in the gait. The net result is an organic motion in which all the parts are orchestrated towards a common goal. The best proof of this comes simply from watching an evolved animat do its thing—as there may be no numerical measure of "style" above and beyond the fitness values used to determine selection in the GA.

3. Physics

The dynamics of motion are created using a qualitative physics model, tailored specifically for this animation system. It incorporates forward dynamics—acceleration is caused by forces exerted internally within the animat by way of deformations in its internal structure (autonomously changing angles of the joints connecting limbs). The model is simple yet produces many of the salient features of interacting objects in the real world, such as gravitational effects, inertia, angular and translational momentum, friction, and dampening. An animat is treated essentially as a rigid body which can deform itself autonomously but cannot be deformed passively by way of outside forces such as collision with the ground surface.

4. The Structure of the Animat

The scheme for expression of morphology from the genetic representation is designed to generate many variations. Previously I had developed a biomorph design with an open-ended structure based on an idiosyncratic scheme for expression of morphology from the genotype. Figure 1 shows three examples from this morphology scheme.

Through a convoluted embryology, genes affect the number of limbs, branch points of each limb, and branching angles for each limb, resulting in asymmetrical forms. This scheme was implemented as an experiment to see if the pressure to affect locomotion would cause symmetrical structures to emerge from a search space of mostly asymmetrical forms. Many trial runs suggested that the search space was too large for the GA to easily converge on symmetry, and that the representation may not have been designed sufficiently for *evolvability*. It did not incorporate a "constrained embryology", in which there are few genes yet each gene controls a more powerful feature of the phenotype (Dawkins, 89). In discussing his own search for a useful biomorph, Dawkins refers to the emergence of segmentation in animal morphology in the natural world as a watershed event— as a likely increase in evolvability for those animals which chanced upon it. No such structural scheme existed in my earlier design.

After exploring variations on this open-ended biomorph design, I settled on a generalized bilateral-symmetric structure, which still allowed for variation, and even occasional asymmetric features, but was characterized by a more realistic embryological scheme—with segmentation. Variations in topological connectivity of limbs and angles in the joints give rise to structures resembling spiders, birds, four-legged mammals, human-like shapes, and, of course, a great variety of monsters.

In this morphology scheme, a few basic elements are always present: a central body functioning as a backbone, having from one to four segments, and opposing pairs of limbs (as many as three), each limb having from one to four segments. Thus, the simplest topology is one body segment with a pair of opposing one-segment limbs (Figure 2a). All segment lengths are set to be of equal length. Angles of limb connectivity are specified with a local (*yaw, pitch, roll*) coordinate system.

Figure 2 shows three typical un-evolved animats. In some cases, asymmetry can arise (Figure 2c). For instance, if the number of body segments is less than the number of limb pairs, extra limb pairs grow out from other limbs. This embryological quirk has been kept in the biomorph design, keeping with the idea that features like this might by chance produce useful strategies for locomotion.

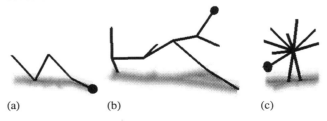

(a)   (b)   (c)

Figure 1 Three examples of un-evolved biomorphs from an earlier scheme for morphology - without symmetry. The black spheres indicate head coordinates.

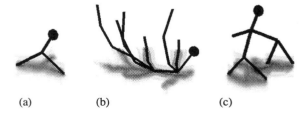

(a)   (b)   (c)

Figure 2 Three un-evolved animats illustrating morphological variety in the search space.

A list of effects of the genes for morphology is given below:

- number of body segments
- *pitch* angle of joints between body segments
- number of opposing limb pairs
- *pitch* angle at the branch point of each limb pair
- *yaw* angle at the branch point of each limb pair
- changes in *pitch* angle at the branch point among consecutive limb pairs
- changes in *yaw* angle at the branch point among consecutive limb pairs
- number of segments in each limb
- *pitch* angle of joints between limb segments
- *yaw* angle of joints between limb segments

5. The Head

Only one body part is differentiated from the rest—the head—which is the coordinate where the first body segment starts. This coordinate is used for some experimental secondary terms in the fitness function, which encourage more realistic behavior in the animats, similar to a strategy described by Ngo and Marks (93). All other body parts are free to evolve to function with no constraints. The secondary fitness terms relating to the head are added to the locomotion fitness term with varying weights associated with them. They are:

1) reward for the head being held high (proportional to the average height of the head throughout the animation)
2) penalty for head movement (proportional to the total length of the trajectory traced by the head throughout the animation—which can be much longer than the total length of the trajectory traced out by the animat's center of mass)
3) penalty for head collision on the ground (proportional to the accumulation of downward impact at which the head collides with the ground—if and when it hits).

6. Motor Control

McKenna (90) has developed a virtual roach which attains locomotion through oscillators in each of the six legs. Coupling of the oscillators, which operate at coordinated frequencies, generates different rhythms determining a set of gaits. In a similar fashion (yet much simpler), this system utilizes a rhythmic motor control program which consists of a combination of sinusoidal angular motions in multiple limbs. The differences from one animat's motor program to another can vary greatly according to genetic variation. An animat's motor program is continually active at all times and cannot change during its lifetime. Unlike the figures modeled by Ngo and Marks (93), and van de Panne and Fiume (93), which incorporate a stimulus/response model for controlling motion (physical-based motor control), these animats use a time-based motor control system.

The genes which affect motion are expressed somewhat differently than in the case of morphology. Attributes of motions within the whole collection of joints are created by one number-series generating function, as a way to coordinate them. This function takes five control parameters (determined by five genes) as input, plus the number of joints, and it generates a periodic series as output. Each number of the series corresponds to one joint angle in the animat. The following pseudo-code example illustrates this function.

```
INPUT:    size, start, step  low, high
          number_of_joints;

integers:      size, start, step,
               number_of_joints;
real numbers: low, high;

BEGIN
number = start;

LOOP from (index=1) to (index=number_of_joints):
 (
   number = number + step;
   output_parameter[index] =
   low + (number MOD size)/size * (high - low);
 )
END;

OUTPUT:  a periodic series of
         real-number parameter values
```

The purpose of this scheme is to allow a small number of genes to determine an indefinitely long series of values to control motions in multiple limbs, where the number of limbs can vary. The series is periodic to encourage regularity, yet with a large number of possible polyrhythms in the total animat's motions. Thus, an animat can move all its limbs either in total synchrony, with various wave motions, in alternating fashions, or irregularly. In embryological expression from genotype to phenotype, the periodic series generator is used to determine six motion attributes (with six associated sets of five genes determining the inputs to the function). The six motion attributes, for the whole series of joints, are:

- amplitude of sine wave motion in the *pitch* angle of the joint
- amplitude of sine wave motion in the *yaw* angle of the joint
- phase offset of the *pitch* angle sine wave
- phase offset of the *yaw* angle sine wave
- on-off switch to enable or disable the joint's *pitch* angle motion
- on-off switch to enable or disable the joint's *yaw* angle motion

In addition to the genes controlling these motions, there is one global *rate* gene which controls the overall frequency of sine motions in all the joints.

7. Evolution

A population evolves towards improved average fitness, where an animat's fitness is proportional to the distance traveled from the location where an animat starts at the beginning of the animation. Direction of travel does not matter. The secondary fitness terms concerning head activity can contribute to the overall fitness. To determine the fitness of an animat, the animation is run for 1000 time steps, in which the animat is essentially dropped into an infinite, flat world. The animat has a limited life span in which to gather distance.

A population of 100 animats is used in each experiment. The population is not updated in parallel as in the standard GA but rather in increments of one individual at a time as shown in the steps below:

Initialize: Set all gene values to be random, then animate and evaluate each animat to determine all initial fitness values

Iterate: Cycle through the following sequence 1000 times:

- Each genotype is ranked by fitness within the population
- Two genotypes are selected randomly with chances of being chosen proportional to ranking, and they mate via crossover.
- They produce one offspring which is subject to some mutation
- The new offspring is animated and evaluated for fitness
- Its genotype replaces the lowest ranking genotype

Limiting reproduction to one mating at a time causes slower convergence than updating in parallel but it has the benefit of inhibiting premature convergence, and it guarantees that the most fit genotypes at any given time are never lost. Mutation rate is set at inverse proportion to population size.

8. Results

While hopes for Olympic performance were somewhat dampened, there was no lack of amusement. Improved locomotion was clearly achieved by all populations, but often at the expense of the head, which in some populations was swung wildly to shift the center of mass, or used like a foot. In all cases, the fitness reward for locomotion was set at a constant weight, while the secondary head-related fitness terms were set at differing weights, for purposes of experimentation. Choosing the best proportions of weights for multiple fitness terms is not trivial, as indicated also in experiments by Fukunaga et al., (94). In these experiments, it is highly dependent on the degrees of freedom in the animat, the duration of the evaluation, and the nature of the physical model.

It has been found that by merely discouraging rapid head movement and collisions via fitness penalties, other body parts of evolving animats will tend to function not only for locomotion, but also to keep the head steady. In tests where animats were rewarded for holding the head up higher, the result was more upright postures, and some pseudo-human forms—but with one distinctive non-human feature—a magnificent kangaroo-like tail which helped balance the animat, tripod-style, as it awkwardly ambled along. Without any stimulus/response system modeled, these animats have no sense of balance so as to adjust their motions accordingly, thus, sustained bipedal walking is near impossible. Figure 3 shows two such animats from populations which evolved through fitness pressures for head height.

Figure 3 Two animats from populations which have evolved through fitness pressures for head height.

The fitness pressures for discouraging head movement and head collisions had to be fine tuned a number of times before satisfactory results came about. If a fitness term's weight was set too high, such as the penalty for head movement, locomotion was inhibited—since locomotion involves at least some head motion—and the population converged on immobility. Once reasonable ranges for these settings were established, desired behavior was more easily achieved.

Resulting gaits were of a large variety. Figure 4 shows the motions of an animat representing a population which evolved with no fitness constraints on the head, and so each animat in the population was free to do with its head whatever was necessary for locomotion. This animat demonstrates the use of the head as a third foot, aiding in balance.

Figure 5 shows an animat representing a population which evolved through pressures for holding the head high. This animat is roughly horse-shaped, except that it has six legs, and it moves like a crab—sideways.

Alternating limb motions emerged in many populations (as in Figures 4 and 5) but quite frequently populations converged on gaits in which all the limbs swung in unison, creating hopping, inching, and other such strategies. Figure 6 shows an animat which moves in this manner. Of note is the fact that this animat moves *backwards* (the direction of movement is not the direction in which the head is aimed). This illustrates the fact that the head is not used to *lead* motion, but only to modify it.

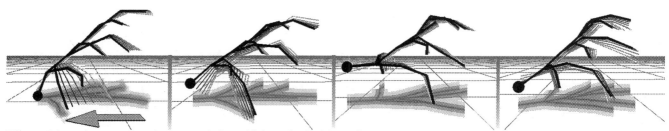

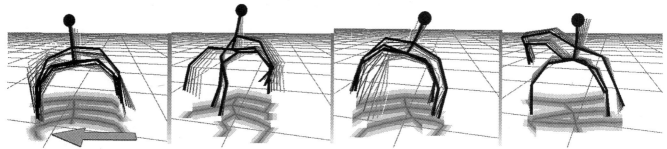

Figure 4 An animat representing a population which evolved with no fitness pressures pertaining to the head. The head is used as a foot. In this depiction (and all which follow), a series of after-images are shown in each frame to indicate the rates and directions of motion in each body part. The arrow indicates direction of travel.

Figure 5 An animat representing a population which evolved with fitness pressures for holding the head high. It might be described as a six-legged horse which moves like a crab—sideways.

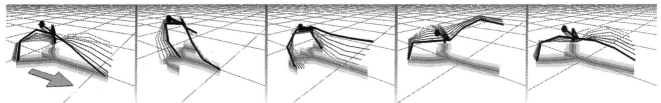

Figure 6 An animat representing a population which has evolved with slight constraints on head motion. This animat does not travel in the direction that the head is aimed, indicating that the head is not used to *lead* motion, only to modify it.

9. Interactive Evolution Overlay

Animats can also be evolved within a separate animation system, called the *Character Evolution Tool*, developed for the purpose of complementing automatic evolution with fitness contributions from a viewer (Ventrella, 94). In this tool, the viewer has the ability to contribute to the evolutionary process by adding fitness values to any one of the animats, as a reward for *style*. This tool resembles other interactive evolution systems in that it displays a number of individuals simultaneously for the viewer to compare and evaluate (Dawkins, 86), (Sims, 91), (Todd and Latham, 92), (Baker and Seltzer, 93). The Character Evolution Tool is designed specifically for evaluation of motion qualities, and it affords the viewer the ability to guide evolution according to such qualities as humor and personality. It complements the system described in this paper in that it offers an additional level of fitness criteria to affect evolution—an aesthetic level. Figure 7 illustrates this tool, with population size set at twelve.

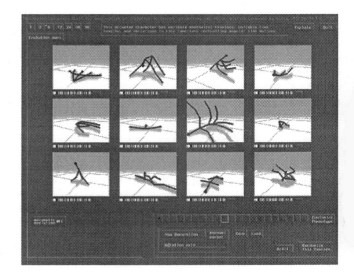

Figure 7 The interface to the Character Evolution Tool, displaying a population of twelve animats.

10. Genotype-Phenotype Design

While spontaneous emergence is an ideal to strive for in artificial life, every experimenter must design a representation on some level. In inventing a new genotype-phenotype representation, the search for an effective representation can become a kind of evolutionary process in itself. In these experiments, before settling on a good anatomy and motion phenotype design, many trials were run, and frequently, ranges of gene effect and embryological constraints were tweaked, along with fitness criteria to accommodate such changes.

These explorations have shown me that careful crafting of a highly evolvable embryological scheme for the development of phenotypes is important, combined with a conceptual compatibility with the domain, such that representation is not too arbitrary. In the case of expressive character animation, representations can be based largely on biological understanding, but can also be derived from traditions in cartoon animation (where physics and motor control may have quite a different meaning—for instance, as *expressive channels*). A genotype-phenotype design methodology which addresses these issues could enrich many areas of engineering, design, and art, and perhaps it could also enrich the field of artificial life, aesthetically.

11. Conclusion

I have demonstrated that morphology is a phenotypic feature which can emerge alongside locomotion behavior, as a result of rewarding animats for covering longer distances during evaluation. I have also shown that adding secondary fitness terms pertaining to motions and positioning of the head can contribute to the emergence of familiar animal forms and motions, as well as some unfamiliar (but funny) characters.

This paper offers a brief look at an artist's use of artificial life techniques and concepts as applied to an expressive medium—character animation. In the Disney tradition, animation is the *illusion of life*. In adopting bottom-up, emergence methodologies, character animation research adds to this the *simulation of life*. The explorations described in this paper are an example of taking this approach towards enriching the art form.

Acknowledgments

This work was produced in the Visible Language Workshop, of the Media Lab at MIT, directed by Muriel Cooper. I would like to acknowledge Joe Marks and Mitch Resnick for their suggestions and advice. Special thanks goes to Ron MacNeil, Suguru Ishizaki, and Louie Weitzman for technical help and many useful ideas.

This work was sponsored in part by PAWS, Inc., USDOT, and News In the Future consortium.

References

Badler, Norman I., Brian A. Barsky, and David Zeltzer. *Making Them Move.* Morgan Kaufman, 1991.

Baker, E., and Margo Seltzer. *Evolving Line Drawings.* Harvard University Center for Research in Computing Technology. Technical Report TR-21-93. 1993.

Dawkins, R. *The Blind Watchmaker.* (software accompanying book by same title) W.W. Norton and Company. New York, London. 1987.

Dawkins, R. *The Evolution of Evolvability.* Artificial Life (proceedings). Addison-Wesley 1989. pages 201-220

Fukunaga, Alex., Lloyd Hsu, Peter Reiss, Andrew Shuman, Jon Christensen, Joe Marks, and J. Thomas Ngo. *Motion-Synthesis Techniques for 2D Articulated Figures.* Harvard University Center for Research in Computing Technology. Technical Report TR-05-94. 1994

Holland, J. H. *Adaptation in Natural and Artificial Systems.* The University of Michigan Press, Ann Arbor. 1975.

Langton, Christopher, editor. *Artificial Life* (proceedings) Addison-Wesley. 1989.

McKenna, Michael. *A Dynamic Model of Locomotion for Computer Animation.* Master's Thesis, MIT. 1990.

Ngo, Thomas J. and Joe Marks. *Spacetime Constraints Revisited.* Computer Graphics (SIGGRAPH conference proceedings). 1993. pages 343-350.

Sims, Karl. *Interactive Evolution for Computer Graphics.* Computer Graphics, vol. 25. number 4. (SIGGRAPH conference proceedings) July, 1991. pages 319-328

Todd S., and W. Latham. *Evolutionary Art and Computers.* Academic Press: Harcourt, Brace, Jovanovich. 1992.

van de Panne, M., and E. Fiume. *Sensor-Actuator Networks.* Computer Graphics SIGGRAPH Proceedings 1993. pages 335-342.

Ventrella, Jeffrey. *Disney Meets Darwin - An Evolution-based Interface for Exploration and Design of Expressive Animated Behavior.* MIT Master's Thesis. 1994.

An Instance of a Parasitic Replicator

Alun Rhys Jones and **Adrian J. West**
Department of Computer Science
University of Manchester
Oxford Road, Manchester, UK.
arj@cs.man.ac.uk, ajw@cs.man.acs.uk

I am a successful parasitic replicator, for although I convey no useful information or aesthetic appeal, I have been able to effectively manipulate the physical world (i.e. the minds of the editors of *Artificial Life IV*) through parasitic action on a host replicator in such a way as to facilitate my own reproduction. The host in question is the paper "An Instance of a Replicator" which appeared in *Artificial Life III*, which I now exploit by reference in a parasitic fashion, in order to convey my methods of reproduction [Schultes, 1994].

References

[Schultes, 1994] Erik Schultes. An Instance of a Replicator. In *Artificial Life III*, edited by Christopher G. Langton. Santa Fe Institute Studies in the Sciences of Complexity, Proc. Vol. XVII, page 1. Reading, MA: Addison-Wesley, 1994.

AUTHOR INDEX

Ackley, David H., 40
Adami, Chris, 269, 377
Angeline, Peter J., 353
Arpin, Eric, 307

Bahm, Alan, 258
Bankes, Steve, 337
Banzhaf, Wolfgang, 109
Batali, John, 160, 343
Beckers, R., 181
Bedau, Mark A., 258
Beer, Randall D., 246
Belew, Richard K., 210
Bersini, Hugues, 382
Bonabeau, Eric, 307
Brogan, David C., 319
Brown, C. Titus, 377

Darley, Vince, 411
Darrell, Trevor, 198
de la Maza, Michael, 325
Dellaert, Frank, 246
Deneubourg, J.L., 181
Detours, Vincent, 382
Doi, Hirofumi, 359
Durand, S., 365

Etxeberria, Arantza, 406

Floreano, Dario, 190
Franklin, Stan, 301
French, Robert M., 277
Furusawa, Mitsuru, 359

Gessler, Nicholas, 430
Godfrey-Smith, Peter, 80
Grzeszczuk, Radek, 17

Hemmi, Hitoshi, 371
Hendriks-Jansen, Horst, 70

Hirayama, Nobumasa, 331
Hodgins, Jessica K., 319
Hogeweg, Pauline, 119
Holland, O.E., 181
Hoshino, Tsutomu, 140, 424
Hosokawa, Kazuo, 172

Inayoshi, Hiroaki, 295

Jakiela, Mark J., 289
Johnson, Michael Patrick, 198
Jones, Alun Rhys, 442

Kajitani, Isamu, 140
Kephart, Jeffrey O., 130
Kitano, Hiroaki, 49
Kitcher, Philip, 343

Levinson, Gene, 90
Littman, Michael L., 40
Lynch, James F., 236

Maes, Pattie, 198
Maley, C.C., 152
Mange, D., 365
Marchal, P., 365
Masujima, Yasuhiro, 331
Menczer, Filippo, 210
Messinger, Adam, 277
Miglino, Orazio, 190
Miura, Hirofumi, 172
Mizoguchi, Jun'ichi, 371
Mondada, Francesco, 190
Moreno, Alvaro, 406

Nade, Toshiaki, 331
Nagayoshi, Masahiro, 331
Nagel, Kai, 222
Nolfi, Stefano, 190

Oliphant, Michael, 349

Paredis, Jan, 102
Peschl, Markus F., 417
Piguet, C., 365

Rasmussen, Steen, 222
Ray, Thomas S., 283
Reynolds, Craig W., 59

Saitou, Kazuhiro, 289
Sardet, Emmanuel, 307
Saruwatari, Takuya, 424
Shanahan, Murray, 388
Shimohara, Katsunori, 3, 371
Shimoyama, Isao, 172
Sims, Karl, 2
Sipper, Moshe, 394
Stauffer, A., 365
Steels, Luc, 8
Stone, James V., 400

Te Boekhorst, Irenaeus J. A., 119
Terzopoulos, Demetri, 17
Thearling, Kurt, 283
Theraulaz, Guy, 307
Toquenaga, Yukihiko, 144, 424
Tu, Xiaoyuan, 17

Umerez, Jon, 406
Unemi, Tatsuo, 331

Vaario, Jari, 313
Ventrella, Jeffrey, 436

Wada, Ken-nosuke, 359
West, Adrian J., 442

Yano, Kiyoshi, 331
Yuret, Deniz, 325

Zhou, Lijia, 301